AF685813

IEMC 96 Proceedings

International Conference on Engineering and Technology Management

Managing Virtual Enterprises: A Convergence of Communications, Computing, and Energy Technologies

August 18 to 20, 1996

Vancouver, British Columbia, Canada

Conference Record IEMC 96
Vancouver, British Columbia, Canada

IEEE Catalogue Number: 96CH35979

ISBN: 0-7803-3552-X
Softbound Edition
0-7803-3553-8
Casebound Edition
0-7803-3554-6
Microfiche Edition

Library of Congress Number: 96-77286

Conference Committee

Preface

Dundar F. Kocaoglu

Program Chair, IEMC'96

Professor and Director, Engineering Management Program
Portland State University, Portland, Oregon 97207-0751, U.S.A.

Every era in history can be remembered with some significant events that establish the color and characteristics of that time period. Some of these events are inventions such as electricity; some are expeditions such as the discovery of the New World; some are exploratory breakthroughs such as the first Transatlantic flight; some are technological breakthroughs such as the development of the computer; some are destructive events such as a major war; some are constructive events such as the flight to the moon. Twentieth century is full of these significant events. Their numbers are constantly increasing. In fact, when we look at the last decade of this century it is difficult to choose a single event as the one which has left its imprint on that period. Rather than a single event, it is more realistic to identify 1990s with the evolution of a new concept; a new way of doing business. That new concept is the emergence of the virtual enterprise. It is the result of three major technologies; communications, computer and energy technologies; getting together. It is the beginning of obsolescence for many of the traditional approaches to planning, organizing, communicating, directing and control. In short, it is the new approach that is bringing sweeping changes to our life and how we conduct it, as well as to technology and how we manage it.

When IEEE Engineering Management Society was choosing the theme for the 1996 conference to reflect on the characteristics of this era, it became obvious that there was nothing more appropriate than the management of virtual enterprises. The conference attracted worldwide attention. A large number of papers was received. About 180 were accepted. This proceedings contains the ones that were submitted in full length. Several others are also included in the conference as presentations without the full length papers being available for publication.

We believe that this volume contains some of the best papers on management of virtual enterprises. We hope it will make a contribution to research, education and implementation of technology management focused on the virtual enterprise.

Table Of Contents

Intelligent Software Agents as Tools for Managing Ethical Issues in Organisations Caused by the Introduction of New Information Technology

Ali Reza Kian Abolfazl
Fax: (+45) 8615 5032

Department of Organisation and Management
The Aarhus School of Business
Haslegaardsvej 10
DK-8210 Aarhus V
Denmark
Email: Kia@hdc.hha.dk

In this paper we reconsider the notion of value in the context of Information Technology (IT), and show how the introduction of new IT can change values in organisations. Subsequently we introduce the notion of Intelligent Software Agents (ISAs) *as* autonomous, social, reactive, proactive *and* subservient *software computer systems. We further look at how the introduction of ISAs to knowledge-intensive industries can help human agents in their understanding of the development of new values while the new IT is taking place in their workplaces, and how the already existing values in the organisation can be dealt with at the same time.*

Introduction

The introduction of new IT, as any other kind of new technology, in organisations usually changes or even destroys many of the values attached to the work done by the employees in the organisations. By values we simply mean the way in which the organisation is defined through the employee and by his/her performed work. In another word, the value here is the way the employee him/herself feels that he/she contributes to the existence of the organisation and consequently defines him/herself. Thus, the introduction of new technology in organisations not only changes the organisations themselves but also the way employees identify themselves with the organisations. Therefore, managers when working on the introduction of any kind of new technology, are confronted with ethical issues. Should they abandon the introduction of IT in order to prevent dissatisfaction of employees, or should they simply ignore the workers' problems and replace those who cannot keep up with the rapid changes of the nature of IT? In the following we shall see how these managers can be helped.

Values

Human beings "are socially constructed. What we mean, how we see the world, our values and valuables are all produced, learnt or acquired by and through interacting with others. Our means of communication such as natural languages cannot be learnt without this kind of interaction. Social norms defining the acceptable or appropriate ways to behave and opinions to hold make the boundaries within which we acquire and develop our behaviour and behavioural patterns. Without these norms, we simply cannot maintain our ways of communication, which are in the core of our existence as species " (Abolfazlian, 1996a). But on the other hand we are the true makers of the same History, culture and society that are supposed to produce us:

> A process going on between man and nature, a process in which man, through his own activity, initiates, regulates, and controls the material reactions between himself and nature. He confronts nature as one of her own forces, setting in motion arms and legs, head and hands, the natural forces of his own body, in

0-7803-3552-X/96/$5.00 © 1996 IEEE

> order to appropriate nature's productions in form adapted to his own wants. By thus acting on the same time changes his own nature. He develops the powers that slumbers within him and subjects them to his own control (Marx1, 1906 - 1909, p. 197).

All human concepts such as values, knowledge etc. are made as a result of interaction with nature and through her own forces. Thus human activity2 is what makes the very same society and its societal rules that consequently draw the boundaries in which human beings will find themselves defined. But these activities are mediated through tools, specially sign systems3. Human beings use a variety of different tools and gadgets to master their mental processes. We call them *psychological tools* or *instruments*:

> Psychological tools are artificial formations. By their nature they are social, not organic or individual. They are directed toward the mastery or control of behavioural processes.....just as technical means are directed toward the control of processes of nature (Vygotsky, 1979).

Natural languages, algebraic symbol systems, schemes, diagrams, mechanical drawings and different aspects of IT are some examples of psychological instruments. By using these instruments and making them part of our behavioural processes, they alter the entire flow and structure of mental functions. They do this by forcing us to re-structure our mental functions and activities in accordance with their structure just as for instance using technical tools determine the form of labour operation and thus alter the process of a natural adaptation. Thus these psychological tools are a defining part of all human concepts, especially values. As mentioned before, human activity is mediated, especially through psychological tools. A human worker's values in and toward an organisation are not only realized and mediated through his/her technical/psychological tools but are also defined by the very same tools. Any change of the mentioned tools, e.g. a secretary's desktop publishing system or a knowledge worker's interface to for instance IT, will change or even destroy the values connected to the work done by the employee. Organisations are introduced to new technology, especially new IT, as a natural consequence of the ever continuous stream of new tasks within the boundaries of organisations. Factors such as *Information anxiety,* i.e. a human agent's negative emotional response to the increasing volume and diversity of available information in our modern Knowledge Societies and organisations are some of the factors that rapidly influence the way we work and our values in the organisations. As a consequence, we need to take these factors into consideration while managing the ethical issues that arise in the organisations because the introduction of new IT changes the aforementioned values. We propose a new kind of specially designed psychological tools, namely Intelligent Software Agents as a useful support system in these managerial situations.

Intelligent Software Agents

Intelligent Software Agents (ISAs) are essentially computer systems whose constituent parts have a special relation to one another and to the environment in which they are realized. What makes the above mentioned relationship special is some especially developed features that these software systems have to possess in order to be identified as ISAs. Although different research groups have different ideas of what exactly these features are and how they should be developed in the agents, there seems to be an

---

[1] In general, I do not agree with Marx's ideas with regard to Economics, and I do not see myself as and am not a proponent of Marxism.

[2] Activity is the nonadditive molar unit of life for the material corporeal subject ... a system with its own structure, its own internal transformations, and its own development.

[3] The interested reader will find a more detailed description of the nature of human activity and its role in human conceptualization in (Abolfazlian, 1996b; Wertsch, 1979).

agreement among almost all of these groups on which features agents, at least, must have. As we have seen in (Abolfazlian, 1996a), they have to be:

- *Autonomous*
- *Social*
- *Reactive*
- *Proactive*
- *Subservient*

An ISA is *autonomous* if it can operate without neither direct human interference nor any interference from other agents working in the same environment. It should also have control over its own actions, plans and internal states. The ISA should be *social* in that it has to be aware of its surroundings. It also has to be able to interact autonomously with other agents, human or non-human, using an interaction tool or gadget. Even though verbal languages are mostly used as tools of interaction by these type of agents, any sign system will suffice for our purpose here. We shall look at some of these systems further on. It is worth emphasizing again that the aforementioned interaction has to be made autonomously by the agent itself otherwise we cannot talk about social interaction with regard to these type of agents.

ISAs should be *reactive*. By reactivity here we simply mean that the agent has to be able to detect changes in its environment and autonomously respond (react) to these changes in the most appropriate way. What appropriate means here is again highly dependent on the surroundings of the agent and specially the other agents in and the societal rules of the society in which the particular agent operates. We should again notice that the respond of the agent should be made autonomously.

ISAs are not just supposed to be reactive as defined above which implies that they have to wait for changes in their environment in order to first detect them and then to react to them autonomously. They must also be capable of taking initiatives and set up sets of actions that will satisfy their goals. In other words, they have to be *proactive*.

ISAs should be *subservient* in that they have to work on behalf of other agents, human or non-human. By defining subservient in this manner we take a position which is very close to the one taken by the Economics when talking about Theory of Agency. Although this feature is not one of the more standard ones we shall see how this will fit within our framework in a very natural way.

Let us now briefly look at an example of these Intelligent Software Agents4.

Softbots

The Softbot (Software Robot) is an ISA running on the internet (Etzioni et.al, 1994). It is a fully implemented Artificial Intelligence (AI) device running on UNIX operating system. "The softbot interacts with its environment, specially different internet resources, through a UNIX shell and the World Wide Web (WWW). As effectors, softbot uses a wide range of file manipulation commands such as 'ftp', 'telnet' and 'mail'. As sensors it uses 'archie', 'gopher', 'netfind' and several other internet facilities. It has a very expressive interface that gives its interaction with the users (human agents) a much more social character. On the other hand it is able to dynamically choose the internet facility to be used in a certain situation. It also is able to autonomously choose the order in which different internet facilities should be used. For instance if it wants to find Ms. Beverly Williams' email address it may choose the 'netfind' facility to start with. But it knows that in order to use the netfind it has to feed it by an institution name or address. Thus it starts by looking in the bibliographic databases that it has access to in order to find an article or technical

[4] For more examples, look at (Abolfazlian, 1996a).

report by Ms. Williams. It knows that if it finds the article it will have the name and/or address of the institution" (Abolfazlian, 1996a).

ISAs as support systems

Now it should be obvious that ISAs such as Softbots can help knowledge workers in organisations when facing problems as Information anxiety (IA). By introducing Softbot, many of the symptoms of the IA such as losing the confidence in him/herself, decreasing in efficiency and so on disappear. Automatic filtering of electronic mail and news, sending routine messages such as meeting reminders or talk announcements, and negotiating a meeting time with others are some other tasks causing IA. ISAs can also help employees to control these situations. In (Abolfazlian, 1996a) we have shown some examples of the ISAs already solving many of these problems. By using ISAs in this manner, the knowledge workers can concentrate on their job, and the value of the labour performed by them is going to be the same even though the subject of the labour is changed.

On the other hand, changes not concerning the subject of the labour but the technical tools, especially Information Technological tools (ITTs), used in order to perform the labour can also be overcome. The problem here is a consequence of rapid changes of ITTs as a result of ever-growing needs in organisations and increasing competition between manufacturers. It should be obvious that the employees, in different levels of organisations, who are introduced to the same new ITTs can build their competence by relying on the collective knowledge distributed among them. A society of ISAs can help in these situations5. Every user gets his/her own agent. Then the agent becomes personalized by relying on the correlations between different users. The agent memorizes its own user's action/reactions in different situations. Then agents start communicating with each other. Each agent finds those agents that are most correlated to itself, i.e. those agents whose users have similar patterns of actions find each other. For example, if A and B are two correlated agents and there is a situation that B's user has reacted on but A's user has not seen or heard of, then A comes back with a recommendation on that particular situation. In this way the ISAs work as mediators of the labour by operating as bridges among the different competences needed to function on different ITTs. When this indirect interaction with new technology is internalized, the workers can operate on the new tool with or without help of ISAs. One could compare the role of ISAs in these situations with that of the supporting wheel of a bicycle. In this way the transition from old to new technology is going to be completed without any loss of the values.

Conclusion

In this paper, we looked at how the introduction of new IT in organisations changes the values in them which consequently arises some ethical issues concerning the management of these new technologies within organistaions. Further on, we introduced ISAs and showed how the introduction of ISAs to workplaces, especially knowledge-intensive industries, can help human agents in their understanding of the development of new values while the new Information Technology is taking place in their workplaces, and how existing values in the organisation can be dealt with in the same time. This resulted in a new way of managing ethical issues concerning new IT.

References

Abolfazlian, A.R.K., "The need for Agents", in Bots and Other Internet Beasties, Joseph Williams (ed.), ISBN #1-57521-016-9, Sams.net, MacMillan Computer Publishing, USA, 1996a.

[5] For a more detailed description of these managerial situations look at (Abolfazlian, 1996c).

Abolfazlian, A.R.K., "On Activity Theory in Cognitive Systems", to appear in the ESSCS - The Journal of European Society for Study of Cognitive Systems, 1996b.

Abolfazlian, A.R.K., "Management of new Information Technologies in Organisations by Intelligent Software Agents", to appear in proceeding of the COST A3 Final Conference on Management and New Technology, Madrid, Spain, 1996c.

Etzioni, O., and D. Weld, "A Softbot-Based Interface to the Internet", Communications of ACM 37(7): 72-76, 1994.

Marx, K., "Capital", Kerr & Company: Chicago, USA, 1906- 1909.

Vygotsky, L.S., "The instrumental method in psychology", in The Concept of Activity in Soviet Psychology, J.V. Wertsch (ed.), 134-143, M.E. Sharpe, Inc. : Armnok, NewYork, USA, 1979.

Wertsch, J.V., "The Concept of Activity Theory in Soviet Psychology: An Introduction", in The Concept of Activity in Soviet Psychology, J.V. Wertsch (ed.), 134-143, M.E. Sharpe, Inc. : Armnok, NewYork, USA, 1979.

Tentative Guidelines for Publications Rights

Imad M. Alatiqi, Suliman Diab,Said Mansour,
Firas Rasoul and Hamdi Shalabi

This paper presents an empirical formula for the calculation of contribution weights of individuals participating in a group research.The objective is to identify the most critical elements that leads to documented results of the investigation.It was found that these elements are: 1 .Initiation of Research Idea. 2. Execution of the research idea. 3.Services of the research idea. 4.Data Analysis. 5.Writeup.These steps of the research process were identified and discussed in the light of their importance for the production of a contribution measure .The difficult problem of giving credit to participants in the investigation is then addressed.A linear mathematical function is adopted to allow the quantification of individuals contribution in every step of the investigation.We then defined a function called the degree of involvement and used a limit of this function that may be used to judge issues like publication rights ,patents rights and the like. The paper presents an example for the calculation of this function.

Introduction

A publication is usually a result of systematic investigation or a record of it. It appears in different forms and shapes. It may appear as a progress report accounting just for the status of an undertaken project or effort. It may appear as a technical report telling the story of a technical or scientific achievement in a project or R&D. It may appear as an article in an annual report or a scientific communication in a conference or a refereed journal summarizing a scientific achievement or a valuable experience. Or it may appear as a patent recording a scientific achievement.

Such systematic investigation includes the following: It includes the initiation of an idea by identifying it and proposing a conceptual solution for it. The initiation can be either requested by a client or innovatively plundered in house. Then implementation and execution of the solution are carried out after detailing the steps. Then certain services are conducted afterwards to complete the picture of the collected data. After that data analysis comes where results are manipulated and interpreted to get conclusions and recommendation. Finally reporting is due to reveal the course of work and type of results and solutions achieved. Reporting is done in one of the forms of publications mentioned above.

Contributions to such work can be divided into two categories (a) authors and (b) participants with acknowledgement.

The systematic steps and contribution to them to establish guidelines for publication rights are detailed below.

1. Initiation of Research Idea

Idea initiator is the person who usually develops the idea for research project and identifies conceptual methodology and resources to carry out the research. If the suggested idea is of a general nature then the assigned person who identifies the conceptual methodology and develops the idea into a research project will have the credit of being the initiator of this idea. This principle can be applied on research ideas resulting from discussion, communication, conferences and exchange of ideas. Such research ideas may also be suggested by various clients and governmental offices through communication channels and transmitted to officer in charge who in turn selects the proper Project Leader to conduct the work related to the preparation

0-7803-3552-X/96/$5.00 © 1996 IEEE

of preproposal and proposal. Therefore, the name of the Project Leader and/or Principal Investigator will appear on these documents and all publications resulting from R&D work, in addition to the idea initiator.

For such an idea the initiator has the right to request his/her name to be considered as co-author on any publications or patents related to the subject even though the R&D work is not carried out by him/her. This is applicable only to innovative and original work. In several cases, the idea initiator and probably Project Leader are the same person and will need advice and feedback from the scientific management in structuring the preproposal and proposaL In general, consultations can be acknowledged, unless they materialize into valuable contribution to work or idea, then a weight function can be applied. If there is a drastic change in the methodology by the new Project Leader and/or Principal Investigator then the initiator's name is referred to and/or acknowledged only. This can be considered for a specific type of idea, i.e., innovative or patentable idea.

II. Execution of the Research Idea

Implementation and execution of a research idea, after it matures into a project takes the following steps:

1. Detailed methodology based on the conceptual idea and the conceptual proposed solution of the initiator(s).
2. Equipment design or equipment acquisition to carry out the planned methodology according to the objectives of the proposed work.
3. Construction, installation and commissioning of the needed equipment to assure precision and reliability of the equipment in carrying out the experiments.
4. Running the experiments detailed in the methodology, supervision and preliminary data evaluation.
5. Data collection and tabulation.

For research carried in social sciences, experiments may be done on human subjects using surveys designed for the specific research purpose.

The above steps are usually carried out by Project Leader and associated scientists, assisted by technicians and professionals of different departments, programs and divisions this is especially true in a matrix oriented organization. Although these steps do not carry the same weight of innovation, yet they are very necessary to implement or execute the overall idea or the work. The degree of contribution that stems from this part is essential, however, when the work is conducted by several individuals a weight function can be applied for evaluating their relative contribution.

III. Services of a Research Idea

Service is another parameter that contributes to the outcome of a research idea. A service is defined as a standard short-term activity in which deliverables are clearly defined prior to commencement. Examples are laboratory tests and measurements, data entry and display.

The role of the authority or participant who provides the service to a project depends on the degree of participation and level of innovation. This can be subdivided into three categories:

1. In the case that the type of service is clearly defined and the procedure is a routine one then the contributor may be acknowledged. The acknowledgement can be addressed to the service unit such as a central research laboratory for conducting

the analysis with reference to the section or the department of the contributor; or it can be addressed to the contributor by name with reference to his/he affiliation.

2. In the case that the purpose of the requested services is known and is given, however, the types and details of such services are not known to the requestor. The needed types of service are then decided upon and conducted by the service unit with the known techniques available, then the contributor can be weighed.

3. The third case is when the contributor form services has to develop the methodology for analysis and/or to participate in data interpretation. In this case, an official request should be made to the service department asking the researcher or the professional to be assigned to that job. The contributor's name then appears in the report and/or articles as an author of that task; in this case the services can be considered an integral part of execution.

The above three rules are applied to any type of technical service whether provided by a laboratory, Computer Center, Electronics Workshop or Mechanical Workshop.

Other types of services such as editing, printing, funding, reviewers, suggestions, etc. can be acknowledged or referred to only.

IV. Data Analysis of a Research Idea

Data analysis is a fourth parameter in the research idea story. It covers the following:

1. Monitoring the execution and checking the produced data against published literature or scientific principles to evaluate the achieved results.

2. Relating and correlating the different parameters to establish the cause-effect relationship.

3. Producing the functional relationships in the form of a graph, a correlation or even a model.

4. Interpreting the achieved results and translating them to pragmatic practices and useful suggestions to the client (conclusions and recommendations) or even to open new avenues for further research and more exploration.

Such analysis is of paramount importance to draw the conclusion for the implemented research idea. The type and degree of involvement and the level of innovation of all points do not carry the same weight; and the analysis can be subdivided into two parts:

a. Analysis while the project is going on and a write up has not been finalized, yet. In this case the analysis is still within the domain of the principle investigator(s) or task leader(s) and the authorship rights can be evaluated using a weight evaluation function.

b. Analysis after the write up has been published and the information has become a public domain. In this case the original authors should be consulted regarding the possibility of contribution to the new analysis and write-up. For the new endeavor the weight evaluation function can be used to determine authorship and name arrangements.

V. Write Up

The final parameter in this series is the write up. This covers reporting the systematic investigation. Such reporting can be subdivided into:

a. Progress Reports:

When different technical tasks are reported by the principle investigators to the project management and line management, and a progress report is produced from the different reported tasks. The person(s) generating such a report is entitled for authorship. Technical reports with the authors' names are kept as an appendix to the produced report, if necessary.

b. Technical Reports:

This type of reports such as project's final reports, semi-annual and annual reports is the responsibility of the task leader, project leader and the participants who carried out the total work. In this case they are entitled for authorship. In case of reports mandated by a contract, the project leader should oversee that the write-up conforms to the project requirements. If the wAter is not an investigator, e.g., a project leader, then either the writer has to be acknowledged as an editor or the write-up has to be modified as in (a), i.e., to be considered as progress report and the technical reports are to be kept as an appendix.

c. Publications:

Publications result from genuine outcome of a project, R&D or part of them published in refereed journals or conferences proceedings. The participants in this work are entitled for authorship. The degree of participation is decided by the task leader/project leader who is responsible for the work leading to such publications taking into account weight criterion, if necessary. The project leader should coordinate publication efforts to ensure fair contribution to team members.

Criterion

The above mentioned activities do compose the body of the work and their completion is necessary for the overall outcome whether a progress report, a technical report, a scientific publication or a patent. However, the participation in these activities and the right to authorship or acknowledgement can be quantified by using some kind of an evaluation function, whenever it is not explicity designated.

The steps of involvement can be detailed at the time of preparation of the project or even initiating the idea and its conceptual solution (proposal or even preproposal period before manpower is assigned). The number of people to be involved can also be stated since it is part of the proposal requirements. Each item in the activities can be assigned some kind of a weight value and a formula as the following can be used.

$$P_J = \Sigma P_I W_U \qquad (1)$$

where

$P_J =$ % weight of each activity or step 1 carried out by a participant.
$W_u =$ weight of participation of jth individual in 1th step
$P_I =$ total number of points for each participant

$P_T = \Sigma P_J$

Where
$P_T =$ the total number of points for all participants in the project or task to be reported.

$\frac{P_J}{P_T} =$ the degree of involvement of each participant.

$\frac{P_J}{P_T}|Min$ is a minimum degree of participants to be pTIML,l agreed upon and it depends on number of participants. Naturally, the

$$\frac{P_J}{P_T}|Min \leq \frac{P_J}{P_T} \qquad (2)$$

For example, if $\frac{P_J}{P_T}|_{participant} \leq \frac{P_J}{P_T}|_{\min imum}$ then

the participant is just only entitled for acknowledgement. If it is more than the minimum then the participant is a co-author and the name appears on the publication as dictated by the degree of involvement. In case of dispute the project leader shall be responsible for resolving it according to the above guidelines. In case such dispute persists then an arbitration committee can be formed to settle the dispute.

Concluding Remarks

The steps described above were discussed at a management meeting at the Kuwait Institute of Scientific Research.It was noted in the meeting that those steps constitue the backbone of any scientific publication and contribution in one or more of these steps should be considered when credit is to be given for publication.Since the guidelines were adopted we have experienced reduction in sensitivity regarding publication of scientific efforts.Although no formal cases were submitted for conflict resolution within the authors division,personal consultation between project leaders and the division director reflected satisfaction that such guidlines have been made available . The degree of involvement criteria were especially useful when marginal contribution was involved.

Acknowledgement:The authors would like to thank the Kuwait Institute for Scientific Research for Permission to publish this article.

Appendix: Example of Weight Calculation

Five persons participated in a research work. The weight value for the completion of the project are as follows while the participants have put time in each step as shown in Table 1.

Steps	Initiation	Execution	Services	Data analysis	Reporting
Weight:	0.3	0.2	0.1	0.25	0.15

Table 1

Participants	Initiation	Execution	Services	Data Analysis	Reporting
J	%	%	%	%	%
1	35	10	-	10	10
2	30	40	-	40 70	
3	20	30	30	20	20
4	10	20	-	5	
5	5	-	70	25	
	100	100	100	100	100

P1 =	(0.35)(0.3)+(0.1)(0.2)+(0)(0.1)+(0.1)(0.25)+(0.1)(0.15) =	0.1650
P2 =	(0.3)(0.3)+(0.4)(0.2)+(0)(0.1)+(0.4)(0.25)+).7(0.5) =	0.3750
P3 =	(0.2)0.3)+(0.3)(0.2)+(0.3)(0.1)+(0.2)(0.25)+(0.2)(0.15) =	0.230
P4 =	(0.1)(0.3)+(0.2)(0.2)+(0)(0.1)+(0.05)(0.25)+(0)(0.15) =	0.0825
P5 =	(0.05)(0.3)+(0)(0.2)+(0.7)(0.1)+(0.25)(0.25)+(0)(0.25) =	0.1475
		1.00

Choice 1

$$if \ \frac{P_J}{P_T}\Big|_{min} = \frac{1}{J_{max}} = \frac{1}{5} = 0.2$$

Then
1=acknowledgment
2=co-authorship
3=co-authorship
4=may be acknowledged
5=acknowledgement

Choice 2

$$if \ \frac{P_J}{P_T}\Big|_{min} = 0.1$$

Then
1=co-authorship
2=co-authorship
3=co-authorship
4=acknowledgement
5=co-authorship

The project leader may select between choices 1 or 2 using common sense on importance of contribution made by the border line participants.

Management Information and Control System for Research Administration

Imad M. Alatiqi*
Ali Mohammad Akbar
Sudhir Padlakar
(965)4811188

Faculty of Engineering ,Kuwait University,
P.O.Box 5969 Safat,13060,Kuwait

This article presents a management control system and a management information system for the problem of research administration in a service organization. We assume a centrally managed process with a decentralized resource and operation. A model control system is developed that emulates the management controls in a simplified fashion. The administration process is assumed to be performed as a cascade system with the management function and project implementation function are in parallel. The research administrators and project investigators are connected through the control process and through the information exchange and personal communication process. The Management Information System is designed to facilitate information flow and improve funds processing from the research administration unit to the investigators. The system was developed on FoxPro platform through a local area network.

A-Introduction:

This article presents an MIS and management control system for a central research administration unit at a service organization .The description that follows pertain to the specific case of Kuwait University. Nonetheless the concept and treatment is generic and can be applied to similar institutions. The research administration unit is responsible for funding, reviewing, monitoring and support of funded research at Kuwait University. The Research Administration (RA) is centrally directed by the vice president for research and graduate studies who also oversees the graduate school and the Scientific Publishing Council.

The MIS has evolved through continuous effort of the RA to upgrade its system and make it transparent to the staff and faculty such that the project approval ,monitoring and
support is done in the most efficient manner. Through office automation and interconnection the management and support staff can service the faculties faster and minimize the waiting period for project approval, equipment procurement and travel procedures.

The purpose of this article is to give a concise description of the procedures that are developed for management of reassert at Kuwait University, and the management information system that has been developed to facilitate the execution of these procedures.

B-The Research Administration Organization:

1. Overview:

The Research Administration (RA) is a funding, support, monitoring and evaluation organization for Kuwait University at large. It operates under the supervision of the Vice Rector for Research and Graduate Studies, assisted by one or more part time directors selected from faculty members. The funding committee is headed by the vice rector for research with membership of 5 or 6 of the most heavily funded faculties. The representation is maintained by the dean of the faculty(college) or the associate dean for research and academic affairs. Traditionally ,the faculties of medical school, engineering ,science, commerce and arts were represented in addition to the RA directors. The funding committee is the highest authority at Kuwait University that is responsible for approval or disapproval of funding for proposed research. It has impressively maintained independent judgment over the years and its

0-7803-3552-X/96/$5.00 © 1996 IEEE

decisions are considered final. Considering that the funding committee meets on the average bimonthly it is inevitable that its decisions require intensive ground work and case preparation. This is one of the most significant duties of the RA. What follows is a brief description of the RA infrastructure. The RA is organized in 3 departments and 2 smaller units, as shown in Figure 1.

Office of the Vice Rector for Research & Graduate Studies

RESEARCH ADMINISTRATION

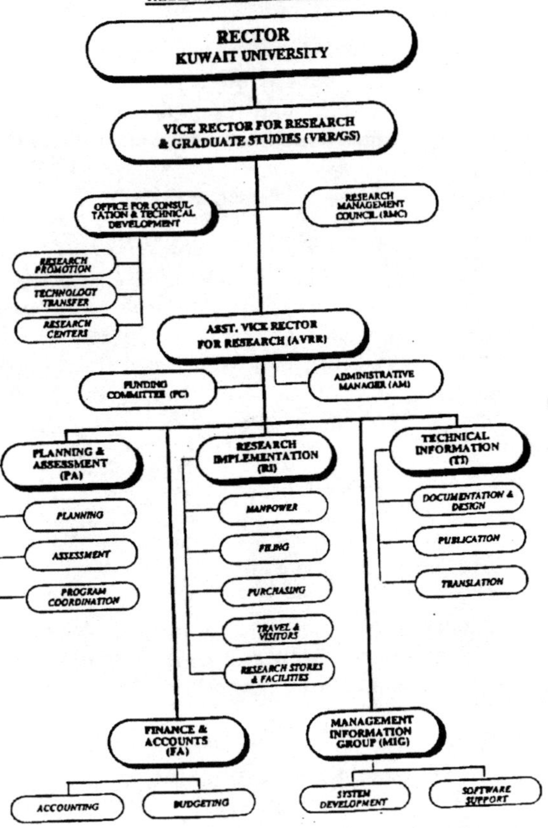

2. Project Planning and Assessment Division(PPA):

This division is responsible of receiving the project proposal from the faculties. The faculties do scrutinize every proposal for technical merit and relevance at the departmental research committee level. The RA only receive proposals that are screened in home department , and then start processing them for further review. The backbone of the RA proposal review and assessment is the ***referee system***. The referee system consists of a confidential data base of over 12,000 specialists in various field which was accumulated over many years and updated periodically(Akbar,et.al,1993). Each proposal is coded by subject area and an analyst matches the project code with the referees data base. The(PPA) division head recommends three reviewers from the list to the RA director who in turn reviews the recommendations .Upon approval the PPA division prepare the mailing document for the referee which includes the proposal, the authors vitae and an evaluation form. Kuwait university follows a point system which includes such criteria as technical merit ,originality, budget soundness , methodology and relevance to national and regional interest. The referee is given two months to submit his report. Upon receipt of all the evaluations the PPA analyst summarizes the referees comments, calculates the average grade and submit the case for review by the upline and hence by the funding committee.

The PPA division also is the secretary for the funding committee and is responsible for further processing of the proposal. This may be preparation of the grant letter for approved project, preparing a rejection letter for the principal investigator(PI), or asking him to modify the proposal according to the committee guidelines. For the most part the PPA department sends the referees evaluation to the PI and ask him to modify the proposal or address the referees comments. The PI's response becomes part of the proposal document and is included in the final analysis. Occasionally the RA would resubmit modified proposals for further evaluation if it expects the revision would enhance the approval chances.

After commencement of the project the PPA also monitor its progress through evaluation of the annual reports and the final report which is refereed by the same reviewers. Upon completion of the project the PPA enters the final project evaluation, resulting publications and other relevant data in the principal investigators file. The PI's file and research history is an integral part of any new proposal case that may be submitted later.

3.Project Implementation Division(PID):

This division takes over from the PPA once the principal investigators signs the grant letters. Its services include manpower supply ,travel arrangement,equipment materials supply .The *manpower* supply involves securing the research assistants and other personnel mentioned in the project plan. The PID receives employment requests from the PI and assist him in finding the required staff either by advertising in the local newspapers or abroad(in limited cases).once the PI identifies a suitable candidate the PID contacts this candidate and help finalizes his employment papers including visa formalities if needed. Occasionally the support staff may be hired on a part time basis and secondement arrangement have to be made.

The *travel* unit is responsible of processing travel arrangement for the project staff which is mainly translate in calculating the travel allowance and securing travel mission approval. Since travel is considered a sensitive budget item,it is handled with extreme care such that the mission can be rendered in the most appropriate and beneficial manner. The travel unit also secures requests for visiting experts travel to Kuwait University which includes visa arrangements ,airfare , accomodation,and honorariums.

The *Purchasing Unit* is responsible on securing equipment and consumable items required by the project ,as given in the project plan. Special attention is given to avoid duplication of equipment in the same department if the equipment is deemed to be of multiservice value. The RA inventory includes major research equipment like scanning microscope , X- ray analyzers ,pilot plants and material testing units. These equipment and
others are deemed primarily for research purpose and the RA does not concern itself with furnishing student laboratories. Other major facilities supplied by the RA include PC's for research purpose,chemicals and supplies and books. The RA also maintains a central store for the frequently used chemicals and materials .

4.Technical Information Division(TID):

This division is the publishing arm of the RA. It is responsible of publishing the research output of Kuwait University supported by the RA. It does not currently publish final reports or scientific papers. Rather,it carries a series of publications for the sake of bibliography relevant to the research effort at the university. Such publications are aimed at informing the research community about the current research trends at KU and providing short but concise references. The most widely circulated publications are the Research Abstracts series and the Research Newsletter. Other publications related to KU research effort are also pursued.

The Technical Information Division operates maintains a Mac desktop publishing capability that includes editing ,formatting,graphic design,proof reading and distribution. It produces camera-ready documents that is later sent to the printers.

5.The Management Information Group(MIG):

This is the group responsible for system development and maintenance within the RA. It consists of three programmers/system developers operating a Novell Network linking 40 workstations. This group is vital to the RA operation and it offers in-house maintenance and troubleshooting,system development and training. The MIG group carried a major load in developing the Management Information System for the RA which will be described in the next part of this paper.

6.The Accounting Section :

This section is responsible on monitoring the projects accounts,processing funds requests for petty cash and maintaining projects budget reports. The accountants also participate in preparing the annual budget plan for the RA. Advance solicitations from the university faculties are filled on forms provided by the RA and processed by the accountant to give an estimate for the required future budgets. The accounting section also prepares periodic project expenditure reports for the principal investigators.

C-The Management Information System

1.The Control Strategy:

A simple definition of a system comprises an input ,a value adding process ,an output and a feedback loop(Belardo and Pazer 1995).Management information systems typically has raw information as input,the MIS software and its implementation as the process .The output could be envisioned as the research results realized from the project. The input is the sum of the principal investigator (PI) and his team efforts and the supporting operations of the RA staff(controllers).The directors decisions and funds allocation is implemented by the controllers through execution and update of MIS. The controllers also perform the Research Administration's(RA) routine operations. The PI receives funds and contract terms from the directors and implement it in his domain trying to match these terms to his specific objectives. The research results are a combined effort from the PI domain and the RA domain. This situation leads to a cascade system model applied to parallel processes and series processes .This "hybrid cascade" control system is shown in figure 2. This scheme is justified largely because the lower rank officers has to take higher frequency actions than the directors. The actions of the assessment,implementation and accounting groups may be reviewed and modified if necessary by the directors. The ability of the PI to communicate with the various administrative levels of the RA has contributed significantly to the efficiency of the RA operations and projects execution.

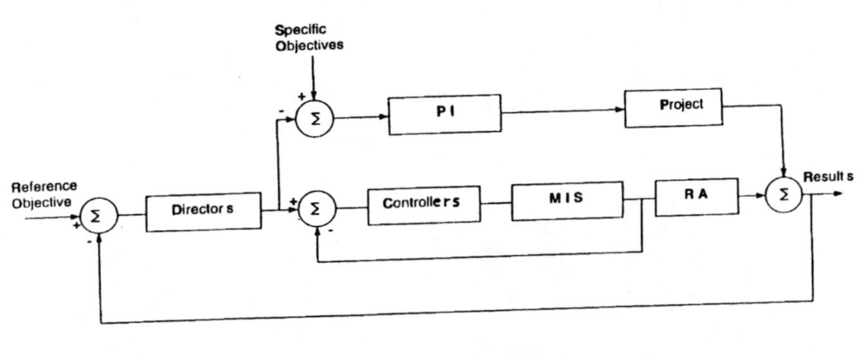

Figure (2) : The Managment Control Strategy

2.The MIS Project:

The Management Information System of the Research Administration was proposed long ago. The suspension of Kuwait University operations in August 2nd 1990 due to the Saddam's forces invasion halted all activities. After the liberation the emphasis was on to rebuild KU infrastructure which includes the RA offices and operations. The university generously granted funds to supply the necessary PC's and other infrastructure equipment. The MIS project commenced in 1992 and was carried out entirely by the staff of the RA. In order to provide incentive for quick completion of the project the RA assigned three staff members from the management information group. The group worked extra hours every working day for 6 months till the project was completed. The RA director assumed project director role and the Vice Rector for Research and Graduate Studies gave the directives and participated in project reviews. The full staff of the RA was aware of the projects through formal introductions and every section was consulted for its specific needs. Early participation of all sections was deemed necessary for the integrity of the project and to keep the motivation level for subsequent application(Globerson et.al.1995).The PC-network available within the RA made it easy to develop the system on the server and the share it by all parties. The project was culminated by the documentation of the MIS User Guide(Alatiqi 1994) after B-testing was rendered in cooperation with the RA staff.

3.The Information Systems:

The MIS covers all the subsystems handled by the RA except technical information department and filing operations which may be developed later(Figure 3).The MIS clusters around the assessment and planning subsystem. Basic data of the PI and his project are entered through this subsystem. Other subsystems cannot accept this data until this basic information is fed to the MIS. The accounting system is responsible for all project expenditure auditing. It passes on expenditure data to assessment. Project expenditure is distributed over 4 heads of accounts: running cost,manpower cost,travel and visitors,and equipment cost.

IMIS DATA FLOW CHART

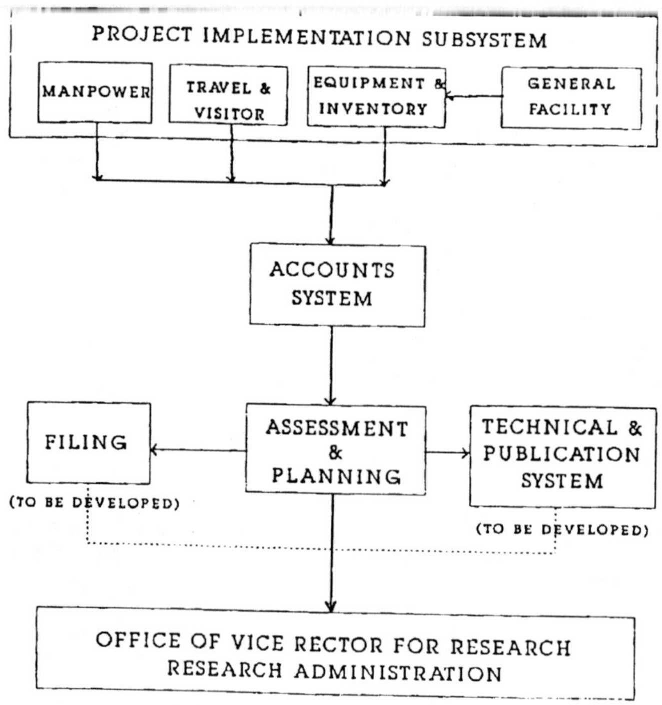

The manpower subsystem and the travel and visitor subsystem,both are part of the project implementation option, passes on their expense information to the accounting subsystem. Inventory is maintained of all consumable items and equipment purchased for the project ,as well as general facility equipment. Expenses of consumable supplied to the PI are passed to the accounting sub system as running cost. The equipment purchases for the PI within the project budget will be posted against equipment budget. The general facility equipment will have a project title and number on its own and its capital cost will be posted against that project. The general facility project in its present form was initiated after the invasion in order to replace major equipment lost to the invaders and provide facility for maintenance and operation of these projects. Because most department have limited manpower allocation, general facility projects have maintenance allocation under running cost(after warranty expiration)and manpower allocation for technicians under manpower cost item.

The user may access any subsystem by entering its activation character,but data modification is restricted to authorized personnel each in his own subsystem.

D-Concluding Remarks:

We tried in this paper to devise a model for the management control system of the research administration process in a service organization with a typical application of a middle size university. This system has been in operation at Kuwait University since 1980. We believe this system has many unique features that has contributed to the remarkable success and expansion of the research efforts at KU. The most significant feature in our view is the ability to centrally fund ,monitor and service the sponsored research at KU using cooperative management and decentralized resources. We also described the newly adopted MIS that has added an important efficiency feature to the RA operations. We believe that the interaction between the faculty and the administration has a significant effect in the success of the research process. The system may be expanded to for-profit organizations by inclusion of additional elements like financial rewards, development processes and commercialization. Many academic institutions have included a developmental style organization were technology,patenting and licensing plays a complimentary role to the research process. No attempt was made in this paper to cover these aspects in order to contain the scope to a focused treatment.

Acknowledgement:

The MIS project has been funded by Kuwait University under grant number RA045. The authors would like to acknowledge technical assistance of Shameem Tayabee,Subrata Basu,and Srinivas Deo.The staff of the RA offered the most helpful comments and support for this project.

References:

Akbar,A.M.; Alatiqi,A.M. and Sharma,P. Research Development at Kuwait University 1980-92.Kuwait University,1993.

Belardo ,S. and Pazer,H.L."A Framework for Analyzing the Information Monitoring and Decision Support System Investment Tradeoff Dilemma."IEEE Trans.Eng.Manage. 42(4):352-359,Nov.1995.

Globerson S.;Shalev,I. and Shenkar, O."The Impact of Technological Change in a Service Organization"IEEE.Trans.Eng.Manage.42(4):382-386,Nov.1995.

Alatiqi,I.M. Integrated Management Information System User Guide.Research Administration(Kuwait University),1994.

Towards a Better Human Resources Management within the Virtual Corporation

Charles-Henri Amherd

Faculté d'éducation
Université de Sherbrooke
Sherbrooke, Québec, J1K 2R1

Zhan Su
(418) 656-2185

SORCIIER
Faculté des sciences de l'administration
Université Laval
Ste-Foy . Québec, G1K 7P4

Diane Poulin
(418) 656-3105

SORCIIER
Faculté des sciences de l'administration
Université Laval
Ste-Foy, Québec, G1K 7P4

This research shows how career development - as far as it contributes to answer fundamental concerns regarding the needs for human resources planning in a virtual corporation; to satisfy them throughout human resources recruitment; to put emphasis on skills during human resources evaluation and to expand those skills with training - represents an alley that should be privileged.

Introduction

The emergence of virtual corporation represents probably the most important change in the organizational universe since the 20's with the model proposed by du Pont and Sloan on giant and decentralized organizations. Amoung the major challenges that will result for management, the one that has to deal with skills and human resources management is certainly the most obvious. In the following pages, we will propose a definition of the virtual corporation with its characteristics and, mentionned the challenges it represents for management and particularely for human resources management.

Definition of the virtual corporation

Since the birth of this concept, virtual corporation has been the object of many definitions. Handy (1990), one of the first authors to write about it, uses the terms «federalism organization» and «shamrock organization» to describe the new links between work and workers. For Davidow and Malone (1992), virtual corporation is like a molecule that is made, broken and remake each time different atoms are cropped up. According to them, virtual corporation is a group formed around a joint project. That group might include a manufacturer, a supplier, a distributor, a sales representative and few customers. Once the project is finished, the group is disolved permitting each member to participate into other projects. These authors talk about an «elactic corporation». As for Poulin, Montreuil and D'Amours (1995), they talk about a network firm who «organizes and manages its operations in a flexible matter by orchestring a synergy between its inside and outside networks to improve its competitivity».

There are two types of associations that we can qualify as virtual corporation: a temporary association of corporations linked by the information technologies towards a common goal; or employees's association that are not constantly linked by a common working place. It's latest situation - intra-corporation - will be discussed in this paper.

0-7803-3552-X/96/$5.00 © 1996 IEEE

Particularities of the virtual corporation

Disruption provoked by the virtual corporation in the world of organizations is total. Even the tendency of the last ten years regarding downsizing and back to core business did not have the same impact. It did not succeed to destroy the vertical caracteristic of management of many groups. The massive lay-off of middle managers have reduced the organizational chart but left the structure intact. For those who favour this concept, joint-ventures and strategic alliances are the first steps that will become a stream of spontaneous partnerships. Whereas the conglomerates based on financial settings, the virtual corporation will be founded on skills e.g. human resources.

According to Byrnes (1993), five key attributes will permit a good functioning in the virtual corporations:

- Excellence: each partner brings its core competence to the effort, to create a best of everything organization. Every function and process could be world-class, something that no single company could achieve.
- Trust: these relationships make companies far more reliant on each other and require far more trust than ever before. They will share a sence of co-destiny, meaning that the fate of each partner is dependent on the other.
- Technology: informational networks will help far-flung companies and entrepreneurs link up and work together from start to finish. The partnership will be based on electronic contracts to keep the lawyers away and speed the linkups.
- Opportunism: partnerships will be less permanent, less formal, and more opportunistic. Companies will band together to meet a specific market opportunity, niches, etc.
- No borders: this new corporate model redefines the traditional boundaries of the company. More cooperation among competitors, suppliers, and customers makes it harder to determine where one company ends and another begins.

Among the fondamental changes that virtual corporation proposes, the one that modifies the verticality of management by abolishing middle managers is probably the most significant. We find two entities or types of partnerships e.g. top management and employees.

- top management keeps as much responsibilities but has less control and privilegies linked to its status. They must learn to live with a contradiction: know more about the company but at the same time exercise less control on its operation.
- employees will have to take on greater decisionmaking.

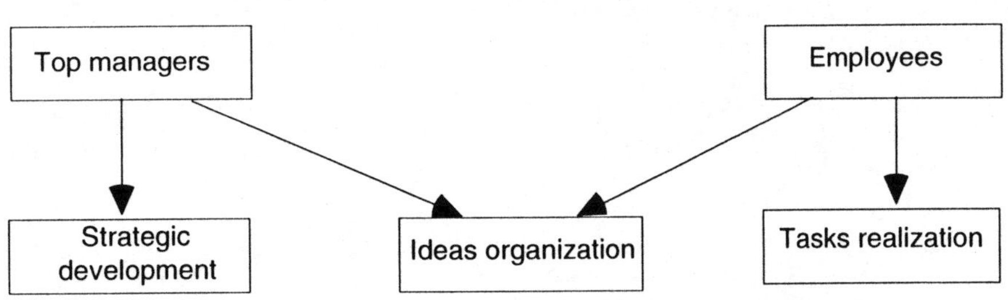

Exhibit 1. Personnel role and responsibility in the virtual corporation.

If the roles of strategic development and tasks realization are respectively devolved upon top managers and employees; the one for ideas organization addresses both.

Challenges of human resources management in virtual corporation

If we admit that the source of virtual corporation constitutes the most important transformation in the world of corporations in the last fifty years, we can measure the extent of challenges that management has to face. In that perspective, each of the five key attributes of the virtual corporation - excellence, trust, technology, opportunism and no borders - will represent a challenge to face. Furthermore, the question of skills will play an important role in virtual corporation; more than the financial aspect found in conglomerates. These challenges will have to be focused before all on human resources management (HRM).

For HRM, these challenges will refer to excellence and trust. They will be hard to face because they have to satisfy two requirements: «catching up» and «leadership».

- catching up means that HRM must be at the same strategic level than all the other functions in the company. In other words, HRM must make up for lost time; where its role was more secondary and human resources considered as a less important resources.
- leadership refers to the fact that human resources hold first place in the virtual corporation.

How to achieve this goal ?

First of all, let's mention the concerns that HRM will have in a virtual corporation oriented towards its skills. Its preoccupation will be:

- anticipating its needs throughout human resources planning;
- establishing how to satisfy them through human resources recruitment;
- locating its skills with human resources evaluation;
- aiming at developing its skills with human resources development and training.

These answers to the four classical preoccupations in HRM (e.g. Sekiou and al., 1992) will become very important in a virtual corporation. In order to achieve this goal, they must not only have strong links together - considering the necessity of integration that is the soul of the new HRM (e.g. Guérin and Wils: 1992,1993) - but they must evolue in order to take into perspective the «catching up» and «leadership» we evoked. To obtain this goal, they must distinguish themselves from traditional HRM by becoming more personalize, anthropocentric and evaluative depending on the terminology borrowed from Guérin and Wils (1993).

Career development: an answer to the challenges facing human resources management

In this perspective, career development (CD) as viewed by Amherdt (1995, 1996) will enable us to contribute towards this answer. Before going any further, lets give a few precisions on this concept.

- Among many definitions on career, we favour the one by Super (1979) who defines a career as the whole and/or sequence of roles and working experiences, paid or not, that a person will have during the course of its life.
- Career development results from interaction (Hall, 1988) between career management centered on organizational needs and career planning focused on individual needs. It constitutes a priviledged premises where converge organizational and individual aspirations.

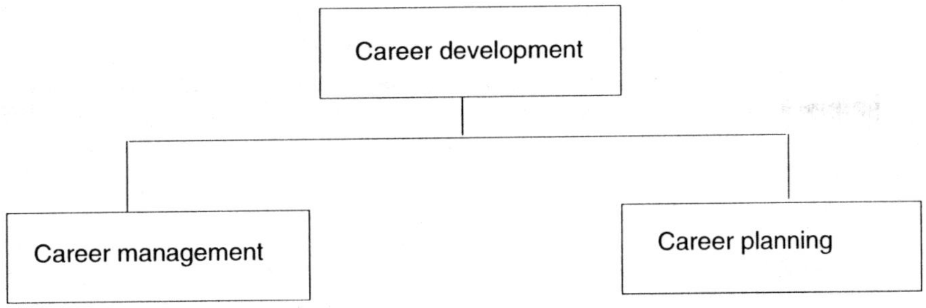

Exhibit 2. Career development and its components

- The application of a CD will take place throughout the implementation of a career development program (CDP) constituted by six steps in its standard form (e.g. Amherdt, 1995). Among these, three are very important in regard of the present problem:

  * step II: identify organizational needs in career development;
  * step III: identify individual needs in career development;
  * step IV: link between organizational and individual needs in career development that will result to the project at hand.

The internal integration of HRM (e.g. Guérin and Wils, 1993) concerns all the informations produced by the CD at steps II, III and IV of the program that will be use by the four operational activities of HRM (planning, selection, evaluation and development of HR).

CD (steps of program)	HRM (activities)
Identify organizational needs in CD	Planning
Identify individual needs in CD	Selection, evaluation
Link between organizational and individual needs in CD	Development

Exhibit 3. Contribution of career development (CD) to human resources management operational activities

Steps of a career development program (CDP):

II) **Identify organizational needs in CD**
Identifying professional roles - also the qualities needed to achieve them correctly, their number and the links they have with each others - that are existing presently inside the corporation also those who are susceptible to accure in a short, medium or long term according to the being of the virtual corporation in question.

III) **Identify individual needs in CD**
Identifying a number of variables proper to the individuals - such as qualification, career status, interest and personal values, strenght and weaknesses, role and responsibility that the person agrees to assume regarding its career development - which influence the career development.

IV) **Link between organizational and individual needs in CD that will result to the project**
Virtual corporation and its people will be able to exchange with each other based on their mutual needs. At this stage, the convergence of each individual interest can be find.

The exhibit 3 shows us that (a) step II of the CDP will give the pertinent informations such as increasing professional roles that will be use in human resources planning, (b) step III of the CDP will contribute to circle individual variables particularely useful when we are interested in human resources selection to fill a function or to evaluate human resources during work process, and (c) step IV of the CDP - the result of the project - will support all the requirements for human resources development by giving them its meaning.

In this matter, excellence that is pursued in virtual corporations can meet the challenge if it is built around people who have been helped to find out who they are and what they want to become. Since mutual trust is so necessary in virtual corporations, it must also be present in every CDP; encouraging such trust is the key element for its preservation and its development.

Conclusion

Mobility and flexibility, outside a person, demanded by all the actors in a virtual corporation is acceptable only if they come within one's knowledge; what the individual wants to become on a personal level and regarding his career development. For each individual who is a part of the virtual corporation, stability even permanence so precious that it is legally claimed, is no longer found like it used to be in the traditional organization but is now within the consciencious needs linked to the career development and the necessity to answer it.

References

Amherdt, C. H. «Le développement de carrière (DC) en tant que stratégie de gestion des ressources humaines (GRH) en amont du développement des ressources humaines (DRH) et qui vient soutenir l'avantage compétitif» in Globalisation et compétitivité: impacts sur les politiques nationales et les stratégies d'entreprises, 1996.

Amherdt, C. H. Rapport final concernant le programme pilote en développement de carrière (PPDC) à l'intention du module maintenance du service des immeubles de la ville de Montréal, Centre de recherche sur l'éducation au travail (CRET), Université de Sherbrooke, 1995, p. 49.

Byrnes, J. «The virtual corporation», Business Week, no. 3304, 1993.

Davidow, W. and Malone, M. «Virtual Corporation», Forbes, December 7 1992, pp. 102-107.

Guérin, G. and Wils, T. «Gestion des ressources humaines: du modèle traditionel au modèle renouvelé», Les Presses de l'Université de Montréal, 1992.

Guérin, G. and Wils, T. «Sept tendances clés de la nouvelle GRH», Gestion, février 1993, vol. 17, no. 1, pp. 22-23.

Hall, D. T. and Associates «Career Development in Organizations», San Francisco, Jossey-Bass Publishers, 1988.

Handy, C. «The age of paradox», Harvard Business School Press, 1990.

Poulin, D. and al. «L'Organisation virtuelle en réseau», in Martel, A. and Oral, M. «Les défis de la compétitivité», Publi-Relais, 1995.

Sekiou, L., Besseyre des Horts, C. H. Chevalier, A. and Fabi, B., «La gestion des ressources humaines», Les Éditions 3 L, Montréal, 1992.

Super, D. «The Babble that is Babel: Basic Glossary for Career Education», Journal of Career Education, vol. 5, 1979, pp. 156-171.

Wexler, J. M. «Ties That Bind», Computer World, June 28, 1993, pp. 97-98.

Human Resource Development Challenges In A Virtual Organization

W. O. Anderson, Jr. P.E.

Bell Communications Research, Inc.
1801 California Street, Suite. 4595
Denver, CO 80202

As a general observation, the United States Regional Bell Operating Companies are aggressively moving to a "virtual enterprise" structure along with many other American corporations. Employees are given new processes and new tools to use as they work in large "mega-centers", remote locations or from their home. All of these changes present a challenge to addressing human resource development and this paper discusses how training will be accomplished in the new virtual organization

INTRODUCTION

"The telecommunications industry will propel all American industry into the 21st century"
Representative Jack Fields (R-Texas)
Western Communications Forum, 2/14/96

The telecommunications industry has gone through major changes in the last twelve years and continues to re-engineer and restructure itself. Dwight L. Allen, Director of Telecom Consulting at Deloitte & Touche stated: " I can't think of any company that has all of their restructuring accomplished or is facing the market in exactly the way they want to. Everyone is going to have to constantly adapt and adjust" (1) . This restructuring is moving many companies to a "Virtual Enterprise" arrangement where many employees will be concentrated in large "mega-centers" while others are scattered throughout a serving region. Those with specific technical expertise are to be coaches to newer employees in the work group. Some people are part time, while others are contract employees or from another company that is a strategic alliance partner.

The last ten years has seen the emergence of Market Units in the former Bell System telephone companies and these personnel are scattered throughout a region. For example, U S WEST Communications has Market Unit personnel in Denver, Minneapolis, Seattle, Albuquerque, Salt Lake, Portland and Omaha. Beginning in 1993, the company collapsed 560 Operations Centers into 26 "mega-centers" with plans to eliminate 6000 jobs. Two major Service Assurance centers monitor the health of hundreds of network elements and send instructions to maintenance personnel scattered throughout their large fourteen state region. In late 1995, AT&T announced that they would eliminate 40,000 jobs over the next three years. Downsizing is nothing new for telecommunications companies. Since 1984, AT&T and the seven Regional Bell Operating Companies (RBOC) have cut over 200,000 jobs. However, independent analysis indicates that up to **3.6 million new jobs** will be created by the 1996 Telecommunications Reform Act.

0-7803-3552-X/96/$5.00 © 1996 IEEE

RBOC Downsizing

	Total Number of Employees		
	1984	**1995**	
Ameritech	77,514	61,144	
Bell Atlantic	85,000	62,000	
BellSouth	98,796	89,544	
Nynex	99,155	68,800	
Pacific Telesis	79,000	50,000 *	
SBC Communications	71,370	58,720	
U S WEST	70,000	61,048	
Total	580,835	451,256	22.3% Reduction

* Spin-off of AirTouch Communications in 1994 reduced 4900 employees, Source: Company Annual Reports

Even when jobs are eliminated and companies re-engineer their processes, human resource development for the remaining employees is still a critical factor for success of this new virtual enterprise! Companies have found that hardware and software are no longer strategic differentiators. Human resources; their knowledge, customer focus and performance will make the difference. If a new software system is utilized to support a new process, employees must be trained to use the features and functions of the new software or process efficiency improvements will not be realized. Many Local Exchange Carriers (LECs) such as U S WEST Communications will also utilize existing operational software systems in their new processes to manage the complex telecommunications network. As employees are reassigned to the new centers, there is a significant need for new and supplemental training on these systems. Other large telephone corporations are going through similar changes. Thomas W. White, President - GTE Telephone Operations stated in his February 13, 1996 Western Communications Forum keynote address that " Work force concerns are one of the most critical issues in our industry"(2).

Tools For The Virtual Organization Employee

Companies are making significant capital outlays to deploy PCs, Work Stations and Laptop computers for employees. The computers come with new software, new word processing and spread sheet programs plus desktop publishing capabilities. Most will have software for accessing corporate databases for e-mail, product information, sales tracking systems and many other capabilities. Employees are expected to learn to operate these new tools "on the fly"; yet, just a small amount of formal training would greatly increase the effectiveness of the users. There are videos and internal instructions that help, but a mandatory one or two day instructor led class usually produces significant improvement. Other tools in the virtual company "tool box" are Fax machines, pagers, voice messaging, ISDN phones, cellular phones and mini-printers for laptop computers. Video conferencing is now an established alternative to business travel for many situations. A February, 1996 article in the *Wall Street Journal* stated that: "Many firms are purchasing the desktop video-conferencing systems to install in small branch offices to allow employees to hook up with other locations" (3).

COMPANIES ALLOWING WORK AT HOME /ROAD VIA COMPUTER & MODEM

Year	Percent
1993	39
1994	42
1995	49

Source: U S A TODAY (4)

Hotels realize that many of their guests need a room that accommodates the new virtual company employee. Marriott is introducing "The Room That Works" in many of their hotels. In cooperation with Steelcase and AT&T, they developed an innovative guest room that "redefines the idea of productive business travel"(5). This new environment includes a redesigned desk, chair, lamp, power sources and telecommunications connections. They also have Fax, photocopying, overnight mail pickup and typing available in the hotel. As we can see, virtual company employees are usually sent out with a "tool box" full of new tools and the environment around them is changing to accommodate this new work style. The question now becomes: "Are they trained and how do we keep them trained?"

Training Challenges

Most companies moving to a "virtual enterprise" structure will notice a significant drop in experience as downsized employees leave the company and the remaining employees learn new functions on new and/or modified support systems. During these changes, company leaders may not realize that there is a human resource development gap, but the supervisors know it immediately! Many supervisors are initiating a Gap Analysis to determine resident skills in their remaining work force. From the results of this analysis, human resource development programs are developed to bring employees to a level of technical competence needed to address the complex, new technologies and processes. For example, in addition to installing and servicing Plain Old Telephone Service (POTS), Local Exchange Carrier technicians are now installing and maintaining ISDN, Frame Relay, ATM and high bit rate services such as DS1 and DS3. In addition to copper facilities, fiber optics are being aggressively deployed and the first Advanced Intelligent Network services are now appearing. Wireless technologies will also join the network. Technicians are scattered throughout a region and delivery of specialized training in a timely and cost effect manner is critical. Even where there are large concentrations of employees in the "mega-centers", supervisors can ill afford to release many employees off of the job for long duration training.

The training organization of U S WEST has worked with many suppliers to develop computer based training and encourage all employees to dial into on-line training or request disks to be mailed to them. They have established a new multimedia training room at the Littleton, Colorado Network Reliability Center that is open 24 hours a day. The room is equipped with 20 personal computers for computer based training (CBT), plus voice, video and CD - ROM equipment. There are six other training rooms in the Littleton facility with training equipment including 31 Sun Workstations, 12 Macintoshes and nine special terminals used by engineers in the Capacity Provisioning center.
If employees can get off the job, the training has to be **focused**. For example, in 1995, Bellcore Learning Services delivered training to 3000 U S WEST Communications employees, with over 2000 attending instructor led sessions "on-site" at a U S WEST location. Many of the courses delivered were "tailor made" addressing the specific needs of the group. Many other employees received additional training through Computer Based Training (CBT) and videos developed by Bellcore, U S WEST Learning Systems and other suppliers. Some of the U S WEST International off shore employees received training in such areas as "Basic Telecommunications" , ISDN, AIN and Broadband technologies via CBTs and Videos provided by Bellcore Training and Education Center (TEC).

The leadership of Pacific Bell is fully aware that their new virtual corporation will not only be re-engineered but will be deploying totally new technologies in its network. Robert Lee, Executive Vice- President discussed "Capability 97" in a February, 1995 article in *Training & Development* magazine. Mr. Lee asked " What good will it be to have new technology by 1997 if our employee skills were developed in the 1980s"(6). In the same interview, he argues that "The training we give people that allows them to sell more hits the cash register right away. The training we give our technicians enables them to install very complex services with fewer defects, and that reduces expenses right away." Similar arguments were raised by Frank Casazza, General Manager - Network Services at Pacific Bell. In a companion article written by Patricia A. Galagan entitled "Business Units Take the Lead in Capability 97", Mr. Casazza concluded that "Training requirements have increased dramatically on account of new technology. At the same time, we are driving expenses out of the business and doing more work with fewer people. If the work isn't done well, or if it's not accurate, you end up costing, not saving money"(7). Both of these executives agree that as a company moves into the virtual company structure, it should not slight training! Pacific Bell is spending millions of dollars deploying a new hybrid / fiber coax network while at the same time re-engineering their operations. These leaders are ensuring that their employees will be ready for this new virtual corporation operation.

The need for continued training is not just a LEC need. The feature article in the February 21, 1996 *USA TODAY* newspaper's Money section addressed job security concerns in America. In a full page story by David J. Lynch and Del Jones entitled "Job Security Redefined in Changing Corporate World", they stated that "To grab the high paid jobs in the future, workers will need the right skills. But given the pace of technological change, those skills are likely to have short shelf lives. Result: a huge need for on-the-job training" (8). They said that companies like Xerox, Corning and United Technologies invest heavily in training programs.

Information Highway Traffic Jam

Virtual companies will set up Local Area Networks and have dial up access for their employees to reach computers day and night, good or bad weather. However, already the vision of instant, quick access to corporate networks is tarnished. In January, 1996, the *Wall Street Journal* reported that during a major snowstorm "many in the Northeast dialed in from home PCs and spent the day faxing, sending electronic mail and surfing the Internet. They also used mobile telephones and pagers"(9). All of this work at home activity had a major impact on the telephone network. Nynex noted a 50 percent increase in telephone line demand in some parts of Massachusetts. This work at home activity created dial tone delay in accessing a line and once a PC was connected to a server, the response time for accessing information was significantly affected. Unfortunately, this scenario will be repeated many times over as other companies embrace the "virtual enterprise" concept. One way to keep a work-at-home employee productive during these times is to supply computer based training on disks and/or CD-ROMs. Another tactic is to require employees to download CBT files from the server periodically to be used in these "rainy day" situations.

Combining A Staff Meeting With A Technology Update

The new "virtual companies" will have employees scattered around the country as well as concentrated in mega-centers. This creates an interesting culture situation. Those scattered will have minimal personal contact with headquarters or mega-center personnel. Most of the communication will be via computer data transfer and/or voice communications which could be a series of voice messaging transfers. To overcome this impersonal business relationship, supervisors will call periodic face to face staff meetings to build teamwork. Information transfer is now done at electronic speeds, but organizations are finding that team building still requires the face to face interaction of people. When people gather, it would be an excellent time for instructor led technology and / or process update sessions.

Training Delivery Options

Even though a considerable amount of human resource development will shift to a self paced, multimedia CBT and video environment, instructor led training will still have a place in the virtual enterprise. Instructor led training can be tailored to the specific needs of a class and the ability to have specific questions answered in "real time" is a significant advantage for this type of training delivery. Companies could chose Distance Learning multi-cast method utilizing satellite systems to inform a large number of employees in disperse locations. Bellcore Learning Services has recognized that there needs to be many delivery mechanisms to effectively support the human resource development needs of its clients as they move into a virtual enterprise environment.

Multiple delivery formats

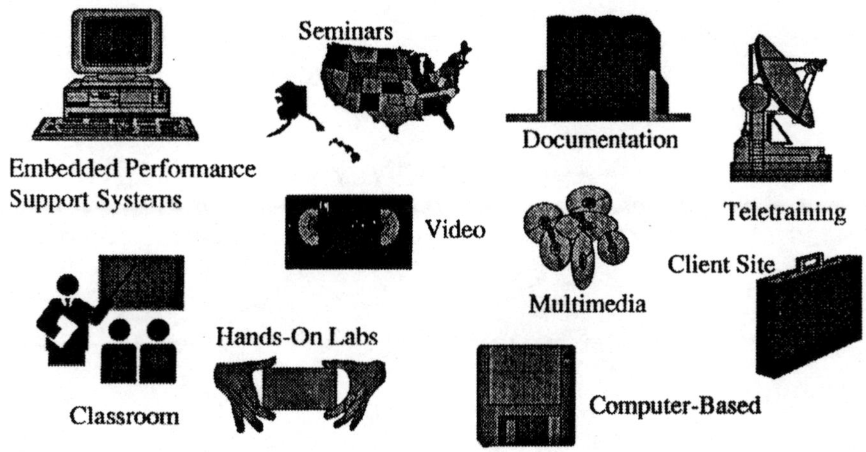

Summary

Telecommunications companies in America seem to be leading the way into a new, downsized, re-engineered "virtual enterprise" structure and with this new environment, a significant human resource development need has been identified. Employees are now expected to utilize all of the new "tools" such voice mail, e-mail, FAX, PCs, laptops and all of the software supporting new processes! The question is whether these employees will have the skills to operate in this new environment? Suppliers of training like Bellcore Learning Services have recognized this need and structured training in multiple delivery formats that is focused and addresses the specific needs of the employee and the supervisor. It is now up to the leadership of these new virtual enterprises!

References

(1) Editor, Business Week, magazine, p.30, January 15, 1996, "The Bloodletting at AT&T is Just the Beginning"

(2) Kleiman, Carol, The Denver Post, newspaper, Section J, p.1, November 5, 1995, "Telecommuting and the 'Virtual Office"

(3) Editor, The Wall Street Journal, New York, NY, newspaper, p.B1, February 27, 1996

(4) Editor, USA TODAY, newspaper, p.B1, December 15, 1995, "Telecommuting on the Rise"

(5) Editor, Marriott Design & Product Services Bulletin # 950281

(6) Editor, Training and Development, magazine, p 34, February, 1995, "The View From the Top"

(7) Galagan, Patrica A., Training and Development, magazine, p.29, February, 1995, "Business Units Take the Lead in Capability 97"

(8) Lynch, David J.; Jones, Del, USA TODAY, newspaper, p.B1, February 21, 1996

Technology Associates of Colorado: An Experimental Approach to Dealing with the Effects of Corporate Downsizing

Dr. Lawrence K. Anderson
719 531-0162

Technology Transfer Associates
2545 Karamy Court
Colorado Springs, CO 80919

Mr. Glenn C. Marshall
719 488-3083

Marshall & Associates
140 Wuthering Heights Drive
Colorado Springs, CO 80921

Technology Associates of Colorado (TAC) is a non-profit, member-based, consortium of consultants with a mission to assist transitioning technical professionals while providing a service to the local community and industry. In this paper we discuss the rationale behind the formation of TAC, its method of operation and some of the strengths and weaknesses of the volunteer-based approach taken.

Introduction

In this paper we describe an experiment in the use of virtual organizational concepts in the area of technical consulting. Technology Associates of Colorado (TAC) is a unique, volunteer consortium of consultants formed in Colorado Springs in 1993 as a response, in part, to corporate and defense downsizing. It provides an umbrella under which individual consultants can share experiences, obtain training, market their expertise and form project teams in response to market opportunities.

Background

The economy of the Greater Colorado Springs area, with a population approaching 400,000 people, is heavily dependent on defense - approximately 50% by most estimates. In the period 1990-1992 Colorado Springs, like most of the United States, was caught up in a combination of general corporate downsizing and defense industry cutbacks. The situation was exacerbated by a general regional economic downturn in the late '80's resulting, in large part, from the crash of the Colorado real estate market. In an ironic good news/bad news twist, the employment picture for engineering professionals was further clouded by the view that Colorado Springs enjoyed an unparalleled "quality of life". This attracted professionals from other parts of the country and provided a disincentive for those already here to relocate. The net result was a very substantial and growing local oversupply of professionals, including technical professionals (Colorado Springs Job Service Center, 1993).

By 1993, the economic revival of the technical segment in Colorado Springs was well underway. But the revival was quite uneven. The relocation of segments of major engineering-based organizations (such as Apple Computer and MCI) and the expansion of existing ones (e.g. Atmel and Symbios Logic) had begun to create a significant number of jobs for production workers and junior engineers. But, for the most part, this had not been translated into a demand for full-time, experienced, management-level engineers.

In 1990, in a tangentially related development, the University of Colorado at Colorado Springs (CU - Colorado Springs) and El Pomar Foundation, the largest philanthropic foundation in Colorado, announced the formation of the Colorado

0-7803-3552-X/96/$5.00 © 1996 IEEE

Institute for Technology Transfer and Implementation (CITTI) at CU - Colorado Springs. CITTI's mission was (and largely remains) the catalysis of technology-driven local economic development, using both University and community resources. For most people, this mission was equated with creating jobs. The local surplus of engineering professionals represented both a challenge and an opportunity when it came to fulfilling this mission.

In early 1993, CITTI became aware of groups of engineers in the Binghampton, NY and Hartford, CT areas, displaced from IBM and United Technologies, respectively, who had each banded together to market their expertise (Gedney, 1993; Dolinger, 1993). At the same time, a number of local displaced engineers came to CITTI looking for employment alternatives and eager to be part of a solution. These engineers had a number of things in common:

- Relatively senior, many with general and/or project management experience
- Displaced from large companies and with little experience in "fending for themselves"
- An almost total lack of marketing skills or experience, except for the highly specialized government sector
- Technical skills which ran largely to systems engineering
- Anxious to try the brave new world of consulting
- Willing and able to invest **time** (but not much **money**) in getting started.

In response to an apparent need, CITTI agreed to incubate a volunteer-led, member-based nonprofit organization (TAC) whose goals were to:

- Serve as a support group for transitioning technical professionals
- Provide training in "real world" business survival tactics
- Serve as a "co-op" under which members could market their services.

In return for providing free office space and general guidance to TAC, CITTI expected, as a *quid pro quo*, that TAC would serve as a resource when individuals or companies came to CITTI seeking business or technical assistance.

Structure

To minimize expense and formality, TAC was set up as an unincorporated project under CITTI, and hence under the University. It has individual members (currently about 40) in two categories (principals, associates). Members pay a small annual fee (currently $350/year) and are expected, in addition, to put in "volunteer hours" in service to TAC (currently 8 hours/month). In return, members have access to training seminars, have their resumes and skill sets entered into a comprehensive, searchable database for marketing purposes (more on that below) and are provided with a number of other services (such as low-cost Internet access). Members also have opportunities for limited reimbursement from TAC's contract and grant income for professional services they render to the community in TAC's name.

TAC is governed by a volunteer Board of Directors consisting of both members and "outside" individuals, including the Director of CITTI as a representative of the University. Day-to-day operations are overseen by an Executive Committee consisting of the volunteer officers [president, vice president and chairpersons of various operating committees (currently Marketing and Process Development)] and the paid General Manager - the only paid position in TAC.

Finances

TAC has an annual operating budget of approximately $40,000, disbursed roughly as follows:

• Salaries (to General Manager)	55 %
• Services and supplies	20%
• Returned to members in the form of "finders fees"	2%
• Contract payments to members	23%

Funding for TAC comes from:

• Member fees	33%
• "Commission" on jobs obtained through TAC	13%
• Contracts and grants	54%

Subsidy from the University (in the form of rent-free use of facilities) and contract and grant income has been absolutely critical to TAC's start-up and survival. In particular, TAC has had a two year TRP contract through the Mid-Continent Technology Transfer Center (MCTTC) to deliver technology transfer and assistance services locally.

Operations

TAC's objective is to match the needs of potential clients with the expertise of TAC members. Requests for help can come in the form of a problem ("We are designing a water-cooled medical device and we can't seem to keep the seals from leaking"), an opportunity ("We're looking for a partner to help us win a global positioning application contract") or a specific personnel need ("We're looking for a software developer with C++ experience as applied to telecommunications").

Clients have had a wide range of needs:

• Information systems development	33%
• Government proposal writing	20%
• Global positioning technology development	15%
• Manufacturing and environmental assistance	11%
• Training assistance (temporary faculty)	9%
• Custom software development	7%
• Database development and maintenance	5%

To serve this broad based clientele, TAC members possess a wide range of expertise:

• Business management	31%
• Information systems management	15%
• Manufacturing and environment	12%
• Government contracting	9%
• Software development	9%
• Hardware design	6%
• High tech marketing	3%
• Other science & technology backgrounds	15%

TAC's principal tool in matching the needs of potential clients to the expertise of members is a comprehensive electronically searchable database of members skills, interests and experience. For each member the database includes a short summary (available at TAC's Web site, *http:/www.usa.net/tac*) as well as a complete résumé.

When a client request for assistance is received by TAC's general manager, she searches the member database looking for matches and queries those members so identified as to their interest and availability. She may also e-mail a summary of the request to the membership at large. Once a member has expressed interest in the job, it is up to that member to negotiate directly with the client, perform the work and remit a portion (currently 15%) of the resulting fee to TAC as a "referral" fee. If more than one member shows interest (not often) or if it appears that a team approach is needed, then one or more TAC officers may be involved in organizing the response.

Strengths and Weaknesses

Strengths

TAC's greatest success has been as a support and networking organization for out-of-work engineers and individual consultants who are seeking to broaden their contacts. Through this networking activity three new *virtual* companies have been formed, two in the area of global positioning satellite applications and another in business systems integration. Participants in these companies have maintained their membership in TAC and their identity as individual consultants while their companies present an image of a broadly based enterprise with in-depth capability.

TAC has developed effective partnerships with CITTI, the MCTTC, the National Consultants Registry, NIST's Mid-America Manufacturing Technology Center and the local Economic Development Corporation and Chamber of Commerce. This has enhanced its ability to help both its clients and its members.

TAC has also been very effective in its retraining mission, offering a series of weekly program meetings in areas of interest to technology-based entrepreneurs. These have covered a wide range of topics in accounting, finance, law, marketing, time management and general business practices crucial to the success of any enterprise but focused on filling the gaps in TAC members' experience.

As a result of the networking and retraining activities, many TAC members have "graduated" to full-time jobs with conventional (*real!*) companies. And, in fact, an informal survey of TAC members suggests that the networking connections that lead to this possibility is viewed as one of the most important benefits of membership.

Weaknesses

TAC has been less successful in positioning itself as a credible supplier of contract services, especially to larger companies and in developing a customer base of established companies. There are several reasons for this:

- As a result of high member turnover (averaging about 50%/year) - the "graduation" mentioned above - it has proved difficult for TAC to establish and maintain a consistent and stable set of capabilities.
- Although TAC members have a diverse set of skills, many are perceived to have management level *breadth* rather than the specific application-oriented *depth* often demanded.
- As an unincorporated project of CITTI, TAC lacks the legal authority to contract in its own right and be accountable for its actions and those of its members.
- As a volunteer-based organization, it has been difficult for TAC to hold its members accountable for specific deliverables.
- Only recently has TAC launched a comprehensive program to market itself coherently to the local corporate technology community. This delay in developing and implementing an effective marketing plan is a direct result of the fact, noted in the introduction, that most of TAC's members have no background and little professional interest in marketing *per se.*

While TAC has had little trouble in attracting technical and other professionals displaced from larger companies, it has been less successful in attracting and retaining already established, successful consultants as members. This may reflect a perceived inequity in the cost/benefit formula. TAC is based on the premise that "the whole is greater than the sum of the parts" - that TAC, marketing itself as a team, can attract more total business than can the individual member consultants. In return for this increased business, TAC members are expected spend time marketing TAC as an entity and pay to TAC dues and a portion of the fees earned from TAC-generated work. TAC has failed to convince some members and many potential members that this is a fair trade.

TAC has also had limited success in positioning itself as covering the entire front range of Colorado from its headquarters in Colorado Springs. To aid in this effort, it set up volunteer-manned offices in Denver and Boulder. The Boulder office has since been closed, although the Denver office, after relocation to a business incubator run by the Colorado Innovation Foundation, has been more successful. The lack of full-time paid staff at the branch locations is the most often cited reason for the difficulty in establishing viable operations there.

Findings

Where TAC has been successful, one can identify several factors which have been instrumental in that success:

- A full-time, paid, professional general manager who can:
  - Gather and maintain a comprehensive database of member skills and interests
  - Coordinate a wide-ranging series of training and informational seminars
  - Translate client problems or resource needs into concise statements and communicate these to the membership in a timely and effective manner.

- A *physical place*, complete with meeting facilities and an office for the general manager, where members can gather.

- Participation in TAC by *service providers* (lawyers, accountants, etc.) as well as technologists, which has allowed TAC to provide a full range of services critical to small businesses and start-ups.

Reasons for TAC's shortcomings can also be postulated:

- Lack of experience and interest in *marketing* among TAC's members - this is clear from the TAC skill set presented above. Engineers as individuals are often uncomfortable in marketing themselves. Aggregating engineers, as in TAC, doesn't seem to change this much!

- Lack of contracting authority as a corporate entity, thus forcing clients to deal with TAC members as individuals. This is most limiting when TAC attempts to assemble a team approach to a client opportunity.

- A perception among some TAC members and some potential clients that TAC is more of a *club* than a serious business entity. This stems in part from the lack of contracting authority and accountability mentioned earlier.

Conclusions

TAC has been in operation for almost three years, during which it has served about 20 clients, ranging from government agencies to technology-based start-ups. It has provided a very valuable local network to assist transitioning technical professionals and served as a catalyst for the formation of new *virtual* technology-based companies. Like many start-ups, it has suffered from a shortage of capital and marketing expertise. To mitigate these problems and fulfill completely its members expectations as a job generating co-op it may have to rethink its non-profit status and incorporate as a contracting entity.

TAC also serves to highlight some of the limitations of *virtuality*. While it is possible for a member of TAC to participate in TAC's networking and consulting activities by a combination of telephone, fax and e-mail alone, experience has shown that the members who see the most action are those who choose to interact face-to-face in TAC's *real* physical facility!

References

Colorado Springs Job Service Center: Data for 1990-91, 91-92 and 92-93.

Dolinger, Sheldon. "Background: Connecticut Technology Associates", P.O. Box 270714, West Hartford, CT. 06107, 1993.

Gedney, Ronald W. "Network 2001", 9 S. Washington Street, Binghampton, NY 13901, 1993.

Multimedia As Value Enabler for Decision Making in Virtual Enterprises

S. Arunkumar
(+91) -22- 578 2545, Ext. 2736
sak@cse.iitb.ernet.in

Dept. of Computer Science & Engineering,
and School of Management and Biomedical Engineering
Indian Institute of Technology, Bombay
Powai, Mumbai - 400 076
India

S. K. Jain

School of Management
Indian Institute of Technology , Bombay
Powai, Mumbai - 400 076
India

Decision making in organizations as contrasted with virtual enterprises is based on physical communicating functional departments. The twin technologies of computing and communication, and the need for management problem solving expertise have made it necessary to have a virtual enabler for effective decision making in enterprise processes. This paper has brought out value addition aspects of multimedia technology in decision making and developed a Multimedia Architecture for Decision Processes(MADP), that can be used at strategic, tactical and operational levels of enterprises. The study was applied to the financial analysis of mergers and acquisition decisions, and equipment diagnosis.

Keywords : Multimedia, MADP, MCU, ICU.

Introduction

Physical location of business units are no longer constraints for effective coordination and management in today's information revolution era. Information as providing value addition in the context of market space has been brought out by (**Rayport, 1994**). Advanced development in technologies allow enterprises to integrate their segregated business units virtually in space. One of the advance technologies that acts as a key to Virtual Enterprises (VE) is **Multimedia**. The twin technologies of multimedia and communication can play crucial role in decision making activities of virtual enterprises.

As is widely recognized, information in a variety of media forms (text, data, graphics, still-images, animation, full-motion video, and speech and non-speech audio) is expanding at an accelerated rate. It is also recognized that the present methods of information processing are not adequate to cope with complexity of decision making process. New designs and synergistic methods of intelligent interactive information management are needed, including fusion of knowledge based systems with various media forms. Multimedia has a promising potential to contribute on this front. Decision making process is vital to enterprises but it involves vast amounts of data and information. Integrated information management is a vital contributor to business goals and objectives (**Framel, 1993**). Effective management of information is a key to leverage the decision making processes in an enterprise and total information management includes video, audio, and text.

There are three levels of managerial tasks in enterprises (**Thierauf, 1988**). First is top level management, which is responsible for strategic decision making. Strategic decisions are of longer term and have to consider the major opportunities and threats facing an enterprise in the long term. These decisions give long range directions that affect the entire organization. Top management directs, formulating the vision and mission of enterprises, which are then translated to tactical and operational levels. They provide guidance, direction and boundaries for tactical and operational management. Second is the middle level management, which use the control information to implement the top level management strategies at various functional levels. Third is the lower level management, which deals with the day to day activity of the enterprise. It is concerned with structured and respective activities that are measurable in terms of specific results. Operational control information allows lower management to comment on how operating standards and policies can be improved to assist day to day operations. A decision support system has to integrates across the levels. For a discussion on decision support system, see (**Bidgoli, 1989**). Multimedia and communication help also to cut through the middle layer.

This paper is organized as follows. This section has brought out management requirements of multimedia based decision support systems. The next section discusses value addition aspects of multimedia in the decision making process. A new architecture is proposed in the third section, and the information flow and processing described in the fourth section. The last section presents implementation aspects and conclusion.

Value Addition in Decision making through Multimedia

The term multimedia implies a new technology based medium where one can fuse data, animation, video with text and graphics through a computer supported base (**Furhi, 1994**). Multimedia can be examined in three parts: media, technology and products. Media include text, audio and video. The technologies that make multimedia possible are optical storage and computer. Multimedia involves concepts from video production, audio recording, animation, and other fields novel to computer experts, system analysts and programmers. Integration of media consists of displaying objects, each embedded in a separate medium, in a window on a screen or, for audio, recorded independently and connected by synchronization. Current authoring systems are largely of this type (**Blattner, 1994**).

Multimedia is a powerful tool which will alter the way we look at knowledge and it provides a new vision of reality. Multimedia systems try to take advantage of human senses to facilitate our interaction. Multimedia combines, audio and video with text materials to enhance communication and enrich its presentation. The potential applications of multimedia are listed and briefly discussed in (**Hedberg, 1993**).

Strategic decision making process involves each and every functional department and several outside factors. Therefore, taking strategic decision in VE, timely and reliable data and information play a critical role. But physically segregated units of VE make it very difficult to get current and relevant information on time and involves high cost. Due to delay in unavailability of information on time, decision making process may get effected. Multimedia technology can help in creating virtual functional departments, from where needed information can be extracted.

Strategic decision making is a complex process. Cognitive perspective of information system support for ill-structured problem domains as obtained particularly in strategic management is the focus of (**Yadav, 1992**). The decision making can be affected by many factors, one of the most important being the way in which information is presented. Multimedia allows users to distill quickly the essence of large amounts of business data to an effective form and reduce greatly the mass of financial reports and other papers. For example, "at a glance" graphics presentation of information allows managers to start their thinking processes quickly. A "picture" may tell managers immediately what they need to know. Information in this visual aid might otherwise be buried in stacks of computer generated reports. Because of the “at-a-glance” presentation power, response time is reduced enormously, thereby increasing the productivity of decision making. Due to the limited size of the computer screen, one can assume that at most four windows could be provided depicting some data in the form of graphs/text. The inclusion of audio to graphical and textual information like graphs, tables and images could be of substantive value significance as this additional modal may provide knowledge units, filtered information and interpreted data with broader aspects of related areas of the problem. This would increase the contextual base of the problem and improve the effectiveness of the decision making process. Audio interactive systems, provide more on-line information which is not subjected to any distortions through textual conversions. The power of multimedia is in accomplishing core-business functions more effectively by plugging into well-worn human input paths (**Kay, 1995**).

Information as an important resource for organizations is brought out in (**King, 1991**). Enterprises are not able to respond to market, not due to lack of information but due to lack of the capability of decision makers to selectively evaluate from much data and information and present in an effective form. Therefore, proper management and presentation of information become crucial in decision making. Multimedia not only provides presentation richness to information but also application domain specific knowledge.

With change in levels of management, features of information element (like video, text, graphics and audio) also change. From top level management to the lower level management, level of interpretation and complexity of information element decreases, whereas the extent of standardization increases. Knowledge and information requirements are deterministic at the bottom, but stochastic and probabilistic at the top management level. Macroscopic knowledge should take precedence, and this is the reason why strategic decision, which focuses on macroscopic views -is so important a prerequisite to other more specific plans. Level of information aggregation increases from operation to strategic level.

Having observed the value addition aspects of Multimedia in decision making, we can conclude that Multimedia leverages the decision making capabilities towards:

- Better problem understanding.
- Decreasing decision making time.
- Enhancement of decision quality.
- Decreasing total involved cost of decision making process.
- Increasing effectiveness of decision.

Multimedia Architecture for Decision Process

The importance of multimedia in the process of decision making was brought out in the last section. This section proposes a new architecture called Multimedia Architecture for Decision Process (**MADP**), to utilize the potential of multimedia technology in decision making.

Figure 1 shows the various modules of the layered architecture for multimedia based decision support system. The top layer has direct interaction to users and represents user interface layer, middle layer has processing part of the architecture and the bottom layer the databases. (**Grosky, 1994**) has proposed a multimedia information system which also can support MADP. A client-server multimedia information system is discussed in (**Papa, 1994**). User interface layer of multimedia system contains the user, the I/O devices and run-time references files, which contains mapping of user actions to internal events.

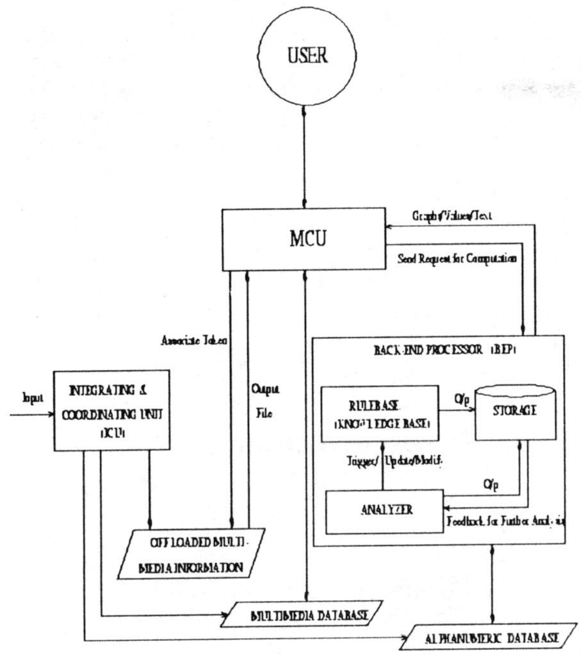

Fig. 1. MULTIMEDIA ARCHITECTURE for DECISION PROCESS[MADP]

Integrating and Coordinating Unit(ICU)

This module coordinates different inputs from the external environment. It is an interface for loading and organizing data in the different databases. The multimedia database provides a vital infrastructure for the Back-End Processor (**BEP**) and subsequent consultation through the Multimedia Coordinating Unit (**MCU**). For each content segments (i.e., portions of image, voice or text) stored, a set of consistent description (called logical structure) is associated with the information element in the database or generated files. ICU has a menu driven interface that guides data/information entry and organizes these media and alphanumeric databases according to processing and presentation requirements. Off-loaded media files comprise those that are externally available as for example details of design and operation of work center supplied by the manufacturer and updated, or standardized company specific pre-formatted multimedia files, as for example responding to an expert financial analyst pertaining to a consultation.

Back-End Processor (BEP)

The BEP primarily provides computation functionality. It does mathematical calculation, generates graphs using alphanumeric database, and passes messages containing reports, queries, and menus to Multimedia Coordinating Unit (**MCU**). There are three modules in BEP, viz., Rule-base, Storage and Analyzer; these are discussed in the following subsections.

Analyzer

This part of the architecture process different data according to the requirements of application. Analyzer activates and/or triggers the different parts of the back-end-process, depending upon the specific application. It transfers values of different variables to the rule base activating corresponding rules. Depending on the predicate and condition, analyzer generates objects and transfers it to MCU for presentation. In a learning mode, it updates/modifies rules.

Rule-Base (Knowledge-Bank)

The primary objective of this unit is to provide qualitative dimension to problem analysis. Knowledge-Bank is a set of rules - crisp, fuzzy or both. It is made up of facts, states and values about the domain to which the system addresses itself. The rule-base is formed by experts in the application domain. The rules to storage are transferred depending on output from the analyzer.

Rules are of three types: those programmed, user updated and machine-learnt (**Arunkumar, 1990 & 1992**). The rule-base can also have management values and priorities provided by management and used in decision making.

Storage

This module stores output from the analyzer as well as from the rule-base. The stored output can be used for subsequent analysis as also to provide feedback.

Multimedia Coordinating Unit (MCU)

This unit provides the unified user-view of the multimedia system. MCU accepts inputs from the various modules of the MADP and integrates them to give effective presentation of information. It also provides navigation control to the user. It identifies particular audio, video and text to form working files for presentation, by a combination of linkages files, which activates media and text files.

The sequence of audio, video and text files have header record. This header specifies the potential records to be linked for offering best presentation. This process is called header file association. This association performs matching and activates the particular files according to the requirements of application on hand. This unit consist of link network of information element, where each information element will have unique identifier. The unit focuses on the mechanism by which link and non-link components are glued together to form the network for presentation system. Presentation specifications might include information on screen location and size of a presentation window, as well as a "mode" for presenting a components. These components contain the chunks of text, graphics, images, and animation, that form the basic content in the multimedia system.

Requirements of Multimedia Coordinating Unit

- This requires some rules (meta rules) on the basis of which, information in different media and from different files are to be integrated. For example in the companies' profile comparison, it provides rules for multi-modal presentation management, which includes text, data, graphics, video and audio. Users might like to see companies' activity graphs at the top portion of the screen simultaneously. Therefore, MCU should have Display Dimension Knowledge base, which is discussed next.

- Meta rule base is updated automatically by machine learning approaches or updated through the ICU, following significant changes in the environment.

- It provides navigation; authoring and following associations are the main interaction approaches used in the system.

Architecture of Multimedia Coordinating Unit (MCU)

The architecture of MCU is shown in figure 2. MCU scans the media files and activates these files on screen to present in a desired manner. It also helps in getting abstract knowledge which is application dependent. The architecture consists of Multimedia User Interface (**MUI**) along with Integrator/Disintegrator and Link Network. Links (**Groanbaek, 1994**) are most fundamental notion of hypertext systems and their function is to define a logical navigation mechanism in a document.

The information at either end of the link is referred as nodes, and the entire node and link structure is referred as a Multimedia Link Network. One of the basic aspects of a link, is that it defines a traversal from one node to another.

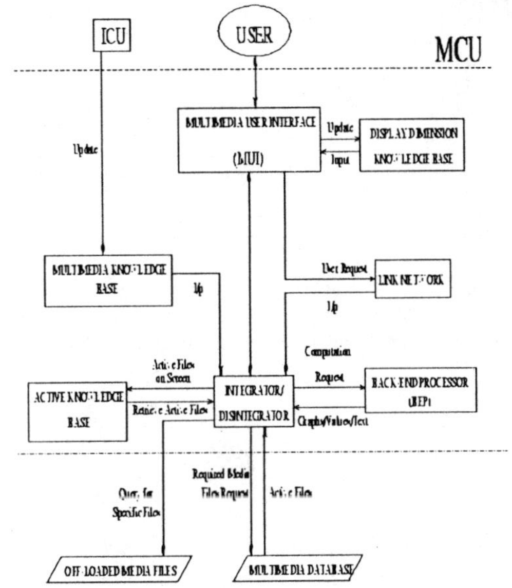

Fig. 1 Architecture of MULTIMEDIA COORDINATING UNIT (MCU)

Multimedia User Interface (MUI) Level

The interface agents are processes that are executed autonomously or semi-autonomously carrying out mission specified by a user. These interface agents help to tailor presentation. They may also can take a input from Dimension Knowledge Base specifying the location of various data forms on the screen.

Method of activation of interface agent

There are different ways by which tasks can be initiated by interface agents as delegated by the user. Three types of methods for activation of multimedia user interface agents have been proposed in the literature (**Sanchez, 1994**):

- User Invoked Method: Under this method of activation, interface agents services are explicitly requested by the user to carry-out a specific task. In order to accomplish this, the user must be able to visualize and interact directly with the available agents.

- Application Invoked Method: With application-invoked method, the user invokes an application which in turn can invoke interface agents that perform related tasks (such as constructing views, translating between different data, etc.). Upon task completion, or before if user input is needed, direct user-agent interaction may occur.

- Event Triggered Method: An interface agent may be activated by pre-specified events. Events may include the occurrence of specific times, data modification, data contents, or relationships among information units.

Integrator/Disintegrator

This module integrates different media and alphanumeric files after taking inputs from the Link Network and is required for presentation. Disintegration is the segmentation of the presented media files in the Active Knowledge Base for later reuse. MCU also dynamically updates data for front end display.

Information Flow and Processing in MADP

MADP starts functioning, either through ICU or MCU. When user wants to enter data and information in different media forms, then ICU comes in its role to coordinate these different media information, by storing and attaching them with associate tokens in a pre-defined manner. It also adds rules to Multimedia Knowledge Base, whenever required. When user starts querying or wants to analyze the information for decision making, then MUI passes a message to the Link network in response to an user action. Link network can determine the system to which action should be next. It transfers signal to Disintegrator/Integrator unit, which first searches user requirements for presentation in Active Knowledge Base, and if not present, passes the signal to BEP for computational work, which transfers graphs, knowledge, and values to the Integrator/ Disintegrator. The latter may also query for some media files such as those associated with graphs, knowledge, and text generated by BEP, from Multimedia Database and Off-loaded files. This unit also takes input from Multimedia Knowledge base, which specifies integration amongst data/information. It also keeps active files (currently displayed) in Active Knowledge Base.

Application and Conclusion

We descibed below the application development environment and further enhancements.

Implementation Environment

The MADP was implemented on a Macintosh Power PC using the MacroMedia Authorware Professional and Symantec C++ compiler. The implementation was guided by (**Vaughn, 1993**).

Application Domain

MADP has been applied to two domains: Mergers and Acquisitions (M & A) decision making and machine diagnosis.

The issues in the strategic analysis of M & A are given (**Bogue, 1986**). Macroeconomic information as the economic indicators of the country and industry were taken from published literature. These included temporal dimension of the economic phase data, employment, inflation, price indices, balance of payments, gross domestic product, agriculture and industrial outputs, and the like. Company specific data include product data and financial variables as net-worth, sales, profit, asset base, return on investment and the likes. Graphs were plotted and their interpretation was part of the off-loaded multimedia files. Sample rule base was constructed for merger and acquisition and the efficacy of the approaches explored.

The approach was also explored for an operational problem of machine diagnosis. Here, the databases consist of multimedia information on faults and rectification. Either for real-time or off-line, the fault features are communicated to the BEP Analyzer for matching and diagnosis, based on which corrective action is initiated.

Conclusion

This paper has proposed a multimedia based decision based support which is especially of import for virtual enterprises. A prototype of the proposed architecture has been implemented and applied to two domains, one at the strategic and other at the operational level. Further refinements to the architecture and extension to client-server networked environments are anticipated.

References

Arunkumar, S. and Salunke, M. P. "Learning from Examples Using Extended Generation Trees with Application to Hemorheology." *Proc. Medinfo'92*, Geneva, Switzerland. Sept. 1992.

Arunkumar, S. and Yegneshwar, S. "Knowledge Acquisition by Hierarchical Learning from Examples." *Proc. Seventh Intl. Machine Learning Conference*, Austin, Texas. June 1990.

Bidgoli, H. *Decision Support Systems : Principles & Practice*. West Publishing company, St.Paul. 1989.

Blattner, M. M. "In our Image: Interface design in 1990s." *IEEE Multimedia.*. Spring: 25-35, 1994.

Bogue, C. M. and Buffa, S. E. *Corporate Strategic Analysis*. The Free Press, A division of McMillan, 1986.

Framel, E. J. "Information Value Management." *Journal of System Management* . December:16-21 & 38-41, 1993.

Furhi, B. "Multimedia System: An Overview." *IEEE Multimedia*. Spring.: 47-59, 1994.

Groanbaek, K. and Trigg R. H.[Eds]. "Hypermedia: System Design Applying the Dexter Model." *Communication of the ACM*. . 37(2) 1994.

Grosky, I. W. "Multimedia Information System." *IEEE Multimedia.* Spring:12-23, (1994).

Hedberg, S. "See, Hear, Learn." *Byte.*. July:119-128, 1993.

Kay, S. A. "The Business Case for Multimedia." *Datamation*. June: 55-60, 1995.

King, W. R. and Grover, V. "The Strategic use of Information Resources: An Exploratory Study." *IEEE Transactions on Engineering Management.* 38(4): 293-305, 1991.

Papa, M. P. "The Development of an Object Oriented Multimedia Information Systems." in *Object Orienred Methodologies and Systems*, E. Bertino and S. Urban [Eds.] Lecture Notes In Computer Science 858, Springer-Verlag,1994.

Rayport, J. F. and Sviokla, J. J. "Managing in the Market Space." *Harvard Business Review*. . Nov.-Dec.: 141-150, 1994.

Sanchez J. A. and Leggett, J. J. "Hyperactive: Extending an Open Hypermedia Architecture to Support Agency." *ACM Transaction on Computer Human Interaction*. 1(4):357-382, 1994.

Thierauf, J. R. *User-Oriented Decision Support Systems.* Prentice Hall:8-13, 1988.

Vaughan, T. *Multimedia: Making it Work.* Osborne McGrawhill, 1993.

Yadav, S. B. and Khazanchi, D. "Subjective Understanding in Strategic Decision Making: An Information Systems Perspective." *Decision Support Systems*. 8:55-71, 1992.

Content Selection for a Course in Total Quality Management

B. Michael Aucoin
(409) 845-6223

Energy Systems Laboratory
Texas A&M University
College Station, Texas 77843-3581

This paper describes the selection of the content for a graduate engineering course in Total Quality Management. Simply defining the subject of TQM is challenging because it encompasses so many disciplines and "methods". The majority of publications on TQM take either a technical approach or a managerial approach. Often missing is a balanced approach to TQM combining both aspects of the subject. Such an approach is crucial considering that many organizations report that their TQM programs have failed. The paper describes the process of selecting the content for the TQM course and the criteria used for selection.

Principles in content selection

This paper describes the process of selecting the content for a graduate course in Total Quality Management (TQM) at Texas A&M University. This course is offered by the College of Engineering under a Systems Engineering curriculum, and is designated SYEN 602. Any graduate student in engineering may take this course. Thirteen students took this course which was offered in the fall semester of 1994 and the author served as the instructor.

The previous instructor, now retired, had chosen an excellent text (Brocka 1992). However, initial research into the literature suggested that a broader approach would be needed to address the controversies which were readily apparent in TQM discussions, and to provide the engineering students with sufficient depth and breadth in the subject to enable them to be successful in quality management.

There was no lack of available information on the subject. A cursory search of the database in the Texas A&M library indicated over 350 books and over 1,500 articles since 1983 with the keywords "total, quality and management".

Many texts deal with TQM with a dogmatic approach, presenting it as the solution to any problem. But many who try to implement TQM are faced with failures. How can they be explained? Is it a failure of TQM or is it a failure of the implementation? It was crucial to address these issues in the course content.

There are two basic decisions to be made in approaching the content to the TQM course. The first involves a choice on whether or not to take a critical approach to TQM. One can present TQM as a wonderful method with no drawbacks, or alternatively investigate the many failures of TQM alongside its successes. The second approach was chosen for several reasons, but primarily because the students must learn how to discern "good" TQM from "bad" TQM, and one way to accomplish this is to make sure that both sides of the debate get air time. For students to be effective with TQM, they must keep their brains engaged, not just blindly follow a set of TQM rules, any one of which could be "broken" for good reason.

The second decision involves whether the content should be directed toward engineering or business, or a mixture of the two. The overwhelming majority of resource materials in TQM tend to emphasize one or the other. The best treatment for this class combines the engineering and managerial aspects in a balanced approach. This approach was chosen for the following reasons.

0-7803-3552-X/96/$5.00 © 1996 IEEE

The course needed a strong foundation in engineering. Because the class consisted of engineering students, most or all would enter careers in which they would primarily be engaged in engineering activities, at least in the early stages of their careers. Considerable time was allocated to the statistical process control and design of experiments topics in TQM. These were also considered important because they often offer quick, objective and significant improvements in quality.

It was also considered important that the students be exposed to key factors in the success of quality strategy. Much of the research and experience with TQM point to management as the single factor which determines the success or failure of TQM programs. It was important that the students get substantial time with discussion of management. However, most of the students in the course would not have had any previous management experience or training, so therefore the discussions would necessarily be related to basic management principles.

With these guiding principles in mind, the chosen course content will now be described.

Foundations of TQM

The first several lectures consisted of an introduction to the subject of TQM and its relation to engineering processes and management.

Introduction to TQM

In the first lecture, an introduction to TQM was provided, highlighting the ongoing debate over the merits of TQM. This discussion provided the framework for a critical evaluation of TQM programs to lead into an examination of effective management later in the course. In this class, a brief overview of the development of management theory over the last 75 years was provided to help the students develop a basic background in management. Considerable discussion was provided on the Hawthorne studies which began the notion of worker participation and productivity. Further discussion was provided on the genesis of the quality movement through Deming and others in the last several decades which led to the recent explosion in TQM programs. Product quality management has progressed from only checking for quality at the end of production to planning for quality at all stages of production.

What is quality?

Some aspects of quality are readily understood and can be agreed upon by many, but in some cases, there can be significant differences of opinion on identifying what constitutes quality, or what degree of quality is appropriate. To a large extent, the improvement of quality relies upon parameters which can be measured, but other facets of quality are philosophical and elusive. Ultimately, the pursuit of quality can become a way of life.

We also proposed that the road to quality often leads through humility. This occurs because improvement often requires an admission of imperfection or even failure. Many people refuse to do this. Likewise, organizations often seem unable to address their problems or admit their shortcomings. This may be the single reason why many TQM programs fail: because some individuals and organizations are unwilling to admit that they need improvement.

Included in this discussion was a treatment on the nature of the course experience. A significant challenge in presenting a course on quality management is to make the subject tangible and "present" to the class. To do this, throughout the course, we used the teaching of the course itself as subject for discussion about quality management. This was to attempt the bring the principles of TQM into the classroom by focusing on the instructor/student relationship, and trying to treat the students as customers. For example, we surveyed the students to determine their topics of interest in TQM to tailor the course curriculum to their interests.

Understanding the masters

We discussed the developments and teachings in TQM provided by Deming, Shewhart, Taguchi and others. Deming is widely recognized as the father of the quality movement, and could arguably be named as the single person most responsible for the course of commerce in the second half of the twentieth century. Each of the other leaders made significant contributions to the subject. The discussion of these teachings showed the breadth of knowledge and wisdom available on TQM and was used to encourage students to seek solutions from many sources.

As part of this topic, we discussed that although there are many "formulas" for implementing a TQM program, it is important to keep an open mind in addressing quality and not simply try to apply a recipe. For example, it is almost considered indisputable that quality involves consistency. However, at least one well-known author recommends that organizations relentlessly pursue *inconsistency* to promote individualized service to customers (Peters 1994). Almost every "rule" in TQM can be challenged in some instance, and so the study of the subject necessarily requires students to think clearly about their objectives.

Quality in technique and analysis

The next major module in the course presented the concepts of TQM which involve measurement and analysis of processes.

Tools and techniques

This segment of the course presented the many tools and techniques which are used for analyzing the quality of products and organizations. These tools and techniques included organizational analysis and statistical analysis. First, the students were presented tools such as cause and effect diagrams, Pareto charts, and flowcharting to use in solving quality problems. For practice with these techniques, the entire class participated in a brainstorming meeting using a cause and effect diagram to solve the problems with parking on campus, a perennial frustration for students.

Statistical process control

Several class sessions were spent on the techniques for performing statistical process control (SPC). As engineers, the students would likely encounter a need for using SPC in their careers, so in-depth treatment of the subject was appropriate. The techniques for SPC included run charts and analysis of sample charts. A discussion was also conducted for natural and unnatural patterns in run charts. The topics of process control and capability were also presented.

Design of experiments

The next major segment of the course concentrated on Design of Experiments (DOE). DOE is a methodology for experimenting with processes to understand and predict their behavior. DOE is considered an important aspect of TQM because it enables adapting parameters for process robustness, and optimization of quality. TQM to some extent can be static and retrospective, as well as involving fine-tuning. DOE is forward-looking, seeking to make great strides in optimization, and is a natural extension of TQM. The students could immediately see the benefits of DOE. This segment of the course included instruction in the basic elements of DOE, and a number of examples involving class participation.

Miscellaneous tools

The tools and techniques section of the course concluded with a presentation on a variety of tools and techniques such as the Plan-Do-Check-Act (PDCA) cycle, the quality house and failure modes and effects analysis.

Quality in strategy and organization

The next major element of the course concentrated on the strategic and organizational aspects of TQM.

Customer focus

Throughout TQM, the customer is considered the most important aspect of business. Again, this principle was related to the delivery of the class, and education in general. We began the class with an informal survey of their level of satisfaction with the course. This exercise served as a springboard for a discussion of the pros and cons of surveys, as well as how to actually implement TQM in the classroom. These experiences provided an excellent means for discussing how we all can become more focused on our "customers".

The lecture included a discussion of a chapter from *The E-Myth* (Gerber 1986). In this chapter, Gerber describes a stay at a hotel which is exceptional at meeting and even anticipating his needs, giving him a truly amazing experience in customer satisfaction. The class on a number of occasions pointed to this discussion as one in which they learned the most valuable lessons about TQM and about becoming focused on customers. The reaction of the class is perhaps indicative that we generally accept mediocre customer service the standard in our society. When we encounter exceptional service, it can transform us.

Unity of purpose

Next we discussed means to develop a unity of purpose in an organization, that is, getting everyone to work on the same problems in the same direction. This subject was presented in light of another chapter from Gerber's book, in which employees of the hotel were invited to make their jobs into a game. The object of the game is to continually better oneself and the substance of the game is to help people recover a lost sense of community and human value through the hotel. This discussion served to show how the pursuit of quality transcends numbers and measurable parameters and can become motivational and fulfilling.

Strategic implementation

The literature is full of articles about quality programs at organizations. Mainly they describe training efforts and committees. We chose a much different approach.

Much of what is published about TQM is written to impress someone, and in so doing leaves the impression that quality is achieved by forming a committee or that quality comes easily. This observation is presented in light of the disturbing statistics on effectiveness of TQM programs: only one-third of TQM programs meet with success. What goes wrong with TQM programs?

What is wrong is that the program becomes the focus. TQM is mostly about good management which is mostly about common sense. It is also about admitting that everyone can improve, and that the first step to improvement is recognizing that things can be improved. As noted earlier, many individuals and organizations have great difficulty with admitting the need for improvement. While many TQM programs can therefore be reduced to mere window-dressing, real TQM has substance.

The key to implementation of TQM is to focus on the customer and on the process, not on the program or the committee.

Included in this presentation were two case studies, each of which analyzed one company with major quality problems. One company had poor methods for dealing with their problems, and their problems persisted, even though they had a quality control function (Leonard 1988). The other followed a path of improvement through a series of learning processes from fixing to prevention to anticipating problems (Sirkin 1990). These cases served to demonstrate that the improvement of quality does indeed involve humility and often pain.

Plant tour and presentation

Representatives of the Westinghouse EAP, a plant for manufacture of electronic systems in College Station, Texas, graciously provided an in-class presentation on TQM and a tour of their plant. The personnel were quite candid in their discussion of the challenges of achieving quality, and offered a real-world perspective to the students.

Course closure

The final lecture consisted of a summary of the course and a discussion of some of the major trends in business which would affect the students (and the issue of quality management) in the coming decade. These issues included: downsizing, outsourcing, the internal customer, and telecommuting.

Evaluations

At the end of the course, students provide formal evaluations to the university for improvement of instruction. These evaluations indicated that the students judged the course content to be worthwhile and rewarding. As evidence of this reaction, all students in the class gave the highest rating of quality to the statements: "The teacher stressed important concepts in the lecture," and "Students gained knowledge and/or competence as a result of the course."

Conclusions

The process of selecting the content for a graduate class in TQM at Texas A&M has been presented. The course content implemented a mix of technical and managerial aspects of TQM. The class presented the successes and failures of TQM programs to indicate that a critical mind is needed to discern the best implementation strategies for TQM. This approach to the course content appears to have been appropriate as indicated by positive evaluations by the students in the course.

References

Brocka, Bruce and Brocka, Suzanne, Quality Management, Implementing the Best Ideas of the Masters, Irwin, Homewood, IL, 1992.

Gerber, Michael, The E-Myth - Why Most Businesses Don't Work and What to Do About It, Harper Business, 1986.

Leonard, Frank S., "The Case of the Quality Crusader", Harvard Business Review, May/June, 1988.

Peters, Tom, The Pursuit of WOW!, Vintage Books, New York, 1994.

Sirkin, Harold, and Stalk, Jr., George, "Fix the Process, Not the Problem", Harvard Business Review, July/August, 1990.

Challenges in Technology Development and Commercialization: A Case Study of Fallen Power Line Detection

B. Michael Aucoin
(409) 845-6223

Energy Systems Laboratory
Texas A&M University
College Station, Texas 77843-3581, USA

B. Don Russell
(409) 845-7912

College of Engineering
Texas A&M University 77843-3126, USA

This paper presents a case study providing the challenges involved in the development and commercialization of technology to detect fallen power lines on electric utility systems. Fallen high voltage conductors are a significant safety hazard to the public and utilities have sought technology to detect fallen lines for decades. This technology represents a significant technical accomplishment, but the success of the project also depended heavily upon innovative solutions to significant marketing, risk management and technology transfer issues. This study provides insights which can be applied to other technologies which involve complex technical, marketing and risk management issues.

Background

Electric distribution systems are designed to efficiently, economically and safely provide power to a large number of customers. The system operates normally the vast majority of the time, but occasionally, the electric current takes an unintended path. This condition is known as a short circuit or fault, and occurs when one conductor contacts another or an object or ground. If the short circuit path is highly conductive, the fault results in a high current which creates enough heat to melt the circuit conductors if not interrupted quickly. Fuses and circuit breakers are installed to protect the circuit by interrupting power when current becomes excessive. This overcurrent protection is common, and until recently, represented the state of the art in fault detection.

If the fault is through a path which is not a good conductor (through a high impedance), the fault current remains low, and is about the same as the current observed for normal service to customers. Conventional overcurrent protection cannot reliably detect and interrupt such faults, and such faults may persist indefinitely.

Often these low current faults occur when a conductor breaks and falls to the ground, as may occur during a storm or as the result of a vehicular accident. It is estimated that there may be several thousand down line faults annually in the U. S. which must be cleared manually. A photograph of down conductor staged fault is provided in Figure 1.

Figure 1. An energized 7200 volt conductor arcs to an automobile during a staged fault test.

0-7803-3552-X/96/$5.00 © 1996 IEEE

When this type fault occurs, energized high voltage conductors may fall within reach of the public. Because arcing often occurs with these faults, a fire hazard also exists. When a downed conductor fault occurs, a fault detector can potentially be used to prevent electrical contact by a bystander.

As an analogy, a high impedance fault can be compared with a minor leak in a water supply system which serves many customers with a strong, steady stream of water. It is easy to identify a major break in the line; it is very difficult to identify a minor leak when it has to be differentiated from normal water usage.

Research history

Texas A&M initiated research on high impedance faults in the mid-1970s. Research continued with a contract from the Electric Power Research Institute (EPRI) in 1978. This research led to the demonstration of a microcomputer-based prototype system in the early 1980s.

The approach taken consists of a system at the distribution substation which monitors electrical conditions on each circuit. The system looks for patterns or signatures which are indicative of a high impedance fault and can automatically de-energize the circuit or issue an alarm if a fault is detected. The principle used is analogous to hearing a crackle on an AM radio during a thunderstorm - it is easy to differentiate lightning from music. Likewise, one can determine that a power line has fallen based upon the electrical arcing which most down lines exhibit.

These initial efforts demonstrated that high impedance faults could be readily detected under most circumstances, but these approaches had some drawbacks. As will be explained shortly, it was also apparent that considerable work was needed for utility acceptance of the technology.

In the mid-1980s, with funding from NASA and NSF, Texas A&M pursued additional sophisticated detection methodologies, including the application of expert systems to the problem. With another EPRI contract starting in 1988, Texas A&M solicited formal utility requirements for a detection device through a workshop, and several utility focus groups and telephone surveys. This market research showed that substantially more sophistication was needed in the fault detection system, and considerable attention would have to be given to formidable marketing and risk management issues.

Market needs and considerations

It was believed at the start of the project in the 1970s that there would be a ready market for a product which detected faults. When the early prototypes did that, it was surprising that few took interest in seeing the technology come to market. Instead, utilities identified many reasons why the technology would not work because they feared the technology would complicate their risk management problems. As a result of these obstacles, researchers were forced to examine the very serious marketing and risk management issues involved with this technology. Five significant market issues were apparent

Legal and risk management issues

Whenever a power line falls, there is the potential for injury or fire, and if these occur, the utility involved may be the target of litigation. In the past, utilities have rightly asserted that the state-of-the-art did not permit the detection of faults in a manner consistent with service continuity. However, any change in the state-of-the-art could have profound effects on the legal and risk environment.

One of the most challenging aspects of high impedance faults concerns how to optimize safety. While one would like to de-energize a conductor if there is any chance of human contact, outages can diminish safety as well. Consider the hazards of a busy intersection when traffic signals are out. Likewise, some customers depend on electricity for medical equipment. Therefore, service continuity is very important.

Given this situation, there is an emotional environment which surrounds the discussion of high impedance faults among utilities. Utilities are dedicated in their desire to reduce the hazards from these faults, but any solution must truly reduce public exposure to all hazards, not complicate matters further. Risk management became the key issue to solve for the success of this technology.

Complex economic incentive

Short circuits with little or no impedance at the point of the fault can melt the conductor if not de-energized quickly. With such faults, there is a direct economic incentive for providing protection. With high impedance, fallen conductor faults, there is little or no circuit damage and no direct economic justification for performing high impedance fault detection. However, if a down line leads to an injury, litigation can result in judgments of several millions of dollars. In essence, the economic consideration of downed conductor detection is highly related to the potential for litigation.

Utility personnel expect that new detection technologies cost approximately the same as conventional protection systems, or about $2,000 - 3,000 per circuit. This cost benchmark places significant constraints on the solution which can be implemented, and implies that only a substation-based system applied one per circuit is justifiable at the present time.

Perfection or substantial improvement

Utilities prefer that a detector identify only those faults which are immediate hazards. An intact, unfaulted line hanging one meter above the ground, is hazardous but cannot be detected because it offers no electrical information to indicate the hazard. It is a significant step for utilities to take to accept a technology which detects many hazardous conditions but not all. To approach "perfection" would require a system with a cost orders of magnitude above what is acceptable to most utilities. To some, such conceptual limitations mean that fault detection technology should be totally rejected, perhaps because of the difficulties for risk management of accepting "imperfect" detection concepts.

Bias toward continuity of service

Electric utilities are judged primarily by how well they keep the lights on. Any technology which is introduced which has the capability to de-energize a circuit must demonstrate that it will almost never operate falsely, a characteristic known as security. The bias of operation of any product is to keep the customers' lights on. The early fault detection techniques had not demonstrated a sufficient level of security.

Fault location

To automatically clear a fault, the substation-based detector requires interrupting the entire circuit for a detection. Then it becomes a difficult matter to patrol the entire circuit to determine the location of the fault, a process which could take from 30 minutes to several hours. Many utility personnel stated that this situation was not desirable. It became necessary to offer an option to either automatically de-energize the circuit, or provide an alarm to an operator. If the operator receives confirming information, such as from a customer call, that a line is down, the operator can remotely de-energize the circuit immediately. Otherwise, a crew is sent to patrol the circuit.

Technology demonstration

The successful development of the technology depended upon the resolution of these key marketing and risk management issues. These issues were all addressed in the research, and the resulting detection algorithms were very sophisticated and ultimately required the use of a high-powered microprocessor. This research led to the demonstration of a prototype beginning in 1991 at five utilities throughout the U. S. The detector successfully detected all downed conductors it was expected to detect, and logged several unit-years without a false operation. Based upon the success of this demonstration, commercialization efforts were undertaken.

Intellectual property

Because of the sophistication of the solutions needed to address the needs of the utilities, eight different algorithms were developed over the course of the research, and each of these formed the basis for a patent application. Because of the duration of the project, and the fact that research contracts originated from several sources, the ownership of intellectual property became complex.

As of the writing of this paper, the technology has resulted in three issued U. S. patents, three patents pending and two applications in the examination process. Patent applications have been filed in a number of countries.

Technology commercialization

It was determined that the most effective method for bringing the technology to the market would be through an organization with a substantial marketing foundation. General Electric expressed considerable interest in the technology and entered negotiations for a license.

The complex intellectual property situation presented a challenge. It was believed to be difficult to have all parties holding intellectual property interests to agree to "simultaneous" licenses. If the parties had all proceeded independently, it is possible that a full agreement would never have been reached. A full agreement was crucial because the technology would be severely limited if any component were missing. It was proposed that all the intellectual property be consolidated in one entity to facilitate licensing. Each party licensed their intellectual property to a company, Power Solutions, which consolidated the intellectual property for an exclusive license to General Electric in 1992.

General Electric decided that based upon their market research, the technology for high impedance fault detection should be augmented with a number of other monitoring functions. With this product strategy, work began in 1992 to perform the commercial design of the product, called the Digital Feeder Monitor (DFM).

Research personnel from Texas A&M were heavily involved with the design engineers during the development process, providing significant input to the design process, and ensuring that the technology was accurately incorporated in the product. This integration of research, development and commercial personnel into one team was crucial to the success of the development effort and the transfer of technology to General Electric.

GE organized a group of utilities into an advisory team for the DFM and a number of them ordered pre-production units for testing at about 50 locations. This process greatly facilitated the modification of design features, and the product was formally introduced to the market in 1995. Since that time, a number of utilities have ordered the product for evaluation or application.

Success factors

The high impedance fault problem evolved into a difficult one to solve, and it is worthwhile to identify the factors which led to the successful solutions developed on this project.

While the development of detection methods proved challenging, it was crucial to focus on the issues involving the implementation of the technology, and particularly, the issues of risk management and security in operation. By listening carefully to the utilities, techniques could be developed to effectively address their needs for implementing the technology.

A number of sophisticated techniques, including expert systems, proved to be necessary for the fault detector. These techniques could only be implemented as digital hardware and software technologies matured at a reasonable cost in the late 1980s to early 1990s.

The research team at Texas A&M included individuals with a strong expertise in technology commercialization. These capabilities proved to be important in developing the commercialization plan and identifying an appropriate commercial partner. During product development, the research team was extensively involved in the design of the DFM, thus facilitating the quick and effective transfer of the technology to General Electric.

Perhaps the most important success factor has been the good relationships and long-term support among the research team at Texas A&M and EPRI, General Electric and the utility partners. All parties involved on the project have been dedicated to the solution of this problem and have persisted in bringing the technology to market.

Conclusions

This paper has presented a study of the development and commercialization of technology to detect fallen power lines which are hazardous to the public. The successful maturation and exploitation of the technology required the creative discernment of utility needs and resolution of their concerns with risk management and security in operation. The success of the project also depended upon innovative approaches to long-term support, licensing of the intellectual property and effective relationships among all parties involved.

References

Aucoin, B. Michael, and Jones, Robert H., "High Impedance Fault Detection Implementation Issues," *IEEE Transactions on Power Delivery,* Vol. 11, No. 1, January, 1996, pp. 139-148.

Aucoin, B. Michael, and Russell, B. Don, "Fallen Conductor Faults: The Challenge to Improve Safety," *Public Utilities Fortnightly,* Volume 129, No. 3, February 1, 1992, pp. 38-40.

Improving Competitive Responsiveness Via The Virtual Environment

Ian Barclay
Jenny Poolton and Zoe Dann
Liverpool John Moores University, England

BICC Cables Professor in Technology Management
School of Technology Management,

The changing nature of the global competitive environment is bringing with it significant changes to both the internal and external activities of companies. One of the key competitiveness drivers of the new environment is a company's degree of "agility" or responsiveness. The need to retain and concentrate on core competencies means that many needed activities have to be resourced externally in dispersed locations. The way in which these internal and external resources are accessed and managed is a key component in defining a company's Responsiveness Capability.

This paper addresses the concept of an organisations Responsiveness Capability, how it can be modelled for practical purposes and addresses the roles the virtual enterprise and IT in improving overall responsiveness.

1. Introduction

The long prevailing economic paradigm of economies of scale, production output driven is giving way to a more customer focused model. At the heart of this emerging model of engineering company activity is the need to cope with increasing product variety, reductions in product life cycles and number of repeat orders and an ever decreasing batch size. Accompanying these changes is the need to constantly reduce "time to market" for product developments as a means of ensuring competitive advantage.

The traditional certainties are giving way to an uncertain and turbulent future environment where a company's capability to respond to events (agility) will be a key aspect of its competitive armoury. This responsiveness (or agility) is a newly emerging and potentially vital issue for companies. We have defined responsiveness as:

> "The ability to react purposefully and within an appropriate timescale, to significant events, opportunities or threats (especially from the external environment) to bring about or maintain competitive advantage".

In order that companies may address the issue of responsiveness, we have developed a model of the component parts of the responsive organisation in an attempt to define Responsiveness Capability. This model has both external and internal elements, with the external focus becoming increasingly important as the market driven economy replaces the provider driven one.

The use of Information Technology (IT) in capturing, manipulating and exchanging data relating to new product developments is a firmly established discipline. Traditionally, this IT capability has primarily related to the internal activities of a company. Increasingly it is the external activities (market forecasting, supply chains etc.) that are being addressed.

This greater emphasis on the external activities has demanded that a Company's "organisation" has to extend beyond its physical and geographical location. With the focus on core competence approach to activities, expertise is being concentrated in specific locations which companies have to access. "Physical" access to dispersed expertise is often uneconomic, time consuming and impractical. This is

0-7803-3552-X/96/$5.00 © 1996 IEEE

where the "Virtual Organisation" or aspects of it, become important to a company. This can range from the non personal exchange of basic product design information (STEP's etc.) or the simulation of face-to-face contact (teleconferencing etc.). Indeed, the two can be combined to provide an interactive, virtual design process and team.

2. The Responsiveness Need

Over the past 10-15 years, the nature of the competitive environment for commercial manufacturing companies has changed from one focused on the optimisation of manufacturing performance around quality and cost objectives to one with multifaceted performance requirements [Dussauge]. This has been brought about by changing customer demands for products and services that are increasingly specific to their needs and wants [Gale]. This has led to a new operational environment for many manufacturing companies where increasing product variety is the order of the day and manufacturing systems have to be capable of producing "one of a kind" [Browne] leading to consequent reductions in product life cycle and numbers of repeat orders and a continual decrease in batch size [Crawford].

The long prevailing economic paradigm of economies of scale (the cost-price driven model) is based on the need for product and component standardisation and the dedicated automation of production to bring about low unit cost and hence reduced sales price. This approach to manufacturing has long held sway and is the basis of the mass production industries of today [Womack]. However, this economy of scale, cost-price driven model of the economic environment is being increasingly questioned as the means of providing future competitive advantage [Ayres]. A new manufacturing paradigm is needed to take advantage of the new requirements of the market .

Competitive advantage for a manufacturing company is becoming increasingly dependent on its ability to react quickly and purposefully to commercial opportunities and threats. This search for responsiveness was described as early as 1978 [Blake]. Subsequently, practical evidence of this change in approach was given in the work of Kanter in her description of the break up of corporate USA into smaller, more responsive operational units which gave a clear indication of the primary corporate attribute needed for success in the 1990's and beyond; that is responsiveness.

At the heart of this transformation in the manufacturing based industries is the concept of "speed to market" [Kupra], with companies organising for reducing product development times to meet quickly changing market needs [Carter & Baker]. The work of Barclay & Poolton has clearly defined these changes in terms of human, organisational and procedural factors, information requirements and the tools and techniques associated with this new environment. The evidence supports the view that the organisation with the ability to respond quickly is also one that is likely to "get it right first time" [Cooper]. This responsiveness is based on a holistic approach, with the customer, design, development, manufacturing and suppliers working together and being supported by appropriate information systems [Keen]. The core concepts of this new responsiveness (or "agility") have been defined [Kidd] as follows:

- A strategic focus on becoming agile.
- A strategy to exploit agility for competitive advantage.
- Integration of organisation, people and technology into a co-ordinated, interdependent system which is a key competitive "weapon".
- An interdisciplinary design method to achieve the integration of organisation, people and technology.

The generic features of this agile organisation are:

- Integrated enterprise.
- Human networking organisation.
- Enterprises based on natural groups
- Increased competencies of all people.
- A core competencies focus.
- Virtual corporations.
- An environment supportive of innovation, experimentation and learning.
- Necessary It support infrastructure

Our work at Liverpool John Moores University has shown that a Company's ability to respond meaningfully to the threats and opportunities that appear is, increasingly, a key competitive weapon. Whilst not being able, at this stage, to actually "measure" this degree of responsiveness, it is quite clear that certain companies are far more capable of responding quickly to a competitive challenge than are others [Slater]. This work has led us to develop a model of the responsive organisation and this is described in Section 3 below.

Organisations need systems and awareness which allow the prediction and assessment of changes in the external environment [Cetron]. The traditional approach to this has been the use of strategic planning processes [Johnson], technological forecasting [Martino] etc.. These approaches are being supplemented by closer collaboration with both customers and suppliers [Beimans]. Indeed, suppliers are increasingly seen as a major source of technological development with companies retaining core skills and releasing non core activities [Wheelwright]

Organisations also need the ability to analyse the potential impact of these changes upon their own internal systems and organisation. For minor changes the current design/manufacturing processes will usually cope. For major changes, the full capability of the design, development and manufacturing expertise will be challenged [Syan]. This means that in a rapidly changing market environment, organisations need to be flexible, innovative and have the ability to constantly re-organise and regenerate themselves [Peters].

The degree of responsiveness, or agility, will depend on the nature of the market, the "newness" of the product's function, upon technological issues such as the "newness" of the product's "structural" changes and the nature of the company's design, development and manufacturing process [Inwood]. Increasingly, companies cannot support development activity solely from their internal capability. The general trend in the reduction of personnel is exacerbated by the need to incorporate an ever increasing range of technologies into products. Thus, companies are forced to selectively use outside expertise as they cannot maintain the breadth of technological knowledge within the company [Prahalad].

For successful product development in a changing environment, a company needs a high degree of responsiveness which is dependent on both internal and external capabilities:. This responsiveness can be split into two aspects:

* **Internal Capability**: The expertise held within the organisation.
* **External Capability**: The expertise of suppliers, customers, agencies and consultants etc. available to the organisation

The way in which these two capabilities are identified, organised, managed and used will be instrumental in achieving success in the product development. The internal development activities must be well understood, organised and managed and supported by an accessible network encompassing the external expertise. The balance of internal and external capability becomes a strategic issue of what expertise has to be retained, and what can be accessed as needed and how the two are organised and managed [Rigby]. At the heart of this management process is the exchange of information.

Both internal and external capability are increasingly reliant on Information Technology (IT) networks which capture and transfer complex data and integrate the design/manufacturing process [Stark]. These IT networks are becoming virtual enterprises and transcend the traditional geographical proximity previously necessary to provide responsiveness. For complex capital goods, the network may even drive design and manufacture at several levels with no specific company or plant having ownership of the product or process [Werther]. (Airbus International may be seen as an example of this).

3. A Model of the Responsive Organisation

3.1. Introduction to the model

This model of the responsive organisation focuses on the new product development process. This focus has been chosen as it encompasses both major and also incremental changes. It has also been chosen as new product development has to be approached holistically, with equal emphasis on markets and customers, the design, development and manufacturing process and systems. It also lends itself well to the consideration of the integration of customers, suppliers, overall logistics and information systems (from initial data capture to market delivery).

The Responsiveness Capability for any company depends upon its:

- Product development strategy and need.
- Internal activities (design, development and manufacturing).
- External linkages and networks (consultants, supply chain etc.).
- Organisational issues of management and personnel capability and culture.
- Technological level, complexity and capability.
- Information needs, systems and their management.

We are also proposing that the level of Responsiveness Capability needed is largely determined by the market sector, its economic determinants and its needs and volatility, company size, the nature of the product and the degree of need to control risk.

Our work in this area has led us to define a six factor responsiveness model that leads to a company being able to assess and improve its responsiveness. This six factor model is shown in Figure 1. and each factor is explained in detail below:

FACTOR	COMMENT
I. Responsiveness Drivers	Mainly external influences.
V	
II. Responsiveness Timescale	Product/market dependant.
V	
III Responsiveness Strategy	Reactive/Predictive.
V	
IV. Responsiveness Focus	Holistic or specific.
V	
V. Responsiveness Providers	Component parts of responsiveness.
V	
VI Responsiveness Capability	Internal/external management.

Figure 1 : The six factor responsiveness model.

3.2. The responsiveness drivers

The Responsiveness Drivers (RD's) are the events or influencing factors to which the company has to respond. They are mainly concerned with changes to the company's external environment, although significant internal events (e.g. loss of a key individual). might be included. They can be broken into two groups:

Company specific	**Global (STEP factors)**
Competitor activity	Social changes
Customer activity	Technological developments
Supplier activity	Economic changes
	Political changes.

This is where the company's strategy has a major determining role on its Responsiveness Capability and relates directly to the Responsiveness Timescales and Focus (described below).

3.3. The responsiveness timescales

Clearly, large, R&D based organisations have a different Responsiveness Timescale (RT.) than a small, development based company. The appropriate timescales are largely determined by the market requirements, the degree of technological difficulty and influences such as approval requirements (e.g. pharmaceutical product testing). Whilst appropriate timescales can be defined, the company's imperative is to consistently reduce them through improving internal efficiency and effectiveness.

3.4. The responsiveness strategy

Within the company's overall strategic framework there is the choice of approach to the market and the development of products and manufacturing and related systems to deliver these products to the market. This approach can range from "new to the world" to "me too" or "Oh Hum" type products. Thus, Responsiveness Strategy (RS) can be viewed as having two extremes:

Reactive RS:	Events (RD's) occur and the company reacts to them.
Predictive RS:	The company predicts events (RD's) and exploits them.

The evidence from our current new product development work is that most companies are trying to move to a more predictive strategy to improve responsiveness and competitiveness.

3.5. The responsiveness focus

Developing and getting a product to market involves many stages and we have found the use of a modified version of Cooper's 13 Step new product development process model (Figure 2.) to be a useful representation of the overall process.

1. Initial screening
V
2. Preliminary market assessment
V
3. Preliminary technical assessment
V
4. Detailed marketing study
V
5. Business & finance analysis
V
6. Product development
V
7. In-house product test
V
8. Customer product test
V
9. Test marketing
V
10. Trial production
V
11. Pre-commercial trial
V
12. Production start-up
V
13. Market launch
V
14. Post launch support

Figure 2 : The 14 step process model

Companies can improve their responsiveness at any of these steps. The nature of the development will determine where the most appropriate emphasis should be placed on the process. The focus may be on the overall process (for a large/new development) or on a select number of steps for smaller changes. This will relate to the nature of the RD's and RS.

3.6. The responsiveness providers

We define the Responsiveness Providers (RP's) as the key component parts that provide the company's ability to be responsive. These are mainly internally controlled and managed and two categorisations are given below:

Agile Manufacturing	**Concurrent Engineering**
(Kidd)	(Clark and Fujimito)
Organisation	Integration
People	Core competencies and skills
Technology	Information
	Innovation

We use the broad definition provided by Kidd and break this down into sub-groups e.g.:

Organisation: Integrated, networked, virtual, natural groups, innovative culture, experimenting, learning, integrated development process.

People Natural groups, innovative, leadership, core competencies and skills.

Technology	Core technologies, IT infrastructure, computer based tools and techniques, core competencies focus.

3.7. The responsiveness capability

Responsiveness Capability (RC) is the "degree of responsiveness" that a company possesses or needs, that is, its ability to react to and/or predict events in order to manage, control and take advantage of them. We envisage two distinct, but integrated types of RC management:

- Internal RC Management: The control of events within the company.
- External RC management: Influencing/exploiting of events outside the company.

4. The Role of "Virtuality" in Responsiveness

4.1. An overview

As we stated in the introduction, the focus on the core competence approach to activities is leading to expertise being concentrated in specific and dispersed locations. "Physical" access to this dispersed expertise is often uneconomic, time consuming and impractical. "Virtual" access becomes important to a company in accessing and utilising this expertise in the most effective and efficient way. The way in which the Virtual Organisation influences responsiveness can be reviewed via three types of interaction, all of which feature prominently in the virtual environment :

Interaction Purpose: "Hard" versus "Soft" interactions with specific purposes.

Interaction Type : Accessing, capturing, processing, monitoring and archiving.

Interaction Level : Within the levels of the responsiveness model.

4.2. Interaction purpose

We have chosen to use the well known "Hard" versus "Soft" analogy for this aspect of our work. The "Hard" interaction is concerned mainly with the capture, processing and manipulation of data used within the organisation. This can range from market and business information to the non personal exchange of basic product design information (STEP's etc.). The design and operation of the IT system is of paramount importance here. Compatibility between both internal and external (customers and suppliers) systems is essential if maximum synergy, and hence responsiveness, is to be achieved. In our experience, this is the main area emphasis for most companies.

The "Soft" interactions address what are commonly known as the interpersonal and cultural considerations of an organisation. Whilst the "Hard" interactions can be viewed mechanistically, Virtual Organisations are complex and are formed to address complex issues in the most practical ways. Handling complexity demands team based operations (Barclay et al) and, in turn, team based operations demand mutual respect, openness and trust, values and qualities that have to be cultivated (Hershock et al) . The cultivation of these values and qualities can be enhanced and cemented by the simulation of face-to-face contact through teleconferencing etc.. This is the "oil" that "lubricates" the whole development process and which improves an organisation's responsiveness.

The importance of the "Soft" interactions is often underestimated in the Virtual development process and the two should be addressed and combined to provide an interactive, virtual design process and team that can respond within desired timescales.

4.3. Interaction type

When considering the interaction type, we are mainly concerned with the "Hard" interactions and the way in which the supportive IT system can promote responsiveness via the Virtual Organisation. A simple overview is presented here :

Interaction type	Influence on responsiveness
Accessing	Open systems, giving easy access to information, reduce lead times and promote responsiveness. Internally, this brings the issue of data security to the fore but this can be addressed with a hierarchy of access approval. The "trick" appears to be to get the authority at the lowest possible level (Jones & Saunders). Externally, this means access is provided to external data bases, either global (eg. patent network) or customer/supplier specific.
Capturing	Accessed information needs to be transferred into a form that is both capable of virtual manipulation and also reusable. Global standards and systems are essential here for increased responsiveness through virtual processing.
Processing	This is an extension of the Capture interaction.
Monitoring	The complexity of the modern product development process has brought about major changes in the style and approach of managers (especially senior ones). In responsive organisations power can no longer be exercised by the withholding information. The new management roles are ones of monitoring and facilitating. Programme tracking and monitoring systems that can be used "virtually" are essential.
Archiving	Once completed, the development activities and data needs to be archived to allow further learning and lead time reduction to take place on the next generation of products. In this way, accessible archives increase responsiveness.

4.4. Interaction level

The interaction level relates directly to our responsiveness model introduced in Section 3.1. above :

Interaction level	Considerations
Responsiveness Drivers	These relate to the strategic processes of the company, both company specific and global. The Virtual Organisation is far more likely to be able to predict events and respond positively to them within the appropriate timescales (Calantone et al).
Responsiveness Timescale	Global, Virtual Organisations can significantly reduce lead times by effectively having "24 hour" R&D operations. For example, Virtuality and IT systems allow European based R&D companies to set up R&D operations in different time zones (e.g. Australia) to allow continuous operations (and half lead times?).
Responsiveness Strategy	Clearly, the Virtual Organisation has more choices in terms of its strategy and approach to its business (Slater).
Responsiveness Focus	Moving toward a more Virtual environment allows a company to focus on the development activities that are critical and have to be retained. At the same time, non critical, but important activities can be released but still be controlled.
Responsiveness Providers	These have been discussed under the type and purpose of the interactions and in Section 3.6. above.
Responsiveness Capability	This covers the development to manufacture process and its relationship to the external environment. This must also be extended beyond the internal systems to the area of supply chain management.

5. The Future of Virtuality and Responsiveness

As markets become more fragmented and difficult to access, getting the right product to the right market in the shortest possible time will become a critical competitive activity. Using IT systems within a virtual framework will become an integral part of this development activity, helping to utilise scarce resources and reducing complexity. As networks such as the Internet develop and IT hardware prices reduce and systems become more user friendly(Miles), the use of the virtual environment must extend to even the smallest of companies as a means of improving their responsiveness in the face of global competition (IEEE. Spectrum).

References

Ayres R U (1990) "CIM, Driving forces and applications", CIM: Revolution in Progress, Proc. final IIASA Conf. CIM Technologies, Organisation and People in Transition (Haywood W ed.) pp 9-26, Laxenburg, IIASA.

Barclay I., Holroyd P & Poolton J. (1994), " A Sphenomorphic Model for the Management of Innovation in a Complex Environment", Leadership and Organisational Development Journal, Vol. 15, No. 7, pp. 33-42.

Barclay I & Poolton J, (1994), "Concurrent Engineering: Concept and Practice", International Journal of Vehicle Design, Vol. 15, No. 3/4/5, pp. 529 - 544.

Beimans W. G. (1990),"The Managerial Implications of Networking", European Management Journal, Vol. 8, No. 4, December, pp 529-540.

Blake S. P. (1978) "Managing for Responsive Research and Development", W. H. Freeman & Co., San Francisco.

Browne J, (1992), "Future integrated manufacturing systems, a business driven approach", Organisations, People and Technology in European Manufacturing (Kidd P. T. ed.), Luxembourg Commission of the European Communities, pp 17-30.

Calantone R. J., Vickery S. K. & Droge C (1995), "Business Performance and Strategic New Product Development Activities: An Empirical Investigation", Journal of Product Innovation Management, Vol 12, No. 1, pp. 224-234.

Carter D. L. & Baker B. S. (1992), "Concurrent Engineering: The Product Development Environment for the 1990's", Adison Wesley Inc., Reading, Mass.

Cetron M., (1994), " An American Renaissance in the Year 2,000: Trends that will affect America's Future - and You", The Futurist.

Clark K. B. & Fujimito T. (1991), "Product Development Performance", Harvard Business School Press, Boston, Mass.

Cooper R. G. & Kleinschmidt E. J. (1995), "Benchmarking the Firm's Critical Success factors in New Product Development", J. of Product Innovation Management, Vol. 12, No. 5, November.

Crawford C. M. (1994) " New Products Management", R. D. Irwin Inc., Boston, Mass..

Dussauge P., Hart S. & Ramanantsoa B., (1994), "Strategic Technology Management", J Wiley & Sons, Chichester.

Gale B. T. & Wood R. C. (1994), "Managing Customer Value, Creating Quality and Service that Customers Can See", The Free Press, New York.

Hershock R. J., Cowman C. D. & Peters D. (1994) "Action Teams that Work", Journal of Product Innovation Management, Vol 11, No. 2, pp. 95-104.

IEEE. Spectrum (1996), "Technology 1996: Analysis and Forecast Issue".

Inwood D. & Hammond J (1993), "Product Development: An Integrated Approach", Kogan Page, Baddock, Herts..

Johnson G. (1988), "The Process of Managing Strategic Change", Management Research News, Vol. 1, No. 4/5, pp. 43 - 46.

Jones J. W. & Saunders C. (1994), "Information Acquisition During Decision making Processes", IEEE. Trans. on Engineering Management, Vol 41, No. 1, pp. 41-49.

Kanter R. M (1992), "When Giants Learn to Dance: Managing the Challenges of strategy, Management and Careers in the 1990's", Routledge, New York.

Keen P. G. W. (1988), "Competing in Time: Using Telecommunications for Competitive Advantage", Balinger, Cambridge, MA.

Kidd P. T. (1994), "Agile Manufacturing", Adison Wesley Ltd., Reading, Mass.

Kupra D. C. (1992), "Time is a Primary Metric", Manufacturing System: Foundation of World Class Practice, National Academic Press, Washington, pp. 166-172.

Martino J. (1983), "Technological Forecasting for Decision making", Elsevier, New York.

Miles I., (1995), "Cutting Edge Technologies: Bringing Computer Power to the Consumer Market), IEEE. Engineering Management Review, Vol.; 23, No. 1, pp. 10-22.

Peters T. & Waterman R (1982), "In Search of Excellence", Harper & Row, New York.

Prahalad C. K. & Hamel G. (1990), "The Core Competencies of the Corporation", Harvard Business Review, Vol. 68, No. 3, pp. 79-91.

Rigby D. K. (1994), "Managing the Management Tools", Planning Review, Sept. - Oct.,

Slater S. F. (1993), "Competing in High Velocity Markets", Industrial Marketing Management, Vol. 22, pp. 255-263.

Stark J. (1992), "Engineering Management Information Systems", Van Nostram Reinhold, New York.

Syan C. S. & Menon U. (1994) "Concurrent Engineering: Concepts, implementation and practice", Chapman Hall, London.

Werther W. B., Berman E. & Vasconcellos E. (1994), "The Future of Technology Management", Organisational Dynamics, Winter

"Wheelwright S. C. & Clark K. ((1992), "Competing through Development Capability in a Manufacturing-Based Organisation", Business Horizons, Vol. 15, No. 4, pp. 29-43.,

Womack J. P., Jones D. T. & Roos D. (1990), "The Machine that Changed the World", Rawson Ass., New York.

Evaluating Quality and Equity Issues in the Management of Hospital Waiting Lists

Judith Barlow
Computer Science and Information Systems Department
College of Arts and Sciences
American University

Rhiannon Tudor Edwards
Department of Public Health
University of Liverpool
Liverpool L693BX England

Waiting lists act as the predominant rationing mechanism for non urgent health care in the British National Health Service (NHS) (Lindsay and Faegenbaum, 1984). Inequalities persist in waiting times between clinical specialities and geographical areas which are in no way related to the benefits that patients can gain from being treated, or the health sacrifices they make as a result of having to wait (Frankel and West,1993). This paper describes a unique yet simple waiting list simulation program which combines activity and health gain data in order to facilitate experiments which explore what happens when patients are selected for treatment according to a range of familiar waiting list criteria which reflect management policy options for the British National Health Service (NHS).

Introduction

Evaluating the quality, equity and effectiveness of alternative policies for allocating scarce resources is difficult. Waiting lists act as the predominant rationing mechanism for allocating non urgent health care resources in the British National Health Service (NHS). Within the new NHS, internal market purchasers face a number of alternative waiting list policy goals, such as

- reducing the number of patients waiting,
- minimising waiting times,
- treating first those who would benefit most from treatment,
- treating first those who have "suffered" the most while waiting for treatment, and
- ideally making sure that waiting time is distributed fairly between patients waiting for treatment of different conditions.

In the face of limited resources (e.g. hospital capacity, medical professionals), simultaneously satisfying these goals may be impossible.

Results of a simulation experiment carried out on a published set of data from Guy's Hospital (London) general surgical waiting list (Gudex et all 1990)are presented. The results demonstrate how different patient selection rules lead to different numbers of patients being treated, different waiting times and different overall health gains and health sacrifices. It is hoped that this paper will help move the waiting list debate beyond the current focus on waiting times, towards the consideration of the as yet unmet health needs that waiting lists represent.

Waiting Lists as a Rationing Mechanism

Waiting lists may well provide a fairer rationing mechanism for non urgent health care than alternative mechanisms such as ability to pay which exist in countries such as the United States. What is crucial is that the criteria used to manage them be open and accountable (Higgins and Ruddle,1991). Day to day decisions about who is

0-7803-3552-X/96/$5.00 © 1996 IEEE

treated next from a waiting list are largely the responsibility of individual doctors and their clerical staff (Royal College of Surgeons of England, 1991). These decisions are often complex and intuitive, compounded by wide evidence of variation in clinical practice (Dowie, 1983). Unravelling this implicit rationing process is a formidable task for both clinical and non clinical researchers.

At a policy level, within the NHS internal market (Holliday, 1993), purchasers decisions about the contracts they set with provider hospital units influence the relative time patients wait for treatment in different clinical specialities and areas of the country.

What Goal for Purchasers?

A number of conflicting waiting list policy goals exist for purchasers. Should they aim to:

i. Set contracts which aim to meet as many patient needs as possible with existing capacity as they arise?

ii. Contract to maximise throughput of patients and minimise waiting times in line with government targets? (Department of Health, 1992)

iii. Contract to maximise the health gains from available resources?

or:

iv. Contract to distribute waiting time fairly - to reflect the relative health gains patients receive from treatment and the health they sacrifice as a result of having to wait.

In this paper we show how questions like these which face every purchaser in the NHS may be quantitatively modelled through the identification of operational criteria by which patients may be selected from a waiting list for treatment.

The Development of the "In-Waiting" Program

The "In-Waiting" program is the result of a collaborative project between a health economist and a computer scientist/operations researcher. The field of health economics has given waiting lists relatively little attention in the past. The literature focuses on (1) the need for explicit criteria or admissions indices for the management of waiting lists (Culyer and Cullis, 1976), (2) Estimation of the cost of time to patients having to wait (Propper,1990), and (3) the modelling of waiting lists as a market clearing mechanism (Lindsay and Faegenbaum, 1984).

More recent research has been focused on the qualitative nature of waiting lists, moving the debate over their management beyond simple waiting times and waiting list sizes. Two authors have calculated the health gains from treating patients on NHS waiting lists in terms of Quality Adjusted Life Years (QALYs) (Gudex et al, 1990), (Haycox, 1991). QALYs are the best known generic measure of health gain developed in the field of health economics. Their calculation involves discounting a stream of added years gained from treatment versus no treatment, adjusted to reflect the quality of life in which these additional years are lived (Gudex and Kind, 1988).

"In-Waiting" Program Design

"In Waiting" is written in C++. An object-oriented design strategy is used to encapsulate the important components in the simulation model: the waiting list object class, patient object class, hospital object class, referral subclass, admission subclass, and discharge subclass. The system clock object class drives the program by keeping track of events (admissions, discharges, referrals) and calling the appropriate objects to simulate events and update simulation statistics.

Object-oriented programming utilises a richer set of data structures which contain not only program data, but also include information on the actions that affect this data as well as procedures which are executed on other objects due to changes in program data. One of the primary data elements used in object-oriented design is the "object." Each object is an "instance" of one of the model's predefined "object classes."

Given data on the number of patients of each treatment type, the average age of patients requiring each treatment, and number of weeks to simulate; a seeded random number generator is used to assign a specific age and treatment type to each patient joining the waiting list. Then a baseline simulation using a "first-come, first-served" selection rule is run at the beginning of each simulation run. The user may then specify other selection rules and the simulation is rerun. Output is generated which allows the user to compare "first-come, first-served" selection criteria to other decision rules.

Waiting List Data and Research Questions

Simulation experiments were carried out on a published set of data (Table 1) which provided both activity data and health gain data necessary, (Gudex et al, 1990).

Table 1: Guy's Hospital General Surgical Waiting List

Condition	Number Of Patients On List	Qaly Gain From Treatment	Bed Days Required
Subcutaneous lumps	325	.28	1.9
Skin lesions	115	.26	0.4
Unilateral varicose veins (female)	107	.29	2.3
Unilateral inguinal hernia (male)	99	1.34	3.5
Bilateral varicose veins (male)	89	.26	2.8
Unilateral varicose veins (male)	50	.22	4.0
Ganglion	37	.39	1.4
Piles	37	.77	6.0
Bilateral varicose veins (female)	28	.41	2.8
Ingrowing toenail	24	.56	1.4
Total	911		

Source: Collated from Gudex et al, 1990.

This data was at the time that the program was being piloted a unique set of data with information available on both patient numbers, waiting times and health gain (QALYs) gained from the treatment of different conditions on the waiting list. The data is by now purely illustrative - demonstrating the data necessary for the "In-Waiting" program.

Five Different Patient Selection Rules

The waiting list management policy options outlined in Section 3 were operationalised in the following five patient selection rules.

i. First Come First Served (FCFS)
ii. Health Gain per bed day maximisation (THG/B)
iii. Claim on resources (CR)
iv. Random (RAND)
v. Suffering (SUFF)

The FCFS rule is most closely represents the status quo and is the most obvious place to begin. The THUG/B rule selects patients in order of their expected health gain from treatment and, given a specified number of beds, will lead to their most efficient use in terms of the total health gain that can be produced. The CR rule operationalises a throughput maximisation policy, selecting patients in inverse order to their bed day requirements. The RAND rule, reflects the view that it is unethical to try to compare the health gains patients get from treatment and that the only truly amoral way of selecting patients is to toss a coin (Harris, 1986). . Finally, the SUFF rule aims to operationalise as closely as possible the clinical decision process. Patients are selected in order of the total health sacrifice they have made to date as a result of having to wait, i.e. their weekly difference between their pre and post treatment health state multiplied by the number of weeks they have waited.

Results

Table 2 shows that the CR rule maximises throughput. Average waiting time for patients treated during the simulation period is greatest under the FCFS rule, as would be expected, and smallest under the THUG/B and CR rule. Each week the model selects patients for treatment and adds new patients to the list. Both the THUG/B and CR rules select patients according to criteria other than how long they have been waiting. In the case of the THUG/B rule patients who join the waiting list in any given week, who can expect a large health gain per bed day from treatment will be admitted immediately. In the case of the CR rule, patients joining the waiting list who require a very short length of stay in hospital will be bumped to the front of the queue and treated immediately -- explaining the small average wait of patients under these two rules. Total health gains from treating patients on the waiting list is greatest as anticipated under the THUG/B patient selection rule. Patients make the greatest health sacrifice as a result of having to wait under the FCFS rule, and far smaller sacrifices under the THUG/B and CR patient selection rules. Recall that these rules had the smallest average waiting time per patient treated of all the rules.

Table 2: Patient Selection Rules

	First Come, First Served (FCFS)	MAXIMIZE Total Health Gain per Bed Day (THG/B)	MINIMIZE Claim on Resources (CR)	Random (RAND)	MINIMIZE Suffering (SUFF)
Total Admissions	1284 (24.226)	1303 (25.456)	1565 (14.565)	1277 (21.101)	1205 (22.888)
Average Waiting Time	31.3 (0.620)	2.56 (0.379)	7.56 (0.195)	22.3 (0.381)	19.7 (0.659)
Average Health Sacrifice Per Patient (QALW)	0.59 (0.013)	0.05 (0.010)	0.13 (0.004)	0.42 (0.008)	0.39 (0.012)
Average Health Gain per Bed Day (QALW)	9.37 (0.175)	15.2 (0.100)	9.6 (0.088)	11.2 (0.142)	7.39 (0.320)

Standard deviations are in parentheses.

Statistical Analysis

One-factor analysis of variance (ANOVA) was used to test the following hypothesis: Null hypothesis (H0): all 6 selection criteria have equal population mean values of outcome variables, versus, alternative hypothesis (HA): some or all selection criteria do not have equal population means for these variables.

The critical value of F, for significance level P=0.05 and with 5 and 594 degrees of freedom for these data, is 2.229: since this was greatly exceeded in all instances here, H0 was rejected in each case in favour of HA. (If any F had not reached this value, the corresponding H0 would not have been rejected.)

Discussion

Different patient selection rules lead to significantly different numbers of patients treated, different waiting times, different levels of health gain achieved from available resources and different levels of health sacrifice as a result of patients having to wait for treatment. While optimum quality in health care is attained when the maximum number of patients are treated, the patients wait a minimum amount of time, each achieves significant health gain from treatment, and there is a minimum level of health sacrifice as a result of waiting time. While the joint optimization of these four criteria can be used as a definition of quality health care, however, none of the patient selection rules simultaneously scored well in all four areas. Further study of equitable trade-offs between the four measurable objectives in the model is needed.

It is hoped that this collaborative project will help to move the waiting list debate beyond waiting times. The "In-Waiting" program facilitates the combination of activity and health gain data for experimental modelling purposes. Since we are not only talking about waiting time but rather and more importantly the rationing of health and possible improvements in health - this paper has demonstrated how collaboration between different fields, and the application of techniques of computer simulation and Operation Research can help to quantify issues of quality and equity in health care.

References

Bailey NTJ. A Study of Queues and Appointment Systems in Hospital Outpatient Departments, With Special Reference to Waiting Time, J.R.STATIST. SOC. 1952; 14: 185-199.

Davies R, Davies T. Using Simulation to Plan Health Service Resources, J. Royal Society of Medicine, 1986; 79: 154-157.

Dowie R. General Practitioners and Consultants: A Study of Outpatient Referrals, King Edwards Hospital Fund for London, 1983.

Drummond MF. Output Measurement for Resource Allocation Decisions in Health Care, Oxford Review of Economic Policy, 1989(b); 5: 59-74.

Edwards R.T. and Barlow J. Rationing Health Care by Waiting List: An Extra-Welfarist Perspective, Centre for Health Economics, University of York, Discussion Paper 114, 1994.

Frankel S, West R. Rationing and Rationality in the National Health Service, The Persistence of Waiting Lists, Economic Issues in Health Care, MacMillan, London, 1993.

Gudex C, Kind P. The QALY Toolkit, Discussion Paper 38, Centre for Health Economics, University of York, 1988.

Haycox A. Constructing a Meaningful Analysis of Waiting Lists: A Case Study, North Western Regional Health Authority, July 1991.

Holliday I. The NHS Transformed, Baseline Books, Nuffield Press, Oxford, 1992.

Higgins J, Ruddle S. Waiting for a Better Alternative, Health Service Journal, 11th July, 1991.

Lindsay CM, Feigenbaum B. Rationing by Waiting Lists, The American Economic Review, 1984; 74: 404-417.

Mullen P. Waiting Lists and the NHS Review: Reality and Myths, HSMC Research Report 29, 1993(a).

Propper C. Contingent Valuation of Time Spent on NHS Waiting Lists, The Economic Journal, 1990; 100: 193-199.

The Royal College of Surgeons of England, Guidelines for the Management of Surgical Waiting Lists, June, 1991.

Taha HA. Operations Research, McMillan Publishing Company, New York, 1989.

Worthington DJ. Queueing Models, for Hospital Waiting Lists, J. Opl. Res. Soc. 1987; 38: 413-422.

Worthington D. Hospital Waiting List Management Models, J. Opl. Res. Soc. 1991; 42: 833-843.

Wright MB. Case Study: The Application of a Surgical Bed Simulation Model, European Journal of Operational Research, 1987; 32: 26-32.

Yates J, Leiu. Damned Lies and Waiting Lists, British Medical Journal, 1991; 303: 303.

Management and Control in the Virtual Supply Chain

Lynne F. Baxter,
James M. Ritchie,
and H.H. Seeto
+44-131-449-5111,

Heriot-Watt University,
Edinburgh, United Kingdom

This paper explores some of the issues arising from the potential implementation of virtual environments (VEs) for the management and control of supply chains. VEs, more commonly known as Virtual Reality (VR), coupled with world-wide telecommunications technology will potentially enable organisations that are physically remote from each other, to operate processes and obtain performance information in other organisations. This paper examines what is theoretically possible with VEs within supply chains and maps this onto research currently being carried out which investigates how manufacturing supply chains operate in the United Kingdom, especially Scotland.

Introduction

Supply chains in the UK are usually dominated by an Original Equipment Manufacturer (OEM), whose recent objectives have been, in theory, to divest themselves of parts of the chain that they do not think that they are capable of adding value to; to replace this they establish partnership relationships with supply networks. Suppliers can be large companies in their own right, however, many are classified as indigenous Small to Medium sized Enterprises (SMEs). The OEM relinquishes day-to-day control of making the part for their product but retains overall power through demanding high performance levels from the supplier in terms of quality delivery and cost.

This paper examines how supply networks have been managed in practice and discusses the implications for this with the future introduction of VEs. The authors use a theoretical framework for the process of introducing technology as a guide in determining how wide area networked VEs might be implemented in the UK, drawing specific conclusions with regard to the management and control of the process. The latter part of the paper investigates how this new form of communications technology might do which will adversely affect the way in which global operations network with their suppliers.

Virtual Possibilities

VEs, coupled with world-wide telecommunications, will revolutionise the way in which products are developed, manufactured and presented in the modern competitive marketplace. This will happen in a number of ways.

Currently, products are delivered to the end customer through activating a chain of processes throughout a network of organisations. The authors contend that globally applied VE technologies will involve simulating scenarios which will result in changing the way new products are made as well as who is involved in the process. For example, through world-wide VE supplier networks OEMs can potentially capture information on the ultimate customer's requirements and not base product design on perceptions of these. One of the most important implications is that the time required for new product development as well as the overall product manufacturing cycle time should reduce substantially. Globalised companies may even have different product development disciplines situated in different parts of the world, (Ritchie et al, 1995); suppliers and customers linking into the chain as they are required in the process. Consequently, this technology will make it appear as if the chain is one organisation but the reality will comprise many organisations and internal company disciplines which will have to link together to manage the process; therefore, this technology is

0-7803-3552-X/96/$5.00 © 1996 IEEE

not likely to be used to its full advantage if its application is not thought out properly and not implemented sensitively.

The 'virtual' possibility considered in this paper is that OEMs will be able to use this technology to test scenarios and even run the manufacturing operations in supplier companies anywhere in the supply chain and that one condition of such a relationship is that suppliers must have VE technology. Whereas previously the OEM negotiated with a supplier as to whether they could make a part to certain performance criteria, they will be able to test whether this is the case and, in theory, link into the supplier's information and manufacturing systems, directly affecting and influencing the making of supplied parts. The consequences of such a situation for suppliers, especially SMEs, are enormous and could interfere considerably with their profitability and their ability to serve other customers.

Supply Networks

Harland (1995) notes that there are a number of ways in which firms can interact with one another and ultimately form a network. The specific ways supply networks are configured can differ, but the main issues concern the breadth of a network and the number of tiers it has between raw material supply and the final customer. Harland also states that telecommunications technology can alter the way a network operates. Most people are now familiar with the VE called the Internet; imagine the communications revolution which will take place when information and ideas can be exchanged across a similar VE only with the full immersion capabilities of the type now being demonstrated in currently marketed immersion-based VR systems.

Nishiguchi (1994) classifies networks three ways:

- vertical integration, where a firm owns most of its supply network;
- strategic dualism, where the firm retains key areas of production but out sources components;
- clustered control, where an assembler deals with a first tier of suppliers who, in turn, manage the second tier and so on.

Using VEs there exists the possibility of managing both clustered control and strategic dualist networks as if they were vertically integrated. Being on an information technology network enables simultaneous communication throughout the system which would simulate vertical integration more effectively.

Many organisations are moving away from vertical integration to network configurations for supply chains. This was seen as a solution to implementing themes identified as being successful in Japanese manufacturing contexts, such as just-in-time manufacturing and total quality management or Western ideas in strategic management, such as 'core competencies' and 'value chain analysis'. These developments imply the disintegration of supply networks in tune with Western thinking but closer co-operation between chain linkages to manage the delivery and quality issues akin to Japanese thinking. Closer links are also desirable if an organisation is to maintain overall control when dispensing activities because this process will in the long term reduce the company's knowledge and ability to control the out sourced activity. VEs applied in the way suggested may see the rebirth of vertical integration as a strategic business approach.

Many writers, for example Saxenian (1994), have pointed out the benefits of having close regional networks to spur innovation and develop new products. These benefits come about through the operations of 'philos' style relationships (Krackhardt, 1992) or strong ties between components of networks. This is because of three related conditions:

- network organisations co-located in one region have the potential for more face-to-face interaction, therefore more information is passed;
- there is a degree of affection between network participants which, in turn, helps performance;
- region based networks are established over time, which helps develop an experience base for other organisations to draw upon.

VEs could either undermine regional networks or be used to shorten the time taken to establish them, depending on the development trajectory chosen by network participants.

Developing closer links is strongly associated with a reduction in the overall supply base and organisations are encouraged to rationalise the numbers they manage and single or dual source components (Lamming, 1993). The effort organisations expend in setting up these arrangements form a powerful barrier to entry for prospective suppliers. The question is: "Will VEs enable the use of a wider network of suppliers for OEMs or enhance the current practice of restricting the number of companies involved in the supply chain?"

Supply Chains in Scotland and the rest of the United Kingdom

In the main as suggested by Nishiguchi (1994), organisations have been moving away from vertical integration to the other two modes. The rhetoric may well be of partnership and co-operation in developing in new relationships (Carlisle and Parker, 1986) but practice suggests that most manufacturing supply chains are reducing the numbers before partnering suppliers. Purchasing departments are culling relationships but maintaining a 'hands-off' approach with everyday operations providing performance levels are achieved (Waters-Fuller, 1996). Western companies in JIT environments also tend to liaise with the layer below them in the supply chain, VEs may enable them to copy the Japanese approach of multi-layer supply chain involvement.

There is little evidence of the type of regional networks developing as a boost to innovation, but there are several initiatives by government agencies, such as Scottish Enterprise, to create and sustain viable supply bases at an operating level to support inward investing OEMs. It would be fair to say that elements of the 'philos' relationship can be seen, especially in a Scottish context, with meetings, visits, sports events etc. providing opportunities for information exchange and bonds to develop. How deep these actually run is open to question.

Currently, the suppliers and supporting organisations would like VEs to be used to develop regional networks but the way the OEMs have been influenced by strategy and operations rhetoric would suggest that it would be implemented to undermine them by providing a way of forming closer ties with more distant suppliers and simulating a regional network on a global scale. Thus, as a consequence of this, the key problem is how to influence events.

Networks for Developing and Implementing Technology

Callon and Latour note that a key factor on whether a technology becomes popularised, and along which trajectory it develops, is dependent on whether the people involved in creating it are skilled at forming coalitions to guard and guide the development forward. This is known as the 'sociology of translation' and has four phases or 'moments'. The first moment is problematization; one group of people try to show another that they have the correct solution to their problem. They have to lead the target group through a series of practices the original group can define and control. These are known as 'obligatory passage points'. The next phase is where momentum is developed through the enrolling group positioning itself between the target group and other groups; in fostering relationships the enrolling group has control over and is known as 'interessement'. The phase after that is known as 'enrolment' and occurs when the alliances and coalitions agree on what they want the outcome of the process to be. The last phase is called mobilisation, where the enrolling agencies control the enrolled to preserve the pattern of interests which is satisfactory to the enrolling agency. Callon and Latour do not expect the process to be successful every time.

Callon (Law, 1986) apply their theory to scallop fishing off the coast of Brittany in the seventies. He noted that a conference was held in the early seventies to discuss the industry, because what was a very lucrative business looked as if it might fritter out because of over fishing. Three scientists managed to enrol all the interested groups in their network and guided the process to develop their prefered solution. A new method for propagating larvae was heralded and funds were assigned to implementing it. The academics convinced the fisherman and economic development agencies to support their preferred solution. They tested and implemented their technology. However the process can be undermined. The scallops did not behave predictably, and the academics were undermined.

Implications for the Development and Implementation of VEs in Supply Networks
The authors would argue that it is possible to use the network translation framework to help project what might happen with VEs in UK. supply chains. There are a number of groups which need to be 'enrolled' if this process is to be managed successfully for either of the likely protagonists. The framework is particularly useful in allowing the technology itself to be seen as a player.

One likely scenario is that OEMs will argue that they are in the position to dictate how the technology is designed and implemented. They might argue that the technology is necessary as a solution to compete with their rivals and should be adopted by the SMEs because they will lose the business elsewhere. They will gain support of the Economic Development agencies, who will be keen for the SMEs to embrace the new technology. The technology could then be used to eliminate steps in the manufacturing process and remove control from the SMEs for their daily operations. This could reinforce the small network-distant ties ethos which exists at the moment and undermine the suppliers profitability, which would lead in the end to the VE being tested locally and operated on a global basis!

A less likely scenario, but one which the SMEs and Economic Development agencies might like to consider, is to develop their knowledge of VEs to 'interest' the OEM in using it in a way which supports philos and region-based networks and enrols them into that process. These technologies are notoriously difficult to make operational and establishing yourself as a knowledgeable actor in this process could yield substantial dividends, including developing a customer base on a global scale.

This highlights the final point, the technology might not collude with either of the likely protagonists' objectives. In order to be managed effectively, it should be treated as an actor in the process in its own right or else none of the organisations discussed will gain the full benefits, rather like the scallop example above. Electronic data interchange (EDI) and other communications technologies are not as widespread in their use as one might believe because simpler methods of communicating are more effective.

Conclusions

In this paper we have discussed some possibilities for and implications of VEs in the supply chains. A framework has been used to underline that there are alternatives in the way that VEs could be adopted and that this is likely to be a very social process. The contradictions in theory and practice are likely to be replicated in the trajectory that VE might take. OEMs may use it to reinforce their position in local networks which, in turn, could lead to their demise. However, the framework enables another trajectory to be projected which could lead to local (regional) networks being reinforced. This contradiction on its own required further investigation.

References

J.M. Ritchie, et al (1995) 'Using Virtual Reality for Knowledge Elicitation in a Mechanical Assembly Planning Environment' 12th Conference of the Irish Manufacturing Committee, Cork, Ireland.

Saxenian, A. (1991) 'The Origins and Dynamics of Production Networks in Silicon Valley', Research Policy, vol 90, no 3.

Nishiguchi, T. (1994) 'Strategic Industrial Sourcing', Oxford University Press, Oxford.

Harland, C. (1995) Networks and Globilisation: A Review of the Research, Warwick Business School, Warwick.

Krackhardt, D. (1992) 'The Strength of Strong Ties' in Nohria N. and Eccles R. 'Networks and Organisations' HBS Press, Boston.

Lamming, R. (1993) Beyond Partnership, Prentice Hall, London.

Carlisle, J and Parker, R. (1986) Beyond Negotiation, Wiley, Chichester.

Waters-Fuller, N. (1996) The Implications of JIT for Suppliers, unpublished PhD dissertation, Napier University, Edinburgh.

Law, J. (1986) Power Action and Belief, a New Sociology of Knowledge? RKP, London.

Virtual Enterprises and the Growth of the Information Sector in Canada

Hadj Benyahia
(514) 987-3497

Université du Québec à Montréal
Département d'informatique
C.P. 8888, Succursale Centre-ville
Montréal, Québec, Canada
H3C 3P8

Virtual enterprises are essentially information enterprises. This paper presents an assessment of the economic impact of such enterprises by an analysis based on an estimation of the number of information workers in Canada, their contribution to the national production and their labor productivity.

Introduction

The objective of this study is to analyze a major aspect of virtual enterprises, namely their informational production. Indeed virtual enterprises are essentially information enterprises (Davidow 1993) as their principal functions consist of creating, processing and disseminating, or of providing an infrastructure of information support. As a result, virtual enterprises form an integral part of a new sector of the economy referred to as "Information Sector " (Miles 1990). The research effort therefore, consists of evaluating, for the first time, the contribution of this sector to Canadian economic growth.

Main Components of the Information Sector

The information sector may be compared to an iceberg, the visible portion of which represents the New Information Technology (N.I.T.) sector, which groups together computer and telecommunication industries (components, hardware and software). These industries only contribute 1 to 2% to national employment and Gross Domestic Product (G.D.P.), but they are of strategic importance as they serve as the engine for the vast information sector. The hidden area of the iceberg represents not only the traditional information industries existing prior to the electronic revolution of the early 1980's (e.g. telephone, radio and television, editing, financial services, education, public administration industries, etc.), but also those industries which utilize informational products and services for the production of non-informational goods and services (e.g. automobile and aeronautical industries). According to the Organization for Economic Cooperation and Development (OECD 1993), information occupations are classified into four main categories, as follows:

Table 1. OECD Classification of Information Occupations

Information Producers	Information Processors	Information Distributors	Information Infrastructure Occupations
- Scientific and Technical - Market search and co-ordination specialists - Information gatherers - Consultative services	- Administrative and management - Process control and supervisory - Clerical and related	- Educators - Communication specialists	- Information machine-workers - Postal and telecommunication

0-7803-3552-X/96/$5.00 © 1996 IEEE

Evaluation of the Contribution of the Information Sector to Canadian Economic Growth

This section presents the results of a comparative analysis concerning the evaluation of employment, production and productivity in the information sector, in the N.I.T. Sector and in the whole Canadian economy for 1980-1990, the period for which data is available for estimation.

The Contribution of Informational Employment to National Employment

Table 2. Evolution of Employment in the Information Sector and in Total Canadian Economy: 1980-90

	1980	1985	1990	1980-85	1985-90
Information Sector					
- Employment	5,500,530	6,043,555	7,112,615	—	—
- Annual growth (%)	—	—	—	1.9	3.3
- % of total economy	44.8	46.2	47.7	—	—
N.I.T. Sector					
- Employment	85,823	91,790	131,995	—	—
- Annual growth (%)	—	—	—	1.5	7.5
- % of total economy	0.7	0.7	0.9	—	—
Total Economy					
- Employment	12,273,255	13,074,460	14,905,395	—	—
- Annual growth	—	—	—	1.3	2.6

Source: Employment in the information sector has been calculated from catalogues no 93-116 and 93-332 of Statistics Canada. Employment in the N.I.T. Sector and in the whole economy was provided by catalogues no 31-203; 56-001; 56-205; 63-222.

Table 2 shows that the information sector is on the threshold of becoming the largest employment sector in Canada by providing almost 50% of jobs in 1990. One can also see that the employment growth rate in this sector is greater than that of the total economy. As a result, informational employment is less vulnerable to periods of economic recession (e.g. 1980-85) than national employment.

The N.I.T. Sector is a new sector that has only begun to develop since the early 1980's with the emergence of the micro-electronic revolution. As a consequence, its employment accounted for less than 1% of national employment in 1980-85 with a modest annual growth of 1.5% during this takeoff period. Conversely, since 1985, the N.I.T. Sector has undergone a rapid employment growth: twice that of the information sector and nearly 3 times that of national employment. We can therefore say that the N.I.T. Sector has its own dynamics of employment since, although its direct jobs are relatively few, this sector is of strategic importance for the creation of indirect jobs in the information sector by providing it with new products and services.

Table 3. Evolution of Information Categories in Canada

	Relative Shares (%)			Annual growth of employment (%)	
	1980	1985	1990	1980-85	1985-90
Information producers	21.7	22.6	23.5	2.7	4.1
Information processors	61.4	60.6	60.5	1.6	3.2
Information distributors	10.6	10.4	10.3	1.4	2.9
Information infrastructure occupations	6.3	6.4	5.7	2.1	1.0
Total Information Sector	100	100	100	1.9	3.3

Table 3 shows that employment in information processing is the most important occupational category of the information sector. Even if the relative share of this category has varied little (about 60%) during the 80's, its internal composition has greatly changed as a result of computerization. Indeed, within this category, many occupations are declining (e.g. bookkeepers, supervisors, inspectors, viewers, testers, typists, etc.) while others are growing (e.g. general managers, managers, correspondence and reporting clerks, word processing operators).

The information producers' category has seen a growth in its relative share due to the rapid rise in scientific and technical personnel (e.g. engineers, systems analysts and computer programmers whose number have more than doubled during the 1980's). As a result, the annual growth rate in this category was greater than that of the information sector as a whole.

The information distributors' category consists mainly of teachers and communication specialists for which the annual growth rate has doubled from 1.4% in 1980-85 to 2.9% in 1985-90.

The last category concerning the information infrastructure occupations presents the lowest relative share and employment growth in the information sector. This situation is caused by the reduction of computer operators, telephone operators, telephone and telegraph linesmen, compositors and typesetters and printing engravers.

The Production of the Information Sector and Its Contribution to National Wealth of Canada

The production of the information sector can be expressed by the following economic relation: Production of the information sector = Labor income + Capital income

$$= \sum_{i=1}^{i=n} (S_i * E_i) + \left(\frac{36}{64}\right) \sum_{i=1}^{i=n} (S_i * E_i)$$

Relationship in which:

n = number of information occupations.
S_i = average employment income of the ith informational occupation.
E_i = number of employees in the ith information occupation
$\left(\frac{36}{64}\right)$ = ratio between capital income share (36%) in G.D.P. and employment income share (64%) in G.D.P.

Table 4. Evolution of the Information Production and G.D.P. in Canada: 1980-1990

	1980	1985	1990	1980-85	1985-90
Information Sector					
Total employment income (constant $ million)	115,271	137,921	173,751		
Average employment income (constant $)	20,956	22,821	24,428		
% of national employment income	49.1	52.3	55.2		
Production (constant $ million)	180,111	215,501	271,485		
Production as % of G.D.P.	47.1	49.2	53.7		
Annual growth (%) of production	—	—	—	3.6	4.7
N.I.T. Sector					
Total information income (constant $ million)	1,263	1,665	2,930		
Average information income (constant $)	25,002	28,426	31,649		
Production (constant $ million)	5,049	6,459	10,331		
Production as % of G.D.P.	1.3	1.4	2.0		
Annual growth (%) of production	—	—	—	5,0	9,8
Total Economy					
Total employment income (constant $ million)	234,827	263,554	314,436		
Average employment income (constant $)	19,133	20,157	21,095		
G.D.P. (constant $ million)	381,992	438,450	504,787		
Annual growth (%) of G.D.P.	—	—	—	2.7	2.8

Note: Use of constant dollars eliminated the artificial overstatement of revenues due to inflation. For this purpose the G.D.P. price index has been utilized, with 1986 as the base year=100; 1985=97.7; 1980=73 and 1990=118.5.

Source: Estimation of employment income and production of the information sector are calculated from Statistics Canada (Catalogues no 93-116 and 93-332).

Table 4 reveals the following main observations:

The information sector has become the largest sector of economic activity in Canada since 1990, accounting for more than half (53.7%) of Canadian production. The annual growth of production in this sector (3.6% and 4.7%) was greater than that of the whole economy (2.7% and 2.8%). This growth is even more significant when we consider the N.I.T. Sector (5.0% and 9.8%).

While employment in the information sector represents 44 to 47% of national employment (Table 2), the income generated accounts for 49 to 55% of employment revenue for the whole economy. This difference indicates that the average salary in the information sector exceeds that of the total economy (24,228$ compared with 21,095$ in 1990). As a result, the information sector exercises a redistribution effect that reduces the disparity in salaries in the whole economy. This favorable effect can be explained by the high percentage of managers, specialists, technicians and educators whose mean annual revenue is greater than the national average. It may thus be anticipated that the continuation of the process of computerization will lead to a better salary equity by reducing the gap between high and low wages in the Canadian economy.

Evolution of Productivity in the Information Sector and in the Whole Canadian Economy: 1980-90

Among the numerous definitions of productivity, the simplest and the most used has been chosen, namely labor productivity. This corresponds to the ratio between national or sectorial production (G.D.P.) and national or sectorial employment.

Table 5: Evolution of Productivity in the Information Sector and in Total Canadian Economy: 1980-1990

	Productivity (constant $)			Annual growth (%)	
	1980	1985	1990	1980-85	1985-90
Information Sector	32,774	35,658	38,169	1.7	1.3
- Information producers	39,475	46,548	49,090	3.3	1.1
- Information processors	29,635	31,554	34,107	1.2	1.5
- Information distributors	40,212	39,360	41,341	-0.4	0.9
- Information infrastructure occupations	27,276	30,102	30,587	1.9	0.3
N.I.T Sector	59,207	70,373	78,294	3.5	2.1
Total Economy	31,124	33,534	33,866	1.5	0.2

Table 5 shows, as expected, that productivity in the information sector exceeds that of the total economy in the 1980's. Productivity in the N.I.T. Sector, in particular, has been approximately double that of the information sector and of the whole economy. It then follows that the information sector contributes to the improvement of productivity gains in the entire economy. Moreover, the N.I.T. sector is identifiable as a high productivity sector.

Among the four professional categories of the information sector, another expected result can be observed, namely, the category of information producers achieves the highest level of productivity. On the other hand, the categories of support and processing show the lowest productivity. The category of information distributors achieves an intermediate productivity level since 1985-90.

Lastly, and chiefly, Table 5 includes an important result which confirms the productivity paradox observed in the United States (Keyes 1995). Indeed, even if the value of productivity in the Canadian information sector increased in the 1980's (32,774$ to 38,169$), its annual growth rate has diminished from 1.7% in 1980-85 to 1.3% in 1985-90. This decline in productivity growth is also seen in the N.I.T. Sector, from 3.5% to 2.1%. This paradox is justified if one considers that it is precisely in the decade of the 80's that American and Canadian enterprises made massive investment in information technology to improve the productivity of their employees. A review of the scientific literature (Dué 1993 and 1994) allows to identify the main causes of this paradox as follows:

- The learning curve: According to this explanation, we are still in a transition period, as the micro-electronic revolution is only 15 years old, and therefore relatively new. Several historical examples from the industrial revolution (e.g. electricity, tool machinery and telephone) demonstrate that there is a considerable delay (20 to 60 years) between the introduction of a new technology and its mass distribution resulting in substantial productivity gains.

- Difference between technical and Organizational Productivity: The fact that technology has a potential impact on users, as well as on organization, does not imply that this potential will evolve into a real impact. The reason is that technology by itself is a necessary but not sufficient condition to achieve productivity gains. Organizational changes must accompany technological advances. For example, the advent of the paper-less office, expected in the 1980's, was not realized and even a contrary effect has resulted, because of mismanagement (lack of strategic planning and lack of project management).

- Problems in quantifying productivity of information technology: According to R. Dué (1993), 95% of U.S. companies have no metrics program to measure the productivity of their investment in information technology. Moreover, cost benefit analyses of information systems often take into account only short-term tangible benefits although a large part of these benefits are intangible and only come in the long term because of the learning curve. Finally, it must be noted, that 80% of investment in information technology is realized in the service sector, whose productivity cannot be accurately measured.

References

Davidow, M.H. and Malone, M.S. "The Virtual Corporation". Harper Collins Publishers, Scranton, PA, 1993.

Dué, R. "The Productivity Paradox". Information Systems Management. Winter: 69-71, 1993.

Dué, R. "The Productitvity Paradox Revisited ". Information Systems Management. Winter: 74-76, 1994.

Keyes, J. "Solving the Productivity Paradox ". McGraw Hill Inc., New-York, 1995.

Miles, I. "Mapping and Measuring the Information Economy ". the British Library Publications. Report 77, Great Britain, 1990.

OECD. "Usage Indicators: New Foundations for Information Technology Policies ". Report 31, Paris-France, 1993.

Statistique Canada. Ottawa, Canada.
Catalogue no 93-116 and 93-332. "Employment Income by Occupation "
Catalogue no 31-203. "Manufacturing Industries of Canada".
Catalogue no 56-001. "Communications".
Catalogue no 56-205. "Cable Television".
Catalogue no 63-222. "Software Development and Computer Service Industry".

Acknowledgment

The author would like to thank the Department Industry Canada for its financial support, Dr. Probst for his helpful comments and A. Belayachi for his work as research assistant.

Profile of the Virtual Employee and Their Office

Dana Mary Bracco
(810) 293-7831

Wayne State University
808 Sunset Lane
St. Clair Shores, MI 48082-1248

Successful virtual enterprises are highly dependent on productive and efficient employees. A mix of technical knowledge, behavior attributes, and leadership skills enable a virtual associate. Enterprises can no longer employ professionals from "cradle to grave." In a competitive, global environment where value added and cost containment are key, the virtual employee reins.

Introduction

Businesses are rapidly changing to meet their number one goal, customer satisfaction. As competitive factors force companies to rethink how they do business, the virtual enterprise represents a way to maximize resources at a minimum cost. This change will have a major impact on the employees who will provide services to the virtual enterprise. The "virtual employee" or "virtual associate" will face many challenges in the rapidly changing business world.

The speed at which change occurs is increasing. The need for continuous learning is required. The methods and knowledge to adapt to technological information exchange is necessary.

Who are these virtual associates? Where do they work? How do they work? How do they keep current? These questions and many more will soon be answered.

Virtual Associate and Their Relationship with Business Enterprises

The term "employee" will disappear as more and more individuals choose to become "virtual associates". Life time employment is in the past. The future will result in a mirage of multiple relationships with the virtual associate responsible for their own destiny. The virtual associate is responsible for managing their career. The virtual associate determines what type of work they want to perform, what skills, knowledge, and experience are required, and who the enterprises are that will utilize their services. Entrepreneurial and "just-in-time" workforce are ways to describe the virtual associate.

The virtual associate will possess certain technical skills, behavioral attributes, leadership and business skills, and continual learning ability that will enable them to survive in this rapidly changing technological environment.

Technical Professionals

Although there are many professionals who can function as a "virtual associate", the focus will be limited to technical professionals. Technical professionals are individuals who have been faced with change for years. The technical professionals are defined to include engineers, designers, architects, computer scientists, and technical support. They can be employed in many fields and provide a service or a product as the end result of their efforts. Technical professionals who have not kept pace with the recent changes in technology, whether it is information or tools, will not be able to keep up in the future. The rapid pace of change will require continuous learning. There are certain skills that a technical professional needs to possess and will need to constantly be aware of what knowledge and skills are required by the virtual enterprises.

0-7803-3552-X/96/$5.00 © 1996 IEEE

Technical Knowledge and Skills

A base knowledge will be required. Academic institutions will need to build more application into the curriculum. There will no longer be time for on-the-job training as it is currently known today. Laboratories and simulation will provide a means to train technical professionals in an academic setting. More partnerships will be formed with business and academia. The focus will be on core competencies supported by business and behavioral skills.

The virtual associate will be required to bring a skill set, knowledge, talents, and experience to the enterprise they will provide services to.

Computer literacy and diversity in application is a requirement. The ability to adapt rapidly to changing software applications and hardware technology will be a part of the virtual associate's continual learning process. Identifying needs of the customer will allow the virtual associate to always be utilized. The virtual associate is responsible for their own productivity and efficiency.

Problem solving skills, concurrent work, modeling and simulation, soft prototyping, and data access capabilities are essential.

A virtual worker, in any specialty or role, faces the challenges of complex, simultaneous work, teammates outside the organization, rapidly evolving specialized knowledge, continuous change, and the need to learn, manage, lead, and work using new technologies. Those who work in virtual environments must be highly competent in all these dimensions. These are highly trained performance people. (Grenier, Metes 1995) Often these virtual associates are referred to as knowledge workers, possessing specialized skills to meet the needs of a virtual enterprise.

As a part of a team, a virtual associate can serve in different capacities. They can serve as a project leader, a virtual associate, or a technical reviewer depending on the specialized knowledge required for the project. The components of each project will vary and the competencies needed to achieve the desired results or outcome need to be completed in a complex, collaborative manner.

Behavioral Attributes

The virtual associates will need to adjust to many changes. No longer will the virtual associate report to a building or an office outside the home. There is psychological and social adaptations that will need to occur, especially for associates who are used to "going to the office." There may be feelings of isolation or loneliness in the transition stage. Associates will be faced with constant change and stress because the work will become more demanding with greater time compression. The high rate of change will provide challenges and require that the virtual associate make decisions on a regular basis and may often face opposing views but will be prepared to provide the facts based on their knowledge.

Self-starter, team oriented, flexibility, collaborative, and leadership qualities are attributes of the virtual associate. These high performance people will not be passive and have the ability to work in multi-functional, cross-cultural situations. The virtual associate is disciplined and has a mission. Other desirable traits include competence, honesty, trustworthy, reliability, and commitment.

Opportunities are more available to the individuals who possess the behavioral attributes necessary to excel in a virtual enterprise. They expect and accept change as the way it is done. They are aware of the demanding environment and thrive in an environment of rapid change. They are able to function as a part of a virtual team to accomplish the project or project(s) they support.

Being a virtual associate is not for everyone. Failure to support the requirements will result in the inability to be employable or earn a healthy income. "Managing career" is the responsibility of the virtual associate, not of a business organization. Leadership and business skills complement the behavior and technical capabilities.

Leadership and Business Skills

"Managing career" is more than making sure the core competencies are up-to-date, it requires investment in continual learning and technological tools to be able to complete the projects to be supported.

The virtual associate must have the ability to market their expertise, to manage their time, to finance their technological investments, and to successfully develop business relationships. In a sense, the virtual associate is an entrepreneur. Developing a plan to position themselves requires established goals and continual reviewing progress towards those career goals.

No longer will the virtual associate receive a paycheck. The virtual associate will be paid a market rate for their capabilities and successful execution requirements on projects. The virtual associate will be responsible for their own benefits and plans for retirement.

The key to success for a virtual associate is to ensure they possess the proper combination of technical, behavioral, and leadership/business skills. This can only be accomplished through a process of continual learning.

Continual Learning

Knowledge of self is another critical area of self-awareness in the virtual work relating to one's own learning style. Some people learn by doing, some through class lectures. Some are receptive to visual information or sound or writing. Some learn sequentially or analytically; others learn by tackling whole concepts. Virtual learning is learner centered. The virtual associate is responsible for accessing information and knowledge suited to the situation and the form that suits their learning style. (Grenier, Metes 1995)

The way the virtual associate learns will change. The traditional classroom and training setting may not be appropriate or available when the skills are needed. Alternative training will be available through video and audio conferencing, virtual courses on the Internet which represents the need for distance learning. Virtual associates can gain the knowledge in a "just-in-time" approach. This will place pressure on successful completion and acquisition of requisite skills.

Increasing complexity, constant changes in processes and technologies, pervasive and shifting needs for boundary, or cross-cultural knowledge, and dependence on shared "knowledge assets," demands an accelerated, breakthrough development of capabilities. (Grenier, Metes, 1995)

Experiential learning, action learning, and interactive computer learning are methods to keep the virtual associate challenged. There is a need for competency not just facts and information. Application is critical. Focus on shared learning and team based approaches will need to be integrated. Use of information technology is a requirement.

Virtual associates must be willing to learn and adapt to changing roles and competencies which is the necessary requirement to move forward.

Constant changes in technology will require skills to be updated on a more frequent basis. Being able to utilize the technology to complete a project or product are important. The virtual associate will be responsible for determining the "tools of the trade" and having the capability to efficiently use the technology.

Technology and Tools

In the past few years, technology has changed at such a rapid pace that today's equipment is out of date tomorrow. It is the responsibility of the virtual associate to purchase the equipment and technology to be able to complete projects. No longer will an enterprise be responsible for supplying the necessary tools to perform the work. The virtual associate will be compensated to enable them to continually invest in the necessary technology.

The ideal situation would be an integrated system of all the technologies: Fax, Telephone, Mobile Phone, Pager, Computer, Imaging Technology, Printer, Videoconferencing, Audioconferencing, E-mail, and Voice Mail. The "complete wireless data solutions" implies having the ability to access the information whenever and wherever the virtual associate needs it.

Changing data requirements and format pose special issues for the virtual associate who may be serving multiple enterprises. Platforms will need to be standardized to be able to communicate effectively and productively.

Integrated Services Digital Network (ISDN) availability will greatly increase the speed of communication which is much faster than modem. Availability of ISDN varies greatly by region within the United States and because of incompatibility of equipment purchases by local phone companies, ISDN subscribers can't always be sure that they'll be able to communicate with ISDN subscribers in other areas. Seamless ISDN communication across the country should be possible by 1996 (Mannes, 1995).

Work Environment

The virtual associate can work virtually anywhere. They are portable. As Bleecker (1994) states, everyone can be a "Road Warrior" who uses mobile communications technologies to break the time and space barriers of the enterprise. The computer-telephone-fax combination allows businesses to operate anytime, anywhere.

Proximity is not important. Virtual enterprises will have greater access to global resources allowing them to have the resources required to complete their projects or products. Technology such as video and audio conferencing become increasingly important to observe nonverbal factors (video) or to listen to the tone or inflection in voice (video and audio). This could eliminate the traditional meetings as they are today. The nonverbal and verbal aspects are lost in pure computer or data interactions.

Virtual offices truly save money, but they require some upfront investment. Virtual associates will be able to contact anyone without really knowing where they are at. No longer will enterprise be concerned with colocating. They will be concerned with codestiny, the fate of each virtual associate is dependent on the other (Byrne, 1993).

Barriers and Adaptation Issues

There will be a loss of power. Throwing out the old ways to run a business and bringing in the new ways of operating an enterprise. Small enterprises will have an advantage of larger enterprises. Although companies have been empowering employees for years, the virtual associate is truly empowered and independent. Resistance to change will result in great frustration and possible loss of work for those who fail to adapt.

No longer is there security. No one will take care of you from cradle to grave. The virtual associate is on their own. There will be shock and fear of the unknown. Stress, demands, time compression, and rapid change will place pressure on the virtual associate.

In this transition, some will enjoy the new found freedom and strive in the new environment. Others will have a difficult time in adapting and feel greater frustration and stress than ever before. This changing paradigm will require changes in structure and operation. There are several benefits for the virtual associate.

Benefits and Opportunities

This new work arrangement will provide different patterns for working. The virtual associate will have great flexibility in the time and place they complete their work as long as they achieve the established deadlines. They will have the ability to select the type of work which best matches their capabilities. The travel time to work will be greatly reduced if not eliminated altogether. The virtual workspace will be designed by the virtual associate to meet their needs.

The virtual associate will have more control over the work they do. They will need to form relationships with several virtual enterprises. This will require that they use a project management approach to scheduling and completing their responsibilities.

Conclusion

As hierarchical structures in business organizations virtually disappears, the terms "company", "manager", and "employee" will be replaced with "enterprise", "leader", and "associate". The business climate has grown globally and rapidly. Keeping up with the changing technology and resource requirements can be very challenging for an enterprise. As customers are willing to pay only for value added products and services, cost containment is an important issue.

The virtual enterprise will use only the resources they require, when they are required. Changing needs will require changing resources. A virtual enterprise can no longer afford to "carry an employee from cradle to grave." It is no longer cost effective, efficient, or productive to operate in this manner.

The virtual enterprise is born. The time has arrived for the virtual associate who operates efficiently and productively to meet the latest resource requirements for the virtual enterprise.

Bibliography

Bjerklie, David and Cole, Patrick E. "Age of the Road Warrior," Time, Spring 1995.

Bleecker, Samuel E. "The Virtual Organization," The Futurist, March/April 1994.

Byrne, John A. "The Virtual Corporation," Business Week, 8 February 1993.

Davidow, William H. and Malone, Michael S. The Virtual Corporation. (New York: Harper Business, 1992).

Grenier, Ray and Metes, George. Going Virtual: Moving Your Organization Into the 21st Century (Upper Saddle River, New Jersey: Prentice Hall P T R, 1995).

Handy, Charles, "Trust and the Virtual Organization," Harvard Business Review, May/June 1995.

Mannes, George. "The Virtual Office," Popular Mechanics, March 1995.

Requirements for Virtual Enterprise Management in Developing Countries

Prof. Dr. Walter Eversheim
+49 241 807405

WZL RWTH Aachen
Steinbachstr. 53B
D-52056 Aachen Germany

Dr. Carlos Frederico Bremer
+55 16 2749237

SEM EESC USP
Av. Dr. Carlos Botelho 1465
13560-250 - São Carlos - SP - Brazil

Jörg Kampmeyer
+49 241 807384

WZL RWTH Aachen
Steinbachstr. 53BSteinbachstr. 53B
D-52056 Aachen Germany

The management of Virtual Enterprises (VE) in developing countries faces different challenges then in industrialized countries. However, using the framework proposed by the Agility Forum as reference, some analysis can be done. In developing countries deficits in infrastructure have to be regarded, yet the cultural behavior in these areas outlines potentials for a successful application of VE. This paper discusses the adequacy of the VE strategy for industries in developing countries, their local infrastructures, and requirements towards the VE management. Finally, the approach of the „GVE testbed" towards an accelerated implementation of VE projects between Brazil and Germany is described.

Introduction

Global competition forces industries all over the world to produce on a world class level. That means, that even industries in developing countries will have to produce with quality, cost and lead time competitive to companies from industrialized countries. However, seeing agility as a necessity for maintaining global competitiveness (Goldman et al. 1994), the Virtual Enterprise (VE) strategy also has to be incorporated in industries from developing countries.

The Virtual Enterprise (VE) is a temporary network of independent companies linked by information technology to share skills, cost and access to another markets. Nowadays, new organizational and technological developments, e.g. more flexible organizations, better and faster telecommunication and improved logistic concepts, are supporting the formation of such networks. However, these new concepts and methodologies may not take into account the specific needs of industries in developing countries. This paper aims to describe an approach towards the integration of companies from developing countries with regard to the specific infrastructure situation in those areas. Therefore, the Agile Virtual Enterprise (AVE) Case Framework proposed by the Agility Forum (Goranson 1995) will be used as a reference system. After some analysis about the infrastructure and the management activities in developing countries an existing project and some of its partial results to establish a VE testbed between Brazil and Germany is described.

Management of virtual enterprises

The management of a VE can be understood as the entire management of a product´s life cycle including all necessary processes and infrastructures. The AVE Case Framework covers this requirement by synchronizing the VE life cycle processes with the product life cycle. However, in the following the management of a VE should be understand as the effective use of the AVE Case Framework.

The more a network organization like the VE becomes a hollow organization, the

0-7803-3552-X/96/$5.00 © 1996 IEEE

more its management has to concentrate on boundary spanning, that means the market oriented bundling of the partners` complementary core competencies and the efficient management of the entire value chain. Therefore, most hollow networks are managed by broker institutions that take care of all coordinating and configurating activities, so that the involved companies can focus only on their own core competencies (Miles/Snow 1986, Ritaine 1990). Regarding to the AVE Case Framework, the broker has to take care, that all necessary infrastructure is available and the single processes of the VE are synchronized to market demands. Thereby, the broker institution itself has to achieve a high level of agility, since its function changes from an architect of a new network to a sort of a lead operator, that configures all resources and operations within the cooperative network. Further on, the broker has to function as a caretaker, that provides the partners with upstream or downstream information throughout the entire supply chain. However, the role of the broker is crucial for the VE`s success. In the following, a strategy will be described that enables an accelerated implementation of the VE strategy within developing countries by an expended use of the broker concept.

Adequacy of the virtual enterprise strategy for developing countries

The application of the VE strategy can be seen as a promising approach, since it offers small and medium sized enterprises (SME) from developing countries the opportunity to globalize their business without loosing their flexibility and their economic independence within traditional long term oriented and very rigid cooperation with industries from industrialized countries.

The successful application of the VE strategy relies on the existence of an appropriate infrastructure. Therefore, the specific situation in developing countries has to be analyzed. Since a generalization of infrastructure in developing countries is difficult to be done, this paper focuses on a more specific analysis of the infrastructure in South America.

Physical and legal infrastructure in South America do not need special attention, since production within international cooperations has become daily business for local industries. However, cultural infrastructure can be considered as an important advantage for companies from developing countries since some of the characteristics of the VE strategy has been already applied to cooperations. The fault of some resources in developing countries traditionally has forced their industries to project and manufacture products in an outstanding creative and innovative manner, using sometimes cooperations with other industries as a solution to get the necessary resources. These cooperations are not always formalized in the sense of an official collaboration. However, in a specific sense the VE strategy represents an already frequently used strategy of companies from developing countries.

Information infrastructure is the critical one for the implementation of the VE strategy in developing countries, since it depends on a high investment in communication technologies. In some countries the cost to get a 64Kb line is very high, in others it just does not exist. As an example, in the state of São Paulo, which is the most industrialized one in Brazil, since January a 64 Kbit public line is available for about 666 US$ per month (Informatica 1996). A more satisfying alternative is the achievement of a private line with a data exchange rate up to 2 Mb. However, the enormous cost of such a line makes it inaccessible for small and medium sized companies. Standards for exchange of data are not wide spread as well. Electronic Data Interchange (EDI) protocols are already used, yet the exchange of product model data using standards like IGES, VDA-FS or STEP is very rare.

Management of virtual enterprises in developing countries

In order to analyze and define the requirements for VE management in developing countries there can be considered four cases of collaboration, which are:

Case 1: Both the broker and the partners within the developing country, i.e. the whole management process occurs inside developing countries.

Case 2: The broker is situated in an industrialized country and some partners in developing countries, i.e. the main activities of management occur in industrialized countries, but the operation is with partners of industries in developing countries so that an efficient integration has to be realized.

Case 3: The broker is located in a developing country and some partners in industrialized countries, i.e. the main activities of management occur in developing countries and, thereby, all processes and infrastructure necessary for the VE management have to be available at the broker's site.

Case 4: Partners are located within both developing and industrialized countries. Brokers are also located in both areas, i.e. distributed brokers take care of their local networks.

Under the aspect of global competition, Case 1 does not have to be taken into special consideration. However, the frequent usage of informal cooperation within developing countries offers great perspectives to an application of the VE model. Yet a significant problem is the lack of an extended information infrastructure that enables an efficient communication between the partners. A reasonable way to improve this situation could be the usage of Internet Services, although it is very limited due to a lack of interactivity. Since information technology in developing countries still does not correspond with VE requirements personal contacts will cover most interorganizational communication processes.

Within Case 2, the broker will have to put enormous effort in the formation activities of the VE. The partner search needs to be done by personal contacts, so that the probability to find the right one on a long distance contact is not assured. The activity of operation in comparison to a VE in industrialized countries has a lot of deficits, mainly due to the restricted communication possibilities. Concerning the information infrastructure, the best possibility is to connect to a private digital line, though the integration is not guaranteed. The usage of Internet services is also possible, yet does not offer all forms of communication needed, e.g. Video/Audio Conferencing or CAD Conferencing.

Case 3 seems to be hardly difficult to realize. Due to the inappropriate information infrastructure, the broker has to develop a very easy way to find, form and operate a VE. The most appropriate chance is when a broker finds an opportunity inside its own country by cooperating with companies from industrialized countries to achieve complementary core competencies. Yet, this case could be a possible business for brokers that have access to high velocity digital lines.

However, none of the three cases fulfills one significant requirement of the VE strategy: the fast communication between all partners throughout the whole VE life cycle. Therefore, a model that corresponds with Case 4 is intended to be implemented between German and Brazilian partners as described below.

Model for virtual enterprises in developing countries

To increase the formation of VE between Brazil and Germany, a scenario has been proposed within a project (Global Virtual Enterprise - GVE) that aims to create a VE between a German first tier supplier of a worldwide operating gear manufacturer and a Brazilian network of second tier suppliers. The goal of this VE is the global development of gear units for the German as well as the Brazilian market.

To guarantee a fast transcontinental communication with high interactivity, a Sub-VE of brokers is defined. This Broker-VE will function as an information backbone between both the Brazilian and the German network partners. The Broker-VE is founded by the Brazilian Factory Model (FIM) of the University of São Paulo, and the Aachen Demonstration Laboratory for Production Technology (ADITEC) at the Machine Tool Laboratory (WZL) of Aachen University of Technology, Germany (Bremer et al. 1995). All transcontinental communication is prevented by an international operating telecommunication carrier that is cooperating within the Broker-VE. All communication necessary for operation between the partners will be carried out on a Virtual Telecommunication Network.

The main idea of this concept is to encourage the establishment of brokers in Brazil, which are integrated with Brokers in Germany, building a global network or a backbone of brokers (Figure 1). Inside the participating countries the broker institutions have to establish their own information infrastructure and management concept. In this way a *Case 4* has been created, that integrates the advantages of the cases mentioned before into a new concept.

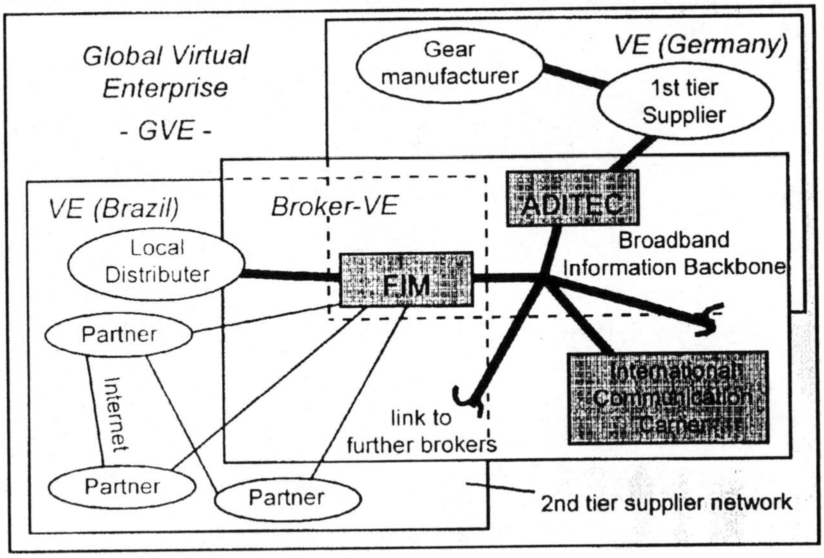

Figure 1.Structure of the Global Virtual Enterprise Network

Although this scenario does not solve the problem of operation among the partners of the VE, it increases the chances of a successful partner search and formation. Additionally, the use of fast information infrastructures becomes available for all partners within the network. The GVE represents a concept with high requirements especially towards process management. However, it represents an approach to make operation of VE more flexible and agile.

The information infrastructure inside the developing countries has already being tested with the use of internet services. The broadband infrastructure is already implemented at ADITEC as the German broker. The backbone between the two brokers is intended to be formed by ISDN services. A process of this cooperation for the case of gear production is already being modeled.

Conclusions

The VE strategy is a reasonable alternative to traditional cooperation for industries in developing countries since it offers an opportunity to them to maintain their traditionally high structural flexibility by achieving a positioning of economics of scale. The requirements to manage a VE in developing countries depends on which of the four cases defined it is inserted. However, the deficits in information infrastructure within developing countries have to be taken into account. Thereby the implementation of the GVE-model can help to overcome some of the restrictions.

The presented model can be an effective way to establish and improve the number of VE in developing countries, since necessary investments in infrastructure can be reduced. Even large enterprises within developing countries, that already have access to high velocity digital lines, may use the structure presented in this paper to create new businesses. A next step in the project would be to extend the broker network to other testbeds in industrialized and developing countries. Such a Global Broker Network is a first crucial step for the spreading of the VE strategy.

References

Bremer, C.F., Erb, M., Kampmeyer, J., and Correa, G. „Global Virtual Enterprise - A Worldwide Network Of Small And Medium Sized Production Companies.“ XV Encontro Nacional de Engenharia de Produção - Brazil First Congress of Industrial Engineering, Universidade Federal de São Carlos, São Carlos, Brasil, Setembro 1995.

N.N.: “A Telesp abre o leque de Serviços Digitais.“ Informatica Exame, Editora Abril, São Paulo, Fevereiro 1996.

Goldman, St., Nagel, R., and Preiss, K. “Agile Competitors and Virtual Organizations - Strategies for Enriching the Customer.“ Van Nostrand Reinhold, New York, 1995.

Goranson, Ted. “Agile Virtual Enterprise - Best Agile Practice Reference Base“, http://absc.amef.lehigh.edu/Ex_Proj/MAVE/Mave.html

Miles, R.E. and Snow, C.C. “Organizations: New concepts for new forms.“ California Management Review, Vol. 18, 1986, No. 3, S. 62-73.

Ritaine, E. “Prato: an extreme case of diffuse industrialization.“ International Studies of Management & Organisation, Vol. 20, 1990, No. 4, pp. 61-76.

Research on the "Agglomeration Economies" model* —Analysis of the development of high technology zone

Li Cai
(412) 621-1352

Dept. of Engineering and Public Policy
Carnegie Mellon University
Pittsburgh, Pennsylvania 15213-3890

Changbo Qui

School of Management
Jilin University of Technology
Changchun,130025 P.R. China

We have presented the "Agglomeration Economies" model using System Dynamics Model, drawing from the practices of "the Beijing Municipality's Experimental Zone for the Development of New Technology Industries" which is called "the Beijing Experimental Zone" for short below. Our "Agglomeration Economies" model describes the "Critical Mass" for high technology zone development. It also captures that when the factors affecting the "Agglomeration Economies" react on one another to reach the "Critical Mass", the zone develops with a high speed. We summarize that the most important factor affecting the "Agglomeration Economies" is the policy guidance of Chinese government in the zone.

Currently, one hundred and twenty high technology zones have been set up either by local governments or the central government in China. But some of them often pay attention to the imput of capital, technology and intelligence superficially, and overlook the best combination essentially among these factors. Therefore, it is necessary for us to research the development of high technology zone by means of the "Agglomeration Economies" model.

I. Definition of " Agglomeration Economies"

Although there are many kinds of high technology zones, all of them essentially are high technology complexes. A high technology complex has at least three factors. The first factor is the radiator of high technology such as universities and research institutes. The second factor is the group of enterprises based on a kind of or a few kinds of high technologies. The third factor is the economic and social conditions related to the first and the second factors. The definition of "Agglomeration Economies" of high technology zone is that a high technology complex can absorb funds and trained personnel from its external environment and can output new products and new enterprises into its environment continuously in order to enhance its economic strength and increase new employment occupations and form a good circle of economic relations.

According to the analysis of information we obtained, we think that many researchers have studied the "Agglomeration Economies", such as the construction and function of high technology complexes. But they scarcely studied the theory and quantitative relationship among the factors affecting the "Agglomeration Economies". On the other hand, most of researchers thought that the formation of "Agglomeration Economies" depends on the relationship among the enterprises in the zone (Oakey 1984), but we think that the most important factor to form the "Agglomeration Economies", besides the factor above, is the policy guidance of government , especially in China.

0-7803-3552-X/96/$5.00 © 1996 IEEE

* The project supported by Huo Ying Tung Education Foundation

II. The "Critical Mass" of "Agglomeration Economies"

A high technology complex consists of many universities, research institutes, enterprises and service networks, which are related to one another. Therefore it is an open system. According to the theory of Synergetics and Dissipative structure, in some conditions subsystems of the high technology complex can be organized and harmonized by themselves to change into macroscopic systematic construction from the macroscopic confused state or to change into advanced systematic construction from the lower systematic state. On the basis of the law of a harmonic and rush theory of high technology evolution (Cai 1995), a high technology complex can develop with a high speed when every factor affecting the development of the system harmonizes with one another. The "Critical Mass" is the measurement synthetically responded by affecting factors of high technology complex. When the degree of synthetic action of these factors reaches and tops the "Critical Mass" which causes the qualitative change in the development of high technology complex , the system will develop to the high level.

III. The "Agglomeration Economies" analysis of "the Beijing Experimental Zone"

A. survey

According to our investigation on "the Beijing Experimental Zone", over 700 high technology enterprises emerged from 1983 to 1989 in this zone. Among these enterprises, the microelectron and information industries were the top industries. The proportion of these industries' sales to the total sales was about 60 percent. "The Beijing Experimental Zone" has become the largest sales market of microcomputers in China. This zone has brought about the "Agglomeration Economies" beyond doubt. Generally speaking, the enterprise agglomeration and the market development as the normal form of high technology complex chiefly depended on the manufacture industry, such as "Silicon Valley" and "Route 128" (Dorfman 1983) in U.S.A. and Shang hai Cao He Jing high technology zone in China. While the economy development in most of enterprises in "the Beijing Experimental Zone" chiefly depended on trade. This zone has developed in the form which is tradeÆresearch and developmentÆmanufactureÆsales. The serious research of the "Agglomeration Economies" in this zone is significant for other high technology zones to use for reference.

B. The factors affecting the "Agglomeration Economies" and the degree of agglomeration

"The Beijing Experimental Zone" emerged in a specific time and environment. The main factors affecting the "Agglomeration Economies" are as follows: 1. The reform of educational, scientific and technical systems. 2. Chinese government called for studying the ways to deal with the new technology revolution in 1983, which stimulated the demand of users in China for microcomputers. 3. "The Beijing Experimental Zone" is close to the location of the Chinese central government, so this zone can get a lot of information very quickly. 4. A high degree of intelligence density. When these factors react on one another to reach the "Critical Mass"(that is the degree of agglomeration defined below), the "Agglomeration Economies" can be formed.

The degree of agglomeration as we called is the measuring degree synthetically by the main affecting factors. When the degree of agglomeration is beyond the "threshold", the "Agglomeration Economies" with a good circle of relations can be formed. The main factors of the "Agglomeration Economies" in quantitative terms in this zone are as follows:

1. Coefficient of science and research achievements.

This is the coefficient of development tendency of science and research achievements of zone. Through simulating the data we collect and analyze, the model is below:

$$r = \begin{cases} 1 + 0.28t & t < 1986 \\ 2.12 + 0.00025t & t > 1986 \end{cases} \qquad \text{-----(1)))}$$

$$n = 0.2372 + 0.20371p + 1.8575g + 0.002264m \qquad \text{------(2)}$$

$$r = r'(n / n0) \qquad \text{------(3)}$$

in this model:
r'—the coefficient affecting science and research achievements of the zone from outside.
p—the number of talented persons in this zone, the unit is a person.
n—the number of science and research achievements in this zone, the unit is the number of projects.
n0æthe number of science and research achievements in this zone in 1983, the unit is the number of projects.
g—the number of international academic exchanges in this zone, the unit is one person multiplied by times of exchange.
m—the investment of R&D in this zone, which consists of the development funds and R&D funds offered from outside of this zone, the unit is RMB10 000.
r—the coefficient of science and research achievements.
t—time, the unit is one year.

2. The "pulling force" of market demand in the zone from outside.

We can use the forecasting data of microelectron products demanded by domestic and foreign markets, and amend these data by means of the tendency of the market demand on this zone, in order to obtain the model of market demand tendency as follows:

$$d = \begin{cases} 0.1 + 0.2t & t < 1986 \\ 0.9 + 0.5t & t > 1986 \end{cases} \qquad \text{-------(4)))}$$

in this model:
d—the coefficient of tendency of market demand .

3. The "pushing force" of technology innovation in China

By means of specialist investigation, according to the results studied by about 30 specialists with good experience, we can put forward the model of "activation coefficient":

$$a = \begin{cases} a0 & t < 1991 \\ 0.225836 + 0.0009t & t > 1991 \end{cases} \qquad \text{------(5)))}$$

in this model:
a—"activation coefficient".
a0—"activation coefficient" before 1991.
By means of the computer simulation, we can obtain the model of a0

$$a0 = (a1cea1t) / (1 - a11cea1t) \qquad \text{------(6)}$$

in this model , a1=0.06414595, c=1.39625, a11=0.17308.

4. The intensity degree of market supply of raw material

In "the Beijing Experimental Zone" the source of raw material of 80 percent of the enterprises can depend on the market supply. Therefore, the degree of market supply of raw material is the main factor affecting the "Agglomeration Economies".

5. The intensity degree of funds.

The speed and quality of industrialization of science and research achievements is greatly related to the imput of funds in the experimental zone. Therefore the imput of funds is the main factor affecting the "Agglomeration Economies".

The degree of agglomeration of the experimental zone is the degree measured by five factors acting synthetically. The model of this degree of agglomeration as follows:

$$D\Delta = (r / ro)dasf \qquad \text{------}(7)$$

in this model

DΔ—the degree of agglomeration.

s—the coefficient of intensity degree of market supply of raw material.

f—the coefficient of intensity degree of funds.

r_0—the basic coefficient of science and research achievements in 1983.

C. The model of the "Agglomeration Economies"

To analyze the "Agglomeration Economies" of this experimental zone, we use System Dynamics Model (Wang 1987). Using System Dynamics Model, first, we analyze the causality and obtain the figure of causality shown in figure 1. In this figure, 5 feedback circles can be produced because of the reaction among all the factors. Second, we need to study the quantitative relationship among all the variables to obtain the flow chart(which is omitted).

Figure 1. The causality of "Agglomeration Economies"

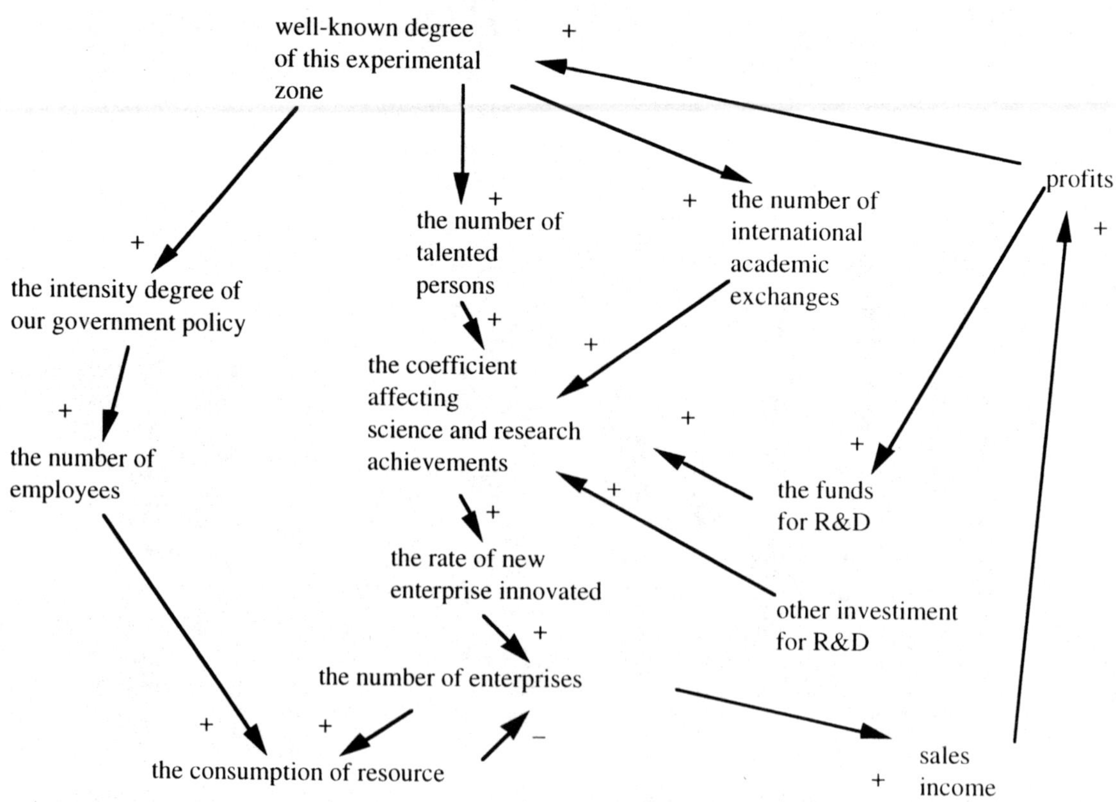

To set up the System Dynamics Model, we can use DYNAMO language to simulate the "Agglomeration Economies" of this experimental zone from 1983 to 1999. The "Agglomeration Economies" can be measured by 2 indicators composed of the number of enterprises and total sales. The results are as follows:

1. In this experimental zone, the number of enterprises develops with the curve of the life cycle, that is, the initial stage is from 1983 to 1985, the development stage is from 1986 to 1989, and its maturity stage is from 1990 to 1999,which is shown in figure 2.

Figure 2. Growth of number of enterprises

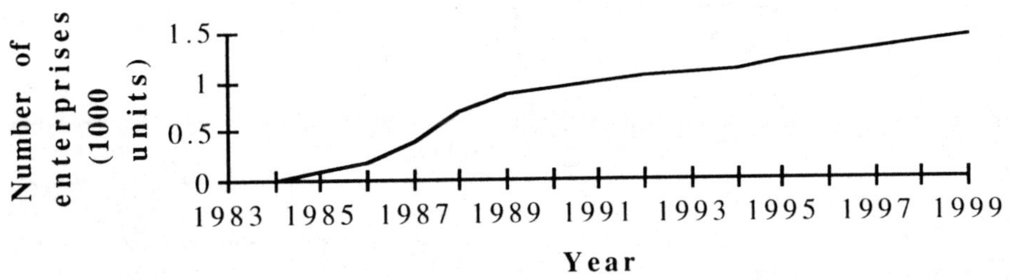

2. In this experimental zone, the trend of total sales is similar to that of the number of enterprises. We know that the initial stage is from 1983 to 1988, the development stage is from 1989 to 1993, and the maturity stage is from 1994 to 1999 from figure 3.

Figure 3. Trend of total sales

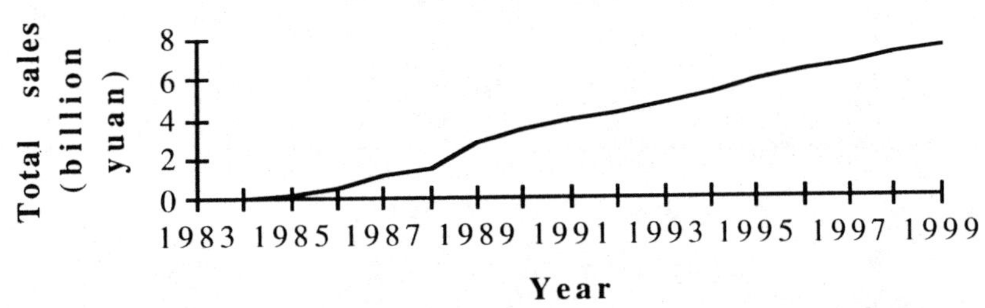

Through further research for this model, we know that the "Agglomeration Economies" of this experimental zone develop with a high speed when all the factors affecting the "Agglomeration Economies" react harmoniously. It needs about five years to develop this experimental zone with the policy guardance of Chinese government. Five years later, this experimental zone will be able to develop in the better conditions of market economy.

References

Oakey, R. P. 1984. High Technology Small Firms: Regional Development in Britain and the United States. New York, N.Y.: St. Martin's Press.

Cai, Li. 1995. High Technology Economics. Changchun, P.R.China: Jilin Science and Technology Press.

Nancy S. Dorfman. 1983. "Route 128: The Development of a Regional High Technology Economy." Research Policy 12(December): 299-316.

Wang, Qifan et al. 1987. Theory and Application of System Dynamics. Beijing, P.R. China: New Times Press.

Technological Learning, Architectural Innovations, And The Virtual Utilily Concept

Elias G. Carayannis, Ph.D.
(202) 994 4702
(202) 994 4705 (Fax)

Science, Technology, and Innovation Program
School of Business and Public Management
George Washington University
Washington, DC 20052
E-mail: caraye@gwis2.circ.gwu.edu

Alistair Preston, Ph.D.
505 277 0335
505 277 ()344 (Fax)

Anderson Schools of Management
University of New Mexico
Albuquerque, NM
E-mail: alistair@unm edu

Shimon Awerbuch, Ph.D.
603 891 1342
603 891 1392 (Fax)

Independent Economist
Nashua, NH
E-mail: awerbuch@aol.com

This paper highlights the potential benefits derived from deploying new methodologies in technology assessment and risk management that drive strategic technological and infrastructural investments by power utilities in response to internal and external contingencies they are confronted with in a rapidly deregulating, increasingly competitive environment.

The role of technological learning in helping to continually re-shape decision making criteria and worldviews is illustrated and leads into formulating some ideas about how to integrate strategic (qualitative) with financial (quantitative) criteria in valuating architectural innovations such as the Virtual Utility (W) concept.

Moreover, the paper explores the nature of the different kinds of technological innovation, architectural, radical, generational, and incremental, and then illustrates -- using examples from diverse technology-driven industries, such as the information technology, pharmaceuticals, and industrial chemicals industries -- how capital budgeting, because of its narrow accounting orientation often fails to capture and qualify strategic benefits associated with emerging technologies, thus impeding the diffusion and adoption of technological innovations.

Introduction: Defining The Concept Of A Virtual Utility

"With the October announcement of our PowerChoice proposal, Niagara Mohawk opted to accelerate the introduction of competition... Through PowerChoice, we have developed a plan to ensure the success of Niagara Mohawk as a regulated energy provider and to position the company to take advantage of emerging unregulated business opportunities.... PowerChoice proposes to split Niagara Mohawk into two parts - a generating company ... and a holding company."
(William Davis, Niagara Mohawk CEO, Annual Report, 1995).

"Rather than squander their energies on developing defensive strategies to resist radical restructuring and protect their markets, executives at all utilities should pour those energies into figuring out how to adapt to the coming new world."
(Navarro, 1996: 120).

US electric utilities are currently undergoing significant regulatory, market, and technological changes. This paper examines how these changes should affect the traditional way for making strategic investment decisions so that utilities can successfuly adapt and respond to the challenges confronting them. We will stress the need for new ways of thinking about strategic investments and the role of technological leaming in shaping emergent worldviews of the decision makers

0-7803-3552-X/96/$5.00 © 1996 IEEE

involved: "Six years ago when PacifiCorp told its coal miners therd have to compete for the right to supply the company's plant here [in Point of Rocks, Wyoming], the mines were skeptical... They became believers when the company trucked in 300,000 tons of coal, $5 million worth, from outside suppliers... Such aggressive tactics are becoming more common among electric utilities, as competition is gradually introduced into the industry." (The Wall Street Journal, 3/2611996).

In short, this paper examines the problems associated with the application of traditional capital budgeting practices to conduct benefit / cost analyses of radically new concepts and technologies in an environment of rapid technological, market, and regulatory change. The paper explores the nature of the different kinds of technological innovation and then illustrates how capital budgeting often fails to identify important benefit categories.

The potential benefits accruing from adopting a new approach to capital budgeting and the positive impact of technological learning (Carayannis, 1994, 1995a, 1995b, 1996; Dodgson, 1991a, 1991b) in the valuation and deployment of new technologies are also examined. The paper concludes with some recommendations for improving the capital budgeting process by making qualitative assessments to better understand the benefit implications of architectural innovations such as the Virtual Utility (W) concept.

One way to conceptualize the Virtual Utility, is as a model for organizing power generation and distribution by minimizing non-value-adding activities, such as excess generation capacity, and providing energy on a Just-In-Time basis. A Virtual Utility is a virtual organizational entity comprised of an alliance of various power generating and distributing entities, such as Non Utility Generators (NUGs) or Independent Power Producers (IPPs) that utilize modular power generation technologies, such as photovoltaic cells (PV). It is a considerably more advanced form of the current evolving business model of power utilities, described by a Baltimore Gas & Electric spokesperson: "there are some inherent difficulties for an outsider entering a competitor's electricity market, ... personal relationships with big customers have been cemented over many years. And on-site services such as power-equipment management for large customers are more practical for the local utility: 'Our position is to stay as close to our customers as possible and offer not only a competitive price but also the types of services that they would not receive from someone outside the service territory'." (The Wall Street Journal, 3126/1996).

Moreover, a Virtual Utility can be defined as "as a producer of high-value-adding intangibles embodied in fully mass-customized electric services that uses passive, capital-intensive, and infinitely durable power generation and telecommunication technologies" (Awerbuch, Carayannis &Preston, 1996).

The Virtual Utility, with its modular generating technologies, its capabilities to deliver specialized new services, and its flexible supply arrangement enabled by new financial instruments, such as energy options and futures contracts, creates a set of benefits that may differ considerably from the traditional direct benefit-cost realm usually examined in capital budgeting. Thus, while particular generating technologies may or may not have lower "busbar" costs, in the case of the Virtual Utility it is the synergism of the organizational structure that has the potential for cost reductions. This is similar to the way that flexible process technology reduced costs in manufacturing, even though direct unit costs were not always lowered. Comparing busbar costs may have been satisfactory in a previous era, when utility resource alternatives were technologically homogeneous, consisting largely of fossil fired options, which had essentially the same mix of operating, overhead, capital, transmission and distribution costs and where output was sold to an essentially captive market (ibid): "Electricity deregulation is posing a potent challenge to the nuclear-power industry, which is struggling to make its costs competitive without compromising safety. With several states making plans that eventually could allow customers to choose among power suppliers, just as they can choose among long-distance phone carriers, cost-control has taken on new urgency for utilities." (The Wall Street Journal, 2/1/1996).

The current integrated resource planning (IRP) environment offers a considerable range of technological options, including capital intensive solar renewables, such as PY, and Demand Side Management (DSM) technologies. These technologies have fundamentally different overhead, operating, and capital transmission and distribution costs than traditional central generating plants. Given these differences it is no longer sufficient to select resource options on the basis of their busbar or direct costs.

The above argument does not invalidate capital budgeting approaches, rather it suggests that we must be much more innovative in conceptualizing and recognizing the novel benefit streams that new, modular energy technologies may produce under a Virtual Utility environment.

In the next section we lay the groundwork for the reconceptualization and the recognition of the novel benefit streams of a Virtual Utility.

Technological Innovation, Technological Learning,And Benefit/Cost Measurement Systems

"The 1992 Energy Policy Act allowed for wholesale competition in the industry and set in motion forces that make retail competition virtually inevitable. In a half dozen years, experts say, many Americans will enjoy choices in electricity similar to those now available in the telephone industry: What equipment to purchase, what level of service to buy and which supplier to use all will be up to the customer." (The Wall Street Journal, 312611996).

Technological Innovation and Organizational / Technological Learning

In this section we will define some of the basic terms and concepts that we use to make the case for a Virtual Utility. These include the four types of technological innovation as well as the related concepts of organizational and technological learning.

An innovation is "an idea, practice, or object that is perceived as new by an individual or other unit of adoption." (Rogers, 1995: 11) thus, a technological innovation is a new idea, practice, or object with a significant technology component.

The literature on technological innovation distinguishes between four generic types of technological innovation: architectural, radical, generational, and incremental (Henderson & Clark, 1990; Morone, 1993; Betz, 1993).

Incremental innovations exploit the potential of established designs, and often reinforce the dominance of established firms (Henderson and Clark, 1990:). They improve the existing functional capabilities of a technology by means of small scale improvements in the technology's value adding attributes such as performance, safety, quality, and cost (Betz, 1993: 20 21). Such innovations lend themselves more directly to traditional valuation procedures.

Generational (Henderson, 1989; Morone, 1993) or next-generation technology (Anderson & Tushman, 1990; Betz, 1993) innovations are incremental innovations that lead to the creation of a new but not radically different system, such as in the case of Cor~ng discussed in Morone (1993), where seven distinct generations of processing technology were developed over the course of fifteen years: "As one generation was being implemented a second was nearing completion, and development of a third was under way, a pattern that is striking in its similarity to the process of generational change in the semiconductor industry." (ibid: 215).

Radical innovations introduce new concepts that depart significantly from past practices and help create products or processes based on a different set of engineering or scientific principles and often open up entirely new markets and potential applications. They provide "a brand-new functional capability which is a discontinuity in the then-current technological capabilities" (Betz, 1993: 20). In the field of electricity generation and delivery, photovoltaics (PV) and similar modular renewables mayrepresent a radical innovation. To the extent that these produce benefit categories not previously understood or conceptualized, they do not directly lend themselves to traditional capital budgeting valuations.

Architectural innovations, a term coined by Henderson and Clark (1990), serve to extend the radical-incremental classification of innovation and introduce the notion of changes in the way in which the components of a product or system are linked together. Betz (1993: 21) using the term systems innovations, illustrates how reconfiguring existing technologies may generate radical innovations that provide new functional capabilities. We consider that the Virtual Utility, defined as a distributed approach to generating and delivering electricity using modular technologies, mayrepresent an architectural or systems innovation in the sense that it alters the traditional components used to manufacture electricity and hence alters the nature of the product in a fundamental manner. Although arguably the product still consists of electrons, their creation, availability, quality and delivery options are sufficiently altered so as to fit the definition of an architectural innovation.

In the case of learning like the case of most other fundamental concepts, there is little consensus as to what learning is and how it takes place. In economics, learning is perceived as tangible, quantifiable improvements in value-adding activities, in management, learning is seen as the source of "sustainable competitive efficiency" (Dodgson, 1993: 376), whereas in the innovation literature, learning is considered as a source of "comparative innovative efficiency" (ibid: 376).

Teece et al (1990) define learning as "a process by which repetition and experimentation enable tasks to be performed better and quicker and new production opportuntities to be identified".

Learning "involves a fundamental shift or movement of the mind" (Senge, 1990: 13) and moreover, "real learning gets to the heart of what it means to be human" (ibid: 14).

Teece et al (1990) also focused on the nature of learning as both, an individual and an orgarl17ational process: "Learning processes are intrinsically social and collective phenomena. Learning occurs not only through the imitation and emulation of individuals, as with teacher student, or master-apprentice, but also because of joint contributions to the understanding of complex problems. Leaming requires common codes of communication and coordinated search procedures."

Organizational learning is the development of insights, knowledge, and associations between past actions, the effectiveness ofthose actions and future actions (Fiol and Lyles, 1985) and by extension, a learning organization is "one skilled at creating, acquiring, and transferring knowledge, and at modif~ing its behavior to refiect new knowledge and insights" (Garvin, 1993).

Argyris and Schon (1987) developed a three-level typology of organizational learning that incorporates single-loop and double-loop learning with deutero-learning: "Organizational leaming involves the detection and correction of error. When the error detected and corrected permits the organization to carry on its present policies or achieve its present objectives, then that error-detection-and-correction-process is single-loop learning. Double-loop leaming occurs when error is detected and corrected in ways that involve the modiflcation of an organization's underlying norms, policies, and objectives." (ibid: 3). As to the third level of organizational leaming, "when an organization engages in deutero-learning, its members learni about previous contexts of learning. They reflect on and inquire into previous episodes of organizational learning, or failure to learn. They discover what they did that facilitated or inhibited learning, they mvent new strategies for learning, they produce these strategies, and they evaluate and generalize what they have produced." (ibid: 27).

Technology is defined as "a design for instrumental action that reduces the uncertainty in the cause-effect relationships involved in achieving a desired outcome. A technology usually has two components: (1) a hardware aspect, consisting of the tool that embodies the technology as a material or physical object, and (2) a software aspect, consisting of the information base for the tool" (Rogers, 1995: 12).

Technological learning is defined as "the process by which a technology-driven firm creates, renews, and upgrades its latent and enacted capabilities based on its stock of explicit and tacit resources. It combines purely technical with purely administrative learning processes and takes place at the individual, intra-, and inter-organizational levels" (CarayaDnis, 1994). At the highest, strategic level, technological learning "serves to "leap-frog" onto a new competit*e realm and to increase the slope of the learning curve as well as the rate by which the slope per se increases by means of enhanced and innovative organizational routines" (ibid: 582-583).

Dodgson also studied the concept of technological learning (1991 a, b; 1992) and he argues that it underlies the "differential ability of firms to learn quickly about technological opportunities" (Dodgson, 1993: 51).

Technological learning, by its definition, encapsulates the processes of adaptiation and renewal of technology-driven organizations such as power generating and distributing utilities. It is thus the most fitting concept of learning to use when examining the concept of the Virtual Utility.
Moreover, the four types of technological innovations, architectural, radical, generational, and incremental, are co-related to successive levels oftechnological learning (Carayannis, 1994, 1 995a, 1996) which is increasingly seen as a critical succes factor affecting the restructuring of complex industries such as power utilities (Carayannis, 1996).
The traditional components ofthe electricity generation and delivery system include i) inflexible, central-station generating facilities with non-zero marginal costs, coupled with ii) inflexible distribution systems marked by high sunk costs. This system, which requires significant organizational support and consumes significant overhead and transactions costs (Wil liamson, 1975), could traditionallynot be operated outside of large, hierarchical org~ni7ations with broad operational support capabilities and agglomeration (scope/scale) economies. New modular technologies may alter this situation just as technological progress has eliminated agglomeration economies in other processes including imaging and health care delivery.
Radical and architectural innovations require new sets of organizational skills and knowledge and thus create problems for existing firms: since it is difficult and costly for firms to readjust their skills and knowledge base, such firms face a distinct disadvantage in the adoption of the new technology (Henderson and Clark, 1990)' which partly explains why radical new technologies are often diffused through new ventures or start-ups.

The failure of established firms to readily adopt radical or architectural innovations may also be explained in part by the failure of traditional capital budgeting tools, (e.g. discounted cash flow or the revenue requirements method) to correctly identify their benefits. Large firms have formal capital budgeting processes which typically require project 'champions' to convince top-level managers that the project will have a positive impact on the bottom line. This requirement places most of the emphasis on accounting numbers and leaves little room for intuition. The entrepreneur, on the other hand, less restricted by the rigid corporate process, may be freer to follow instinct.

Historical Evolution of Utility Capital Budgeting Methods

The basic Revenue requirements Method (RRM) and EPRI-TAG concepts can be traced back to the begitming of this century, although the procedures did not mature until the post World War II era (Awerbuch, Mouck, et. al., 1996). These procedures were conceived in a different technological era, and, as applied, no longer work well. Generally, discounted cash flow procedures (and their utility derivatives MM and TAG) were developed for manufacturing and generating technologies with the following specific characteristics:

i) They were active—with a clear distinction between periods when the equipment is 'in-use' and when it is 'off' or 'idle', at which point costs are essentially zero. This simplifies the process of allocating cost to output.

ii) Theywere expense-intensive—cost consisted primarily of direct, operating outlays. This is an important distinction; it is usually easier to compute and allocate direct operating costs than sunk capital costs. Capital costs are allocated as arbitrary periodic depreciation charges under the accounting "matching" principal. Since sunk capital costs are small for expense-intensive technology (by def~nition) the procedure by which they are allocated to each unit produced, no matter how arbitrary, will not significantly alter total unit costs. This implies that costs and revenues can be more easily "matched" by accountants and are hence less ambiguous or problematic.

iii) They posses f nite durabilit~ each day of using a machine brings the required maintenance or replacement of some component that much closer. This leads to a more direct conceptual linkage between usage and cost hence forms the basis for various accounting procedures, including the arbitrary allocation of sunk cost to each unit produced through such fabrications as accounting depreciation.

iv) Theywere marked by relatively lowrates of technological progress-- this has several imp]ications. First, depreciation on particular asset can be conceptualized as being driven by the rate at which that asset wears out. Since technology remains constant, an asset with low wear on it will be worth a relatively high proportion of its original value. Thisdoes not hold when rates of change are high-- for example with computers-- where the value of a used asset has more to do with the rates of technological change than with how much "wear" it has had (e.g. how manyhouts it has been used) (Awerbuch, Mouck, etal, 1996).

Low rates of technological change suggest that a new machine will have only incremental improvements to it: for example: while the new milling machine may have better speed control and somewhat higher slew rates, it is essentially unchanged so that it will not require major changes in the way the firm produces its product. As a result, overhead and other support costs are not likely to change much from what they were with the old machine. This illustrates that the replacement decision, and, for utilities, the capacity addition decision, was at one time well supported by the existing accounting. Indeed there was little need for intuition: "intangible" costs and benefits, if any, were small.

The attributes of activeness, expense intensity and f nite durability could be found in most traditional, process-oriented technologies, whether the product was screws, or kilowatt-hours. However, the nature of technology has changed: new process technology, whether CIM or PV/renewables, has attributes that differ considerably, as follows:

i) It is frequently capital intensive-- not expense-intensive; fixed costs are not as easily dealt with by the traditional accounting model.

ii) It is often passive-- i.e. there is little distinction between its state of being 'in-use' versus its state of being 'idle' or 'off; costs are virtually the same. This suggests a margi~al cost functionradicallydifferent fromwhat utility economists are used to seeing. It also makes it harder to allocate costs to output: the personal computer, for example, loses value not in relation to the hours it is used, but as a consequence of technological progress in the industry.

iii) It possess essentially inf nite durability—its actual use contributes little to its "wearing out;" Indeed, given today's shorter manufacturing life-cycles and high rates of technological change, technology becomes functionally obsolete before "wearing out" in the traditional sense (ibid).

The investment decision for passive, capital intensive technologies is considerably more complicated, and accounting information no longer provides decision support. This issue affects the valuation of the W which involves the use of passive, capital intensive generating and telecommunications technology which radically alters the manner in which electricity is produced and delivered.

Towards An Enhanced Framework For Valuating A Virtual Ltitllty: Technological Learning And Strategic Capability Options

The RRM and similar accounting-based techniques suggest that passive renewables cost more than conventional power generation. But these techniques have a dismal record in identifyingpromising new technologies. They failed to see the important benefits of such major innovations as computers, CIM, computer-aided design (CAD) and robotics (See: Awerbuch 1993a). The benefits of new process technologies, whether in manufacturing or electricity generation can be fully understood only through the use of more robust measurement concepts.

Emergent Technologies and Their Atypical Cost / Benefit Attributes

A fundamental lesson regarding capital budgeting to be gained from the experience in manufacturing is that new technologies provide an indivisible bundle of numerous benef t-cost attributes which invanably will differfrom those of the conventional technology. This was not generally the case in previous technological eras, where, given low rates of technological progress, a new machine may have differed only incrementally from the previous vintage. And although the differences between new and existing technology choices are generally multidimensional, traditional analyses tends to focus on only one attribute—direct cost—both initial and operating. It is easy to see why this happens. Most of the important attributes of new process technologies have no direct accounting measure and are therefore easy to ignore or to relegate to the "intangibles" category (Awerbuch, Carayannis & Preston, 1996).

The traditional focus on cost ignores other important technology attributes such as throughput, and quality which in some cases are the important cost drivers. Other attributes which drive cost relate to the way in which new technologies affect project risk, reversibility and flexibility, all of which are bynow widely recognized. For example: We may be able to determine a set of expected manufacturing costs for a capital intensive CIM (Computer Integrated Manufacturing) process and compare them to the expected costs under the current labor-intens*e process although the two are not directly comparable: a labor-intens*e process is subject to the vagaries ofthe labor market while the capital intens*e process presents a set of known, up-front outlays coupled with very low annual operating costs. Barring implementation problems the costs of capital-intens*e process are known with near-certainty (ibid).

Flexibility/Reversibility

The flexibility attributes likewise play an important role in electricity generation. Conventional, fuel-intens*e technology is quite inflexible: it requires the firm to commit to the construction large, irreversible, central-station plant, such as coal, some eight to ten years prior to completion of the plant. The construction period is fraught with uncertainty—environmental and siting permits, various local, state and federal approvals, etc. Assuming the project survives these hurdles, it must operate per the original forecast, by now, easily ten to fifteen years old.

This option-set is created by the modular capability of new generating alternatives including small gas turbines, PV and other renewables. This capability can be seen in several different ways. First, it means that these technologies can be installed in small increments as load grows, much the same way the local telephone company installs central-office equipment. It also means that projects can be scaled back or moved geographically, should conditions change. The rapid response to changing load and other conditions reduces uncertainty associated with the long lead-times oftraditional, lumpy central-station technologies (Vogelsang, 1988). It also greatly reduces the enormous costs of planning and implementing lumpy central-station investments. Flexibility/Reversibility also means that PV/renewable installations tend to have lower sunk costs and hence can be salvaged at some reasonable fraction of their original cost. This is in comparison to inflexible technology, say a coal plant, where the sunk costs of, for example, engineering design and (literally) bricks and mortar mean that once completed, such plants will generally have a negligible or even negat*e salvage value. The ability to move or scale-back such projects reduces risk, and, moreover, enhances the likelihood of demand-supply equilibrium (ibid): "[Niagara Mohawk's strategic initiat*e] PowerChoice envisions an open generation market in which, eventually, all customers would choose their suppliers. Customers would be able to purchase from the market, or from marketers, brokers, or energy service companies, under any terms and conditions they could negotiate... Pending approval of other aspects of PowerChoice, direct access by retail customers would begin as soon as the competit*e generation market is operational." (Niagara Mohawk Annual Report, 1995).

Strategic/Capability Options and Environmental Attributes

The set of strategic/capability options provided by new technology suggests that by adopting a new process technology, firms develop capabilities in the sense of Baldwin and Clark (1992), who define capabilities as "groups of expenditures, which when taken as a whole, represent an investment for the firm.". New technology generally also provides a different set of environmental attributes which are often more benign. Our tools for measuring these benefits are poor. Even poorer is our ability to value the results inter-temporally, where theoretical as well as practical impediments exist (ibid).

The Need for a New Accounting "Language of Thought": The Lean[1] Accounting and Budgeting Paradigm

In characterizing the state of accounting for governmental entities, Anthony (1988) concludes that "We do not have the accounting concepts we need." Similarly, we cannot fully understand the value of VU and distributed generation concepts without first enhancing our vocabulary for expressing and measuring its potentially non-traditional benefits. Lacking more comprehensive accounting concepts, we tend to value new technologies using a vocabulary that worked for the previous vintage technology. For example, the busbar cost concept, $/kWh, is based on the idea that short-run cost rises linearly with each additional kWh produced so the incremental kWh is the cost driver. This is clearly not the case with pass*e technologies such as PV, or, most likely, even for fossil generators.[2]~ We are thus attempting to evaluate new generation alternatives using inappropriate concepts belonging to an earlier vintage of active, expense-intensive technologies.[3]

The search for appropriate accounting vocabulary is not a new endeavor. In the last decade several radically new accounting concepts were developed in other fields notably the concept of the diagnostic related group (DRG) to replace patient-bed-days as an output measure for health care and, to give a second example, the relatively novel notion (or "cost construction") that there is an avoidable cost for manufacturing defective products, or for that matter, to keeping parts inventories. These idea have direct applicability in electricity generation, where the current generation/delivery system incurs a variety of costs, such as the cost of excess capacity, that are specifically identified by a cost-accounting entry. This is not unlike the situation in manufacturing, where the cost of producing defectives was also ignored until the accounting vocabulary was enriched. Once recognized, the principle gave rise to a widespread "quality" movement in American industry.[4]

Similarly, it is reasonable to expect that the development of enhanced cost measurement concepts in electricity will give rise to new ways of organizing the electricity generation process as well. For example, While utility planners and engineers generally find it impossible, for example,

In other words, the manner in which cost and cost-causation is conceptualized can have a significant impact on the firm's organization. For example: the operating cost notion in transportation has traditionally been the $/mile concept, or for goods movement, $/ton-mile, which suggests that the further goods are moved the higher the cost. This measure suggests that we generally want to m1nimize the total distance that goods are moved, which is traditionally operationalized by using warehouses and other trans-modal shipment points. Yet consider the package express business, as illustrated by Federal Express, which picks up packages all over the continental US and moves them to a central location, so that a package sent from, say, Boston and destined for Burlington, Vermont will get there via Tennessee. The existence of this particular organizational structure suggests that $/mile is not the major cost driver for this industry. Rather, it suggests that cost is driven more by the number of times a package is handled, sorted, stored in remote trans-shipment points.

Utilities perform numerous functions related to keeping large centralized power generation stations operating. Some of these costs involve the high level managerial resources required to plan, undertake and supervise the construction of major lumpy generating facilities requiring investments that are quite sizable relative to firm's equity. Others are in the form of transaction costs (WIlliamson, 1975) which reflect the exchange of material and information. Examples are: ordering, executing and confirming ffiel and material movements; performing production schedules and forecasting inventory needs. Still others require additional working capital or overheads-- for example -- the costs of ffiel inventories and ffiel transfer/holding facilities.

The Role of Technological Learning in the Cost / Benefit Valuation of Strategic Capability Options

The concept of creating strategic technological capability options implies that by investing in given technologies, f~lms create opportunities or 'options' for themselves, to make still additional investments in the ffiture (e.g.: Bierman and Smidt, 1988; Baldwin, 1989; Clark, 1994; and Henderson, 1990). The idea is more commonly understood in terms of "staying on the learning curve". It is nicelyillustrated inthe example byKaplan (1986, 1990) who observed that machine tool companies that adopted numerically-controlled production technology in the early 1970's, found it easier to switch to integrated computer-based manufacturing a decade later.

Baldwin (1989) and Baldwin and Clark (1992) explainthe process quite plainly by observing that these firms developed capability options, which could be "exercised" to reduce the cost of adopting subsequent, more advanced technology vintages. In this context, the capabilities developed in the first technology transition facilitated a subsequent transition to the more advanced technology. Thus part of the pay-off of the original move to numerically controlled equipment was the capability the firm developed to deal with automation in the manufacturing process. This implies that part ofthe benefit the original investment was the capability to work with automated processing equi~ment, a capability that translated more easily into the adoption of CIM technology the 1980's .[5]

Carayannis and Stokes (1995b: 10) observed a similar phenomenon in the case of at BASF Corporation where lessons learned from a major reengineering project faci litated the conversion to ISO 9000 standards: "Reengineering proved to be a useful "tool" for ISO 9000 Certification. The Vision '96 teams discovered that their redesign of core business processes facilitated going through the changes needed for ISO 9000: 'Vision '96 will provide a system for capturing the ISO 9000 requirements in a straightforward manner"' (ibid.). Implementing major business process reengineering projects and instituting ISO 9000 standards helped BASF "stay on the learn1ng curve": "The ISO 9000 Certification programs serve as "tools" for institutionalizing, within the organization, a culture of introspection and Root Cause Analysis" (Carayannis and Stokes, l995b: 10).

Such *first-hand* technological learning (CarayanDis, 1994, 1995a, 1995b, 1996) within a firmwill ultimately enable a particular utility to adapt distributed VU concepts to its own needs; indeed the evidence suggests that o~anizational learning through f rst-hand experimentation can have a significant impact on productivity even in industries with relatively homogeneou6s product. For example: by designing their production process around new Bessemer technology[6] in the 1860's, American steel mills were able to double their output relative to established English f~ms that adopted the technology without fully integrating it to their processes (Clark 1987: 67). In these older mills, floor plans, job classifications and organizational structure were all grounded in the earlier crucible process, and while this basic Bessemer idea was "clear enough from the very begi~ning," (Clark 1987: 67), it apparently could not be fully exploited by these mills without first hand technology-driven organizational learning.

Carayannis and Stokes (1995b) identified similar instances and patterns of first hand technological learning in a longitudinal study of BASF AG from 1865 to 1993: "BASF fosters its corporate culture (and in the process of doing so) technological lea~ng, in a variety of ways: a) corporate training and development / retraining shems; apprenticeship programs programs; written rules for work...b) a high status on its corporate R&D laboratories and personnel, c) the firm's managers also carefully cultivate new generations of management personnel by giving them a variety of assignments, not only in different operations, but also in different plants".

Furthemore, there is also strong evidence and in particular in the case of the nuclear power industry (Carayannis and Madlifassi 1992; Carayannis, 1996), that utilities are realizing worldwide the need to become increasingly adaptive and proactive in their decision making and resource allocation in order to survive and prosper in an increasingly competitive environment: '~uclear power success stories, across different countries, have been characterized by a tendency for openness and an emphasis on facilitating individual and organizational learning. The re engineering of core business processes, currently a popular management tool across many different industries, is exactly and instance of a practical and tangible way for encouraging both individuals and the organization as a whole, to radically improve their learning practices. Namely, prog~ams such as the Core Process Redesign (CPR) as it is called at Niagara Mohawk, aim to assist employees to learn better and faster how to do things more safely, effectively, and efficiently" (Carayannis, 1996).
That implies that they have not only to "stay on the learning curve" but to actually "drive the learning curve" by engaging in higher order technological learning such as technological learning-ho~to-learn "through assimilating and standardizing explicit best practices, industry wide and worldwide" (Carayannis, 1996) and "through learning "by analogy" based on industry wide and worldwide benchmarking of tacit best practices" (ibid.).

The concept of the Virtual Utility would further foster the institutionalization and practice of higher order technological learning processes both, within and across utilities, thanks to its underlying philosophy of "leanness" and its emphasis on creating and leveraging off strategic technological capability options.

Conclusions: Insights Gained And Lessons Learned

The established accounting theory and existing capital budgeting procedures are inappropriate and even misleading for the valuation of technological innovations of an architectural caliber, including the Virtual Utility concept.

This has historically proven to be the case with several technologies such as CAD, CIM, robotics, steel mini-mills, information technology and biotechnology. The lessons learned however, should help us create a family of promising approaches to better valuate thepotential pay-off from the Virtual Utility innovation. In principle, a qualitative assessment is in order to pinpoint where the benefits of such an architectural innovation may lie and what form they will take. This procedure, which is aimed at helping to identify and define potential benefits, consists of evaluating the following issues and threshold questions (Awerbuch, CarayaDnis & Preston, 1996):

" a) Are anticipated benefits largely cash-flowrelated?

b) What are the "non-cash " benefits? The experience in manufacturing suggests that these will take the form of enhancements in quality, reliability and flexibility. While these ultimately do affect cashflow, it is hard to conceive of them in this fashion. Obviously this question requires some understanding of what the quality-reliability-flexibility concepts mean in the case of electric delivery.

c) Does the new technology enhance capabilities or strategic options for the f rm? This evaluation necessarily entails an assessing the VU in terms of its ability to yield new capability and strategic options for the f rm to serve new customers and new markets. These capability/strategic options may come in the form of increased ability to serve specialized needs ormarkets, or to provide newproducts and services that consumers may value. Flexible organizations, using flexible technologies coupled with innovative f nancial arrangements may be able to better serve emerging customers and market. Part of the valuation of the VU necessarily depends on an assessment of these possibilities.

d) Does thef rm really have a choice about adopting new technologies or organizational structures such as W? One line of argument suggests thatf rms do not really have choices regarding the adoption of broadly based innovation when competitors are embracing them. Indeed there are numerous examples off rms in electronics and other industries who did not stay on the "learning curve" and subsequently lost considerable market preeminence. The threshold question is whether failure to adopt VU and related innovations will place the f rm at a strategic disadvantage in the future.

e) Does the VUyield complementary overhead and other cost reductions?
A complentarity of CAD, for example, is that it enables direct input to computerized production equipment thus reducing the cost of product changes. Considerable analysis, coupled with experience will be required to evaluate the extent to which the VU concept provides such benef ts. However, this much is certain -- simplistic analyses that compare busbar costs will not capture the needed decision-mahng information".

---

[1]The work of Womack et al (1991) is what inspired our view ofthe emerging accounting paradigm as lean. Womack et al outlined the principles of definition of lean production as teamwork, communication, eff cient use of recources and elimination of waste, and continuous improvement: "The lean producer ... combines the advantages of craft and mass production, while avoiding the high cost of the former and the rigidity of the latter...Lean production is "lean" because it uses less of everything compared with mass production." (ibid: 13). By that token, lean accounting focuses on value adding actitivities especially those formerly classified as "intangibles".

[2] Short-run marginal costs of PV-based electricity are essentially zero. Fossil generation may involve short-run costs that are driven by activities other than incremental kWh generation. This implies that once these costs are understood, electricity may be sold using different pricing mechanisms, which may lead to significantly higher usage

as the result of new markets and applications. The effects of such innovative pricing, like the impact of the word-processor or the enormous growth in demand for telephony over the last decade, are virtually impossible to predict.

[3]The idea may be roughly comparable using standard cost procedures to compare, say, autos and horse-drawn carriages at the turn of the century. People did not routinely travel great distances by horse and carriago marginal cost for travel beyond a few miles was too high (e.g. the slow pace requires overnight stays, riding in cold carriages, etc.). In fact, it is doubtful whether anyone evaluated carriage ownership on the basis of cost per mile driven, a measure that makes much more sense for a car, where, with sufficient usage, much of the operating cost becomes variable. Interestingly a PV-powered car may make costs largely fixed again for most urban drivers, thus making the "$/mile" cost concept again inapplicable.

[4]Although the ideas were forcefully expressed quite early by some experts including G. Edwards Deming.

[5]This evidence suggests that under certain situations, process technology should be adopted even where the present values are relathely low, because the investment is needed "to stay on the learning curve," i.e.: to create the options for getting onto some subsequent technology in the future. This point was first illustrated over twenty years ago by Bierman and Haas (1973) and subsequently by Howell and Soucy (1987). Traditional valuation processes in electricity cannot reflect such options, even qualitatively, yet clearly they exist: excursions into DSM, PV installedfor grid support, all of these investments produce capability options which may become valuable in a future, competitive market. Arguably such options may not be as valuable as they are in manufacturing, since electricity maybe a somewhat more homogeneous product whose process technology has traditionally been more readily transferable from firm to finn witness the existence of the Electric Power Research Institute, whose aim it is to disseminate research results throughout the industry. Restructuring howcver seems to be eliminating such industry cooperationwith fmns recognizing the need to achieve competitive advantages.

[6]The Bessemer process produced batches of steel almost continuously, every 15 minutes, as opposed to every few days under the old crucible process;

Technological Life Cycles: Causes and Effects

Mario W. Cardullo, P.E.
(703) 698-6019

Department of Industrial and Systems Engineering
Virginia Polytechnic Institute and State University
Northern Virginia Graduate Center
2990 Telestar Court
Falls Church, VA 22042

All technological developments progress through a series of life cycles. Managers of technology must be fully aware of how these life cycles interact. Technology life cycles can have significant impacts on society. These impacts arise from interaction of a myriad of forces which may be represented by a model, using the Pearl equation of sigmoidal curves. Technologies impact the productivity of industries and in turn, affects the number of workers. The introduction of new technology leads to higher productivity and could eventually cause economic growth requiring more workers. However, new workers will require technological skills that the displaced workers did not possess. This paper presents the results of an initial analysis of the interaction of technological life cycles in terms of socioeconomic factors.

Technological life cycles

Technology must be translated into a useful product, process or service, otherwise it will just remain a technological curiosity, i.e., fusion power or high temperature superconductivity. The development of new technology is an acquisition process, either internally or externally. The acquisition process of new technology, either internally or externally, is a dynamic process, with a well defined sequence of actions. These are not random actions, but proceed in expected patterns. All technologies progress through a series of life stages or life cycle. The acquisition process is the first stage of the life cycles of any technology.

There are various types of life cycles which must be considered in managing technology. Life cycles express the dynamic nature of various systems (Cleland and King 1983). The technology which is proposed as the solution to a particular problem moves from state to state as it evolves from a creative solution to final implementation and then to phase out. A manager of technology must consider the following life cycles:

- Technology life cycle
- Product/process/service life cycle
- System life cycle
- Project life cycle
- Organizational life cycle
- Business life cycle

Table 1 shows a summary of the various stages of each of the cycles that a manager of technology should consider during the total life cycle process.

0-7803-3552-X/96/$5.00 © 1996 IEEE

General	Technology (Martino 1993)	Product/Process/ Service (Cleland and King 1983)	System (Blanchard and Fabrycky 1990)	Project (Cleland and King 1983)	Organizational (Cleland and King 1983)	Business (Howard Jr. and Guile 1992)
Birth	Scientific Findings	Establishment	Acquisition Conception	Preparation or Initiation	Creativity	Emergence
Early Growth	Laboratory Feasibility	Growth	Design and Development	Implementation	Directive	
Later Growth	Operating Prototype		Production/ Construction		Delegation	Growth
Stable Maturity	Commercial/ Operational Use	Maturation	Utilization Use	Operation	Coordinate	Maturity
Unstable Maturity	Diffusion to other areas				Collaborative	
Phase-out	Social and Economic Impact	Declining Use	Phase-out		Dissolution/ Acquired/ Merged	Possible Decline
			Disposal			

Table 1. Summary of Life Cycle Phases

At any one time, each of these life cycles interact. The manager of technology will find the:

- technology in the middle of the research and development portion of the technology life cycle; while the
- product in which the technology is embedded is in the growth stage, e.g., the technology may be replacing another technology, i.e., 32 MB for 16 MB DRAM chips; while the
- overall system in which the product serves as one of the components is in the utilization stage; while the
- project is in the operational stage; while the
- organization is in a collaborative stage facing a series of complex problems; and finally the
- overall business life cycle is in decline.

Due to this complexity, the technology manager faces problems and opportunities. As in all technological complex systems, the underlying non-linearity conditions exist which can cause a chaotic and potential destabilization of all the cycles.

Life cycle interactions

Technology life cycles can have a significant impact on society. According to the Chairman of the Federal Reserve, Alan Greenspan: "*Human skills are subject to obsolescence at a rate perhaps unprecedented in American history....the fallout from rapidly changing technology.*"(WSJ 1996). Since computers became commonplace millions of positions have been altered. Other technologies have caused the elimination of many positions by replacing them with advanced technologies. The remaining positions have required the worker to be more technically competent. These new technology positions usually result in higher wages.

The following is a simplified analytical scenario of the interaction of technology, productivity and worker requirements. This is presented to illustrate that life cycles can have unexpected results in the socioeconomic arena.

The assumed scenario begins when a new technology is introduced into an industry. Due to a number of factors, this new technology finds initial resistance when it is introduced. Since the new technology has a definite productivity value, i.e., it allows workers to produce more units in a similar time, the benefits eventually lead to the new technology becoming well established in the industry. This higher productivity allows the industry to meet existing demands with less workers or produce more with the same number of workers. As the technology makes further penetration into the industry, more enterprises employ the technology. As an industry grows, possibly due to decreased cost due to increased productivity, the number of workers will either remain the same, producing more, or decrease to further decrease cost. However, trends of employment indicates that the number of the original workers will decrease, such as is happening in the current "right-sizing" of organizations. The displaced workers will be replaced by a smaller number of workers who have up to date technological skills.

The introduction of electronic typesetting resulted in the displacement of type setters is an example of this effect. These displaced type setters were replaced by a smaller number of computer operators and programmers. The following is a simplified model of this situation using a "S" shaped growth model based on the Pearl[1] equation.

If we assign:

u = units in use at time "t"
U = upper limit of units which can be placed in use
t = time

Then:

$$u = \frac{U}{1 + a_u e^{-b_u t}}$$

If the productivity of workers is a function of the number of new technologies in use displaced by a learning factor, assuming perfect learning and p = productivity at time "t" and is given by a productivity factor "A":

$$p = Au = \frac{AU}{1 + a_u e^{-b_u t}}$$

If we assume the requirement in an industry using the technology is "r" at time "t" and is also given by a Pearl type equation, then:

$$r = \frac{R}{1 + a_r e^{-b_r t}}$$

where "R" is the limit of the requirements. Then the number of workers ("w") required is given by:

$$w = \frac{r}{p} = \left(\frac{R}{AU}\right)\left(\frac{1 + a_u e^{-b_u t}}{1 + a_r e^{-b_r t}}\right)$$

Then for a constant requirement "r" $a_r = 0$ and the number of workers working in industry using the technology is represented by:

$$w = \left(\frac{R}{AU}\right)\left(1 + a_u e^{-b_u t}\right)$$

Figure 1 shows a resulting reverse "S" shaped curve. If the requirements follow an "S" shaped curve we find a situation depending on the constant factors where the number of workers can either decrease or increase over time. History seems to indicate a trend which consists of a cyclical curve around an increasing slop as new technologies stimulate new users but also increase productivity. The difference between the cyclic motion and the fact that the worker levels increase and then may have to decrease as the new productivity reduces the need for workers (See Figure 2).

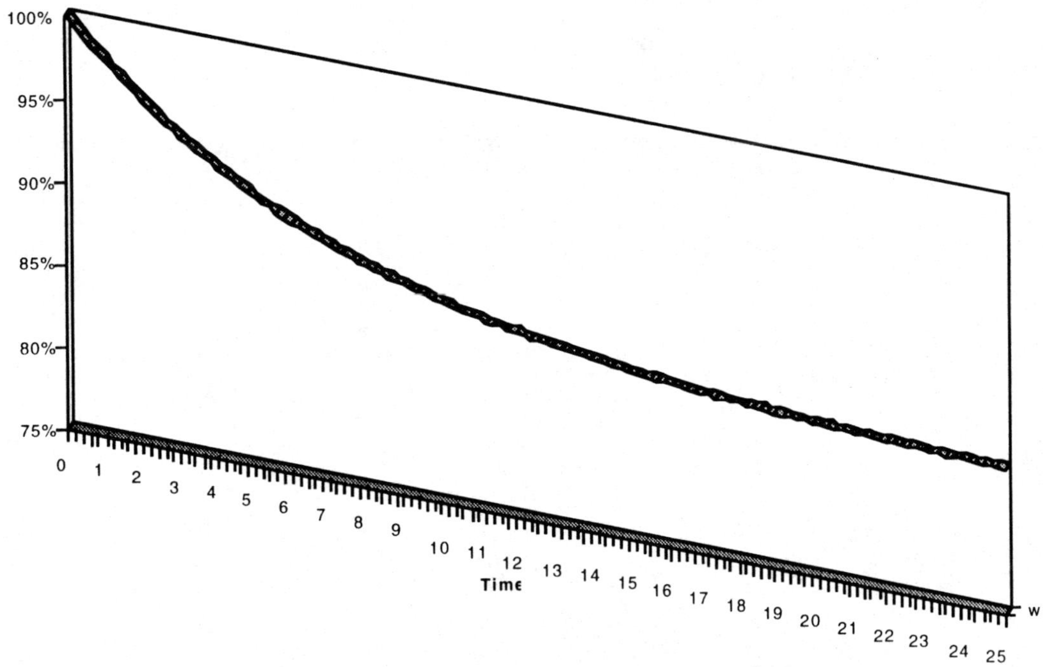

Figure 1 Impact of Introduction of Technology on Worker Requiren

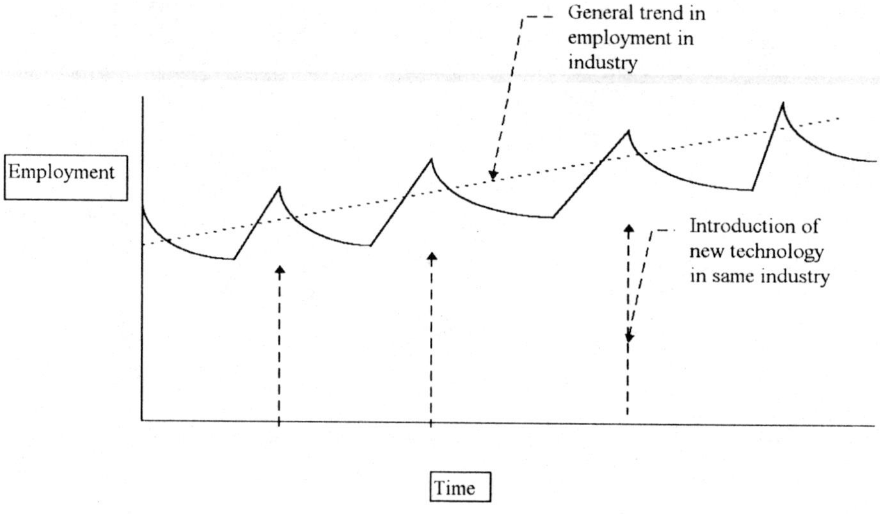

Figure 2 Potential Impact of Introduction of a New Technology

The cyclic nature of these interactions can lead to situations under certain conditions where the non-linearity cause highly chaotic conditions. In the early ninetieth century the displacement of skilled workers by mechanical looms caused serious social strife as seen by the Luddite[2] movement. There is some evidence that such anti-social behaviors are again rising.

As *The Wall Street Journal* (1996) stated "*The anticipated boom in productivity, however, could well spawn more labor upheavals.*" It appears that 18 million U.S. works may be at risk due to technology. It is important that we develop a better scientific understanding of how technological cycles impact not only the management of technology, but the underlying socioeconomic system upon which they are based.

References

Blanchard, B. S., and Fabrycky, W. J. (1990). *Systems Engineering and Analysis*, Prentice-Hall, Englewood Cliffs, NJ.

Cleland, D. I., and King, W. R. (1983). *Systems Analysis and Project Management*, McGraw-Hill, New York, NY.

Howard Jr., W. G., and Guile, B. R. (1992). *Profiting from Innovation: The Report of the Three-Year Study from the National Academy of Engineering*, The Free Press, New York, NY.

Martino, J. P. (1993). *Technological Forecasting for Decision Making*, McGraw-Hill, Inc., New York, NY

WSJ (1996). Outlook, *The Wall Street Journal*, April 22, 1996, p. 1

[1] Pearl's (1924/25) work dealt with the rate of increase of fruit flies within a bottle; rate of increase of yeast cells in an environment; and the rate of cell increase in white rates.

[2] A group of British workers who between 1811 and 1816 rioted and destroyed laborsaving textile machinery in the belief that such machinery would diminish employment. This is also used for individuals who oppose technical or technological change and are named after Ned *Ludd,* an English laborer who was supposed to have destroyed weaving machinery around 1779.

World Class Manufacturing: Techniques for Implementation for Small & Medium Sized Enterprises

Hassan Farsijani
(44) 1274-384264

Department of Industrial Technology
University of Bradford, BD7 1DP, UK

Andrew Carruthers
(44) 1274-384264

Department of Industrial Technology
University of Bradford, BD7 1DP, UK

The range of techniques associated with competitive manufacturing has expanded rapidly since the inception of MRP in the 1960s. Schonberger (1986) integrated these techniques in to the generic term world class manufacturing (WCM). The range and sophistication of these techniques places WCM status beyond the aspirations and competence of many small and medium sized enterprises (SMEs). The paper explores how the concept of WCM can be made relevant to SMEs, through the example of two case studies based on small to medium sized manufacturing companies experiencing both rapid growth and increasing international competition.

The Importance of the SME

While SMEs have always played a key role in the economies of all major industrial societies to both GDP and employment levels, it was not until the 1980's that this role gained wide appreciation in the UK. ACOST (1990) has reported that in UK manufacturing, 98% of all companies employ less than 500 people and account for more than one third of manufacturing output and 40% of total manufacturing employment. Rothwell (1989) found that out of 4400 significant innovations made by British companies between 1945-1983, 37% came from companies with less than 500 employees and 23% from companies with less than 100 employees. Between 1980-1983, 50% of the significant innovations were made by companies with less than 500 employees. For these reasons, support for SMEs is high on the agenda of all the main political parties in the UK.

While the definition of what constitutes a SME varies, it is generally based on the number of employees and financial turnover. In practice SMEs are usually characterised by simple organisational structures which facilitate rapid decision making and often display a high degree of innovation. What is often apparent in many SMEs is that the management techniques and operating structures appropriate to their roots in the "garden shed" have been carried forward into much larger scale production. The management techniques and operating structures employed are one way of identifying the maturity of the organisation.

The Concept of World Class Manufacturing

Schonberger (1986) coined the term "World Class Manufacturing" to cover the many techniques and technologies designed to enable a company to match its best competitors. What is readily apparent from the literature is that the concepts were developed principally in relation to the needs of larger scale manufacturers. Relatively little attention has been paid to the needs of the SME and there exists scope for framing the ideas implicit in WCM for the SME seeking to improve its competitive position. It is not sufficient, however, to attempt to apply techniques appropriate to large organisations to the operations of a small scale manufacturer. The techniques covered by WCM must be made recognisable and relevant to the SME if there is to be any possibility of them being adopted.

0-7803-3552-X/96/$5.00

When Schonberger (1986) first introduced the concept of "World Class Manufacturing" into popular parlance, the term was seen to embrace the techniques and factors as listed in figure 1. However, techniques of competitive manufacturing were well established before 1986 and would include statistical process control as one of the early manifestations. The substantial increase in techniques can be related in part to the growing influence of the manufacturing philosophies and economic success of Japanese manufacturers from the 1960s onwards.

What is particularly interesting from a review of the literature, is that while there is a degree of overlap in some of the techniques, it is clear that relative to the elements that were seen as constituting WCM in 1986, the term has evolved considerably. In recent years, it has expanded most notably to include simultaneous engineering, bench marking and increasing emphasis on issues relating to manufacturing strategy. Clearly, the management of a company seeking world class status would appear to be faced with a far more complex task than was the case previously.

1980	1986	1989	1991	1996
JIT/Kanban	JIT/Kanban	JIT/Kanban	JIT/Kanban	JIT/Kanban
TQC	TQC	TQM	TQC	TQC
TPM	TPM	TPM	TPM	TPM
TQM	TQM	TQM	TQM	TQM
Cellular Manuf	Cellular Manuf	Cellular Manuf	Cellular Manufacturing	Cellular Manufacturing
MRP II	MRP II	MRP II	MRP II	MRP II
	CIM	CIM	CIM	CIM
	Empowerment	Empowerment	Empowerment of Staff	Empowerment of Staff
	Kaizen	Kaizen	Kaizen	Kaizen
		Technology Mangt	Technology Management	Technology Management
		Systems Mangt	Systems Management	Systems Management
		Training	Training	Training
		Total Material Flow	Total Material Flow	Total Material Flow
		Design Mangt	Design Management	Design Management
		Manuf Strategy	Manufacturing Strategy	Manufacturing Strategy
		Perform Measures	Performance Measures	Performance Measures
			Benchmarking	Benchmarking
			Structure & Culture	Structure & Culture
			Innovation Strategy	Innovation Strategy
			Corporate Strategy	Corporate Strategy
			Group Technology	Group Technology
			Batch Size Reduction	Batch Size Reduction
			Set up Reduction	Set up Reduction
			Non-Financial Measures	Non-Financial Measures
				Simultaneous Engineering
				Business Process Re-Eng
				Vendor Quality
				Supplier Partnership

Figure 1
The growth of techniques and factors associated with the concept of World Class Manufacturing

Gap Analysis

One of the main reasons for the inappropriate use of advanced manufacturing technologies and techniques in many companies arises from an inadequate understanding of their production problems and the integrated nature of modern technology (Carruthers 1990). All too often, technological solutions are imposed which necessitate the company to engage in an organisational metamorphosis to effectively employ them. This can often produce sub optimal results. Ideally, the reverse process should occur, where the company progresses from a detailed understanding of its problems which ensures that a particular technology or technique is adapted to meet the needs of the company. This process of adaptation should also take into account the size and skills available in the management and workforce. WCM needs to be framed in terms of the needs of the SME rather than the other way round. For many SME's, becoming "world class" does not always mean implementing the most advanced technologies, instead its competitiveness may arise from the flexibility and skills of its workforce, or a unique market niche and business strategy.

A useful framework for analysing the deficiencies of the manufacturing operations is to identify "gaps" in the production and business processes that lead to inefficiencies and compare these to its own model of what constitutes "world class" in its field as shown in figure 2. For example, the company may employ modern production technologies, but as a result of poor factory layout and organisation of materials it is not using these to their full potential and so a general "flow gap" exists. By applying an iterative process and identifying "gaps" in its performance, the company can assess the suitability of potential solutions at a level appropriate to the requirements and resources of an SME, and gradually move nearer to its model of "world class" (Farsijani and Carruthers 1996).

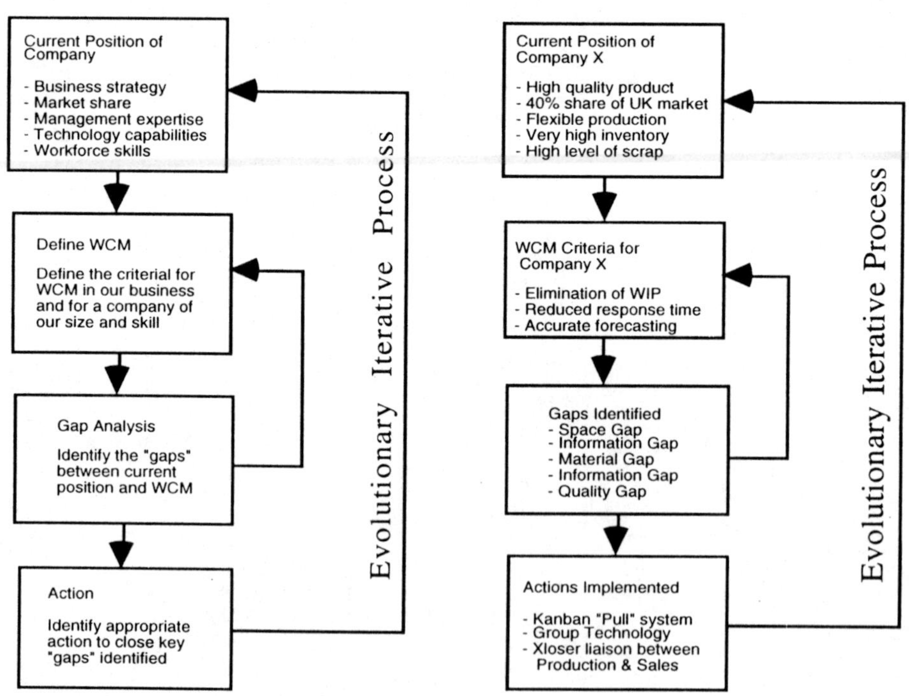

Figure 2

Gap analysis model and its application

Case Studies

A key aspect of testing the concepts developed in the research programme has been the collaboration of several SME's. A brief comparison will be made between two SMEs. Company A is a contract electronics manufacturer. It has sales worth several million pounds a year and has doubled in size on an almost annual basis over the five years or so of its existence, so that it has about 130 employees at present. It employed a small group of senior managers from the outset who were experienced in the industry and who could attract sufficient capital to enable the company to invest in advanced manufacturing equipment. The nature of their market is such that a company has to be World Class however large, or small it is. As a consequence Company A has not just invested in the most advanced equipment, but in training for its employees, in a management information system, in product development and in a programme of continuous improvement that (independently) parallels the Gap Analysis outlined above. Company A has the clear objective of maintaining rapid growth so that it can seek a listing on the London Stock Exchange.

Company B, although in electronics manufacture, does not have the capital resources to invest in the latest production technology. Its business is focused on the design of printed circuits boards and on the manufacture of electronics used in more 'robust' applications. These circuits use discrete components that can be assembled and soldered by hand. Most of the company's capital is tied up in working capital so that growth is limited to organic growth derived from profits. The company is seeking to broaden its company base so as to reduce its dependence on particular contracts. This dependence has been illustrated in dramatic form recently, as a drop in orders has led to a reduction in the size of the workforce of almost 50%. The company is not able to fund the equipment necessary to meet current legislative requirements for new products, so product development (other than circuit boards) has ceased. The company competes on the price for a given volume of order combined with speed of response. Its management is unaware of the concepts included in the term World Class Manufacturing, but it is acutely aware of the need to reduce costs whilst maintaining quality and speed of response. It is small enough, at 16 employees to enjoy good communications internally and to be able to implement Gap Analysis in its production and material control functions.

The greatest contrast between the two companies lies in their relations with their suppliers and customers. Company A has achieved a size and importance that enables it to build long-term relationships with its customers (the companies that order the manufacture of their own product lines) to ensure the lowest cost of components from its suppliers and delivery direct to the end-customer. It is able to implement a limited form of Just-in-Time, whereas Company B is too small to be able to dictate terms to suppliers or customers, so many aspects of Just-in-Time are inappropriate to its circumstances.

Conclusion

While the concept of what constitutes "World Class Manufacturing" has received considerable attention in the academic literature, it has been developed principally in relation to the needs of larger scale manufacturing organisations. Given that small and medium size enterprises make up a significant proportion of the manufacturing capacity in industrial economies there is a clear need for a rethink of the concept in order to make it relevant to their specific needs. Faced with a growing list of the elements needed for "world class" manufacturing the management of SME's appear faced with a daunting task. Focusing on the manufacturing operation in an SME using the "gap analysis framework" has proved its utility in assisting the management of one SME to make progress towards a model of WCM that is applicable to their circumstances.

References

Advisory Council on Science and Technology. The Enterprise Challenge: Overcoming Barriers to Growth in Small Firms. HMSO: London, 1990.

Carruthers, A. Business Implications of Manufacturing Innovation: The Experience of Increasing Automation in the Small to Medium Sized Company. PhD Thesis. Aston Business School: The University of Aston in Birmingham, UK, 1990.

Farsijani, H., and Carruthers, A. "World Class Manufacturing and the Small and Medium Sized Enterprise." Proceedings of the International Symposium Advanced Manufacturing Processes, Systems and Technologies (AMPST 96). University of Bradford: Bradford, UK, 1996.

Rothwell, R. "Small firms, Innovation and Industrial Change." Small Business Economics. 2(1), 1989.

Schonberger, R.J. World Class Manufacturing: The Lessons of Simplicity Applied. Macmillan: New York, 1986.

The Virtual Enterprise Framework and Toolbox

John W. Center
612-482-9828

Center Associates
Post Office Box 26173
St. Paul, Minnesota 55126

Joyce A. Thompsen
612-921-5871

Zenger Miller
3601 Minnesota Drive, Suite 880
Minneapolis, Minnesota 55435

Optimizing the performance of a virtual enterprise requires the creation and application of a unique framework and set of tools that serve the confluence of both technological and social aspects of the new business imperative. The authors present a composite of complex challenges to the virtual enterprise, a range of alternatives that respond to those challenges, and deeper exploration of a leading edge framework and its attributes. Also, a tool box approach is used to describe the practical application of the proposed framework.

Introduction

Incremental improvement, long held as an acceptable and predictable source of measurable progress, is now being portrayed as a vastly insufficient solution to the long-term sustenance of any business enterprise in today's complex world. This condition is complicated by the exponential acceleration of change in every dimension, greater degrees of uncertainty. A further complication comes from mandates for new definitions of environmental and human stewardship, and new strategic and political rules for identifying and capitalizing upon any potential opportunity.

The context

So, what form of enterprise can be sufficiently responsive not only to survive in this environment, but also has a good chance of thriving? In 1991, Lehigh University's Iacocca Institute published a report entitled *21st Century Manufacturing Enterprise Strategy: An Industry-Led View*. The study was in response to a Congressional request for identifying requirements for American industries to return to a state of global manufacturing competitiveness. (Goldman, 1995, p. xxi).

A new "agile competitive environment" was defined as driven by "the customer-perceived value of information and services embedded in physical products and delivered over a period through continuing relationships between vendor and customer." A synthesizing system of "rapidly maturing, computer-based production, information, and communications technologies" and new approaches to the human, physical, and intellectual assets among and within organizations were described as significant factors contributing to a "fundamental redefinition of industrial production." (Goldman, 1995, pp. xxi-xxii).

This confluence of both technological and social factors creates a dynamic platform for organizations to coordinate geographically and institutionally dispersed capabilities into a singular virtual enterprise. The platform also presents an opportunity to achieve a powerful competitive advantage in today's complex business environment.

The agile and virtual enterprise is posited as the strategic and organizational model necessary to create and assemble productive resources quickly, frequently, and concurrently in the face of complexity and uncertainty. (Goldman, 1995, pp. 201-201).

0-7803-3552-X/96/$5.00 © 1996 IEEE

In doing so, the virtual enterprise becomes dependent upon the availability and accessibility to communications, computer, energy, and transportation technologies. Also, the virtual enterprise must use new mental models within the traditional management disciplines. Social factors rising from the impact of the emerging intricacies of globalization, and the growing consciousness of respect for the rights of individuals, combine to produce the imperative for an applications-intensive, peresciential organization.

We have experienced two centuries of various forms of industrialization. Those two centuries of experience have given us a set of management processes:

- The English system centered on craftsmanship.
- The American system with interchangeable parts.
- The Taylor Scientific Management focus on efficiency of labor.
- Statistical Process Control.
- Numerical control and the onset of information processing.
- Computer-integrated manufacturing and flexible systems.

Our experience is being challenged by the new interdependencies of technology and social factors within the virtual enterprise. (Davidow, 1992 pp. 26-30).

Challenges of the virtual enterprise

The complex challenges facing the virtual enterprise include crossing global boundaries established by distance, time, language, political systems and related regulation, new rules for economics, and the emergence of new consumer and institutional markets.

The virtual enterprise must function simultaneously as an interactive and learning organization. It must include multidirectional communications paths. It must host new forms or adaptations of familiar control, feedback, and measurement tracking as well.

Processes must be designed for simultaneous information access (Grenier, 1995, p. 9). This is a significant departure from reliance on only the typical serial information flow. Timely and easy access to accurate information is essential to serve as a real-time, future-oriented, generative stream of critical knowledge bases and genuine return on intellect.

All technology choices for production and services within and without organizations must be astutely adjusted to the most urgent and important needs of the enterprise and its unique dynamics.

The life force of the agile and virtual organization is founded on absolute trust. Every member will be informed, informing, and performing as promised. No unknown or unidentified weak links are acceptable.

The typical void of a shared history must be quickly displaced by a practical shared vision that can be measured both in behavioral and economic terms. Communications nuances that are normally nurtured in face-to-face encounters must be acquired through other means that can effectively overcome physical separation. Both subtle and obvious individual and group differences must be identified and valued.

Leadership presents a paradox. Leadership may have the appearance of invisibility by virtue of its embodiment in <u>every</u> member. The keeper of the critical knowledge area, and its application for results, will become a momentary leader.

Framework alternatives

Current alternatives to a framework for the virtual enterprise might be presented in a continuum. It spans from work collaboration within or among groupings of people to classic teams, then to super teams, and eventually to virtual super teams. Groupings have long manifested themselves in the natural work group, focus groups, cross-functional teams, project teams, the matrix unit, and task forces of assorted types.
Virtual teams differ significantly from traditional or classic teams. The virtual team members may come from different organizations, must demonstrate selected competencies. They are expected to perform according to the situation rather than a prescribed role, expect to enjoy a trusting environment, execute work processes that are designed to be flexible and adaptive. They also derive authority from knowledge, function in a continuously changing environment, work collaboratively with others across distances, and tend to be technologically networked (Grenier, 1995, p. 218).

Virtual teams are designed to effectively manage the realities of "time compression, distributed resources, increasing dependency on knowledge-based input, the premium on flexibility and adaptability, and that fact that most of the information we use today is in electronic form" (Grenier, 1995, p. 217).

The Virtual Super Team (VST) is described by Terence Burton and our colleague Jack Moran as "the next generation of lateral management employee involvement teams." The VST functions as a "de facto small business and manage themselves as a self-contained value center." Organizational robustness is required as the foundation for the breakthrough results and high performance potential of the VST. The following characteristics must be intact: "boundarylessness, seamlessness, flat, fluid, virtual, one box called the customer success function, hierarchy replaced by dynamic, geometric patterns that represent critical customer processes, pools of customer needs, or fractals of resources assembled, reorganized, or redeployed on demand to maximize customer success, and extremely structured and disciplined but capable of changing instantaneously and automatically." (Burton, 1995, pp. 145-149).

Selected organizational philosophies and values must also be intact: awareness rather than unconsciousness, permissive instead of dogmatism, universality, not exclusion, education rather than training, integration as a substitute for isolation, full participation, beyond mere involvement, and holistic rather than targeted to displace alienation and distrust (Burton, 1995, pp. 150-153).

The continuum of work groupings actually extends past those just discussed. The far edge of the continuum reaches a Quantum Age organizational structure that reflects a natural force. This is the *non-structure* of a Swarm, a moving mass with a purpose.

The swarm

This leading edge framework is familiar and presents vivid images. Consider these: bees, wasps, bats, gnats, termites, schools of fish, butterflies, salmon, locust storms. One can look deeper into nature and consider the electron cloud. Visualize the budding "vesicles that emerge as needed from one set of intracellular organs to deliver their payload at the right destination" (Rothman, 1996). Consider the mission-based space shuttle team constituted of both the orbiting elect and the earthbound trackers with their highly interactive, integrated, and interdependent infrastructures.

Mandelbrot's fractal geometry, while appearing unstructured and random, actually produces multiple iterations and self-symmetry in its fractal curves, an advanced order beyond the realm of equilibrium. Applying the principles of fractals and chaos theory to organizations, we have the opportunity to achieve elegant motion, *unpredictable* order, optimized flexibility and simultaneous natural efficiency (Burton, 1995, pp. x-xvii).

Looking for images among people, consider the 100th Boston Marathon with its 37,000 runners. Consider the traffic flow on the LA freeways. Watch the women's sailing team competing for the America's Cup, where they use simple headnods to signal the assumption of another teammates' function and performing totally in synchrony.

Attributes of the swarm

The Swarm is more than an image. It is more than a name for a non-structure. A Swarm also represents a set of major characteristics and attributes. The names of those attributes can be easily remembered because their first letters spell SWARM. The SWARM is:

- Self-organizing
- Willful
- Agile
- Reliant
- Moving

Self-organizing

The Self-Organizing descriptor represents the SWARM's inherent capability for self-sustenance, achieving self-order, affiliation by member choice. It also represents a tendency and capacity for generative, progenitive, even perescientia| (pervasive knowledge) behavior, peregrination to accommodate shifting boundaries, high degrees of freedom, instantaneous cohesion when signaled. Also included is self-leadership, and comfort with blurred responsibilities.

Willful

The Willful attribute represents mission-focused, accountable, responsible, autonomous, comfortably directed within an ambiguous environment, self-directing, and self-coaching.

Agile

The Agile characteristic means instantaneously reconfigurable, innovative, temporal, capable of a swift focus, and can be wispy or perhaps illusionary in surface appearance while capable of high tensile performance.

Reliant

The Reliant descriptor captures access to competencies and resources, autonomous behavior, highly creative and innovative yet results-oriented, possessive of an elite set of competencies, unified in a purpose, demonstrating professional integrity.

Moving

The Moving category refers to the activity level, mobility, speed, intensity of aggression, high level of energy, and tendency toward only the most necessary yet temporary hovering on any issue.

Applications

Images come to mind including the menacing Africanized bees that practice intense defensiveness, absconding, endurance, and frequent swarming for action (Rinderer, 1993). Another SWARM image is the halo nuclei phenomenon with neutrons orbiting in a "misty quantum" cloud (Austin, 1995).

The SWARM can seem free form, able to absorb high energy pulses or transients, accommodating the elements of time, configuration, density, and speed to behave as necessary in its alternative steady states. It invokes the behavior of circuit theory, fluid analysis and thermodynamics, complex statics, chemical reactions, and systems interdependencies in an environment of anticipatory synthesis.

The environments

The SWARM concept can be applied in any of three broad scenarios or environments:

- Green-field or genesis processes.
- Retrofit or conversion environments.
- Ambient or focus process.

Retrofit

Within the retrofit situation, a conversion or transformation must occur. It requires displacing the plethora of seemingly intractable silos, pseudo-teams, departments, project assignments, and classic matrices.

Green-field

Within the green-field situation, the genesis process presents the single-time opportunity to create a SWARM to fit the central business purpose. Close examination of the compelling business case for the green-field will reveal the degree of fit of the SWARM model to produce the optimal performance.

Ambient

The ambient situation will already host an existing or near-SWARM, yet it will not likely have been identified as such. An example might include a globally-seeded technical sales, engineering, and support team behaving as a SWAT-team to identify, create, and service customers. A fraction of the benefits of the SWARM model will have been realized. The desired increased yield can be achieved with focused energy and concentration on the model's attributes.

The toolbox

Achieving optimal results from the SWARM model is dependent upon the practical application of a variety of tools. The toolbox has compartments for sets or collections of tools. Let us give the compartments the labels:

- Design.
- Process.
- Measurement.
- Sustenance.

Within each compartment, an array of tools exists for application to the specific attributes of the SWARM model and the three broad scenarios.

Design

Design tools will serve as diagnostics that can accommodate the survey method, specifications inherent in the scenario, desired structuring, component determination, and prototype work. An opportunity or problem appears on the horizon. Design tools help focus the efforts to identify the situation and its characteristics. They also help focus the effort to determine the needs, to put limits and requirements on the operation, and to define or identify the critical knowledge areas and competencies necessary for optimal performance. These tools also help determine in the SWARM concept applies to the situation. Examples include various forms of business and organizational analysis techniques used to establish specifications to establish the purpose for the SWARM or describe a desired future state.

Process

Process tools are categorized to fit the requirements of pilot lines, procurement, preparation, assembly, and distribution. Examples might include the creation of a bill of materials of critical knowledge areas to fit the business purpose. A pocket of the organization may be identified for an advanced learning laboratory of the applicability of the SWARM, especially in a retrofit scenario. Process tools will also focus on the acquisition, selection, and development of the critical knowledge areas, competencies, innate abilities, and motivations required for optimal performance. Issues of information access and flow, business process redesign, communications strategies, and all systems or infrastructure aspects will be examined, reconstructed, or aligned to feed the SWARM model capability.

Measurement

Measurement tools are used for testing, control, inventory, and gauging quality. The measurement tools will focus on input/throughput/output tracking, load factors or use, achievement of potential or capacity, asset application, and gauge immediacy, accuracy, efficiency, and absolute or relative yield.

Sustaining

Sustaining tools can be equated with strapping, packaging, protection, repair and modification functions. These protection tools include such things as circuit breakers or strapping tools for stress or strength management, repair or intervention tools. They also include tools for environmental management or scrutiny, and for sponsoring roles to affirm trust and mutual respect, celebration, circular feedback loops, and the foundation for the peresciential environment.

Conclusion

The practical application of the SWARM model as a framework for the virtual enterprise, and its accompanying toolbox, is still underway. Results will be duly compared and contrasted with those of the earlier framework attempts of groupings, classic teams, virtual, and super virtual teams. Two observations are already clear. The very nature of the SWARM model dictates that it is not an immediate or precise fit for the traditional or hierarchical enterprise. The SWARM is a strange attractor within the virtual enterprise and can produce extraordinary results in its organic life-form.

References

Austin, Sam M., and Bertsch, George F. "Halo Nuclei." *Scientific American*, June 1995, Volume 272, Number 6, pp. 90-95.

Burton, Terence T., and Moran, John W. *The Future Focused Organization: Complete Organizational Alignment for Breakthrough Results*. Prentice-Hall, Inc.: Englewood Cliffs, New Jersey. 1995.

Davidow, William H., and Malone, Michael S. *The Virtual Corporation: Structuring and Revitalizing the Corporation for the 21st Century*. HarperBusiness: New York, New York. 1992.

Goldman, Steven L., Nagel, Roger N., Preiss, Kenneth. *Agile Competitors and Virtual Organizations: Strategies for Enriching the Customer.* Van Nostrand Reinhold: New York, New York. 1995.

Grenier, Ray, and Metes, George. *Going Virtual: Moving Your Organization into the 21st Century*. Prentice-Hall, Inc.: Upper Saddle River, New Jersey. 1995.

Rinderer, Thomas E., Oldroyd, Benjamin P., and Sheppard, Walter S. "Africanized Bees in the U.S." *Scientific American*, December 1993, Volume 269, Number 6, pp. 84-90.

Rothman, James E., and Orci, Lelio. "Budding Vesicles in Living Cells." *Scientific American*, March 1996, Volume 274, Number 3, pp. 70-75.

Control of Electronic Business Transactions (EBT) in the Banking Industry

H. Michael Chung
hmchung@csulb.edu

Department of Information Systems
College of Business Administration
California State University, Long Beach
Long Beach, CA 90840-8506

Kristen Russell

Andersen Consulting
Houston, TX 77084

Electronic Business Transactions(EBT) are the interorganizational exchange of business data and monetary transactions in a standard format. This study explores control issues to facilitate secure business transactions in the banking industry. The paper details the risks and analyze them.

Introduction

The banking system in America has changed dramatically in recent years due to the rapid advancements in computer and communications technology. Since financial businesses are heavily dependent on real-time information, information technology plays a vital role in day-to-day operations and the price and mix of services offered. Financial transactions can occur more frequently and more rapidly while relative costs of information processing decrease.

Banks are no longer primarily concerned with lending money, but they are emphasizing noncredit services as well. Services such as funds transfers, Automatic Teller Machines (ATM), and cash management techniques are just a few of the new services and products being offered to customers at both retail and wholesale levels.

However, as changes in information technology revolutionize the financial services area, significant issues have come to light with regard to the present control mechanisms regulating the financial services industry. Although fundamental policy objectives, such as fostering institutional stability and integrity or protecting consumers, remain the same as in the past, the focus of the current regulation involves access to services, system security, and the effects of changes in telecommunications industry (Bhatt, 1995). With today's technologies, Electronic Business Transactions (EBT) have created new risks, security problems, policy concerns, and service performance issues that are critical to the financial services industry.

The next section describes today's EBT. The third section discusses the EBT issues followed by their analysis.

Current Status of Electronic Business Transactions (EBT)

Information technology has allowed corporations, financial institutions, government agencies, and consumers to use electronic payments systems to efficiently complete financial transactions. The two principal wire transfer systems used for electronic payments, Fedwire and CHIPS (the Clearing House Interbank Payments System), together process more than 80% of all dollars moved between financial institutions.

Traditional payments means, such as cash and checks, account for the vast majority of smaller transactions. These traditional payment mechanisms used to purchase goods and services cover most of the volume of transactions, well over 90%. More specifically, consumers used currency in payment for 73 percent of transactions, checks 17 percent and the rest with the other electronic means as detailed in Table 1 (Department of Treasury, 1995).

0-7803-3552-X/96/$5.00 © 1996 IEEE

Table 1. Means of Payment

Transaction Type	Percentage
Cash	73
Check	17
Credit Card	5
Automatic Clearing House	3
Debit Card	2

But, the value of all transactions completed by non-electronic means provides only 17% of the total value of all transaction types. By the year 2000, about 2 billion transactions that toady would be made with cash will instead be made with cards, reaching up to over $ 6 trillion (Egner, 1991). EBTs are the primary means of making the small volume, large-dollar payments.

A majority of transactions on Fedwire are transfers of federal funds between depository institutions and purchases and sales of government securities. Each Fedwire payment instruction is immediately debited to the sending bank's Federal Reserve account and credited to the receiving bank's account. At peak processing time, an average transaction takes less than 120 seconds from receipt to settlement (Federal Reserve, 1990). CHIPS is a private network operated by the Clearing House Association. The primary role of CHIPS is to facilitate the movement of dollar payments between financial institutions around the world. Currently, CHIPS is estimated to handle more than 90% of such payments. The size of transactions on the two networks is very large. The average size of a CHIPS transfer was about $5.2 million and Fedwire payments averaged about $3.0 million.

On the consumer side, ATM, direct debit at the point of sale, and telephone based services have been steadily growing with an exception of home banking. ATM becomes a competitive necessity for a bank and approximately 10 million ATM transactions are made per hour every day. ATM transaction prices are around 10 cents per transaction range and it is expected to drop below 10 cents (Egner, 1991).

The infrastructure for home banking was not adequately developed in the 1980's and the expectation for home banking to be a profit making service was not realized due to the differing perception of consumers.

Electronic Business Transactions (EBT) Issues

The dollar volume of transactions that move through the two major EBT networks is quite large along with a big potential in electronic consumer banking. However, there appears to have been a compromise between system access security and control versus convenient customer access and service price (VanHoose and Sellon, Jr., 1989; Deane and Pringle, 1995). In this section, the risks inherent in participating EBT are discussed.

The main risks associated with EBT include a settlement risk and fraud. The Federal Reserve requires all depository institutions to have a positive balance in their reserve account at the close of the business day. Daylight overdrafts are the deficient amounts that reserve accounts may have prior to final settlement at the end of the day. Daylight overdrafts can be the result of intentional or unintentional mismatching of payments and receipts. Unintentional overdrafts often result from poor planning, inadequate communication, or computer problems. On the other hand, intentional overdrafts might involve borrowing federal funds from another institution in order to maintain a positive end-of-day balance.

A risk associated with daylight overdrafts is that of default. Since these overdrafts are viewed as unsecured loans, they expose the Federal Reserve and other financial institutions to potentially serious financial loss. This problem is defined as the possibility that default by one institution on a private wire system could lead to additional defaults by other institutions, threatening the stability of the entire payments system (VanHoose and Sellon, Jr., 1989). The primary method of reducing daylight overdrafts places an upper limit on the amount of intraday credit available in individual institutions.

Another risk to the large-dollar EBTs is fraudulent transfers. Because most payment orders are transmitted electronically, a bank may not know whether an EBT has been authorized by its customer. Internal collusion is also a big threat for financial institutions.

While safeguards related to physical access control are widely implemented, it is necessary to consider critical aspects of encryption and workflow controls. Encryption involves the scrambling of a message sent from one party to another to protect the confidentiality and integrity of electronic transactions. For example, a digital envelope can be created by the sender encrypting the message and the other party decrypts the message. A digital signature allows cryptographic digest of the message and attaching this for checking by the recipient (Welch, 1994). Therefore the choice of a suitable encryption method is of critical importance.

Workflow controls refer to the employment practices, separation of duties, and software development standards in place by an organization. For example, callbacks can be used by employees to verify transfer requests. An algorithm may be implemented to reduce errors or to impose transfer limits by limiting the maximum amount of any transfer without non-electronic approval. Also, the beneficiaries to which funds can be transferred can be restricted.

Another issue is related to erroneous payment orders. If a customer and a bank have agreed on a security device to detect erroneous payment orders and the customer can show that the bank failed to comply with the system, liability falls on the bank.

Analysis of Issues

The increased use of electronic payments has improved the efficiency of financial markets by lowering the costs and increasing the speed of transactions. But, there has been a clear tradeoff between reducing payments system risk and promoting payments system liquidity. Overdraft policies and security devices are the two main factors that can affect the performance of payments systems. Without regulated overdrafts, the obvious advantage is the resulting increase in speed at which payments can be sent or received. But, with the implementation of overdraft caps and pricing of overdrafts, payments processing will be slower because institutions will need to synchronize electronic funds flow to avoid unnecessary expenses. As a result, the gains from reducing payments system risks against the costs of reduced payments system liquidity must be balanced (VanHoose and Sellon, Jr., 1989).

As far as security measures are concerned, the more secure the transfer system is, the more delays in transaction processing will occur. Often customers may adopt less secure procedures because of the costs associated with the actual security devices and the costs associated with the transfer delays resulting from higher level security. However, by implementing less secure procedures, the customer is assuming more risk. Again, tradeoffs must be made between security, costs, convenience, and risk. The main public concern over the payments system is in its safety and integrity, followed by an interest in the efficiency of the system (Welch, 1994).

Trends in banking, such as decentralization of operations, networking, and open systems will promote the electronic utility of delivering enhanced banking and information services to customers. Another important trend is the globalization of banking networks. With the availability of information on a constant real-time basis, financial services markets face the events of 24-hour trading, payments over electronic transaction systems in different countries made in different currencies, and difficulties in establishing global network regulations.

Conclusion

The paper described the issues associated with EBT: payments systems, electronic transactions and recordkeeping, and the institutional structure within which EBT operate. Such digital business environment demands more attention on control mechanisms involving encryption and workflow control.

References

Bhatt, V., "Financial Systems, Innovation and Development," Sage, Thousand Oaks, California, 1995.

Deane, M and Pringle, R, "The Central Bank," Penguin Books, New York, New York, 1995.

Department of Treasury, "Exploring the World of Cyberpayments," September, 1995.

Egner, F., "The Electronic Future of Banking," Financial Sourcebooks, Napcrville, Illinois, 1991.

Federal Reserve Bank, "The Federal Reserve in the Payments System." *Federal Reserve Bulletin*, 76, 293-298, May 1990.

VanHoose, D. and Sellon, Jr., G. "Daylight Overdrafts, Payments System Risk, and Public Policy," *Economic Review*, 74, 9-29, September-October 1989.

Welch, B., "Electronic Banking and Security," Blackwell Publishers, Oxford, England, 1994.

Design of an Information System for a Networked Industry

Sophie D'Amours [1,2],
Pierre Lefrançois [1] and
Benoit Montreuil [1]

[1] SORCIIER, Faculté des sciences de l'administration,
Université Laval, Ste-Foy (Québec), G1K 7P4
[2] GRAP, Faculté des sciences et génie,
Université Laval, Ste-Foy (Québec), G1K 7P4

The industrial network is composed of active firms and public organizations evolving in a specific industry and of an industrial network manager. The Information System for a Networked Industry (NetIndSI) is constructed to transfer information between firms of a business network. The firms are modeled as processors *in a network linked together to manufacture* products *or deliver* services. *These* processors *can perform many tasks on the NetIndSI such as bidding on a project, requesting bids for a project, analyzing strategic information, developing bidding strategies, developing strategic alliances, managing the firm operations and contracts, and archiving pertinent information. The industrial network manager is modeled to maximize the network contribution to its member's. Its role is (i) to manage membership, (ii) to answer requests from the network members, (ii) to ensure information flow between the network members, (iii) to distribute decision supporting tools to the network members and (iv) to develop new business opportunities for the members.*

Introduction

The new economy, the global market, the emerging power of technology and information, are among factors changing the rules of competitiveness in this last decade of the twentieth century. Theses economical and technological changes are unpredictable, fast and brutal. To survive, a firm has no other choices than to question itself and to restructure, reengineer and reconceptualize its business and processes (Vallerand *et al.* (1995)). To survive, the firm, whether it is small or large, needs to count on a strong innovation capacity and a flexible organization (Poulin *et al.* 1995). The firm now focuses on developing its core competencies (Quinn 1994) and creates short, mid and long term partnerships in order to respond efficiently to market expectations. The firm is now deeply dependent upon its relation with others, as it evolves through series of business networks.

The firm that configures and manages its operations with flexibility by the orchestration of the synergy between its internal facilities and its business network is called the Networked Enterprise (Poulin *et al.* (1994), Beaudry (1995), Paché et Paraponaris (1993)). The Networked Enterprise is a business form as well as an organizational system where economical and organizational processes highly depend on the business inter-firm relations characterizing it (Butera (1991)). In fact, the Networked Enterprise relies upon dense, fluid and flexible partnering relationships taking place inside and outside the firm (Nohria and Eclles (1992)).

The configuration of these networked enterprises and business networks have been studied by many authors. Miles and Snow (1992) have discussed stable, internal and dynamic networks; Artzen *et al.* (1994) have addressed logistic networks, also named production/distribution networks; Montreuil *et al.* 1992 have proposed the concept of Symbiotic Manufacturing Networks and Poulin *et al.* 1994 have rigorously classified industrial business networks. These networked enterprises and business networks are characterized by a set of nodes (economical units) and links (flow and relations), through which the nodes exchange valuable information to exploit the synergy of their gathering.

Traditionally they were built based on pure make or buy transactional considerations. These transactional considerations have led to the establishment of two well known forms of government : the markets and the hierarchies (Willianson (1994)). Today,

0-7803-3552-X/96/$5.00 © 1996 IEEE

however, many authors are questioning the fundamental element of this analysis : the transaction cost. For example, Ring and Van de Ven (1992) believe that the analysis of transaction costs is imperfect to support make or buy decisions since the dynamic aspect of the inter-firm relations, the frequency and the periodicity of the inter-firm relations as well as the impact of risk and trust are not considered. In that sense, Poulin *et al.* (1994) propose that the actual pertinent decision should take an extended form which is : make, buy, make-together or not-make. These decisions are now part of the Networked Enterprise strategic thinking.

The Networked Enterprise is facing the complex problems of configuring its network by selecting its partners and by establishing partnership, and of orchestrating them to realize its mission, according to the needs and goals of its partners (D'Amours (1995)). The Networked Enterprise is then forced to develop new capabilities in terms of planning of distributed operations, project scheduling, control of distributed operations as well as research, evaluation and selection of new partners (Martel *et al.* (1995)).

To support the networked enterprise in taking its decisions we propose a blackboard based information system which is built on the model of personalized electronic market introduced by Malone et al. (1987). We present an information based model of the networked enterprise and the networked industry, as well as a prototype of a Networked Industry Information System (NetIndIS).

We have defined three basic characteristics required from a NetIndIS:

(1) It is perceived as a seamless integrated system by the various partners of an industry, although it permits the exploitation of the locally most adequate information technologies, depending upon the configuration of the network and of its partners.

(2) It has an up-to-date comprehensive product-process-processor model of the industrial network supported. We call this the *Network Object Domain.*

(3) It provides a network of inner-enterprise and outer-enterprise manufacturing decision support tools, to support, for example, the processes of product, process, network design and the planning and control activities of the network. We call this network the *Network Agent Domain* which consist of the *Networked Firm Agent Domain and the Industry Network Manager Agent Domain.*

Exploiting the object/agent domains: The blackboard-based information Web

In order to design a NetIndIS, we adopt a vision where the industry network is seen as a system of interacting blackboards to which are connected planning and control agents. These agents realize networked internal and external planning and control activities for a networked firm and organization and are designed to post and retrieve data on their domain blackboards, to interpret such data and, through reasoning, execute behaviors that permit networking with other members of the network (networked manufacturing, strategic alliances, product development ...).

To do so, the agent-based approach builds around four major concepts: 1) inner-firm/organization planning and control (P&C) agents, (2) networked planning and control (NP&C) agents, (3) inner-firm/organization blackboards used as the core of the information system of a network node, and (4) network blackboards used as the core of the information system of a manufacturing network. **Figure 1** presents a simplified illustration of an agent-based industry network information system. The information transferred within the NetIndIS pertains to what we call its *Network Object Domain* which relies on three basic concepts: *processors*, *products* and *processes*.

The firms of a network are modeled as *processors*, acting within *processes* to transform *products*. We consider that the network processors are enabled to perform tasks such as preparing a bid to realize a process or a sub-process, requesting bids for a process or sub-process, analyzing information flowing within the network in order to adapt its process bidding strategies and finally for managing its own processes. These task performing capabilities are distributed within the NP&C agents of the networked firm and organization.

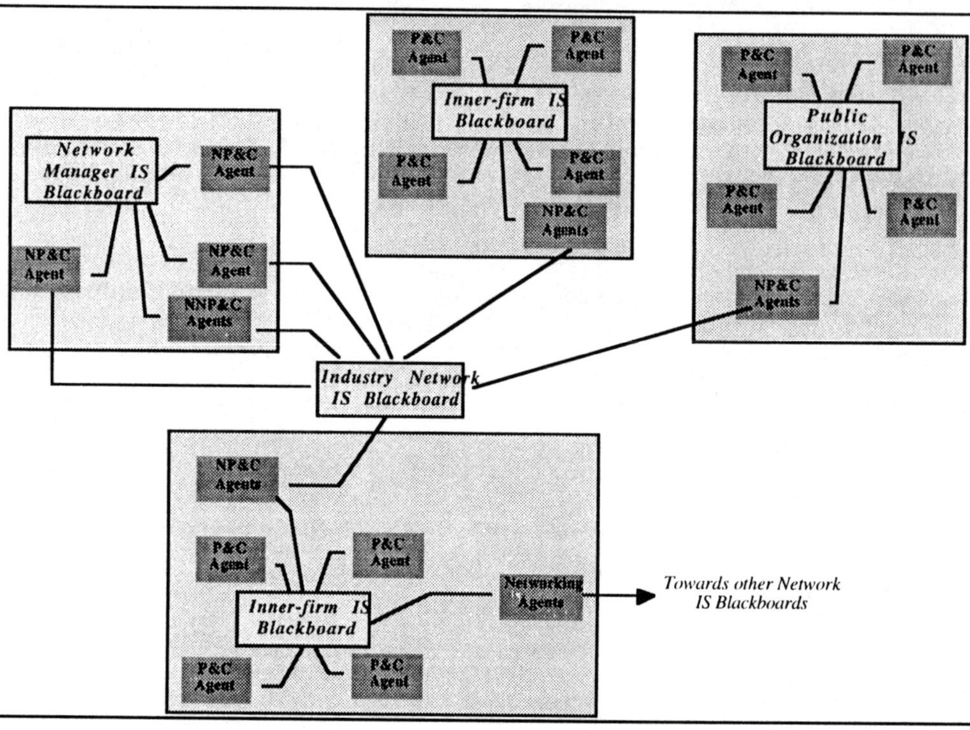

Figure 1. An industry network information system: *The BB-based NetIndIS.*

Inner-firm/organization IS blackboards connect the set of agents of a networked firm such that they can communicate within the firm. Also connected to the blackboard are networking agents who, as their name implies, permit flow information between the inner-firm IS blackboard and different network IS blackboards. These inner-firm/organization agents form what we call the *Networked Firm Agent Domain.* Their role is to coordinate the operation of the network, eventually offering information on the connected processors, their processing capabilities and their product offering. Such agents however, only have a passive role within the networking system, since all the activity is generated through the interaction taking place between the various agents of the networked agent domain.
To develop our prototype we have concentrated on modeling the networked firm agent domain and developing though the NetIndIS their mode of action.

Figure 2 summarizes the roles of the key agents within this domain.

Inner-firm/organization agent domain	Networking task
Intelligence Agent	Networking agent whose task is to analyze the information (bids and bid requests) posted on a network IS blackboard. This agent is connected to a network IS blackboard.
Bid Analyzer	Agent whose task is to analyze the bids received for the realization of a process or sub-process and to find the time-dependent processor-process assignment.
Data Manager	Agent whose task is to archive and retrieve for later use information.
Bid-Taker	Networking agent whose task is to release processing orders and to seek for inner and outer firm bids for processes or sub-processes to be realized. This agent is connected to a network IS blackboard.
Contract Manager	Agent whose task is to control contractual terms.
Process Manager	Agent whose task is to plan and control the realization of a process or sub-process once the time-dependent processor-process assignment has been made and an inner processor has been selected.
Production Planner	Agent whose task is to establish production programs, which means to decompose the process into sub-process and to define delays in order to respect demand.
Bid-Maker	Networking agent whose task is to prepare a bid for the realization of a process or a sub-process. This agent is connected to a network IS blackboard.
Strategic Manager	Agent whose task is to develop strategic plans for the networked firm.

Figure 2. The roles of the key members of the *Networked Firm Agent Domain.*

The Industrial Network Manager Agent Domain

Also connected to the NetIndIS blackboard is the industrial network manager agent domain. The industrial network manager is modeled to maximize the network contribution to its members. Its role is (i) to manage membership and search for new subscriptions, (ii) to answer requests from the network members, (ii) to ensure information flow between the network members, (iii) to distribute decision supporting tools to the network members and (iv) to develop new business opportunities for the members. **Figure 3** describes the roles of the key members of the industrial network manager agent domain.

Industrial manager agent domain	Networking task
Membership Manager	Networking agent whose task is to manage membership issues such as payment, business ethic, marketing and technical support.
Information Provider	Networking agent whose task is to generate and diffuse pertinent information on the industry (price, market, technology, capacity..) and to analyse performance of the industry members.
Message Manager	Networking agent whose task is to transfer information between members of the NetIndIS.
Decision support tools provider	Networking agent whose task is to analyze the needs of the network members in terms of decision support tools and to provide these tools on the NetIndIS.
Business opportunities Analyzer	Networking agent whose task is to seek for business opportunities by developping partnership with other industry network manager and by brokering major projects accross the NetIndIS.

Figure 3. The roles of the key members of the *Industry Network Manager Agent Domain.*

Conclusion

Information integration through different information systems such as CIM and CIM in the extended enterprise is actually at the center of many research program. Its positive impact has been shown under different perspectives in terms of organization, structure, cost and delivery performances in D'Amours et al. [1995], Venkatraman [1995] and in Srinivason et al. [1994]. However, when we extend this concept to the industry, we predict new benefits. First at the firm level, the NetIndIS should permit a rapid acces to a large set of buyers or providers, a better understanding of the industry, a reduced commercialization effort, a better utilization of resource and an increase in business opportunities. Second, at the industry level, the NetIndIS should permit a better and more structured market organization, a greater availability of dedicated knowledge, a greater availability of planning and control systems, a continous improvment, an increase in the number of transactions through the industry, as well as a technological positioning throughout the world. We intend to pursue the development of the NetIndIS prototype, to test its real impact, to develop more dedicated decision support tools and to extend the industrial network agent domain.

References

Arntzen, B.C., Brown, G.G., Harrison, T.P. et Trafton, L.L. (1994). Global Supply Chain Management at Digital Equipment Corporation, Interfaces, Vol.25, no.1, p.69-93.

Baudry, B. (1995). L'économie des relations interentreprises, Éditions la découverte, Paris, 125 p.

Butera, f. (1991). La métamorphose de l'organisation, du château au réseau, Les Éditions d'Organisation, Paris, 245 p.

D'Amours, S, Montreuil, B. and Lefrançois, P. (1995). Networked Manufacturing: The impact of Information Sharing on Scheduling Performance, Working Paper, SORCIIER, Faculté des sciences de l'administration, Université Laval.

D'Amours, S. (1995). La planification des opérations dans les réseaux manufacturiers symbiotiques, Ph.D. Thesis, École Polytechnique de Montréal, Université de Montréal.

Malone, T.W., Yales, J. and Benjamin, R.I. (1987). Electronis Markets and Electronic Hierarchies, Communication of the ACM, Vol.30, No.6, p.484-497.

Martel, A. and Oral, M. (1995). Les défis de la compétitivité : vision et stratégies, Publi-Relais, Montréal, 287 p.

Miles, R.E. et Snow, C.C. (1992). Causes of Failure in Network Organizations, California Management Review, Vol.34, No.4, p.53-71.

Montreuil, B., Lefrançois, P., Ramudhin, A. and D'Amours, S. (1992). A Conceptual Introduction to Symbiotic Manufacturing Netwoks, Proceedings of CEMIT 92/CECOIA 3Conference, Tokyo, Japon, p.505-508.

Noria, N. and Eccles, R.G. (1992). Face-to-Face: Making Network Organizations Work, Networks and Organizations, Structure, Form and Action, Harvard Business School Press, p. 288-308.

Paché, G. and Paraponaris, C. (1993). L'entreprise en réseau, Presses Universitaires de France, Paris, 127 p.

Poulin, D., Montreuil, B. and Gauvin, S. (1994). L'entreprise-Réseau: Bâtir aujourd'hui l'organisation de demain, Publi-Relais, Montréal, 335 p.

Quinn, J.B. and Hilmer, F.G. (1994). Strategic Outsourcing, Sloan Management Review, Été, p.43-55.

Ring, P.S. and Van de Ven, A.H. (1992). Structuring Cooperative Relationships Between Organizations, Strategic Management Journal, Vol.13, p.483-498.

Shaw, M.J., J.J. Solberg and T.C. Woo (1992). System Integration in Intelligent Manufacturing, IIE Transactions, Vol. 24, p. 2-6.

Scrinivasav K., Kekre S., Mukhopahyay T. (1994). Impact of Electronic Data Interchange Technology on JIT Shipments, Management Science, Vol.40, No. 10, p.1291-1304.

Vallerand, J. and Montreuil, B. (1996). L'OR^3 : Une méthodologie de développement stratégique de l'Organisation Réseau, Proceedings of Colloque International de Managemetn Stratégique, IAE de l'Université de Lille, France.

Venkatramam, N. (1995). Reconfigurations d'entreprises provoquées par les technologies de l'information, L'entreprise compétitive au futur, Les Éditions de l'Organisation, Paris, 349 p.

Williamson, O. (1994). Les institutions de l'économie, Inter Éditions, Paris, 402 p.

Review of Organization Structures of Concurrent Teams Responsible for Product Development in the Electronics Industry

Tugrul Daim
tugrul@emp.pdx.edu

Portland State University, Engineering Management Program
PO Box 751 Portland, OR 97207-0751

This paper explores the characteristics of the organization structures in concurrent product development in today's competitive markets. To be able to extract the most effective implementations the paper is sampling from the electronics industry in which product development cycles have been reduced significantly in the recent years. Several case studies from the literature, and interviews with product managers from the electronics industry have been used.

Concurrent product development (CPD)

Concurrent Product Development or as usually referred Concurrent Engineering has been the key to success in product development and time to market reduction (Rosenau, Shina, Shunk, Carter, Rochford, Scheiber, Tolhurst, Waddel, Smith, Prasad).

The idea of concurrent product development became visible in 1988 (Barkan, Evans, Stauffer, Winner, Henke, McDonough). Institute for Defense Analyses (Winner) defined it as a systematic approach to the integrated , concurrent design of products and their related processes, including manufacture and support. This approach is intended to cause the developers, from the outset, to consider all elements of the product life cycle from concept through disposal, including quality, cost, schedule, and user requirements. Many (Henke, McDonough, Durand, Calantone) described it as a concept of product development that emphasizes the response to customer expectations. According to the authors this concept embodies team values of cooperation, trust and sharing in such a manner that decision making proceeds with large intervals of parallel working by all life-cycle perspectives early in the process, synchronized by comparatively brief exchanges to produce consensus. (Figure 1)

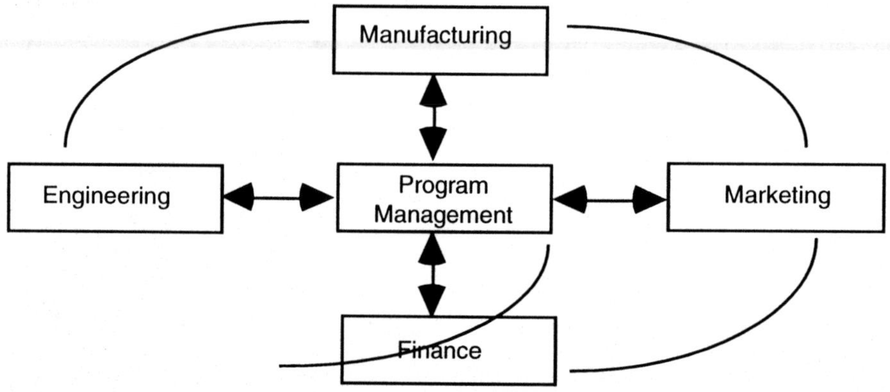

Figure 1. Common Elements of CPD

Basic principles of CPD are summarized as early problem discovery, early decision making, work structuring, teamwork affinity, knowledge leveraging, common understanding, ownership, and constancy of purpose. (Rosenau, Shina, Shunk, Carter, Rochford, Scheiber, Tolhurst, Waddel, Smith, Prasad)

0-7803-3552-X/96/$5.00 © 1996 IEEE

Concurrent product development in electronics industry

Electronics industry has been the solid proof of the fact that concurrent product development works for organizations who are eager to reduce the time to market for their new products. The significant results have been observed in Printed Circuit Fabrication (Scheiber, Tolhurst, Baumgartner, Hurt, Kovarsky, Osborne) and semiconductor industry (Parker, Hildebrand).

This section will explore case studies from the electronics industry. The facts about the cases were collected from the literature (Shina, Shunk,Ettlie) The problem area before CPD was implemented, the particular element of CPD that was implemented, and the results after implementation are presented.

Company/ Product	What was the Problem	What was the Solution	What were the Results
HP/Multi-Meter	Market competition	QFD, DFM	Reduced time to market
Chipcom/ Switching Hubs	High Growth, Less Resources	CAD/CAM Integration	Reduced time to market
Sun / Client Servers	High Costs at Product Development	DFM	Reduced Development Costs
Mercury/ High End Computers	Difficult to Manufacture Complex Design	Concurrent analysis at all functions for the complex design	Successful and Systematic New Product Introduction
Northern Telecom/ Telecom. Equipment	Market Competition in time to market	DFM Gateways during New Product Introduction	Reduced time to market
Digital/ Printed Board Assembly	Inefficient Production Lines	Employee Involvement Site Feedback	More efficient Production Lines

Table 1. Examples of Concurrent Product Development

The way the project is handled matters a lot in terms of the end results (Wright, Dome). In order to highlight the method how the new tools were implemented, the following table was derived. It shows how the organization was structured and how the projects were driven. In all of the cases the project champions (Beatty), who were acting as a glue in bringing different functions together are also highlighted.

Organization Structure	Drivers of the Implementation (Project Champions)		
	Top Mgmt	Project Mgr	Cross Functional Team
Project			
Functional			
Matrix	////	////	//////

Table 2. Organization Structure and the Project Champions

Organization structures vary among three major structures (Nicholas, Meredith): Pure Functional, Pure Project and Matrix type of organizations. In all of the cases studied the structure was definitely a matrix type, which was allowing cross-functional teams function effectively. In all of the cases the team was the main driver. In some cases top management was driving the projects with the team and in some their participation was through the project manager. In some cases both types of top management involvement have been seen.

Cross functional teams

To understand the findings out of the cases studied above, program managers in the local area (Portland, Oregon) have been interviewed. They were asked to describe how their teams are organized during new product development. A total of 12 managers from the electronics industry were interviewed. Both interviews with program managers and literature supported the success achieved by utilizing cross-functional teams in a matrix type of organization. Figure 2 describes this organization structure.

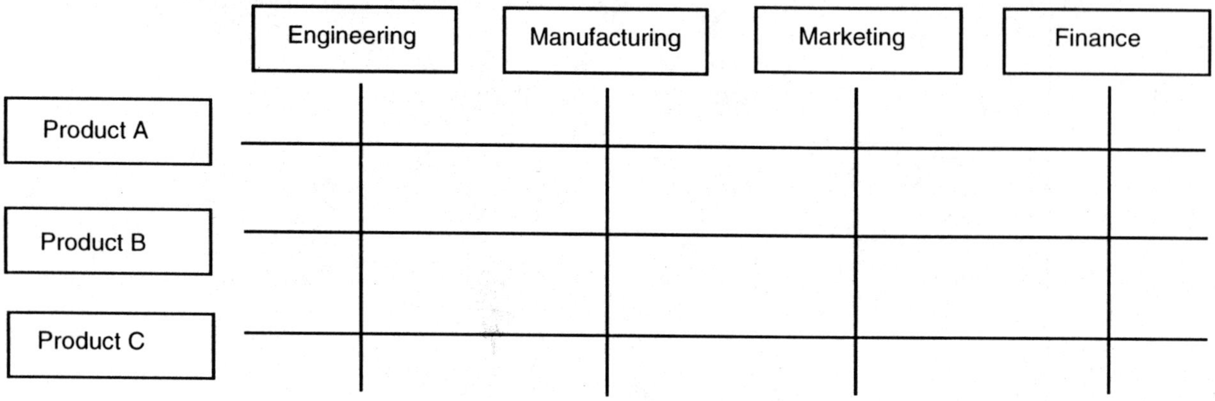

Figure 2. Matrix Organization that facilitates the cross functional teams

Matrix organization has been studied extensively in the literature (Meredith, Nicholas). The intersection points in the structure represents a team member from the corresponding function. The program managers interviewed pointed out that team members work on multiple projects at a time.

All the teams under study had at least three major sections:

- Engineering
- Manufacturing
- Marketing

In many companies all functions are involved throughout the project. Engineering is more involved in the beginning when product is developed. The phases of a product development program can include the following:

- Ideas
- Design
- Prototype
- Production

The program managers who were interviewed stated that involving all the three functions (Engineering, Manufacturing, and Marketing) throughout the program definitely reduces time to market for the reasons below:

- Engineering and Marketing will work together to design a sellable product
- Engineering and Manufacturing will work together to design a manufacturable product
- Marketing and Manufacturing will work together to set the production schedule

Design for Manufacture (DFM) and Quality Function Deployment (QFD) are two types of practices that are utilized significantly in the electronics industry, and that provides the interaction among different functions listed above.

QFD is described as a method for structured product planning and development that enables a development team to specify clearly the customer's wants and needs, and then to evaluate each proposed product or service capability systematically in terms of its impact on meeting those needs. QFD is well known with the matrices that are constructed during the process and referred as "House of Quality". DFM on the other hand is described as the full range of policies, techniques, practices, and attitudes that cause a product to be designed for the optimum manufacturing cost, the optimum achievement of manufactured quality, and the optimum achievement of life cycle support. From the descriptions we understand both methodologies should be used to produce a manufacturable product with a customer ready to buy it.

Conclusions

The general conclusions that should be drawn from this study are:

1. Empowered cross-functional teams are the main key success factor in concurrent product development in the electronics industry.

2. Matrix type of organization is the best organization structure that facilitates cross-functional teams responsible for new product development in the electronics industry.

3. New product development can only be successful through concurrent involvement of different functions in the electronics industry.

References

Barkan P. "Simultaneous Engineering" Design News, 44:A30 (March 1988)

Baumgartner D., "Quick-Turn Design and DFM", Printed Circuit Design, 19(3):18-22, (March 1996)

Beatty C. "Preaching the Gospel: The Evangelists of New Technology" IEEE Engineering Management Review, 20(3):15-25 (1992)

Calantone R., Di Benedetto A. "An Integrative Model of the New Product Development Process" The Journal of Product Innovation Management, 5:201-215 (1988)

Carter D. Concurrent Engineering, Addison Wesley Publishing Co,, 1992

Domb E., King B. Tate K., "Systematic Innovation: A Method for Resolving Design and Manufacturing Problems in New Product Development", PM Network, 10(3):42-43, (March 1996)

Durand T. "Concurrent Engineering and Interfunctional Project Groups" *International Journal of Technology Management* 10(1):67-78 (1995)

Ettlie J., Stoll H. Managing The Design-Manufacturing Process, McGraw-Hill Inc. 1990

Evans B. "Simultaneous Engineering" Mechanical Engineering, 110(2):38-39 (Feb 1988)

Henke J., Krachenberg R., Lyons T. "Cross Functional Teams: Good Concept, Poor Implementation", The Journal of Product Innovation Management, 10(3):216-229 (June 1993)

Hildebrand C., "Fast Company", CIO, Aug 1994, 66-67

Hildebrand C., "Widget Racers", CIO, Sep 1995, 69-76

Hurt L., "DFM for Test and the Customer", Printed Circuit Design, 19(3):24-26, (March 1996)

Kovarsky M., "Tools for the Trade", Printed Circuit Design, 19(3):28-32, (March 1996)

Mcdonough III E. "Faster New Product Development: Investigating the Effects of Technology and Characteristics of the Project Leader and Team" The Journal of Product Innovation Management, 10(3):241-250 (June 1993)

Meredith J., Mantel S., Project Management, John Wiley & Sons, Inc, 1995

Nicholas J., Managing Business and Engineering Projects, Perentice Hall, New Jersey, 1990

Osborne S., "Flex Circuit Manufacturability", Printed Circuit Design, 19(3):34-38, (March 1996)

Parker K., "What the Microprocessor Sowed", Manufacturing Systems, March 1994, 20-32

Prasad B. Concurrent Engineering Fundamentals, Prentice-Hall, New Jersey, 1996

Rochford L. , Rudelius W. "How Involving More Functional Areas Within a Firm Affects the New Product Process", The Journal of Product Innovation Management, 7(4):287-299 (December 1992)

Rosenau M. "Faster New Product Development", The Journal of Product Innovation Management, 5:150-153 (1988)

Scheiber S. "System Houses Pursue Many Paths To Ease Test" Electronic Business Buyer, 21(7):87-93 (July 1995)

Shina S. Successful Implementation of Concurrent Engineering Products and Processes, Van Nostrand Reinhold, New York NY, 1994

Shunk D. Integrated Process Design and Development, Richard D. Irwin Inc. 1992

Smith J., Tomasek R., Jin M., Wang P. "Integrated System Simulation", Printed Circuit Design 12(2):19-22 (Feb 1995)

Stauffer, R.N. "Simultaneous Engineering - What it is" Manufacturing Engineering, 101(3), (Sep 1988)

Tolhurst B. "Timely Success Strategies", Printed Circuit Design 12(3):60-62 (March 1995)

Waddell P. "Systems Design", Printed Circuit Design 12(2):13-15 (Feb 1995)

Winner R, Pennel J, Bertrend H., Slusarczuk M, The Role of Concurrent Engineering in Weapons System Acquisition, IDA Report R-338, Alexandria, VA, Institute for Defense Analyses

Wright C., Schacht N., "Where the Rubber Meets the Product Development Road: Institutionalizing Project Management in IBM's Development Community", PM Network, 10(3):17-19, (March 1996)

Designing the Virtual Enterprise

Richard C. Dorf
(916) 752-7395

Graduate School of Management and College of Engineering
University of California, Davis
Davis, CA 95616-8609
rcdorf@ucdavis.edu

The virtual enterprise integrates business components located in multiple locations using technological aids for business communication. All enterprises seek to optimize the use of business resources, including communication technologies, in order to achieve maximum total utility for the enterprise. A business design process is proposed which may enable the enterprise to attain the optimum use of resources in the pursuit of the virtual enterprise.

Introduction

The virtual enterprise integrates business components located in multiple locations using technological aids for business communications. Examples of communication aids include fax, e-mail, voice mail, audio conferencing, and video conferencing. The permanent collocation of a corporate workforce is transitioning towards a distributed workforce. As the enterprise further organizes around the flow of information, a strict requirement for collocation of workers decreases. An additional force for change is the trend towards outsourcing and contracting for many tasks formerly accomplished by the employees of the enterprise. As teams are formed for specific projects and later dissolved and reformed anew, a flexibility is achieved utilizing distributed people with many skills that may be realigned to fit the requirements of the organization. The virtual enterprise is said to overcome bureaucracy and the cumbersome nature of large, integrated enterprises thereby resulting in a shift to a nimble, responsive organization with higher incentiveness for the workers (Byrne 1996).

The virtual enterprise may be also termed the network organization since it is built upon the electronic network. The benefits to the corporation include reduced office space requirements and the ability to engage independent contractors without concern for their location. The benefits to the worker include reduced commuting, no requirement to relocate for the job, and flexible work hours and situation.

Requirements for Success

The worker in the virtual enterprise should be a worker who can function fairly independently and is competent in accomplishing the required tasks. This may disadvantage the beginning worker in training. Furthermore, supervision and management of the remote worker may be more difficult than in a collocated work place. The mediation of conflicts between workers may be less effective in a network organization.

The communication of tacit knowledge usually requires face to face interaction. The elements that may be lost are those ideas and concepts which are difficult to transmit such as emotion, rapport, passion, group synergy, and sharing. The culture of an organization is built on shared values and beliefs as well as team spirit, loyalty and a sense of community. Brainstorming, task delegation, conflict and problem resolution may require in-person meetings.

Specific types of enterprises may be successful as virtual business. For example, small firms focused on computer programming or technical editing services could successfully utilize an Intranet—a private network that spans geographic regions and provides databases, e-mail, and collaboration software (groupware).

0-7803-3552-X/96/$5.00 © 1996 IEEE

It is critical to retain and control the firms core competencies, tacit knowledge, and critical parts of the value chain in order to capture the rewards of the market. To rely on others for core competencies exposes an enterprise to significant hazards (Chesbrough 1996).

Workers in a virtual enterprise must possess skills with electronic tools and be capable of continuous sharing of incomplete work via a network. The role of the collocated core group will be to maintain and communicate the vision and strategy and to engage and manage the network relationships.

Workers find rewards and motivation from the physical aspects of a sense of place with an organization. Trust is a key feature of any organization. Small work groups facilitate bonding and trust. Membership in a group will require a sense of community to replace the sense of place. The idea of the virtual organization as a "box of contracts" is flawed (Handy 1996). The virtual organization will be challenged to extend trust to its members in order create community.

Innovation and positive change within an organization requires significant organizational stability based on a strong widely shared culture, community, vision, and strategy. Actions taken by the enterprise leadership must reinforce shared performance measures and people practices. The network organization will require the clear alignment of control, structure, performance measures, and rewards across a network of non-collocated people.

The Business Design of the Virtual Enterprise

Any business, virtual or collocated, requires a design based on a full understanding and description of the market, customer, and technologies as well as the business mission or purpose and the business core competencies (Drucker 1995). The business design should maximize utility for the customer and other stakeholders. First, an organization defines, or redefines, its tasks and then decides which it will perform and which it will outsource. The organization then configures its resources; arranges its culture and incentives; and sets a strategy for capturing profit (Slywotzky 1996).

The design process will generally follow the following sequential steps.

- describe the business process and the subtasks
- write specifications for each process
- assess capabilities and core competencies
- design the business system of processes
- assign processes as collocated, networked, or outsourced
- reevaluate and iterate the steps required

The decision to collocate or disperse certain process or tasks can be based on the following impact pattern:

1.	Competency required:	Peripheral	— Significant	— Core
2.	Nature of task:	Common	— Important	— Creative
3.	Stability of task:	Stable	— Evolutionary	— Dynamic
4.	Effect of task or activity:	Autonomous	— loosely connected	— systemic

Examining a given process in the context of the four items will enable one to determine if it should be a) outsourced (contracted), b) networked with employees, or c) collocated centrally. If the profile of the task lies in the right hand most column, then, the task should most likely be collocated centrally. The profile of a task aligned with the middle column calls for a network of employees. Finally, the profile of a task or process aligned with the first (left most) column will call for the outsourcing of the task to a supplier.

Case Examples — New Airlines and Book Publishing

Thirty new airlines have been certified in the United States during the period January 1994 to March 31, 1996 (McCartney 1996). In the past, the starting of an airline was a daunting process requiring capital, airplanes, spare parts, facilities, mechanics, reservation systems, ticketing, boarding staff among others. The success of Valujet Airlines, a 1993 startup that has grown from two planes in 1993 to 45 planes in 1996 has been widely reported (McCartney 1996). Suppliers are available to handle many of the functions and planes are readily available for lease. Sterling One Airlines started with 5 employees and subcontracted the pilots, ticketing, maintenance, and most functions. These new airlines offer low priced fares. Examining the impact pattern of the preceding section one can envision that many of the tasks will fall in the first (left-hand) column and lend themselves to outsourcing. However, an airline should possess some core competencies (marketing for example) and will require some centrally collocated staff.

Many readers of this paper may have, or plan to, publish a book. This writer has 30 years of experience in the publishing industry and has experienced an evolution towards outsourcing the manufacturing, inventory and fulfillment functions while networking the copy editing and review processes and retaining centrally only the acquisitions editor and the marketing function. The sales function is often outsourced, but many publishing companies utilize a non-collocated network sales force.

Conclusion

The virtual enterprise integrates business functions located in multiple locations using technological aids for business communication. A design method for the virtual business enterprise is proposed so that the collocated functions are based on the core competencies and include the dynamic, creative, and systemic functions. Less critical, but important, functions are networked. Finally, relatively common, stable, peripheral and autonomous functions can be outsourced to suppliers.

References

Byrne, J.A., "Has Outsourcing Gone Too Far?" Business Week, April 1, 1996, 26-28.

Chesbrough, H.W. and Teece, D.J., "When is Virtual Virtuous?" Harvard Business Review, January 1996, 65-73.

Drucker, P.F., Managing in a Time of Great Change, Dutton, New York, 1995.

Grenier, R. and Metes, G., Going Virtual, Prentice Hall, Upper Saddle River, NJ, 1996.

Handy, C., "Trust and the Virtual Organization," Harvard Business Review, May 1996, 42-46.

McCartney, S., "Conditions are Ideal for Starting an Airline," Wall Street Journal, April 1, 1996, A1.

Price Waterhouse Change Integration Team, The Paradox Principles, Irwin, Chicago, 1996.

Slywotzky, A.J., Value Migration, Harvard Business School Press, Boston, MA, 1996.

A Virtual Framework For Technology Acquisition

CJ Broadfoot*, JC Greenwood*, S Forbes*,
SK Banerjee**, TS Durrani*
Tel: 0141-552 4400

University of Strathclyde
*Department of Electronic and Electrical Engineering
**Department of Design, Manufacture and Engineering Management
204 George Street, Glasgow, Scotland
G1 1XW

This paper presents results of work conducted on "Re-engineering the Technology Acquisition Process" - a project sponsored by the UK Engineering and Physical Science Research Council under its 'Management of Technology' initiative.

The research has focused on the management of technology acquisition in Scottish-based, global electronics corporations, and has established global communications and virtual presence as key elements of the technology acquisition process.

Greater flexibility in work practices is being encouraged by companies in response to evolving global markets. One such approach being increasingly adopted is remote and distributed work groups. This has led to the need for a globalised (shared) technology information resource to assist rapid product development. To facilitate the co-ordination of information throughout an extended organisation, the University of Strathclyde is developing a software environment which will allow access to technology information by users at distributed sites and will act as a management tool to support technology acquisition decisions.

Introduction

The evolution of global markets has been characterised by increased competition and rapid change. Product life cycles are becoming ever shorter and companies are placing greater attention on time to market considerations. Globalisation of markets has lowered barriers to the transfer of technology across geographical boundaries. Technological innovation and incorporation of new technologies into products have become key competitive factors. In response, companies have sought to improve and refine various aspects of their technology management activities. Thus, identification and evaluation of technologies from a variety of sources now plays an important role in the effective management of technological resources. As the technology content of 'Global Products' varies widely depending on the market, the management of information underpins the ability of the global corporation to utilise technology as a strategic lever.

In this context, one solution is to create a virtual environment in which business units are linked together by networks which share market, product and technology information. Recent advances in communication technologies now offer companies the potential for real time exchange of information world-wide. Flexibility and performance are key features in such an exchange system unconstrained by geographical boundaries.

A key benefit for companies operating in a virtual environment is that business unit synergies can be exploited through the sharing of knowledge and experience. The virtual environment can also facilitate staff empowerment by providing the information resource they need to make informed decisions; thus enabling the company to respond more quickly to the requirements of different markets.

The University of Strathclyde has conducted extensive research on the Scottish electronics sector. This has highlighted the emphasis placed by many companies on improvements to their product development and process innovation. The studies have shown that, in general, the technology acquisition process has not been formalised to the same degree or significant benefits achieved by globalisation of this process. To provide an effective vehicle for the acquisition of technology in a distributed organisation, operating in a global context, a Virtual Technology Acquisition Framework (VTAF) has been developed at Strathclyde. This framework exploits the advantages of a centralised regime, coupled with the advantages of a decentralised enterprise.

Defining the Virtual Framework

Technology acquisition in global corporations

In many cases, global corporations have evolved in response to geographical market opportunities, but may not have fully adapted their existing technology acquisition practices. Technology acquisition in these companies needs to be co-ordinated with other technology management activities. As Hammer (1990) has stated, companies should "use the power of modern information technology to radically redesign business processes in order to achieve dramatic improvements in their performance". Thus, integrating the technology acquisition process in global companies can provide the benefits of streamlined technology management systems, as well as allowing access to world-wide technology information.

The virtual environment

By formalising the shared experience and technological knowledge held throughout an organisation, capabilities in key technologies can be nurtured. Chesborough (1996) supports this view by asserting that the most successful virtual companies are ones at the centre of a network which leverages their own capabilities. The virtual environment can also help companies to develop a wider perspective regarding technology applications. For example, a technology which may be inappropriate in one market, perhaps due to economic or cultural considerations, could be applicable in other geographical markets.

The virtual technology acquisition framework (VTAF)

Initially, a technology acquisition model was developed to represent activities that support technology acquisition decisions at the business unit level. However, research identified the need for a solution that fully utilises communication tools and supports a co-operative virtual environment. This concurs with the views of De Meyer (1993) who asserts that, "to be effective, one has to create mechanisms on an international scale to diffuse, validate and integrate new knowledge across the whole network." The proposed VTAF, shown in Figure 1, provides such a mechanism.

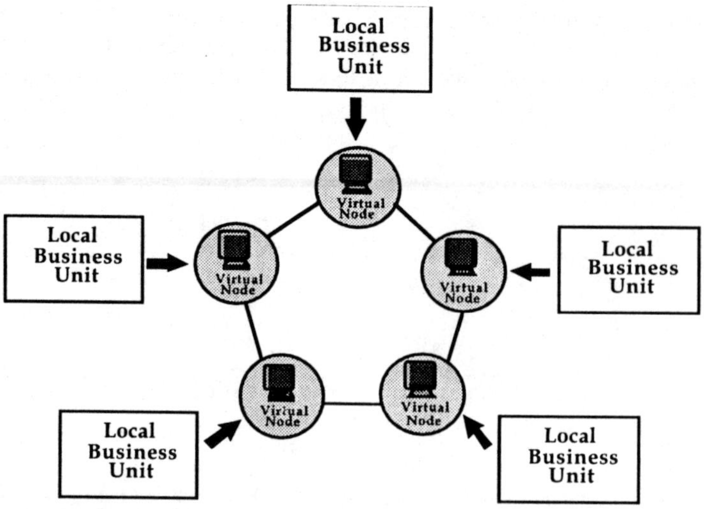

Figure 1. The virtual technology acquisition framework

The virtual framework comprises a series of Local Business Units (LBUs) which have responsibility for the development of their associated market-places. The VTAF establishes inter-connection between the LBUs via virtual nodes. The sharing of information resources can be managed through a well documented company procedure.

The VTAF supports the technology acquisition process by providing a structured information resource which allows companies to evaluate key technologies, both at the corporate and the business unit level. By enabling local business units to share information, the VTAF raises awareness of new technological opportunities. This may encourage collaborative technology developments, allowing the company to gain competitive advantage in key technologies.

Studies at the University of Strathclyde (Broadfoot, 1996) have shown that there is clearly a need for a holistic approach to technology management; one that encourages the free flow of technological ideas and information between decentralised business units throughout the company. The VTAF, which has been developed to address this need, possesses the following attributes:

- Facilitates virtual interaction
  The VTAF provides a platform for the exchange of information and viewpoints relating to technology acquisition. It enables information to be disseminated across distributed sites in the organisation.

- Acts as a global knowledge base
  The VTAF consolidates knowledge held across a number of operating units into a unified corporate framework, thus co-ordinating the company's capabilities more effectively. The creation of an information resource available to relevant employees in the company helps to codify technical knowledge which might otherwise remain tacit.

- Supports world-wide benchmarking
  The virtual framework can facilitate world-wide benchmarking for product development and technology acquisition performance. This is an important consideration for those companies that aspire to world class manufacturing status.

- Supports decision making at a corporate level
  Technology, product and market information held on the VTAF supports the formulation of a world-wide technology management strategy. The model

facilitates communication throughout the organisation, enabling relevant personnel to contribute to the strategy formulation process.

- Supports operational decision making at the business unit level
  The information which is collated, evaluated and disseminated via the VTAF supports operational technology acquisition decision making. Although these decisions generally focus on technology and customer issues, the VTAF ensures the integration of these issues across business units.

The VTAF is supported at each local business unit by a technology acquisition model.

Local business unit (LBU) environment

The LBU environment, shown in Figure 2, comprises three main elements which contribute to a technology acquisition model: (a) technology specialist activities, (b) strategic management activities and (c) core team activities.

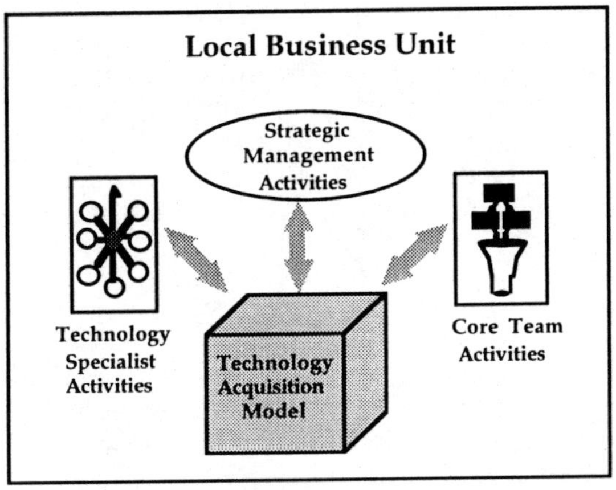

Figure 2. The local business unit environment

The technology acquisition model (TAM)
Each LBU uses a Technology Acquisition Model (TAM) as a support tool to identify the most appropriate technologies from the most suitable source of acquisition. The TAM captures technology information and decisions produced by the local business managers, technology specialists, and core product development teams (Durrani, 1996).

A principal user and owner of the TAM is the core product development team. This group uses the model to receive and store technology information associated with the products under development. The core team can supply information obtained from customers, suppliers and various functions throughout the business unit.

Local business managers contribute information about the strategic plans of the business unit, including market, technology, human resource, manufacturing and general business strategies.

Technology specialists provide expert information about the current state of technologies, including technology scanning and forecasting. This group may provide information on R&D activities, R&D resource plans and legislative needs.

The TAM captures information for technology acquisition decisions around three main areas: (a) the customer/market view, (b) the competitive abilities of the company and (c) the choice of acquisition route available. The model is broadly based on the initial stages of the QFD approach (Akao, 1990). Technology is, therefore, matched to market-place requirements and linked to the appropriate source of acquisition.

A range of other technology, market and product information can be accessed from the model, which acts as an information exchange. The model is designed to allow LBUs to access any relevant information that may exist within the company's global environment.

The TAM supports re-engineering activity and offers a significant competitive advantage by exploiting virtual networks to transfer technology solutions across global markets. The primary focus for the model is to assist core product development teams to evaluate technology acquisition options for a class of products.

The technology acquisition process (TAP)

The research team at Strathclyde has formulated a technology acquisition process (TAP) that utilises the information resource of TAM and supports decision making, both at business unit and corporate levels.

The TAP underpins a five stage evaluation process:

(i) Identify market-place requirements
Market-place requirements define individual customer needs that could be included in future product variants. Identification and classification of all market-place requirements for a particular product is the first step in a process to determine the most appropriate technology option. These are assessed using qualitative judgement of a range of criteria, e.g. product price, functionality, market share. This facilitates their prioritisation and classification into one of the following categories: (a) essential, (b) valued or (c) desirable.

(ii) Identify technology solutions
For each market-place requirement, all feasible technical solutions are identified. The information is primarily of a technical and engineering nature and is collated by the core team, with expert inputs from Research & Development, Technology Development Group, designers, engineering, and manufacturing.

(iii) Classify technology solutions
The technical solutions identified in the previous stage are assessed to determine their value to the company's competitive position. Each technology solution is classified to establish the technology/market position on the technology acquisition model. The technologies are classified into one of three categories: (a) basic, (b) core or (c) future core (McGrath, 1992).

(iv) Identify sources of technology acquisition
Technology can be acquired from three principal sources: (a) internally through research and development, (b) externally through licensing or purchase, or (c) through alliances with other organisations. When choosing the most appropriate source of technology acquisition, the company seeks to optimise internal resource utilisation. External sources of technology may be considered in order to overcome resource deficiencies in the development of a new product.

(v) Make the technology acquisition decision
Information collected during the previous stages of the process provides a number of technology acquisition options. The implications of adopting each option is assessed and then comparisons are made between them. Decision Webs, and other graphical tools have been designed at Strathclyde, are used to promote discussion and quantify views about the available options (Greenwood, 1996). This enables the most appropriate technology acquisition option to be selected.

Implementation aspects

The software has been implemented in a windows environment as a distributed database with user friendly interfaces. To facilitate use within a multi-user/virtual network, two approaches are being studied: (a) exploiting Groupware, with Lotus Notes as the preferred medium, and (b) employing the Internet, with the software being implemented using HTML. While the Internet offers the obvious infrastructure for the VTAF, issues of access and security are still to be resolved. However, fire-walls can be constructed to overcome this.

The TAM offers ease of implementation in a virtual environment. While specific core teams have ownership of the model, information can be distributed and accessed throughout the virtual environment. In particular, tasks at stages (ii) and (iv) of the TAP, i.e. identifying technology solutions and sources of technology, can be carried anywhere in this environment, and relevant information mounted on the TAM for future use, or exchanged throughout the organisation.

Conclusion

The Strathclyde research has established a virtual framework that utilises company-wide information to support technology acquisition decisions at the business unit level. The VTAF combines the advantages of a centralised technology management system together with the benefits of a decentralised market interface. The opportunities provided by the emerging communications technologies allow global companies to manage and develop their technological capabilities across disparate markets.

References

Akao, Y, Quality Function Deployment: Integrating Customer Requirements into Product Design, Productivity Press: Cambridge, Massachusetts, 1990.

Broadfoot, CJ, Greenwood, JC, Forbes, S, Banerjee, SK and Durrani, TS, "Technology Management: Dealing with Change" Working Paper, University of Strathclyde, 1996.

Chesborough, HW and Teece, DJ, "Organising for Innovation", Harvard Business Review, pp65-73, January-February, 1996.

De Meyer, A, "Internationalising R&D Improves A Firm's Technical Learning", Research-Technology Management, pp42-49, July-August, 1993.

Durrani, TS, Forbes, S, Greenwood, JC, Broadfoot, CJ and Banerjee, SK, "Defining the Technology Acquisition Model", Working Paper, University of Strathclyde, 1996.

Greenwood, JC, Broadfoot, CJ, Forbes, S, Banerjee, SK and Durrani, TS, "Strategic Management of Technology Acquisition" Working Paper, University of Strathclyde, 1996.

Hammer, M, "Reengineering Work: Don't Automate, Obliterate", Harvard Business Review, pp104-112, July-August, 1990.

McGrath, ME, Anthony, MT and Shapiro, AR, Product Development: Success Through Product and Cycle-time Excellence, Butterworth-Heinemann, 1992.

An Appraisal of R&D Management Research Literature as the Output of a Virtual Organisation

Leslie T Falkingham
Tel 44 1788 542121

Technical Director
GEC ALSTHOM T&D Vacuum Equipment Ltd
Rugby, CV21 1BD, UK.

Richard Reeves
Tel. 44 1234 754773

R&D Management Centre
Cranfield University
Bedford MK43 0AL, UK.

We have created a database of information about papers in a body of research publications on R&D Management, and used this to examine the structure of research activity in the field. We found that different types of author tended to produce papers with different types of outcome and also that different authors appeared to subscribe to different schools of thought about the R&D management process. Researchers generally subscribed to different schools of thought from researching practitioners, which raises a question about the utility of this body of research for those who might be thought to be its intended beneficiaries.

Introduction

There is a large body of research publications on the topic of management of industrial research and development. As is normal, this body of knowledge has arisen piecemeal from the world research community without any overall plan or deliberate structure. This research community could be considered to be a virtual organisation, which is clearly stable and productive, but we think that the nature and usefulness of its output is worth examining.

There is also a large body of people who actually practice the management of R&D as their primary occupations. These practitioners are presumably the intended recipients of the output of all this research effort.

We have set out to appraise the body of research by studying its product, i.e. its body of publications. This was done by constructing a database of information about the publications and then analysing it.

Definition of the body of research studied

Most publications on R&D address corporate level concerns, but in a large organisation the need to know how to manage is also felt strongly at the level of the section or unit, which often has considerable autonomy. Our concern was with papers which dealt wholly or in part with a topic defined as "R&D management concerns of the R&D section". This is a subset of publications on all topics in R&D management.

Our method of research

When setting out to review literature in the field just defined, it was decided to record reference details of all reviewed publications in a database. This proved to be unexpectedly fruitful, and led on to our research method, which consists of compiling a database of information about all the papers in the field, and then examining the database to look for summary information and relationships.

0-7803-3552-X/96/$5.00 © 1996 IEEE

An example of the method - author type and output type

It was noticed during initial reading of the papers that authors might be placed into the following categories of *author type*:

- *academics*
- *journalists*
- *consultants*: individuals or employees of consultancy organisations
- *practitioners*: practice or directly supervise R&D
- *corporate*: managers above practitioner level who write on R&D

Academics dominated, which is not surprising, but the significant contribution made by journalists was not expected by us.

It also proved possible to classify the *output type* of the paper into the following classes:

- rules
- strategy
- tactics
- model
- comment only
- data
- measurement method

While reading the papers, the impression arose that the type of author was significant in predicting the type of output of a paper. To test this, it was decided to create columns in the database for recording these two attributes of a paper. *Author type* was gleaned from the authors' biographical notes or addresses in their papers, and was deducible for all but one paper. *Output type* was assigned by the reviewer, and this proved practicable for 100% of papers. The following table was then extracted.

OUTPUT TYPE	Academic	Journalist	Consultant	Practitioner	Corporate
Rules	80	3	22	6	23
Strategy	67	17	13	11	68
Tactics	60	20	24	26	20
Model	59	0	4	6	12
Theory	19	0	2	2	6
Comment only	0	8	8	4	7
Data	18	0	0	0	3
Measurement method	11	0	2	0	4
Totals	314	52	77	55	144

Table 1. Variation of Output Type of papers with Author Type

Without enquiring into what interpretation should be put on the above results, it was decided to conduct further work along the same lines.

Construction of the database

Selection of attributes

The database was then expanded to record the following *attributes* of each paper in addition to *title, author, reference*, *author type* and *output type*:

1. The research methodology used.
2. Nationality of authors.
3. Affiliations of authors.

4. How formal the work was; whether it reports:
   - opinion
   - experience
   - a research result
5. The subject matter of the paper:
   - strategy
   - tactics
   - R&D management
   - general management
   - organisational structure
6. The contribution of the paper to theory;
   - describes theory
   - tests theory
   - is not relevant to theory
7. Degree of relevance to R&D management
8. A grading of our opinion of the importance of the paper
9. A grading of the quality of writing of the paper
10. The school of thought underlying the work; (this concept will be explained below).

The list above was modified and extended slightly in the light of experience with the papers. Once this was done there was no particular difficulty in categorising all the papers.

The publications included

By April 1996, 676 papers relevant to the subject of R&D management at the unit level, published since 1975, had been entered. The bulk were refereed journal papers, which were found by the usual methods of computer searching of indexes, scanning of identified key journals, and some citation analysis. The list of papers is thought to be more or less comprehensive for the restricted field in question, and comprises about 20% of papers on all R&D Management topics. The decision was made not to do searches for key authors as this might bias the database towards those who feature a number of times, and lead to neglect of rarer authors.

Categorising the papers

All papers included were read by the reviewer, and were categorised according to his opinion. The list of attributes above was modified and extended slightly in the light of experience with the papers, and then there was no particular difficulty in categorising all of them.

Some results from the database

The publications were distributed among 119 journals as shown in the league table below.

PUBLICATION	PAPERS
Research Technology Management	152
IEEE Engineering Management Review	62
Research Management	51
R&D Management	51
Engineering Management Journal	19
Management Science	19
Professional Engineering	7
5 journals	5
7 journals	3
12 journals	2
70 journals	1

Table 2. Where papers in this field were published

The table shows that publications are quite widely dispersed in this field with 22% of papers found in Research Technology Management, and 52% of papers found in the top 6 journals. The last 10% of papers are dispersed individually among 70 journals.

We extracted a league table showing which countries papers originated from. A Japanese researcher working in France and publishing in the USA would be classed as French.

COUNTRY	PUBLICATIONS
USA	438
UK	114
Canada	45
Japan	14
Germany	8
Switzerland	6
Holland	5
France	4

Table 3. National origins of the publications

The database concerns solely papers written in English, which may explain the small number from Japan. One author has significantly distorted the results, by elevating Canada to third position almost by himself. Professor R. J. Cooper of McMaster University and his co-authors have written 26 papers within the field of study.

We examined how many papers came from each organisation. Work in this field is very widely dispersed, with the most prolific organisation producing only 26 papers, 4% of the whole. The next most prolific produced only 10 papers and, 125 organisations produced 1 paper each.

We defined four levels of *formality of research* as follows:

- *Formal Study*, which is the reporting of a research project.
- *Review,* which is a review of existing information.
- *Theory*, which represents a theoretical discussion or analysis.
- *Comment*, which is a report of the author's views only, with no formal research.

	Journalist	Academic	Consultant	Corporate	Practitioner
Formal	2	65	4	15	6
Review	7	85	13	43	11
Theory	6	80	24	38	9
Comment	37	92	22	52	30

Table 4. Research Formality and Author Type

Table 4 shows that surprisingly few papers were classed as formal research.

Schools of thought in R&D management

While recording a case study in an industrial R&D laboratory, which was the working environment of one of us, it was noticed that different managers, when apparently faced with the same information, drew quite different conclusions and favoured different courses of action. We formed the hypothesis that these different managers think in different ways or perhaps subscribe, consciously or unconsciously, to different forms of reasoning about management activity. The publications on R&D management also appeared to reveal the presence of these same schools of thought underlying the approaches to research on the topic. The *schools of thought*, as formulated by us, were as follows:

1) *Cookbook* - Empirical rules can be found that apply most of the time
2) *Deterministic* - R&D can be managed by logic and reason

3) *Chaotic* - R&D is complex and variable, and there are limits to manageability
4) *Biological* - R&D management must adapt continuously to change

We allocated a school of thought to each paper, and produced a table to show how the schools of thought are distributed among the papers of the different researcher types.

	Journalist	Academic	Consultant	Corporate	Practitioner
Cookbook	30	189	55	46	11
Deterministic	9	89	11	92	12
Chaotic	7	26	8	2	13
Biological	5	10	3	7	21

Table 5. Variation of School of Thought with Author Type

Practitioners were strongest on the biological school, which was least favoured by the other author types. This raises a question about the match between the mind sets of the majority of those who supply research papers and of those who presumably would be expected to benefit from them.

Verification by authors' own opinions

All of the above relies on our opinions in classifying the papers. To investigate whether the authors of the papers would agree with our classifications, a questionnaire was developed and sent to 296 of the more recent authors. This asked for the same information as is contained in the database, and to date 115 responses have been received, a response rate of 39%. The responses indicate good correlation with the researcher's opinions and appear to validate our findings.

We in effect asked authors to classify each paper according to its school of thought by asking the following question:

> "Which of the following most closely describes the conceptual framework in which your paper is set?"

and authors were asked to tick one of the four definitions shown under Schools of Thought above. All but one author answered this without any apparent difficulty. We take this as an indication that the Schools of Thought hypothesis has validity.

Discussion

We believe that the type of analysis which has been presented here shows promise as a means of providing information about the total effort in a research field. The database can be interrogated remarkably rapidly to ask new questions within its scope, but the number of correlations possible is too large to be conveniently reported.
No one controls the total research effort in the field examined, and it is not subject to market forces. It is perfectly possible that it does not supply knowledge to the presumed end users efficiently. One of us (LTF) is a practising R&D manager and he found no publications at all concerning his core problem, which was how to decide strategy at the unit level.

Most practitioners must obviously be non-publishers, and these have played no part in this study. A preliminary trial of non-publishing practitioners has indicated that these are mainly of the biological school of thought, quite unlike most of the researchers, but more in line with researchers who are practitioners. We intend to carry out a further study of practitioners.

Investment in Facility Flexibility Under Uncertain Competitor Market Entry

Richard M. Franza
(513) 255-7777 ext. 3380

Graduate School of Logistics and Acquisition Management
Air Force Institute of Technology (AFIT/LAS)
2950 P Street, Bldg. 641, Rm. 302D
Wright-Patterson AFB, OH 45433-7765

Cheryl Gaimon
(404) 894-4934

School of Management
Georgia Institute of Technology
Atlanta, GA 30332-0520

For firms competing in markets characterized by short life cycles, quick changeover of manufacturing processes from one product generation to the next is critical. Longer changeovers delay a firm's market entry. As a firm's market entry time is delayed, its opportunity to accumulate demand is diminished. The extent to which demand is lost reflects the market entry time of the firm's competitor. A model is introduced to explicitly examine the effect of uncertainty associated with a competitor's market entry time on both manufacturing and marketing strategies. The firm's objective is to maximize expected profit obtained from a series of high volume products manufactured on a single facility.

1. Introduction

Today's marketplace is characterized by technological change and product innovation. The introduction of a product innovation severely limits the ability of the previous product generation to realize demand and revenue. As a result, product life cycles are short and the time it takes to bring a new product to market significantly impacts on a firm's competitive position, (Blackburn 1991, Qualls et al 1981). It follows that a firm must have the ability to quickly changeover for the manufacture of successive product generations. In addition, empirical studies indicate that product profitability is strongly related to a firm's market entry time, (Griffin 1993, Kurawarwala and Matsuo, 1993). This marketing phenomena increases the pressure on manufacturing to be capable of rapid product changeovers.

Readying a facility for the production of a new product is particularly challenging for high volume manufacturers. Traditionally, such firms invest in dedicated technologies to achieve consistent quality and unit cost efficiency. Unfortunately, dedicated facilities are not conducive to rapid changeovers. Alternatively, high volume manufacturers can invest in manufacturing facilities allowing more rapid and less costly changeovers. Such facilities are said to exhibit changeover flexibility. However, achieving this flexibility for high volume production entails trade offs. First, flexible facilities require a significantly higher initial investment. Second, flexible facilities sacrifice some of the cost efficiencies derived from dedicated technology. In conclusion, the investment in facility flexibility (manufacturing decision) must be analyzed in the context of the competitive (marketing) impact.

The purpose of this research is to determine the optimal level of changeover flexibility for a firm producing a series of high volume products on the single facility. A stochastic model is introduced to capture uncertainty regarding the time a competitor enters a new product market. In addition, we analyze how the level of uncertainty associated with the competitor's market entry time affects the firm's manufacturing and marketing strategies.

This paper builds on earlier work of Gaimon and Singhal (1992) and Franza and Gaimon (1994). In contrast to the earlier work in which competition was implicitly

0-7803-3552-X/96/$5.00 © 1996 IEEE

considered, this paper explicitly models the behavior of a firm's competitor. In addition, unlike the deterministic approaches previously taken, this model captures the stochastic nature of the competitor's entry time.

2. The Model

Corresponding to the competitive environment motivated in the previous section, we introduce a stochastic decision making model.

2.1 Time-Related Parameters

The firm invests in a manufacturing facility with a known useful life denoted by **L**. The useful life is the length of time the firm will use the facility for sequentially manufacturing a series of high volume products. The facility's life reflects the natural deterioration of manufacturing systems over time.

The length of the average life cycle for the products manufactured on the facility is represented by **T**. The value of **T** can be estimated from forecasts of product and process innovations. The start of each new product's life cycle is referred to as time zero. Time **T** of one product is time zero for the next product. At time zero, the new product or process technology is available to the two firms in the market and demand for the new product begins.

Based on the useful life of the facility and the length of the average product life cycle, the number of products successively manufactured on the single facility, **n**, is determined by **L/T**, rounded to the nearest integer. Since **L** and **T** are forecast (and are exogenous), the number of products manufactured on the facility is exogenous, as well.

Next, consider the relationship between a firm's facility flexibility and its market entry time. The firm's initial investment in flexibility dictates the speed with which facility changeovers are realized. Furthermore, the time required for facility changeover determines the firm's market entry time. Therefore, a firm's optimal level of facility flexibility is a proxy for its optimal market entry time. Let **m** denote the time the firm enters the market for each successive product following the start of each product's life cycle at time zero. Therefore, **m** represents the time required to complete each facility changeover, and **T-m** is the length of the production cycle for each of the **n** products manufactured on the facility.

A high (low) level of facility flexibility is associated with an early (late) market entry time and a small (large) value of **m**. It should be noted that $\mathbf{m} \varepsilon [0,\mathbf{T}]$; where $\mathbf{m}=0$ represents a facility of maximum flexibility (instantaneous changeover) and $\mathbf{m} = \mathbf{T}$ represents a facility incapable of production changeover during the product's life cycle.

The firm's competitor faces an analogous relationship between facility flexibility and market entry time. The competitor's initial investment in facility flexibility determines its market entry times and production cycles. Let δ denote the random variable representing the competitor's market entry time following the start of each product's life cycle, $\delta\varepsilon[0,\mathbf{T}]$. The probability density function and cumulative distribution of the random variable are denoted by $g(\delta)$ and $G(\delta)$, respectively. The time required by the competitor to complete each high volume product changeover is δ; the production cycle for each product is $\mathbf{T}$-δ.

2.2 Flexibility / Cost Relationships

I(m) denotes the initial investment cost for the facility, including manufacturing and materials handling equipment, tools, and other start up expenses. It is assumed that the initial investment cost increases at an increasing rate as the level of flexibility increases so that **I(m)** decreases at a decreasing rate relative to **m**. Therefore, $\mathbf{I_m}<0$ and $\mathbf{I_{mm}}>0$.

A(m) denotes the average cost of adapting the facility from one high volume product to the next and is expressed as a function of the changeover time, **m**. A less flexible facility incurs a greater changeover cost due to the purchase of new manufacturing and materials handling equipment, tools, re-organization of the shop floor, and the disposal of some portion of the existing. It follows that **A(m)** increases at an increasing rate relative to **m**. Therefore, $\mathbf{A_m}>0$ and $\mathbf{A_{mm}}>0$.

The average unit production cost incurred for each of the **n** high volume products is denoted by **C(m)**. A more flexible facility has more complex operations and materials handling technologies, and lower production efficiency. Therefore, the average unit cost decreases at a decreasing rate relative to **m,** giving us $\mathbf{C_m}<0$ and $\mathbf{C_{mm}}>0$.

2.3 Revenues / Market Entry Time Relationships

An important element of the competitive framework introduced is captured by the mathematical form of the revenue function. The magnitude of revenue earned by the firm is driven by the difference between its market entry time and that of its competitor. Numerous empirical studies (Biggadike 1976, Urban et al 1986, Robinson 1988, Carpenter and Nakamoto 1989) have shown there are long-term market share benefits due to early market entry. Additionally, Huff and Robinson (1993) show that as the difference between the entry times increases, the earlier entrant has a greater market performance advantage. Therefore, it is assumed that as the value of δ-**m** increases, the firm's revenue increases.

The revenue earned by the firm for each of the **n** successive products manufactured on the single facility is denoted by Π(**m**,δ) and is comprised of two elements. The competitive revenue potential, R(**m**,δ), is a function of the difference between the market entry times of the firm and its competitor, (δ-**m**). The second element of revenue reflects the length of the firm's production cycle, (**T**-**m**). Therefore, the shorter the firm's production cycle, the less revenue earned. The functional form of Π(**m**,δ) is defined below.

$$\Pi(\mathbf{m}, \delta) = R(\mathbf{m}, \delta) * S(\mathbf{m})$$

We know that as (δ-**m**) increases, the firm's revenue increases. However, the rate at which revenue increase is unclear. Empirical evidence does not exist indicating whether revenues increase at an increasing or decreasing rate as (δ-**m**) increases. Therefore, we make no assumption regarding the sign of the rate of change in R(**m**,δ), and we solve the model for both possibilities.

Next, the second element that impacts on a firm's ability to earn revenue is considered. The function S(**m**) captures the effect on revenue due to the length of the firm's production cycle. Note that S(**m**) lies in the [0,1] range and can be interpreted as a percentage. As facility flexibility increases, the percent revenue potential realized increases at a decreasing rate because the volume of demand is lower earlier in a product's life cycle. Therefore, $\mathbf{S_m<0}$ **and** $\mathbf{S_{mm}<0}$.

2.4 Optimal Market Entry Time (Facility Flexibility)

As described above, the costs incurred due to the initial facility investment, changeover, and production are independent of the effect of competition. Therefore, only the revenue earned by the firm is affected by the uncertainty associated with the competitor's market entry time. The firm's objective is to determine the optimal level of facility flexibility (as indicated by **m**) to maximize expected revenue minus the costs incurred over the life of the facility.

The facility investment cost is incurred once. The revenue earned and the production cost incurred are realized **n** times, once for each product. It is assumed that the changeover cost is also incurred **n** times. Therefore, the investment cost does not include the cost of the initial setup to ready the facility for the manufacture of the first product. This gives us the following objective function.

Maximize

$$\mathbf{n}\, E[\Pi(\mathbf{m}, \delta)] - I(\mathbf{m}) - \mathbf{n}\, A(\mathbf{m}) - \mathbf{n}\, C(\mathbf{m})$$

2.5 Competitor's Market Entry Time

Our primary focus in this research is: (i) to examine the effect of a competitor's new product market entry time on a firm's investment in changeover flexibility, and (ii) to evaluate how the stochastic nature of th competitor's market entry time impacts on the firm's optimal facility flexibility decision.

In the next section, we summarize results obtained in which g(δ) satisfies a uniform distribution. The uniform distribution fits environments in which a firm estimates that its competitor is (approximately) equally likely to enter a new product market at any time within a specified interval. Analysis is conducted to examine the impact of various specifications of the time interval over which the uniform distribution is defined. This facilitates comparisons of scenarios in which different degrees of uncertainty exist regarding a competitor's market entry time. In addition, comparisons are made of the optimal investment in facility flexibility for a firm that anticipates its competitor's market entry will occur early versus late in the product's life cycle.

3. Results

Analysis of the model offers rather intriguing results. First, consider the results (analytic and numerical) when the competitive revenue potential function increases at an increasing rate as (δ-**m**) increases. As the expected market entry time of the competitor increases (decreases), the firm's optimal market entry time decreases (increases). Initially, these results might appear counter intuitive. However, consider the following interpretation. First, suppose a firm expects its competitor to enter a new product market early in the product's life cycle. If the firm wishes to realize the revenue advantage associated with entering the new product market prior to its competitor, then it must make a substantial investment in facility flexibility. However, under this scenario, the potential revenue advantage that the firm could attain is relatively small and insufficient to offset the higher costs of investing in the higher level of flexibility. In conclusion, these results indicate that firms optimally avoid head-to-head competition for early market entry under the given conditions.

Conversely, suppose the firm's revenue potential increases at a decreasing rate as (δ-**m**) increases. As the expected value of the competitor's entry time decreases (earlier entry), the firm's optimal market entry time decreases (earlier entry). Therefore, in this scenario, firms are more likely to compete head-to-head for increased revenues. In conclusion, the assumption regarding the rate of change in revenue is critical to a firm's strategic investment in changeover flexibility and its market entry strategy.

Other results indicate that as the competitor's market entry time becomes more uncertain (the variance increases), the firm optimally invests in more changeover flexibility. This is consistent with conventional thought that flexibility is a hedge for uncertainty. In this case, changeover flexibility is a hedge for uncertainty regarding competitor market entry. Furthermore, these results are obtained regardless of whether revenue increases at an increasing or decreasing rate as (δ-**m**) increases.

For more comprehensive discussion of the model and results, see Franza and Gaimon (1996).

4. References

Biggadike, R. E., *Entry Strategy and Performance*, Harvard University Press, Cambridge, MA, (1976).

Blackburn, J. D., *Time-Based Competition: The Next Battleground in American Manufacturing*, Business One Irwin, (1991).

Carpenter, G.S. and K. Nakamoto, "Consumer Preference Formation and Pioneering Advantage, *Journal of Marketing Research*, 26 (August 1989), 285-298.

Huff, L. C. and Robinson, W. T., "The Impact of Leadtime and Market Age on Sustainable Market Advantages," Working Paper, University of Michigan, May 1993.

Franza, R. M., and C. Gaimon, "Flexibility, Pricing, and Learning for High Volume Products with Short Life Cycles," Working Paper, Georgia Tech, School of Management, April 1994.

Franza, R.M. and C. Gaimon, "The Effect of Uncertainty of Competitor Market Entry Time on Changeover Flexibility Investment," Working Paper, Georgia Tech, School of Management, April 1996.

Gaimon, C. and V. Singhal, "Flexibility and the Choice of Manufacturing Facilities Under Short Product Life Cycles, *European Journal of Operational Research*, 60 (1992), 211-223.

Griffin, A., "Metrics for Measuring Product Development Cycle Time," *Journal of Product Innovation Management*, , 10 (1993), 112-125.

Kurawarwala, A. A., and H. Matsuo, "Cost of Delay in Time to Market and Capacity Restriction," Working Paper 93-01-02 , University of Texas at Austin, 1993.

Qualls, W., R. W. Olshavsky, and R. E. Michaels, "Shortening of the PLC - An Empirical Test," *Journal of Marketing*, 45 (1981), 76-80.

Robinson, W. T., "Sources of Market Pioneer Advantages: The Case of Industrial Goods Industries," *Journal of Marketing Research*, 25 (1988), 87-94,

Urban, G.L., T. Carter, S. Gaskin, and Z. Mucha, "Market Share Rewards to Pioneering Brands: An Empirical Analysis and Strategic Implications," *Management Science*, 32 (1986), 645-659.

Facility Flexibility: A Manufacturing and Marketing Interface

Cheryl Gaimon and Alysse Rosewater
(404) 894-4934
Atlanta, GA 30332-0520 USA

School of Management
Georgia Institute of Technology

The strategic interface between manufacturing and marketing is analyzed for a firm investing in a single facility to manufacture a sequence of high volume products with short life cycles. The extent of investment in facility flexibility and the dynamic price charged over each product's life cycle are key decision variables linking the manufacturing and marketing domains. The investment in facility flexibility impacts on a several factors contributing to the dynamic profit maximizing objective. For example, the investment in facility flexibility impacts on: the firm's market entry times, dynamic production costs, and optimal pricing strategies over the lifetime of each product generation. Several dynamic elements are captured in the decision making process including price competition, pioneering advantage, facility efficiency effects, and volume based learning.

1. Introduction

Product life cycles are shortening due to a combination of changing technologies and increased competition to meet consumer needs. The effect of short product life cycles on a firm's ability to generate profit is profound. First, as product life cycles are reduced, the available window of opportunity to realize demand for each product generation shortens. Furthermore, as the time to market for the next product generation is delayed, increased competition is likely, thereby driving down unit revenues and demand opportunity. Lastly, early market entry affords the firm learning curve benefits which lower the production cost and serve as a barrier to competitive entry.

With short product life cycles, a firm must have the ability to rapidly specify and develop new product concepts. Integration with manufacturing is a fundamental requirement in order to establish the means of production for the new products. Typically, a lack of integration with manufacturing results in a significant delay in market entry, higher unit costs, and inferior performance of quality related attributes. In the context of dynamic market forces, rapidly changing technologies, and international competition, firms cannot afford to ignore the benefits of integrating manufacturing with R&D and marketing decisions.

1.1. Focus of Paper

In this paper, a model is introduced to integrate strategic decisions in the manufacturing and marketing domains for firm operating in a high volume production environment characterized by short product life cycles. The firm's objective is to maximize the profit obtained over the life of the facility. In particular, the dynamic profit reflects the initial investment in the facility, the changeover costs incurred to ready the facility for the manufacture of successive products, and the revenue and cost streams for each product.

The extent of investment in facility flexibility is a decision variable impacting on several dimensions of the model. A high level of facility flexibility enables the firm rapid changeover of its manufacturing processes and early market entry. To capture this relationship, the investment in changeover flexibility is a proxy for the firm's market entry time for each high volume product. One facet of the strategic manufacturing and marketing interaction is thereby established.

0-7803-3552-X/96/$5.00 © 1996 IEEE

The investment in facility flexibility has implications on the dynamic production cost incurred. The initial unit cost for each high volume product to be manufactured reflects the underlying efficiency of the facility. A dedicated facility is more efficient and thus has a lower initial unit production cost. Alternatively, to achieve early market entry, a firm investing in a flexible facility incurs a high initial unit production cost. However, early market entry also affords the firm a greater volume of cumulative demand over time. If volume based learning exists, then early market entry is associated with greater reductions in the unit cost over the production cycle. In conclusion, the impact of facility flexibility on the dynamic unit production cost is complex and embodies opposing forces.

Lastly, the investment in facility flexibility has implications on two other costs incurred by the firm. First, as the initial level of facility flexibility increases, the changeover cost incurred to convert from one high volume product generation to another decreases. Second, the initial investment cost in a facility increases as the level of facility flexibility increases. In other words, to realize more rapid market entry for each high volume product, the firm incurs a higher initial investment cost for facility flexibility.

So far, we have characterized the internal costs incurred by the firm (investment, changeover, and production) associated with the technology choice decision. Next, we consider the ability of the firm to generate revenue for each of the high volume products manufactured over the life of the facility. Clearly, the revenue earned at each instant of time is defined as the product of the dynamic price and demand. The dynamic demand function reflects several external market forces which drive the manufacturing based decision. First, the level of demand opportunity is expressed as a function of the firm's market entry time. Therefore, the potential benefits on the volume of demand due to pioneering advantage are included. Second, since the model is dynamic, we can accommodate the possibility that price competition increases over time. As a consequence of the dynamics of both the internal production cost and external market forces, the optimal pricing policy is obtained over each product's life cycle.

To summarize, a model is introduced to analyze the manufacturing and marketing interface for a firm producing a series of high volume products with short product life cycles on a single facility. The strategic issues analyzed in the model include the optimal market entry time for each product generation (investment in facility flexibility), as well as the optimal pricing strategy over the lifetime of each product generation. The model incorporates the trade offs of manufacturing efficiency and volume based learning on the dynamic unit production cost. The model captures other manufacturing related expenses including the initial investment in facility flexibility and the changeover costs realized between product generations. From a strategic marketing perspective, the model embodies dynamic forces such as price competition and pioneering advantage.

1.2. Related Literature

The advantage of early market entry is the subject of a large body of literature in the field of marketing (see Kerin et al, 1992, for a review). A firm entering the marketplace first is referred to as the pioneer. The pioneer realizes benefits in the marketplace such as consumer recognition and manufacturing benefits such as volume based learning. In contrast, to distinguish itself in the marketplace, a later entrant must charge a lower price, offer additional product attributes, or spend heavily on advertising.

Relatively little literature exists modeling the link between pioneering market advantage and manufacturing decisions regarding investment in flexible technologies. Gaimon and Singhal (1991) evaluate the benefits of investing in flexible facilities for a series of high volume products with short product life cycles. Franza and Gaimon (1994) extend the model to include a pricing decision which impacts on demand. The demand and facility flexibility impact on production costs through efficiency and learning effects. However, in contrast to the formulation introduced in this paper, these models are static in nature. Therefore, the effect of time based competition, pioneering advantage, dynamic costs and pricing strategies are not easily analyzed.

2. The mathematical model

In this section, the dynamic profit maximizing model is introduced. The profits obtained from a series of high volume products manufactured on a single facility are maximized over the lifetime of the facility. The length of each product's life cycle is an exogenous parameter which can be forecast based on factors including the rate of technological change and competitive activity in the marketplace. While the length of the life cycles of the products manufactured on the single facility may differ, our model employs an estimate denoting the average product life cycle. This and other time related parameters are defined below. In addition, the notation required to define the mathematical model is given in Section 2.1.

2.1. Model Parameters and Functions

Time Related Exogenous Parameters

- t — Time measured over a product's life cycle from the perspective of demand in the marketplace. Zero denotes the time demand starts for each product's life cycle.
- T — Average length of each product's life cycle. Therefore, when one product's life cycle is complete (time T), the next product's life cycle begins (time zero).
- L — Length of the facility life.
- n — Number of high volume products manufactured over facility life; $n = L/T$ rounded down to the nearest integer.

Decision Variables

- m — Market entry time for each successive product; proxy for facility flexibility. During the time intervals $[0,m]$, the facility changeovers occur. Production cycles to meet each new product's demand occur during the time intervals $[m,T]$.
  Clearly, $m \in (0,T)$.
- $p(t)$ — Price charged per unit sold at time t; expressed as the average over the n high volume products at the same stage in their life cycles; *control variable*.
- $D[m,p(t),t]$ — Demand at time t expressed as a function of market entry time, price, and time; assumed to be separable in m and p.
- $\delta(t)$ — Accumulated demand through time t; *state variable*.
- $c[m,\delta(t),t]$ — Unit production cost at time t expressed as a function of flexibility (m), learning from cumulative demand ($\delta(t)$), and time.

Exogenous Functions for Demand and Cost

- $d_0(m,t)$ — Portion of demand at time t impacted by market entry time, m, and independent from price; i.e, "demand opportunity" at time t corresponding to market entry time m.
- $d_1[p(t),t]$ — Function indicating the effect of price on demand at time t; independent from market entry time. May reflect increasing or decreasing competition in the marketplace.
- $I(m)$ — Investment cost in facility flexibility enabling market entry at time m.
- $A(m)$ — Facility changeover cost from one high volume product to the next, expressed as a function of flexibility (i.e., market entry time).

2.2. Demand

The demand function introduced is motivated by several assumptions. (i) Demand is separable into two components, $d_0(m,t)$ and $d_1[p(t),t]$, as defined above. (ii) As market entry time is delayed, demand opportunity decreases at a non increasing rate so that $d_{0m} \leq 0$ and $d_{0mm} \geq 0$. (iii) Demand is linearly impacted by price such that $d_1[p(t),t] = d_1(t)p(t)$. (iv) The extent of price competition is a non decreasing function of time reflecting the likelihood of competitor market entry, ($d_{1t} \geq 0$)). (v) The nature of demand for each high volume product follows the same pattern over time.

Given the above discussion, the equation representing the volume of demand for each high volume product is written as follows:

$$D[m,p(t),t] = d_0(m,t) - d_1(t)\, p(t) \qquad (1)$$

Cumulative demand at time t reflects the demand met during each production cycle. Equation (2) defines the volume of cumulative demand for each high volume product.

$$\delta(t) = \int_m^t D[m,p(\tau),\tau]d\tau,\ \delta(m) = 0 \qquad (2)$$

2.3. The Production Cost

The effect of facility flexibility on the production cost is twofold. First, to achieve early market entry, a firm may employ flexible manufacturing equipment. However, higher flexibility results in lower efficiency of the production processes. As a consequence, the firm incurs a higher initial unit production cost for each product generation ($c_m < 0$; $c_{mm} > 0$). Second, as cumulative demand ($c_\delta < 0$) increases over the production cycle, the unit cost for each product generation is reduced over time due to learning. The impact of learning on reducing the unit production cost decreases as cumulative demand increases ($c_{\delta\delta} > 0$), (diminishing returns). Lastly, we assume that $c(m,\delta(t),t)$ represents the average facility changeover cost at time t for each high volume product.

2.4. The Investment and Changeover Costs

Two other costs are incurred by the firm which are independent of demand. As facility flexibility is increased (m is reduced), the initial investment cost increases at a decreasing rate, ($I_m < 0$ and $I_{mm} > 0$). Clearly, the facility investment cost is incurred once over the planning horizon. In addition, the cost of facility changeover from one high volume product generation to the next is incurred n times. The magnitude of each changeover cost decreases at a decreasing rate as facility flexibility increases, ($A_m > 0$ and $A_{mm} > 0$). We assume that $A(m)$ represents the average facility changeover cost for all n high volume products.

2.5. The Profit Maximizing Objective

The firm's profit at each instant of time during the production cycle of a single product generation is determined by the product of demand and the unit price less the unit production cost. In addition, the profit earned over the life of the facility includes the initial investment and n changeover costs. Since the functions representing the volume of demand, the unit production cost, and the changeover cost are expressed as averages over all n products (not averages over time), the calculation of profit simplifies to the following equation.

$$\text{Maximize} \quad \underbrace{n\int_m^t \{p(t) - c(m,\delta(t),t)\}D[m,p(t),t]dt}_{(a)} - \underbrace{n\,A(m)}_{(b)} - \underbrace{I(m)}_{(c)} \qquad (3)$$

The term given in (a) denotes the total revenue minus production cost obtained from all n product generations. Term (b) represents the total cost incurred to change over the facility for the production of the n product generations. Term (c) is the cost of investment in facility flexibility.

3. Analysis of solution

For a complete discussion of the model presented in Section 2, or a detailed analysis of the solution, see Gaimon and Rosewater (1996). Due to space limitations, highlights of the solutions obtained follows.

3.1. Optimal Market Entry Time (Facility Flexibility Level)

Analytic results show the manner in which the optimal market entry time is impacted by both manufacturing and marketing forces. The optimal solution indicates that earlier market entry is advocated for each product generation if: (i) the average length of the life cycle is reduced (number of product generations is increased), (ii) the magnitude of demand opportunity is increased, (iii) the degree of price competition is reduced, (iv) the rate of learning based production is increased, (v) facility flexibility has less impact on lowering manufacturing efficiency, (vi) the initial cost of investing in facility flexibility is reduced, or (vii) the facility changeover cost is increased.

3.2. Optimal Pricing Policy

Similarly, results are obtained showing the manner in which the optimal pricing policy is influenced by both manufacturing and marketing variables. The optimal price is the average of the maximum price permitted and an adjusted value of the unit production cost. The unit cost is adjusted down by the marginal value of an additional unit of cumulative demand. Therefore, the optimal price charged at each instant of time is reduced in order to increase cumulative demand. This affords the firm the opportunity to obtain further unit cost reductions from volume based learning. Lastly, it is shown that in response to either an increase in the level of price competition or a reduction in the magnitude of demand opportunity, demand is increased and the optimal price is reduced over the entire manufacturing cycle.

4. References

Franza, Richard M., and Cheryl Gaimon, "Flexibility and Pricing Decision for High-Volume Products with Short Life Cycles," WP, Georgia Institute of Technology, (1994).

Gaimon, Cheryl, and Alysse Rosewater, "Facility Flexibility: Impact on Market Entry and Dynamic Pricing," WP, Georgia Institute of Technology, (1996).

Gaimon, Cheryl, and Vinod Singhal, "Flexibility and the Choice of Manufacturing Facilities Under Short Product Life Cycles," *European Journal of Operational Research*, 60 (1992), 211 - 223.

Kerin, Roger A., Rajan Varadarajan, and Robert A. Peterson, "First-Mover Advantage: A Synthesis, Conceptual Framework, and Research Propositions," *Journal of Marketing*, (1992), 33-52.

Outsourcing: Focusing on Core Competencies by Leveraging Resources

Mr. James C. Gordon
Mr. Aidan C. Gordon
(604) 324-2441

Gordon Crane & Hoist Inc.
104 - 8475 Ontario Street
Vancouver, BC V5X 3E8

This paper presents an overview of outsourcing as a means of performing tasks beyond the scope of the permanent core staff of your organization, by having a wide variety of companies at your disposal to work with you, supplying their specific expertise on a job-byjob or project-by project basis. This allows your company to stay focused and have the freedom to operate and maintain specialized business pursuits without carrying the heavy burden of large overhead during downturns in the business cycle. The paper concludes with a specific example of a lesser known application in outsourcing with respect to heavy engineering.

Definition

The term "Outsourcing" has been bandied around in the business world for a few years now; however, it is a concept which seems to have many different interpretations.

The expression "Outsourcing" can be used to describe the setting up of off-shore factories using cheap labour and resources or alternatively, the laying-off of full time staff and hiring them back as contract labour. The current favoured trend is to contract out certain peripheral functions of a core business to companies specializing in that particular field. These specialists are referred to as resource companies or "resources". Essentially the strength of the core business is the sum of its resources. A situation where the sum is significantly more powerful than its individual parts.

An Old Concept with a New Name

Outsourcing is being heralded by some as a wave of the future which is revolutionizing the business world. However, the concept of outsourcing is certainly not new, it is just being applied in new areas. Even within the classic model of a "traditional business", there are several common examples of outsourcing.

Certainly all but the larger businesses outsource their accounting and legal requirements. It is not the average 10 man company that can justify professionals such as a lawyer or accountant on full time staff. Rather we utilize them as outside resources, often quite informally.

There is, however, a critical aspect to outsourcing that differentiates an outsourced service from a standard Client-Contractor relationship.

For example, if you were considering having the of fice roof replaced, it is doubtful that you would just phone up a roofer and tell him to replace the roof by the end of the week. In reality what would happen would be a process of scope analysis, vendor selection, short iisting, tendering, review, negotiation and finally there would be a price agreed upon, a contract signed and the roofer would perform the work.

Yet on the other hand your company probably engages the services of an accountant to review the quarterly or annual records, and expects additional services whenever requested. If there is a particularly onerous request from Revenue Canada requiring a written response, it is certainly not in your best interests to go out to tender to three accountants to see who is going to write the cheapest letter.

0-7803-3552-X/96/$5.00 © 1996 IEEE

The relationship between the accountant and the business owner is one of understanding, trust and co-operation in maintaining the financial records of the company.

In a well outsourced business that type of relationship prevails. There is no room for traditional contract "values" of presenting a very low bid and then sticking it to the client on the extras. These are long term relationships based to a high degree on trust, mutual understanding and co operation.

Common Areas for Outsourcing within the Traditional Business Model

- **Legal Services**
- **Accounting Services**
- **Janitorial Services**
- **Payroll Services**
- **Transport and Courier Services**

It is generally accepted that the proliferation of outsourcing developed in response to the economic downturns in the 1980's. Many companies were fighting for their lives and outsourcing was considered a last resort in reducing overheads to a bare minimum. What followed in the 1990's was an interesting development as other companies began to see the strong economic advantages to farming out functions which were not viewed as a part of the core business. Larger companies in particular began to outsource their Information Services (IS), and Logistics and Transportation.

It is not uncommon for the outsourcers to subcontract for one another, which in fact can get rather convoluted. For example, one of the clients of Octel Network Services, a Dallas based firm that operates a vast number of electronic voice "mailboxes", is the colossal computer services company Electronic Data Systems. EDS was, of course, one of the originators of information-technology (IT) outsourcing and has a multi billion dollar contract to run the computer and telecommunication networks for Xerox. However, Xerox provides invoicing and billing services for Motorola, which in turn designs and manufactures parts of Octel's voice-mail systems.

The concept is not restricted to the service industry of course. Other companies, particularly small businesses, began to outsource key areas of their business, including manufacturing of the product itself.

The interesting extrapolation is what happens when more and more of the routine business functions are outsourced? The answer in its extreme is the Virtual Business. Essentially a one man of fice that functions like a large business. All aspects of the business are outsourced, including the sales, advertising and promotion, purchasing, and manufacturing. The business then essentially becomes akin to an exercise in Project Management. It is the responsibility of the business owner to ensure all pieces of the jigsaw fit and that no pieces are missing.

Why Outsource?

In a Coopers & Lybrand study, 81% of the fastest growing companies in the US relied on the services of temporary, part time or contract employees to help them grow.

Non-traditional employment now accounts for over 25% of the US work-force and almost one third of the Canadian workforce. But more startling is that almost half of the net new jobs created on both sides of the border in the last decade are non-traditional jobs. (B.Helm Assoc., 1995).

The top five reasons given for outsourcing are:
1. Outside providers are more efficient (70%).
2. To keep focus on own products and growth (45%).

3. To save cost of benefits/administration (42%).
4. To reduce overhead investment or debt needed (41%).
5. To ease the burden of regulatory compliance (21%).

If we look at outsourcing in terms of Project Management principles, it is easier to see the advantages and drawbacks.

Time

The management of several resource companies requires good planning skills and a firm grip on the overall production schedule. One late supplier can cause hardship for all other resources if the project as a whole is unduly delayed.

In this respect it is important to have back-up suppliers in key areas. If one major resource you depend on heavily is not available when you need to schedule them for a large project, there should always be a second choice ready to step in like the stand-in for a theatrical production. This may be their big break. Another reason to choose your suppliers carefully, and not just on prlce.

Resources

In this case, of course, the term "resources" refers to human resources. This is the. essence of the outsourcing concept. You must manage your own resources and utilize them where they are most beneficial and let your outsourced suppliers concern themselves with their own manpower.

This is also an area where there are negative aspects to outsourcing. In particular, there are certainly social implications of having increased numbers of independent contractors, perhaps self-employed, who have no access to traditional benefit programs or even to unemployment insurance.

In addition, job security is a relic from an earlier age. One of the advantages of outsourcing is undoubtedly the increased flexibility for companies to bring in more resources when business is good or alternatively simply not renew contracts when times get tough. However, since this in effect side-steps the relationship between employer and employee, adequate notice and redundancy pay are no longer required. But, how does a worker plan for this downturn when one of his major contracts is terminated at short notice?

Cost

Hiring back on contract may seem that you're paying considerably more for the same work, but that is not the case. The hidden costs of direct employment such as UI premiums, pension plan, vacation time and medical allowances all add up. The secondary benefits result from smaller premises for fewer staff and less equipment in the shop or of fice. Contract employees tend to be more efficient and their results can be more easily tracked.

Savings in the construction industry are estimated at 20% of payroll costs through avoidance of the related benefits and expenses (B.Helm Assoc., 1995).

Risk

Although the ultimate risk is always with the core company a good portion of the risk can be passed on to the resources by ensuring they are aware of their accountability for their portion of the work. Similarly, as the project is already broken down into smaller resources, so then is the risk reduced.

However the risks are real. Clients often can't and in most cases shouldn't need to differentiate who did what when things go wrong. When several big American and

Japanese manufacturers were forced to recall millions of vehicles earlier this year, it was, and rightly so, the car makers who took the heat. Few customers blamed the outsourcer in Japan that had supplied the defective seat-belt assemblies.

Scope

The scope of each resource is defined at the outset; it is unwise to wait until each resource comes into play. Although it is tempting to get complacent with familiar resources, it is important to give as much notice as possible to ensure they are able to perform when scheduled. Even if they have a small part to play in the final installation, they're informed from the very start - and then there are no surprises.

Contract / Procurement

Your Resource Group should be made up from set suppliers whose quality is known through past experience and whose expertise is well established. There is a degree of trust along with a level of proprietary knowledge that is passed on, and it's important that it remains in your control..

Communication

This cannot be over stressed. Outsourcing only works if everyone knows what js expected from them and what they expect of others. This is a critical area that can cause problems if the lines of communications aren't constantly open.

Quality

The quality of the overall project is the sum of the individual efforts of each resource working cooperatively. Quality is the goal and everyone in the system must be aware of how valuable their individual contribution is.

Outsourcing in Heavy Industry

The nature of the business of Gordon Crane & Hoist is, strangely enough, the sales, design, manufacture and service of Electric Overhead Travelling Cranes & Hoists

These are the large, yellow steel structures which run on overhead rails in factories, warehouses, pulp mills, heavy equipment repair shops etc.

Gordon Crane & Hoist emerged some 12 years ago when the present owners bought the Western Canada operation of a large international crane company which was downsizing. Having been in the employ of the vendor prior to the downsizing the new owners were well aware of the reasons for the downsizing, namely the cyclic market conditions traditionally in the heavy equipment business, i.e. the first thing to go in a declining economy is capital equipment.

We love that question in the crane industry "How's Business?" - "Oh same as usual, up and down!" For the new owners, being aware of the problem was only half-way to a solution, not a solution in itself. With a product that was greedy in manufacturing space and which required a highly trained workforce, one could try to survive by saturating the market with a very inexpensive product on the premise that if engineering departments were being downsized the Bean Counters would control purchasing. The "spec." would be secondary and price would rule. Low Dollars x High Volume = Reasonable Profit.

But what if the government screws up the energy policy and at the same time the bottom drops out of the pulp and paper market. Alberta and BC become basket cases. Low Dollars x Low Volume = Unreasonable Losses. Has this happened before? "Yes, 10 to the 6 times" - and the prognosis? Well, discounting the intervention of Lotto 6/49, the bank calls the loan and you are gone!

What was that profound insight of Karl Marx? With the advent of the factory, the worker no longer owned the tools and without the tools he would remain a servant and never again be master of his own destiny. Really? We have just discovered outsourcing by osmosis. Using outsourcing the tools of production are between our ears and will always be there as long as the Tower of London days don't come back.

Our core staff has the expertise in the market knowledge and in the application of correctly applied engineering. Our resources offer the manufacturing capability, the engineering services, the field service, and much more. We have manufacturing facilities at our disposal in Pennsylvania, Montreal, Toronto, Central Alberta, and here in Vancouver. The resource we choose is dependant on the nature of the equipment and the geographic location of the project.

Our four principle outsource suppliers are referred to as our Strategic Partners. We do the marketing, sales, and project co-ordination in our territory, they supply the equipment as either components or as a finished product. Two recently completed projects illustrate the co-ordination of these resources.

We quoted a Vancouver consultant on 11 cranes for an Ontario based project, had the equipment built at our facility closest to the final destination to minimize transportation and. had the cranes installed by a specialty crew from Alberta.

For a recent 85 ton crane supplied to the paper industry, we utilized hoisting machinery manufactured in Norway. The project was engineered and several key components were manufactured in Pennsylvania, and the main structure was manufactured and assembled in Alberta. The project was delivered to a paper mill in BC on time, under budget, and performed impeccably on first push of the button.

On both these projects, quality control included trips to all North American points of manufacture for periodic inspection, co-ordination of shipping for just-in-time deliveries, supervision of the final assembly and in shop testing prior to delivery to the customer.

The small business which extensively outsources can be analogized to a wheel in which the core staff are at the hub and each resource is a spoke. At Gordon Crane & Hoist our flexible network of resources produces a wheel which can have up to 20 resource spokes. A wheel that can carry heavy loads over great distances, at high speeds, and yet can still turn on a dime when required.

Are Gordon Crane & Hoist near that elusive futuristic goal of the Virtual Office? No not really. But when we are called to give an opinion on a crane upgrade and have to climb 50 ft in the air to crawl over a dust covered crane at the wet end of a paper machine where the temperature is 140°F and the humidity is 100%, it could be said that a virtual office in the Bahamas does have some appeal, even if it does mean moving there and having to give up the privilege of paying Canadian taxes.

References

B.Helm Associates Ltd. "The Impact of Outsourcing on Service Industries and the Workforce" Industry Canada Business Services Industries Branch, March 1995

Buss, Dale D. "Growing More by Doing Less" Nation's Business, Dec 1995 v83 nl2 pl8(5)

"Eight in Ten Growth Firms Use Outside Workers, Coopers & Lybrand Survey Finds." Business Wire, Jan 24, 1996 p 1241147

"The Outing of Outsourcing" The Economist, Nov 25, 1995 v337 n7942 p57(2)

"Virtual Companies, Real Profits" Nation's Business, Dec 1995 v83 nl2 p21(1)

Improving the National Science and Technology Policy Development Process[1]

James Gover, Fellow
505/284-3627
Sandia National Laboratories
Albuquerque, NM 87185-0103

Elias G. Carayannis
202/994-4702
George Washington University,
Washington, DC

Mark Peterson
505/271-5104
University of New Mexico,
Albuquerque, NM 87131-1121

The constituent-driven political process, where research and development performers provide the most vocal advocacy for federal R&D, is unable to create programs that can survive the scrutiny of critics and maintain bipartisan support during extreme budget pressure. The political process can be supplemented during program conception by introducing models that relate funding inputs to desired public outcomes. Game technologies developed by companies for strategic planning and the military for war simulation can be used to predict behaviors of those affected by science and technology policies.

Introduction

We review the government's response to the US industrial competitiveness problem and conclude that the policy development process led to programs whose public outcomes were perceived to be unacceptable by Congress. We propose a science and technology policy development process that we believe is both more likely to yield outcomes that meet public needs and that will be perceived by the public, Congress, and the President as having done so.

The Problem

Table I is a partial list of products and technologies that are generally regarded to be of economic importance.[2] In practically every instance where US industry has lost market share to foreign competitors, competition was dominated by the need to rapidly make cyclic improvements to existing products and their manufacturing processes. (Gomory 1992) Competitive difficulties have stemmed from:

- Lack of emphasis on manufacturing equipment and processes, especially: (1) the application of quality principles to manufacturing, (2) inattention to process R&D and production tool R&D in comparison to product R&D, (3) weak linkage between science and technology innovation and production processes, and (4) failure to make modeling-intensive, deterministic design methodology part of our nation's technology infrastructure[3];

- Preference for niche products with high profit margins accompanied by abandonment of commodity products with low profit margins that foreign competitors are sometimes willing to sell at prices below their manufacturing cost, and

[1] This work was supported by the United States Department of Energy under Contract DE-AC04-94AL85000.
[2] Data were compiled while Gover served as an IEEE Competitiveness Fellow in the Office of Senator William V. Roth, Jr.
[3] Arguments for the fourth point have been convincingly made to Gover by Gib Marguth, Sandia National Laboratories.

- A national regulatory environment that is perceived to disadvantage the US as a site for manufacturing commodity products that results in US companies either abandoning these products or moving their manufacturing off-shore.

Industry Response to the Problem

The intensity of international competition has caused US companies to introduce many changes in their organization, management, and structuring of R&D. These changes reflect the progress that companies are making in learning the strategic management of technology. (Carayannis 1995) Major changes include:

- Central corporate research laboratories (CCRLs) have been downsized in companies that compete in industry sectors where commercialization is driven by the need to improve existing products and processes. R&D is being moved to manufacturing facilities in companies' business units. Formal technology transfer processes are being used to shorten transfer time. (Bosomworth 1995)

- R&D projects are designed to support companies' business strategies by nurturing those core technologies that differentiate them from competitors.

- Preference is being given to R&D that can be captured by the company.

- R&D laboratories are being reengineered and restructured to be compatible with the ways that each company commercializes its products. Companies are particularly improving their ability to compete in sectors where commercialization is product and process improvement driven.

- Companies look everywhere for innovation with partnering, strategic alliances (Carayannis 1996), collaboration, and consortia used to gain access to new technology. (Gover 1995)

US companies have improved their competitiveness. Even though major US firms increased sales by 4.4% or $63B between 1991 and 1992, they reduced their employee count by 3% or 233,000.(BW 1992) In 1993, US multinational companies announced nearly 600,000 layoffs, 25 percent more than in 1992 and 10 percent more than during the 1991 recession.(The Economist 1995) Therefore, while US companies have made progress in solving their competitiveness problems, public issues related to employment (Kapstein 1996) remain, as do the trade deficit.

Government Response to the Problem

Congress took action during the late 1980s and early 1990s to address the "competitiveness problem". Pressure on Congress came from constituents seeking better employment opportunities and constituents that would directly benefit (this usually means receive public money) from these programs. To emphasize that industry, in contrast to the public, was their customer, federal agencies often stressed that they were "industry-led" rather than "government-led". These programs include:

- Technology Transfer Initiative
- Advanced Technology Program
- Manufacturing Extension Project
- Technology Reinvestment Project
- Defense Research Projects Agency became ARPA
- SEMATECH and Other Consortia Supported
- Small Business Innovative Research Program Increased Awards

Analysis of the Government Response

Throughout the evolution of these programs little was invested in determining:

- how much the US public would benefit from each project and whether public benefits outweighed public costs;
- how public return from projects should be measured;
- whether the free market had actually failed;
- why the free market had failed and how each R&D project addressed the root cause of the market failure;
- the trade-off of public institution versus private institution performers in maximizing public return;
- the importance of direct R&D assistance to industry, valued at less than $5 billion, in comparison to alternative R&D options including:
    - optimizing regulations, reputably an annual $500 billion to $800 billion cost to companies,
    - reducing health care costs, reputably a $200 billion annual cost to industry and a $1 trillion annual cost to the Nation,
    - reducing energy costs, reputably a $100 billion annual cost to industry and a $500 billion annual cost to the Nation;
    - reducing crime, reputably a $40 billion annual cost to industry and a $425 billion cost to the Nation;
    - improving education productivity, reputably a $50 billion dollar cost to companies and $500 billion annual cost to the Nation;
- how program costs should be shared between the public and companies to maximize the ratio of public benefit to public cost;
- how an entire industry sector can be strengthened without distorting domestic competition; and
- how intellectual property ownership should be distributed to maximize public benefit.
-
- Not surprisingly, in view of
- the weak intellectual framework for most of these programs, particularly their lack of attention to economic models and their inability to clearly articulate public benefit;
- the bureaucratic practice of treating program critics as ideologues and ignoring their concerns4;
- the public perception that many programs are industrial subsidies of low public value;
- the low political rewards of federal R&D not conducted at a university;
- the high political risks associated with cutting entitlement accounts;
- the rapid growth in health care costs; and
- the perception that the public wants the federal budget balanced;

many of these programs are being curtailed by Congress, just as has been done with this class of programs since they were first introduced during the early 1960s. (Fusfeld 1994)

Recommendations

The problems associated with federal support of commercial technology stem from the inability to develop an intellectually-credible, publicly-defensible federal S&T policy. As illustrated in Figure 1, Congress, by design intent, is responsive to constituencies and the most vocal science and technology constituencies are R&D performers that directly benefit from a program. Federal commercial technology policies must be influenced by public policy experts well-versed in technology issues.

[4] This criticism is not always valid. See, for example, (Rogers 1996)

The following processes will introduce an increased level of rational rethinking into federal science and technology policy for economic development. (Drucker 1995)

Much of the dispute over commercial technology programs stems from a disagreement over the tacit models each advocate and critic possesses and the compatibility of a particular program with their model. The convergence of these conflicting models to a common intersection will allow identification of acceptable roles for government to play in commercial technology R&D. As a first step the process must identify those events and conditions of market failure which warrant intervention in economic sectors. This, in turn, should lead to selection of programs with the greatest potential for successful implementation. Finally, convictions with regard to program effectiveness will be borne out of rational assessment rather than self-interested assertion.

Commercially-relevant technology policy is driven by a wide array of models including: free market model, market failure model, commercialization model, competitiveness model, infrastructure model, innovation models, long-range/short-range R&D model, etc. Programs have attempted to implement specific models under a variety of conditions, but little or no effort has been made to do systematic analysis. Furthermore, different models work best in different environments and create different outcomes. Application of commercialization models, for example, suggests avenues for technology transfer from federal laboratories that differ from those paths sometimes pursued by laboratories. (Gover 1994)

A National committee of public policy experts, distinguished by their research on National technology policy, should be established to rationalize those models that are relevant to National economic growth. This committee should be chaired by a Nationally recognized figure from the science and technology community of the stature of Ralph Gomory, President of the Sloan Foundation. This committee would be tasked to evolve a model for how the federal government should best support science and technology in ways that stimulate the National economy to increase its annual growth rate from 2.3 percent to 3.5 percent.

A report of the findings and recommendations of this blue-ribbon committee would include the definition of a national program to support economic development. Roles for universities, federal laboratories, states, and companies in executing this program would be a major component of this report as would an assessment of how current programs could be optimized. Recommendations would include the agency infrastructure and governance system required to manage these programs. Upon completion of this report, the next step would be to assess the response of stakeholders affected by these recommendations. Game theory provides the vehicle for this.

Game theory, extensively used by military planners, has recently been offered as a tool for strategic planning by businesses. (Brandenburger 1995) It has demonstrated utility for policy development. Prosperity Games (Berman 1995), a high-level interactive forum for exploring complex issues in a gaming environment, could be reconfigured to test the response of companies, federal agencies, the Congress, federal laboratories, universities, and states to the recommendations of this committee. Prosperity Games are an adaptation and extension of move/countermove and seminar War Games. They have already been successfully used to address issues in electronics manufacturing, industry-government interactions, and international competitiveness issues. Prosperity Games provide opportunities for exploring current realities while simultaneously creating and studying other revolutionary alternatives. Prosperity Games are not a replacement for the political process; rather, they complement it by facilitating the political process on a compressed time scale.

A report including the findings and recommendations of this committee and the response of the legislative and executive branches of the federal government, industry, federal laboratories, states, and universities to these recommendations gathered at the

Prosperity Games exercise would be delivered to the President, Speaker of the US House of Representatives, and the Majority Leader of the US Senate. The recommended new process is illustrated in Figure 2.

References

(Berman 1995) Berman, Marshall, Ivan Berry, and J. Pace VanDevender, Prosperity Game for the National Electronics Manufacturing Initiative, Sandia National Laboratories Report, SAND95-0724.UC-900, May 1995.

(Bosomworth 1995)Bosomworth, Charles E. and Burton H. Sage, Jr., "How 26 Companies Manage Their Central Research", Research Technology Management, May-June, 1995, pp.32-40.

(Brandenburger 1995) Brandenburger, Adam M. and Barry J. Nalebuff, "The Right Game: Use Game Theory to Shape Strategy", Harvard Business Review, July-August, 1995, p.p 57-71.

(BW 1992) Business Week , "R&D Scoreboard", June 28, 1993, and June 29, 1992. Business Week obtained their data from Standard & Poor's Compustat Services.

(Carayannis 1996) Carayannis, Elias G. et al, "Strategic Alliances as a Source of Early-Stage Seed Capital in Technology-Based, Entrepreneurial Firms", HICSS, January 3-6, 1966.

(Carayannis 1995) Carayannis, Elias G. and Raymond Stokes, "The Strategic Management of Technological Learning: Fadische Anilin und Soda Fabrik (BASF) AG", Journal of Focus on Change Management, Dec. 1995-Jan. 1996, pp 18-22.

(Drucker 1995) Drucker, Peter F. "Really Reinventing Government", The Atlantic Monthly, February, 1995, pp 49-61.

(The Economist 1995) The Economist, "A Survey of Multinationals", June 24, 1995, p 5.

(Fusfeld 1994) Fusfeld, Herbert I. Industry's Future: Changing Patterns of Industrial Research, American Chemical Society, 1994, p. 220.

(Gomory 1992) Gomory, Ralph "The Technology-Product Relationship: Early and Late Stages", in Technology and the Wealth of Nations, edited by Nathan Rosenberg, Ralph Landau, and David C. Mowery, Stanford University Press, 1992.

(Gover 1995) Gover, James "Corporate Management of R&D - Lessons for the U.S. Government", Research Technology Management, March-April, 1995, p 35.

(Gover) Gover, James "Optimizing Federal Technology Transfer to Promote Commercialization", Journal of Technology Transfer, Dec., 1994, p.p. 36-50.

(Kapstein 1996) Kapstein, Ethan B., "Workers and the World Economy", Foreign Affairs, vol 75, no 3, May/June, 1996.

(Rogers 1996) Rogers, Everett, Elias G. Carayannis, et al, "The Anatomy of the CRADA Process in New Mexico and a Comparison with Japan", HICSS-29, January 3-6, 1996.

Table I: Nations and continents originating breakthrough R&D compared to those first commercializing products from this research and compared to those leading today's market for these products.

Technology or Product	R&D Leader	First Product	Current Leader
Advanced Composite Materials	US	US	Japan/US.
Anti-Skid Brakes for Autos	Europe	Europe	Japan/Europe
Automatic Focus Cameras	US	US	Japan
Automobiles	Europe	Europe	Japan
Biotechnology	US	US	US
Commercial Jet Aircraft	Europe	US	US
Communications Electronics	US/Europe	US	US
Compact Disk Players	Europe	Europe	Japan
Computer Aided Design	US	US	US
Consumer Electronics	US	US	Japan
Copiers	US	US	Japan/US.
Desk Top Computers	US	US	Japan/US.
Digital Watches	Europe	US	Japan
DRAM Memory	US	US	Japan
Drugs and Pharmaceuticals	US	US	US
Electronic Fuel Injection	US	US	Europe/Japan
Facsimile Machines	US/Europe	US	Japan
Fiber Optics	US	US	US
Flat Panel Displays	US	US	Japan
Fuzzy Logic	US	Japan	Japan
Hand-Held Calculators	US	US	Japan
High Temperature Superconductors	Europe/US.	Japan/US	?
Integrated Circuit Test Equipment	US	US	Japan
Jet Engines	Europe	Europe	US
Lithography Steppers	US	US	Japan
Medical Imaging Technology	Europe	US	US
Microprocessors	US	US	US
Military Radars	Europe	Europe	US
Notebook and Desk Top Computers	US	US	US/Japan
Numerical Control Machine Tools	US	US	Japan
Rocket Propulsion Technology	Europe	Europe	US
Robotics	US	US	Japan
Semiconducting Lasers	US	US	Japan
Software	US	US	US
Super computers	US	US	US
Television Sets	US	US	Japan
Video Games	US	US	Japan
VCRs	US	US	Japan

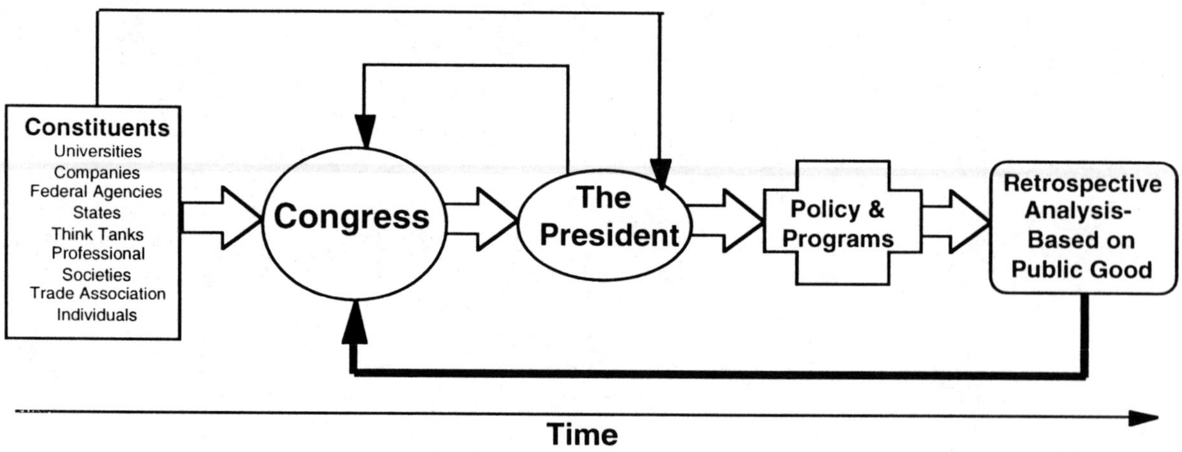

Figure 1: The Policy Development Process. Except for health care issues, most of the advocacy for federal science and technology policy derives from R&D performers. Most analyses are performed retrospectively, often after program are nearing completion. Programs that fail to achieve their public outcome, poison the wells for future federal R&D programs.

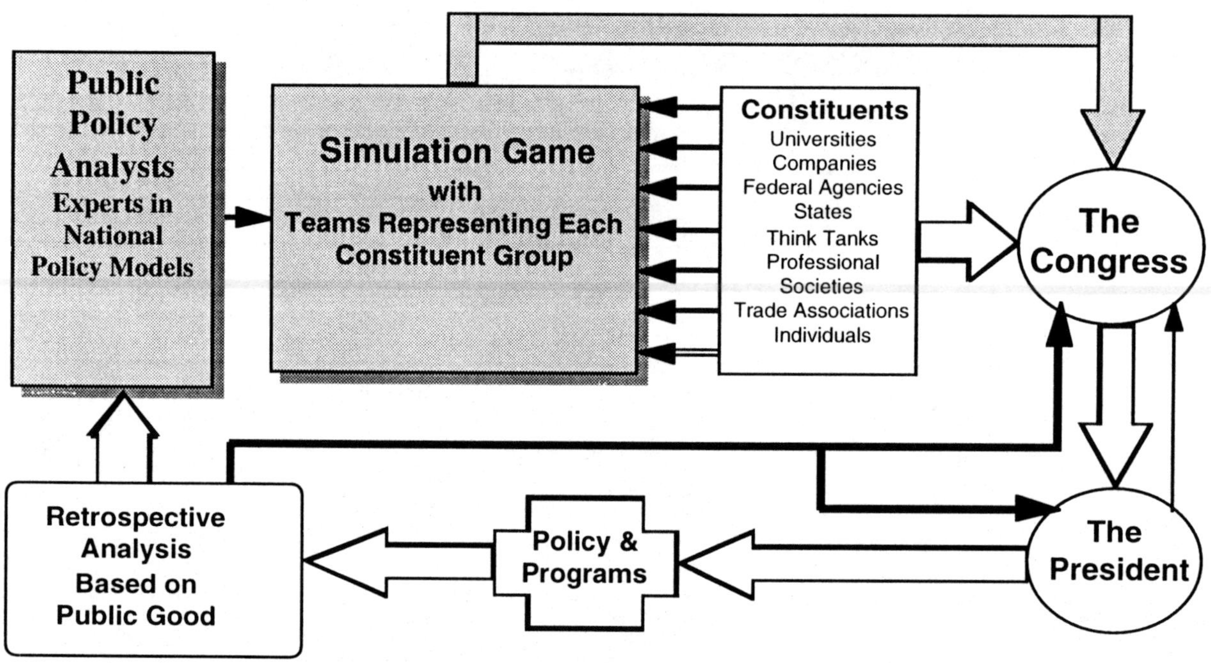

Figure 2: Public policy experts known for their expertise in (1)models relevant to commercial technology and/or (2) retrospective program analysis would recommend programs. Stakeholder response and proposed modifications to these programs would be determined through use of a simulation game.

Engaging Engineers in Science and Technology Policy Development[1]

James Gover, Fellow
505/284-3627

Sandia National Laboratories
Albuquerque
NM 87185-0103

Mark Peterson
505/271-5104

University of New Mexico
Albuquerque
NM 87131-1121

To engineers the democratic political process is emotional, irrational, and offers only superficial adherence to systems approaches. Yet, the political process is profoundly affecting our livelihoods, future prospects for technology advances, and nations' ability to maintain economic growth in an increasingly complex global environment. The engineering approach to problem-solving can be useful in the political process, but it requires the engagement of engineers in that process. In leaving policy development to lawyers, engineers are implicitly permitting rational problem-solving approaches to be avoided because vigorous advocacy in their behalf has been absent. There are numerous ways for IEEE to use its resources to engage engineers in policy development.

Introduction

The democratic political process is virtually the antithesis of the engineering process. The political process requires the building of consensus, and the satisfying of demands in ways that have nothing to do with objective rationality and everything to do with subjective expectations and a concept of power not related to ergs, watts, horsepower, octane, or BTUs. We argue that the rationality of the engineering approach to problem solving can be an effective voice in the political process. For too long engineers, and engineering organizations have been reluctant to engage in the public policy debate. In leaving the field to others, rational problem-solving approaches have been avoided.

Engineers have within their tool kit of technology the ability to analyze problems and logically and objectively arrive at systemic solutions that have high knowledge intensity. Fresh, creative perspectives for policy synthesis can emerge from the minds of engineers. Engineers can provide advocacy for high quality, defensible policy positions that are supported by thorough, unbiased, timely research. However, most engineers are inexperienced in using these skills to deal with "soft issues" such as policy development. When engineers address "soft issues" without bringing to bear their analytic skills, their contributions differ little from the "man in the street". That is, the engineers' contributions are not differentiating.

If one assumes that every US member of IEEE becomes only weakly involved in the national political process and that each member can influence the position of 10 family members and friends, then our members have the political power of approximately 1 percent of the population. If, however, their policy positions are knowledge intensive, well researched and easily shown to be in the best interests of the American public, this differentiating capability further increases their political strength. If our members are then willing to move from an intellectually involved state to a politically active state in which they contact the offices of their Senators and Representatives three to five times a year and write thoughtful, knowledge-intensive

0-7803-3552-X/96/$5.00 © 1996 IEEE

1 This work was supported by the United States Department of Energy under Contract DE-AC04-94AL85000

letters to members of Congress, their political influence can become equivalent to that of an even larger percent of the population. If they can get local and national press coverage on behalf of their issues, their influence can be further multiplied.

For IEEE-USA to achieve its political potential, we recommend that our primary interface with the political system be at the local level through the chapters and sections and that emphasis at the national level be focused on providing knowledge-intensive data to chapters and sections keeping them up-to-date on policy related issues, creating opportunities for our members to present their knowledge-intensive positions for public scrutiny, and providing lobbying training to these small groups.

The Changing National Environment

The extraordinary budget pressure brought on by the perception that the public wants the budget balanced coupled with the political risks of cutting social security, Medicare, and Medicaid, has changed the environment such that we predict Federal R&D, like other discretionary accounts, will be increasingly examined under a microscope. We believe that funding pressures will demand thorough analysis and unfounded, qualitative assertion is nearing the end of its useful life. Furthermore, to be funded, the path to public good must be clear and defensible. We are entering a period of unprecedented opportunity for IEEE-USA, if, and only if, we are able to make the transition to a knowledge-intensive institution.

Congress is not just looking for institutions to come by and advise them on what to do. In fact, that happens many times every day, but the recommendations are often in conflict and only support special interests. Rather, during this period of unprecedented budget pressure, many members of Congress are looking for well-researched, intellectually-defensible proposals that are politically practical. These allow Congress to move beyond the debate of special interests and justify their position to their constituents.

There is an almost endless amount of data needed to make better policy decisions and IEEE could play a major role in the development and publication of that data. It goes without saying that management by data can't be done in the absence of data. Despite that, government rolls on making decisions of monumental importance in the absence of data. Congress can be persuaded to take action on most any matter if its members are handed objective analyses that have been socialized to the public. Our objective should be to become the principal source of those objective analyses.

At the local level, with a conscious well-organized effort to put the same objective data into the hands of our membership as we provide to members of Congress we have the opportunity to influence individual votes of members in very effective ways. Political scientists refer to this as the "Utah Plant Manager Effect." While lobbyists may be able to provide detailed and accurate information about policy impacts on the District's populace, nothing sells the point as effectively as when Senator Hatch receives a call from the Brigham City, Morton Thiokol production manager telling him with great specificity what jobs will be lost or benefits derived from his vote on the new round of technology transfer funding before Congress.

Increasing the Policy Role of IEEE

It is well known that many members of IEEE, including US members, are completely insulated from S&T policy issues that determine their future. Many of these issues extend well beyond the borders of the US and are of interest to our international members. It is proposed that the IEEE pioneer the development of a program that is designed to increase the awareness of US and international members of S&T policy issues, inform IEEE members of the trade-off involved in these issues, inform the general public of S&T policy issues, and make state and national governments aware of these issues. Worthwhile activities include, but are not limited to:

- A session on S&T policy should be added to each IEEE international conference. The policy sessions would focus on S&T policy issues related to the technical subject of the conference. For example, a session on energy policy could be included at conferences sponsored by the Power Engineering Society, a session on communications and information policy could be included at conferences sponsored by the Broadcast Technology Society, a session on aerospace policy could be included at conferences sponsored by the Aerospace and Electronics Systems Society, etc.

- IEEE should develop a tutorial publication, perhaps a book, that includes an articulation in layman's language of the technical problems, the costs, the relevant social and political issues, etc. of various technologies and identifies which of these options are being pursued within various countries. Such a publication could be coordinated by the Social Implications of Technology Society and worked in conjunction with other relevant technical societies. The Grand Challenges identified by the TAB New Technology Directions Committees and the related TAB society evaluation of emerging technologies could be useful resources for such a publication.[2]

- IEEE Transactions editors should institute an editorial practice of encouraging at least one policy paper in each edition. The policy paper would be relevant to the specific area of the Transactions and subject to the usual rigorous review normally used for technical papers in the Transactions. IEEE Transactions on Engineering Management has successfully followed this practice for several years.

- IEEE-USA entity positions should be transmitted to IEEE members on the Internet with invitations to comment extended.

- Each IEEE-USA Technology Policy Council committee should develop a relationship with the science and technology editor of the New York Times, The Wallstreet Journal, the Washington Post, and The Economist and provide them with tutorial materials that can be used for special policy-focused articles, or in the case of The Economist, used for their surveys, for example, "A Survey of Nations' Industrial Policy" or "A Survey of International Energy Policy", or "A Survey of Telecommunications Policy".

- In every Proceedings of The IEEE we should encourage or even require the authors to write a major section that addresses policy matters of importance to the topic under discussion.

- The Washington Watch section of Spectrum should be expanded to include international S&T policy issues and more depth on US S&T policy issues.

- We should create a Science and Technology Policy Review Journal, and encourage collaboration between IEEE members and social scientists in the submission of articles to this journal. The policy analytic profession today is dominated principally by lawyers and academics who understand the political process very well, but know little about the actual technical aspects of the policy solutions many carry around ready made to be applied to the first available problem There is a large body of policy research journals that publish papers closely related to science and technology policy that are read by only a small group of IEEE members. Again, imitating the practice of a journal of the IEEE Engineering Management Society, IEEE Engineering Management Review, it is recommended that IEEE-USA publish a review journal of science and technology policy papers. This journal would select papers from policy journals published around the world and print those judged to be the best.

---

[2] Chris Brantley, IEEE-USA, personal communication.

- With the exception of a few science and technology writers like Michael Schrage and G. Pascal Zachary, newspapers and television typically avoid articles or interviews of substance that deal with S&T. The news media seeks breakthroughs, S&T delivers incremental advancements. It is well known among Congressional staff that S&T isn't sexy, a euphemism meaning that it doesn't get press for the boss. Thus, S&T activism has few political rewards for members of Congress. It is time to raise the ante! Publishing a scorecard on Congress to see how each member votes on S&T issues is an important first step. Such a document could record how many new S&T bills each member of Congress introduces, rank members as supporters, antagonists, etc., of S&T issues and single out the antagonists for "hard-ball" political attacks. For example, we could bring press attention to those who actively support S&T and attack those critical of S&T. There is much that can be done in this area. We could bring press to our members that run in political races and encourage our membership to make contributions to their campaign. We could offer an annual award to the member of Congress that most supported S&T issues. By sponsoring well-publicized policy conferences that can attract several hundred people from outside the beltway, we can attract more interest among members of Congress.

- We should create an IEEE National Policy Think Tank that both conducts and synthesizes authoritative studies of public policy issues of importance to our customers. Because the ideological preferences of those funding think tanks must be satisfied, most of the time the policy recommendations of think tanks can be predicted. We propose that IEEE-USA develop a think-tank that has an expert researcher for each of the committees of the Technology Policy Council and that we use our journals as vehicles for publishing the studies of this think tank.

- We should conduct a case study of the four or five chapters and sections that are most successful in developing politically active groups. The purpose of this study would be to identify what works for the benefit of other chapters and sections. Several chapters and sections, generally those in the largest cities, have been particularly successful in developing politically active groups that have been successful in policy development at the state level. The Power Engineering Society has established a Public Affairs Council to manage the policy affairs of that society.[3]

- Many companies have found their retirees to be an extraordinary source of political activists. To assure that their political agendas include that preferred by the company, retiree meetings are held and the company passes on their political concerns and preferences for national policy. In addition, the retirees may be provided with guidelines on how to lobby, telephone lists of members of Congress and their legislative interests, and information on political issues.[4] It is appropriate, in particular, that IEEE-USA grassroots policy work make maximum use of retired IEEE members.

- To stimulate grassroots political activism, we recommend that each chapter and section create an officers position, Government Liaison Officer, that is part of the progression to section or chapter president. This officer would be in charge of establishing a political action group in each chapter.

Conclusion

Science and technology has enjoyed a long run of funding plenitude in the post World War II era. The allocation of resources to S&T activities was generous enough that blind alleys, and Golden Fleece award winners had very little impact on its general direction or bounty. The media and the political public relations machines churned out endless stories of science fiction rather than science fact, and the public

---

3 Ann Hartfiel, Manager IEEE-USA Professional Programs, personal communication.
4 Sandra Donovan, consultant, personal communication.

became increasingly ignorant and cynical about the meaning and role of science and technology. For a variety of reasons having to do with cynicism, huge budget deficits, and an increasing public sense of "something being wrong," the nation's public policy priorities have begun to shift in ways that threaten to undo the real benefits and progress afforded society by the application of science and technology to mankind's existence. A failure to respond proactively and effectively to this challenge by one of the most influential of representational groups in this field, IEEE, would represent an abject abdication of responsibility. Poor Richard wrote, "God helps them that helps themselves." In our political system the clash of interests is the process by which policy is made. It is not precise. It is not scientifically analytic, nor is it objectively efficient. It strives for the maximization of utility within the constraints of a system which must achieve pluralities of support among disparate interests. Well focused, particularized interests often prevail over much broader, but also more amorphous and less deeply felt interests. The worst political defeats come to those who lose by default. If we don't suit up, we'll never get in the game, and if we don't get in the game, we can't possibly win.

Social Networks and the Development of Two Oil Technologies in Norway

Arent Greve
Tel. (int.) 4755 959 453

The Norwegian School of Economics and Business Administration
Breiviken 2, 5035 Bergen-Sandviken, Norway
arent.greve@nhh.no

Julia Harkola
Tel. (415) 354-1170

Stanford University Department of Civil Engineering
Stanford, CA 94035-4020 USA
julia@ce.stanford.edu

This study investigates how social networks were used in the development of two related oil technologies, STL/STP, during the period 1992-1994. The network spanned across several firms and constituted a professional community that began in the late 1960's. The most important relations were established by 1985. The theoretical framework is founded on how human and social capital contribute to the innovations through distributed cognition in social networks. The network enabled mobilization of complementary resources during the development process through activating the connections. This indicates that the social relations of technology developers are a crucial asset.

Introduction

This study focuses on the role of social networks in the development of two technologies in the Norwegian oil industry. We investigate how social relations and networks facilitate the combination of various resources to accomplish complex tasks. The subject technologies, STL/STP (Submerged Turret Loading/Submerged Turret Production), are systems for off-shore buoy loading and oil production from underwater installations. The STL-buoy was developed by APL (Advanced Production & Loading Systems) and STP by Framo Engineering. Statoil coordinated both projects. The development of STL started between 1990 and 1992, and APL was founded early in 1993. STP was developed in 1994.

Theory

Most theories related to innovation focus on the firm and try to identify attributes, structures, or personnel that can produce successful innovations. Absorptive capacity indicates a firm's ability to develop or adopt innovations; it is measured by R&D costs (Cohen & Levinthal 1990). One problem is that it assumes that the expenditure of R&D resources has the same effect across firms and it disregards the effects of external networks. Firms develop communication skills that allow them to understand externally developed knowledge, so cognitive theories can be applicable. Cognition is primarily an individual activity, however, research has been done on collective cognitions (Walsh 1995). These are based on the development of cognitive categories and common understanding of the environment through interaction.

We use a social network perspective where norms, values, and social system properties influence behavior (Granovetter 1985). Social networks may be used to get information and resources (Granovetter 1982). Innovation development requires a combination of own assets and complementary assets from others (Teece 1987), including human resources, information, and tangible assets. A social network is effective for asset mobilization to the extent that one may use social capital with complementary knowledge through distributed cognition. *Social capital* is assets an actor shares with social relations (Burt 1992). Through *distributed cognition*, each expert in a group deploys his/her expertise to solve specific problems enabling a joint solution (Hutchins 1991). A shared frame of reference is developed that enables coordination of cognition through social interaction.

0-7803-3552-X/96/$5.00 © 1996 IEEE

Social networks are defined as a set of relations among actors (Wasserman & Faust 1994). Social network location determines an individual's access to information and other actors. Social positions are associated with actors' different cognitive representations (Pattison 1994) and determine what resources an actor is able to mobilize in innovative activities (Håkanson 1989).

Methods

Case study method

This descriptive study traces social network establishment among the core group of ship engineers that became crucial to the development of STL/STP. It combines qualitative and quantitative methods with the advantage of attention to detail and generalizability of the results referenced to theory.

Interviews were conducted with STL/STP development participants at APL, Statoil, and Framo Engineering. We designed a name list with empty space for additions. For each listed person, we asked the respondent the relationship duration, communication frequency, and communication content. To analyze networks over time, we used the data on how long a connection had been in existence.

Network measures

Network measures can be used for individuals and their position, and properties of the whole network. Block models combine these and show how the network can be divided into blocks of structurally equivalent actors (Wasserman & Faust 1994). *Structural equivalence* defines actors' similar network positions in terms of connections to others. Structurally equivalent actors may or may not be connected to each other.

Network *density* indicates the number of existing relations, compared to the maximum possible number. *Degree* counts an actor's direct contacts and represents social capital. *Closeness centrality* measures the reciprocal distance to all others in the network (Freeman 1979) and indicates connectivity. Degree and closeness centrality indicate ability to get information and influence the social environment.

Betweenness centrality is computed as the number of optimal paths passing through each node. People located between clusters have high betweenness centrality and can control information flow. *Flow betweenness centrality* is based on all paths among the actors in a network. It allows actors to choose information paths. *Information centrality* is determined by the total flow of information between two actors with information loss based on the distance between them. It is based on the theory of statistical estimation (Stephenson & Zelen 1991).

We use the CONCOR method for dividing the network into blocks of structurally equivalent actors (Borgatti, et al. 1992). Structurally equivalent actors' position leads them to share communication sources and, therefore, they have similar social capital (Burt & Talmud 1993).

Results

Network nodes were assigned identification numbers with the rightmost three digits for the individual and the leftmost two digits identifying the firm (place of employment in 1995). The most important firms are: APL (1), Framo Engineering (2), and Statoil (3).

By 1970 there were two clusters: a clique composed of 21001, 3012, and 3005, and a star shape with 1013 in the middle connecting 1023, 1004, and 3017. During the 1970's, a relation between 21001 and 1013 connected the two clusters and more people entered the network. 1004 was central in STL concept development and 1013 became the most central person for development work. The STP concept originated among 3012 and 3005 while they worked on STL during 1993; prototype was developed about one year later.

In 1985 the network contained 33 people, including the most important for STL development . The most central people (normalized degree) in the network were: 1013 (79%), 3005 (45%), 3011 (38%), 3012 (22%), 3009 (38%), and 1004 (20%). A professional community was established; the members communicated and exchanged technical information. Since most worked in different organizations, they also developed some complementarity that would prove useful for later development. In 1985, 1013 was a university engineering department faculty member.

More people joined the network and existing members established new connections. There were 57 people in 1990, 67 in 1992, 91 in 1993, and 111 people in 1994; the maximum was 113 people early in 1995. During 1995 the network was reduced to 108 people.

Centrality measures (Table 1) indicate the extent to which the network was centralized or if there were critical nodes through which information flowed (betweenness centrality). Normalized centrality measures are comparable across years. Degree increased because the most central people added more relations, particularly between 1992 and 1994 when STL/STP was developed. Network size and the number of central actors' contacts increased, but the mean number of contacts was stable since more actors with fewer connections were added.

Year:	**1985**	**1990**	**1992**	**1994**	**1995**	**1995a**
Measures:						
Degree, %	70.0	58.0	58.0	85.8	88.9	82.5
mean, normalized	13.8	10.3	9.9	10.9	11.2	11.1
Closeness centrality, %	9.1	9.5	51.4	67.2	69.0	69.5
mean, normalized %	15.6	16.3	40.2	44.9	45.6	46.2
Betweenness centrality, %	38.0	42.6	58.9	49.7	49.2	49.1
mean, normalized	4.7	3.5	2.4	1.16	1.11	1.16
Flow betweenness centrality, %	54.3	38.9	35.9	20.4	19.4	20.1
mean, normalized	5.7	2.8	2.13	1.02	1.00	1.03
Information centrality, mean	0.28	0.40	0.34	1.26	1.16	1.11
Density	0.13	0.10	0.09	0.10	0.10	0.10
n, people in the network	33	57	67	111	113	108

Table 1. Network centrality measures

Closeness centrality results also indicate that more actors were connected to each other, providing better access to the network. A few high centrality people connected a large number of others and served as important communication bridges early in the observation period. Later, network betweenness centrality was reduced as more members became inter-connected. Comparing betweenness centrality and flow betweenness centrality measures shows that initially there were fewer alternative paths, but as development started more people were connected and network centralization decreased. This also indicates that the network became less dependent on a few very central people. The increase in information centrality indicates that there were also more alternative information paths.

A structural equivalence analysis using CONCOR was made for 1995 with two partitions; the four blocks can be interpreted as activity niches. Block 1 consists of 15 firms with 46 individuals that were engaged in STL development, fabrication, and installation (including consultants and users). Block 2 (5 firms and 21 individuals) has most of the people working with STP, which had not yet been installed at any oil field, most of the higher level managers, and two potential users. Block 3 includes 14 firms with 22 individuals that deal with oil field supplies and marine consultants that are required for STL/STP installation and use. Block 4 (17 individuals) contains three oil fields, five firms working with field installations, and two ship owners that are potential users of STL/STP.

The diagonal in Table 2 is the density within each block; the remaining cells indicate the density between blocks. The density of the entire network was 0.10. Between blocks 2 and 4 there was a higher density than for the whole network. This shows that the managers had most of their contacts among themselves and with the oil fields. There were few relations between blocks 2 and 3, probably because STP was not being installed at any oil fields during data collection.

Table 2 shows the density within and between blocks of structurally equivalent actors. The density of the entire network was 0.10.

	Block 1	Block 2	Block 3	Block 4
Block 1	**0.18**			
Block 2	0.04	**0.46**		
Block 3	0.05	0.01	**0.14**	
Block 4	0.07	**0.13**	0.05	**0.24**

Table 2. Density within and between blocks

Discussion

Prior research used cognitive structures to explain organizational behavior (Walsh 1995). With firms or work groups as the unit of analysis, researchers may not be able to detect collective cognition among individuals in professional communities existing beyond firm boundaries. Development of complex technologies requires the combination of complementary resources, the suppliers of which must be able to communicate with each other. This study shows that a professional community was developed over several years. The social relations where the ideas for STL/STP originated were established by 1970, and the most important relations for STL/STP development had been formed before 1985. The social structure represents social capital that was built up over time. The structure includes individuals with different competencies and the interaction history that made cooperation possible. The complexity of the innovations is reflected in the distributed cognition represented in the network.

Structural equivalence analysis indicated that firms with common innovation roles had similar communication structures. Members of structural equivalence blocks develop a common understanding of the environment and their tasks, and share norms for appropriate behavior (Galaskiewicz & Burt 1991). Structurally equivalent actors may share cognitive structures. The distributed cognition necessary for accomplishing complex tasks may, therefore, take place across structurally equivalent positions.

References

Borgatti, Everett, and Freeman. 1994. *UCINET IV*. Version 1.40/X Colombia: Analytic Technologies.

Burt, Ronald S. 1992. *Structural Holes: The Social Structure of Competition.* Cambridge, MA: Harvard University Press.

Burt, Ronald S. and Talmud, Ilan. 1993. Market Niche. *Social Networks.* 15: 133-149.

Cohen, Wesley M. and Levinthal, Daniel A. 1990. Absorptive Capacity : A New Perspective on Learning and Innovation. *Administrative Science Quarterly.* 35: 128-152.

Freeman, Linton C. 1979. Centrality in Social Networks: Conceptual clarification. *Social Networks*. 1: 215-239.

Galaskiewicz, Joseph and Burt, Ronald S. 1991. Interorganization Contagion in Corporate Philanthropy. *Administrative Science Quarterly*, 36, 88-105.

Granovetter, Mark.1982. The strength of weak ties: a network theory revisited. In Peter V. Marsden and Nan Lin (eds.), *Social Structure and Network Analysis*. Beverly Hills, CA: Sage Publications, 105-130.

Granovetter, Mark. 1985. Economic action and social structure: the problem of embeddedness. *American Journal of Sociology,* 91: 481-510.

Hutchins, Edward. 1991. The social organization of distributed cognition. In L. B. Resnick, J. M. Levine & S. D. Teasley (ed), *Perspectives on Socially Shared Cognition*. Washington DC: American Psychological Association.

Håkanson, Håkan. 1989. *Corporate Technological Behaviour. Co-operation and Networks*. London: Routledge.

Pattison, Philippa. 1994. Social cognition in context: Some Applications of Social Network Analysis. In Stanley Wasserman and Joseph Galaskiewicz (eds.), *Advances in Social Network Analysis: Research in the Social and Behavioral Sciences*. Thousand Oaks, CA: Sage Publications, 79-109.

Stephenson, K. and Zelen, M. 1989. Rethinking centrality: Methods and applications. *Social Networks*. 11: 1-37.

Teece, David J. 1987. Profiting from technological innovation: implications for integration, collaboration, licencing, and public policy. In D.J. Teece (ed.), *The Competitive Challenge*. Cambridge, MA: Ballinger Publishing Company.

Walsh, James P. 1995. Managerial and Organizational Cognition. *Organization Science*. 6: 280-321.

Wasserman, Stanley and Faust, Katherine. 1994. *Social Network Analysis: Methods and Applications*. Cambridge: Cambridge University Press.

Virtual Teams - Constrained by Technology or Culture?

Dr. Francis Hartman

NSERC/SSHRC Chair
Project Management Specialization
The University of Calgary
2500 University Dr. NW
Calgary, Alberta
T2N 1N4

Connie Guss

Ph.D. Student
Project Management Specialization
The University of Calgary
2500 University Dr. NW
Calgary, Alberta
T2N 1N4

This paper provides a preliminary view of a new era of organizational investigation into virtual organizations and their functional units, virtual teams. The question posed is whether a shift to virtual organization is constrained more by technology or by corporate culture. This preliminary analysis of existing literature is a catalyst for future empirical research in virtual organizations and teams, as defined in current literature and research projects. Discussion of key factors for success and known technical and cultural challenges provide some practical ideas for making virtual teams work. A preliminary conclusion on the basis of a literature review suggests that the social and corporate cultural barriers are more significant than technological barriers in promoting the growth of virtual teams. The paper concludes with some issues that may be faced and resolved over the next decade.

Virtual Organizations and Teams

Electronic networking and cyber-communication is fueling unique organizational cultures that are now being discussed in the literature. These new ad hoc forms are often described by their organizational structures under such labels as spiderís webs (Reich, 1991); New Organizational Forms (NF) and Management Technologies (MT) (Tucker, Meyer and Westerman, 1996); multi-divisional or M-Forms (Bartlett and Ghoshal, 1996); modulars, clusters, learning networks, perpetual matrix, spinouts or virtual corporations (Bartlett and Ghoshal, 1989; Miles and Snow, 1986; Quinn, 1992; and Rodal and Wright, 1993). Crossroads (1996) defines virtual organizations as those whose workflow communications are conducted across publicly available media to allow geographically independent operation of all aspects of its internal affairs. At the heart of each of these constructs is a common theme - a virtual organization consists of individuals working toward a common goal, but without centralized office buildings, physical plant, or other characteristics of a traditional industrial age corporation (Rodal and Wright, 1993). A virtual corporation can marshal resources that include customers, suppliers, competitors and independent entrepreneurs to quickly exploit brief windows of opportunity.

The functional units with virtual organizations are flexible teams that quickly and professionally execute multiple projects, anywhere and anytime. These virtual alliances have many labels (Jessup, 1996; Mezias and Glynn, 1993; Knoll and Jarvenpaa, 1996). Guss (1996) describes virtual teams as a temporary network of independent professionals, separated by geographic, temporal and psychological distance, whose use of telecommunications tools for business communication is interdependent, to satisfy the business requirement of sharing skills and working to meet a common goal. The common element in definitions and discussion of virtual teams is that there is no traditional context to participate and evaluate social

0-7803-3552-X/96/$5.00 © 1996 IEEE

interaction and communication of project team members and their organizations. The emphasis of research needs to shift to investigate the nature of the social relationships of organizational cultures rather than the enabling technologies to determine communication flows and barriers in virtual settings.

Culture and Technology

The literature pushing appropriate technological capability to solve organizational problems far outweighs discussion of cultural barriers in developing virtual organizations and virtual teams. As early as the 1960s researchers such as Chandler noted that there were ìpsychological hazards of adjusting to new ways, new tempos, and new dutiesî (1962, p. 320). However, Flower (1995) notes that traditional emphasis is still on overcoming technological barriers to bring about swift, clean and managed shifts to virtual teams and alternate organizational forms. However, the process of letting go, stopping what is done, or doing something completely different, has not made it into the vocabulary of corporate management (Flower, 1995).

Schonís (1967) explanation of organizations as biological organisms that are actively struggling to maintain stability through ëconservative dynamismí is an excellent model for todayís virtual structures. This organic model clarifies the reasoning behind the need for paradoxes in shifting corporate culture to facilitate technological change, at the same time as causing destruction to existing organizational states. Traditional organizations are not committed to the radical change needed to embrace the paradox, and they remain in a vulnerable position of being resistant to change. Corporations have to learn that chaotic states afford the greatest avenue for change so that organizations can continually reach what Flower (1995)calls ìlocal optimumsî, or false peaks, where they can not grow without losing, changing or transforming what they already have. The effectiveness of ìorganic organizationsî for project teams has been demonstrated through testing of a new approach to project management developed at The University of Calgary (Hartman 1995).

At a project level, project managers seek technical capability and software rather than answers about what processes of communication help expose and overcome limitations and inefficiencies on the job. Technology is important in enabling a virtual team to communicate, but Demingís teachings clearly indicate that improvement in the quality of projects demand improvements in processes. One of the key process improvements that the engineering and construction industry needs to make is the area of communication to facilitate the transfer of knowledge and needs between teams and projects. In the future, the industry will find it more difficult to rely on tacit knowledge (on the job know-how) of organizations or individuals in virtual environments.

Techno-Social Challenges

Some techno-social and cultural challenges are presented in the following section. The topic is vast, and this paper only provides a glimpse of some of the more prominent problems. Virtual organizations provide unique opportunities to significantly reduce cost structures and utilize the publicly accessible infrastructure of the global networks (Crossroads, 1996). However, barriers do exist, and one of the most significant is perception of the effectiveness of communication technology. Newer theories are beginning to test the logic that computer-mediated communications reduce personal influence. Recent research by Walther found that anticipation of future interaction (a team knew if it would work with a client again) accounted for the differences between use of virtual communication and face to face (f+f) interaction on ìthe immediacy, similarity, composure and receptivity of group membersî (1996, pg. 12). As well, researchers are also beginning to find that electronic media can facilitate ëhyperpersonalî (better than f+f) communication (Walther, 1996, Reich, 1991; Knoll and Jarvenpaa, 1996). Technical limitations of software and videoconferencing are briefly described below to provide a glimpse at some common problems.

One example of a socio-technical limitation is that traditional Group Decision Support System (GDSS) software is developed for same place, same time teams and research is lacking in its effectiveness in virtual teams (Jessup, 1996). Jessup suggests that traditional GDSS tools come close, but are not ideal for virtual teams. These tools need to be better suited for distributed group member use, and they need to be facilitator-less and easily accessible.

Another example of socio-technological limitation seen in public videoconferencing is that salient cues (those that stand out from others) are harder to pick up than in f+f communication because the communication is ëhyperí formal. That is, responses are brief, structured and very formal (Storck and Sproull, 1995). As a result, participants tend to rely less on task competence and more on communication competence (Storck and Sproull, 1995). As well, technological limitations such as low bandwidth (capability to transfer amounts of information), slow video frame rate, and poor resolution contribute to difficulties in application of existing technology to virtual teams. Interestingly, Hudson (1995) reports that userís perception (rather than actual visual quality) of communication quality is most closely linked to frame rate.

Similarly, desktop videoconferencing suffers from problems, with bandwidth as a large bottleneck. Bandwidth limitation is a serious technological factor because it takes about 17 minutes to transfer a megabyte file over a telephone line (Hudson, 1995). Desktop videoconferencing is an option that offers use of Integrated Services Digital Networks (ISDN) or Ethernet channels. The former is a circuit switched bandwidth not shared by all users who want to access it, and the later is packet switched whereby the bandwidth shared by all users. Although ISDN is better for videoconferencing, it can only handle a limited number of frames per second, giving lower quality ëfreeze frameí communication.

Cultural Challenges

An organization is primarily a social structure that is motivated by external economic objectives and whose culture is created by a set of interlocking behaviors that, repeated over time, become expected norms (Bartlett and Ghoshal, 1996). In a virtual setting, there is no organizational context, so that the centrifugal forces that drive entrepreneurial units may result in team fragmentation, isolation and inter-unit competition, and blockages in the flow of knowledge (Bartlett and Ghoshal, 1996).

Because of its importance in organizations, teams are the key focus for research on the impacts of technology on communication (Poole and Holmes, 1995). Although numerous communication technologies exist, few guidelines are targeted at building virtual teams. Knoll and Jarvenpaa (1996) found that recruiting is difficult, and long lead times are needed to create virtual teams. Once virtual teams form, researchers find that participation is the most important factor in collaboration and learning (Knoll and Jarvenpaa, 1996); Leidner and Jarvenpaa, 1994). This is critical in a culture of reciprocity and information sharing characteristic of the Internet (Norton, 1994).

To participate in accomplishing goals, virtual teams must learn the social and technical aspects of communicating by means other than f+f interaction. Teams tend to resist communication by new and unfamiliar means in cultures that are conservative and risk aversive. However, Reinsch and Shelbyís (1996) results found that only 5 out of 344 responses concerned a need for more or better information to communicate effectively, and 51 responses pointed to the need for more self-confidence to improve communication.

Another important factor in virtual teams is cohesion. Knoll and Jarvenpaa (1996) found that virtual teams working toward a common goal (that was also challenging) showed more satisfaction with the process and the final product than those with no common goal. Also, teams with no common goal experienced more conflict, (although all levels of conflict were all very low). Interestingly, conflict concerned process issues (how the team functioned) but the teams attributed the conflict to product issues. Recent work by Hartman (1995) on self-managing teams with Amoco and PanCanadian found that focusing an organic team on common objectives accelerated team formation and significantly reduced conflict. Four factors were deliberately introduced to the team culture to encourage this. These were: a propensity to take risk, openness of communication, ownership of oneís job (or responsibilities) and through those three, trust. The result was the effective elimination of the ìstormingî phase of team building.

Even in practice in real time projects, team difficulties are often social, rather than strictly technical in nature. A recent study with one of Mobil Oil Canadaís facilities engineering projects in Calgary, Alta. supports this finding. The data show that of the 203 individual written comments collected as part of the post project assessment, communication and team processes were the root causes of more than half of the problems, although initially the team thought the technical problems were the primary cause (Wells, Solinger, Guss and Dilger, 1995).

Virtuality in the Future

Global trends to outsource work and downsize corporate employee pools combined with the widespread availability of telecommunications devices continues to push organizations into ëvirtual employmentí. In the future, a shift to virtual teams will be a consequence of the need for high speed communication of new ideas among experts world-wide, for a competitive edge. Corporate metaphysicians can argue that successful virtual enterprises like Visa International have almost no tangible assets, but its transactions still total 1/3 of a trillion dollars per year (Stewart, 1995).
In the future, internal capabilities to transfer communication loads (information) through communication tools (load transferring devices) will overtake the practice of configuring resources and capital for competitive advantage. Knowledge management will refer to use of loads, and the flow and stocks of knowledge where a stock can be altered by derivation and acquisition of knowledge (Holsapple, Johnson, and Waldron, 1996). The computer will then be considered a knowledge worker that possesses its own ability to represent and process knowledge.

Summary

Value creating intangibles such as knowledge and speed of information transactions are still ignored by financial institutions, and many other industries and businesses. Virtual organizations provide a unique opportunity to advance geographically independent entities in a web of interconnected industries, saving on physical plant and eliminating unwanted and unnecessary travel. Known social and cultural barriers rather than technological barriers appear to more significant in slowing down the use and function of virtual organizations and teams. However, a shift in thinking to view an organization as a living organism will build a better understanding of how corporations may have many cultures in a virtual environment. What is more certain is that while inheriting many cultural elements and problems from existing organizational forms, virtual enterprises will support their own distinct cultures (Norton, 1994).

References

Bartlett, C., and S. Ghoshal. Managing Across Borders: The Transnational solution. Harvard Business School Press: Boston, 1989.

Bartlett, C. and S. Ghoshal. Beyond the M-Form: Toward a Management Theory of the Firm. http://www.gsia.cmu.edu/bosch/bart.html. 1996.

Chandler., A. D. j.r. Strategy and Structure. The M.I.T. Press, Cambridge, MA, 1962.

Crossroads. About Virtual Companies. http://www.idiscover.co.uk/rodz/virt_co.htm. 1996.

Flower, Joe. ìThe Structure of Organizational Change: a conversation with Kevin Kelly.î http://dazzelm.com/is/constantfluxwhitepaper.html. 1995.

Guss, Connie L. Communication Flows in Virtual Teams that Enhance the Success of Projects in the Telecommunications Industry. Research Proposal, The University of Calgary Project Management Specialization, 1996 April.

Hartman, Francis. ìA Case Study of the application of Self-Managed Project team principlesî. Proceedings, PMI Symposium, New Orleans, 1995.

Holsapple, Clyde, Johnson, Linda Ellis, and Vincent. R. Waldron. ìA Formal Model for the Study of Communication Support Systemsî. Human Communication Research, 22(30): 1996.

Hudson, Rhett. DT-5: Enabling Technologies. Desktop Video Conferencing. http://fiddle.ee.vt.edu/succeed/videoconf.html. 1995.

Jessup, Leonard M. (ed). Pushing the GSS Envelope: distributed collaboration for virtual teams on the World Wide Web. http://ezinfo.ucs.indiana.edu/~ljessup/gwcent3.html. 1996.

Knoll, Kathleen, and Sirkka L. Jarvenpaa. Leaning to Work in Distributed Global Teams. http://uts.cc.utexas.edu/~bgac313/hicss.html. 1996.

Leidner, D. E. and Jarvenpaa, S.L. The Use of Information Technology to Enhance Management School Education: A theoretical view. Working paper, Bayor University, 1994.

Mezias, S.J., and M. A. Glynn. ìThe Three Faces of Corporate Renewal: Institution, revolution, evolutionî. Strategic Management Journal, 14: 1993.

Miles, R.E., and C.C. Snow. ìOrganizations: New concepts for new formsî. California Management Review, 28: 1996.

Newhagen, John. and Sheizaf Rafaeli. Why communication researchers should study the Internet: a dialogue. http://shum.huji.ac.il/jcmc.vol1/issue4/rafaeli.html. 1996.

Norton, Tim. ìThe Internet and the Use of Global Computer Networks: an investigation of their culture and its effects on new usersî. http: //foo/curtin.edu.au/Thesis/Chap6.html. 1994.

Poole, Marshall Scott, and Michael Holmes. ìDecision Development in Computer-Assisted Group Decision Makingî. Human Communication Research, 22: 1995.

Quinn, J. B. Intelligent Enterprise. Free Press: New York, 1992.

Reich, R.R. The Work of Nations. Vintage Books: New York, 1991.

Reinsch, Lamar, and Annette. N. Shelby. ìCommunication Challenges and Needs: Perceptions of MBA Studentsî. Business Communication Quarterly, 59 (1): 1996.

Rodal, Alti, and David Wright. ìA Dossier on Partnershipsî. The OPTIMUM: Journal of Public Sector Management, 24 (3): 1993.

Schon, Donald A. Technology and Change: The New Heraclitus. Pergamon Press, 1967.

Stewart, T. A. (1995). Trying to Grasp the Intangible. http://pthfinder.com/@@NeV6pxAu7AQGTo/fortune/magazine. 1995.

Storck, John and Lee Sproull. ìThrough a Glass Darkly: what do people learn in videoconferences?î Human Communication Research, 22(2): 1995.

Tucker, Mary L., Meyer, Dale, G., and James W. Westerman. ìOrganizational Communication: Development of Internal Strategic Competitive Advantageî. The Journal of Business Communication, 33 (1): 1996.

Walther, Joseph B. ìComputer-Mediated Communication: Impersonal, Interpersonal and Hyperpersonal Interactionî. Communication Research, 23 (1): 1996.

Quality Management in Software Development Process: An Empirical Model

József Györkös
Ivan Rozman
Romana Vajde Horvat
Marjan Hericko

University of Maribor
Faculty of Electrical Engineering and Computer Science
Smetanova 17
SI-2000 Maribor
Slovenia
e-mail: gyorkos@uni-mb.si

A methodology/courseware for introducing such a quality system in software development organizations is described in this paper. The authors used the ISO standardization (ISO 9000-3: Standard for QA in design/development, production, installation and servicing) as the basis of the methodology and refined it with well known related process models. It is not limited to the standard but also concerns human factors regarding the continual improvements toward the quality system within training programs and group-work. The methodology is supported by an integrated software product which enables software process assessments and goal definitions and offers a hypertext based software-process-improvement knowledge base.

Introduction

The PROCESSUS project (Software Process Improvement: Assessment and Introduction of a Quality System) has been initiated by the cooperation of the research group Laboratory for Informatics and by the Slovenian local industry. A heterogeneous group of researchers supported also by governmental funds is not aimed only to satisfy the needs of the participating organizations but also to provide the community with a methodology that can be applied ("reused") for the wide range of potential organizations seeking for quality of software process. The methodology should be supported by a software tool that can give an interactive hypertext-based learning and advisory support for those involved in software process improvement.

The partners from the local industry participating in PROCESSUS project can be divided in three groups (large scale software development, large in-house development, small scale software development) each with its own specific concerns. The final methodology should be flexible enough to fulfill the needs and ambitions of all three different types of organizations. The methodology allows group-work of the participating organizations not only because of the resource optimization but also because of the positive effect of experience exchange between the participants.

Why not apply the existing approach for the software process quality improvement? In related works ((Haase, 1995) and (Paulk, 1995)) we can find excellent guidelines and suggestions but they are without personal contacts with experts for particular method only of a minor value. Common deficiencies attributes of the related and internationally available approaches are:

Language. In non-English speaking countries English is not applicable as a communication language in the scope of the organization's quality system. However, for a few very narrowly specialized software groups it is partially acceptable to use English but in the broader view this is not sufficient - the quality system always fits in the organization as a whole with all the personnel.

ISO 9001. The final goal of the participating organizations and of the entire Slovenian community is to follow the ISO standardization. Authors of existing quality

0-7803-3552-X/96/$5.00 © 1996 IEEE

models began to proclaim the ISO-compatibility only in the last few years (Bramford, 1993) - the PROCESSUS project was initiated two and a half years ago (Rozman, 1994). An exception is the TickIT model (TickIT, 1992) which is built on the basis of ISO 9001.

Costs. Results of a cost-benefit study and a SWOT (Strengths, Weaknesses, Opportunities and Treats) analysis showed mid- and long- term benefits of establishment of a methodology for quality system introduction at the national level. Comparability with the related works ensures wider acceptance of the proposed methodology.

Quality Models - The Quality Manual

The Quality Manual comprises the organization from all three levels of the quality system (MSQH, 1991): the top level (policy), the middle level (tactical issues) and lowest level (operational issues). The purpose of the Quality Manual and the documented quality system is twofold (MSQH, 1991): (a) to instruct the employees and (b) to inform the customers. It is not necessary, even more it is impossible to keep all the operational details in the Quality Manual. First, the operational details may vary from project to project and, not at last, the company's know-how can also be incorporated in them. However, in the Quality Manual there must be a clear set of references to the appropriate documents, such as procedure guides, in-house standards or detailed work instructions.

PROCESSUS Methodology

In Figure 1 the PROCESSUS Algorithm is shown. The aim of the process as a whole is to maintain continuous contacts between the consultant and the organization seeking for quality improvement. We can imagine the importance of this contacts knowing that the support of the management has the crucial role in the commitment of the quality systems issues.

According to our experience (Györkös, 1994) the first iteration takes 18 months - the first assessment is made and the Quality System is established at the declarative level.

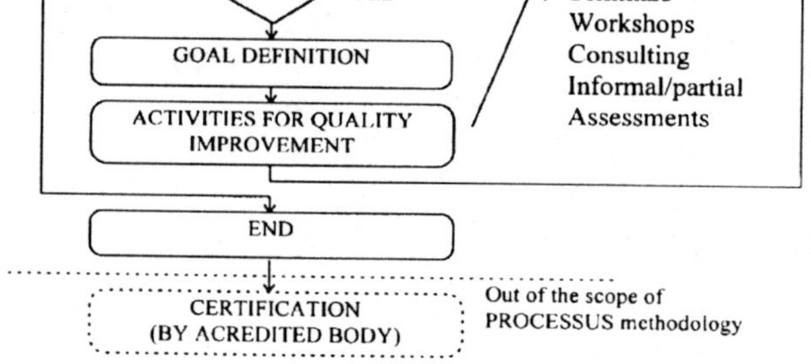

Figure 1 The PROCESSUS Algorithm

Assessments

The first principle of software engineering (out of 30) given by Davis (Davis, 1994) is: make quality a number. Similar requirements for measurable quality are presented in experience based references literature (Gilb, 1993): Among different approaches to quality measurements which are usually concerned with the quality of the final product (low development cost, elegant code, good response time etc.) we choose the quality of the software process expressed with the level of maturity - accomplishment of requirements contributed by different quality models.

The assessment is supported by a software tool which has the following basic functionality: (a) questionnaire based on ISO, SEI CMM and on the combination of both of them, (b) follow up questions for each basic question, (c) quantitative assessment, (d) outline for descriptive assessment, (e) goal definition support. The tool can be used for self assessment or by the auditing groups (Györkös, 1994).

Seminars

Seminars are a means of acquiring basic knowledge about the quality system related topics. Regardless of their previous experience and knowledge participants are provided with systematic issues which enhance their understanding of the essence of the quality system and for the stepwise building of it. Titles of the seminars are depicted in the first row of Table 1. The subsequent schemata of key process areas in SEI Capability Maturity Model inspired the content and the schedule of the seminars to be held within PROCESSUS.

Another reason for the seminars at the "ex-cathedra" level lies in the heterogeneous auditorium - more or less experienced practitioners, engineers or managers which are hard to be pulled away from the need for immediate writing of the quality manual. The seminars utilize the group and introduce an inevitable discipline needed for further work at the workshops. At the seminars the creativity of the participants can be expressed during the short "my view" and "my experience" breaks. The outcome of these breaks is subject to discussion by all participants.

Workshops

In the first workshop the assessment method is introduced to the participants. The method of work is rehearsal. Reading questions from the questionnaires, follow-ups, explanatory comments and understanding of the quality system structure are the aims of this workshop. Next four workshops are aimed towards the development of the first draft of the Quality Manual (QM). The method of work at these workshops (WS2-WS5) is as follows: (a) one week before the workshop participants are provided with a written Quality Manual outline; (b) at the workshop participants are provided with the text-processor template of the already given outline of the Quality Manual and they are asked to write the first draft of the document, (c) required comments and suggestions are given and an extensive experience interchange appears between the participants from different organizations, (d) if needed, brainstorming-like techniques are introduced to solve ambiguous problems.

Table 1 Titles of seminars and workshops

ID	Seminars (SE)	ID	Workshops (WS)
SE1	Introduction: Quality Models	WS1	Quality Assessment Method
SE2	ISO 9001 and ISO 9000-3 Standards	WS2	QM (part 1): The Quality System
SE3	Requirements Management	WS3	QM (part 2): Lifecycle Activities 1
SE4	Project Management	WS4	QM (part 3): Lifecycle Activities 2
SE5	Design	WS5	QM (part 4): Supporting Activites
SE6	Testing, Validation and Verification	WS6	Review of QM's
SE7	Configuration Management		
SE8	Metrics		
SE9	Interpersonal Communication and Workflow		
SE10	Object Oriented Techniques (AS)		

AS: Advance Seminar; QM: Quality Manual

Consulting

After attending the workshops participants finish their Quality manuals, prepare the supplementary documents and stepwise introduce the quality system among all members of the organization. The consultancies are given in three directions:

Refinement of the Quality Manual. However, there are alternatives in the QM templates given to the participants, but all details can never be accomplished. On-site consulting and discussions are needed to optimize the content of the Quality manual and find the balance between the level of common and formal issues. The latter enables the Quality Manual to be used as a reference.

Preparation of the Supporting Documents. Regardless of the organization's maturity level there are usually more documents available which can serve as a part of the "Supporting Documents" appendix of the Quality Manual (usually these documents are templates for: requirements specification, black-box testing, failure reports, configuration management etc.). These documents have to be updated and adopted, and the redundancy should be avoided. CASE tools are also a great source of standardized documents - by introducing the quality system the existing CASE-generated documents can be retained and incorporated into the documentation. A case study (Györkös, 1995) shows that the degree of compliance between the observed CASE tool and the Quality system was full in 39%, partial in 29% and none in 20% of CASE-generated documents. Where possible, computer aided support for documents management should be introduced and maintained.

Introduction of the Quality System. Our experience with the introduction of the Quality System at all levels of organization are limited to the actual projects. Awareness of the Quality System should be increased by motivational activities of the top management (comparations, common interests, sustainable effort distribution). At the operational level the Quality System is introduced by disciplined use of documents (Quality Manual and Supporting Documents) and by the Quality System incessant adoption and improvement. If software engineering is to be real-engineering discipline, the workflow for software process is a target that must be achieved.

Conclusions

The PROCESSUSS Methodology is verified at the level of the first iteration of the model depicted in Figure 1. The declarative level of the Quality System is achieved by 4 out of 5 partners. Important steps toward implementation have been done as parallel activity by the preparation of supporting documents for the Quality Systems, for which reason the expected length of the second iteration in the model should be not longer than 8 months. After this period the organization is fully prepared to apply for the official certification by authorized institution.

We are now performing the second iteration of the PROCESSUS project. Activities are being updated regarding the lessons learned from the first iteration and first comparations in quality improvement process can be done. The compatibility of the proposed methodology with some well known directions and approaches (e.g. SPICE (Kujava, 1995)) is an important issue to its use in the future. Our realistic but still optimistic view is that the PROCESSUS methodology has also to mature.

References

Bramford, R.C., Deibler, W.J., "Comparing, contrasting ISO 9001 and the SEI capability maturity model", IEEE Computer, October 1993.

Davis, A.M., "Fifteen Principles of Software Engineering, IEEE Software", Volume 11, November 1994.

Gilb, T. and Graham, D., Software Inspection, Addison-Wesley Publishing Company, 1993.

Györkös, J., "A Support for Auditing the Software Process", Austrian-Hungarian Seminar on Software Engineering, University of Klagenfurt, , pp. 7.1-7.4, April 1994.

Györkös J., Rozman I., Vajde Horvat R., Herièko M., "Compliance of a CASE Environment and of ISO 9000-3", ISCO - 3; IFIP WG 8.1 Conference, Marburg, 1995.

Haase, V.H., "Measurement of Software Process Quality: Assessments and Self-Assessments", Proceedings of International Conference on Software Quality ICSQ'95, University of Maribor, 1995.

Kujava, P., "SPICE: The Software Process Assessment Model", Proceedings of ESI-ISCN'95: Measurement and Training Based Process Improvement, European Software Institute, Vienna, 1995.

MSQH - Modeling a Software Quality Handbook, STRI TS2 - Icelandic Council for Standardization, Reykjavik, Iceland, 1991.

Paulk, M.C. et al, The Capability Maturity Model: Guidelines for Improving the Software Process, Addison-Wesley Publishing Company, 1995.

Rozman, I., Vajde-Horvat, R., Györkös, J., "United View on ISO 9000 Model and SEI CMM", IEEE International Engineering Management Conference, Dayton North, IEEE Press, 1994.

TickIT, Guide to Software Quality Management System Construction and Certification, Department of Trade and Industry and the British Computer Society, Issue 2.0, 1992.

Identifying Critical Factors Impacting Virtual Work Group Performance

M.E. Hacker and B.M. Kleiner
(540) 231-4596

Department of Industrial and Systems Engineering
Virginia Polytechnic Institute and State University
Blacksburg, VA
24061-0118

Virtual teams are an increasingly frequent response to the new competitive environment. Technological advances and global competition have moved virtual teaming from the "good idea" stage to a critical strategy for many organizations. Our understanding of critical factors relative to virtual teams is limited as very few studies have been completed. Virtual teams are characterized as having high social complexity. Jointly optimizing the social and the technical organizational subsystems has been demonstrated by sociotechnical theory to improve overall performance. Consequently, technical system interventions may improve the virtual team's ability to leverage its social system complexity. This paper explores a sociotechnical approach to investigating technical system interventions to enhance virtual team performance.

Introduction

Technological advances have made it possible for anyone, globally to become your competition overnight. The new competitive environment is being called 'agile' where agility refers to the ability to thrive in an environment of continuous and unanticipated change, (Goldman, 1994). Previous unidimensional strategies of low cost, high quality or fast cycle times are insufficient. Now, even performing competitively in all of these areas simultaneously does not guarantee success. In the new agile environment we see highly individualized products, products with platform upgrades, short market life cycles, producing to order and prices being driven by consumer perceived value (Goldman, Nagel and Preiss, 1995). Our traditional organizations are struggling to change with this dynamic environment. Traditional organizations were structured for long production runs of identical products within stable environments. This is no longer the most likely scenario. Today, "companies must organize their operations around the fact that in more and more markets customer requirements are changing more rapidly and more unpredictably than ever before (Goldman, 1994 p.2)".

A cornerstone to being competitive in an agile environment remains leveraging people and information (Goldman, 1994). What is different though, is *how* people and information are leveraged. The new structure is called the 'virtual enterprise'. "The virtual enterprise is a reconfiguable, computer-networked, customer solutions delivery system (Kovac, 1994 p.4)". The term 'virtual' is derived from computer systems to describe 'virtual memory'- which is computer memory that shifts and varies depending on the situation (Kristof, Brown, Sims and Smith, 1995). This same characteristic is now required of organizations, thus these of 'virtual' to describe organizations. The building block of a virtual enterprise is the virtual work group or team. "Virtual teaming is an optimal way to work in the current environment of time compression, distributed resources, increasing dependence on knowledge-based input, premium on flexibility and adaptability and availability of electronic information and communication through networks (Grenier and Metes, 1995 p.8)". Unfortunately, most organizations still do not know how create effective local work groups (Tompkins and Associates, 1995; Kepner-Tregoe, 1996). There is a growing need to understand the critical factors in virtual work group performance. This paper presents a conceptual model as a starting point for identifying these factors.

0-7803-3552-X/96/$5.00 © 1996 IEEE

Sociotechnical systems theory and virtual teams

Our current understanding of effective collocated work group effectiveness will not greatly enhance our understanding of virtual work groups. This is because most report disappointment with their work groups' performance (Tompkins et al., 1995) and most employees' experiences in work groups are unsatisfying (Kepner-Tregoe, 1996). Current scholarly research is limited so little is actually known about how effective work groups operate (McGrath and Kravitz, 1982).

A useful framework for studying virtually work groups is derived from sociotechnical systems theory. Sociotechnical research has demonstrated in the field that jointly optimizing the social and technical subsystems of an organization leads to best overall performance results (Pasmore, 1988). "The sociotechnical approach seeks to utilize those capabilities of the social system which are distinctively human, for example, the ability to be flexible and adaptive, to make judgments and decisions. Only by matching these human capabilities with the complementary characteristics of the technology can the best overall results be obtained (Hill and Emery, 1993 p.279)". Virtual teams by their very nature are the organization's response to increasing its flexibility and adaptivity. Since virtual work group members accomplish their work through technology linkages, the sociotechnical framework, with its emphasis on both technical and social subsystems, is a nice fit for research on virtual work groups.

The social subsystem

The social subsystem is defined as the people in the work group and their relationship to each other and others within the broader organization (Pasmore, 1988). Virtual work groups are temporary, self-managed teams with members from separate corporate organizations which are linked through technology to respond to a specific business opportunity (Kristof, 1995). Although a simple concept to understand, it is not a simple concept to implement. An inherent characteristic of virtual teams is their complexity. From sociotechnical research we understand that group complexity can impact group performance. Complexity is the degree of differentiation and integration (e.g., structural communication, control and coordination mechanisms) the work group experiences (Hendrick, 1986). Differentiation is further classified as: horizontal- having multiple organizations in the group, vertical- having different management levels represented and spatial- having members that are geographically dispersed. Additionally, the ability for team members to be located internationally increases the likelihood of higher levels of complexity due to individual differences such as ethnic and cultural differences. Virtual work groups are high in complexity since each type of differentiation typically exists.

Work groups have two demands: task or production demands and social or emotional demands. Complexity makes it difficult for work groups to satisfy either demand. The multiple perspectives often strain the social demand, creating intragroup dynamics that make achieving the task demand difficult (McGrath and Hollingshead, 1994). Intragroup dynamics include behaviors that create inefficient meeting processes and impact output effectiveness (e.g., nonparticipation, inattention, domineering behaviors, lack of cooperation, lack of trust, etc.). Virtual work groups can only be successful where these behaviors are severely limited (Goldman, 1995). Addressing these type of behaviors in collocated groups has not been easy. For virtual work groups it will be even more difficult due to the higher inherent complexity.

The task demand is further challenged by the virtual aspect of the environment itself. Most people are not accustomed to working in close collaboration with people with whom they are not collocated. For many, the new methods of communicating and collaborating are uncomfortable. "Many people are reluctant to use the electronic infrastructure to do their work, which may be highly communicative, knowledge intensive and ego-involved. They communicate best face-to-face, like to persuade over the phone, are resistant to wide-spread electronic information sharing and don't like the planning and formality that are required to make virtual processes work (Grenier, 1995 p.211)." The virtual environment represents a unique challenge to the social subsystem. To identify critical factors impacting virtual work group performance we must evaluate factors that enable the group to excel. Determining the optimal level of complexity is not the issue. The issue for researchers and business leaders is to identify the critical elements which enable complex groups to optimize performance. Sociotechnical theory directs us (in this situation) to the technical subsystem. Sociotechnical theory guides us to technical subsystem interventions to enable virtual organizations to leverage their inherent complexity of the social subsystem and optimize the system.

The technical subsystem

The technical subsystem includes more than just technology, it also includes the rules, procedures and policies used by organizational members to transform inputs into outputs (Pasmore, 1988). Perhaps, collocated work groups are not meeting performance expectations and team members are dissatisfied with their group experiences because of the lack of attention to relevant sociotechnical variables. Although virtual teams require the use of information technology to exist, little is known about how the technology can be configured to optimize work group performance. Virtual work groups use computer-based group decision support systems as a mechanism for structuring group processes. Structure is necessary for coordinating, integrating and controlling the activities of the various organizations working together (Grenier, 1995). There is little empirical evidence indicating how much or what type of team process structure is optimal.

Conceptual model

McGrath (1991) has defined a framework, defined by modes of the processes that all work groups engage in:

Mode I	inception and acceptance of a project (goal choice)
Mode II	solution of technical issues (means choice)
Mode III	resolution of conflict (policy choice) and
Mode IV	execution of the performance requirements of the project (goal attainment)

Implementation methodologies link modes together in a systematic manner through defining and structuring the activities within each mode. Little research exists about implementation methodologies. Most scholarly attention has been limited to Mode II in the form of problem solving and decision making research (McGrath, 1991). Most of these studies support the premise that formalizing group processes improves the quality of decisions (Pavitt, 1993). Studies evaluating factors impacting cross functional work groups have found that formalized team rules and procedures are significant to cooperation and performance (Pinto, Pinto and Prescott, 1993). Computer augmented meeting studies have demonstrated that procedures that simplify the handling of information, organize group processes and procedures that enable the group to deal with internal group dynamics whether computer aided or not, improve team performance (DeSanctis and Gallupe, 1987). Anecdotal case studies support the formalization of the implementation methodology as a mechanism for increasing group and organizational performance (Melum and Collect, 1995). The evidence supports the notion that formalizing group processes is critical to improving team interactions and increasing group performance. The benefit of formalization may be most pronounced for complex work groups such as virtual teams. There are few specifications about how much formalization leads to optimal performance. We propose that in order to understand the critical performance factors for virtual teams, we must identify the optimal level of formalization for each of the four phases or modes of team processes.

A conceptual model to guide future research involving virtual work groups is presented in Figure 1. The model includes the team activities and their impact on team interactions. Unlike most small group conceptual models the proposed model does not include input elements (e.g., individual, team, work and, task characteristics). In this model, the complexity dimension is the only defining input characteristic of virtual work groups since by definition, virtual work groups have high complexity across all the traditional input characteristics. This model suggests that formalizing virtual team processes improves team interactions which increases team performance.

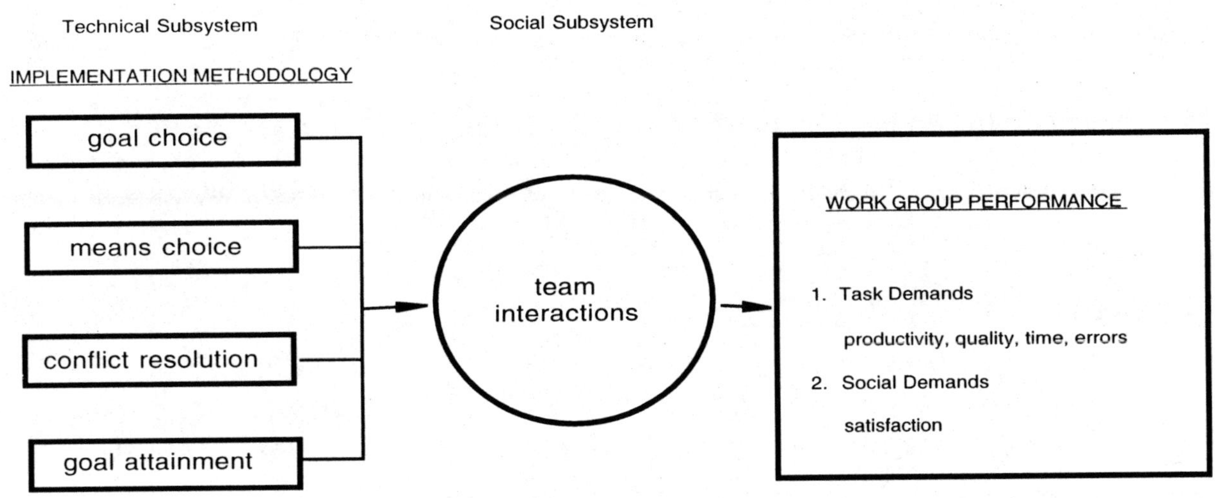

Figure 1. Conceptual Model

Summary

There is a growing need to identify the critical factors for virtual work group performance. The existing body of knowledge about work groups is extensive but studies of virtual teams and field studies are scarce. This paper presented a conceptual model as a starting point for studying virtual work groups. The conceptual model was an extension of the literature which documents the positive impact formalizing of team processes. Since virtual work groups represent high complexity, it was suggested that formalization may represent a key factor in performance and warrants attention as the next step to understanding the critical factors impacting virtual work team performance. This study represents just one of several studies about virtual teams being performed in the Group Decision Systems Laboratory.

References

1. DeSanctis, G. and Gallupe, R.B. "A Foundation for the Study of Group Decision Support Systems". *Management Science.* 33: 589-609, 1987.

2. Goldman, S.L. *Agile Competition- The Next Business Paradigm.* Lehigh University, Bethlehem, PA, 1994.

3. Goldman, S.L. *An Agility Primer.* Lehigh University, Bethlehem, PA, 1994.

4. Goldman, S.L., Nagel, R.N. and Preiss, K. *Agile Competitors and Virtual Organizations: Strategies for Enriching the Customer.* Nan Nostrand Reinhold: New York, 1995.

5. Grenier, R. and Metes, G. *Going Virtual.* Prentice Hall: Upper Saddle River, NJ, 1995.

6. Hendrick, H. "Macroergonomics: A Conceptual Model for Integrating Human Factors with Organizational Design". in *Human Factors in Organizational Design and Management III*, Hendrick, H. Ed. North Holland: New York 1986.

7. Hill, P. and Emery, F. "Toward a New Philosophy of Management". in *The Social Engagement of Social Science: A Tavistock Anthology*, Trist, E. and Huge,M. Ed. University of Pennsylvania Press: Philadelphia, III, 1993.

8. Kepner-Tregoe. *Worker Performance*. Kepner-Tregoe, Inc.: Princeton, NJ, 1996.

9. Kovac, F.J. *Virtual Enterprise Strategy*. The Goodyear Tire & Rubber Company, Bethlehem, PA, 1994.

10. Kristof, A.L., Brown, K.G., Sims, H.P.J. and Smith, K.A. "The Virtual Team: A Case Study and Inductive Model". in *Advances in Interdisciplinary Studies of Work Teams: Knowledge Teams the Creativ the Edge II.*, Beyerlein, M. and Johnson, D. JAI Press: Greenwich, CT, 1995.

11. McGrath, J. and Kravitz, D. "Group Research". *Annual Review of Psychology,* 33: 195-230, 1982.

12. McGrath, J.E. "Time, Interaction and Performance (TIP): A Theory of Groups". *Small Group Research,* 22: 147-174, 1991.

13. McGrath, J.E. and Hollingshead, A.B. *Groups Interacting with Technology.* SAGE Publications: Thousand Oaks, CA, 1994.

14. Melum, M.M. and Collect, C. *Breakthrough Leadership: Achieving Organizational Alignment Through Hoshin Planning.* American Hospital Publishing, Inc.: Chicago, 1995.

15. Pasmore, W.A. *Designing Effective Organizations: The Sociotechnical Systems Perspective.* Wiley & Sons: New York, 1988.

16. Pavitt, C. "What (Little) We Know About Formal Group Discussion Procedures (A Review of Relevant Research)". *Small Group Research,* 24: 217-35, 1993.

17. Pinto, M.B., Pinto, J.K. and Prescott, J.E. "Antecedents and Consequences of Project Team Cross-functional Cooperation". *Management Science,* 39: 1281-1297, 1993.

18. Tompkins and Associates. *Teaming for Excellence Survey.* Tompkins Associates Incorporated, Raleigh, NC, 1995.

Virtual Organizations - an Opportunity for Learning

Dr. Francis Hartman

NSERC/SSHRC Chair
Project Management Specialization
The University of Calgary
2500 University Dr. NW
Calgary, Alberta
Canada T2N 1N4

Eur Ing Dr. Rafi Ashrafi

Project Management Specialization
The University of Calgary
2500 University Dr. NW
Calgary, Alberta
Canada T2N 1N4

As the idea that new enterprises may be formed and operated successfully without the normal physical constraints and trappings of office buildings, incorporation organization charts and so on takes hold, so other opportunities will arise. One of these is the opportunity to learn and improve through the adoption of better practices from hitherto alien cultures. This paper presents just some of the intriguing notions that arise from the findings of a recent pilot study on project management practices in seven different industries. The industries investigated were product development, utilities, oil and gas, entertainment, infrastructure (traditionally government), systems development and construction.

The findings include a number of potential areas for significant process improvement through the adoption of ideas and practices of one industry by another. The opportunities for such improvements will be enhanced through virtual organizations that span different industries and cultures. This is just one more area where competitiveness may be enhanced.

Background

Virtual teams and organizations are with us to stay. The organizations that will benefit most from this shift in how we work will be those that learn. And those are the ones that will overcome the barriers to learning. Before we discuss the understanding of the barriers in learning organizations, let us first define what learning organizations are. A very comprehensive definition of a learning organization is given by Galvin (1993) in which he stated that "A learning organization is an organization skilled at creating, acquiring and transferring knowledge and at modifying its behaviour to reflect new knowledge and insights." As rightly recognized and emphasized by Galvin, acquiring knowledge and wisdom is all well and good, but it needs to be disseminated and translated into action and changed behaviour before it does anybody any good.

Kiernan (1995) recognized that in the 21st century, successful companies will be effective learning organizations. Barriers to learning in a company come from all directions up and down, inside and out. Kiernan identified three types of organizational barriers. First, there are horizontal barriers, that so often prevent collaboration between different departments and functional specialists. Second, there are vertical/hierarchical barriers, which separate senior executives from rank and file employees at the coal face. And the most intractable of all, the corporate boundary itself, that psychological Great wall of China that separates the entire company from other key players in its business ecosystem: customer, suppliers and competitors.

In the current changing global environment, other barriers include geography, ideology and mindset. All these barriers prevent organizations from identifying and

0-7803-3552-X/96/$5.00

using these opportunities for company leverage. To understand what and how we can leverage the learning experiences of virtual teams, we need to understand the main barriers to such learning. Three of the more important of these are described below. These are language, culture and technology. Barriers to learning may be overcome in part through the alignment of team members in a virtual organization.

Alignment of the objectives, needs and processes used on projects is an essential part of achieving success. The term "alignment" in this text means identifying common objectives and using these as a focus. It also means identifying and negotiating compromises on conflicting elements, so that expectations of all parties are brought together with minimal conflict and potential for subsequent misunderstanding or disappointment. To achieve effective alignment requires a high degree of open and effective communication between the participants of a virtual team. This can be tedious and apparently repetitive since all messages need to be fed back to the originator. The principle is the same as that of a double control feedback loop, since without feedback, we cannot be sure that the correct signal has been sent. As more cultures become involved in the team, so the need for hard work on communication grows.

For the purposes of this paper, it is important to understand what is meant by "culture". We refer to the large or subtle differences between the perceptions, experience, knowledge and language of individuals. Some examples of this include cultural differences within an organization (e.g., marketing and engineering), between organizations (e.g., customer, supplier and subcontractor) or between regions, time zones, countries, etcetera.

An interesting side-effect of alignment is the improved understanding of common objectives, development (to some degree) of common language and team norms. Team norms draw from all of the involved cultures, though there is a clear tendency to adopt those of the more dominant (or noisy, or aggressive) ones! The opportunity to learn lies in the adoption of norms as well as in the process of their adoption. The low hanging fruit is the permanent adoption by the real organization of virtual team norms that lead to better performance than current practices. The other area for learning is in the review of options for the selection of the norm. If we identify a better solution, we should adopt it and improve our corporate - or personal - performance as a result.

Examples of cross industry learnings

A study at The University of Calgary has created an opportunity to look at best practices in project management in several quite different industries. Listed below are the industries, what the "different" practice is and how other industries may benefit from it.

INDUSTRY	DIFFERENCE	LEARNING OPPORTUNITY
Live Entertainment	Strong culture and traditions, focus on creativity blended with public deadlines (the show must go on!).	Routinely on time without complex and costly Scheduling (e.g. CPM)! Focus on key issues and let experts take responsibility for their own part.
Oil and Gas Capital Projects	In the North Sea, Owners are presenting their engineers with problems to solve, not solutions to solve - they are using performance specifications, not detailed ones.	Use of industry standards and harnessing the knowledge of practitioners and experts leads to substantial savings.
Commercial Construction	Companies are turning away from confrontation in contracting, building trust as the foundation of working relationships.	Partnerships and alliances are reducing time and cost for project delivery. The CII Cost/Trust relationship demonstrates the effect clearly.
Software Development	Software project managers have developed skills in managing uncertainty.	All projects are uncertain. This is rarely adequately identified at the time of funding approval. Uncertainty analysis and representation can improve decision making on projects.
Engineering Design	Specialization, learning from experience, developing specialist skills and coordination to match.	Software developers (e.g., programmers) rarely work on the same project type more than once. Learnings are lost, experience value is reduced. Specialization can turn this around.
New Product Development	Stage gates are established to ensure that product development is orderly and inappropriate products are weeded out early.	Management of a portfolio of projects when the project investment options are increasing, is easier if predetermined investment criteria can be defined and used to eliminate projects as early as possible to avoid spending on those that will not proceed.

Are learnings transferable? If so, how?

The following simple steps may be used to help determine whether a possible learning is transferable and is worth transferring.

A: What is the underlying principle?
A first step is to identify the issue being addressed and how this is being done.

B: Acid test - would we do this at home?
This may sound crazy, but it is a wonderful test of the efficacy of a management principle. It is easiest to explain by example.

Example A:
Industry practice: use of exculpatory clauses in contracts. Home equivalent: would I buy insurance (risk being transferred by clause) without knowing what the premium is (how much the contractor has added to the cost as a result)?

Example B:
Industry practice: project measurement and control of the right things. Home equivalent: would we use the fuel gauge (cost to date) to determine our speed (schedule performance - earned value)?

C. Does the principle apply in our environment?
Be careful with this one, the easy answer is NO, but in reality it is often YES (but we do not want to admit to it).

D. How do we adapt it to adopt it?
This is where there is no simple answer. But frankly, if we see the advantage of the new or different process, we are likely to find a way!

Barriers in Virtual Team Learnings

There are many barriers to learning. For the purposes of this paper we will address just three of them, although they are probably the most significant. These three are language, culture and technologies.

Language: This is one of the important barriers. Some of the major issues related to language barriers are as follows.
- We do not recognize a communication breakdown. Half the solution to a problem is knowing that it exists.
- We misinterpret words that have (a very or subtly) different meaning in another culture.
- We do not know that we have misunderstood someone else.
- Emotive issues skew our perception of reality.
- Differences in priorities: focus or expectations put emphasis on different issues, resulting in differing interpretations of messages sent or received.

Culture: As we move towards the use of more virtual teams, the understanding of cultural differences and how to harvest rewards from them becomes more important than ever before. Some of the key issues related to corporate culture follow:
- A blame culture: assignment of blame (responsibility) for errors and their punishment override learning and suppress innovation and risk taking
- The well-known Not Invented Here (NIH) syndrome: if we did not invent it, it could not possibly be any good. External ideas are rejected.
- Competition: if we believe we are better than our competition, why should we adopt any of their ideas?
- Implication of poor performance: if your idea is better than ours, it implies we are wrong or performing poorly. Ego: our idea must be better!
- Risk taking: some cultures encourage risk taking while other cultures avoid it, and may even punish it. Nobody has ever been fired for following company policy.

Technologies: Rapidly developing information and communication technologies are breaking a number of barriers; however some of the technology related barriers include the following.
- The level of sophistication of the organization: the cost of upgrading or changing technologies to adopt better methods or processes may be prohibitive.
- Legacy technologies: with rapidly changing software and hardware, the cost of change and associated resistance is an increasingly important issue, both for small and large organizations.
- Training: not only the direct cost of technological change (real and consequential) but also the associated cost of training personnel needs to considered. The total cost may be prohibitive. Alternatively, training may be skipped and the technology never properly utilized.

Conclusions

The authors believe that cross-cultural learnings are one of the keys to maintaining a competitive edge. After all, we are adopting better ideas that have already been tested by others. This MUST be one of the most efficient (high return, low risk) ways of improving team performance.

The fluid and relatively raw environment in which virtual teams operate is a healthy place for the exchange of ideas, the selection of the best and the transfer of results (osmosis) between virtual team members. The opportunities created by the effective development of a virtual team are real and ready for exploitation. Conscious awareness of this opportunity will allow us to harvest the best in "proven improvements" from our project partners.

References:

Galvin, David (1993) Building a Learning Organization, Harvard Business Review, July- August 1993, p.80.

Kiernan, Matthew J. (1995) Get Innovative or Get Dead!, Douglas & McIntyre, Vancouver/Toronto.

Paradoxes in Corporate Theories of Technological Innovation*

Claus Holm
clh@hdc.hha.dk

Research Fellow
The Aarhus School of Business
DENMARK

John Parm Ulhøi
jpu@hdc.hha.dk

Associate Research Professor
The Aarhus School of Business
DENMARK

* Funds for this research project was made available by a grant from The Danish Social Science Research Council and The Danish Technical Science Research Council.

The aim of this paper is to analyse some key paradoxes or dualisms in corporate theories of technological innovation. In recent years, the understanding of corporate innovation and technological change seems to have caught a Sculla of paradoxes. Poole and Van de Ven (1989) draw attention to the notion of paradox and the necessity of illuminating theoretical tensions and contradictions in order to stimulate a more broadly-based theory building. We argue that the paradoxes presented here are among the most important challenges of future research and that they cannot be dealt with without a substantial rethinking of theories of technological innovation. In the first section, two major paradoxes in the field are presented and analysed. In the next section, a structurationist *framework of technological innovation based on Giddens´ work (1976; 1979; 1984) is proposed as a promising research avenue. Further, a new view of the notion and role of planning is presented, based on the ideas developed in the paper. Finally, some implications for future research are discussed.*

Introduction

The study of technology and innovation in organisations has grown enormously over the past few decades (Ulhøi, 1993), producing a vast literature. Despite this, our understanding of technological innovation and the innovative process in organisations remains inconsistent and undeveloped (Wolfe, 1994) and is fraught with definitional ambiguity and dissension (Bamberger, 1991). In addition, the field is challenged by a growing stream of social constructivist research, drawing attention to the notion of meaning in shaping technological innovations (Orlikowski, 1992) and criticising more objectivistic and rationalistic positions.

We argue in this paper that *paradoxes* (Poole & Van de Ven, 1989) will be a key issue in future research. A main aim of this paper is to propose a *structurationist* framework of technological innovation based on Giddens´ work (1976; 1979; 1984), which is argued to be an interesting possibility for dealing with paradoxes.

We have only limited knowledge of how technology planning/strategy interact with innovation projects. It is hoped that the proposed framework can produce some new insights here. A second important aim of this paper is to reinterpret the role of planning in technological innovation based on structurationist theory.

Key paradoxes in technological innovation

Technological innovation as objective vs. subjective phenomenon

Social science has been dominated by a fundamental division and *metodenstreit* between objective and subjective positions (Schutz, 1970; Burrell & Morgan, 1979). In a similar manner, this dualism has pervaded the study of technology and innovation in an organizational context.

0-7803-3552-X/96/$5.00 © 1996 IEEE

The objective view was common and widespread in earlier (economic-based) studies of technological innovation (Orlikowski, 1992). Thus, a "hardware" view of technology was a major assumption of this line of thought. Most past research on technological innovation has had a strong hardware orientation (Van de Ven & Rogers, 1988). Later, a "social technology" view, still defining technology in objective terms, but more broadly as including hardware, knowledge and tasks, gained prominence (Thompson, 1967).

Against this image, scholars within the "strategic choice" view (Burrell & Morgan, 1979) emphasise that technology and innovation are not deterministic or objective phenomena. Rather, they are a manifestation of ongoing human action, choice and negotiation. Examples of research within this tradition are socio-technical and social constructivist studies (Orlikowski, 1992). The emerging social constructivist tradition (e.g Woolgar, 1985; Bijker, 1987) emphasises how shared interpreations and meaning develop around artefacts.

Technological innovation as planned vs. emergent/chaotic process

One of the major paradoxes in the literature is that the innovation process is, on the one hand, portrayed as a planned and approximate linear process, while, on the other hand, scholars continue to emphasise its chaotic and intuitive nature.

The planning view has prevailed and flourished in the "technology strategy" literature (e.g. Adler, 1989; Holt, 1991) and in certain technological innovation process theories. In the former, the widespread attention to (normative) techniques, frameworks and tools as regards choice of technology strategy and the integration of technology and business strategy epitomises this planning and control approach. In sum, the planning view in innovation research manifests itself in a domination of rational choice models (Abrahamson, 1991) and prescriptive guidelines (Ulhøi, 1993).

Paradoxically, the innovation process is also being portrayed as a non-linear or emergent process rather than a planned one. The non-linear, chaotic and unstable nature of the innovative process are described by scholars like e.g. Rosenberg (1982), Van de Ven & Angle (1989), and Mintzberg (1994).

Similarly, social constructivist studies of technology and innovation emphasise the emergent and non-linear aspects of technological innovation. The context of innovation is portrayed as one of multiple realities and equivocality (Weick & Ashford, forthcoming). The latter describes contradictory interpretations of the situation, disagreement, and lack of understanding. Technological innovation implies the social construction of a (new) reality, the reduction of equivocality, closure, enactment and processes of organising (Weick, 1979). In this context, "organising" can be seen as the opposite metaphor of "planning and control".

Bridging paradoxes: A structurationist framework of technological innovation

This paper proposes a *structurationist* model of technological innovation for bridging paradoxes, based on Giddens´ work in social theory (1976; 1979; 1984). Organisational researchers are becoming increasingly inspired by structurationist theory, a major example being Orlikowski (1992). We argue that research in and conceptualisation of technology and innovation from a structurationist perspective appears to be an interesting research avenue. Furthermore, it tends to offer two major advantages. Firstly, a solution to the subjective-objective dilemma is offered, inasmuch as the structurationist perspective allows one to embrace and incorporate both. Secondly, the theory makes possible a deeper investigation into the role and institutionalisation of meaning in innovative activities.

In essence, structuration involves the reciprocal interaction of human actors and structural features of the organisation, e.g a planning system. Through human interaction, three substantial elements, *meaning*, *power* and *norms* (Giddens, 1976) are continuously created and recreated. The theory revolves around the "duality of structure" argument. Structures have a dual nature, they are both the medium and outcome of action. Structures only exist if they are continuously produced and reproduced.

The main components of the structurationist framework proposed here are (i) human agents - knowledgeable and reflexive actors, e.g designers, users and decision makers (ii) technological innovations - material artefacts, and (iii) institutional properties of organisations - e.g. structural arrangements, strategies, culture and control mechanisms. The key premise of the framework is that technological innovation takes place through reciprocal interaction of (i) human action/the way actors socially construct and attach meanings to artefacts and (ii) the structural properties of the organisation. Actors participating in innovative activities are enabled and constrained by structures (e.g. planning systems, resources), yet these structures are the result of previous actions.

Two major propositions follow from this. Firstly, a view of technological innovation as objective force *or* socially constructed product is rejected. Secondly, technological innovations are interpretive flexible. The outlook for these is a function of the meanings different actors and communities-of-practice (Brown & Duguid, 1991) attach to its development and use. The meaning of meaning in the paper is close to the definition of Gray et al. (1985), who consider meaning as having cognitive (concepts, maps) and deeper, less easily changeable properties (values, ideology).

A reinterpretation of planning in technological innovation

Following a structurationist approach, planning/technology strategy is seen as a structural feature of the organisation and the institutional context of action and innovative activities, continuously shaping and being shaped by these. Thus, planning is not an independent "objective" structural feature, but is continously being reaffirmed by the practices of actors. Accordingly, planning is the product of institutionalised (collective) meaning and practice that, conversely, guide future action and meaning creation.

The institutional context of technology planning acts as a boundary on meaning attachment and organising processes (Weick, 1979). The linkage between innovation projects and planning implies the reinforcement or change of the latter, i.e whether actors conform with the "meaning boundary" or choose to "act otherwise" (Orlikowski, 1992) thus changing the structural features.

The choatic/planning paradox reflects this ongoing interaction between meaning attachment and iterations in earlier phases of innovative activities and the institutionalisation and assuming reification of meaning in planning and structural features. A visual description of the interaction between planning/strategy and innovation projects is shown in figure 1 below.

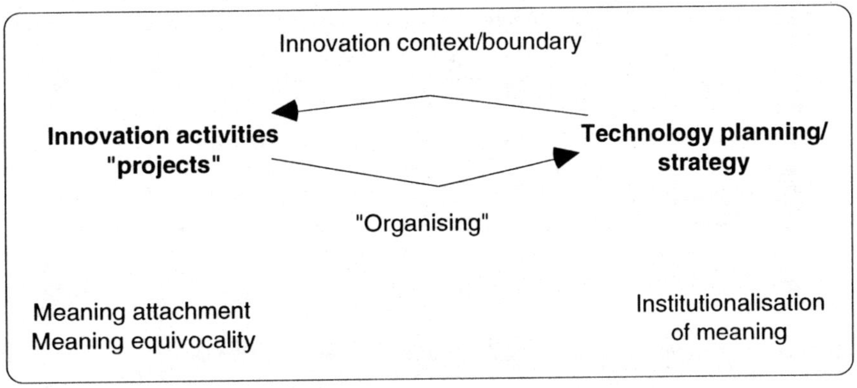

Figure 1. The interaction between innovation projects and technology planning/strategy - A structurationist view

A further implication is the acknowledgment of the fundamental conservative nature of planning, as convincingly argued by Mintzberg (1994), and the importance of distinguishing between planning on the one hand and "visionising" and "strategising" on the other (Mintzberg, 1994). Whereas planning embodies institutionalised and conserving features, visionising and strategising is a continuous more long-term process of readjustment and redirection, based on more informal learning and coupled to the socially constructed and dominant gestalts created and defended by various members of the organization, notably those who have the power to engage in this fundamentally creative process.

Implications and future research

Contemporary theories of innovation and organisation are beset with paradoxes. This paper has proposed a structurationist framework of technological innovation as a way of bridging major paradoxes in the field and to underscore the duality, complexity and inherently dialectic nature of technological change and innovation in an organisational context.

Technological innovation is more a dialetical and continuously evolving process rather than a linear or rational-analytic process as traditionally assumed in mainstream strategic literature. It is a process that involves the "interpretive flexibility" of technological innovations, and with it, processes of social construction and meaning attachment as argued by social constructivists, though within institutional and structural boundaries.

Research is needed on the dialetic between action, e.g. innovative projects, and structural properties of organisations, and, in this respect, the limits and contours of the interpretive flexibility of actors. In particular, more investigation is needed into the role of communities-of-practice, and how situated arrangements affect the structuration process.

References

Abrahamsson, E. (1991). Managerial Fads and Fashions: The Diffusion and Rejection of Innovations. *Academy of Management Review*, **16**, 586-612.

Adler, P.S. (1989). Technology Strategy: A Guide to the Literatures. In: R.S. Rosenbloom & R.A Burgelman (Eds.), *Research in Technological Innovation, Management and Policy*, vol. 4: Jai Press, 25-151.

Bamberger, P. (1991) Reinventing Innovation Theory: Critical Issues in the Conceptualization, Measurement, and Analysis of Technological Innovation. *Research in the Sociology of Organizations,* **9**, 1-35.

Bijker, W.E. (1987). The Social Construction of Bakelite: Toward a Theory of Invention. In: Bijker, W.E., Hughes, T.P., and T. Pinch (Eds.), *The Social Construction of Technological Systems.* Cambridge, Ma: MIT Press, 159-187.

Brown, J.S., & Paul Duguid (1991). Organizational Learning and Communities-of-Practice: Toward a Unified View of Working, Learning and Innovation. *Organization Science*, **2**, 40-57.

Burell, G., & Gareth Morgan (1979). *Sociological Paradigms and Organizational Analysis.* London: Heinemann.

Coombs, R., & Albert Richards (1991). Technologies, Products and Firm´s Strategies - A Framework for Analysis. *Technology Analysis & Strategic Management*, **3**, 77-86.

Giddens, A. (1976). *New Rules of Sociological Method.* New York: Basic Books.

Giddens, A. (1979). *Central Problems in Social Theory: Action, Structure and Contradiction in Social Analysis.* Berkeley, CA: University of California Press.

Giddens, A. (1984). *The Constitution of Society: Outline of the Theory of Structure.* Berkeley, CA: University of California Press.

Gray, B., Bougon, M.G., & Anne Donnellon (1985). Organizations as Constructions and Destructions of Meaning. *Journal of management*, **11**: 83-98.

Holt, K. (1990). Technology Strategy - Is There a Need for It. In: Khalil, T. & B. Bayraktar (Eds.), 1990, *Management of Technology II. The Key to Global Competitiveness.* Norcross, Georgia: Industrial Engineering and Management Press, 407-416.

Maidique, M.A., & A.L. Frevola (1988). Technology Strategy. In: Burgelman, R.A. and M.A. Madique (Eds.), *Strategic Management of Technology and Innovation.* Homewood, Illinois: Irwin.

Mintzberg, H. (1994). *The Rise and Fall of Strategic Planning.* London: Prentice-Hall.

Orlikowski, W.J. (1992). The Duality of Technology: Rethinking the Concept of Technology in Organizations. *Organization Science*, **3**, 398-427.

Poole, M.S., & A.H Van de Ven. (1989). Using Paradox to Build Management and Organization Theories. *Academy of Management Review*, **14**, 562-578.

Rosenberg, N. (1982). *Inside the Black Box: Technology and Economics.* Cambridge: Cambridge University Press.

Schumpeter, J.A. (1961). *The Theory of Economic Development.* Cambridge, Mass.: Harward University Press.

Schutz, A. (1970). Concept and Theory Formation in the Social Sciences. In: Emmet & McIntyre (eEds.), *Sociological Theory and Philosophical Analysis.* New York: Macmillan.

Thompson, J.D. (1967). *Organizations in Action.* New York: McGraw-Hill.

Ulhøi, J.P. (1993). Towards a Theory and Methodology of Management of Technology. In: *Proceedings of The International Conference on Technology Management*, April 4-7 1993, Denver, Colorado, USA.

Van de Ven, A.H., & E.M Rogers (1988). Innovations and Organizations - Critical Perspectives. *Communication Research* , **15**, 632-651.

Van de Ven, A.H., & H.L. Angle (1989). An Introduction to the Minnesota Innovation Research Program. In: Van de Ven, A.H., Angle, H.L., & M.S. Poole (Eds.), *Research on the Management of Innovation: The Minnesota Studies*. New York: Harper & Row, 3-30.

Weick, K.E. (1979). *The Social Psychology of Organizing*. Reading, Mass.: Addison-Wesley.

Weick, K. E., & S.J. Ashford. Communication and Learning in Organizations. Draft Version. Forthcoming in Jakli & Putnam (Eds.), *The New Handbook of Organizational Communication*. Sage: 1996.

Wolfe, R.A. (1994). Organizational Innovation: Review, Critique and Suggested Research Directions. *Journal of Management Studies*, **31**, 406-431.

Woolgaar, S. (1985). Why Not a Sociology of Machines? The Case of Sociology and Artificial Intelligence. *Sociology*, **19**, 147-167.

New Trends In Space Program Management

Edward G. Howard
Steve Kirkner
(301) 713-0098

U.S. Dept. of Commerce/NOAA
1315 East-West Highway/SAO4
Silver Spring, MD 20910

The next generation of weather satellites at geosynchronous orbit may be smaller, less costly, and developed in less time than current space programs. Working closely with NASA in teams like their New Millennium Program (NMP), we at NOAA are trying to find new ways to design, manage and implement the next generation of GOES (Geosynchronous Operational Environmental Satellites). In this paper we describe three trends: trade on requirements early using small teams, infusion of technology, and early focus on lowering costs.

Trade on Requirements Early

Technical Requirements with Small Teams

A lot has been written about getting the right requirements defined early in a mission to save costs. (Shishko 1995, Wertz 1996) Even though we are part of a large government organization, we have broken down our early studies into small expert teams--which seem to be working well. (Howard 1995) Small teams can be useful in providing a broad vision, fit, and rationale for new technical programs. They can then be used to "explore" several design levels to narrow in on technical requirements.

Why is the small team element so important? A small group of knowledgeable experts are highly motivated, have flexibility, can anticipate changes, can develop common insight and can do rapid iteration and prototyping. These are some of the same favorable characteristics of virtual teams which we discuss later.

In our first example, we know the imager on our present GOES satellite can be improved for new products and services to the nation. Rather than going directly to the highest performance option of design, we arrive after discussion with experts at four design options.

Table 1. GOES Imager (Storm Sat) Design Options

Option 1: 2 channels
VIS at 0.51 - 0.71 um
IR at 10.2 - 11.2 um

Option 2: 5 channels (add)
IR at 3.7 - 4.1 um
IR at 6.9 - 7.9 um
IR at 11.5 - 12.8 um

Above still at 1 km spatial resolution, VIS
4 km spatial resolution, IR

Option 3: 5 channels above, plus
0.5 km VIS resolution
2.0 km IR resolution
CONUS scan coverage @ 5 min.

0-7803-3552-X/96/$5.00 © 1996 IEEE

Option 4: All of the above plus:
Add IR channel at 5.4 - 7.2 um
NEdT of 0.1 C at scene temp.
bit depth of 12
Dynamic range to Tmax of 320K
Tmax of 345 SIRW

CONUS coverage @ 30 sec.
Encircled energy 80% from 1.5 fov dia.
User data access via digital TV reception

Option 1 is a "bare bones" option that pegs the lowest cost for a new imager design. Option 2 adds IR channels. Option 3 increases temporal and spatial resolution performance. Finally, option 4 might represent the highest performance option, but at a much higher anticipated cost.

Our point is that there are design to cost options and low cost options that need to be explored in early stages of a new space program to satisfy, if not to optimize, performance. Lower cost may have to prevail over optimal performance and small teams have the flexibility to look at many early design approaches.

Our second example is a small expert team studying design trade-offs for a new passive microwave instrument for GOES.

Table 2.GOES Passive Microwave Design Concepts

Antenna Size: 2 meter

Ground Resolution 66 km at 118 GHz
42 km at 183 GHz
19 km at 400 GHz

Scan 5000 km square in 3 hours

Instrument Noise of 0.03 K at 118 GHz
0.38 K at 400 GHz

Noise reduction require factor of 2 over todays levels

Moving antenna issue 0.14 o/sec to be studied

Stronger science/product uses to be stated.
Future use in numerical weather models to be stated.

Dr. Dave Staelin of MIT/LL has Chaired this small team to find a lower cost, workable design for microwave. Microwaves can penetrate non-raining clouds and explore storm interesting regions along with visual and other IR sounder instruments.

Note that a modest 2 meter antenna at a few microwave frequencies appears to satisfy many observing needs. Our goal is to explore lower cost means to significantly improve weather needs via microwave at geo orbit. Even with approaching technology of materials and large antennas that may become possible, we seek to reduce antenna size and associated cost and risks.

Program Requirements

In addition to the use of small teams for technical requirements, we seek better ways to manage space programs. We are cooperating with NASA's New Millennium Program which seeks to shorten development time and risk. (Casani 1995)

They are suggesting new methods of concurrent engineering that include:

- get a lead time start on production out of design
- resolve conflicts fast by mutual understanding
- use high communication skills/tools
- keep teams small at no more than 5 to 20
- use virtual teams via computer tools
- seek integration among sub-problems, and
- decompose large complex tasks into small simple ones.

In the past few years the literature is filled, also, with sound articles on "becoming a virtual enterprise." (Harris 1996, Chesbrough 1996, Mitzberg 1995, Dess 1996) We believe the government can learn from virtual organization principles--especially virtual information technology to make faster, smoother decisions with less controls. However, government needs the ability to set goals, to ensure long term continuity of service, to settle conflicts early, and to coordinate complex or systematic innovation/ technology. Some of these latter needs require elements of centralization which are at opposite ends to a virtual company. It may be that spacecraft alliances and new ventures offer an intermediate possibility for some future spacecraft data attainment.

Technology Infusion

We believe that new technology will allow us to perform better spaceflight missions at lower cost. However, we advocate evolutionary technology instead of aggressive technology. Instead of conventional spacecraft "block changes" when satellite, instruments and ground systems get changed all at once, we may introduce technology first into spacecraft, then into instruments, then into ground systems. We think this lowers risk and gives us more time to study the real future requirements, as well.

It may be necessary to flight validate some of the technologies for later missions and to cooperate with others in technology needs. Our example in technology is to reference our work with NASA's New Millennium Program, again. (Howard 1995)

Table 3. GOES Technology Needs Shared with NASA's New Millennium Program

Instruments

- Microwave radiometer
- Advanced imager or sounder
- Lightning mapper
- Ocean color imager
- Advanced solar instruments

Spacecraft

- Advanced solar cells
- Stability and control systems
- Thermal designs needed
- Greater on-board calibration
- Multiplex of high data sensors
- Data compression

Operations

- SEP for station keeping/transit to geo
- Limitation on sun angles for operations
- Data fusion techniques
- Use of configuration flying
- Cross-calibration of sensors on different spacecraft

As the table shows, we have listed future technology needs for geo syn spacecraft under three headings: new instruments, spacecraft improvements, and operations. A number of key technologies have been targeted for more study by both NOAA and NASA in the next few years.

Lowering Mission Costs

Our final section of this brief paper focuses on methods for lowering costs. We offer several conclusions first and then select several guidelines that may be useful.

We believe it is possible to lower cost in many different ways--some of which work for some groups, some for others. i.e. there is no one best formula for cost reduction. Some groups believe in a highly structured approach to design; others prefer small, free wheeling teams. Some groups insist on doing most of the work in house, others prefer the virtual approach with nearly all subsystems contracted out.

We suggest that each progam be examined closely for its unique requirements, worker, resources and timing needs. But there are some good guidelines that may be useful to the readers. (Wertz 1996)

In the above text by Wertz the authors provide discussion, case studies and lists that can be used in advanced mission design.
For example in chapter 10, Dr. Wertz suggest actions the government can take to create an environment that reduces costs. Five of our favorites are: force trading on requirements, strong mission engineering, rewarding low costs, decentralization or streamline procurement, and sponsor technology infusion or R&D to reduce future mission costs. In the main body of this paper we have stressed several of these elements--especially trade on requirements, technology infusion, and lower costs.

Summary

There are signs on the horizon that better methods of spacecraft design and procurement are happening: reinventing government is a priority, saving money is a congressional priority, change is happening at NASA to smaller spacecraft, the FAR allows more flexibility, and other agencies have shown success with small spacecraft and shorter development cycles with smaller teams.

We at NOAA are adapting to these changes with a positive and proactive spirit. We have given examples of three new trends in R&D management via future spacecraft: trading on requirements early with small team experts in both technical and program requirements, incremental technology infusion to enhance performance, and early focus on lowering costs.

References

Casani, E. Kane and Wilson, Barbara, "The New Millennium Program: Technology Development for the 21st Century," AIAA, Reno Meeting Papers, Jan 17, 1996.

Chesbrough, Henry W. and Teece, David J., "When is Virtual Virtuous? Organizing for Innovation," Harvard Business Review, Jan-Feb, 1996, pp. 65-73.

Dess, Gregory G. et. al., The New Corporate Architecture," IEEE, Engineering Management Review, Issue on Virtual Enterprise, Vol. 24, No. 2, Summer, 1996, pp. 20-28.

Dittberner, Gerald et. al., "NOAA's Future Earth Observations From Advanced GOES Satellites," AIAA and IAF Paper 95-b.2.01, 46th IAF Congress, Oslo, Norway, Oct. 2-6, 1996.

Harris, R. Clark, et. al., "The Virtual R&D Laboratory," Research Technology Management, Mar-Apr, 1996, pp. 32-36.

Howard, Edward and Gunson, Mike, "New Millennium Program, Instrument Technology Workshops, GOES Program, Presentation Material, Nov. 7, 1995, JPL and NOAA.

Mitzberg, Henry, "Managing Government, Governing Management," Harvard Business Review, May-Jun, 1996, pp. 79-83.

Shishko, Robert, NASA Systems Engineering Handbook, SP-6105, June, 1995, JPL/NASA.

Staelin, David and Kerekes, John, Minutes of the First GMSW (Geosynchronous Microwave Sounder Working Group), Oct. 5, 1995 and subsequent minutes by MIT/LL for NOAA, Washington, D.C.

Wertz, James R. and Larson, Wiley J., Reducing Space Mission Costs, Kluwer Academic Publishers and Microcoslm, Inc. May, 1996.

A conceptual framework for quality information systems

Syohei Ishizu

Aoyama Gakuin University
6-16-1 Chitosedai, Setagaya-ku, Tokyo, Japan 157

According to product life cycle; e.g., product planning, product design, process design, manufacturing, and sales & services; many kinds of information systems which relate to quality of products are developed, and quality information systems are configured. Indeed, some systems have quality data base, process the quality information, and support total quality management. But the concepts of quality information systems are not clear, and how to develop quality information systems is also vague. In order to clarify the concepts of quality information systems, we first characterize the properties of quality information. And then we configure the quality information systems, and classify the types of quality information systems. Quality information is also important in the following area: evaluation of products, organization of the other information such as cost or technological information, and organization or re-design of management activities.

1. Introduction

In corporations, many kinds of activities are performed in order to offer the products with high quality. In order to perform these activities efficiently, many types of information systems are developed. Most of the advanced information systems, such as re-engineering or flexible manufacturing systems, are related to quality information of products or index of customer satisfaction. In this sense, quality information becomes one of key concepts in management information systems, and developing quality information systems (QIS for short) becomes very important. But the concepts of quality information and the methodology for developing QIS are not clear, since the properties of quality information are not clear.

Main objectives of this paper is to propose a conceptual framework for quality information systems based on the characteristics of quality information, and presents developing tools which are used in TQM. In order to introduce a framework for quality information systems, we follow the next steps.

1 We first characterize the properties of quality information systems from a standpoint of goal seeking systems.
2 We next discuss about the aims of quality information systems, and we classify the quality information systems according to the properties.
3 We finally shows that the methodologies which help to develop of quality information systems based on the total quality management tools.

2. Quality information systems as advanced information systems

One of the main aims of quality information systems is of course to develop advanced TQM based on IT power, such as knowledge processing, multi-media, networks, etc. In TQM, some methodologies and tools are developed and they are also useful to develop QIS. QIS may also contribute to information technology (IT for short) as shown in Fig. 1. QIS may be useful for the reasonable evaluation of IT or information systems, since TQM usually introduce evaluation of products and management systems. And QIS may be useful for the differentiation of IT or information systems.

The directions of advanced TQM are studied by the research group of QIS1,2 and summarized as follows.

1 Reliable quality assurance systems.

0-7803-3552-X/96/$5.00 © 1996 IEEE

2 Quick response to the customer satisfaction.
3 Effective use of the knowledge of the engineering technologies.
4 Simple or lean quality assurance systems.
5 Development of application packages of the tools of TQM.
6 Diffusion of end user computing.

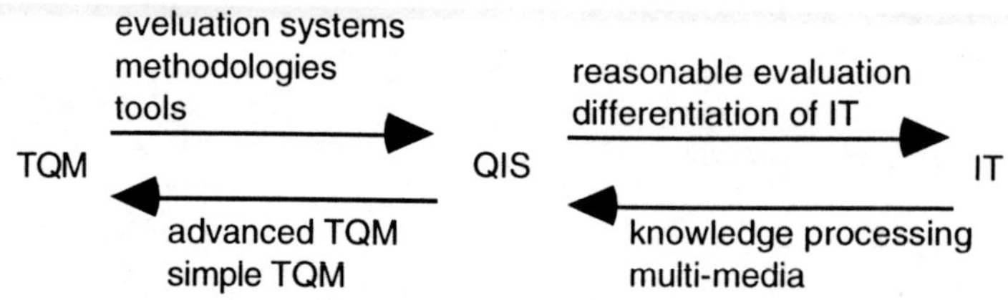

Fig.1 Aims of quality information systems

3. Basic properties of quality information systems

The properties of quality information may be summarized as follows.

1 Uncertain information
Quality information naturally involves external environmental systems such as customers symbol 146 \f "Times New Roman" needs, and quality information systems treat uncertainty environmental information. This means that QIS has uncertainty and is naturally open systems.

2 Invisible information
Quality information is different among products. We must define quality for each product, and we must develop the way to measure the quality of product.

3 Multi-levels information
Quality information is recognize as the multi-levels information; e.g., planning quality, design quality, process design quality, quality of manufacturing, and quality of sales & services. This means that the quality information may be represented by many types of information.

4 Complex information
One product may be divided into and represented by many kinds of quality items. Quality information of product has many quality items which closely connected each other even in the same level. This means that quality information itself is regarded as a complex system.

Information systems which support goal seeking may be classified into recording systems, evaluation systems, and goal seeking systems. Recording systems may record the information of process units, in order to recognize the process mechanism. Evaluation systems may be consists of the evaluation of inputs and outputs information. Goal seeking systems may consists of information of decision making and the decision premises.

We mainly discussed about quality information systems as product life cycle. But there are some other kinds of management cycle in corporation according to the range of cycles.

1 Production cycle
In manufacturing, each product is produced repeatedly. This cycle is production cycle, and the inspected data are gathered to understand the process mechanism in this cycle. Mainly focused information in this cycle may be quality information of manufacturing.

2 Product cycle
This cycle is new product development cycle according to the product life cycle, and we mainly discussed about this cycle before. In this cycle, process mechanism consists of planning quality, design quality etc. The overall activities of this cycle are usually recognized as quality assurance activities of new product. Mainly focused information may be quality information of the product and technological information.

3 Management cycle
This cycle is a cycle of system design in which the structures of activities are re-designed. Process mechanism involves many different planning quality in this cycle. Mainly focused information may be market information.

4. Methodologies for developing quality information systems

From a standpoint of goal seeking systems, quality information systems are classified as shown in Fig.2. From the aspects of management cycles, QIS are classified by the production cycle, product cycle, and management cycle. From the aspects of supporting levels of information systems, QIS are classified into recording systems, evaluation systems, and goal seeking systems.

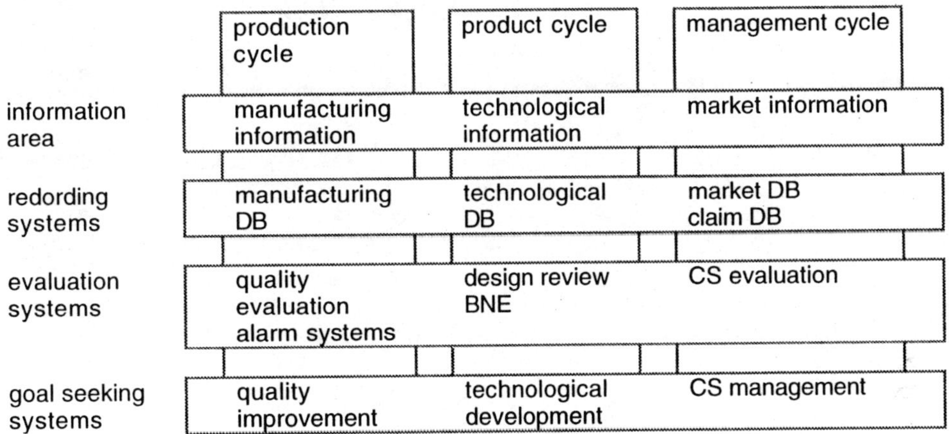

Fig. 2 Types of quality informatio systems

In TQM, many to types of charts are used to clarify the quality information and quality assurance activities in TQM. These charts are also useful to develop QIS. Many types of quality analysis are performed to solve quality problems and quality goal seeking. Table 3 shows that the useful charts and analysis as tools to developing QIS. These tools are listed in Table 3 according to the information areas and supporting levels.

Table 3. Main task and the tools for developing quality information systems

	main task	production cycle	product cycle	management cycle
recording systems	activities analysis data synthesis data recording	process chart QC process chaart	list of quality problems list of QA activities systems chart of QA AFD	marketing analysis products matrix portfolio analysis
systems evaluation	evaluation by quality data finding of quality problems	QA chart list of inspections	design review inspection	evaluation of QAS CS evaluation
goal seeking systems	seeking of the causal factors analysis and coordinate problems solving creativity	statistical analysis engineering analysis resolution of quality problems	statistical analysis engineering analysis resolution of BNE R&D	marketing analysis CS analysis CS management market creation

5. Summary

Quality information becomes a key concepts in management information systems, and developing quality information systems becomes very important to survive in the changing environments. Main objectives of this paper is to propose a conceptual framework for QIS based on the characteristics of quality information, and to present tools which are used in TQM and useful for the development of QIS. In order to introduce a framework for quality information systems, we first characterize the properties of quality information systems from a stand point of goal seeking systems. Next we discuss about the aims of quality information systems. And we classify the quality information systems according to the properties. Finally we show that the methodologies which help to develop of quality information systems based on the tools for TQM.

References

1 Furukawa, O., Development of quality information systems and the research activities of academic and business sectors,Trans. Quality Information Systems Symposium, 1994 (in Japanese).

2 Ishizu, S., The state of the art of quality information systems in Japanese companies,Trans. Quality Information Systems Symposium, 1994 (in Japanese).

3 Ishizu, S., A methodology of coordination in Japanese company-wide quality control, Int. J. General Systems, Vol.14, pp. 317-330, 1988.

4 Mizuno, S. and Akao, Y., Quality functional deployment, JUSE, 1978 (in Japanese).

5 Shiba, S., Graham, A., and Walden D., A new American TM - Four practical revolutions in management, Productivity Press, 1993.

6 Tannock, J.D.T., Automating Quality Systems, Chapman & Hall, 1992.

Electronic Messaging - The Lifeline of the Global Enterprise

Kai Jakobs
+44-131-650-5165

University of Edinburgh
Department of Computer Science
Edinburgh EH9 3JZ, Scotland

Rob Procter
+44-131-650-5177

University of Edinburgh
Department of Computer Science
Edinburgh EH9 3JZ, Scotland

Robin Williams
+44-131-650-6387

University of Edinburgh
Research Centre for Social Sciences
Edinburgh EH1 10Z, Scotland

This paper raises some of the issues related to large corporate e-mail systems. A brief discussion of whether or not current functionality suffices is given first. The result of this discussion is then related to the development of corporate e-mail systems. We show that a number of problems typically experienced stem from a lack of early strategic thinking. Subsequently, end-user problems of messaging services are discussed, mcluding user training and support and mechanisms in place allowing end-users to forward complaints and suggestions. This is followed by a brief analysis of e-mail's organizational impact.

Motivation

Electronic mail has made its way into the offices of virtually all large international organizations. However, for distributed IT systems such as e-mail, two kinds of barriers to successful implementation may be particularly important. The one most commonly recognized is caused by differences between various proprietary solutions or different generations of technology, which may mean that systems cannot interoperate or that some functions cannot be shared. Another, potentially more significant barrier arises from the commitment of end-users to their own locally-chosen systems - which may represent a substantial investment made by large numbers of people in learning how to use a system and to apply its functionality to their working activities.

We explore these and other, related issues through a number of case studies on the adoption and development of e-mail services in large, international organizations in a variety of different business sectors including finance, chemistry, aviation, and oil.

Does Functionality Matter?

When attempting to identify the constituents of a usable, useful and therefore successful corporate e-mail system, functionality is one of the potential contributors that springs to mind. This is supported by the frequent complaints about the insufficient number of users participating in the standardization process (cf eg. (ETSI 1992) or (Hawkins 1995)). Standards - and thus systems - that cannot survive in the market since they do not reflect users' needs are often feared to be the result of this reluctance.

Primarily the European companies represented in our studies did not come up with a big list of additional functionalities they would like. The list of requirements most prominently featured more user-friendly addressing, which was unanimously declared a 'pain in the neck', followed by notification services, enhanced security features, understandable and uniform error messages, multimedia messaging, and simpler editing of messages.

0-7803-3552-X/96/$5.00 © 1996 IEEE

Given that X.400 is the corporate e-mail backbone of choice in Europe, the only item from this list that is neither actually met by the standards' specification (as eg 'notification') or rather beyond the scope of a standard (as eg 'error messages') is 'addressing'. It may therefore be stated that:

- the list of additional functionalities required is surprisingly short,
- most problems appear to stem from inadequate implementations, and thus
- the standard in question is not that bad after all.

This result may come as a (pleasant) surprise. The explanation, however, qualifies the amazement. So far, e-mail has been little more than a convenient interpersonal communication system. In particular, it has not yet become part of any business processes; and there are virtually no mail-enabled applications in use. Judging by the outcome of studies of US-based organizations this will change in the near future. This change, in turn, is likely to yield a considerable number of additional requirements (Jakobs et al. 1996). Therefore, our recommendations for corporate users are

- try and push service providers/implementors to provide full implementations of the standards,
- participate in standardization activities to ensure that future demands will be met.

Trying to explain why apparently nobody has cared about future developments so far, we note that for quite some time other, more down-to-earth e-mail related issues had to have higher priorities (like eg. providing reasonably smooth interworking between different systems). This is primarily due to the development of e-mail in multinational organizations. The next section will elaborate on these issues.

Introduction Strategies

The issue of corporate IT strategies has been one of growing concern in recent years as IT has become increasingly identified with competitive advantage (Fincham et al, 1994). Boynton and Zmud (Boynton and Zmud, 1987) summarised the options available to corporate managers as follows:

> "Will these strategies be driven by environmental forces, ... by chance events? ... or by a coordinated sequence of events? Will these strategies be initiated by a proactive IS function? ... by senior management? ... by subunit management? ... or by a carefully orchestrated network of influential and IT-knowledgeable managers located throughout the organization ? "

We will outline and discuss the different approaches employed and hope to contribute to a better understanding of a crucial part of an overall IT strategy - the approaches adopted for introducing technological innovations into the company.

The Top-Down Approach

This is a straightforward approach with the advantage that once decided upon, the introduction of the service is backed by senior management, and thus progresses rather smoothly. Compatibility issues can be solved more easily and a solution providing homogeneous services throughout the whole organization will be much more cost-effective. However, one problem with top-down strategies is that they may be subject to resistance from within the organization. This was the experience of one of our case study organizations, where the introduction of e-mail was initially confounded by people's resistance to change, itself linked to the fact that at that time e-mail's benefits were not really understood. This has changed dramatically over time. It should, however, be noted that the 'top-down' approach rarely occurs in multinationals.

The Bottom-Up Approach

Our study suggests that overall strategic decisions concerning messaging services are not normally made at the very beginning of e-mail penetrating the company. Rather, we found that the initial IT-related decisions were made at departmental or site level. This result may partly be explained by the structure of the case study organizations, the majority of which are subdivided into a number of almost autonomous companies or branches, located around the globe.

The following introduction 'strategy' holds for about two thirds of the organizations within the case study. A group of employees obtains a messaging tool, maybe to fulfil a specific work requirement, or perhaps bundled in with other software. The new service soon becomes popular. Slowly, mainly by word of mouth, information about benefits provided spread throughout the department. The number of users increases steadily, though still within the department or site, rather than at the organizational level. However, at the same time very similar developments take place at many sites, resulting in an extremely heterogeneous environment.

Subsequently, at some point, a central entity takes over and tries to harmonize the different services with management backing. This stage also marks the transition from an almost purely 'technical' approach (ie largely concerned with providing a solution to day-to-day problems) to a higher degree of strategic thinking (ie the realization of the corporate importance of an organization-wide electronic mail system). However, technical problems such as eg. providing and maintaining systems interconnection and smooth interworking still prevail.

Also at this stage most organizations start looking at more flexible and feature rich systems. In contrast to the initial service introduction, this move has always been part of an overall IT strategy. However, some more strategic thinking during these earlier stages of the development would have saved considerable trouble.

It should be noted here that this approach very much resembles the way LANs were introduced a few years earlier. To some extent this justifies the fear that similar mistakes will be made in the future if and when other new 'distributed' IT systems are to be introduced.

The third step, which many of the organizations interviewed are currently pursuing, is characterized by the introduction of a uniform local e-mail environment, interconnected through a messaging backbone which also offers access to the outside world. Completion of this step means that a homogeneous service will be available for most, if not all, users, and that the number of different gateways will be minimized.

> "Until we've got to the stage we've got an entire user population on one e-mail system we're going to have a degree of user annoyance."

End-user Related Topics

This section discusses two issues that appear to be crucial for success of failure of a corporate e-mail system.

User Training And Support Facilities

Attitudes varied with respect to user training. The range of comments included

> "Full user support on technical issues is provided on site, including user training"

> "User support enjoys comparably low priority".

The latter is quite surprising, given the fact that the importance of user training has long been recognized, cf eg (Nelson and Cheney, 1987).

The majority of the organizations considered (re)training of users as being of principal importance. In most cases, a substantial amount of time and money has gone into these issues, which include initial training courses for management and support staff; 'training the trainers' being a popular first step.

One company, which insists that staff follow an introductory training course, considers e-mail as a major (positive) part of today's corporate culture, a view with which staff apparently concur. On the other hand, two companies said that user training is low on their respective priority lists; one of which sees itself as being "still at an early stage of e-mail use" (after about eight years), with the other admitting that "despite our efforts e-mail is not widely used"

Some of the case study organizations have realized that forcing users to move from one e-mail system to another is not a good policy. Rather, they have tried to 'persuade' their users to move in the desired direction, by offering, for example, migration tools and better support facilities for the new service, yet whilst retaining (for some interim period) interconnection to the old one.

Despite the different attitudes towards training, the case study organizations are unanimous in their belief that help-desks are crucial to success. In fact, every of them offers access to a help-desk facility to its users.

Reacting to User Comments

The organizations show similarly diverse approaches towards compilation of and reaction to - user reports, suggestions and complaints. Whereas some do have sophisticated mechanisms in place for automatic logging of such user input, and have formally established bodies to analyze and react upon these reports, others simply leave it to the help-desk staff. The quality and volume of user input itself seems to reflect the respective approaches: users who know that their comments will be considered seriously appear to be much more willing to actually submit them (Jakobs et al. 1995).

Overall, we conclude that users within the case study organizations have generally responded positively to the introduction of e-mail services, but note that:

- e-mail services will lose their appeal if people do not receive 'enough' messages and
- It is important that progress continues to be made in terms of enhanced servlces.

Organization Issues

This section very briefly addresses general organizational issues related to e-mail.

Some of the interviewees acknowledged that e-mail had at least contributed to organizational changes. The most prominent effect reported was that it had helped to create a flatter hierarchy, with complete management levels having vanished following its introduction. It should, however, be noted that e-mail was seen as being only a facilitator of this trend, and not its prime cause Interviewees generally agreed that success in achieving a flatter hierarchy could be largely be attributed to the newly gained ease of communication with senior management. They also agreed that organizational downsizing had increased the importance of e-mail.

It was reported by more than one interviewee that e-mail had also been an important factor behind decisions to establish spin-off companies. E-mail was seen as the means to reconcile the goal of tackling niche markets more efficiently whilst maintaining close (communication) links between the spin-off company and its parent.

On the more external side, e-mail was said to have provided more efficient communication with business partners, and even to have helped forge new business partnerships.

Finally, numerous case study organizations envisaged e-mail enable teleworking to a greater extent, a fact that may well contribute to extremely significant organizational changes in the future.

Summary and Concluding Remarks

Electronic mail has been introduced by an ever growing number of organizations. Whilst this does not necessarily guarantee any improvements for the company in itself, it provides a means to generate real business benefits. Usefulness will be raised further once e-mail has become an integral part of the business process. Once this has happened, it will imply a range of additional functional requirements to current standards specifications.

End-user reactions to e-mail seems generally to be very positive, even though it is widely perceived as an important facilitator of corporate restructuring. This aspect of corporate e-mail services may yet lead to a less favourable reaction amongst end-users if downsizing trends continue.

An introduction strategy that incorporates requirements analysis, training of users well in advance, and appropriately reacting to user feedback are major cornerstones of a successful corporate e-mail service. An e-mail service which is not at odds with established working procedures and which is to be used by well trained employees, is most likely to leave both, employer and staff satisfied. Above all, adequate training appears to be crucial for a successful uptake of e-mail services.

Finally, to answer the question of how companies have chosen to approach the introduction of e-mail services, we found that it is largely subunit management that triggered the adoption of e-mail. Only in a very few cases have we found that the development was initiated by senior management. Subsequent stages of e mail use reveal the use of a more centrally planned approach as its strategic value becomes evident.

References

Boynton, A.C. and Zmud, R.W. "Information technology Planning in the 1990s: Directions for Practice and Research", MIS Quarterly 11(1), 59 - 71, 1987.

ETSI Directorate "Progress on User Participation in the Standardization Process" Presented at the 3rd Interregional Telecommunications Standards Conference, Tokyo 1992.

Fincham, R.; Fleck, J.; Procter, R.; Scarbrough, H.; Tierney, M. and Williams, R. "Expertise and Innovation: Information technology Strategies in The Financial Services Sector", Oxfoed University Press, 1994.

Hawkins, R.W. "Enhancing the User Role in the Development of Technical Standards for Telecommunications", Technology Analysis and Strategy Management, 7 (1), 1995.

Jakobs, K. and Fichtner, M. "Introducing Electronic Mail in Large Organizations Lessons Learned from Case Studies", Proc. 1st Int. Conf. on Telecommunications and Information Markets, 137 - 143, 1995.

Jakobs, K.; Procter, R.; Williams, R. and Fichtner, M. "Corporate E-Mail in Europe - Requirements, Usage and Ways Ahead", Proc. 4th Int. Conf. on Telecommunication Systems, Modelling and Analysis, 1996.

Nelson, R.R. and Cheney, P.H. "Training end users: An exploratory study" MIS Quarterly 11 (4), 547 - 559, 1987.

Fuzzy Multiobjective Linear-Programming-Based Justification Of Advanced Manufacturing Systems

Cengiz KAHRAMAN
Ziya ULUKAN

Istanbul Technical University
Department of Industrial Engineering
80680, Macka-Istanbul-Turkey
E-mail: cengiz2@ayasofya.isl.itu.edu.tr
E-mail: ziya@ayasofya.isl.itu.edu.tr
Tel: 90-212-293 13 13/2073 Fax: 90-212-240 72 60

Ethem TOLGA

Galatasaray University
Faculty of Engineering and Technology
80840 Ortaköy-Istanbul-Turkey
Tel: 90- 212 - 227 44 80 / 192

Many of the objectives of the advanced manufacturing tehnologies (AMTs) lie in strategic areas such as better quality, flexibility, and shorter lead times. We propose a fuzzy multiobjective linear-programming which considers intagible benefits in AMTs and expands the constraints by adding tolerances. The transition from vagueness to quantification is performed by applying the fuzzy set theory. The approach also considers the vagueness in the objective functions by using the membership functions. The main advantage of the fuzzy LP, compared to the unfuzzy problem formulation, is the fact that the decision maker is not forced into a precise formulation because of mathematical reasons. A case example is given in the paper.

Introduction

Many of the advantages of the advanced manufacturing technologies (AMTs) lie not in the area of cost reduction but rather in strategic areas such as better quality, flexibility and shorter lead times. Two characteristics of these technologies are the flexibility that maintains the value of the equipment over the long run and their synergy when linked together (Meredith and Suresh,1986). The capital investment in AMT proposals is generally very large and cannot be implemented in a piecemeal manner. Many of the significant benefits from AMT proposals are not easily quantifiable in cash flow terms. Application of the traditional capital budgeting techniques to AMT proposals without recognizing the benefits of enhanced flexibility, quality etc., generally results in an undervaluation and their rejection (Usher,1993).

Using LP model for the justification of AMTs is not a new approach. Park and Son (1988) formulate a multistage investment decision model that has two objective functions. One is to maximize capital performance and the other is to maximize production performance. Son (1994) attempts to solve the inherent problems with productivity including the productivity paradox by an unconventional use of a LP method. Chandra and Schall (1988) use the Leontief input-output model and linear programming to determine performance measures of a flexible manufacturing system for the economic justification of the system. Madu and Madu (1993) use a LP model for the allocation of resources for the transfer of mutually dependent technologies. Kuei et al. (1994) use a LP technique to assist decision makers in allocating company money to automation technology.

To deal with the vagueness of human thought, Zadeh (1965) first introduced the fuzzy set theory, which was oriented to the rationality of uncertainty due to imprecision or vagueness. The fuzzy set theory is primarily concerned with quantifying the vagueness in human thoughts and perceptions. Linguistic terms can be properly represented by the approximate reasoning of the fuzzy set theory.

Wilhelm and Parsaei (1991) define two fuzzy linguistic variables to implementing a strategy for computer integrated manufacturing: The importance associated with each of a set of goals common to all enabling technologies, and the capability of each technology to meet the strategic CIM goals of the organization. Kahraman and Tolga (1996) use a multi-attribute fuzzy scoring model to select the best alternative among the technologies. Kahraman et al. (1995) develop a fuzzy flexibility analysis for AMTs. They accept the numerical values of flexibility parameters as triangular fuzzy numbers and use a fuzzy present worth analysis. Kahraman et al. (1996) propose a model based on fuzzy benefit-cost (B/C) ratio for the justification of AMTs. Their goal is to capture the high level of risk and uncertainty in AMT projects by using the capability of representing vague knowledge of the fuzzy set theory.

Fuzzy Multiobjective Linear Programming

The main advantage, compared to the unfuzzy problem formulation, is the fact that the decision maker is not forced into a precise formulation because of mathematical reasons even though he / she might only be able or willing to describe his / her in fuzzy terms. Linear membership functions are obviously only a very rough approximation. Membership functions which monotically increase or decrease, respectively, can also be handled quite easily (Zimmermann, 1994). The role of the constraints can be different from that in classical linear programming where the violation of any single constraint by any amount renders the solution infeasable. Fuzzy linear programming offers a number of ways to allow for all those types of vagueness. By contrast to classical LP, "fuzzy LP" is not a uniquely defined type of model but that many variations are possible, depending on the assumptions or features of the real situation to be modelled.

Korhonen et al. (1989) propose a general approach to semi-structured decision-making, which makes it possible to consider multiple objectives (flexible goals), "hard" constraints (inflexible goals), and "soft" constraints (fuzzy goals) within the same framework. Rommelfanger et al. (1989) present a new method called "alpha-level related pair formation" for solving linear programming problems with fuzzy parameters in the objective function. Tanaka and Asai (1984) formulate a fuzzy linear programming problem whose elements are subjectively perceived. This formulation is performed by the fuzzy function. Slowinski and Tenghem (1993) compare the methods FLIP and STRANGE developed by themselves, respectively. FLIP: Fuzzy LInear Programming and STRANGE: STRAtegy for Nuclear Generation of Electricity.Zimmermann (1993) surveys major models and theories in mathematical programming and offers some indication on future developments which can be expected. Turtle et al. (1994) show how fuzzy logic can be employed using straightforward LP tools. Julien (1994) investigates the application of fuzzy set and possibility theories for the representation of imprecise information in water quality management problems. Fuller and Fedrizzi (1994) explore stability analysis in possibilistic programming by extending previous research results to possibilistic linear programs with multiple objective functions. They use multiobjective possibilistic linear programs with continuous fuzzy number coefficients.Nakahara and Gen (1994) propose a quantitative formulation of LP problems with fuzzy number coefficients, by using the ranking criteria proposed by themselves, and show an algorithm for solving the formulated problems in some cases. The optimization of an objective function with fuzzy number coefficients is formulated as the user oriented extension of the optimization of an objective function with real coefficients by the proposed ranking criteria. Carlsson and Fuller (1995) introduce measures of interdependence between the objectives in order to provide for a better understanding of the decision problems and to find effective and more correct solutions to multiple criteria decision making problems. Sakawa et al. (1995) show that large-scale fuzzy LP problems can be reduced to a number of independent linear subproblems and the overall satisfying solution for the decision maker is directly obtained just only solving the subproblems. Herrera and Verdegay (1995) study some models for dealing with fuzzy integer LP problems which have a certain lack of precision of a vague nature in their formulation and present methods to solve them with either fuzzy constraints or fuzzy numbers in the objective function or fuzzy numbers defining the set of constraints.

The multiobjective linear optimization problem may be given as (Zimmermann, 1994)

$$\text{Max.} \quad Z(x) = z_v(x) = c_v x \qquad v=1,2,\ldots,k \tag{3}$$
$$\text{s.t.} \quad Ax \leq b, \quad x \geq 0$$

where x:n-dimensional decision variable vector. The problem consist of n decision variables, m constraints, and k objectives.

Fuzzy linear programming uses fuzzy set theory within the framework of mathematical programming to solve the multiple objective linear programming problem.

Let c_i and c_i be the upper bound and the lower bound for the attainment of the objectives respectively. The membership function of the fuzzy objective function i, $\mu_{oi}(x)$ is monotonically increasing from 0 to 1, that is,

$$\mu_{oi}(x) = \begin{cases} 0, & \text{if } c_i^T x \leq \bar{c}_i \\ (c_i^T x - \hat{c}_i)/(\bar{c}_i - \hat{c}_i), & \text{if } \hat{c}_i \leq c_i^T x \leq \bar{c}_i \\ 1, & \text{if } c_i^T x \geq \bar{c}_i \end{cases} \tag{4}$$

The constraints can be hard or soft. If soft, the right hand side b can be exceeded by the amount p, which is under the discretion of the decision maker. The membership function of the fuzzy constraint j, $\mu_c(x)$ is monotonically decreasing from 1 to 0 over the tolerance interval (b_j, b_j+p_j), that is,

$$\mu_{cj}(x) = \begin{cases} 1 & \text{if } (Ax)_j \leq b_j \\ 1-((Ax)_j - b_j)/p_j & \text{if } b_j \leq (Ax)_j \leq b_j + p_j \\ 0 & \text{if } (Ax)_j \geq b_j + p_j \end{cases} \tag{5}$$

The membership function of the decision set $\mu_E(x)$, is given by

$$\mu_E(x) = \min_{ij} \{\mu_{oi}(x), \mu_{cj}(x)\} \qquad i=1,\ldots,k \quad j=1,\ldots,m \tag{6}$$

for all xεX.

The min-operator is used to model the intersection of the fuzzy sets of objectives and constraints. Since the decision maker wants to have a crisp decision proposal, the maximizing decision will correspond to the value of x, x_{max}, that has the highest degree of membership in the decision set (Martinson, 1993):

$$\mu_E(x_{max}) = \max_{x\geq 0} \min_{ij} \{\mu_{oi}(x), \mu_{cj}(x)\} \qquad i=i,\ldots,k \quad j=1,\ldots,m \tag{7}$$

Introducing one new variable, λ, which corresponds essentially to Eq. 2, this problem is equivalent to solving the following crisp LP problem

$$\begin{aligned} & \text{Max } \lambda \\ \text{s.t.} \quad & \lambda\,(c\text{-}c)_i - c^T x \leq -c_i \qquad i=1,2,\ldots,k \\ & \lambda p_j + (Ax)_j \leq b_j + p_j \qquad j=1,2,\ldots,m \\ & x \geq 0 \end{aligned} \tag{8}$$

Weights can be easily incorporated into fuzzy programming by linearly weighting the membership functions. The maximizing solution can then be expressed as

$$\mu_E(x_{max}) = \max_{x\geq 0}\ \min_{ij}\ \{w_{oi}\mu_{oi},\ w_{cj}\,\mu_{cj}\} \qquad \begin{matrix} i=1,\ldots,k \\ j=1,\ldots,m \end{matrix} \tag{9}$$

The weighting of the membership functions for the objectives is done inversely proportional to the weights proposed for each objective and the following linear programming formulation is obtained,

$$\begin{aligned} & \text{Max } \lambda \\ \text{s.t.} \quad & \lambda w_{oi}\,(c\text{-}c)_i - c^T x \leq -c_i \qquad i=1,2,\ldots,k \\ & \lambda w_{cj} p_j + (Ax)_j \leq b_j + p_j \qquad j=1,2,\ldots,m \\ & x \geq 0 \end{aligned} \tag{10}$$

Fuzzy Multiobjective Linear Programming-Based Justification of Advanced Manufacturing Systems

In a product-mix LP model, the objective function is to maximize profits or minimize costs; constraints consist of various production input resources such as labour, capital, materials, and energy. A product-mix LP problem may be given as in formulation (11).

$$\begin{aligned} & \text{Max } Z = s_1x_1 + s_2x_2 + \ldots + s_nx_n \\ & \text{s.t.} \\ & \text{Labour:} \quad a_{11}x_1 + a_{12}x_2 + \ldots + a_{1n}x_n \leq b_1 \\ & \text{Capital:} \quad a_{21}x_1 + a_{22}x_2 + \ldots + a_{2n}x_n \leq b_2 \\ & \quad\vdots \qquad\qquad \vdots \\ & \text{Energy:} \quad a_{m1}x_1 + a_{m2}x_2 + \ldots + a_{mn}x_n \leq b_m \\ & \qquad \text{all} x_j \geq 0 \quad \text{for } j=1,2,\ldots,n \end{aligned} \tag{11}$$

where s_j: the market price of unit product j; x_j: the amout of product j; a_{ij} $(i=1,2,\ldots,m)$: the cost of input resource i used for making a unit or product j; b_j:the maximum operating budget for resource i available in the firm.

Son (1994) suggets a simple LP model which includes the intangibles as in formulation (12).

$$\begin{aligned} & \text{Max } Z_1 = s_1x_1 + s_2x_2 + \ldots + s_nx_n \\ \text{s.t.} \quad & a_{p1}x_1 + a_{p2}x_2 + \ldots + a_{pn}x_n \leq b_p \\ & a_{q1}x_1 + a_{q2}x_2 + \ldots + a_{qn}x_n \leq b_q \\ & a_{f1}x_1 + a_{f2}x_2 + \ldots + a_{fn}x_n \leq b_f \\ & \text{all } x_j \geq 0 \quad \text{for } j=1,2,\ldots,n \end{aligned} \tag{12}$$

where b_p: the operating budget allocated to productivity-related resources and aggregates all b_1, b_2,...,b_m in the LP model (11) above; b_q: quality-related aggregate resources; b_f: flexibility-related aggregate resources.

For Son's (1994) LP model, another objective function, Z_2 may be to minimize inventory costs in process as in Eq. 13.

$$\text{Min } Z_2 = r_1x_1+r_2x_2+...+r_nx_n \tag{13}$$

or

$$\text{Max } Z_2 = -r_1x_1 -r_2x_2- ...-r_nx_n \tag{14}$$

where r_j (j=1,2,...,n): the inventory cost in process of unit product j. Weights on the objectives Z_1 and Z_2 are w_1 and w_2 respectively. Weights on the costraints p,q, and f are w_p, w_q and w_f respectively. Then, fuzzy multiobjective linear programming is given as in formulation (15).

$$\text{Max } \lambda$$

s.t.

$$\begin{aligned}
&w_1(c_1-c_1)l - (s_1x_1+s_2x_2+...+s_nx_n)\leq -c_1\\
&w_2(c_2-c_1)l + (r_1x_1+r_2x_2+...+r_nx_n)\leq -c_2\\
&w_pp_pl \quad + (a_{p1}x_1+a_{p2}x_2+...+a_{pn}x_n)\leq b_p+p_p\\
&w_qp_ql \quad + (a_{q1}x_1+a_{q2}x_2+...+a_{qn}x_n)\leq b_q+p_q\\
&w_fp_fl \quad + (a_{f1}x_1+a_{f2}x_2+...+a_{fn}x_n)\leq b_f+p_f
\end{aligned} \tag{15}$$

$$\text{all} \quad x_j\geq 0 \quad \text{for } j=1,2,...,n$$

Many of the parametric values of the fuzzy multiobjective LP model developed above are readily available or estimated at the accounting department or other departments in a company. For example, s_j is generally determined by the market; a_{ij} is easily measured by cost accountants. The intangible parametric values and resources are difficult to obtain; however, they are not impossible to estimate. For example, the cost of flexibility-related aggregate resource for making a unit or product j, a_{fj} (j=1,2,...,n), can be obtained by calculating the unit cost of set-up, waiting, idle and inventory costs (Son, 1991).

Conclusion

The LP model used in this paper integrates productivity, quality, flexibility, and reduced lead time. Fuzzy LP permits to expand the resources such as quality and flexibility-related aggregate resources and the decision maker may assign different weights to the various objectives (and/or constraints). The fuzzy multiobjective LP provides a great flexibility to make the estimates of the problem parameters. The main advantage of the fuzzy LP, compared to the unfuzzy problem formulation, is the fact that the decision maker is not forced into a precise formulation because of mathematical reasons.

References

Carlsson, C., Fuller, R., Multiple Criteria Decision Making: The Case for Interdependence, Computers & Operations Research,Vol. 22, No. 3, pp. 251-260, 1995.

Chandra, J., Schall, S.O., Economic Justification of Flexible Manufacturing Systems Using the Leontief Input-Output Model, The Engineering Economist, Vol.34, No. 1, pp. 27-50, Fall 1988.

Fuller, R., Fedrizzi, M., Stability in Multiobjective Possibilistic Linear Programs, EJOR, Vol. 74, No. 1, pp. 179-187, 1994.

Herrera, F., Verdegay, J.L., Three Models of Fuzzy Integer Linear Programming, EJOR, Vol. 83, No.3, pp. 581-593,1995.

Julien, B., Water Quality Management with Imprecise Information, European Journal of Operational Research, Vol. 76, No. 1, pp.15-27, 1994.

Kahraman, C., Tolga, E., A Multi-Attribute Fuzzy Scoring Model to Select the Best Technology, in Proceedings of the Fifth International Conference on Management of Technology, Miami, Feb. 27-March 1, 1996.

Kahraman, C., Tolga, E., Ulukan, Z., Fuzzy Flexibility Analysis in Automated Manufacturing Systems, in Proceedings of INRIA/IEEE Conference on Emerging Technologies and Factory Automation, Paris France, pp. 299-307, October 10-13, 1995.

Kahraman, C., Tolga, E., Ulukan, Z., Fuzzy Justification of Advanced Manufacturing Technologies, in Pre-Prints of the Ninth International Working Seminar on Production Economics, Igls/Innsbruck, Austria, Feb. 19-23, 1996.

Korhonen, P., Wallenius, J., Duckstein, L., Multiple Objective Linear Programming over a Fuzzy Feasible Set, in Applications of Fuzzy Set Methodologies in Industrial Engng, G.W.Evans, W.Karwowski,M.R. Wilhelm (Editors) ,Elsevier, Inc., pp. 225-235, 1989.

Kuei, C.H., Lin, C., Aheto, J., Madu, C.N., A Strategic Decision Model for the Selection of Advanced Technology, Int. J. Prod. Res., Vol.32, No. 9, pp. 2117-2130, 1994.

Madu, C.N., Madu, A.A., A Systems Approach to the Transfer of Mutually Dependent Technologies, Socio-Econ. Plann. Sci., Vol.27, No.4, pp. 269-287, 1993.

Martinson, F.K., Fuzzy vs. Minmax Weighted Multiobjective Linear Programming Illustrative Comparisons, Decision Sciences, Vol. 24, No. 4, pp. 809-924, July/August 1993.

Meredith, J.R., Suresh, N.C., Justification Techniques for Advanced Manufacturing Technologies, Int. J. Prod. Res., Vol. 24, No. 5, pp. 1043-1057, 1986.

Nakahara, Y., Gen, M., Formulation and Analysis of Fuzzy Linear Programming Problems by User Oriented Ranking Criteria, Computers & Industrial Engineering, Vol. 27, No. 1-4, pp. 457-460, 1994.

Park, C.S., Son, Y.K., An Economic Evaluation Model for Advanced Manufacturing Systems, The Engineering Economist, Vol.34, No. 1, pp. 1-26, Fall 1988.

Rommelfanger, H., Hanuscheck, R., Wolf, J., Linear Programming with Fuzzy Objectives, Fuzzy Sets and Systems, Vol. 29, pp. 31-48, 1989.

Sakawa, M., Sawada, K., Inuiguchi, M., A Fuzzy Satisficing Method for Large-Scale Linear Programming Problems with Block Angular Sturucture, European Journal of Operational Research, Vol. 81, No. 2, pp. 399-409, 1995.

Slowinski, R., Teghem, J., Fuzzy Versus Stochastic Approaches to Multicriteria Linear Programming under Uncertainty, in Readings in Fuzzy Sets for Intelligent Systems, D. Dubois, H. Prade, R.R. Yager (Editors), Morgan Kaufmann Publishers, Inc., pp. 810-821, 1993.

Son, Y.K., A Cost Estimation Model for Advanced Manufacturing Systems, Int. J. Prod. Res., Vol. 29, No. 3, pp. 441-442, 1991.

Son, Y.K., Linear-Programming-Based Productivity and Business Stratagy Analyses in the Factory Automation Era, Int. J. Prod. Res., Vol. 32, No. 2, pp. 443-455, 1994.

Tanaka, H., Asai, K., Fuzzy Linear Programming Problems with Fuzzy Numbers, Fuzzy Sets and Systems, Vol. 13, pp. 1-10, 1984.

Turtle, H., Bector, C.R., Gill, A., Using Fuzzy Logic in Corporate Finance: An Example of a Multinational Cash Flow Netting Problem, Managerial Finance, Vol. 20, No. 8, pp.36-53, 1994.

Usher, J.S., Whitfield, G.M., Evaluation of Used- System Life Cycle Costs Using Fuzzy Set Theory, IIE Transactions, Vol.25, No.6, pp. 84-88, November 1993.

Wilhelm, M.R., Parsaei, H.R., A Fuzzy Linguistic Approach to Implementing a Strategy for Computer Integrated Manufacturing, Fuzzy Sets and Systems, Vol. 42, pp. 191-204, 1991.

Zadeh, L.A., Fuzzy Sets, Information and Control, Vol.8, pp. 338-353, 1965.

Zimmermann, H. -J., Applications of Fuzzy Set Theory to Mathematical Programming, in Readings in Fuzzy Sets for Intelligent Systems, D. Dubois, H. Prade, R.R. Yager (Editors), Morgan Kaufmann Publishers, Inc., pp. 764-809, 1993.

Zimmermann, H.J., Fuzzy Set Theory and Its Applications, Second, Revised Edition, Seventh Printing, Kluwer Academic Publishers, 1994.

The Effect of OO Life Cycle on Software Project Management

Emdad H. Khan and Moheb R. Girgis
Tel: (+973) 777 336
Fax: (+973) 682 582
E-mail: (emdad, mohib) @ cs.uob.bh

Department of Computer Science,
University of Bahrain,
P.O.Box 32038, Isa Town, Bahrain

The successful development of a software system must consider appropriate management practices, regardless of the software life-cycle adopted: object-oriented or structured. The structured analysis and design (SAD) and object-oriented (OO) software development life-cycles are introduced. The impact of OO life-cycle model on various software project management activities, such as planning, resource allocation, scheduling, estimation, and staffing, is discussed. The management factors that are adaptable from SAD with little or no modification, and the new ones specific to OO are addressed. For example, management needs to introduce changes in the project team roles and activities, and create new staff such as class designer and implementor, and class library manager to maintain and manage reusable classes. The issues that the management must consider before adopting OO Technology as the basis for future development of the application systems are highlighted.

Introduction

Software project management is responsible for planning and scheduling development activities, and monitoring progress to ensure that the project activities are carried out to the required standards. Rapid technological changes in development methods and tools outdate previous management experience. Lessons learned from that experience may not be transferable to new projects.

Software have been developed for the last two decades using the traditional SAD approach or its variations. Each variation created a new set of problems for the project management. Now in the 1990s, there is a radical shift in the development approach, most notable is the emergence of the object-orientation. The OO paradigm has impacts on many aspects of project management such as the structure of software development process, the roles, evaluation and reward of the development team, the analysis and design techniques, and methods that are applied.

First, we review the SAD approach followed by the OO paradigm, and the project management activities. Then, we discuss various challenges imposed by this new paradigm, and the associated changes needed in the management activities.

Traditional SAD Approach

In the SAD approach, software is developed using a top-down functional decomposition strategy, starting with a high-level view and progressively refining this view into a more detailed design. In traditional software development several approaches have been followed, some of the well known approaches are waterfall, incremental delivery, exploratory, prototyping, fourth-generation techniques, and

0-7803-3552-X/96/$5.00 © 1996 IEEE

spiral model. The life-cycle activities comprises three main phases, regardless of the development approach used: analysis, design, and implementation, as shown in figure 1.

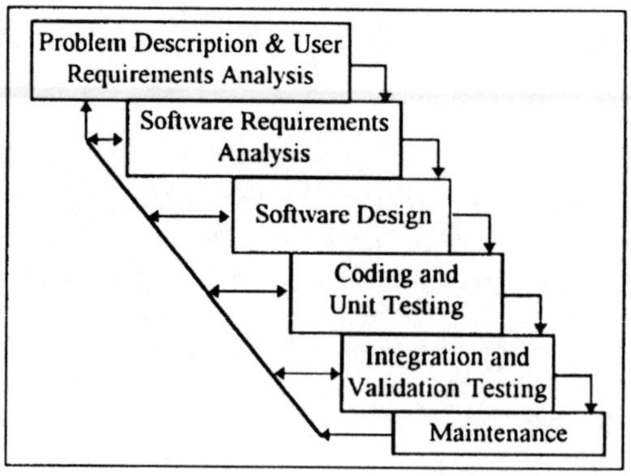

Figure 1. Traditional SAD Life Cycle

OO Approach

The OO approach centers around modeling real world in terms of objects which encapsulates data and operations. The basic OO concepts such as: objects, encapsulation, inheritance, polymorphism, and class relationships are not discussed here. The OO life-cycle supports the new programming styles of exploration, refinement, factoring, and prototyping. The building of prototypes for classes/applications and bottom-up construction of application are key differences in the structure of an OO project compared with the SAD approach.

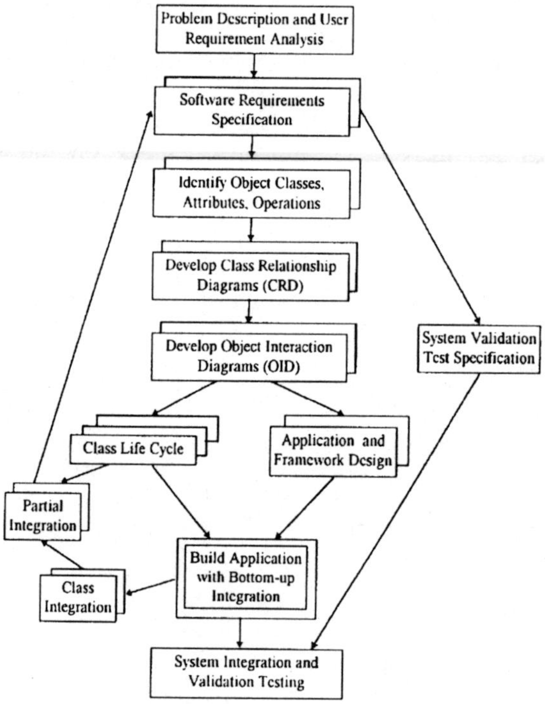

Figure 2. Object-Oriented Life Cycle

The OO life-cycle comprises the same three phases: analysis, design, and implementation, as shown in figure 2, and at each phase the same object model is used. It is more incremental and iterative in nature compared with the SAD life-cycle. It is iterative in nature as we do refinement and iterate from design to analysis and from analysis through design to implementation. It is incremental as we gradually increase the integration level by adding new classes, and deriving new subclasses from existing classes via inheritance.

OO Life Cycle Steps

- Problem description and user requirement analysis.
- Software requirements specification.
- **OOA:** the real world problem domain is analyzed to develop the system's desired behavior model, as shown in figure 2 and 3. The activities are:
    - identify objects/classes
    - identify their attributes and operations
    - develop class hierarchies in the form of class relationship diagrams showing specialization/generalization and whole/parts relationships

- **OOD:** the system's architecture is created through class and application design activities. Class design involves both bottom-up and top-down procedures. Before designing any new class, the designer should check the organization's class library or the OO language built-in class library to identify existing reusable classes. The class design activities are:
    - refine individual classes identified during OOA
    - refine class hierarchies through abstraction of common attributes and operations - creating more generic classes
    - develop object interaction diagrams showing the objects' dynamic behavior and how objects communicate by message passing
    - determine the semantics (life-cycle) of each object from creation to destruction
    - add new classes to manage user interfaces, and transactions

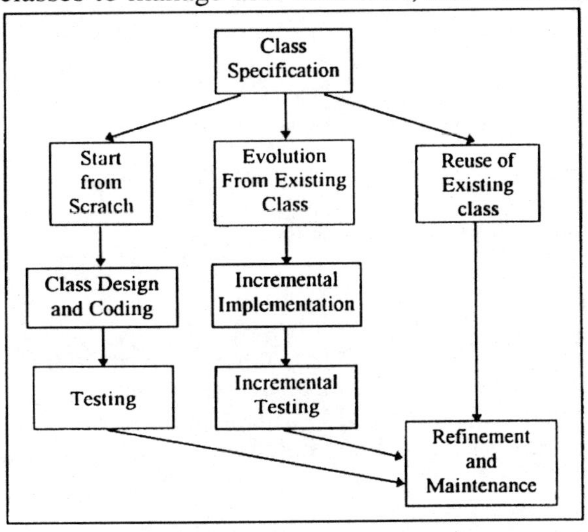

Figure 4. Class Life Cycle

Application design is achieved bottom-up after the constituent class design is complete. Within the OO life-cycle there is a sublife cycle for each class as shown in figure 4. The class life-cycle may be initiated at many different times in the application development process. The application design activities are:

- assemble existing or newly designed classes to carry out the functions of the desired system
- uncover additional classes to enhance system's functionality, which are refined via class design

- **OOI:** The designed classes are implemented using OOP languages or ODBMS. Coding activities are:
    - declare the class structure consisting of specification of its attributes and methods (operations)
    - write the code for each method of each class

- **Testing:** Testing an OO system is a continuous process and involves the same basic approach as testing SAD-based systems. Testing activities are:
    - unit testing: involves testing individual classes and mechanisms. Both black-box and glass-box testing may be required for each method to ensure that objects behave as specified
    - subsystem testing: involves testing a complete class category or subsystem, including the interrelations of classes. Each object of the subsystem is tested for dynamic behavior in connection with message passing with other cooperating objects
    - system testing: involves testing the system as a whole when assembling new releases

Software Project Management Activities

The successful development of a software system must consider appropriate management practices, regardless of the life-cycle adopted: object-oriented or structured. Khan (Khan, 1994) discussed the effects of the SAD methods on the project management and showed the benefits of adherence to formal management practices. Booch (Booch, 1994), Jacobson (Jacobson, 92), and (Pittman, 93) discussed management issues associated with OO projects. They noted that the basic SAD management practices, such as task planning and walkthroughs, are unaffected by OO technology, but in OO projects the tasks being scheduled and the products being reviewed are different.
Project management activities include: planning, estimation, monitoring, and scheduling. It also encompasses selection of a development methodology, CASE tools and a programming language, and maintenance of effective communication with user.

Effects of OO Paradigm on Management Activities

OO projects require a different work-breakdown structure than is typical. As mentioned earlier, it includes: identifying classes, their attributes and operations, developing class hierarchies, and object interactions. Thus, scheduling the various types of resources and activities also differs from SAD life-cycle. The blurred boundaries between OOA, OOD and OOI presents problem for project management. Some form of artificial boundary should be made between OOA and OOD. In fact, the output of OOA are considered as a preliminary design, while the output of OOD as the further refinement of classes with additional design detail. The boundary between OOD and OOI is in two parts: class design and implementation, and application design and implementation.

Estimation and metrics

Pittman (Pittman, 93) discussed three approaches to estimating OO projects: estimating by analogy, with metrics, and with methods. He discussed how to use Boehm's COCOMO model in OO development. Traditional metrics may to some extent also be used in OO software. For example, the McCabe cyclomatic complexity measures could be used to give a complexity measure for the object interaction diagrams, and to generate a cyclomatic metric per class. Chidamber and Kemerer (Chidamber, 93) proposed six formal metrics for OOD: weighted methods per class, depth of inheritance tree, number of children (subclasses), coupling between classes, response for a class, and lack of cohesion in methods. Jacobson discussed similar and other OO metrics. Booch suggested two measures of progress: number of completed and working classes/modules, and stability of key interfaces.

Many traditional software quality assurance (SQA) measures such as review (formal review, inspection, and walkthrough), testing and metrics are directly applicable to OO systems. For example, the defect-discovery rate, and defect density are two relevant quality measures that are applicable to OO systems.

Project team structure

OO projects are staffed differently than SAD projects. We present the development roles that are applicable to both OO and SAD projects, followed by the roles relevant to an OO project.

Roles for both SAD and OO project

The roles which are applicable to both OO and SAD include:

- *Project manager*: manages the project's deliverable, tasks, resources, and schedules.
- *Problem domain analyzer*: extracts and interprets the user's requirements from the real word problem domain.
- *Quality assurance group*: measures both the product and the process to develop high quality product.
- *Integration manager*: assembles compatible versions of released class categories and subsystems in order to form a deliverable release.
- *Documenter*: produces user documentation of the product and its architecture.
- *Toolsmith*: creates and adapts software tools to facilitate the production of project's deliverables.
- *Prototypers*: investigates different solutions at early stages in order to prepare for later development.
- *User interface coordinator:* designs user interfaces.

New roles for an OO project

Booch, Jacobson, Wilkie, and Hares and Smart (Hares, 1994) proposed various alternatives to a development team structure for an OO project. Based on these, we identified the following main OO roles:

- *System architect*: responsible for evolving and maintaining the overall systems architecture.
- *Methodologist*: an expert in the OO method used and supports the development team in applying the method.
- *Class designer*: designs the class structure of a system/subsystem and groups of cooperating objects.
- *Class implementor and tester*: implements and performs unit testing for classes.
- *Application designer*: designs and develops a specific application having knowledge of business analysis and using reusable software components.
- *Application implementor*: assembles designed classes to carry out the functions of the system.
- *Reuse library manager*: responsible for acquiring, producing, and adopting components for general use within a project or the entire organization.
- *Framework team*: develops generic, reusable components for use in various projects in the organization.

Reuse Management

The purpose of reuse is to achieve rapid development of robust systems to satisfy new requirements by using previously developed and tested classes that exit in a class library. The OO paradigm must provide explicit support for integrating reuse of analysis, design and code (class) with the new development. Reuse cost resources in the short term, but pays off in the long term.

To achieve reuse and reap benefits, organizations must encourage design, development and utilization of reusable components. This requires managerial and technical support, as well as mutual cooperation between the development teams and the team members. In particular, managers need to adjust the manner in which they supervise, review and compensate software developers. The biases of the development staff against reusing components developed by others should be overcome. Managers should encourage changes in attitudes by modifying the reward structures for developers so that reuse becomes a requirement: writing new code when appropriate reusable components exist should be rejected, while the developer of components with a high level of reuse should be rewarded.

Resource allocation

For OO analysis, resource requirements do not differ much from SAD approach. During OO design, fewer resources are required, because it refines classes identified earlier during analysis. Testing may also require fewer resources, because adding new functionality to a class is achieved by modifying a previously tested component. Integration usually requires fewer resources, because it happens incrementally throughout the development life-cycle. Thus, the net human resources required for OO development is less than that required for traditional approaches.

Costs and Benefits

OO projects may incur higher start-up costs than normal due to the need for training, language and OOAD tools. Unlike SAD, in OO approach the costs/risks profile is higher during analysis and design, but lower during implementation. The real significant benefit comes during maintenance and upgrading of applications, which is very low compared to SAD software. The need to design new applications from scratch is progressively reduced as the reuse class library grows. The risk of reused code is zero as they have been successfully tested and used before.

Guidelines for Introducing OO Technology

The issue of introducing OO technology received wide attention (Booch, 1994, Hares, 1994, Jacobson, 1992, McGregor, 1992, Pappas, 1995, Wilkie, 1993). The introduction of OO technology into an organization must be done with great care and stepwise, and supported by the upper management. Following are some points to consider before adopting OO technology:

- Select an integrated OO life-cycle from a number of successful OOAD methods such as Rambaugh's OMT, Booch OOAD, Coad/Yourdon OOAD, Jacobson OOSE, Shlaer and Mellor OOSAD (Shlaer, 1992), and Wirfs-Brock RDD/CRC (Wirfs-Brock, 1990). Wilkie described criteria for selecting an appropriate OO Method. Alternatively, select a hybrid life-cycle mixing OO with SAD to safe guard old investment in software, SAD tools and techniques during a transition period.
- Introduce the new method smoothly through an appropriate pilot project.
- Get every one proper training: programmers, analysts, managers, marketing, SQA and documentation personnel including senior management.
- Employ mentors to start the first OO project and to help in building in-house OO experts.
- During training, use examples from the actual user organization.
- Provide tools compatible with the selected method. Booch (Booch, 1994) and Honiden et al (Honiden, 93) proposed different kinds of OO CASE tools and facilities, while Pappas identified evaluation criteria for OO tools.
- Perform risk analysis before starting the project, and take steps to manage the risks.
- Not to have high expectations of the first project, especially with respect to reuse. Reuse benefits occur after three or four project cycles
- OO development is not programming using a C++ compiler. Developers have to think in terms of OO concepts.
- Select an appropriate OO language from hybrid languages such as C++ or a pure OO language such as Smalltalk.
- Select a suitable DBMS from specialized ODBMS such as Gemstone/Orion or a hybrid ORDBMS from established relational vendors such as Oracle/Informix.

Our Experience in Managing an OO Project

We completed a typical bank account problem (Khan, 1995) as our pilot OO project. Here, we briefly outline our experience and the lessons learned:

- We selected Coad/Yourdon OOAD method, because of well defined process and simplified notations.
- Having no access to any CASE tool, MS-Word was used to draw the required diagrams.
- We selected a hybrid language, Borland C++4.5 under Windows, which provided a gradual introduction to OO.
- Training program for team members was useful.
- The team must orient their thinking in OO terms and thus make a cultural change in their attitude during the development process.
- Use of an experienced OO designer/programmer as a mentor was helpful.
- The service of a problem domain expert with the analyst was useful.
- Iteration with prototyping during development of classes and application was essential.
- Programming by refinement with bottom-up integration was followed.
- A class hierarchy was developed through generalization/ specialization and implemented using programming by factoring.
- Difficulties were faced in implementing multiple inheritance. Each method was tested separately and added incrementally to its class. Finally inheritance testing was performed.
- Being the first OO project, the completion time was doubled compared to a similar size traditional project.

Conclusion

We showed how the OO paradigm affects the structure of software development process, the roles, evaluation and reward of the development team, analysis and design techniques, methods, and tools. We observed that some traditional estimation models and metrics could be modified to be used in OO projects, while new OO metrics are required to handle the different OO features and work-breakdown structure. We described the development roles that are applicable to both OO and SAD projects, and the roles relevant to an OO project. To achieve reuse and reap benefits, organizations must encourage design, development and utilization of reusable components. We also presented some points to consider before adopting OO technology. The short term benefits of introducing OO technology are in delivering quality software. The long term benefits come during maintenance and upgrading of applications. Finally, we outlined our experience and the lessons learned in our first OO project.

Reference

Booch, G., *Object-Oriented Analysis and Design with Applications.* 2nd ed. The Benjamin/Cummings Publishing Company, Inc, California., 1994.

Hares, J. S. and Smart, J. D., *Object Orientation: Technology, Techniques, Management and Migration.* John Wiley, 1994.

Honiden, S., Kotaka, N., and Kishimoto, Y., "Formalizing Specification Modeling in OOA". *IEEE Software.* 10(1):54-66, 1993.

Jacobson, I, *Object-Oriented Software Engineering.* Addison Wesley, 1992.

Khan, E. H., "Software Project Management: The Effect of Development Methodologies and Environment", *Modeling, Measurement & Control, D.* AMSE Press, 9(1):37-52, 1994.

Khan, E. H., Al-A'ali, M. and Girgis, M. R., "Object-Oriented Programming for Structured Procedural Programmers". *IEEE Computer*, 28(10):48-57, 1995.

McGregor, J. D. and Sykes, D. A., *Object-Oriented Software Development: Engineering Software for Reuse*. Van Nostrand Reinhold, NewYork, 1992.

Pappas, T. L., "Object-Oriented Technology: PC-Based OOAD Tools". *IEEE Software* 12(3):92-95, 1995.

Pittman, M., "Lessons Learned in Managing Object-Oriented Development". *IEEE Software*. 12(1): 43-53,1993.

Rambaugh, J, Blaha M., Premerlani W., *Object-Oriented Modeling and Design*. Prentice Hall, Englewoods Cliffs, NJ, 1991.

Shlaer, S. and Mellor, S.J., *Object Lifetime: Modeling the World in States*. Yourdon Press Computing Series, New York, 1992.

Wilkie, G, *Object-Oriented Software Engineering*. Addison Wesley, 1993.

Wirfs-Brock, R., Wilkerson, B, and Wiener L., *Designing Object-Oriented Software*. Prentice Hall, Englewood Cliffs, N.J, 1990.

Yourdon, E., *Object-Oriented Systems Design: An Integrated Approach*. PrenticeHall, Englewood Cliffs, NJ, 1994.

The Future of Information Systems - Using Social Systems to Create Protocols for the Virtual Environment. (Systems Analysis through Social Analysis.)

Chris Kimble
+44 (0)1904 433380
Department of Computer Science
The University of York Heslington YO1 5DD UK

Tina Conkar
+44 (0)1904 433243
Department of Computer Science
The University of York Heslington YO1 5DD UK

Information is the medium for communication, power-play, politics, and the building block for knowledge systems. It is associated with social interaction, and can be mediated by technology use.

This paper argues that the key to understanding the impact of future technologies lies in the interaction between the social and technical environment. It suggests that future technologies such as Virtual Reality make necessary a move away from traditional methods of systems analysis and design. The interactive nature of such technology requires a validation in the social environment.

The paper proposes the creation of protocols (a set of universally applicable standards) for the Virtual Environment. It suggests that Information Systems are split into three Protocols: Physical, Learning, and Cultural Protocols. Finally it illustrates that their influence over each other can be understood by applying Structuration Theory.

Introduction

"Because of its potential to change us as humans, as communities, as democracies, we need to try to understand the nature of CMC, cyberspace, and virtual communities in every important context - politically, economically, socially, cognitively". (Rheingold, 1995)

The World Wide Web, the Internet, and technologies such as Virtual Reality+ , represent the future of Information Technology (IT); their interactive nature, make possible social contact across the boundaries of space and time. Daily, we hear of more new terms springing from this technology: 'Virtual Communities', 'Cyber Punks', and 'Surfing the Net'. New terminologies for a new society.

Virtual technologies are reminiscent of philosopher Baudrillard's "Simulacrum", a universal reality where the border between artefact and reality has vanished. As such these technologies require a new approach to understanding and building the systems needed to support, what is essentially, a new social phenomenon.

Current analysis and design methods, based on specified logical and sequential processing, lack the flexibility and dynamics for understanding this technology. This paper builds on social theory as a basis for developing a new systems and design methodology for such technologies. It argues that the key to achieving this lies in an understanding of the relevance of information in both the social, and the technical, system.

Background: The business of virtual reality

Tenbruck (Tenbruck, 1989) suggests that today's society is constantly in the process of "producing a new and partial reality" that forces people to keep up with the flow of cultural production. These new realities are the result of media manipulation and aggressive marketing of information that encourage people towards a consumer society. Virtual systems are being hailed as the new tool in the evolution of the

0-7803-3552-X/96/$5.00 © 1996 IEEE

modern business organisation and its associated information systems.

This increased interest in new technology has resulted from a paradigm shift over the last two decades that has emphasised the use and manipulation of technology within the commercial environment. Sociologist Bruno Latour (Latour, 1995) suggests that "It is no longer clear if a computer system is a limited form of organisation, or if an organisation is an expanded form of computer system".

Another stimulus for the creation of 'The Virtual Organisation' is the increasing recognition that, for certain types of work, the Fordist principles of mass production have resulted in a decrease in employee satisfaction and efficiency - a phenomenon described by Peter Drucker as far back as 1951 (Drucker, 1951).

Drucker stipulates that industrialisation destroys the prestige of individuals in traditional work. It has long been argued that new forms of communications technologies have the potential to reverse this trend. For example, Iyenger (Iyenger, 1972) states that technology is "causing an implosion forcing people back together in tribal unity".

McHale (McHale, 1972) provides another perspective on how technology has affected the social environment. He states "(besides the enlargement of the physical world now available to our experience, these media (newspapers, television, and radio), virtually extend our physical images of the world for our daily appraisal. They provide psychical mobility for the greater mass of our citizens. Through these devices we can telescope time, move through history, and span the world in a great variety of unprecedented ways". McLuhan's vision of the "Global Village" (McLuhan, 1989) is now, for some, a reality thanks to the information highways of the Internet.

Information: The gate between two worlds

Information interaction between the social and the technological world can be represented by the image of a mirror. We shall call this the Socio-Technical mirror. Information interchange in both environments share many identical traits. One may be considered to be a mirror image of the other, although there are subtle differences that exist in these opposing systems.

At the most basic level, both environments use various tools to represent information (such as language, symbols, pictures, icons, etc). These are tools that reflect or model real life events and environments.

Information exchange in both worlds results in synergy, where the whole is more than the sum of the parts. In the social system this is achieved through group interaction and the use of shared knowledge. In the technical system, this occurs when data is aggregated and transformed to new data that is then networked and exchanged.

Technical systems have never been able to mirror the social world accurately or completely. The technical system is simply a repository and converter of data that is only given form and meaning through the user. It is people who create knowledge systems (Walsh, 1995) and information from that data. Technical environments lack the framework, based upon a socially derived consensus, required to give the data the form and meaning that it has in the social world.

Virtual information environments (such as E-mail, and the Internet) may represent a departure from this. They have shown themselves to be capable of evolving beyond their original technical specification. ARPANET, the ancestor of the Internet was never conceived as a medium for virtual communities, and yet, today the Internet has grown into an environment that support MUDS and MOOS*.

Traditional systems analysis and design methods were never designed to support the 'soft' issues of information that relate to the social environment, such as power, meaning and control.

How can we address these issues? The answer is with information itself. Information is neither objective (as traditionally extolled in Systems Design) nor completely subjective. Information is instead inter-subjective: common meaning can be extracted through shared action and culture (Mingers, 1995) as happens with virtual communities.

Socially formed information systems (knowledge), and IT systems such as Virtual Information Systems, share the commonality of information. Socially based knowledge systems are created by social interaction, but can often be influenced by the technology that mediates this process. Conversely, Virtual Information Systems, such as virtual communities, are created in artificial environments but are influenced by the social norms that mediate social interaction.

Using social systems to create protocols for the virtual environment

Traditional systems analysis and design methods such as SSADM, IE, Yourdon and other Structured Methods, take into consideration some of the social or 'soft' issues involved in new systems development. However, they more than often fail to adequately make the connection between the technical and social aspects of information, its role and its subsequent use. These methods follow techniques that eliminate or distance soft aspects of information, in a bid to delayer 'fuzzy' issues for clear and systematic analysis. Gibbons et al (Gibbons, 1994) provide an appropriate summary.

"Systems models imply greater stability in the relationship between actors than is justified given what we know of distributed knowledge production. More helpful descriptions could be worked out by trying to develop models that incorporate the evolution of patterns of interconnections; the ability to establish, on a recurrent basis, new modes of exchange _"

This section provides, in outline, a methodology that moves away from traditional systems analysis and design methods. As with 'soft systems' methods there is a shift in emphasis towards social context to achieve systems optimisation. This methodology reflects a structure that places equal weighting on social and technical structure. It recognises the inter-subjective nature of information as the dominant form of information exchange in the organisation and the environment at large.

Previously, this paper identified that information systems can be categorised with both social and technical systems. It has suggested that the new forum for this convergence is fast becoming the Virtual Information System. Therefore, it follows, that a medium is required to facilitate the transition of information systems between the technical and social worlds. The authors suggest that this may take the form of Protocols based on social theories. Protocols, are used in the sense of 'fixed rules of accepted behaviour used by agents'.

It is proposed that Information Systems are split into three Protocols: Physical Protocols, Learning Protocols, and Cultural Protocols. We believe that classic approaches, do not go far enough in taking into consideration the various social and psychological factors that make up the decision making process of individuals. Understanding these factors will make for a better 'fit' or interface between systems design and human interface.

The diagram below illustrates some of the factors that may be included under these headings. The list is by no means complete.

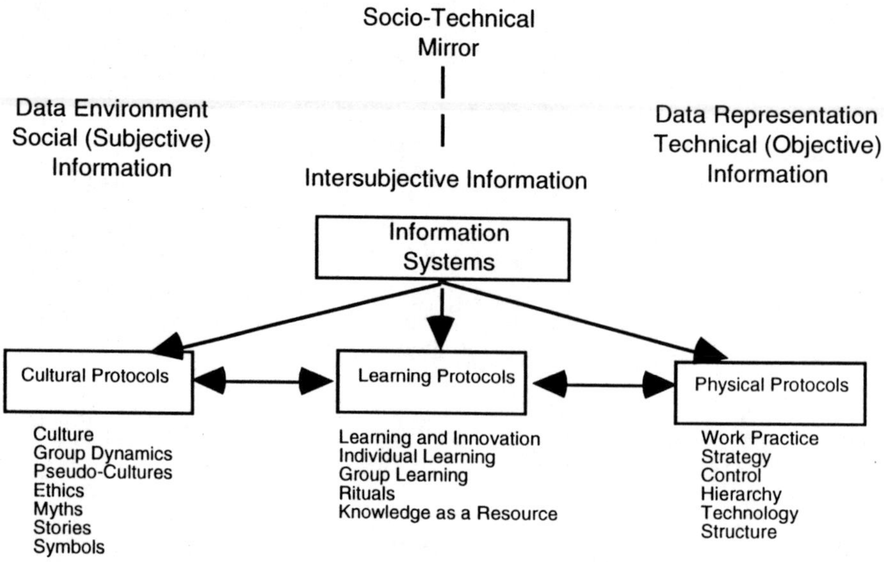

A Mirror Representation of Information Systems in the Social and Technical Environment

These protocols are clearly inter-related both on a micro and macro level. Their influence on each other can be understood by applying Structuration Theory (Giddens, 1984), (Orlikowski, 1992).

There are several reasons for using Structuration theory. Firstly it enables the flexible interaction of social theories with other theoretical approaches. It builds on the relationship between human interaction, social action and social structure, and illustrates the relationships and cause/effects that these factors have on one another.

In this sense, Structuration Theory has proved to be a useful medium for understanding Socio-technical systems. It has been applied to Information systems interpretation by a number of authors (De Sanctis and Poole, 1994), (Orlikowski, 1992), (Walsham, 1990), (Barley, 1986), (Boland, 1985).

Orlikowski's treatment of Structuration Theory in particular takes into account the fluid and dynamic nature of VIS. Orlikowski states "rather than positing design and use of disconnected moments or stages in a technology's life-cycle, the structurational model of technology posits artefacts as potentially modifiable throughout their existence. In attempting to understand technology as continually socially and physically constructed, it is useful to discriminate analytically between human action that affects technology and that which is affected by technology" (Orlikowski, 1992).

Conclusions

Virtual environments have a greater potential to narrow the divide that currently exists between Individual and machine. It is a technology that is offering a new reality. It represents the coming together of two opposing worlds: that of the technical and the social.

Virtual Technologies will, in the same way as the personal computer, play out their future social position in the organisation. As company structures change and adapt in a quest for perfection, profit and globalisation it may prove that this technology is the key to achieving these aims. Virtual systems however are in their infancy and as such require a framework for application and use. This can only be achieved through appreciating the nature of information and the interaction between the social and the technical world.

It involves recognising information is more than an interface for conveying meaning: that it has significances attributed by the users of the information and that these come about through shared action and culture.

The use of a set of universally applicable standards, or protocols that are formed from critical aspects of the social and technical worlds may be the answer to achieving greater integration between the systems. Unlike traditional systems analysis and design methodologies, these protocols will achieve more understanding of the cause and effect of human interaction, action and social structure in relation to technologically mediated information systems. We suggest that using Orlikowski's structuration model (Orlikowski, 1992) may be one way of achieving this.

We believe that such a methodology will make possible the creation of an analytical framework that is compatible with both the technological and societal domains. One that will bridge the gap between these two worlds and help create a culture that can be constantly reconfigured through continual interaction with information.

References

(Barley, 1986) Barley S R, "Technology as an occasion for Structuring, Evidence for Observations of CT Scanners and the Social Order of Radiology Departments", Administrative Science Quarterly, Volume 31, p78- 108, March 1986

(Boland, 1985) Boland R J, "Accounting and the Reproduction of Culture: Budgets and the process of Structuration", Working Paper No 1168, College of Commerce and Business Administration, University of Illinois, Urbana, 1985

(De Sanctis and Poole, 1994) De Sanctis G, and Scott Poole M, "Capturing the Complexity in Advanced Technology Use: Adaptive Structuration Theory", Organisation Science 5(2), p121-147, May 1994

(Drucker, 1951) Drucker P, "The New Society, The Anatomy of Industrial Order", Heinemann, 1951

(Gibbons, 1994) Gibbons M et Al, "The New Production of Knowledge - The Dynamics of Science and Research in Contempory Societies", Sage Publications, 1994

(Giddens, 1984) Giddens A, "The Constitution of Society: Outline of the Theory of Structure",University of California Press Berkley, California, 1984

(Iyenger, 1972) Iyenger M S, "Can we Transform a post industrial society?" from The Futurists, edited by Toffler A, Random House 1972

(Latour, 1995) Latour B, "Social Theory and the Study of Computerised Work Sites", p302, from Orlikowski et Al (Eds.), Information Technology and Changes in Organisational Work, Chapman and Hall, 1995

(McHale, 1972) McHale J, "The Plastic Parthenon", p52, from The Futurists, edited by Toffler A, Random House 1972

(McLuhan, 1989) McLuhan M and Powers B R, "The Global Village, Transformations in World Life and Media in the 21st Century", Oxford University Press, 1989

(Mingers, 1995) Mingers J C, "Information and Meaning: Foundations for an Inter-subjective Account", Information Systems, (5), p285-306, June 1995

(Orlikowski, 1992) Orlikowski W, The Duality of Technology, "Rethinking the concept of Technology in Organisations", Organisation Science 3(3), p398- 427, 1992

(Rheingold, 1995) Rheingold H, "The Virtual Community", Minerva, 1995, p15.

(Tenbruck, 1989) Tenbruck F, "The Cultural Foundations of Society", p29, in Social Structure and Culture, De Gruyter, 1989

(Walsh, 1995) Walsh J P, "Managerial and Organisational Cognition: Notes from a trip down Memory Lane", p281, Organization Science Vol 16, No 3, May/June 1995

(Walsham, 1990) Walsham G, Han CK, "Structuration Theory and Information Systems Research", 10th Internal Conference on Information Systems, p53- 59, 1990

+ Virtual Reality Systems, including Virtual Information Systems (VIS) such as E-mail, increase interaction. They are a medium that allows more flexibility and makes possible representation of real life systems. For example, E-mail is an enhanced conversational medium that approximates the informality of everyday speech. Theoretically Virtual environments have the potential to enable people to interact within virtual worlds, outside the confines of time, space and social conformity.

* MUDs - Multi-User Dungeons, and MOOs - Multi User Object Orientated - textually created virtual worlds that simulate social communities.

Processes and Tools to Support Knowledge Management in a Virtual Organization

Timothy G. Kotnour, Ph.d.
(407) 823-5645

Department of Industrial Engineering and Management Systems
University of Central Florida
Orlando, Fl 32816-2450

Michael Proctor, Ph.d.
(407) 823-5296

Department of Industrial Engineering and Management Systems
University of Central Florida
Orlando, Fl 32816-2450

This paper's contribution is a description of a knowledge management process and tools for a virtual organization. A virtual organization is described as an interconnected set of management systems. The first connection is conceptual in nature and is seen through a project which defines the need for knowledge management. The second connection is informational and is experienced through electronic management tools. Knowledge management is composed of knowledge creation, assimilation, and dissemination. Management tools connect management systems for improved information flow and rapid knowledge growth to effect project execution. These concepts are demonstrated using a project view of a virtual organization's tasks.

Introduction

Organizations are moving to a virtual organization structure due to the competitive advantages gained by advances in electronic communication/information technology and competitive pressures that force a focus on core products. A virtual organization is viewed as an organization in which workers are not physically but electronically connected (Fulk & Desanctis 1995). Grenier and Metes (1995) described a virtual organization as a disperse (in geography and familiarity with other members of the group) set of groups and individuals organized and coordinated based on their core knowledge needed to complete a project (i.e., a virtual project). After the project is completed the team is disbanded and the virtual work group disintegrates to pursue other opportunities. A virtual work group includes members from the "parent" organization as well as partnering organizations. Supporting the virtual organization and work group is an evolving foray of essential tools and capabilities that enable near instantaneous communications and rapid yet knowledgeable responses for communication, coordination, and learning. Technical, product or directional knowledge become key elements in executing a "virtual" project that address customer needs. Partnerships of knowledge workers predominate in virtual relationships. Their productivity is based on system knowledge and integration. Knowledge growth and synthesis is essential to continued performance for a virtual organization (Boland & Tenkasi 1995). A productive organization utilizes the virtual organization knowledge management process and tools in a way that creates knowledge growth and enables synthesis to generate constructive decisions.

A MSM View of a Virtual Organization

A conceptual model called the management system model (MSM) (Kurstedt 1994) is adapted to describe the virtual organization as an interconnected set of management systems. The MSM contains three components ("who manages", "what is managed", and "what is used to manage" or "management tools") and three interfaces ("decision to action", "measurement to data", and "information portrayal and perception"). The "who manages" represents a virtual worker who makes a decision leading to action for completing a task on a virtual project. The "what is managed" component represents what the virtual worker is responsible and

0-7803-3552-X/96/$5.00 © 1996 IEEE

accountable for on the project. A "who manages" component acts on the "what is managed" component through the "decision to action" interface. The "measurement to data" interface collects data from the "what is managed" component. The "what is used to manage" component uses the data to produce and portray information. "Information portrayal to perception" is the third interface. The portrayed information is perceived by the "who manages" component to make decisions. The "what is used to manage" component includes the management tools we discuss later for knowledge management. We use this model to develop two connections between virtual workers.

The first connection is made through the "what is managed" component or tasks on the virtual project. We discuss the second connection through the "what is used to manage" component when discussing the role of management tools for knowledge management. As shown in Figure 1, management domains A and B are connected through their shared responsibilities for "what is managed". One of the primary resources in any organization and especially the virtual organization is the knowledge used in executing a virtual project. In a virtual organization, individuals are selected based on their knowledge for the project tasks. These individuals are connected through electronic channels for information and knowledge management.

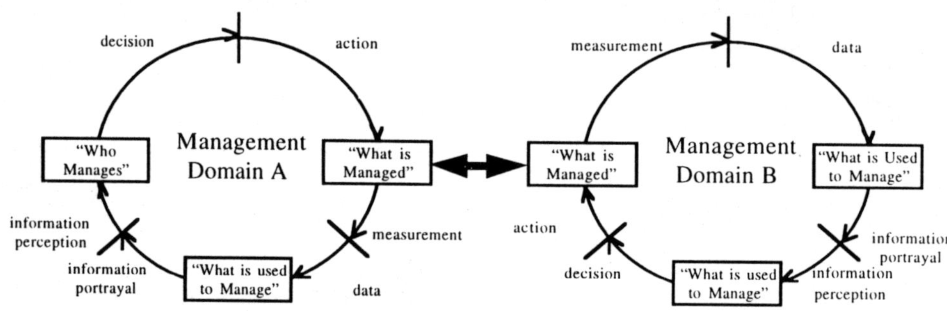

Figure 1: Workers Connected by Shared Responsibilities for "What is Managed".

Knowledge Management for Organizational Learning

Virtual organizations, as all organizations, need to continually produce and share knowledge including using knowledge gained from past projects to solve current projects (Cohen & Sproull 1991). The knowledge is placed in and taken from organizational memory. Organizational memory is the data, information, and knowledge from the past that can be used on present problems (Walsh & Ungson 1991). The process for updating and using organizational memory is organizational learning (Huber 1991). Organizational learning is the naturally occurring process, where as knowledge management is the proactive and purposeful management of the organizational learning process. Knowledge management functions of creation, assimilation, and dissemination are the proactive methods for updating and using organizational memory. For knowledge creation, the organization must have a sub-process to connect the planned actions with the results to learn from the project. Knowledge assimilation is the sub-process to collect the created knowledge and store and refine the knowledge with past organizational knowledge. This knowledge assimilation sub-process adds support to best practices and decreases the support for less effective practices. Knowledge dissemination is the retrieval and distribution of the knowledge to use on another project. The organizational learning process which knowledge management supports is described along two dimensions: modes and timing.

Nonaka (1994) described four modes for organizational learning. Three of the modes (socialization, externalization, and internalization) emphasize direct human-to-human interaction. Socialization is the process for transferring tacit knowledge to tacit knowledge between individuals and includes culture or shared experiences, mentors, and on the job training. Externalization is the mode by which an individual shares his or her tacit knowledge with others by making the knowledge explicit. Internalization transfers explicit to tacit knowledge through learning by the individual. The fourth mode, combination, relies on formal electronic communication methods for sharing explicit knowledge between individuals.

Taking a project view of organizational actions, organizational learning takes place at two time periods: intra- and inter-project. Intra-project learning is the learning existing team members complete on the tasks within the project. Inter-project learning is the sharing of the knowledge between projects. The intra-project learning serves as the foundation for the inter-project learning by providing the experiences to be assimilated for the end of the project learning.

The intra-project organizational learning process focuses on the tasks within a single project. The creation of organizational memory during intra-project takes place during the project planning, execution, and close-out phases. Knowledge is created in 1) the definition of the project plan and 2) the comparison of the actual results to the plan. Explanations for both performance variances and non-variances offer opportunities for learning. Innovative approaches for completing a task should be documented as well as any major problems encountered and the approaches used to overcome the problems. Knowledge creation produces a lesson learned (Juran 1988) which includes the following components: plan (goals, conditions, and steps or actions); results; and recommendations (what worked and why, and what didn't work and why) (Kotnour 1995). Knowledge assimilation occurs through the continuous capture and storage of the lessons learned. Knowledge dissemination occurs through the virtual workers sharing amongst themselves what has been learned on their given tasks and sharing the gained knowledge for inter-project learning. The intra-learning focuses on the tasks within a project.

Inter-project learning focuses on taking the knowledge created, assimilated, and disseminated from the intra-project learning and assimilating and applying appropriately new knowledge with past project knowledge to support knowledge dissemination to and creation in other projects. The inter-project knowledge creation is the drawing of lessons learned from a single project (i.e., lessons learned for the project). The knowledge assimilation is the combining of lessons learned across projects to develop new knowledge. The dissemination function is the sharing of lessons learned across projects to support intra-project learning for the next project. The cycle begins anew for each project.

The characteristics of the virtual organization pose possible barriers and enhancements to the organizational learning modes and timing. We hypothesize that the socialization, externalization, and internalization modes are most exhibited by the high levels of informal communications exhibited in the physical co-location of workers in communities of practice (Brown & Duguid 1991). Traditional balances and expressions between these three modes may be modified for sharing knowledge in the virtual organization because the very nature of the virtual organization relies on electronic communication; that is, during the project the communities of practice may be primarily electronic communities. An analogy may be drawn between the dynamic changes that occurred in the actors when the entertainment industry shifted from radio to television or silent films to talking films. Immediately the dynamics changed the qualities needed for success. These three modes may be modified for inter-project learning because workers within a virtual organization will be consistently changing from project to project. Therefore, the combination mode of sharing explicit to explicit knowledge may be a primary source for sharing knowledge within and between projects. Based on the characteristics of the virtual organization and the need for organizational learning in the virtual organization, we offer the following hypotheses to base our description of knowledge management tools to support organizational learning in a virtual organization:

- Learning takes place within and between projects.
- Virtual communities of practice are dispersed and ever changing.

- The combination mode of Nonaka (1994) will be the primary mode for sharing knowledge within and between projects in the virtual organization because the physically located "communities of practice" of the traditional organization which provide the mechanisms for the sharing of knowledge between workers will not exist. Instead, presence will be experienced in a virtual manner.

Knowledge Management Processes and Tools

In the virtual organization, there are now electronic communities of practice and the primary means of knowledge sharing is through the "combination" mode. The second connection of virtual workers in a virtual organization is through management tools (i.e., the electronic workflow and communications tools). The role of tools in this connection is for the sharing of knowledge between the management systems (Kotnour 1995). This sharing may take place intra-project as the project is being completed or inter-project between projects after they are completed. The notation {management system, cycle} is used to distinguish the components and interfaces of the two different systems and cycles. The concept of Figure 2 is illustrated with two cycles. The first cycle begins with a decision and ends with the update of organizational memory. The first worker {A} during the first cycle {c1} makes a decision {A, c1} that leads to an action {A, c1} on a project {A, c1} all within the first cycle {c1}. Measurement {A, c1} on the task performance produces data {A, c1} for the worker to use in creating a lesson learned. The second cycle {c2} begins with the updating of organizational memory with a lesson learned.

The contents of organizational memory at the beginning of cycle {c2} are shared between management domains {A} and {B}. The information portrayal of both domains {A} and {B} at the beginning of the second cycle {c2} is a set of lessons learned in domain {A} to be shared with domain {B} and, therefore, is the same for each domain. The second cycle begins when the second worker {B} uses this information {B, c2} to make a decision {B, c2} leading to an action {B, c2}. The action on the project {B, c2} produces a lesson learned through measurement and data. For intra-project learning, the c1 and c2 occur within the given project. For the inter-project learning c1 and c2 are used to denote separate projects.

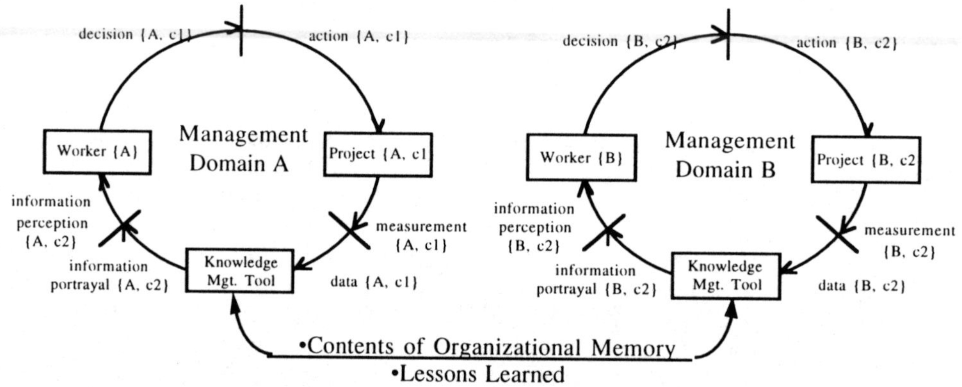

Figure_2: Workers Connected by Knowledge Management Tools.

To support this sharing across management domains, three groups of functions of a management tool need to be defined: creation, assimilation, and dissemination. Creation for intra-project learning focuses on guiding the worker through the construction of a plan, collection and analysis of the actual results, and creation of recommendations (i.e., creating a lesson learned). The assimilation function is the integration of the lesson learned across all tasks within a given project. Dissemination is the function in which the lessons learned are communicated with the other workers. For inter-project learning, the creation function is the development of project-focused lessons learned based on the assimilation of the task-focused lessons learned. The assimilation function is the integration of a project lesson learned across projects and can take place using artificial intelligence techniques. The dissemination functions focus on supporting the worker working on a new project by sharing the lessons learned. The outputs of the tool support the worker in planning a project by providing answers to the following questions:

- What are the actions or set of steps to take?
- What are the alternative sets of steps?
- What are the expected results?
- How do I change the past plans to fit the current situation?
- What should I do that others have done and found to work?
- What should I not do that others have done and found not to work?

As more projects are created information overflow becomes a major problem. Intelligent advisors and search tools can be used to help disseminate the knowledge (Ragusa 1994). The tool should provide a complete set of cases which represent both positive and negative examples.

Conclusions

Given the nature of the virtual organization and the increasing importance of knowledge, organizations need to be proactive in the management of their knowledge. We have offered a description of two related knowledge management processes and a functional description of management tools within these processes. However, there are further virtual knowledge management research issues which must be addressed including: understanding the factors which support and act as barriers to people sharing knowledge; ensuring the quality of the knowledge shared; and understanding how the four modes and timing dimensions of knowledge management differ for a traditional versus virtual organization. The users of a virtual organization need to be aware of the processes for knowledge management and the need to deal with the barriers of electronic communication for knowledge management. This paper provided a framework to understand knowledge management when developing electronic communication technologies for virtual organizations.

References

Boland, R. J. and Tenkasi, R. V. "Perspective making and perspective taking in communities of knowing." *Organization Science*. 6(4): 350-372, 1995.

Brown, J. S. & Duguid, P. "Organizational learning and communities of practice: Toward a unified view of working, learning, and innovation." *Organization Science*. 2(1): 40-57, 1991.

Cohen, M. D., & Sproull, L. S. Editor's introduction. *Organization Science*. 2(1): i-iii, 1991.

Fulk, J. and Desanctis, G. Electronic communication and changing organizational forms. *Organization Science*. 6(4): 337-349, 1995.

Grenier, R. and Metes, G. Going virtual: Moving your organization into the 21st century. Prentice Hall, Inc.: Upper Saddle River, New Jersey, 1995.

Huber, G. P. Organizational learning: The contributing processes and the literatures. *Organization Science*. 2(1): 88-115, 1991.

Juran, J. M. *Juran on planning for quality*. The Free Press: New York, 1988.

Kotnour, T. G. *The effect of lessons-learned sharing processes for organizational learning on decision-making performance*. Unpublished dissertation, Virginia Polytechnic Institute and State University, Blacksburg, Virginia, 1995.

Kurstedt, H. A. *Management systems theory, application, and design*. Author: Blacksburg, Virginia, 1994

Nonaka, I. "A dynamic theory of organizational knowledge creation." *Organization Science*. 5(1): 14-37, 1994.

Ragusa, J. M. "Models and Applications of Multimedia, Hypermedia, and Intellimedia Integration With Expert Systems." *Expert Systems With Applications*. 7(3): 407-426, 1994.

Walsh, J. P. and Ungson, G. R. "Organizational memory." *Academy of Management Review*. 16(1): 57-91, 1991.

Management Of Software Development Processes

Marjan Krasna, Ivan Rozman

University of Maribor
Faculty of Electrical Engineering and Computer Science
Smetanova 17, SI-2000
Maribor, Slovenia
e-mail: marjan.krasna@uni-mb.si

Project do act as a process carefully combined by lower level processes no matter how project ends. Disturbances can cause this processes to act unpredictable. All of the reasons that initiate disturbances can not be foreseen. In the article the object model of project organization is presented. The adoptable object model can be used to simulate software engineering project management plan. The simulation can give us significant insight into dynamics of processes and inter-process relations. It can also highlite some problems we can encounter during execution of the software engineering project plan.

Introduction

Software engineering project management is specific project management problem. Lack of the historic data influence on risk while building project plan. Project plan consists of activities and tasks had to be performed to achieve project goal(s). The author of the project plan can make a complete work breakdown structure needed, but estimation of the effort to complete tasks is still high risk area. In the literature articles can be found explaining fall of the quality projects (Reifer, 1993). These problems inspired us to start the research of processes during project execution. The question appears: is the project in execution phase stable or not? If it is stable then with the perfect plan there are no disturbances. Implication: No profound control needed. On the other hand it may be unstable and control loop is needed to ensure stability. Many authors use statistics to identify processes that significantly interrelate with stability of the project in execution phase and all of them omit interrelations among processes. Now we claim that processes are not stable they are interrelated and worst of all they are even hyper-sensitive. Hyper-sensation means that a small change of one process can cause huge change of one or many of other processes. Since this is not necessarily true each time that a change occurs, statistics fail to foresee this kind of relations. With that statement in mind we can explain why some of projects carefully planned and controlled were not completed successfully. The aim of this article is to open a new dimension to understanding disturbances during execution of software engineering project plan and how to manage software development processes.

In the literature many aspects of software engineering project management can be found. It is not our intention to describe things we already know. More important are things that we can not find, or even if we can find them they are not clear enough. One interesting aspect of the software engineering project management is area of motivation. B.W.Boehm has written an astonishing article (Boehm, 1989) stating that most important object in information system development is individual and it can not be treated like a machine. He also said that expectation of personnel is highest motivation activator. His revised golden rule says “Do unto others as you would have others do unto you - if you were like them”. This rule implies understanding how people want to win and then match people’s tasks to their win condition. In this article it is seen that motivation is very important in the project but it does not say anything about implication into opposite side to make equivalence. With the other words even if we pays enough attention to the motivation this can not be a sufficient precondition for success of a project.

0-7803-3552-X/96/$5.00 © 1996 IEEE

The aim of our on-going research is to find stability factors during execution of software engineering project management plan. In Walston-Felix (Walston, 1977) study of productivity there are 29 factors that affect development process of information systems. None of them were related to human motivation. Boehm's Cocomo (Boehm, 1981) model has 15 factors grouped into 4 groups. Five of factors are related to personnel attributes but only to experiences and capability.

P.J.Shank (Shank, 1993) was the first who finds that motivation is important. Since his research was made for US Aerospace Defense Organization where personnel are carefully selected before they become hired his research was much more narrow than for an ordinary project organization.

All factors in different studies were selected from large list of potential factors and with help of statistics (factorial analysis, group correlation) (Conte, 1986) they combine factors with high correlation into one factor. What is a result of this approach if individual factors that were unified together have hyper-sensitive relation with each other and/or with other factors? The whole statistical theory may work if we have enough measures but are that measures available in sufficient amount?

Another problem is historic data. The computer science is a very fast developing area of science (Frame, 1994). That means many of historic data are out of date (for example, different approach in software development life cycle and programming techniques). Question not without relevance is whether we can adopt data from another organization or not and level of risk involving adoption. Our opinion is we can not adopt data from other organization because we are not them. At least our environment, our personnel and our inter-personal relations are different. Methodologies how to adopt historic data from other organization are still not available nor are available methodologies how to treat our own historic data if we have changed development approach.

We found many areas in software engineering project management specified sufficiently enough to make a good project plan (IEEE, 1994). The starting point of our research was the question weather is possible to make a perfect project plan. We tried to deny previous statement. Assuming that perfect project plan exists we searched the factors that would crash the plan. We put this problem on many project organizations to see if project organization could be stability factor. The most interesting result was that inter-personal relation, and motivation are the highest risk for stability of the project. The explanation is simple. Individual is the crucial factor in software development. Mental activities during execution of software engineering project plan can not be planned (at the present state). It becomes obvious that psychology and sociology would become necessary in software development processes.

Presentation of project model

Model of processes during execution of the software engineering project plan was made. In the model we put all objects (project resources) that are involved in software development. Our object model (Rumbaugh, 1991) is presented in figure 1.

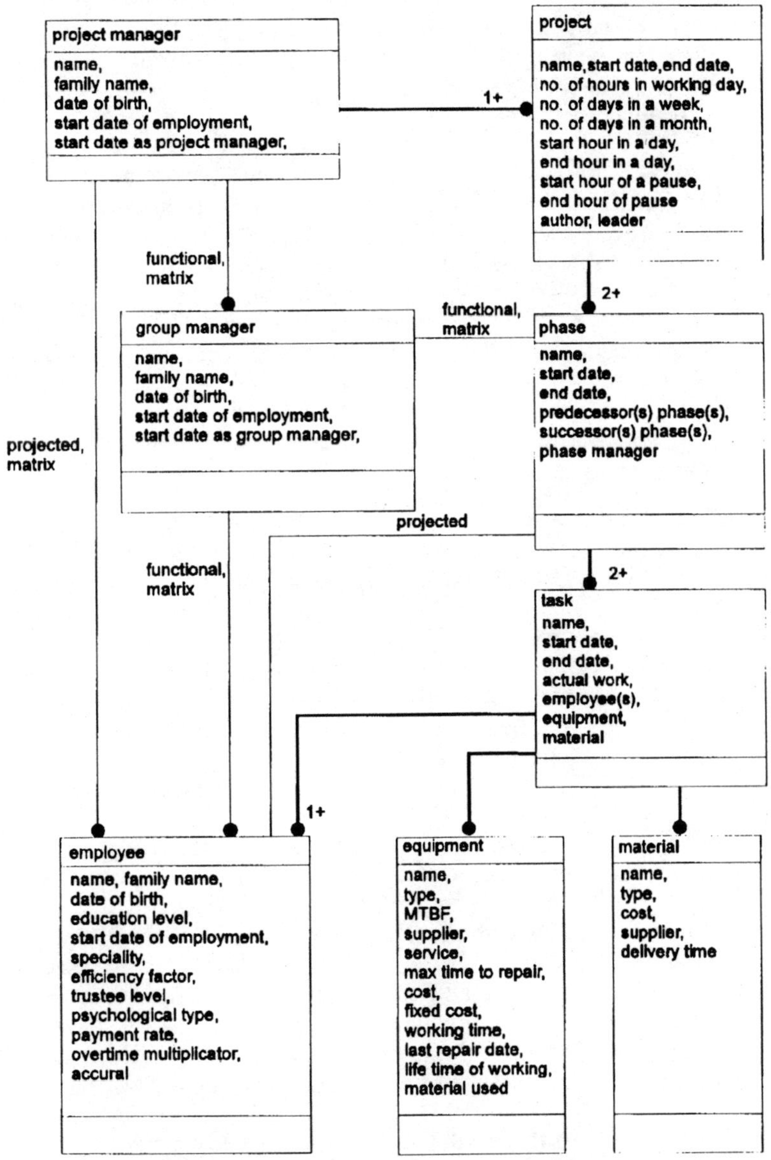

In the model we have information about all objects involved in project execution. For the clarity we join three different models into one model. Most of the model is correct for every known type of project organization. Group manager is presented only if project organisation is functional project organization or matrix project organization and it is not presented on the projected project organization. Connections are varying depending on type of the organization. Thick line works with every project organization, thin line are optional and they depend on type of the project organization (projected, functional or matrix project organization).

Now we must define processes during execution of software engineering project plan. Processes can be seen as multi-level according to WBS (Working Break-down Structure). Highest level consists of single process and this is execution of project plan (object project; see figure 1). Each phase (phase object; see figure 1) in the software engineering project plan is process while it is in the state of execution. Presented task-processes (task object; see figure 1) can be further decomposed to lower level processes depending on project resources assigned to specific task (employee object, equipment object and material object; see figure 1).

At this point we have project model and processes attached to project model. The only things remaining unclear are inter process connections.
There are two types of inter process connections:
1. trivial connections and
2. nontrivial connections.

Trivial connections are connections that can be seen on the figure 1 (denotes type of organization and WBS). Connections that denotes project network diagram are also part of trivial connections. Nontrivial connections are those that can not be seen in the object model and not even in the project plan. These connections are dynamic connections. To simplify the model we present a new object called environment. The environment object is connected to every other object dynamically whenever any change occurs. With this object we can accumulate influences from inside and outside environmental disturbances, and any other disturbances predictable or unpredictable that can occur while project is simulated.

Execution of simulation

Now we are in the position to explore presented model of processes had to be performed to achieve software engineering project goal(s). Exploration of model means use of model to perform simulation that denotes how this model acts in time.

Every object in our model has two distinct methods: evaluation function and state transition function. All of steps through time and any disturbance will cause a model to evaluate itself. That means to signal all of its objects to perform evaluation function and eventually change its own state. If at least one object changes its state then whole model has to perform an update. Every object has to perform its state transition function. This may cause a whole model to perturbate.

We start the simulation with all available data (project management plan, type of project organization, and attributes of objects in our object model). In fixed steps through a time we generate possible disturbances and calculate the results of those disturbances. Disturbances can be created by hand (we put time when disturbance start, type of disturbance and affected object) or we can select a scale of randomly created disturbances (the scale can be from little and unsignificant disturbances up too a lot and very significant disturbances). Since all disturbances are logged we can with multiple simulations with the same preconditions find the crucial disturbances that eventually will cause a project to collapse.

We can put to the model also a corrective disturbances that will start if some predefined conditions are achieved. So we can test the stability of our plan in advance and prepare a necessary corrective actions that we could not even think off before.

Conclusion

Model presented in this article is a result of an on-going research and has potential to help project managers in better understanding what is happening or might happend during execution of software engineering project plan. With simulation where we can vary states of objects and state transition functions attached to that objects it is possible to foresee dynamics of processes that had to be performed. It is possible to identify stability factors and potential dangerous situations that may cause different losses. We can change objects in that model to present any specific organization and project plan. Simulation can predict response of whole system to any interaction with the model. It is possible to foresee how corrective actions taken to guide misguided project influence to the dynamics of processes. For the end we must state that this model is capable to highlight hyper-sensation of processes during execution of software engineering project plan.

Literature

Boehm, B.W. "Software Engineering Economics". Englewood Cliffs, NJ: Prentice Hall, 1981

Boehm, B.W. "Theory-W Software Project Management: Priciples and Examples", IEEE Transaction on Software Engineering, Vol. 15, No. 7, July 1989, pp. 902-916.

Conte, S.D., Dunsmore, H.E., Shen, V.Y. "Software Engineering Metrics and Models", Benjamin/Cummnings Publication Company, 1986

Frame, J.D. "The New Project Management: Tools for an Age of Rapid Change, Corporate Reengineering, Other Business Realities", Jossey-Bass Publishers, San Francisco 1994

IEEE Standards Collection, Software Engineering, 1994 Edition

Rumbaugh, J., Blaha, M., Premerlani, W., Eddy, F, Lorensen, W. "Object-Oriented Modeling and Design", Prentice Hall, Engelwood Cliffs, New Jersey, 1991

Reifer, D.J. "Software Management", Fourth Edition, IEEE Computer Society Press, Los Alamitos, California, 1993.

Shank, P.J. "Modeling as a Tool in Defense Aerospace Software Project Management", Claremont Graduate School, 1993

Walston, C.E., and C.P. Felix. "A Method of programming measurement and estimation". IBM System Journal 16, 1 (1977): 54-73

Managing Emerging Technology Development Through the Network Organization

Arun Kumaraswamy
(212) 533-6556

New York University
Leonard N. Stern School of Business
7-150 Management Education Center
44 West Fourth Street
New York NY 10012

Raghu Garud
(212) 998-0255

New York University
Leonard N. Stern School of Business
7-13 Tisch Hall
40 West Fourth Street
New York NY 10012

Ajit Prabhu
(212) 436-3597

Deloitte & Touche Management Consulting Group
2 World Financial Center, 8th Floor
New York NY 10281-1420

Firms interact in a "lattice-like" structure within an overall regulatory environment to develop emerging technologies in many high-technology industries. We propose that organizational issues at various levels of such a network organization need to be addressed for successful technology development to occur. These levels include the intra-firm level (within individual members), the inter-firm level (among members), and the institutional (overall regulatory framework) level.

Introduction

High-technology industries are characterized by continual change and ever-increasing technological complexity. Continual change makes it incumbent on firms to invest heavily in R&D, while contracting product life-cycles offer smaller windows of opportunity within which to recoup these investments. Additionally, convergence of several technologies makes R&D an increasingly complex task.

Under such conditions, monetary, organizational and regulatory constraints preclude any one firm from possessing all the competencies required to create or benefit from emerging technologies. Even if this were possible, considerable uncertainty exists over technological feasibility, market potential and regulatory regimes, thereby making it difficult for firms to assess alternative approaches and choose the "right" technologies to develop. Furthermore, the absence of clearly (or tightly) specified intellectual property rights makes it difficult for firms to trade competencies using arms-length transactions in a market.

Therefore, firms are better off focusing on their core competencies and relying on others for complementary technologies (Langlois 1992; Prahalad and Hamel 1990). The web of relationships that emerges -- a "lattice-like structure" of interdependent firms within an overall regulatory environment (Powell and Brantley 1992) – represents the network organization.

Analysts have studied manifestations of the network organization such as "keiretsus", "virtual corporations" and dynamic networks (Abbeglen and Stalk 1985; Davidow and Malone 1992; Miles and Snow 1986). Common to this literature is an appreciation that organizing relationships under the network organization allows us to see the whole and its parts. Specifically, each part of the network represents a core competency that, together with competencies possessed by other parts, hastens technological convergence. It is no wonder that emerging-technology development is

0-7803-3552-X/96/$5.00 © 1996 IEEE

accomplished using the network organization in industries including computers and information technologies, biotechnology and automobiles (Fombrun and Astley 1982; Kreiner and Schultz 1993; Womack, Jones and Roos 1990).

Emerging-technology development and the network organization

For successful emerging-technology development, the network organization must possess the following properties (see Garud Kumaraswamy and Prabhu 1995):

- Diversity of membership, that allows members to probe different technological approaches and combinations, and to experience a wide range of stimuli;
- Overlapping competencies and assets required for technology development, and, given the uncertainty regarding future importance of competencies, a redundancy of capabilities;
- Continual evolution and reconfiguration, as learning occurs among members regarding technological and market potential and the relative importance of competencies;
- Implicit control or self-regulation, that fosters decentralized innovation and alternative technological approaches, thereby preventing a single firm from gaining premature control over emerging standards.

Whereas it is easy to create a network with sufficient diversity and redundant capabilities, it is much harder to create one that is continually evolving and self-regulating. For continual evolution and reconfiguration, extensive knowledge sharing and flexible relationships are required; for self-regulation, management of paradoxical organizational pressures is required.

Management challenges within the network organization

These management challenges translate into key issues at three levels of the network organization – intra-firm (within individual members), inter-firm (among members), and institutional (overall regulatory framework) levels.

Issues at the intra-firm level

For successful technology development within the network organization, each member firm needs to develop, absorb and transfer competencies. Accordingly, at the intra-firm level, the primary issue is internal sharing of knowledge and information within each member firm, thereby enabling it to benefit from participation in the network organization.

Traditional hierarchical structures – particularly the SBU structure – constrain the flow of information within a firm, thereby inhibiting knowledge-sharing. Because SBUs are profit centers and resource-allocation within a firm is based on their contribution to overall profitability, SBUs frequently act as rivals and are reluctant to share knowledge among themselves.

Instead, firms are better off organizing their internal technology development efforts around core competencies (Prahalad and Hamel 1990). This prevents "knowledge hoarding" on the part of SBUs that share a competency. Furthermore, firms may enhance knowledge-sharing among centers of competencies and individual SBUs by cataloging development activities carefully, making these data-bases available to all employees, and promoting periodic interactions among researchers and other functional units.

Innovations do not occur in a chronologically convenient manner. Therefore, absent immediate market potential and applications, firms may have to shelve technologies for later use. To ensure that shelved technologies are not lost for ever, firms may catalog all shelved technologies and invest resources in maintaining them (Garud and Nayyar 1994). Additionally, firms may re-use technologies extensively across products and generations; otherwise, the repetitive destruction and creation of knowledge during each new development effort will increase R&D investments to levels that cannot be recouped within short product life cycles (Garud and Kumaraswamy 1995).

Issues at the inter-firm level

As member firms focus only on core competencies and depend on others for complementary competencies, they implicitly "partition" the technology development effort into discrete modules. To ensure eventual integration, such modular development presupposes agreement on interface technologies. In the case of emerging technologies, however, agreements on these interfaces must themselves be flexible and evolve as development progresses.

Such flexible coordination is difficult to accomplish using traditional contracting mechanisms. This is because traditional contingent claims contracts create a rigid framework for relationships by attempting to pre-specify performance under all likely contingencies (Macneil 1980). Furthermore, a zero-sum mentality which prescribes that firms view most relationships as being antagonistic also inhibits flexibility (Garud 1994).

Instead, member firms may adopt relational contracting under which relationships are defined broadly, allowing enough latitude for evolution over time (Macneil 1980). Once relationships are established, continual interactions (including the exchange of strategic information, personnel and knowledge) among member firms create an environment that engenders trust and mutual accommodation. Indeed, as Morgan (1986) points out, the broadest agreement would be one in which only those eventualities that definitely must be avoided (noxiants) are specified.

While relational contracting is prevalent in Japan (Abegglen and Stalk 1985) and parts of Europe (Best 1990), it is a comparatively recent trend within the US (Fortune 1994). Even otherwise, forging relational contracts is difficult in the case of emerging-technology development, because, "....(t)o willingly put the core of your R&D thrust outside your own control and share it with competitors is basically unattractive" (quotation in Murphy and Bower 1983).

The primary issue, then, is the creation of trust among firms that traditionally have had antagonistic relationships. Such trust, in turn, allows the effective and efficient distribution of knowledge among members even though intellectual property rights are not well defined. As a first step, member firms may familiarize themselves with their respective practices, structures and norms for compatibility (Ring 1995). Furthermore, member firms may establish ground rules and resolve differences on the following issues:

- Scope and duration of the joint development effort -- preferably, limiting it to research in fundamental technologies, with development of actual products done individually by members;
- Organization of the joint development effort – preferably, in several small jointly-funded laboratories, each focusing on specific competencies and staffed predominantly by member firms that possess these competencies;
- Transfer of know-how to member firms – preferably, through frequent announcements, presentations, visits and discussions;
- Assignment of intellectual property rights – preferably, vesting rights to technologies arising from joint development in the collective, and controlling/monitoring visits to and from member firms to avoid disputes.

Even as technology development progresses, frequent interactions may be required to resolve technical, cultural and social issues before norms regarding fruitful collaboration and selective disclosure evolve, and knowledge-sharing occurs.

Issues at the institutional level

For "the functional integration of modules as part of a coherent system, an overarching 'framework' of planning and coordination would be necessary" (Astley and Brahm 1989: 258). Such an institutional framework minimizes the costs of controlling the technology development effort.

Self-regulation and open standards

The advent of new information technologies such as design automation and electronic data exchange over EDI networks (Malone, Yates and Benjamin 1987) and the Internet enable connections among member firms and promote integrated development. But, these alone are not sufficient. Rather, a more implicit and self-regulating mechanism that promotes decentralized innovation is required.

Open technological standards provide one such implicit control mechanism[1]. Open standards allow member firms developing complementary technologies or components to coordinate among one another tacitly. Compatibility is assured as long as member firms conform to the open standard, thereby promoting decentralized innovation. To the extent that open standards are broadly defined and conformance is certified by impartial and autonomous bodies, new technologies may be integrated as innovations occur. Thus, the open standards themselves evolve over time.

Such reliance on open standards allows member firms to "trade" knowledge encapsulated in modules or reusable components. At the same time, opportunistic behavior is mitigated because no one firm can change the open standards unilaterally. Therefore, open standards enable the forging of relational contracts among member firms. Furthermore, by minimizing context-specific knowledge (through explicit codification of specifications), and proprietary knowledge (through unrestricted access), open standards promote knowledge-sharing without the threat of intellectual property piracy. Finally, open standards endow the network organization with memory, thereby minimizing its reliance on the original creators.

Increasingly, we are witnessing the emergence of open standards as an implicit control mechanism in networks. For instance, consider the information technologies market – specially, the emerging Internet applications -- where even dominant firms such as Microsoft have had to adopt (and jointly set) industry-wide open standards.

Socio-political and legal institutions

Institutions exogenous to the network such as the socio-political system (e.g., regulatory bodies) and the legal system also affect emerging-technology development. For instance, in the early eighties, there was considerable uncertainty over the legal status of R&D consortia until the Congress passed the National Cooperative Research Act of 1984. More recently, in the emerging cyberspace market, antitrust concerns and uncertainty over the actions of regulators have hindered technology development despite the use of networks (Garud, Kumaraswamy and Prabhu 1995).

Another important cause for concern – specially for knowledge sharing -- are intellectual property laws. Innovative mechanisms are required to ensure intellectual property protection while at the same time enabling knowledge distribution. Such mechanisms include prepaid lump-sum access fees in biotechnology, widespread use

[1] Hierarchical control mechanisms exist where a dominant firm maintains control over the overall technological architecture and delegates design of components to a core group of partners after prescribing interface specifications (for instance, Japanese keiretsus). However, these mechanisms are not appropriate for emerging-technology development because unilateral selection of the technological architecture by the dominant firm is tantamount to the premature adoption of a single technological approach.

of the trade secrets law through confidentiality agreements and non-disclosure laws, and copyright laws that prohibit software piracy but allow programmers to use algorithms or techniques contained in an existing program (David and Foray 1995).

Conclusion

In summary, this paper discussed the role of the network organization in managing emerging-technology development, and issues that need to be resolved for successful technology development. Essentially, what is required is a network organization that enables constituent members to work independently and in unison, even as they evolve over time.

However, the network organization is characterized by a "syndrome of attributes that bear a supporting relationship to one another" (Williamson 1991:271). Addressing issues pertaining to only one attribute (level) will make the relationship between these attributes inconsistent, thereby causing failure of the entire organizational form. Rather, issues at various levels need to be resolved simultaneously. The experiences at Kaleida, a joint multimedia software development venture between IBM and Apple (The Wall Street Journal 1995), and General Magic, an international consortium of firms to develop intelligent-agent technology for use in computer networks (The Wall Street Journal 1993), offer enough justification for adopting such a systemic approach to managing emerging-technology development within the network organization.

Selected references

Abegglen, J. and Stalk, Jr., G. *Kaisha: The Japanese Corporation.* Basic Books Inc., Publishers, New York, 1985.

Astley, W.G. and Brahm, R. "Organizational designs for post-industrial strategies: The role of interorganizational collaboration". In C. Snow (ed.), *Strategy, Organization Design, and Human Resources Management.* JAI Press, Greenwich, CT, pp. 233-270, 1989.

Best, M. *The New Competition: Institutions of Industrial Restructuring*. Harvard University Press, Cambridge, MA, 1990.

David, P.A. and Foray, D. "Accessing and expanding the science and technology knowledge-base". *STI Review,* 16, pp. 13-68, 1995.

Davidow, W. and Malone, M. *The Virtual Corporation.* Harper Business, New York, 1992.

Fombrun, C. and Astley, W. "The telecommunications community: An institutional overview". *Journal of Communications.* 32, pp. 56-68, 1982.

Fortune. "The new golden rule of business", pp. 60-64, February 21 1994.

Garud, R. "Cooperative and competitive behaviors during the process of creative destruction". *Research Policy.* 23, #4. 385-394, 1994.

Garud, R. and Kumaraswamy, A. "Technological and organizational designs for realizing economies of substitution". *Strategic Management Journal, Special Issue.* 16, pp. 93-109, 1995.

Garud, R. and Nayyar, P. "Transformative capacity: Continual structuring by inter-temporal technology transfer". *Strategic Management Journal*, 15, pp. 365-385, 1994.

Garud, R., Kumaraswamy, A. and Prabhu, A. "Networking for success in cyberspace". *Proceedings of the International Conference on Multimedia Computing and Systems.* Washington, D.C., pp. 335-340, May 1995.

Kreiner, K. and Schultz, M. "Informal Collaboration in R&D. The formation of networks across organizations", *Organization Studies*, 14 (2), pp. 189-209, 1993.

Langlois, R. "External economies and economic progress: The case of the microcomputer industry". *Business History Review,* 66, pp. 1-50, 1992.

Macneil, I. *The New Social Contract: An Inquiry into Modern Contractual Relations.* Yalc University Press, New Haven, 1980.

Malone, T., Yates, J. and Benjamin, R. "Electronic markets and electronic hierarchies". *Communications of the ACM*, 30 (6), pp.484-497, 1987.

Miles, R. and Snow, C. "Organizations: New concepts for new forms". *California Management Review*, Spring, pp. 62-73, 1986.

Morgan, G. *Images of Organization.* Sage, Beverly Hills, CA, 1986.

Murphy, W. and Bower, J.L. "The Microelectronics and Computer Technology Corporation". Harvard Business School Case # 9-383-067, 1983.

Powell, W.W. and Brantley, P. "Competitive cooperation in biotechnology: Learning through networks". In N. Nohria and R.G. Eccles (eds.) *Networks and Organizations: Structure, Form and Action.* Harvard Business School Press, Boston, pp. 366-394, 1992.

Prahalad, C. K. and Hamel, G. "The core competence of the corporation". *Harvard Business Review.* May/June, pp. 79-91, 1990.

Ring, P.S. "Transacting in the state of union: A case study of exchange governed by convergent interests". Working paper. Management Department, Loyala Marymount College, Los Angeles, CA, 1995.

The Wall Street Journal. "How do joint ventures go wrong? Ask Kaleida". pp. A1+, November 22 1995.

The Wall Street Journal. "Vague new world: Digital media business takes form as a battle of complex alliances". pp. A1+, July 14 1993.

Williamson, O.E. "Comparative economic organization: The analysis of discrete structural alternatives", *Administrative Science Quarterly,* 36, pp. 269-296, 1991.

Womack, J., Jones, D., and Roos, D. *The Machine That Changed the World.* Rawson Associates, New York, 1990.

Integration With Human Resources Management: Key to Sustained Quality Improvement

Li Chung Shih
+852-2358-7744

Department of Information and Systems Management
The Hong Kong University of Science and Technology
Clear Water Bay, Kowloon, Hong Kong

In this paper, we analyze and suggest the integration of Total Quality Control activities with the Human Resource Management System. We believe that this is the most effective way to create ongoing and long lasting quality improvement activities. A quality improvement system which can renew and revitalize itself continuously avoiding premature death.

Introduction

During the last decade, we have been observing many firms in varied countries trying to implement Total Quality Management Systems-TQMS modeled after the Japanese Total Quality Control System. These systems in some companies fail in short period of less than one year, and in some others even have very promising start and encouraging initial results, but die down after two to three years. It seems that a poor quality program shows its weakness in 3 months and a good one saturates, at longest, in three years. For example, we have observed the subsidiary of a large US cosmetic maker investing a large amount to acquire the quality management package advocated by a globally famous quality guru just to see the enthusiasm and effect last for less than three years.

The reasons for the fast failures are most frequently attributed to: 1)Lack of top management commitment; 2)Poor response or even sabotage from the middle managers in the factory floor; 3)Low motivation of workers and; 4)Inability in working in teams. The last three factors are most frequently explained through cultural differences (Chang 1989, Flynn 1989, Sours 1989, Takamiya 1970, Tasugi 1969) between Japan and the countries where the firms are located.

The longer lasting programs normally die quietly with the feeling of mission accomplished. It is even difficult to notice that it is dead because everything is settled in a new quality practice routine and at a higher overall quality level. You only notice that it is dead because there is not any new challenge that ignites the enthusiasm of the entire company and the pace of quality improvement is almost unnoticeable.

Contrasting with this observation, the best managed Japanese firms can still maintain a high level of enthusiasm, commitment, and accomplishment even after 30 years of quality journey. Though, many of these companies have also faced downturns in their quality programs, they, somehow, were able to revitalize them.

In this paper, we present our analysis and suggestions for a long lasting TQMS. This analysis is mostly based on our observation of the implementation of quality programs in the firms, and on a detailed study of the key managerial factors behind the longevity of the quality programs in best run Japanese companies. Most of these key factors points to the direction of integrating TQC with Human Resources Management System. We believe that this integrated system may be effective in differing cultural settings to help firms achieve sustained and continuous quality improvements.

In Section II, we present a brief description of the main reasons leading to failures in

0-7803-3552-X/96/$5.00 © 1996 IEEE

the implementation of TQC. In Section III, we present the suggestions for integrating TQC with the human resources management system, and in Section IV, the conclusion of the paper.

Main Problems For a Successful TQC

In our understanding, the main problem leading to fast failure in the implementation of TQC programs in the firms is more related to lack of thorough application of Company-Wide Quality Control-CWQC activities, rather than cultural differences. The CWQC activities encloses top-down, lateral, and bottom-up quality control activities (Mizuno 1988). Following, we present the relationship between the most frequently cited problems and lack of a thorough and careful application of CWQC activities.

- Lack or insufficient commitment by top-management as perceived by the managers and employees.
  When top-managers do not apply careful and thoroughly the top-down CWQC activities -- the policy deployment and quality audit activities, -- they may easily fall preys of incoherence or give the false impression that they are only doing lip service, even if they really mean it with their quality initiatives. The incoherence appears especially under conflicting situations. For example, there may be times when the quality and delivery interests are conflicting, and if the managers are not careful enough, they may destroy completely the message of Quality-First they try to convey under the TQC program.
  Another very frequently committed error resulting from it is the abandonment. Frequently the top management name a quality manager (normally a junior manager) and under the guise of good management, delegate all the responsibilities and authority to him. By doing so, they neglect the fact that most of the quality problems are cross-functional and the real authority and power to mediate the conflicting interests is difficult to be actually delegated. The real intention behind the insulation from the process is actually self protection. Any failure will be blamed on the incompetence of the quality manager. When this happens, the delegation becomes abandonment.
- Lack of cooperation by middle-management.
  This problem happens most frequently due to the middle-managers mis-perception of power losing threat. They are told that there will be more bottom-up activities with workers gaining more say. This is automatically and understandably interpreted as a power losing threat. As nobody likes to lose the wielded power, they try by every means to sabotage the program and guarantee their authority. This mis-perception happens due to insufficient understanding of CWQC activities. They do not perceive that the reach and limits of each group of activities are clearly defined. In Japanese companies this power losing threat never materialized because the middle management understood that the bottom-up activities are mostly centered on the improvement of work methods and do not affect the delicate authority issues. The supervisors at the Japanese factory floors never had any doubt of their continued authority on task assignment, performance evaluation, and promotion issues.
- Lack of motivation and teamwork spirit by blue collar workers.
  When the bottom-up activities are not well organized it often results in loss of interest by blue collar workers. For example, when the problem solving tools needed by Quality Control Circles-QCC are not properly taught and supported, it will most probably result in less significant improvements or even failures which dampens any initial enthusiasm. Another example may be seen from the management of suggestion system. If the suggestions are not perceived as being fairly evaluated and properly recognized, or if the replies take too long time the workers motivation in making suggestions will be reduced. This item will be discussed in more detail in next section.

Integrating TQC With Human Resources Management

In order to have a long lasting quality improvement system, it is essential to integrate it with the mainstream human resources management. Most normally the TQC programs are run as independent programs. As most of independent programs in firms are centered on specific tasks and are normally disbanded once the goal is achieved, the quality programs by being independent do not escape the same stigma. Through the integration with human resources management systems of the firms, the quality program may be constantly renewed and revitalized. The most important aspects of the integration are briefly described in the following:

- Integrated definition and promotion of quality campaigns (year, biennial, or triennial) emphasizing specific improvement topics (quality, productivity, safety, etc.) These campaigns help renew the challenges to the company and reaffirm the commitment of top management;

- Link the employees' personal interests with that of the firm. Through the combined application of profit sharing, closed career, job-stability, fringe benefits (e.g. housing, transport, and social club), and company symbols (e.g. uniform, flag, and anthem) to foster the sense of belonging;

- Provide a clear career perspective to all levels of employees and link it with the skills acquired and performance achieved in the quality improvement activities. The career perspective for blue collar workers in most of the western countries are much less visible than in Japan (Koike 1988). Without a clear career perspective and a weak linkage to the skills acquired and performance demonstrated in TQC is difficult to motivate the employees. The employees' motivation is the single most important factor for a sustained quality system;

- Use of the skills acquisition process and performance in TQC as performance evaluation parameters. These parameters will help the firms to implement a more objective performance evaluation system substituting the mostly subjective parameters adopted presently. These parameters may also focus more on the contribution to team performance instead of focusing exclusively on individual performance to foster teamwork spirit (i.e. you will only be promoted if you are well evaluated as a teamworker;)

- Creation and maintenance of the virtuous circle -- Challenge-Support-Achieve-Recognize-Challenge. -- That is, continuously challenge every employee to achieve; support him through timely training and other encouragement; and recognize and reward him properly for the well spent effort. This circle leads to increased self-esteem and fulfillment, which will result in renewed enthusiasm for the improvement activities.

Conclusion

We have been recommending to many firms about the relevance of integrating TQC and their existing human resources management systems, and the response has been more than encouraging. The major problem we have been facing is to make the two groups speak the same language. Most of the HR personnel do not understand well about the methodologies of TQC, and, conversely, most of the TQC personnel do not understand well the limitations and needs of HR functions. In the companies that have successfully put two groups of people to interact and integrate the results have been more than satisfactory. We expect that most of these firms will be able to develop a sustainable quality improvement system and justify through deed rather than words that the quality improvement is an ongoing process.

References

Chang, C. S., Individualism in the Japanese Management System, in Lee and Schwendiman (ed.) Japanese Management--Cultural and Environmental Considerations, Praeger, 82--88, 1989.

Flynn, D. M., Japanese Values and Management Processes, in Lee and Schwendiman (ed.) Japanese Management--Cultural and Environmental Considerations, Praeger, 72--81, 1989.

Koike, K., Understanding Industrial Relations in Modern Japan, St. Martins Press, New York, 1988.

Sours M. H., The Influence of Japanese Culture on the Japanese Management System, in Lee and Schwendiman (ed.) Japanese Management--Cultural and Environmental Considerations, Praeger, 27--38, 1989.

Takamiya, S., Nihon Keiei no Tokushitsu to Saikin no Doko (Japanese Management Characteristics and Current Trend), Sosiki Kagaku, 4(1}: 4--18, 1970

Tasugi, K., Keiei Rinen no Ruijisei to Soi (International Comparison of Business Philosophy--Similarities and Differences), Keizai Ronshu, 69(6): 19--41, 1969

Movement Toward a Virtual Engineering Management Program

E. Ray Ladd, Ph.D.
509-335-7653
FAX 509-335-3818
Home 208-289-4694

Associate Professor
Engineering Management
College of Engineering and Architecture
Dana 101
Pullman, WA 99164-2700
e-mail address: ladd@wsu.edu

The Engineering Management program (E M) at Washington State University is a state wide program currently serving Tri-Cities, Vancouver, Spokane, Wenatchee, and Boeing-Seattle. There is a constant demand to expand the program to other locations with the major limitation being resources (faculty, class room, communications). The program is delivered via closed circuit, live, two-way interactive television. With the advent of virtual technologies the WSU Engineering Management program is moving toward a virtual education program. In addition, WSU's five-year plan wants to establish itself as a virtual university. This paper will explore the design, ramifications and educational effectiveness of such an industrial based Engineering Management program.

Engineering Management Program Overview

The vision of the Engineering Management program at Washington State University is to develop an innovative university/industry partnership for training tomorrow's technical leaders. Engineering Management is a graduate degree program designed to meet the educational needs of working engineers in a manufacturing and non-manufacturing environment. The major thrust of the program has been project management, manufacturing management, quality engineering and management, systems engineering and environmental management. With judicious choice of electives, the student in Engineering Management can carve out a program of study which provides an emphasis in world class management of technology. The flexibility exists to add other courses to the management of technology emphasis, in order to enhance technical knowledge.

The Washington Higher Education Telecommunication System (WHETS) is extensively used in the delivery of the Engineering Management program at Washington State University. WHETS is a fully integrated closed circuit digital microwave system. This 34 credit hour degree consists of a minimum of 22 hours of core requirements and 12 hours of engineering and management electives. See Table 1 for Engineering Management program curriculum.

The Master of Engineering Management degree awarded through WSU Spokane, WSU Tri-Cities, WSU Vancouver, and at the Boeing Company Seattle may be completed in 2-3 years, although 6 years is allowed. The program is delivered via live television with two-way interactive audio/video, and may originate from any of the four sites including the main campus in Pullman Washington.

0-7803-3552-X/96/$5.00 © 1996 IEEE

Table 1 Engineering Management Program
Delivered from a team building, concurrent engineering emphasis

CORE REQUIREMENTS: 22 credits

Prefix	Topics	Credit
E M 505	Finance Management for Engineers	3
E M 540	Operations Research for Decision Analysis	3
E M 564	Project Management	3
Mgt 501/584	Human Resources and Team Building	3
Stat 430	Statistical Methods in Engineering	3
E M 591	Strategic Management of Technology & Innovation	3
E M 702	Comprehensive Final Exam/Project	4

SUGGESTED ELECTIVES: 12 credits

Prefix	Topics	Credit
E M 526	Constraints Management	3
E M 545	Decision Analysis	3
E M 560	Manufacturing and Operations Design and Strategy	3
E M 570	Quality Management	3
E M 575	Performance Management in Technical Organizations	3
E M 580	Quality Control and Reliability	3
E M 585	Quality Engineering Using Experimental Design	3
E M 590	Design for Manufacturability	3
C E 547	Fundamentals of Environmental Engineering	3
ES/RP 511	Environmental Law	3
ES/RP 544	Environmental Assessment	3
ES/RP 545	Hazardous Waste Management	3
M E 574	Advances in Manufacturing Science	3
M E 575	Computer Integrated Manufacturing	3
E M 517	Systems Simulation	3

FUTURE ELECTIVES

Prefix	Topics	Credit
E M 5XX	Systems Engineering	3
E M 5XX	Information Technology	3
E M 5XX	Decision Support Systems	3
	Total Required Core Credits	22
	Total Required Elective Credits	12
	Total Required Program Credits	34

Once a week, via a two-way, voice activated, audio/video system, students join other industry professionals in Spokane, Tri-Cities, Vancouver, and Seattle from companies like Boeing, Hewlett-Packard and Westinghouse. The evening classes are designed to teach students to analyze and improve the performance of their organizations.

Students in the program are normally full-time working engineers from a variety of industry throughout the area. To this end, courses are offered at times convenient to the working adult in a part-time format. WSU is the only educational institution in Washington State granted authority to offer a Masters of Engineering Management degree. Students registered in engineering management are all professionals, with varying engineering backgrounds and job descriptions. Both men and women engineers bring a significant amount of experience with them into the academic arena. Companies spanning from government contractors to self-employed, private groups to state and county agencies are represented by the students in the Engineering Management program. The average age of Engineering Management students is 35, but range from 22-60. Work experience on average is ten years.

The 34 credit hour Engineering Management program conforms to almost the same requirements as other engineering management programs with six core E M courses complemented by 12 hours of electives at the graduate level, and a project or comprehensive exam. Depending on a student's background, additional units may be required. Students in the program generally have a strong engineering or technical background. Table 1 shows the program curriculum.

The Engineering Management program was initially offered in 1982 with a cohort of 28 students. Although the curriculum has evolved over the last 14 years, the program objectives remain intact. These include our goal to prepare engineers to better participate in the management of technology and technology-based companies, to help engineers become better managers, to provide working engineers the opportunity to maintain technical competence, and to introduce and strengthen technical managers in the areas of quality and productivity.

In 1992, the Boeing Company, through the WSU Foundation, made a $7 million gift to WSU designating part of the funds be used to link Seattle to WHETS. This gave Boeing employees the opportunity to pursue a Masters of Engineering Management degree, and will ultimately provide Seattle with the distance learning opportunities for other placebound students via WHETS.

The difficulty that many of these adults face in acquiring the additional skills they need to meet their growing responsibilities are logistical. As full time employees, their professional schedules rarely allow them to attend traditional daytime classes at sites far from their workplace.

With a statewide mission of service to industry and the public, Washington State University has become a leader in the delivery of distance learning. Through its investment in the creation of the WHETS, the university has demonstrated its ability to deliver quality education to students throughout the state who would otherwise be unable to pursue opportunities for advanced learning.

The Engineering Management program is a prime example of the success of this strategy. The course structure is designed to accommodate its students, adults working full-time jobs. These classes are conveniently timed and located, available to employees in classrooms at their work locations or nearby learning centers and are scheduled one night per week, in three-hour blocks, after the normal working day.

In this unique learning atmosphere, students operate in two distinct group environments, one being local and one being network based. First, personal interaction and teamwork occurs with their colleagues at the shared learning site. Secondly, the WHETS environment offers simultaneous, interactive participation from students and faculty throughout the network. This style of learning, which mirrors the growing use of teleconferencing and other two-way and multiple-way communications systems now being utilized in industry, offers participants a variety of collaborative and team experiences through personal and electronic interaction. Since students in these classes come from a variety of professional engineering backgrounds they bring a variety of real-life experiences, and they learn from each other as well as from the instructor and course material.

The program addresses the following areas and concerns:

Define and Produce Product Development and Manufacturing Integration

- integrated product development teams
- understanding and translating customer requirements
- concurrent engineering - design, planning, tool design, etc.
- process development
- design for manufacturing
- digital definition
- variability control in process and product
- risk identification and mitigation

Lean Manufacturing for World Class Competitiveness

- strategic planning
- strategic management of technology and innovation
- policy deployment
- bench-marking
- process improvements
- flow-time reduction
- variability reduction
- right-sizing people and facilities
- team building
- make/buy decision
- sub-contract management
- kaizen and re-engineering principles
- quality in design

Production Systems and Information

- transition of requirements from macro process define to produce
- modeling and simulation
- manufacturing and procurement (produce) requirements generation, ordering and central systems
- world class operations management
- quality improvement and management systems
- project management

Capstone Projects

- team based company sponsored projects

Virtual Engineering Management Considerations

It is believed that in this environment the teleconferencing approach is in itself virtual in nature and that other virtual considerations must be carefully chosen to augment this approach, as well as, the nature of the materials being taught.

The university's five-year plan has indicated a desire to establish itself as a virtual university. The following points from this plan are also relevant to the Engineering Management program.

Situation

- The base of knowledge is growing exponentially requiring sophisticated technology to manage well.

- It is increasingly difficult for students to complete their education in a set time frame and in one location.
- The rate of change necessitates that learning take place not only at a University but also throughout one's life.
- Governmental projections indicate that funding for higher education will not keep pace with demand, so new learning approaches are required.
- We now have a base of experience which indicates that multimedia computer technology lends itself to individual learning styles and facilitates learning.

Target - for the year 2000

- WSU course-work will be designed to be more scaleable and effective for non-residential campus students and for all individuals who want to continue their education.
- Virtual education will be mainstreamed into 100% of the student experience by the year 2000.
- WSU be among the top 5 in the nation in virtual education by the year 2000.

Proposal

- Emphasize support for whole department projects whose curriculum impacts large numbers of students, and whose faculty are ready to move now.
- Establish a detailed business plan from which to promote and secure necessary internal and external funding.

To date virtual WSU has a course inventory of nearly 50 departments, approximately 110 faculty and well over 200 courses.

In this setting, Engineering Management is exploring and identifying ways which additional virtual dimensions can be added to the telecommunication based courses.

The purpose of adding virtual dimension to Engineering Management courses are to enhance the learning in terms of effectiveness and to some extent efficiency experienced by students and faculty.

Categories for virtual considerations via: www, e-mail, desktop video-conferencing and application software for courses include:

- Syllabus/class schedule/important updates
- Special class information
- Contact information
  - -Instructor
  - -Local experts
- Instructor/vita
- Team projects
  - -Assignment
  - -Example paper
- Research projects
  - -Assignment
  - -Example paper
- Grades
- WWW resources
  - -Broad based
  - -Class specific
  - -Useful
  - -Of interest
- Hypertext assignments
  - -Case studies
- Homework
  - -Assignment
  - -Solutions
  - -Examples of student homework
- Lecture notes
- Computer tutorials/assignments
- Student examples of exercises/projects
- Courses using the www in Engineering Management

- WWW seminar/tutorial — -Notes; -Exercises; -Resources
- Student group communication
- Electronic paper — -Turn in; -Grading
- Class roster — -Information on fellow students
- Bulletin board for class information
- Information searches
- Student advice about course
- Class office hours — -WHETS time allowed; -Desktop time established
- Course evaluation

For program materials include:

- Course descriptions
- Course syllabi
- Contact information
- Faculty vita
- Bulletin board
- Long range course schedule
- Short range registration information
- Thesis/project format and expectations
- Example thesis project
- Committee for project report (thesis)
- Student advice on program
- Historical timeline
- Calendar of events and deadlines
- College of Engineering and Architecture Homepage
- Advisors list
- Special articles referring to Engineering Management

Improvements that work well as class interface:

- WHETS — -Two-way audio/video; -Office hours for questions/answers
- WWW — -Class structure; -Handouts; -Instructions
- E-mail — -Paper hand in; -Small communication
- Application software — -Class specific

Summary Insights

- Effectiveness: The effectiveness seems to be as good if not better than the traditional class via student class evaluations, faculty evaluations of WHETS and WHETS evaluations filled out by students.

- Efficiency: The efficiency seems to be considerably less. Several internal studies indicates that WHETS as a media takes twice the class preparation time. And that many of the internet activities take substantially more time than conventional methods, often requiring extensive class assistance to be available.

- Course Strategy: Do not over extend a course via internet. Rather develop it incrementally over time as a course to course development.

- Personal Satisfaction: On a personal, instructional, satisfaction level because WHETS instruction allows travel 2-3 times per location within a semester, personal contact with students is still maintained and satisfaction remains high. If this personal contact were removed then satisfaction would be less. I have had students who have hidden in the electronic class with no personal contact and little personal satisfaction.

Current and Near Future Directions:

- Increasing need for technical education at off campus, remote sights, in order to improve working adults skills and knowledge, therefore increasing need for flexible access to education.
- Increasing interest in private and public organizations to provide educational access to employees.
- Increased flexibility and access to two-way audio, video and data communication via teleconferencing at a lower cost and increased capacity.

WWW Reference Addresses

Virtual WSU Course Inventory
http://www.wsu.edu:8001/vwsu/resources/course-inventory.html

Engineering Management Program - Vancouver
http://vancouver.wsu.edu/fac/holt/emprog/emprog.html

Engineering Management Constraint Management Course
http://vancouver.wsu.edu/fac/holt/em596cm.html

Washington State University Engineering Management - Spokane
http://spookey.sirti.org/~rumsey/enginman.html#stuff

Engineering Management (Spokane Faculty)
http://spookey.sirti.org:80/~rumsey/

Washington State University Business Courses on the WWW
http://w2.cbe.wsu.edu/courses/courses.html

Washington State University College of Business
http://w2.cbe.wsu.edu/

Mgt 585 Homepage
http://vancouver.wsu.edu/fac/tripp/mgt585/mgt585hp.html

Mgt 584 Homepage
http://vancouver.wsu.edu/fac/tripp/mgt584/mgt584hp.html

WSU's Homepage
http://www.wsu.edu

Washington State University Spokane
http://www.spokane.wsu.edu/

Washington State University at Tri-Cities
http://www.tricity.wsu.edu/

IT UNIX Systems
http://www.wsu.edu/UNIX_Systems

WSU Libraries' Homepage
http://www.wsulibs.wsu.edu

WSU Summer Session Catalog
http://www.summer.wsu.edu

First, Fast, and On-Time. The Path to Success. Or Is It?

Denis Lambert
(719)590-2515

Hewlett-Packard Co.
Colorado Springs Division
PO Box 2197
Colorado Springs, CO 80901

Stan Slater, Ph.D
(719)593-3580

University of Colorado at Colorado Springs
PO Box 7150
Colorado Springs, CO 80933-7150

This paper provides a critical examination of the three most significant and popular product development time performance measures: first-to-market, fast product development cycle time, and on-time schedule performance. The data will show that none of these measures are very robust as guidelines for product success and profitability. To address their deficiencies three new, superior guidelines will be proposed: first-to-mindshare, effective market introduction timing, and managed responsiveness. These new guidelines do not eliminate first, fast, and on-time as potentially valid strategies. They are merely saying that these strategies are no where near as universally effective as the product development literature and management gurus would lead you to believe. There are other, equally valid strategies, depending on the type of business environment and industry you are operating within. The three new tenets take this into consideration and create a framework that is more robust in that it allows for more diversity in your strategic options, and provides guidance for determining your optimal strategy.

First-to-Market

It is generally believed that those who get to market first achieve higher than average profits and greater market share. However, upon closer examination this relationship does not seem to be as strong as the advertising suggests.

Seven of thirteen studies reviewed by Kerin, Varadarajan, and Peterson (1992) showed a significant, positive relationship between first-to-market, profitability and market share. However, three of the seven had sample sizes of 22 products or less, and the remaining four were based on the PIMS database, which raises some concerns over the results of these studies. First, the PIMS database is a "self reporting" database, where 60 to 72% of the PIMS companies have classified themselves as pioneers. Furthermore, the PIMS database contains only survivors, therefore studies using the PIMS database as their source run the risk of introducing significant "survivor bias." Even many of the non-PIMS studies have omitted the non-survivors, biasing their results due to an inadequate adjustment for risk (Golder and Tellis, 1993).

Even considering the PIMS database survivor bias, some PIMS studies have shown either a weak or inverse relationship. Slater (1994) suggests that pioneers achieve slightly higher profits, but in doing so they take on greater risk than early followers. Miller et al. (1989) found an inverse relationship between order of entry and market share. Robinson (1988) found that order of entry only explained about 8.9% of the variance in market share. Moreover, using different analysis techniques, Moore, Boulding, and Goodstein (1991), and Van Honacker, and Day (1987) found that superior managerial skills and resources all but eliminated the relationship between entry order and market share.

Of the non-PIMS studies reviewed by Kerin, Varadarajan, and Peterson (1992), Urban et al. (1986) found advertising and market positioning to be much more significant

0-7803-3552-X/96/$5.00 © 1996 IEEE

drivers of market share than order of entry or the timing between market entries. Lilien and Yoon (1990,1989) found that the third through fifth entries were more successful than the first and second entries. Furthermore, Mitchell (1991) discovered that newcomers (those trying to break into a market) benefit from earlier entry, while market incumbents did better with later entries.

By tracking the market share of the main competitors in 36 product categories, starting with the initial market entry, Golder and Tellis' (1993) have attempted to remove the survivor bias and the problems of self-categorization as a pioneer. Their results show that the average market share for pioneers is approximately 10%. This compares to the average market share of 30% for pioneers in the PIMS database. Also, the market pioneers were the current leaders in only 4 of the 36 categories. Moreover, Golder and Tellis found that early leaders, those that enter the market during the early product-market growth phase, conform more closely to the PIMS results, which may support Lilien and Yoon's (1990) finding that the third through fifth entrants did the best. Early leaders entered the market approximately five years after the pioneer and they averaged a healthy 28% market share.

The research reviewed here challenges the traditional view of first to market wins. It appears that many other factors such as managerial skills, superior product design, superior resources, and customer understanding, in addition to order of entry, that drive product success.

Fast Product Development Cycle Times

Customers must improve at certain rates to remain competitive, but at the same time, due to learning curves, implementation expenses, infrastructure restrictions, and capital investments, they can change only so fast. Therefore, at any given time, there is some natural rhythm to the marketplace; one that is fast enough to remain competitive, but slow enough to recoup your investment. Organizations that are significantly slower than this natural cycle are definitely uncompetitive, and if it isn't obvious to them early on, it will be when their faster competitors begin trouncing them (Meyer, 1993; Patterson, 1993; Reinertsen, 1983; Smith and Reinertsen, 1991; Stalk and Hout, 1990; Wheelwright and Clark, 1992).

There is some evidence to back up the notion of natural cycle times, and contrary to conventional wisdom, they may be getting longer in some major markets. Bayus (1994) analyzed the IBM PC market lifetimes and determined that they are not getting shorter. He found the same results when looking at Compaq, Apple, and Hewlett-Packard. Bayus (1994) reported that, "there is no statistical relationship between PC model lifetime and the date of introduction [italics added]." In addition, when one looks at the introduction frequency of Microsoft Windows and Word, it can be seen that the length of time between introductions has increased.

Many organizations forget to ask the questions: what is the cycle time that the customers need and feel comfortable with, and how important is this compared to other customer needs? In fact some of the time-based competition leaders are now asking these questions and a few of them have consciously chosen to lengthen their development cycle times in order to improve another dimension of product success (e.g., customer understanding, quality, cost) (Stalk and Weber, 1993).

Finally, Cooper's (1995) analysis of 103 new products in the pharmaceutical industry found that time efficiency only accounted for 18% of the variability in profit. Although 18% is a respectable number, rephrased it says that 82% of the variation in a new product's profitability is determined by factors other than time.

On-Time Schedule Performance

The dominant belief is that financial success of development projects is very schedule sensitive. Schedule slips translate to large revenue and profit losses (a six month schedule slip reduces profit by 33% (Reinertsen, 1983)). However, the empirical evidence tells a very different story. Cooper (1995) found that the correlation between timeliness and profitability was extremely low (r2 = 0.09, n=103). The Manufacturing Vision Group had similar results, producing an r2 of 0.05 (n=22) (Bowen et. al., 1994). Data within one Hewlett-Packard division showed almost no relationship between hitting your schedule and financial performance (r2 = 0.03).

Rethinking First, Fast, and Predictable

What should be clear from the above analysis is that the goals of first-to-market, fast cycle times, and schedule predictability have several weaknesses and potential short-comings. What is needed are more general, overarching principles that incorporate first, fast, and on-time, along with other options to create competitive advantage. The proposed new tenets are as follows:

First-to-Mindshare (FTM)

Mindshare results when the vast majority of the target market think of your products first, and believe that your company offers the solutions that the other companies must measure up to. Purchasing from any other company represents a greater risk to the customer, especially when considering future competitiveness, compatibility, and support. *Order of entry of the product is no where near as important as order of entry of the dominant position in the customer's mind, and that requires much more than just having the first product.* The following table highlights how your focus changes when striving for first-to-mindshare as opposed to first-to-market.

First-to-Market	First-to-Mindshare
• Product focused	• Portfolio, company, industry focus.
• First product wins the war	• First product wins a battle, best portfolio and series of products, services, and company actions wins the war.
• Physical product	• Complete extended product.
• Short-term product-customer relationship	• Long-term supplier-customer relationship.
• Measure: When was the product introduced relative to the competition?	• Measures: How did the product contribute to future mindshare? How much do we influence standards and customer evaluation criteria?

Effective Market Introduction Timing (EMIT)

Effective Market Introduction Timing (EMIT) looks at customer adoption rates, how much control you have over the window of opportunity, your ability to forecast trends and sales accurately, market discontinuities, your infrastructure capacity (service, support, sales channels, etc.), your strategy for maintaining market visibility, technology and partner trends, and competitive pressures to determine what the optimal introduction cycle time is for your products These eight items are referred to as the EMIT drivers. Within the EMIT framework, faster is not always better, but instead a well thought out market-driven timing strategy is what's important.

Managed Responsiveness (MR)

Managed Responsiveness (MR) looks at the eight EMIT drivers, long-term business objectives, short-term cash flow needs, customer needs, and competitive offerings to provide a framework for making effective project management decisions that are responsive to both the internal needs and external market conditions. It is not blindly responding to events with simple decision rules (such as "hit the schedule"), but rather managing the response to these events in a rational, well-thought out manner, using the three levers of control within the project environment: schedule, development costs, and the scope of the work.

Market-Driven Project Management

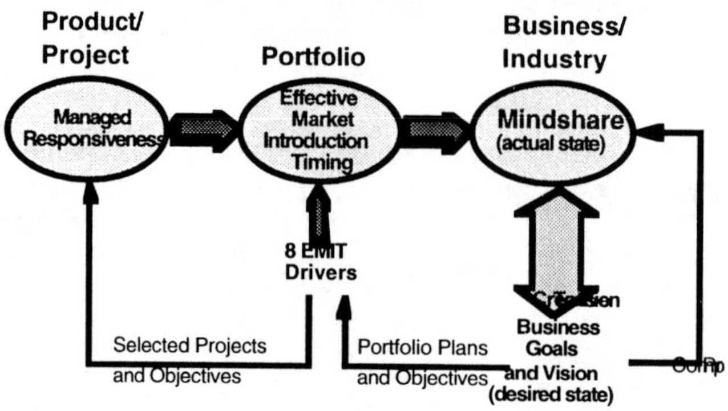

Figure 1

The integration of the three tenets proposed here form the basis for market-driven project management. As shown in figure 1, the fundamental goal is to build mindshare through the use of Effective Market Introduction Timing (EMIT), and Managed Responsiveness (MR). Mindshare requires a steady stream of products and therefore depends on the organization's ability to deliver products based on the eight EMIT drivers. However, in order to deliver the portfolio with the proper market introduction timing and frequency, the organization must develop excellent managed responsiveness capabilities at the project level. In other words, the business-level builds on, and is dependent on the portfolio-level capabilities, which in turn are dependent on effective project-level capabilities.

Concluding Thoughts

The three new guidelines presented here do not eliminate first, fast, and on-time as potentially valid strategies. They are merely saying that these strategies are not as universally effective as the product development literature and management gurus would lead you to believe. There are other, equally valid strategies depending on the type of business environment and industry you are operating within. The three new tenets take this into consideration and create a framework that is more robust in that it allows for more diversity in your strategic options and provides guidance for determining your optimal strategy. The other major advantage of first-to-mindshare, effective market introduction timing, and managed responsiveness over first-to-market, fast cycle time, and on-time schedule performance is that they take a portfolio view versus an individual product view.

Bibliography

Bayus, Barry, "Are Product Life Cycles Really Getting Shorter?" Journal of Product Innovation Management, 11, 1994, 300-308.

Bowen, H. Kent, Kim B. Clark, Charles A. Holloway, and Steven C. Wheelwright, "Development Projects: The Engine of Renewal," Harvard Business Review, September-October 1994, 110-120.

Cooper, Robert G., "Developing New Products On Time, In Time," Research-Technology Management, September-October 1995, 49-57.

Golder, Peter N., and Gerard J. Tellis, "Pioneer Advantage: Marketing Logic or Marketing Legend," Journal of Marketing Research, Vol. 30, May 1993, 158-170.

Kerin, Roger A., P. Rajan Varadarajan, and Robert A. Peterson, "First-Mover Advantage: A Synthesis, Conceptual Framework, and Research Propositions," Journal of Marketing, 56 (October 1992), 33-52

Lilien, Gary L., and Eunsang Yoon, "The Timing of Competitive Market Entry: An Exploratory Study of New Industrial Products," Management Science, 36, May 1990, 568-85.

Lilien, Gary L., and Eunsang Yoon, "Determinants of New Industrial Product Performance: A Strategic Reexamination of the Empirical Literature," IEEE Transactions on Engineering Management, Vol. 36, No. 1, February 1989, 3-10.

Meyer, Chris, Fast-Cycle Time, The Free Press, New York, NY, 1993

Mitchell, Will, "Dual Clocks: Entry Order Influences on Incumbent and Newcomer Market Share and Survival When Specialized Assets Retain Their Value," Strategic Management Journal, 11, January-February 1991, 85-100.

Miller, Alex, William B. Gardner, and Robert Wilson, "Entry Order, Market Share, and Competitive Advantage: A Study of Their Relationships in New Corporate Ventures," Journal of Business Venturing, 4, May 1989, 197-209.

Moore, Michael J., William Boulding, and Ronald C. Goodstein, "Pioneering and Market Share: Is Entry Timing Endogenous and Does It Matter?" Journal of Marketing Research, 28, February 1991, 97-104.

Patterson, Marvin L., Accelerating Innovation, Van Nostrand Reinhold, New York, NY, 1993

Reinertsen, Donald G., "Whodunit? The search for the new-product killers," McKinsey Quarterly, July 1983

Robinson, William T., "Sources of Market Pioneer Advantages: The Case of Industrial Goods Industries," Journal of Marketing Research, 25, February, 1988, 87-94.

Slater, Stanley F., "Competing in High Velocity Markets," IEEE Engineering Management Review, Vol. 22, No. 2, Summer 1994, 24-29.

Smith, Preston G. and Donald G. Reinertsen, Developing Products In Half the Time, Van Nostrand Reinhold, New York, NY, 1991

Stalk, George Jr., "Time – The Next Source of Competitive Advantage," Harvard Business Review, July-August 1988, 41-53.

Stalk, George Jr., and Thomas M. Hout, Competing Against Time: How Time-Based Competition is Reshaping Global Markets, The Free Press, New York, NY, 1990

Stalk, George Jr., and Alan M. Webber, "Japan's Dark Side of Time," Harvard Business Review, July-August 1993, 93-102.

Urban, Glen L., Theresa Carter, Steven Gaskin, and Zofia Mucha, "Market Share Rewards to Pioneering Brands: An Empirical Analysis and Strategic Implications," Management Science, 32, June 1986, 645-59.

Van Honacker, Wilfried R. and Diana Day, "Cross-Sectional Estimation in Marketing: Direct Versus Reverse Regression," Marketing Science, 6, Summer 1987, 254-67.

Wheelwright and Clark, Revolutionizing Product Development, The Free Press, New York, NY, 1992

Information Management for Knowledge Amplification in Virtual Enterprise

Jun Numata
(83) 45 338-5194 134

Information System Center, Sony Co
Goudo-Cho, Hodogaya-Ku, Yokohama
240 Japan

Bangyu Lei and Yukinori Iwashita
(83) 3 5448-5799 3-13-1

Sony Systems Design Co.
Takanawa, Minato-Ku, Tokyo
108 Japan

Based on an analysis of Sony's practice, the core of innovation in product development is viewed, in this paper, as Knowledge Amplification, which depends on an iterative exploration cycle and information emergence through the cycle by capturing both articulable and tacit knowledge. An information system for Knowledge Amplification in Virtual Enterprise is proposed, which includes a symbiotic network for information capturing and sharing through WWW and multimedia emails; an integrated product data manager for creation, evaluation, decision and archiving; an activity scheduler for workflow coordination, and an object-oriented database as shared information server for product, process and organization data.

Introduction

Innovative product development in Sony usually depends on a passion of a project team, including a strong project leader with clear product targets given from top management, and key persons from multi-disciplines. Many of these project teams was organized as so-called matrix-structure, which is a balance between two or more bases of grouping by the creation of a dual-authority structure.

Recently, the demand for unique, high quality and lower cost products with shorter development lead-time for dynamic global market is forcing industry to focus on new product development strategies such as Virtual Enterprise (VE). Actually, consumer electronics industry has practiced the VE concept a little early, driving by the relative shorter life-cycle and complexer multi-displinary technologies of its products. For example, to explore CD business, a fast-changing opportunity in the begining of 1980's, Sony organized the CD project in a form similar to VE. 4 independent groups of Sony and and other exter companies such as Philip were involved for technical design, business design, laser-based drivers design and semiconductor design, in a temporay consorium headed by Mr. Nakajima (Nakajima 1989). In particular, when developing CD's LSIs in Semiconductor Group, another exter-company was involved for virtual cooperation implicitly.

Apparently, to realize innovation in a competitive speed and cost is very important for a virtual enterprise because of VE's characteristics of product uniqueness and cost effectiveness. To support effective and efficient innovation in VE, information issues, i.e., the ability to capture, manage and share structured information, is the obvious area of interest, besides the physical, legal and cultural issues, since VE involves entities with different organization size, geographic location, computing environments, technology deployed, or process implemented.

In next section, we analyze Sony's product development process and consider the core of innovation in product development as Knowledge Amplification (KA), which depends on an iterative exploration cycle and information emergence through the cycle by capturing both articulable and tacit knowledge. The third section lists requirements to an information system for KA in Virtual Enterprise. Then, a

0-7803-3552-X/96/$5.00 © 1996 IEEE

architecture of this information system is presented in section 4. Discussions of the current application state and desired future developments of the proposed system, and conclusions follow.

Knowledge Amplification

A recurring theme through consumer electronics industry's history is the importance of innovation.

Sony mainly focuses on the design and manufacture of consumer electronics products from radio, tape recorders, video camera recorders etc., to CD, Floppy/Hard Disk and CD-ROMs in recent years, that contain mechanics, optics, electronics, electro magnetic, and sophisticated human interfaces (Whitney 1994). This is an area that does not have a single technical focus or well developed engineering models. A project team deals in many technologies and make heavy use of their experience to cover gaps in basic engineering knowledge, CAD tools, and multi-disciplinary techniques. Key elements of consumer electronic products typically carry out multiple functions that include logic, geometric compatibility, and energy transactions in multiple media, such as light, liquid, gas, rotating metal, chemistry and combustion, simultaneously. This results in that product design usually cannot be separated into distinct phases. Instead, components and assemblies are designed together. Extremely strong interest on mobility and smaller size impels consumer electronic products to be as light and compact as possible, and that their components are packed closely together or interact strongly with each other, and distributed in very limited space. Because of this complexity, products are usually designed with simultaneous considerations about manufacturing or assembling methods.

Therefore, our new product development process is very opportunistic, explorative, and incremental. It is not performed using a fix set of events applied in an ordered way. No one person has all the knowledge and skills needed to design and produce new products alone. It is a process wherein various activities occur in an opportunistic manner, either top-down or bottom-up. It can be viewed as exploration process because knowledge about the space of possible solutions has to be obtained before goals can be well formulated. Typically, the initial description of the solution is incomplete and/or ambiguous and/or inconsistent. Reversing some previous decisions either for refinement or for resolution of conflicts with other decisions is inevitable At a coarse level of granularity, it can be divided into 8 subprocesses, i.e., product planning, design planning, conceptual design, primary design, detailed design, manual prototyping, mold prototyping, and mass production, although these tracks are not unique to any particular product, and steps may differ from product to product. The staggering of their start points and overlaps are indicative of partial information sharing and their orders are indicative of their precedence. The degree of dependency that may exist between any two consecutive tasks of product development is much higher than what were used to be, so that the process is almost not possible to be task driven, rather than, at fine granularity levels, information-driven or conflict-driven. In this sense, we epitomize the core of new product development practices as "Knowledge Amplification".

The knowledge amplification in product development means knowledge enhancement as information emergence and syntheses arising from the maturity of a product in related organizations, the product development process in a project team, and the problem-solving process of each individual. product development process

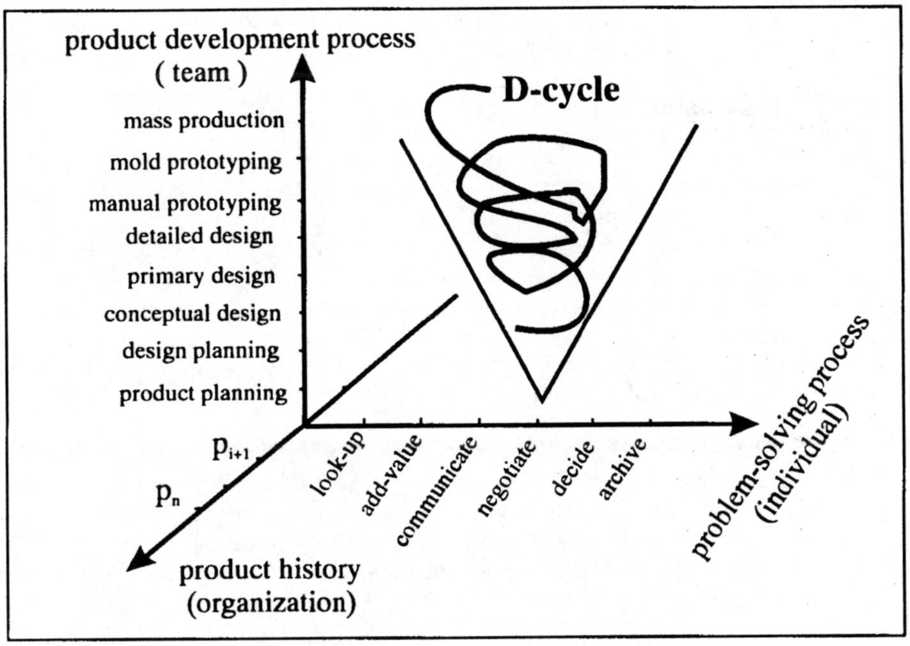

Figure 1: Spiral of Knowledge Amplification

As shown in Figure 1, the ontological dimension of KA can be represented by problem-solving processes of individuals, product development process of related team, and existing enterprise/product data of related organizations. The three elements form a information space, in any point of which KA may occur. The epistemological dimension of KA can be described from tacit knowledge to articulable knowledge, based on the organizational knowledge creation (Nonaka 1993). The process of KA, based on an analysis of Sony's product development, has been proposed as a D-cycle by one of the authors (Numata 1996) as follows:

- confirmation of the status of technological levels among engineers;
- collection of related knowledge and information;
- creation of solution concepts
- convergence of knowledge and information into a solution archetype;
- verification of the solution archetypes

Requirements to information management

To support knowledge amplification in virtual enterprise, the need for overall coordination, consistency, control, and integrity of product data, engineering knowledge, design ideas, and design rationale are critical. The information generated by each designer must be viewed in the context of information generated by other related designers, the enterprise historical product data, and the organization as a whole. In particular, complex electro-mechanical products are growing in complexity as they shrink in size and perform more functions. Getting a product rapidly to market, will depend on countless, highly interdependent decisions by numerous individuals in different disciplines, from senior executives to production engineers, spread worldwide around many engineering centers and manufacturing facilities. Both the way that engineering information is managed by insufficient computerized, paper intensive acquisitions, and the way a project team cooperate together by meetings and "nominications" (communications over drinking after work) will become ineffective, inefficient, increasingly difficult or even impossible in future virtual enterprises. In summary, based on the three elements of KA's information space, an information system for KA should: • manage total enterprise information globally and in real-time; • support the whole product development process across organizations, in particular, the early stages of product development such as functional and conceptual design • support cooperative team work in product development both within and across organizations concurrently, for example, the information collection (look-up), communication, and negotiation.

Architecture of an information system for KA

To transcend the barriers of distance, time, and heterogeneity in computer equipment of virtual enterprises, an information system for KA in VE which satisfied the above mentioned requirements has to be implemented by deploying new technologies. Some core technologies are already moving into place: Internet and related communications facilities and services; OMG (Object Managment Group) and related object technology; and, STEP (STandard for the Exchange of Product model data) and related information modeling technologies. An integrated application of them provides for the technical basis for our system. In addition, Advances in databases, multimedia, and web-based programming is taken into account in our system too.

As shown in Figure 2, this system can be viewed as three functional layers, that is, Symbiotic Network, Knowledge Bases, Communication Basis. From the viewpoint of functionality, the Symbiotic Network is a optimally tailored cooperative working environment, which supports effectively team members sharing information and knowledge, exchanging ideas, presenting/solving questions, and developing new solutions by using information technology in a symbiotic way. In fact, it is an Internet based environment integrating each team member him/her self, their on-line individual computer tools such as electronic circuit simulators, and KA Tools. These k~4 Tools support each member to locate and use information from the Knowledge Bases, to exchange ideas and information each other, and to perceive/confirm tacit knowledge through the Multimedia Navigator during real-time video conferences. Conversely, information created in the symbiotic network will be formalized and stored into the Knowledge Bases.

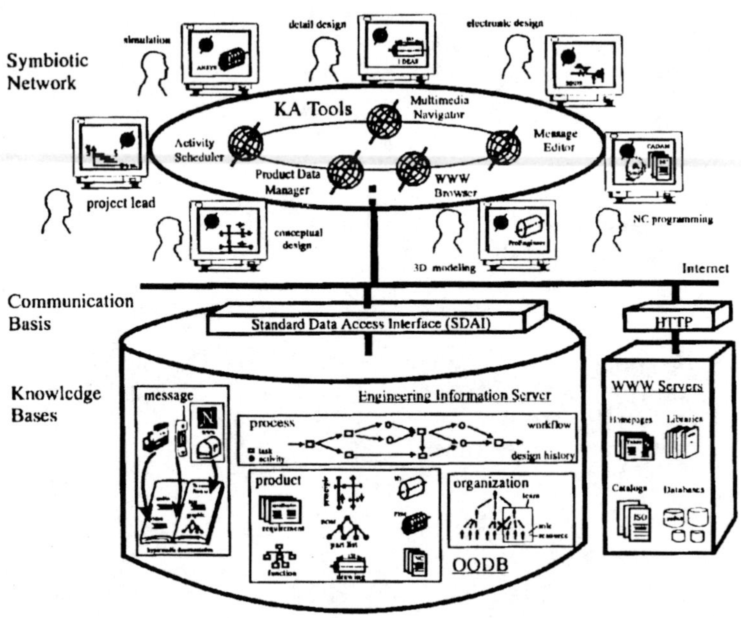

Figure 2: Architecture of Information System

Despite of various knowledge taxonomy, our information system handles knowledge as two parts, tacit and articulable one. The tacit knowledge is perceived and confirmed in the Symbiotic Network, while the articulable knowledge is represented in the Knowledge Bases. After one engineer confirms other engineer's tacit knowledge he/she has perceived, it can be re-organized to generate new product information, in other word, new articulable knowledge. On the contrary, the articulable knowledge can be refereed directly when acquisition of tacit knowledge, either perception or confirmation, occurred in the Symbiotic Network. The Knowledge Bases is composed of an Engineering Information Server and WWW Servers, providing information sources for the Symbiotic Network. We develop the Engineering Information Server to manage product data, process data, organization data, and messages; while we support the team members to access information from other external WWW servers.

The Communication Basis is an Internet-based communication channel, which provides communication services for all information flow, giving access to its naming, directory, and security services.

Conclusions

The proposed system is been developing in Sony Systems Design Co. by some pilot projects. The programming platform is complete Web-based, by using C++ for object-oriented database, Java, and HTML/VRML browsers. Taking the design of printers as example, related product data models has been implemented in an object oriented database.

The proposed architecture will be improved by developing the Engineering Information server into various Internet-based generic information services directly. It is the author's belief that to implement the proposed information system by integrating the modern Internet communication technology, object technology, and information modeling technology will not only result in a revolution in engineering management and product development, but also provide for an solution for information management issues in virtual enterprises.

Reference

Nakajima, H. "The Age of New Audio", Sony's Innovation in management Series, Vol. 4, 1989

Nonaka, I. Management to Promote Knowledge Creation, Nippon Economy Press, Tokyo, 1993

Numata, J., and Taura, T. "Case study: a network system for knowledge amplification in product development process", IEEE Transaction of Engineering Management, 1996

Whitney, D. and Nevins, J., and Fazio, T. and Gustavson, R. "Problems and Issues in Design and manufacture of Complex Electro-Mechanical System", CSDL-R-2577 The Charles Stark Draper Laboratory, Inc., 1994

A Corporate Business Excellence Process

Richard J. Leo

Vice President and General Manager
Xerox Quality Services
Rochester, New York 14644

This document will trace the evolution of Leadership Through Quality (the Xerox approach to TQM) and the development of a company-wide self-assessment and business results diagnosis methodology at Xerox. It will also show how quality assessment is linked to the annual operational planning and management process. It describes the structure and management process by which Xerox managers use facts and data to conduct a company wide assessment of TQM practices, and to diagnose and eliminate the root causes of gaps in business performance to produce process-based continuous quality improvement.

Historical Highlights And A Crisis Of Opportunity

After 15 years of Leadership Through Quality at Xerox, one might ask why there is a continuing need for quality. Early on, David T. Kearns, then Chairman and CEO of Xerox, offered some insight into this question when he said "This is a race without a finish line." While Xerox has made great strides, the world we live in continues to change at an accelerating pace, and we can succeed only if we continue to focus on quality and use our quality strategy to improve our business results.

Paul A. Allaire, current Xerox Chairman and CEO, has recently pinpointed the challenge: "We are facing a 'crisis of opportunity.' On the one hand, we see attractive markets, and we have superior technology. On the other hand, we won't be able to take advantage of this situation unless we can overcome cumbersome, functionally driven bureaucracy and use our quality process to become more productive."

The approach that Xerox has taken to realize the vision inherent in our "crisis of opportunity" was to reengineer our operational definition and management process for the implementation of Total Quality Management company-wide. Today the Xerox 2000 Leadership Through Quality Strategy and Management Model are the means by which a corporate business excellence process is implemented annually across the corporation on a worldwide basis. Our objective is to use a company-wide quality assessment and certification process along with our Policy Deployment Process called "Managing For Results," to achieve world-class productivity as the key enabler to capture market opportunities.

Leadership Through Quality Evolution

The beginnings of the quality movement at Xerox can be traced to our Japanese subsidiary, Fuji Xerox. In 1976, Yotaro "Tony" Kobyashi (then president of Fuji Xerox) began the quality journey by introducing the principles and tools of total quality management, advocated by the Japanese Union of Scientists and Engineers (JUSE), to help the people of Fuji Xerox improve the company's competitive position. This effort resulted in the people of Fuji Xerox receiving the Deming Prize in 1980 and in making many improvements in its business. Examples of these improvements included a doubling of product reliability, a one-half reduction in machine cost, a one-half reduction in time-to-market, a three-quarter reduction in machine size, and a dramatic increase in copy quality as perceived by the customer.

Meanwhile, business problems in the rest of the Xerox family were becoming

0-7803-3552-X/96/$5.00 © 1996 IEEE

apparent on a worldwide basis. The early 1980s was a time of intense business pressure. Market Share and Return On Assets (ROA) declined. Over 20,000 jobs were eliminated, and employee satisfaction dropped as people began to lose confidence in senior management.

During 1982, CEO David Kearns had been witnessing firsthand the implementation of Total Quality Management (TQM) at Fuji Xerox. Kearns believed three key factors made the Japanese better than their U.S. competitors: cost, quality, and expectations. Kearns noticed that the Japanese set much higher expectations for their performance and output than the United States operations did.

At about the same time, several of Kearns' key advisors suggested the use of a TQM approach to address our fundamental business problems. To put this recommendation into action, Kearns commissioned 11 senior managers to explore and outline a TQM approach for Xerox. This design team would draft a "vision book" that would project how the company would look if TQM efforts were successful. The intent was to construct a strategy and implement a plan whose impact would be broad enough and deep enough to turn the competitive tide.

As a result of this effort, the management team decided in 1983 to implement the Xerox approach to TQM, Leadership Through Quality, on a worldwide basis. It took four years to train over 100,000 people worldwide to ensure a common understanding and approach to quality at all Xerox facilities.

The Leadership Through Quality approach was endorsed and quickly made operational in a strong partnership with our Labor Union Management team and line workers. Annual events to recognize the achievements of quality teams began to grow in popularity and participation, eventually involving thousands of employees on a worldwide basis. The use of meeting management, interactive skills, problem solving, quality improvement, benchmarking and other TQM practices became commonplace across the company.

During the four-year deployment period, Xerox had to weather continued downward trends and business pressures, but the effort paid off. In addition to being a strategy for change within Xerox, Leadership Through Quality was a tool for measuring progress in business improvement. By 1987, the people of Xerox had reversed the decline in market share and

Company-Wide Quality Assessment

Xerox came to understand the importance of periodic assessments in order to provide an opportunity for reflection and to stimulate future initiatives. The key to business success, after all, was in the reinforcement of both the commitment and the pursuit of continuous improvement.

The first quality assessment at Xerox was conducted in 1987, initiated by Paul Allaire after he became president of the company. The purpose of this assessment was to evaluate the progress in deployment of Leadership Through Quality, and to compare the quality results against the goals established in 1983. The objective was to measure and define gaps in quality implementation against our strategy and plan and to gain insight into why progress was not being achieved at a faster pace.

The assessment results were used by the management team to help focus on quality improvement opportunities in all areas of the company. This early work on quality assessment set the stage for the later use of the more comprehensive assessment criteria contained in the Malcolm Baldrige National Quality Award and the European Foundation for Quality Management (EFQM) criteria.

The Role Of Quality Awards

The Xerox involvement with the world's premier quality awards has helped us formulate our approach to internal quality assessment. In order to DO things differently, we must SEE things in a different way. The assessment criteria from world-class quality awards help a company see itself from the customer point of view and to organize itself and its priorities for improvement from that point of view.

Through experience, Xerox managers have learned that periodic interventions are required to stimulate and refocus an organization engaged in a race without a finish line. Management learned that in applying for quality awards and responding to each part of the award criteria, Xerox could use the insights gained to address the gaps identified in the 1987 quality assessment. The assessment process, as specified by the various criteria for the different quality awards, proved to be an excellent business improvement tool. For instance, applying for the Deming Prize helped Fuji Xerox gain critical insights it would not have perceived otherwise. Other Xerox units, including the Netherlands, France, Britain (two), Canada, Mexico, and Australia, have similarly benefited.

Integrating Tqm With Business Operations

The quality award application process itself would provide a vehicle necessary for continuous improvement. By the middle of 1991, some of the senior executives at Xerox observed that the quality agenda, while influenced by the findings of various assessment activities and founded upon sound management principles, looked more like an unorganized "laundry list" than a well-structured, focused quality strategy. One member of the Allaire team suggested that Xerox needed an operational definition of Leadership Through Quality in order to put communicable meaning into the Leadership Through Quality concept.

It was also suggested that an internal worldwide Leadership Through Quality annual assessment using this definition should be integrated into the annual planning process to support and measure continuous improvement. The idea behind this suggestion was to link the cycle of quality assessment and diagnosis with the resource-allocating process found within the annual operational planning process.

This linkage between policy deployment for cross-functional work process improvement activities and current business issues is the foundation of the Xerox approach to Business Excellence. The methodology allows teams of Xerox people and managers to identify the root causes for gaps in business performance and target opportunities for improvement. This identification process, in turn, guides the allocation of resources to develop and implement work process improvements.

Xerox Tqm Principles

Thus, Xerox decided to create a Management Model and internal Assessment System for Leadership Through Quality, and to position it as part of the management process that is used in the development of the annual operating plan for each unit. The following five principles guided the design of the management model and assessment methodology.

1. Fully satisfy customer-defined requirements.
2. Top management leads by example.
3. Involve, empower, and motivate employees.
4. Improve results through continuous process improvements by reducing:
    - A Variability in work process outputs
    - B Cycle time through process improvement
    - C Costs flowing from improvements in variability and cycle time

5. Integrate continuous quality improvement into the daily management of the business.

Xerox Operational Definition Of Quality

In order to bring structure to all of the learning produced from both internal and external quality assessments, Xerox moved back to the basics and designed a model that specifies desired state performance; assesses current state performance; and then identifies the gap between these two-focusing on the critical few actions necessary to close the gap.

Work began on this company-wide quality assessment model, formally called the Xerox Management Model, in 1990. This model was designed to become the operational definition of Leadership Through Quality at Xerox. The definition of the Xerox Management Model is based on the quality principles and practices of leading experts, the Green Book (an early Xerox quality strategy document), the Baldrige Award criteria, the European Quality Award criteria, ISO 9000 criteria, and the lessons learned from the Xerox quality assessment (see Figure 1, Leadership Through Quality: The Synthesis of Its Operational Definition).

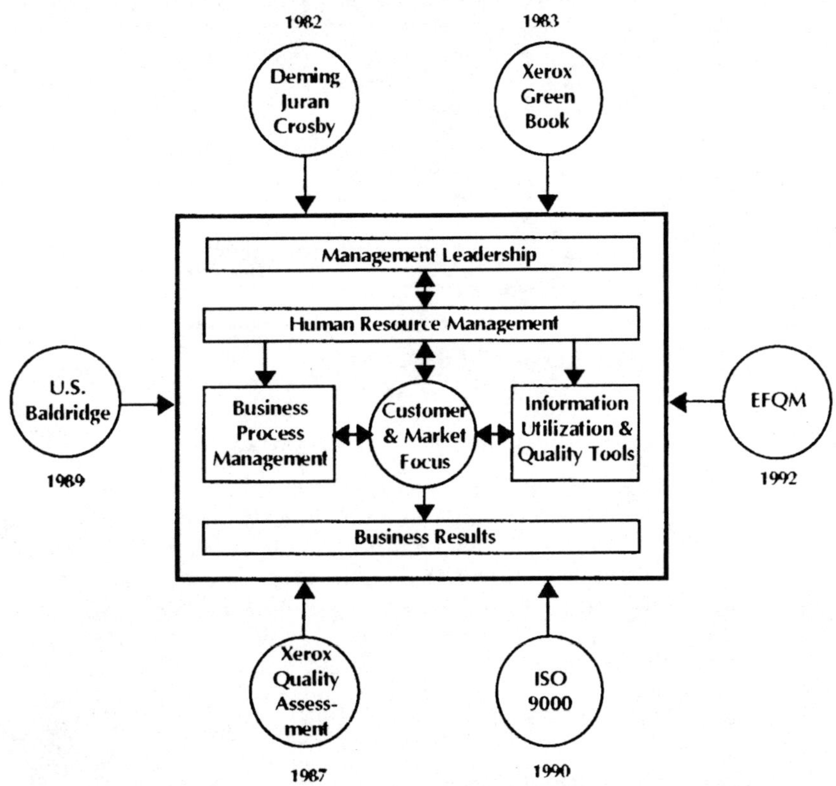

Figure 1. Leadership Through Quality: The Synthesis of Its Operational Definition

Xerox 2000 Leadership Through Quality Strategy And Management Model

The Xerox 2000 Leadership Through Quality strategy is designed to support new organizational and managerial paradigms and to embrace the concept of change as a powerful tool to achieve productivity breakthroughs. Its implementation relies on work process improvement by empowered people to create and increase value for the benefit of our customers.

To ensure ease of understanding and total integration on a company-wide basis, we have created the holistic Xerox Management Model. (see Figure 2, The Xerox

Management Model). The model gives a logical framework that integrates and connects the TQM practices necessary to produce desired business results. The model focuses on the "HOW" of implementation and provides a common company-wide "LANGUAGE," which facilitates a uniform implementation of our management process. The model is designed to be flexible in terms of the unique measurements required to assess TQM improvement across business units with different operational characteristics. It will also evolve in the future to reflect our ever-changing, dynamic business environment.

The Xerox TQM Model identifies the six major categories of our business and how they interact with each other. At the center of the model is the customer and the markets in which customers are found. The elements above and alongside the customer and market focus depict how we organize and manage ourselves to work with our customers. Our ability to focus on our customers is measured by the business results that we achieve.

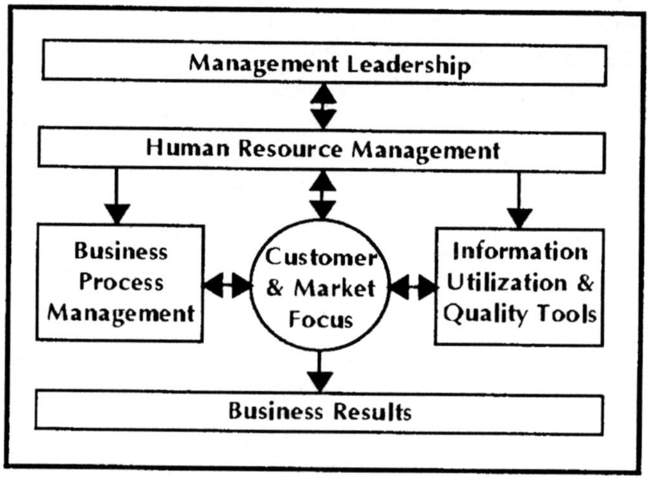

Figure 2. The Xerox Management Model

The Vision Of Xerox As A World-Class Company

The model describes a vision of strong, innovative managers, who lead empowered, motivated, and challenged employees. Both managers and employees focus on the customer and continuous improvement of business processes, using the appropriate quality support, information, and tools. Their efforts produce sustained and improved business results.

Each of the six categories of this Leadership Through Quality operational definition has a desired state, which is defined in terms of the vision of Xerox as a world-class company.

The six major categories in the Xerox Management Model and their desired states are described below:

1. Management Leadership

Xerox management displays a customer focus, exhibits "role model" behavior, establishes clear long-term goals and annual objectives, establishes strategic boundaries, and provides an empowered environment to achieve world-class productivity and business results.

2. Human Resource Management

Xerox management leads, motivates, develops, and empowers people to realize their full potential. All employees are personally responsible for continuous learning and the acquisition of competencies required to achieve business objectives and to continuously improve productivity for customers and Xerox.

3. **Business Process Management**

Business processes are designed to be customer-driven, cross-functional, and value-based. They create knowledge, eliminate waste, and abandon unproductive work, yielding world class productivity and higher perceived service levels for our customers.

4. **Customer and Market Focus**

Current, past, and potential customers define our business. We recognize and create markets by seeing patterns of customer requirements. Anticipating and fully satisfying those requirements through the creation of customer value achieves Xerox business results.

5. **Information Utilization and Quality Tools**

Fact-based management is led by line management; it is achieved through accurate and timely information and by the disciplined application and widespread use of Quality tools.

6. **Business Results**

The Document Company Xerox is the largest and one of the most productive and profitable companies in the global documents market because:

A. Our customers' expectations are exceeded and we achieve 100% satisfaction.

B. Our people are motivated and challenged to achieve superior results for their customers.

C. We achieve optimal Market Share positions across our range of business, resulting in a premium Return On Assets.

D. We grow revenue profitably while effectively managing our cost base to achieve growth in shareholder value.

There are 42 elements of the model that further divide into 35 TQM Practices and 7 Business Results Goals that support this vision. The Business Results Goals are organized into the six categories (this structure is informally called the "Six by Forty two" model). Each TQM practice and business goal has an identified process-based measure that is used to assess progress toward the desired state.

Tracking Quality Improvement

This working definition of Leadership Through Quality is our guideline for measuring quality improvement from one planning cycle to another. Each of the 4~ elements are evaluated on a seven-point scale. The scale ranges from a low end at level one equivalent to no demonstrable competence to the high end at level seven, representing world-class achievement. The gap between the current state and the desired state for each of these practices is used to establish priorities and to identify where effort should be placed for improvement during the next planning cycle. Specific emphasis in the operating planning process is placed on the top priorities identified during this assessment. These priorities are called the "vital few." Year-to-year progress is tracked by using the same assessment mechanism and comparing progress achieved over the prior assessments.

Figure 3, Tracking Progress in Quality Improvement, illustrates the Xerox company-wide assessment scheme as it evolved from the initial application of Leadership Through Quality, through the quality award assessment, to the self-assessment administered by each unit in the company on a worldwide basis.

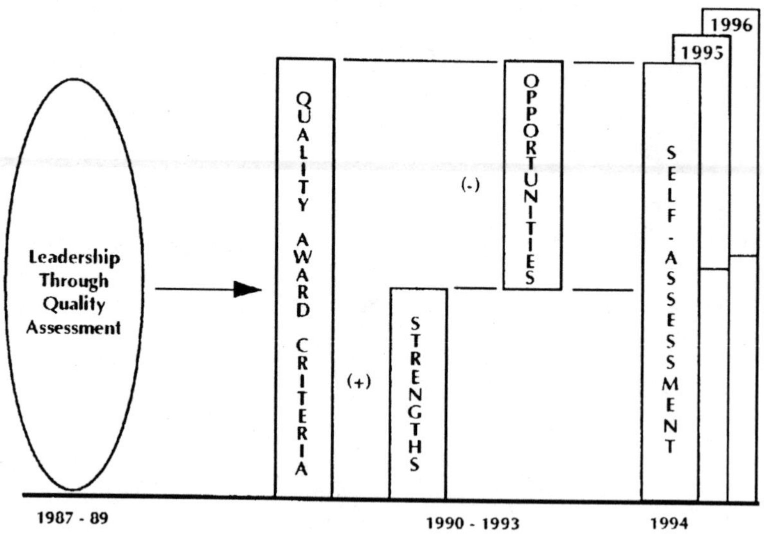

Figure 3. Tracking Progress in Quality Improvement

Managing For Results

Xerox learned that a closed-loop planning process is essential to assuring that identified improvement goals are actually achieved. Figure 4, Results Diagnosis and Process Improvement Cycle, shows in more detail how the self-assessment and diagnosis methodology is used to measure progress. In-process measures for each of the 35 Quality Practices are correlated with the 7 Business Results Goals. The diagnosis of gaps between predicted and actual performance for the 7 business results indicates which of the 35 practices needs attention to help close the gap and to improve business performance in the following planning cycle. These key practices are targeted as the "vital few" for attention when units form their operating plans for the next year

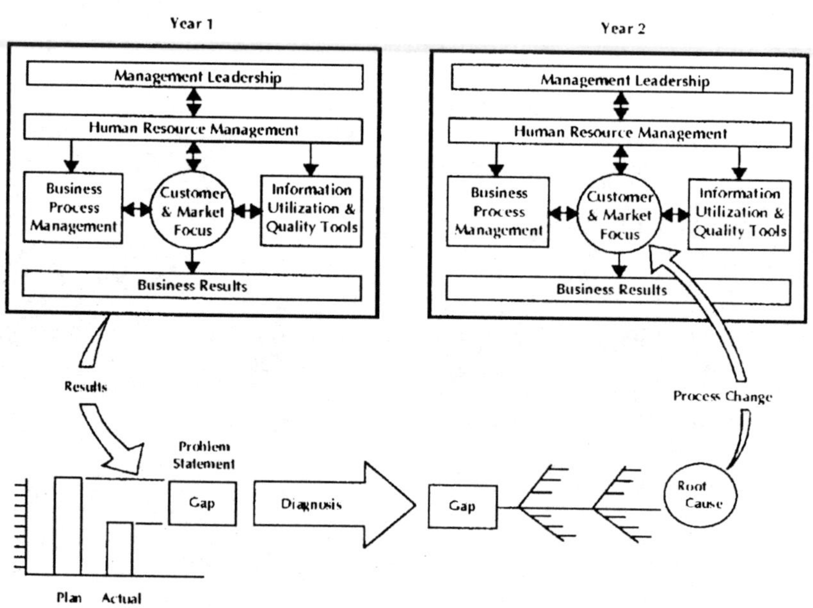

Figure 4. Flesults Diagnosis and Process Improvement Cycle

The policy deployment process provides the linkage between the initial diagnosis and identification of root causes for gaps, the setting or organizational objectives, and the deployment of them to the employee level. This process is called "Managing For

Results" in the United States and "Policy Deployment" in both Rank Xerox and Fuji Xerox. Figure 5, Managing for Results Annual Process, illustrates the major steps in this deployment process and the linkage from the corporate direction to the annual objectives of individuals. This process enhanced the Leadership Through Quality operational definition by establishing linkage between the measurement of process performance, diagnosis of performance gaps, and definition of responsibility for action to improve the process. Congruence in these steps of the planning process assures improved alignment between the strategic direction of the corporation and the actions of Xerox teams and individual employees. It also focuses the corporation's limited resources on the vital few strategic practices targeted for improvement.

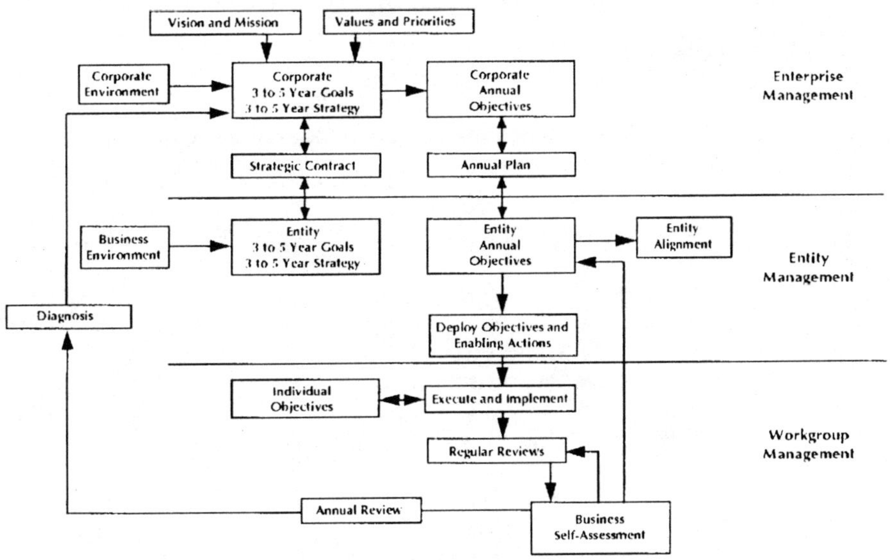

Figure 5. Managing For Results Annual Process

Company-Wide Quality Assessment And Certification

In 1990, the United States Marketing Group (USMG) component of the Xerox Business Products and Systems organization decided to implement the assessment process throughout its 32,000 employee organization. During the application process for the Baldrige award, only a small subset of the total U.S. operation was examined by the Baldrige site visit team. The decision to certify all operating entities in the U.S. was made to involve all employees in the quality assessment process. The approach used was to certify each operating entity within USMG using quality assessment criteria to validate its position. To be certified, an operating unit had to achieve a minimum level of performance.

More than 140 operating entities were assessed by a team of 50 internal examiners in a series of site visits that took place from April 1990 through June 1991. These internal examiners were members of the USMG Senior Management Team and the Corporate Quality Office. Thus, examiners became more knowledgeable and committed to TQM. The examinations revealed a range of TQM competencies. Six units achieved world-class levels of performance, while five units did not initially qualify. Units that did not certify used the assessment results to initiate improvement actions and were given follow-up site visits. Eventually, all entities were certified. Today, this baseline serves as a standard for assessing those critical few areas that need improvement within an operating unit and for measuring improvement over time.

Rank Xerox further refined this assessment methodology in 1992 for its Business Excellence Certification (BEC) Process, which is highlighted in its application for the European Quality Award.

The BEC management process enables each unit of the company to compare its own performance in all of its quality practices against an agreed set of minimum standards from the Xerox Management Model, and to make the necessary improvements to meet them over time. The results are judged by managers from other units of Rank Xerox. Achievement of the standards is formally recognized by certification. (See Figure 6, Business Excellence Certification.)

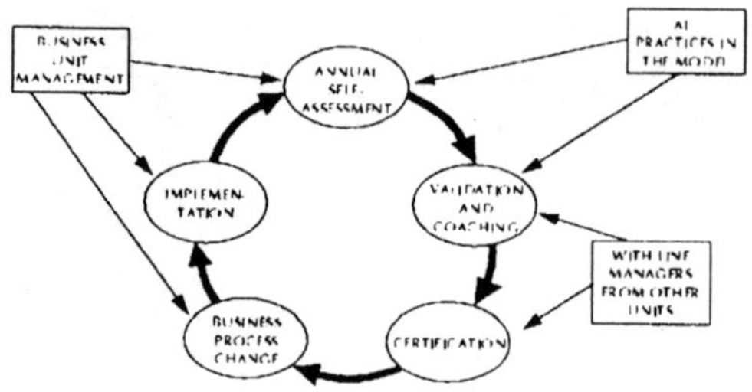

Figure 6. Business Excellence Certification

Each unit in the company is asked to assess its own performance in over 40 quality practices, which are organized into the six categories of the Xerox Management Model. Each of these practices is measured against a desired state. he resulting information is communicated to the unit and then within Rank Xerox. The management team therefore understands how each unit in the organization measures up to our standards for quality performance, and how the company as a whole is doing. The total company view is very useful to identify the systemic issues that can be targeted for breakthrough improvement.

This process has been implemented in more than 40 European, middle eastern, and African countries; 18 manufacturing plants; the Rank Xerox Technical Center; and the International Headquarters location. It involved 250 senior managers trained as examiners and 500 senior managers trained in self-assessment.

Self-assessment is followed by a validation step: a simulated site examination that includes guidance and coaching by line managers, who are qualified examiners from other units. After validation comes a formal site visit and examination. If the minimum criteria are met, certification is awarded. The examination is also conducted by qualified examiners from other units, but not by the same units that were involved in the validation and coaching step.

The results of the assessment help shape the following year's operating plan. The BEC management process provides a detailed picture of where each unit stands, which processes need to be improved, and insights into the root causes that must be eliminated to produce improvement. It produces important customer-focused and process-based information that complements the other sources of data that go into the Policy Deployment prioritization. Another advantage of the BEC process is that it enables comparison of best practices among units. Managers are able to examine the operations of other units from across Europe and use that information to continuously improve the way they do business.

Organizational Reflection And Learning

During 1995, all Xerox units will participate in the annual assessment and validation process. An assessment of the 35 TQM Practices and diagnosis of the 7 Business Results will be used to fine tune priorities and investments for the 1996 Operating Plan cycle.

The intensive learning gained from the Baldrige and other quality award experiences was refined into a cohesive process for assessment and improvement (Figure 7, Continuous Improvement: Year-After-Year). This process yields continuous improvement in a defined set of vital few areas in every Xerox unit, each aligned to defined corporate goals and objectives. The corporate goals themselves are aligned to serve the customers in the markets where we compete.

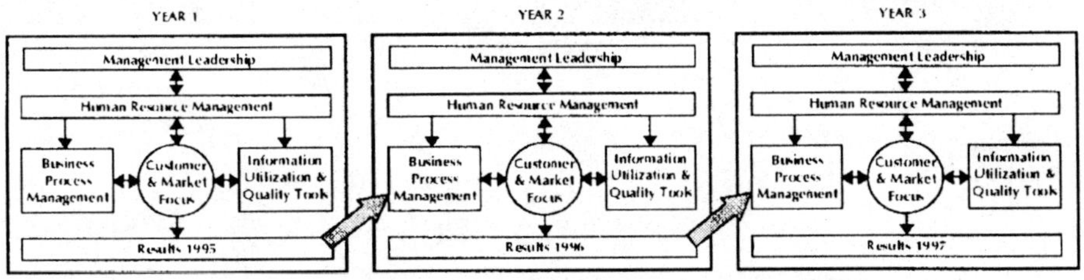

Figure 7. Continuous Improvement: Year-After-Year

We have used what we have learned over the past 15 years about Total Quality Management to position ourselves as The Document Company Xerox, with the goal of becoming one of the most productive and profitable companies in the global document market.

Is Design Quality a Function of Time?

Alan Leong
(206)685-4343

Industrial Engineering
University of Washington
Seattle, WA 98195
alanl@eskimo.com

Robert P. Smith
(206)543-4605

Box 352650
Industrial Engineering
University of Washington
Seattle, WA 98195
robsmith@u.washington.edu

This proposed study extends the work of Smith & Tjandra (1996) and Smith, Eppinger, & Gopal (1995). Using a group design exercise, it investigates the extent to which the final design quality is a function of the number of design iterations. Some recent models of engineering design prescribe ways to compress the time to design a product. These models imply that the total number of design iterations is not related to final quality. This study tests this assumption through a videotape analysis of a simulated design environment.

Keywords: *design research, group design research*

This study is an extension of Smith & Tjandra (1996) and Smith, Eppinger, & Gopal (1995). Design teams composed of senior level engineering students are videotaped performing a design exercise. In this, subjects work within a group design environment called the Delta Design Exercise (Bucciarelli, Goldschmidt, & Whipple 1991). Objective measures of design quality are embedded in the task and made explicit to the subjects as part of their written task instructions. Photographs are made of the final designs to analyze for measures of final design quality. The videotapes are analyzed for the number of iterations, duration of iterations, and the total design time.

Smith & Tjandra (1996)

The procedures in this proposed study replicates a prior experiment in Smith & Tjandra (1996) and incorporates their test subject data as part of this study's current sample. The intent of Smith & Tjandra (1996) was to make observations about the iteration process as it related to the group process of design. In this, subjects work within a group design environment called the Delta Design Exercise (Bucciarelli et. al, 1991). Another prior study also used the Delta Design Exercise to examine the appropriateness of design iteration modes as well as examine the impact of pre-specified design strategies on design quality (Smith, Eppinger, & Gopal, 1995).

The original intent of the Delta exercise was as a teaching aid for showing the necessity for cooperation in design, and for demonstrating the effect of non-quantifiable constraints (such as aesthetics) on the design process. The goal of the exercise is to create a two-dimensional structure in an artificial world which meets a number of constraints and objectives.

The simulation game participants must build their structure using equilateral triangles called Deltas. Deltas come in two colors, red and blue, which have different thermal and cost properties. Each participant is given a distinct role. There are four roles: architect, thermal engineer, structural engineer, and a project manager (cost estimator). The exercise begins with a period of training where each individual

0-7803-3552-X/96/$5.00

participant becomes familiar with their own role, followed by a design phase where the participants come together to negotiate a complete design artifact which meets all of the objectives.

The tasks are all classified as either being analysis or synthesis tasks. An analysis task is when the participants are calculating, either using pencil and paper or using a calculator. The synthesis tasks are when the participants are physically manipulating the pieces. There is also observed idle time. These occur because a player has no calculations left to make on the existing structure, or because their minds wander briefly. The analysis phases constitute the bulk of the time spend during the iterative portion of the design process. The synthesis phase typically involve making incremental changes to the structure to improve performance against whatever criteria were deemed out of compliance.

Current Study

Small design groups composed of engineering students and professional engineers are videotaped performing the Delta Design Exercise. Design activities are coded as to whether the participant is engaged in analysis or synthesis. Each iteration is composed of an analyses and a synthesis phase. The final designs for each design team are photographed for analysis and evaluation of design quality.

Not only will this study analyze videotapes of student design, but teams composed of working engineers will also be videotaped. Their inclusion increases the external validity of the research by including professional engineers as a comparison to the groups composed entirely of students.

It is expected that the total number of design iterations will not be correlated with measures of design quality. It is also expected that the total number of iterations will be correlated with the total design time.

No design process is imposed on the experimental subjects, rather they chose what process to use to determine an appropriate design result. There was no interference by the experimenters in the process after the beginning of the design phase, except in the case of answering a specific technical question, such as how to interpret the meaning of a phrase in the instruction.

This study makes no claims about causality, e.g. that the number of design iterations causes design quality. It also makes no claims about how to improve design quality. This study simply examines the relationship between the number of design iterations and final design quality within design teams. If for example, the number of design iterations is related to design quality, this sets the stage for further research that examines how iteration-related variables impact design quality. The study of how iteration is related to design time and quality is a central concern of the certain design models.

Why Iteration?

In this study, the focus is on measures of design iterations and design quality, as well as on design time. It is clear why a project on design might focus on length of time and design quality. After all, faster product cycles and higher design quality are sources of competitive advantage. However, what makes *iteration* a special focus? And if iteration is important, what design models exist which are relevant to this proposed study?

In most design tasks, a final design is developed through a series of incremental steps in which the emerging design is evaluated at each step to provide information for subsequent steps (Nevill, 1988; Nukala, Eppinger & Whitney, 1995; Wiste & Cannon, 1995; Dwarakanath & Wallace, 1995). At each of these incremental steps, or iterations, activities are repeated because the design fails to meet some yardsticks, or because newly information changes the design situation (Smith & Eppinger, in press a).

What then is the empirical evidence for the role of iteration on design?

Osborne (1993), for example, found that iteration is the main cause of variability in the lead times of projects at a major semiconductor manufacturing company. Using a computer simulation to design simple mechanical tools, Ramachandran, Langrana, Steinberg & Jamalabad (1992) found that better initial starting points resulted much fewer iterations and less overall design completion times. By giving the experimental groups an optimal ordering of design tasks, Smith, Eppinger & Gopal (1995), were able to shorten design times without sacrificing quality as compared to a control group which were given no such strategy. The Smith, et. al. (1995) study should be underscored as a precursor to this current proposed study.

Iteration thus plays a fundamental role in design. As a consequence, well-formulated engineering design models which incorporate iteration are especially relevant to this study.

Iteration and Design Models

Not surprisingly, there are some recently proposed *prescriptive* models of product design utilize iteration as a central parameter (e.g. Ha & Porteus, 1995, Smith & Eppinger, 1994b). They propose strategies to speed up design by reducing the cycle time of each iteration in the process, and reducing the number of iterations (Ulrich & Eppinger, 1995). It is this idea which is being tested, that the quality of the final design is not related to measures of overall time (or iteration).

One such model is the sequential iteration model (Smith & Eppinger, in press a). This uses a probabilistic iteration prediction with a sequential structure to the design process to find a schedule of design tasks which minimizes expected completion time. In Ha & Porteus (1995), such is accomplished by manipulating when design reviews occur. Other models involve ordering tasks so that the probability of rework is minimized (Smith & Eppinger, in press b; Smith, 1995). These models assume that final design quality is unaffected by the number of iterations or overall time to completion.

There also exist prescriptive engineering design models that do not place iteration in such a central spotlight. For example, some propose dealing with imprecision and incomplete information as the central parameter for increasing design quality or compressing time (Haroud, Boulanger, Grelle & Smith, 1995; D'Ambrosio & Birmingham, 1995; Antonsson & Otto, 1995; Wood, Antonsson & Beck, 1990). While others focus on improving the information exchange process (Christian & Seering, 1995). However most of these do assume that design occurs iteratively.

Discussion

The problem domain used in this experiment is artificial. It is a two-dimensional representation of an unrealistic physical domain. The main advantage of this is that all experimental subjects have a similar exposure to the domain, so the experiment is repeatable. Furthermore, it is possible to attribute observations made to design behavior, rather than to prior knowledge. The exercise is well constructed in that the work involved compares to real-world engineering analysis and synthesis. The disadvantages include that all participants are learning about their world as they progress, which is a more naive form of design than would be typically encountered in real-world engineering. Furthermore, the participants know of the artificiality of the situation and so may not have motivation to perform as well as they might in an industrial situation.

The design problem is small, requiring only a few hours to define and determine an acceptable result. This is useful so that videotape records can be kept, and to make observations throughout the design process. It is difficult to extrapolate observations to larger design problems which last several months or years.

Although the research question appears simple, the implications are rich with content. For example, what if the total number of design iterations is unrelated to time but is related to design quality? However, it is not being claimed that the observations will hold directly for more realistic domains, but this method is an appropriate for early steps towards understanding the role of iteration in the design process, and will indicate how future research should be focused.

References

Abernathy, W.J. & Rosenbloom, R.S. (1968). Parallel and sequential R & D Strategies: Application of a simple model. *IEEE Transactions on Engineering Management,* EM-15(1), 2-10.

Bucciarelli, L.L., Goldschmidt, G., & Whipple, B.C. (1991). Delta - A Design Exercise. MIT Science, Technology, and Society Program.

Christian, A.D. & Seering, W.P. (1995). In O.M. Jadaan (Ed.), Proceedings of the 1995 Design Engineering Technical Conferences: Volume 2. New York: The American Society of Mechanical Engineers.

D'Ambrosio, J.G. & Birmingham, W.P. (1995). Preference-directed design. *Artificial Intelligence for Engineering Design, Analysis and Manufacturing,* 9, 219-230.

Dwarakanath, S. & Wallace, K.M. (1995). Decision-making in engineering design: Observations from design experiments. *Journal of Engineering Design,* 6(3), 191-205

Ha, A.Y. & Porteus, E.L. (1995). Optimal timing of reviews in concurrent design for manufacturability. *Management Science,* 41, 1431-1447.

Haroud, D., Boulanger, S., Gelle, E. & Smith, I. (1995). Management of conflict for preliminary engineering design tasks. *Artificial Intelligence for Engineering Design, Analysis and Manufacturing,* 9, 313-323.

Nevill, Jr., G.E. (1988). On research methodology towards a scientific theory of engineering design. In S.L. Newsome, W.R. Spillers, & S. Finger (Eds.) *Design Theory '88: Proceedings of the 1988 NSF Grantee Workshop on Design Theory and Methodology.* New York: Springer-Verlag.

Nukala, M.V., Eppinger, S.D., & Whitney, D.E. (1995). Generalized models of design iteration using signal flow graphs. In O.M. Jadaan (Ed.), Proceedings of the 1995 Design Engineering Technical Conferences: Volume 2. New York: The American Society of Mechanical Engineers.

Osborne, S. (1993). Product Development Cycle Time characterization Through Modeling of Process Iteration. Masters Thesis, MIT

Ramachandran, N., Langrana, N.A., Steinberg, L.I. & Jamalabad, V.R. (1992). Initial design strategies for iterative design. Research in Engineering Design, 4, 159-169.

Smith, R.P. (1995). *Managing risk by reordering tasks in engineering design.* University of Washington, Industrial Engineering Program Working Paper, revised January1995.

Smith, R.P. & Eppinger, S.D. (in press a). A predictive model of sequential iteration in engineering design. *Management Science.*

Smith, R.P. & Eppinger, S.D. (in press b). Identifying controlling features of engineering design iteration. *Management Science.*

Smith, R.P., Eppinger, S.D. & Gopal, A. (1995). Testing an engineering design iteration model in an experimental setting. Proceedings of the 4th International Conference on Design Theory and Methodology. Scottsdale, AZ: ASME.

Smith, R.P. & Tjandra, P. (1995). Experimental observations of iteration in Engineering design. University of Washington, Industrial Engineering Program Working Paper, October 1995.

Ulrich, K.T. & Eppinger, S.D. (1995). Product Design and Development. New York: McGraw-Hill.

Viste, M.J. & Cannon, D.M. (1995). Firmware design capture. In O.M. Jadaan (Ed.), Proceedings of the 1995 Design Engineering Technical Conferences: Volume 2. New York: The American Society of Mechanical Engineers.

Wood, K.L., Antonsson, E.K. & Beck, J.L. (1990). Representing imprecision in engineering design; Comparing fuzzy and probability calculus. *Research in Engineering Design,* 1, 187-203.

Systems Engineering: The Application of Processes and Tool in the Development of Complex Information Technology Solutions

Regina S. Lightfoot, Ph.D
(703) 974-1817

Bell Atlantic Corporation
1310 N. Courthouse Road
Arlington, Virginia 22201 USA

This paper examines the procedures, tool, and standards which comprise the field of systems engineering. The design, development, and implementation of integrated information and telecommunication solutions is a complex and extensive process requiring the involvement of numerous subject matter experts throughout all phases of the process. These experts represent a variety of disciplines, including software engineering, configuration management, network architecture, facility design, operations, procurement, software design, development, and coding, and database management.

The discipline of systems engineering provides a systematic approach for applying procedures, tools, and standards to an information-oriented problem for the purpose of defining, designing, managing, and implementing an effective integrated information technology solution. This paper identifies the major phases of the system engineering process and it highlights the tools which enable large disparate organizations to work effectively in the design, development and implementation of complex integrated information and telecommunication solutions.

Introduction

The design, development, and implementation of integrated information and telecommunication solutions is a complex and extensive process requiring the involvement of numerous subject matter experts throughout all phases of the process. These experts represent a variety of disciplines including software engineering, configuration management, network architecture, facility design, operations, procurement, software design, development, and coding, and database management. Such expertise is rarely located in a cohesive group and often the expertise of employees is supplemented by that of consultants and vendors who may take on a number of these responsibilities. In addition, the complexity of the process is often compounded by aggressive development schedules thus making the efficient involvement of all experts and resources critical to the success of the project.

Definition and Purpose of Systems Engineering

Systems engineering is the orderly process of applying procedures, tools, and standards to a problem for the purpose of defining, designing, developing, managing, and implementing an effective integrated solution which meets a specific need. When applied to the discipline of information technology, systems engineering uses this orderly application of procedures, tools, and standards in the development of information technology solutions which meet a specific information-oriented need. The purpose of systems engineering is to enact the successful design and development of an integrated solution and to prevent the development of a solution which fails to meet the stated need. This is accomplished through the proactive and systematic completion of well defined activities, the expert use of tools to manage these activities, and the implementation of procedures to ensure that the objective of the project (i.e. to meet the specific need) is effectively and thoroughly addressed.

System Engineering Activities

The activities which comprise the systems engineering process are designed to address all of the major events required to advance a project from problem definition through

0-7803-3552-X/96/$5.00 © 1996 IEEE

design and specification to implementation. These activities are performed by the systems engineering and design team. They include the following:

1. Definition of the Need/Problem:

This is a complete description of the functional or performance features desired. This description is presented in qualitative and quantitative terms. These desired features are generally driven by a deficiency with the current process/system. This is stated from the perspective of the user(s).

2. Identification of Requirements:

This is a formal statement of what the system must be able to do. This activity results in the development of the functional requirements document. It describes "what" the system must do not "how" it should be accomplished. It addresses issues such as a) the functions which the developed solution must be able to perform, b) the timeframe in which the system must be operational, c) the environment in which the system will operate, d) the constraints to which the system must adhere, e) the performance characteristics which the system must meet, f) the operational and maintenance requirements the system must incorporate. The Functional Requirements Document can also include a state diagram which highlights the manner in which the system transitions from one state to the next based upon the inputs it receives. A state diagram is an effective means for depicting the various data options which a system might encounter and how the system should react to these inputs.

3. Conceptual Design:

This activity results in a high level object-oriented diagram which defines the major functional elements necessary to meet the identified requirements. The conceptual design is the initial attempt to define how the identified requirements will be met. It outlines the functional boundaries of the elements and identifies the relationship of the elements to each other. Examples of functional elements which are commonly found in complex information technology solutions are:

a. End User Interface or Delivery Subsystem
b. Control Subsystem
c. Access Subsystem
d. Operation and Support Subsystem
e. Transport Subsystem (if extensive telecommunication facilities are used)
f. Processing Subsystem (stated in terms of the specific processing functions to be performed)

4. System Functional Analysis:

This is the process of allocating the system level requirements (from Activity 2) into subsystem components, using the functional elements identified in the conceptual design (Activity 3) as a guide. This process breaks requirements "down from the system level, to the subsystem, and as far down the hierarchical structure as necessary to identify input design criteria and constraints for the various components of the system". (Blanchard, 1991)

5.Design Options:

This process identifies the possible hardware/software options for each subsystem. This process also identifies the interface options for the system including protocols and network elements.

6. Trade-off Analysis:

This is the process of evaluating the various design options according to a predefined set of criteria for the purpose of recommending a hardware/software configuration for each of the system components.

7.Subsystem/Component Design and Specifications:

This is the detailed description of the functions to be performed by each component and the specific procedures and processes each component will execute to accomplish these functions. This document provides the detailed description of each function that the component performs and the specific steps it completes to accomplish these functions. It also defines the interface specifications for each set of components that interface directly with each other. This document specifies "how" each subsystem will accomplish the various "whats" for which it is responsible.

8. Prototype Development:

This activity results in the development of a model of the component/subsystem for the purpose of clarifying requirements and identifying design constraints and limitatoins. A prototype often attempts to illustrate the user interfaces as a means of determining if the design approach will meet the specific needs described in Activity 1. However, a prototype can also demonstrate a technical approach for accomplishing a specific set of tasks. Prototypes which demonstrate technical capabilities are oftencalled brassboards and they help illustrate capability of the design as well as limitations and constraints of the design.

9. Subsystem/Component Development:

This is the activity in which the subsystem/component is constructed according to the defined specificatons. This activity includes the acquisition and/or construction of hareware, the final detailed design and coding of software, and the integration of hardware and software elements to create a single complete subsystem.

10. Subsystem/Component Testing:

This activity verifies that the subsystem/component completely meets the defined specificatons established in Activity 7.

11. System Integration:

This is the process of linking the various components together via the defined interface elements into the final system solution. The activity of integration requires the execution of a series of interface tests to ensure that each interface within the system is functional and operates as defined. These tests examine the manner and format in which data flows between the components.

12. System Testing and Acceptance Testing:

This activity verifies that all functional and performance requirements are met by the system solution. The system and acceptance tests are derived primarily from the requirements identified and documented in Activity 2.

13. Deployment/Operational Use:

This is the process of deploying the system for general use. It is at this point that the system is accessible by the user to meet the general needs for which it was designed and developed.

The activities provided in this list are commonly used in most information technology solutions. The process of systems engineering is customizable to the specific needs of the project and the matter in which any one activity or a series of activities is performed can be tailored to the specific need. The process defined by the systems engineer may include all of these activities in a single sequential logical process or it may require the repetition of several specific activities in the process. At times some Activities may occur concurrently. At other times, the process may include several iterative phases in which certain Activities are repeated in a number of times for the purpose of gathering additional information prior to a critical decision point. Reilly (1993) identifies several different paradigns which a systems engineering process could follow. These include:

a. The Early Prototype in which a prototype of the system is developed first and then the information gathered from the review of this prototype is used in the identification of system requirements.

b. The Spiral which incorporates a series of prototypes, one after each major definition and design phase. The prototypes are refined as additional information is gathered. This process helps the user confirm that the design team has an accurate understanding of the needs.

c. The Rapid Development process brings the final solution to fruition through a series of deliveries. Each delivery is a deployable system. This approach allows the user to experience the system in an operational environment, thereby giving the user a more accurate perspective of how the system meets or fails to meet the problem.

These paradigms are depicted in figure 1 and are provided to illustrate that the systems engineering process can and should be tailored to the specific challenges of the problem being solved.

The Management of the Systems Engineering Process

The activities described above are managed in part through the use of tools which contribute to the orderly execution and thorough completion of the activities. These tools include:

1. The Work Breakdown Structure (WBS):

This is a list of activities and deliverables organized by system which identifies all work elements which must be accomplished in order to deploy the solution and make it operational. It is a top-down process in which each successive level of detail consists of the specific activities which must be completed in order to achieve the previous layer. A WBS is not an organizationchart but can be included in the WBS. The WBS can be depicted in a tree structure or it can be provided in an outline format.

2. Gantt and PERT Charts:

A Gantt chart outlines all major activities and milestones on a timeline such that the elapse time for each activity is evident. The activities which occur concurrently are also evident from a Gantt chart. The detailed activities in the WBS can be used as input in the construction of the Gantt chart. A PERT chart identifies the dependencies between major activities and is useful for identifying the critical path on a project.

3. Configuration Management Plan:

This document identifies the bseline characteristics of a system by specifying the functional and physical characteristics of each item utilized as part of the initial system solution. This includes the specific features and release versions of the hardware, operating system software, database systems, storage devices, protocols and standards, and purchased (off-the-shelf) software systems. This document also describes the process for evaluating, implementing, and tracking changes to the elements which comprise the baseline.

4. Integrated Logistic Support (ILS) Plan:

This document describes all activities which are required in support of the initial acquisition, testing, and sustained maintenance of the system. These activities include facility preparation, procurement and acquisition of test equipment, transportation, delivery, and handing issues, personnel requirements, training requirements, and technical data and maintenance requirements. The degree to which the system integration, system test, and deployment activities are successul is directly related to the thoroughness of the ILS Plan. If the system design and development process is successful but the deployment, support, and maintenance activities encounter problems, it is generally due to omissions and errors in the ILS Plan.

5. System Test Plan:

This document specifies the tests that are required to verify how the system performs. it includes test procedures which "provides the detailed account of each step to be carried out for each test" (Reilly, 1993) including the objective of the test, the facility to be tested, the conditions under which the test is to be performed, the resources which are required, and the expected results.

6. Acceptance Test Plan:

This document specifies the tests that are required to verify that the system meets the functional requirements identified by the user/customer. Similar to the System Test Plan, the Acceptance Test Plan also includes test procedures which provide a detailed account of each step to be carried out during the test.

7. The System Engineering Management Plan (SEMP):

This document identifies all major management activities to be performed in association with the design, development, implementation, and delivery of the solution. The SEMP is a progressive document which can be updated as information and needs change. The SEMP describes the technical activities which must be performed to fulfill the project objectives, the specific engineering paradign which will be followed for this project, and the resources which will be required. As discussed earlier, the paradigm is derived by determining which systems engineering activities will be applied to this project and in what order. The WBS, Gantt charts, PERT charts, the configuration management plan, and the ILS plan are generally included in the SEMP. In addition, the SEMP can include cost projections, measurement indicators, and references associated with the standards which will be employed on the project.

Major System Engineering Milestones

The system engineering process should include a minimum of four major check points of milestones. These milestones are points in the process in which the progress of the project can be reviewed and the decisions can be made concerning whether the project should proceed to the next phase of development. These milestones enable the concerned parties to determine if the work performed to date is consistent with the users's expectations. The four check points are as follows:

1. The System Requirements Review:

This is also called the Functional Requirements Review. This review occurs upon completion of the Functional Requirements Document. Its purpose is to verify to the user that the systems engineering team understands what the solution must be able to do. It provides the user with the opportunity to provide additional input and feedback to the team prior to the start of the design activities.

2. Preliminary Design Review:

This review occurs upon completion of the trade-off analysis. It demonstrates the design structure which will be employed in the development of the system. It also provides initial cost estimates (derived from the design options and trade-off analysis) associated with the development of the solution. It addresses the feasibility of the solution (again derived from the design options and trade-off analysis) regarding the availability of technology to meet the system requirements.

3. Critical Design Review:

This review occurs prior to the start of the Subsystem/Component Development activity. The purpose of this review is to demonstrate that all issues of design have been addressed and "that the design is mature and ready to implement". (Reilly, 1993)

4. System Integration Review:

This review occurs after the integration and testing activities have been completed but before the system is deployed. This review presents the engineering team with the opportunity to demonstrate that the system is ready for deployment. It includes a review of the acceptance test as well as a review of all documentation, procedures, facilities, conversion teams, support teams, and deployment plans.

Summary

The activities, tools, and procedures described in this paper provide the systems engineer with an approach for moving a project from problem definition through design to implementation in an effective and orderly manner. The complexity of information technology solutions requires the effective application of activities and resources tailored to the specific project. The challenge of systems engineering isto develop an approach which matches the required ctivities to the needs of the project in an orderly manner such that the resources are used effectively in the development of a solution which meets the user needs.

References

Blanchard, Benjamin S. Systems Engineering Management. John Wiley & Sons, Inc. New York, 1991.

Reilly, Norman B. Successful Systems Engineering for Engineers and Managers. Van Nostrand Reinhold, New York, 1993.

Figure 1: EXAMPLES OF SYSTEM ENGINEERING PARADIGMS

Early Prototype

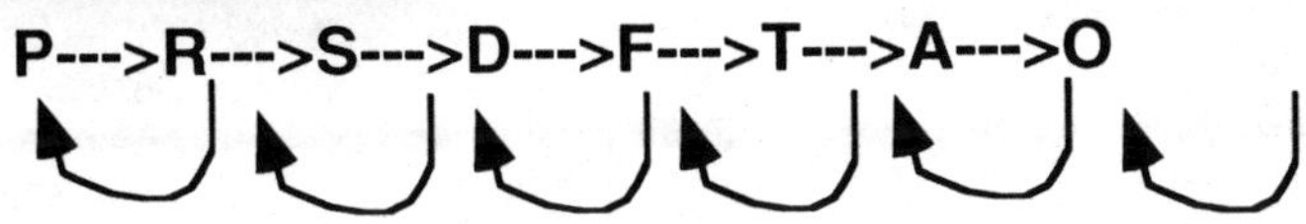

Spiral

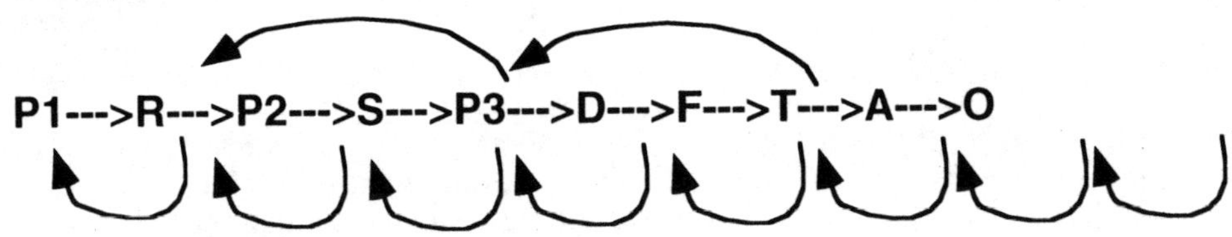

Rapid Development

RSDFTAO-->RSDFTAO-->RSDFTAO....RSDFTAO (final)

Legend: R=Requirements, S=Specifications, D=Design, F=Fabrication, T=Integration & Test, A=Acceptance, O=Operations, P=Prototype

From: Reilly, 1993

Information Technology Support of Strategic Management Activities

Richard G. Donnelly, Ph.D
(202) 994-7155

The George Washington University
2115 G Street, N.W.
Washington, DC 20052 USA

Regina S. Lightfoot, Ph.D
(703) 974-1817

Bell Atlantic Corporation
1310 N. Courthouse Road
Arlington, Virginia 22201 USA

This study explores the relationship between an organization's strategic management process and its use of information and communication technologies. Effective strategic management especially under conditions of changing markets, competition, technology and other environmental factors, appears to rely upon insights from experts. The use of information technologies, the fundamental tools for the information age, could be expected to augment the access to experts and their insights and information and thereby enhance strategic decision making and strategy implementation .

The purpose of this study is to identify the characteristics of the information technologies which are used during the various phases of the strategic management process and to determine the sufficiency of the information technologies in meeting the needs of strategic management activities. The development of a technology-strategic management model illustrates the manner in which information technologies currently support the activities which comprise strategic management. The model enables the researcher to highlight the characteristics about the information technologies and about the strategic management activities which they support.

Introduction

This study explores the relationship between an organization's strategic management process and its use of information and communication technologies. A firm's goal to be competitive in complex and dynamic environments causes it to make strategic decisions in an attempt to build the competencies needed to dominate future markets (Hamel and Prahalad, 1994). Effective strategic management (i.e., strategy formulation and implementation) especially under conditions of changing markets, competition, technology and other environmental factors, appears to rely upon insights from experts on trends in technology, demographics, regulation, and lifestyle (Hamel and Prahalad, 1994). The use of information technologies, the fundamental tools for the information age, could be expected to augment the access to experts. Access to the insights and information of experts could thereby enhance strategic decision making and strategy implementation.

Background

The research question which led to this study was, "Given that information and communication technologies impact the use and communication of information, and given that effective strategy formulation depends upon the manner in which information is shared and communicated, then does an organization's use of information and communication technologies have an impact on its process of strategy formulation, and if so how?" Comparative case studies were utilized to address this question. A series of R&D-dependent, high growth companies were reviewed to determine their detailed strategic decision making processes, and their processes were compared among firms utilizing networked desk-top computers to either a high or low degree (Donnelly and Lightfoot, 1996).

The findings from the study are that networked desk-top computers were utilized in a noteworthy manner in strategy formulation and implementation process in only one of the sample-base companies. In other sample-base firms, networked desk-top computers were not utilized to a noteworthy degree in strategic management.

0-7803-3552-X/96/$5.00 © 1996 IEEE

Networked desk-top computers in the case study companies were used for functions such as E-mail, file sharing, real-time shared document edition, data gathering, and information storing via databases but such use did not have a significant impact on the essential strategic management processes of any of these firms.

Purpose & Objective

The absence of significant evidence concerning the impact of information technologies on the strategic management processes in the sample firms led to a desire to better understand the manner in which information technologies support or fail to support strategic management activities. The purpose of this study is to identify the characteristics of the information technologies which are used during the various phases of the strategic management process and to determine the sufficiency of the information technologies in meeting the needs of strategic management activities.

Approach

This study utilized a four-step approach. The activities which comprise the strategic management process were identified and briefly defined. Second, the categories of networked desk-top information technologies currently available were identified which have the potential to support at least one of the identified strategic management activities. The technology categories were mapped to the strategic management activities to indicate the areas in which the technology is being utilized. Finally, the characteristics of theinformation technology-strategic management matches were examined and distinctions were highlighted. In highlighting the distinctions, the common characteristics among the strategic management activities which were supported by information technologies were identified. The study also identified the common characteristics among the strategic management activities which were not supported by information technologies.

The Activities which comprise Strategic Management

In addressing strategic management in this work, 11 major activities which comprise any strategic management process are considered. They are defined as follows

1. Issue Identification -The process of defining the situation and its characteristics, which together describe a matter to be resolved.

2. Idea Generation - The process of identifying potential solutions which address a specific issue or set of issues.

3. Information Gathering - The process of searching various information sources for data and information which is relevant to the issue under consideration or to the ideas generated in association with the issue.

4. Assessment - The process of analyzing the information which was collected during the Information Gathering process to determine its applicability and significance to the issues to be resolved.

5. Decision Making - The process of making choices between alternatives in light of a specific issue(s) and in consideration of the ideas generated, the information gathered, and the analytical results.

6. Communication - The process of informing relevant persons of 1) the issue(s) to be resolved, 2) the strategy, 3) the process by which the strategy will be accomplished, and 4) the progress of the strategy implementation process.

7. Resource Allocation and Staffing - The process of identifying the required resources (human, capital, equipment, etc.) necessary to implement the strategy and allocating them as needed.

8. Project Planning and Organization - The process of identifying the tasks which must be performed to complete a project or activity. It also includes identifying the interdependencies associated with the tasks and their time and resource requirements.

9. Project Tracking and Controlling - The process of monitoring the progress of a project, including identifying, reporting, and following the progress of problems and issues.

10. Corporate Visioning - The process of developing a picture of what an organization could become and capturing that picture in words such that by reading the words, others envision the same picture.

11. Leadership - The process of conveying a direction to relevant persons and guiding an organization toward that direction by way of example, instruction, and steerage.

Current Desk-top Information Technologies

Each technology category identified meets four criteria which qualify it for consideration. The technology must be an information technology. It must either currently support or have the potential (in its current implementation) to support one of the strategic management activities listed above. In other words, each technology, in its current implementation must be able to provide value (albeit small, in some cases) in at least one of the strategic management activities. The technology must be accessible from a desk-top computer. Finally, there must be a well-established common conceptual understanding of what each technology constitutes.

Technology Category
Artificial Intelligence
Bulletin Boards
Databases
Decision Support Systems
EMail and File Sharing
Expert Systems and Idea Generators
Groupware
Presentation Software
Project Management Software
Statistics and Spreadsheets
Word Processing Software

Mapping Network Desk-Top Information Technologies to Strategic Management Activities

Figure 1 illustrates how the identified information technologies support the various activities which comprise the strategic management process. The technology-capability ratings (E, S, PS, PE) used in this model are defined as follows:

E- Enabling
This technology embodies sufficient capabilities that its use changes the manner in which the activity is conducted or accomplished. This definition suggests that the application of the technology changes the way in which the essential elements of the activity are performed.

S- Supporting
The use of the technology contributes sufficiently to the activity such that it adds value to the process. For the purposes addressed in this model, a technology can only be considered to be supporting when it is able to contribute to one or more essential elements of the process itself.

PE - Potentially Enabling
The technology as currently defined has the potential to be an enabling factor in the manner in which the activity is conducted, however, the technology has not matured sufficiently for this factor to be fully realized.

PS - Potentially Supporting
The technology as currently defined has the potential to support and add value to the process. However, the technology has not mature sufficiently for this support to be fully realized.

Figure 1:
MAPPING INFORMATION AND COMMUNICATION TECHNOLOGIES ON STRATEGIC MANAGEMENT ACTIVITIES

Complexity vs. Sufficiency

		Word Process'g	Present'n Software	EMail and File Share	Statistics Spreadsheet	Bulletin Boards	Data Base	Proj. Mgmt	Groupware	Decision Sup't Sys	Expert Sys. Idea Generat'r	Artific'l Intellig
Most	Leadership											
Complex	Idea Generat'n			S		S	S		PS		PE	
	Corp. Vision'g										PS	PE
	Issue Ident.			S		S	S		PS		PE	
	Info. Gathering			S		S	E		PS			
	Communication	E	E	E		S		S	PS			
	Assessment				E							
	Proj. Control							E				
	Decision Making				S				PS	E		
Least	Proj. Planning							E				
Complex	Staffing							E				
		High Sufficiency									*Low Sufficiency*	

LEGEND:
E - Enabling
S - Supporting
PE - Potentially Enabling
PS - Potentially Supporting

The strategic management activities on the vertical axis are in the order of decreasing complexity of the activity, in which the most complex activity is at the top of the axis and the least complex activity is at the bottom of the axis. The information and communication technologies on the horizontal axis are in the order of sufficiency, in which the most sufficient technologies are on the left end of the axis and the least sufficient technologies are on the right end of the axis. Sufficiency, for the purpose of this model, is defined as the characteristic of possessing enough functionality to meet the need of the process for which it is enabling. For example, word processing software is generally considered sufficient for meeting the need of text/document creation in a communication process.

Note that simply because a technology is enabling does not necessarily mean that it is sufficiently enabling. At its current stage of development, a technology may not possess all of the required characteristics to sufficiently meet the need for which it is enabling. For example, in Figure 1, Decision Support Systems (DSS) are an enabling technology for Decision Making, yet most would hold the view that DSS need additional development before they would be considered to be sufficient.

Technology and Strategic Management Alignments

As the model in Figure 1 illustrates the four most complex strategic management activities (Leadership, Corporate Visioning, Idea Generation, and Issue Identification) are not supported by a sufficiently enabling technology. There are information and communication technologies such as groupware, expert systems, and artificial intelligence which have the potential to be enabling. However in their current state they have not been developed to the level in which they can be considered to be sufficiently enabling. Information and communication technologies such as email, bulletin boards, and databases support these four complex strategic management activities, but these technologies do not meet the expressed needs of these activities. All of the other strategic management activities have at least one enabling information technology to support them.

Convergent vs Divergent Thinking Activities and Supporting Technologies

Another way of ordering the strategic management activities on the vertical axis is by the tendency of the activity to utilize convergent thinking versus divergent thinking. Convergent thinking activities are activities which focus on reducing the various options under consideration with the objective of identifying a single best solution. The child's game in which geometric shapes are matched to the appropriate geometric opening is an example of a convergent thinking activity. In a business environment, the process of selecting a single vendor which best meets the stated requirements at the lowest price is another example of a convergent thinking activity.

Divergent thinking activities are activities which focus on identifying all of the possible solutions to a problem and the opportunities and implications provided by such a solution. Figure 2 is the result of ordering the strategic management activities along the vertical axis by the type of thinking upon which the activity relies. The strategic management activities which tend to rely upon divergent thinking are at the top of the axis and the strategic management activities which tend to rely upon convergent thinking are at the bottom of the axis.

Figure 2:
MAPPING INFORMATION AND COMMUNICATION TECHNOLOGIES ON STRATEGIC MANAGEMENT ACTIVITIES

Divergent/Convergent vs. Sufficiency

		Word Process'g	Present'n Software	EMail and File Share	Statistics Spreadsheet	Bulletin Boards	Data Base	Proj. Mgmt	Groupware	Decision Sup't Sys	Expert Sys. Idea Generat'r	Artific'l Intellig
Diverg.	Leadership											
	Idea Generat'n			S		S	S		PS		PE	
	Corp. Vision'g										PS	PE
	Issue Ident.			S		S	S		PS		PE	
	Info. Gathering			S		S	E		PS			
	Communication	E	E	E		S		S	PS			
	Assessment				E							
	Proj. Control							E				
	Decision Making				S				PS	E		
	Proj. Planning							E				
Converg.	Staffing							E				
		High Sufficiency									*Low Sufficiency*	

LEGEND:
E - Enabling
S - Supporting
PE - Potentially Enabling
PS - Potentially Supporting

One purpose for reordering the entries along a particular axis is to consider how the dispersement of enabling, supporting, potentially enabling, and potentially supporting technologies changes in the matrix as the order along the axis changes. In comparing Figure 1 to Figure 2, the order of the activities along the vertical axis changed slightly. At the same time, the manner in which the technology-capability ratings clustered changed little, if at all. It appears generally that those activities which can be characterized as divergent in nature are also the activities which tend to be more complex. In addition, the activities which have the highest degree of divergent thinking are not currently supported by sufficiently enabling technologies. By contrast, the activities which rely upon convergent thinking tend to be supported by enabling technologies.

Characteristics of Technologies which are Enablers

In considering both Figure 1 and Figure 2, is there anything inherent in the activities which are less complex and rely upon convergent thinking that promotes the development of enabling technologies for these activities?

The list of technologies which are enablers for at least one of the identified strategic management activities are as follows:

Word Processing
Presentation Software
Project Management Software
Databases
Statistics and Spreadsheet Software
EMail and File Sharing
Decision Support Systems

Each of these technologies actually models or simulates a human activity. For example, databases simulate the manner in which people make files and indexes. EMail and file sharing model the communication process of a person writing a letter, attaching a document to it, making the appropriate number of copies, placing them in envelope, addressing them, and placing them in the U.S. Mail for delivery. The foundation of these technologies which simulate or model human activities is mathematics. Computers and the software programs which run on them are founded upon the properties of mathematics. Mathematics, by its nature, is a convergent form of study. Consequently, the reason why strategic management activities, which tend to be convergent in nature, have a greater tendency to be supported by enabling technologies may be that they are easier to model and simulate -- therefore making it easier to create an enabling technology.

Characteristics of Strategic Management Activities which are Enabled by Information Technologies

The following is a list of strategic management activities which are enabled by one or more currently available technologies:

Information Gathering
Decision Making
Communication
Assessment
Project Control
Project Planning
Staffing

These activities embody certain characteristics which facilitates their ability to be modeled in part or in total. Each of these activities are made up of processes which are repetitive in nature. The process to complete a statistical analysis follows the same pattern each time it is performed. Gantt charts, which are produced using project management software and are used to depict the various tasks of a project schedule, are always created using the same set of basic rules. The input tends to be well structured and predictable. It is the repetitive nature of the activities and the well structured input, which tend to make the activities easy to model, either in part or in total. Statistical software and spreadsheets model the part of the assessment process which processes data through statistical formulas. Databases model a well constructed filing system which could be of great value during an information search activity. Project management software contains tools which model the process of scheduling tasks and tracking their progress. Each of these strategic management activities is relatively easy to simulate and model.

Characteristics of Strategic Management Activities which are not Enabled by Information Technologies

The remaining set of strategic management activities --- those which are not enabled currently by a technology, are the following:

Leadership
Corporate Visioning
Idea Generation
Issue Identification

These not only tend to be more divergent and more complex, but they also have other common properties. They are activities which result in producing original thought. The end product of these activities are often new ideas or concepts which are not currently captured or reflected anywhere else. The other strategic management activities (Information Gathering, Assessment, Project Control, etc.) take existing ideas and concepts and apply processes to those ideas and concepts in order to convey them in a different manner. Also, these activities, which are not enabled currently by a technology, have less repetitive properties. For example, the process used to develop a corporate vision does not always follow the same approach and is often more reflective of the personalities involved.

Each of these characteristics (divergent thinking, complexity, production of original thought, and non-repetitive properties) are difficult to model. This is especially true for the characteristic of non-repetitiveness. Computer technology has proven that it adds the most value and it operates the most efficiently when it is conducting processes which are repetitive in nature. It is inefficient to model an activity which is conducted in a different manner every time it is performed.

It is a challenge to model an activity which possesses just one of these characteristics. When these characteristics (divergent thinking, complexity, production of original thought, and non-repetitive properties) are considered together, it is easy to understand why the activities which possess them are a challenge to model. It is simply more difficult to model activities with these properties.

Interpreting the Findings of the Technology-Capability Matrix

The four strategic management activities, which are not currently supported by an enabling technology [Leadership, Corporate Visioning, Idea Generation, and Issue Identification], along with Decision Making are the activities which have the greatest impact on the formulation of the strategies. These activities define the scope of the problem to be addressed. They determine the range of options which will be considered. The manner in which these activities are performed determines the degree to which to the formulated strategy meets the needs of the organization. Remove these activities from the list, and strategic formulation fails to exist. The activity of decision making is supported by an enabling technology, however as indicated in the discussion associated with Figure 1, decision support systems are far from being sufficiently enabling.

The fact that key strategic management activities are not sufficiently supported by enabling technologies appears to indicate the technology does not yet exist which embodies sufficient capabilities such that using the technology changes the manner in which the activity is accomplished. As a result, the impact of the existing technologies on the strategic management processes is small or non-existent and therefore non discernable. This is why the results of the previous study revealed no discernable difference in the strategy formulation processes which could be attributed to the use of networked desk-top computers in a company.

Conclusions

To realize a true impact of technology on the strategic management processes of companies that use them, the technologies must be able to add value to the activities which are central to the strategy formulation process. These are the activities which determine the degree to which to the formulated strategy meets the needs of the organization. In addition, it is important to note that these activities [Leadership, Corporate Visioning, Idea Generation, Issue Identification, and Decision Making] are not isolated to simply the strategic management functions within an organization.

They are part of and essential to many business processes and are used on a daily basis. Consequently, the value that an enabling technology would bring to any one of these activities has a much larger impact than simply the strategic management process of that organization.

If information and communication technologies are to provide true added value to management processes within organizations, they must move beyond the current paradigm and be able to support and enable activities which use divergent thinking processes, which can produce original thought, and which can apply and utilize new information (i.e. Iearn) efficiently.

References

Donnelly, Richard G. And Regina S. Lightfoot. "The Impact of Information and Communication Technologies on the Strategic Decision Making Process". 1996

Hamel, Gary and C. K. Prahalad. Competing for the Future. Harvard Business School Press, Boston, Mass. 1994.

The Impact of Information and Communciation Technologies on the Strategic Decision Making Process

Richard G. Donnelly, Ph.D
(202) 994-7155

The George Washington University
2115 G Street,N.W.
Washington, DC 20052 USA

Regina S. Lightfoot, Ph.D
(703) 974- 1817

Bell Atlantic Corporation
1310 N. Courthouse Road
Arlington, Virginia 22201 USA

This study examines the relationship between an organization's decision making process and its use of information and communication technologies. The research question addressed in this paper is, "Given that information and communication technologies impact the use and communication of information, and given that effective strategy formulation depends upon the manner in which information is shared and communicated, then does an organization's use of information and communication technologies have an impact on its process of strategy formulation, and if so how?"

Comparative case studies were utilized to address this question. A series of R&D-dependent, high growth companies were reviewed to determine their detailed strategic decision making processes, and their processes were compared among firms utilizing networked desk-top computers to either a high or low degree. The findings from the case study are that networked desk-top computers were utilized in a noteworthy manner in strategy formulation and implementation process in only one of the same base companies. In other sample base firms, networked desk-top computers were not utilized to a noteworthy degree in the strategic management process.

Introduction

This study examines the relationship between an organization's decision making process and its use of information and communication technologies. A firm's goal to be competitive in complex and dynamic environments causes it to make strategic decisions in an attempt to position itself as a strong leader and competitor in its industry. Effective strategic decision making, especially under conditions of changing markets, competition, technology and other environmental factors, appears to rely upon inputs from experts. The use of information and communication technologies, the fundamental tools for the information age, could be expected to augment the access to experts and their insights and information, thereby to enhance strategic decision making.

A Discussion on Strategic Management

Strategy is defined as the way in which a corporation endeavors to differentiate itself positively from its competitors through the use of its relative corporate strengths to better satisfy customer needs. [Ohmae, 1982] Strategic management uses the competitive environment to help it define its strategy and then it restructures the organization (its tasks, people, technologies, reward systems) [Pearce and Robinson, 1982] so that it is able to enact its strategy more effectively. The art of Strategic Management emphasizes the utilization and application of resources in new ways to build advantages [Hamel and Prahalad, 1989]. This emphasis requires the strategic management process to be proactive. The organization does not simply react to changes in the environment and reposition itself accordingly. Rather, it proactively seeks out its competitive advantage by redefining the environment to its advantage.

The activities which comprise the strategic management process can be grouped into two disciplines -- strategy formulation and strategy implementation.

0-7803-3552-X/96/$5.00 © 1996 IEEE

The activities included within the discipline of strategy formulation are[1]:

Issue Identification
Idea Generation
Information Gathering
Assessment
Decision Making
Communication

The activities included within the discipline of strategy implementation are:

Communication
Resource Allocation and Staffing
Project Planning and Organization
Project Tracking and Control

Information-oriented activities, such as idea generation and information gathering, are essential to the strategy formulation portion of the strategic management process because the strategies which have the greatest positive impact for an organization and enable to organization to place itself in the best competitive position are those which expertly utilize environmental (both internal and external) information [Porter, 1982].

The Role of Information and Communication Technologies

Information and communication technologies can affect the manner in which we share information and participate in strategy formulation. Communication and information technology enables the reduction of the number of layers through which information must pass in order to reach the people who need it. It thereby reduces "information float" by eliminating the many layers between information capture/generation and information receipt by the user [Huff, 1989; Hammer, 1987]. Furthermore, communication technology enables real-time and near real-time communication of voluminous, complex information over geographical barriers.

The Research Question

The research question addressed in this paper is, "Given that information and communication technologies impact the use and communication of information, and given that effective strategy formulation depends upon the manner in which information is shared and communicated, then does an organization's use of information and communication technologies have an impact on its process of strategy formulation, and if so how?"

The Methodology

This research utilized a comparative case study approach in which the strategy formulation processes utilized by high growth companies that perform research and development and have a high level of networked desk-top computer usage were compared to the strategy formulation processes utilized by high growth companies that perform research and development and have a low level of networked desk-top computer usage. The level of networked desk-top computer usage within a company was assessed by considering the percent of employees which utilize networked desk top computers. The level of networked desk-top computer usage was quantified so that the line of demarcation between high and low levels of networked desk-top computer usage could be defined.

Interviews were conducted to determine the level of strategic management activity within each company. The interviews provided information concerning the level of intensity of the strategic management efforts underway.

The companies selected for the case study were those which were assessed to have a significant amount of strategic management experience. These companies were clustered in groups of high/low networked desk-top computer usage. Each company selected for the study was assessed concerning their strategy formulation activities and any impact that networked desk-top computers had on these activities. The information derived from the case studies was compared to determine if similarities existed among the companies which participated in the case study relating to the level of networked desk-top computer usage in the company and the strategy formulation process which it utilized.

The Case Study Companies

The four companies which participated in the case study are addressed as Companies A through D. The profiles of the four companies are as follows:

Company A
Number of Employees - 15,000
Location - Mid-West
Industry - Aerospace

Company B
Number of Employees - 2,900
Location - North East
Industry- Instruments

Company C
Number of Employees - 2,400
Location - North West
Industry- Instruments

Company D
Number of Employees - 3,200
Location - Mid-Atlantic
Industry - Computers

Analysis

An interactive interview style was utilized during the case study. Information from the interviews was segmented and categorized so that it could more easily be analyzed and formatted into matrices.

Matrix Related to Strategic Management Activities

Figure 1 depicts the involvement of various levels of management among the case study participants for specific issues associated with the strategic management process. If a company label appears in the block for a specific issue, it indicates that the management level associated with that block had a moderate-to-high degree of representation and involvement within the company for that specific issue. The specific strategic management related issues addressed in this matrix are:

Column 1. the people responsible for implementation were part of the strategy formulation process.
Column 2. the management levels participating in the information gathering process for strategic decision making.
Column 3. the number of managers participating in the strategy formulation process.
Column 4. knowledge of the strategy(s) prior to implementation.
Column 5. cross-functional or cross disciplines represented at each level during the strategy formulation process.

Column 6. personal performance objectives and appraisals linked to the strategy(s).
Column 7. awareness of the progress and the status of the implementation of the strategy.

	1. Implementors are part of planning process	2. Involved in gathering information	3. Participation in planning process	4. Knew of strategy prior to implementation	5. Cross-functional planning	6. Performance linked to strategy	7. Implementation status awareness
Officers/ Executives	*Company A* *Company B* **Company C** *Company D*	*Company B* **Company C** *Company D*	**Company A** **Company B** **Company C** **Company D**	**Company A** **Company B** **Company C** **Company D**	**Company B** **Company C** **Company D**	**Company B** **Company C** *Company D*	*Company A* **Company B** **Company C** **Company D**
Division Heads	**Company A** **Company B** **Company C**	*Company A* **Company B** **Company C**	**Company A** **Company B** **Company C**	**Company A** **Company B** **Company C**	**Company A** **Company B** (some div.) **Company C**	*Company A* **Company B** **Company C**	**Company A** **Company B** **Company C**
Mid-Level Managers	**Company A** **Company B** **Company C**	**Company A** **Company D**	**Company A** **Company C** **Company D**	**Company A** **Company C** **Company D**	**Company A** **Company C** **Company D**	*Company A* **Company C** *Company D*	**Company A** **Company C** **Company D**
Lower Level Managers	*Company A* (some div.)	*Company A* (some div.)	*Company A* (some div.)	**Company A** *Company B* (some div.) **Company C** **Company d**		*Company B* (some div.) **Company C**	*Company A* *Company B* (some div.) **Company C**

Figure 1. How various employee levels participate in the strategic management process in their company

If the company label is in bold type, it is an indication of a high level of involvement by the indicated management level. If the company label is in italic type, it is an indication of a medium level of involvement by the indicated management level. If a company's label does not appear in a block, then the managers at the indicated level have either no involvement or have a low level of involvement associated with the activity. For example, it is appropriate to conclude that officers and executives in Company A are not involved in the information gathering process (see Column 2) while officers and executives in Company C are highly involved and officers and executives in Companies B and D are moderately involved.

Matrix related to Network Desk-Top Computer Usage

As a part of this research, it was important to understand the manner in which networked desk-top computers are used in each company. Such understanding assisted in making certain that the study questions are answered from the intended perspective. A simple categorization of the networked desk-top computer usage in each company would be:

Company A- Low networked desk-top computer usage
Company B- Moderate networked desk-top computer usage
Company C- Moderate to High networked desk-top computer usage
Company D- High networked desk-top computer usage

However, these categories do not speak to the manner in which networked desk-top computers are used within the companies. Figure 2 is a matrix which depicts the level of networked desk-top computer usage in each company by function. The functions include EMail, File Sharing, Real-Time Shared Document Editing, Data Gathering, and Databases.

The Impact of Networked Desk-top Computers on the Communication Processes within the Case Study Companies

As depicted in Figure 2, networked desk-top computers are used for a variety of functions within the companies studied including email, file sharing, data consolidation and reporting, and information databases. Of all of the companies, Company D had the highest degree of interconnections among its employees. In fact almost 80% of its employee body, including employees that were based internationally, were utilizing networked desk-top computers. Yet, Company D used little of its networked desk-top computer capabilities for its strategic management activities.

Company C used its networked desk-top computer capabilities for strategic management to a greater degree. Company C maintained an extensive database of customer information. It used this information for both strategy formulation and implementation. The database had an advanced search capability which allowed for the data to be formatted and reported in various ways. Company C used the library search function to search for information concerning industry, market, and technology trends and they encouraged teams charged with the development of new products to use the library search capabilities for ideas on how to enhance its current line of products through improvements and new product development.

Company B, which had moderate levels of networked desk-top computer usage, used their network capabilities to transfer, via modems, financial data to the corporate headquarters for consolidation. This was the only contribution that the networked desk-top computers made to the strategic management processes within this company.

Company C's network was not as extensive as Company D's. Yet, Company C did have the most progressive use of its resources among the four companies. Company C's employees were encouraged to use to consumer database and the library search capabilities as part of their normal assignments. This may have increased the potential for their use during the strategic management process. The employees had become comfortable with the resource and therefore found additional ways to utilize it over and above those initially recommended. Company D referred to its networked capabilities as an email and the employee look-up system. Although the employee look-up system is a database, it is limited to employee information and does not enjoy the degree of cross-functional use that Company C's consumer database has.

One of the groups in Company D installed a real-time conferencing and editing capability in which two or more people can simultaneously view and edit the same document. This group used the real time conferencing and editing capability to address the document creation needs of his group. However, this capability was restricted to only that group and the functions of this group tend to be somewhat specialized. Therefore, the primary focus of Company D's network use was one of passing files and messages and creating and editing documents.

Company C encouraged its employees to use the network as an information resource where they could go to find information. Since the need for information crosses functional boundaries, it was much easier for the employees Company C to use their network across a wider range of functions. This in turn increased the value of the network to the company. The employees saw it as a resource for information instead of simply a tool for creating documents and passing messages.

Company A, like Company C, collected information from its customers for use in the strategic management process. However, their distribution of this information was limited.

The case study companies did not demonstrate a direct relationship between the amount of the networked desk-top computer use within the company and the manner in which networked desk-top computers are used in the strategic management process. Instead the degree to which they are used in support of strategic management activities appears to be a reflection of how they are used in the execution of daily business. Still, a company must have some networked desk-top computer capability to realize any impact on its strategic management processes or any other processes, for that matter. To that end, the contributions that the networked desk-top computers made to the strategic management processes in Company C and even the limited impact that they made in Companies B and D is a reflection of their existence and acceptance within those companies.

	E:Mail	File Sharing	Real-Time Shared Document Editing	Data Gathering & Consolidation	Information Database
High	Company D				
Moderate		Company C Company D		Company D	Company C Company D
Specialized			Company D		
Low				Company B	

Figure 2. How the case study companies use their networked desk-top computers

Results

The purpose of this research was to explore if the strategic management decision making process differs between organizations which use networked desk-top computers and organizations which do not and, if so, to determine how it differs. This was explored by addressing the question, Does the use of information and communication technologies have an impact on the strategy formulation process in organizations? The information gathered during this research revealed that the strategic management decision making process among the four case study companies did not differ between those companies which used networked desk-top computers and those which did not. The strategic decision making and implementation processes of the case study companies did not have any distinctions which could be attributed to the existence of networked desk-top computers or lack thereof.

However, the use of networked desk-top computers in one case study company did have an impact on both the strategy formulation process and the implementation process. This impact was not sufficient enough to distinguish the strategic management process in this company from the strategic management processes in the other companies. Still, the fact that an impact did exist, albeit small, is an indication that networked desk-top computers can be and are potentially beneficial in the strategic management process. There were two prominent distinguishing factors between the company in which networked desk-top computer usage facilitated the strategic management process and those in which networked desk-top computers did not. These distinguishing factors are:

1. The practice of using the networked desk-top computers as an information resource rather than only a messaging and document creation tool.
2. The encouragement of management within the company to use networked desk-top computers in this manner.

Within this case study company, there was evidence of the benefJts of networked desk-top computer usage to the daily work processes of the company in addition to the strategic management process of the company. This one company had found networked desk-top computers to be a valuable information resource in their product planning and development process, and they had become comfortable with the use of this technology. Thus, it was easier for this company to utilize and find value in utilizing networked desk-top computers in their strategic management processes.

Summary

The findings from the case study are that networked desk-top computers were utilized in a noteworthy manner in strategy formulation and implementation process in only one of the same base companies. In other sample base firms, networked desk top computers were not utilized to a noteworthy degree in the strategic management process. Networked desk-top computers in the case study companies were used for functions such as E-mail, file sharing, real-time shared document edition, data gathering, and information storing via databases but such use had a significant impact on the firm's essential strategic management processes in only one case study company.

References

Hamel, Gary and C.K. Prahalad. "Strategic Intent", Harvard Business Review, Vol 67, No. 3, May-June, 1989

Hammer, Michael and Glen E. Mangurion. "The Changing Value of Communication Technology", Sloan Management Review, Vol. 28,1ss. 2, Winter 1987.

Huff, Sid L. "Information Technology and the Future Organization", Business Quarterly, Vol.54, Num. 1, Summer 1989.

Ohmae, K. The Mind of the Strategist: Business Planning for Competitive Advantage. McGraw Hill, New York. 1982.

Pearce, John A. lll and Richard B. Robinson. Strategic Management. Richard D. Irwin, Homewood, IL. 1982.

[1]These activities were compiled by the researcher as a result of reviewing numerous sources on the practice of strtegic management.

ScoreCarding: A Case Study in Effective Management of Virtual Enterprises

Bob Littlewood
(619) 942-1078
Fax: (619) 436-7742

The Strategic Resource Group
P.O. Box 9427
Rancho Santa Fe, CA 92067

To compete, successful companies must often integrate technical expertise or disciplines from outside their traditional areas of operation. Virtual organizations are formed, combining individuals and companies with widely divergent viewpoints and methods of operation. By their very nature, these teams face challenges in operating effectively and staying focused on overall goals. This paper presents the process used by a virtual team comprised of employees from Genta and Procter & Gamble (P&G) to establish and develop a successful long distance business partnership.

The Virtual Partners

Genta is a small pharmaceutical research firm of approximately 160 employees which recently teamed with the P&G department responsible for R&D in the healthcare sector. When they partnered, they confronted several challenges common to virtual organizations, not the least of which was the big company/little company culture clash. A software-supported process known as ScoreCarding was used to successfully facilitate the relationship between the two organizations.

Partnering Is Not a Natural Act

In today's corporate world, many acknowledge the value of partnering with customers and suppliers, and increasingly, of selectively collaborating with those who were previously regarded purely as competitors.

To operate effectively, teams must learn to develop or acquire methods to overcome the five basic obstacles that confront nearly every teaming situation:

- Diversity and cultural differences between companies, organizations and even departments within the same organization. In this case study, the big company/small company cultural differences presented one major obstacle.
- Misunderstandings and conflicts about roles and goals challenge team effectiveness from the beginning. Over long periods of time, changes in personnel and emphasis can create major misunderstandings in these areas.
- Other obligations. Priorities compete between the demands of the parent organizations, which expect individuals to do what they have always done, and the needs of the team.
- Dispersion of team members because of geography or organizational assignments complicates and retards team development and performance. Dispersion is often the default condition of virtual structures.
- A lack of trust. Teams are often made up of strangers whose trustworthiness has not been clearly established. They interact infrequently, and these interactions may not be in person. Participation in teams may not be voluntary. Establishing trust can be particularly challenging when virtual teams are established by management not involved in team development.

0-7803-3552-X/96/$5.00 © 1996 IEEE

Given these problems, Five important capabilities must be developed if the partnership is to survive and succeed. Virtual partners must:

1. Continue to reclarify and adjust the goals, objectives and strategies of the alliance. Without clearly articulated common goals for the virtual team, the "old" goals of the individual components and individuals may set the team operating at cross-purposes and certainly erode effectiveness.

2. Integrate "softside" issues into the process of managing performance. These are often subjective, unspoken, qualitative expectations about the performance of others. Team members find them uncomfortable topics of discussion, but customer satisfaction surveys show that 65% to 85% of the factors that matter most are "softside" issues such as responding in a timely manner, owning up to problems, and maintaining the same level of commitment after the sale as before the sale.

3. Create permission to "do it right the second time". The "zero defects" concept runs the risk of ruling out a second chance to get "closer to right". Teams need tools that continually recalibrate the gaps between satisfaction and performance. Teams need processes that permit them to interpret causes and develop and test solutions.

4. Engage dissatisfactions comfortably and use them positively. Everyone in every relationship is dissatisfied about something at some time or other. When the relationship is working well these dissatisfactions accumulate as unspoken irritants, but sooner or later they reach a flashpoint that can damage the relationship beyond repair.

5. Defuse emotions. Teams need several clear strategies for defusing emotions. For example, a means of establishing reasonable discourse--a forum for asking tough questions and sharing concerns--must be in place to defuse ticking emotional time bombs.

Genta and Procter & Gamble: The Challenges They Faced

Genta is a company of 160 people which specializes in the discovery and development of compounds with medical applications. They teamed with the Procter & Gamble department responsible for research and development in the health care sector. The motivating factor behind the formation of this partnership was the exploration of new opportunities in pharmaceutical development.

More specifically, this joint effort involved research about ways to use DNA to create new compounds for intervening in diseases. Such upstream research entails high risk and great uncertainty, particularly because significant sums of money are usually required. From the start, it was known that the probability of success was low because of various discoveries that would have to be made along the way, but that if it everything worked, the payoff would be substantial. As one participant noted, this type of partnership "is an adventure, but creates a lot of tension."

It is no secret that large corporations have radically different approaches to conducting business than their smaller counterparts. While large organizations are usually blessed with more substantial resources, they also tend to be burdened with more cumbersome procedures and multiple layers of hierarchy. Small organizations, on the other hand, have the advantages of speed and flexibility, but often operate with limited resources. Cultural differences extend to decision making process, communication styles, even dress codes. Such was the nature of the little/big firm partnership that was formed between Genta and Procter & Gamble.

In addition to the pressure to perform and the differing corporate cultures, several other potentially challenging factors were at play. The two organizations were geographically separated with the Procter & Gamble office located in Ohio and the Genta office in California. In addition, they were blending two teams, one of which was multicultural with scientists from the U.S., Taiwan, Iran, Germany and other countries, while the other was somewhat more homogeneous.

Finally, the fact that the participants were primarily scientists meant that they, not surprisingly, wanted to concentrate on the science, and were considerably less interested in management techniques. Chuck McOsker, Procter & Gamble's Section Head for Molecular and Cell Biology and their project manager for this relationship, admits that he and some of the other scientists went in with a skeptical attitude that ScoreCarding was "just another touchy-feely management tool that makes everyone feel warm and fuzzy, but nothing will come out of it." Three years later, he says that it proved far more useful than they expected and wouldn't hesitate to use it again.

"What's Easy for One Isn't for the Other"

Of all of the challenges that Genta and P&G faced, the one of most concern was clearly the little company/big company element. Because both organizations had teamed with others in such research efforts before, they anticipated the problem of integrating a large firm culture with the style of a small firm, and decided that some means of bridging this gap should be introduced from the start. According to McOsker of P&G, both parties were aware that there was "ample opportunity for misunderstanding because what's easy for one isn't for the other." Genta had tried a methodology called ScoreCarding with another partner earlier and had found it helpful. They suggested this approach and Procter & Gamble agreed.

How ScoreCarding Works

ScoreCarding has been used in more than 100 corporate alliances to:

- Shape valuable relationships between internal operating units, or the corporation and its customers or suppliers.
- Provide a guided, disciplined way to manage on-going relationships.
- Measure progress against agreed-upon goals and objectives.
- Create actions that enhance effectiveness and propel the group toward its goals.

ScoreCarding has three major stages--Planning, Installation and Implementation.

Planning

The Planning stage involves:

- Evaluating the value of partnerships, the risks they face, and the contribution ScoreCarding is likely to add.
- Diagnosing selected relationships to determine what's working and what is not, then identifying mutual expectations and necessary changes.
- Establishing goals and objectives for relationships, examining challenges, and selecting appropriate ScoreCarding strategies.

These activities are needed only once for each relationship and analysis may take from 30 minutes to a few hours. Only a few individuals are involved in the host organization. There is little to no involvement with the partner at this stage.

Installation

This stage involves:

- Gaining the partners' agreement to start or experiment with ScoreCarding.
- Negotiating the joint behavior contract, incorporating the seeds for changes needed to enhance the relationships.

- Negotiating mutual expectations and obligations and establishing a mutually acceptable performance benchmark to use in assessing performance and making changes. This, too, is a "once per relationship" activity. Both sides are involved, but only for a few hours.

Implementation

ScoreCarding is a cyclical process. Once under way, team members need to interact on a regular schedule to assess problems and take steps toward resolving them. A cycle is typically a quarter, but can be more frequent if necessary. Each cycle at this stage of the process involves:

- Scoring.
- Collecting scores and comments from each partner to measure performance. This takes about an hour in the first cycle and about 30 minutes thereafter.
- Facilitating agreements between partners in "Agree to Agree Meetings", where performance gaps are identified and analyzed, future issues are identified, and action contracts are negotiated. The first of these sessions may take several hours. Subsequent sessions are much shorter, often less than an hour.
- Preparing for the next cycle by adding agreements to ScoreCards, producing and distributing performance reports that analyze results and trends.

This stage involves every member of the virtual team and, occasionally, outsiders.

The First Steps Are the Hardest

It would be an exaggeration to claim that the relationship between Genta and Procter & Gamble flowed smoothly from the very start. One of the first challenges was determining which issues would be key to the success of the relationship. Bob Klem, Vice President of Research and Development for Genta, admits that some of the earlier quarterly meetings were a little tense. As both partners laid out their expectations, the differences in their corporate cultures quickly revealed themselves. Nowhere was this more clear than with the sharing of information. While Procter & Gamble wanted information to be shared as quickly as possible, Genta scientists preferred to exercise caution in sharing information until they were certain of its implications. Some of the Problems and Solutions:

Problem: At certain stages, P&G felt Genta was not fulfilling promises.
Solution: ScoreCarding identified this issue early, and helped participants develop a case that convinced management to add equipment and resources to the team.

Problem: P&G team members felt they were passing good ideas on to Genta, but they were being ignored because they never got any feedback. Though Genta felt many of the ideas were good, they wanted to reserve comment until each suggestion could be investigated.
Solution: ScoreCarding brought the issue to the surface and helped the team create a protocol
that satisfied both sides.

Problem: Genta felt P&G was not allowing enough time to accomplish tasks P&G had communicated to their management, and that the time lines were not negotiable.
Solution: ScoreCarding identified that P&G continually refreshed their time projections and that these estimates would not be used to measure project effectiveness during these early stages.

Problem: P&G felt it was not getting adequate information on things Genta was doing in scientific areas outside the immediate scope of their project--particularly advanced chemistry research--that could impact the long-term success of the project. Although Genta's progress was positive, the silence on these projects raised concerns at P&G that the research was not going well.

Solution: ScoreCarding alerted Genta to the hidden issue, and they were delighted to schedule updates.

Problem: New people were introduced to the team, bringing their own assumptions and interpretations about its activities and objectives.
Solution: ScoreCarding made it easy for newcomers to understand and adopt the culture the team had developed. New members avoided a trial and error approach.

Problem: How results of the work was communicated to the outside world was an important issue to Genta. As a small company, these communications could have a major impact on the company's stock. It was important for Genta to review and edit even internal announcements in advance. It was a large issue for Genta, but had little or no impact on large P&G.
Solution: ScoreCarding brought the issue to the forefront early, before it caused major problems.

Through the ScoreCarding process, expectations were surfaced and addressed early in the relationship before they had the chance to accumulate and cause serious misunderstandings. Trust was not eroded. In some cases, it was a matter of providing more resources. Participants concur that the early discomfort of discussing expectations and issues up-front was a small price for the on-going benefits of a smoothly functioning relationship. By contrast, many partnerships start with a "being nice" phase that's difficult to get past and ultimately makes it difficult to ever express concerns with any degree of candor.

According to John Alden, Senior Manager for Human Resources in Procter & Gamble's R&D health care sector, "differences of philosophy were resolved in 40 minutes." He speaks of a chain reaction where getting expectations out explicitly allows both partners to focus on the work and meet commitments. The process also provides a way to demonstrate that progress is being made toward commitments. This, in turn, builds trust and makes it easier to solve problems when they do crop up. In short, Alden says that the process "helps build teams where it counts. This collaboration was a model."

Partnerships Require On-Going Maintenance

A common myth with business partnerships is that they eventually get to the point where they become more or less self-sustaining. Unfortunately, this is wishful thinking as partnerships -- and particularly virtual teams -- require considerable care and maintenance. In this case, Genta and P&G representatives met every three months. After early successes, some of the participants questioned whether this was still necessary, but they learned that getting off to a good start is not enough. Klem of Genta notes that, "ScoreCarding gets easy after a year, but it continues to be necessary." The on-going meetings are essential to maintain a good sense of camaraderie, to stop potential problems in their tracks and to integrate newcomers. For this reason, the ScoreCarding methodology incorporates periodic team meetings for as long as the team exists.

Concluding Remarks

Three years into their virtual relationship, both Genta and P&G are very pleased with the research successes that the relationship has produced. Ignoring the widespread belief that technology alone can make virtual organizations work, they applied a systematic methodology to the people side of the equation. In doing so, they overcame the challenges presented by distance, big firm/little firm ways of doing business, scientist vs. business manager perspectives, and multiculturalism, any of which could have derailed the relationship.

Structuring A Virtual Team Working Coordination System (Vtwcs) In The Management Of Projects

Xin Liu
Phone: 0044-141-552 4400 Ext 3056

CAD Centre
Dept. of Design, Manufacturing and Engineering management
University of Strathclyde, Glasgow, G1 1XJ, Scotland, UK
Email: xin@cad.strath.ac.uk, Fax: 0044-141-552 3148

The intention of this paper is to develop a system model and to propose a virtual environment or a Team Working Coordination System (VTWCs) for the computer supported design coordination (DC) towards managing a project and multidisciplinary design teams. To this end, the paper develops a structure of DC which can be used for modelling an appropriate computer support environment. Based on this framework, a virtual system called VTWCs is designed as a computer system for project management support. The system can be seen to act as assistant for the project manager, where the communication and coordination is achieved through a common communication space.

Design Coordination as a Means to Effective Project Management

Design coordination (DC) is defined as an approach to product development in which different design activities are carried out in an integrated, harmonious way, with concurrent views of different aspects of design. DC may be also seen as the dynamic activity and relations which involve the co-ordinated organisation of multi-disciplinary groups, design activities, and product information [Duffy 93]. The most recent work on DC has been to develop a framework of models [Andreason 94]. This framework is an attempt to represent the reality of the design process, with emphasis on the various types of entities which exist in DC. It does not , of itself, describe DC but rather identifies the things which have to be co-ordinated, and the resources available to do it.

From an examination of this framework , it is possible to identify three types of entity; those to do with the product representation, those to do with the activities of design, and those to do with the people or agents who will carry out the design. The corresponding activities associated with these entity groups are modelling, planning/scheduling, and resource grouping, or "teaming", respectively. Thus, DC can be viewed as three basic design activities with respect to three important design factors (see Table 1).

Table 1: Aspects and relations of DC

activities factor	teaming	planning/ scheduling	modelling
product aspects			*
activity		*	
people	*		

From the above view of DC, the main aspects related to DC have been presented as:

- *Product;*
- *Design activities;*
- *Design team;*

The process of design and the activity of DC can be considered as a relationship between these three aspects. From the design process view, the project manager, given

0-7803-3552-X/96/$5.00 © 1996 IEEE

the customer need as the first part of the product aspect, the overall goal to design a product as the activity aspect, and available resources in the team aspect, will break down the goal into sub-activities, such as building the product design specification (PDS) or generating concepts, and then allocate resources or teams to these sub-activities. The result of these allocations will be to modify the product aspect, leading to further development of the planned activities. At this level, coordination includes selecting the team members for a task, and planning and scheduling of activities. A lower internal coordination takes place in each of the three aspects; negotiating in the team aspect, product structuring in the product aspect, and strategic planning in the activity aspect.

Product aspects can be regarded as product specifications or characteristics, properties, functions and requirements. In DC the product aspects can be decomposed into:

a) The physical phenomena (involved in product versions, behaviour/performance, etc.);
b) The types of these phenomena (e.g. electrical, optical, mechanical, etc.);
c) The product life-cycle issues (e.g. manufacturablity, total costs, design for environment, etc.);
d) The technology alternatives for product implementation (e.g. hydraulic vs. electrical drives);
e) Goal structure (decomposition into subgoals).

Activities in product design especially in a DC environment are mainly concerned with the design goals, tasks, and procedures (process). Therefore the design activity model can be decomposed into:

a) Design task definition;
b) Plans and schedules (when to do it, how many stages for a task);
c) Design Procedures.

The design team can be decomposed into:

a) Design resources organisation and acquisition;
b) Available domain skills and knowledge;
c) External resources available to teams.

From the above description, it is clear that both the design activity structure and product structure are dynamic. At the start of the design process, the set of activities is empty and the product contains only requirements. After completion, the product aspects contains all the information defining the product from activities of the design teams.

Modelling DC Structures in Project Management

Design coordination entails integrating or linking together different but interdependent parts of an organisation as they work together to accomplish organisational goals. Research has identified effective coordination as an important factor which differentiates successful from unsuccessful projects. Traditionally in a project management process, once the objectives for a project have been defined, developing a project schedule involves the following steps: producing a work breakdown structure of activities to be performed, allocating the activities to individuals and groups, identifying the interrelationships between activities and sequencing them, estimating activity duration and costs, reconciling the schedule with available project timing constraints and available resources, and finally reviewing and updating the schedule as the project proceeds. However, the major scheduling approaches used in the project management (e.g. Gantt charts, CPM and PERT) are based on events and activities. They are difficult to make visible in a meaningful way to the individual and groups involved in a project and so fail to provide the basis for the development of a shared view necessary for effective project management.

In fact, a manager can best develop an understanding of a complex project through an analysis of its design coordination structure - a system level view of the entire set of interrelated interdependencies between all of the individuals and groups that are involved in the project. Specifying the coordination structure of the project prior to specifying the actions required to manage these interdependencies, and arriving at a shared view of this structure on the part of the project participants, can reduce the structural problem of managing part-whole relationships [Van de en 1986].

The ability of individuals and groups to coordinate has been linked with their ability to perceive and manipulate the same common objects [Crowston 1991] or their ability to come to have similar understanding of the objects that they share [Dougherty 1992]. Therefore the coordination structure can e conceptualised using three elements: project teams/actors, shared objects/tasks, and relations/links between teams/actors and shared objects/tasks. Teams or actors are the individuals and groups involved in a project. Iterdependencies between actors are specified by their relations or links to a set of shared objects or tasks. Conceived in this way, the coordination structure of a project can be made visible using diagrams.

Figure 1 shows the design coordination ensemble for a product design stage of the project management process. It capture the view of a project manager of the generic coordination structure for one stage of a product design project. Actors are shown as ovals and shared objects as rectangles. Verbs specify the nature of the links between actors and shared objects.

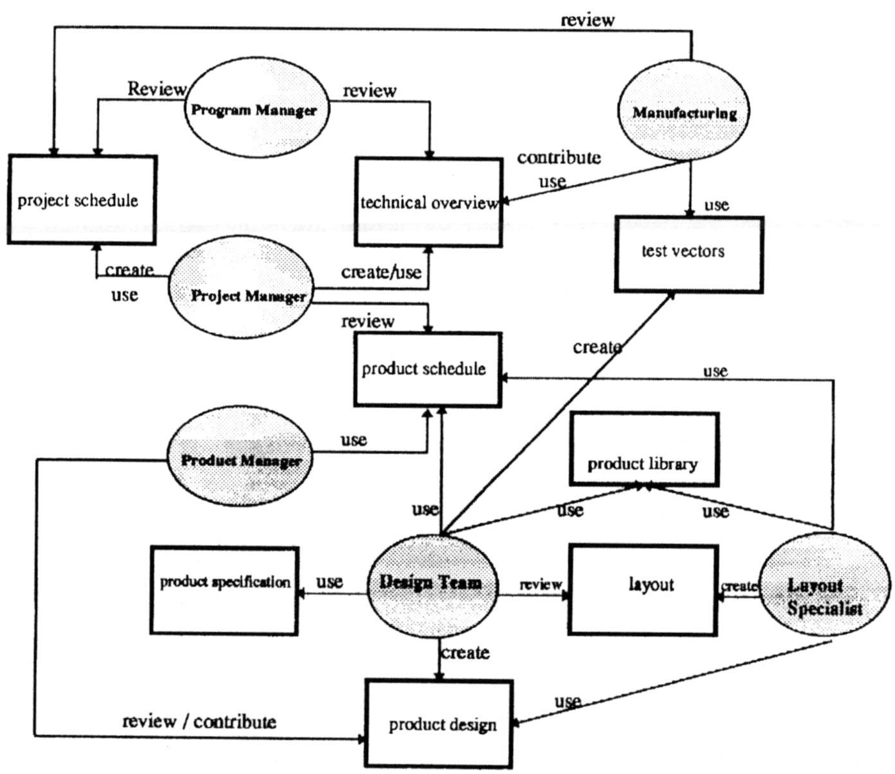

Figure 1: Design coordination ensemble for a product design stage

The design team, generally a self-organising group of 3 to 4 designers, is charged with using the product specification to produce a layout usable for product manufacture. It coordinates with a variety of actors including managers, a layout specialist, and manufacturing around a variety of shared objects. The technical overview contains the technical demands which will be placed on manufacturing given the current state of the product design and is used to coordinate design and manufacturing during the design stage.

In this coordination ensemble, each actor and shared object has a responsibility and attributes or describing characteristics. Links are defined by their actor/shared object endpoints and a full description of how the actor affects or is affected by the shared object. Actor/shared object links specify the power that actors have over the evolution and use of shared objects. It is useful when creating coordination ensemble to create a data dictionary specifying the responsibilities and attributes of all the actors and hared objects and full descriptions of links in the ensemble.

System Architecture of TWCs for DC Support to Project Management

A system architecture for design coordination support to project management - the Virtual Team Working Coordination System has been proposed by using the IDCS [MacCallum 1994]. The system architecture (Figure 2), is based on a blackboard architecture and concepts from distributed artificial intelligence (DAI) [Engelmore 88, and Lesser 87]. In this, design co-ordination is modelled by multi-agents which means that different agents work in their own area but towards the same goal, so that all work can be coordinated together. An agent reasons about its own local actions, and about the actions of the other agents to try and ensure the community acts in a coherent manner. The idea of intelligent agents is used to represent team capabilities with negotiation in a team blackboard. It is assumed that the user of such a system is a project manager who has the responsibility for a project, carrying out the planning and scheduling, and building teams for specific tasks to ensure that the project can be completed in the most effective way.

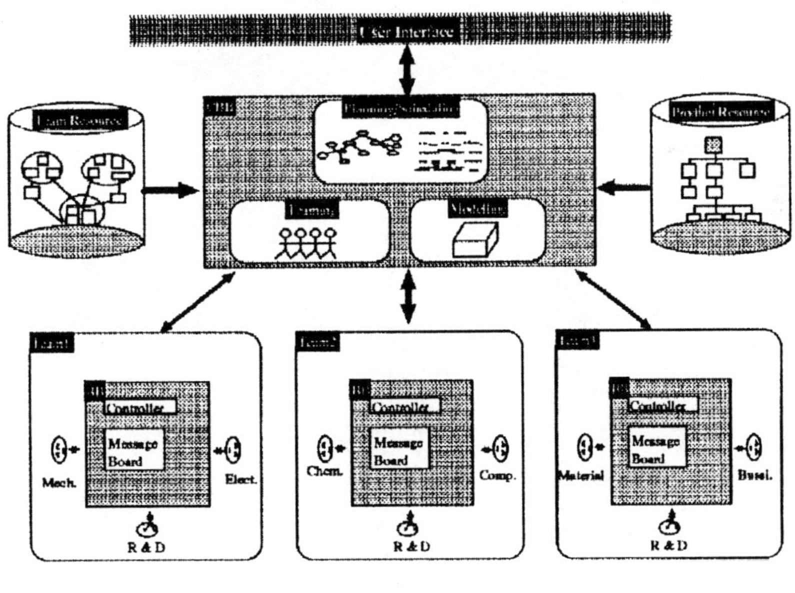

Fig. 2 System architecture for design coordination support to project management

The architecture consists of a user interface to a control blackboard which represents the planning workspace. Through access to available team and product resources, multiple teams can be created and allocated to activities. Each team then has its own negotiation blackboard for operating on its part of the product workspace.

The *Control Blackboard* (CBB) is the medium through which all high level communication takes place. This blackboard is divided into 3 workspaces: *Planning/Scheduling, Teaming* and *Modelling*. *Planning/Scheduling* contains the current plans and schedule of design activities; *Teaming* contains current team structures; and *Modelling* contains the product structure and model information. A control mechanism provides support for the design co-ordination manager. It will assign activities to teams, evaluate actions taken by a particular team; check the consistency of design steps; and assist the negotiation process to support final decision making.

A *Knowledge Resource* (KR) can be viewed as a collection of available design knowledge resources for activities. In IDCS the KRs are grouped into the following two categories:

.*Product Resource:* The knowledge resource for product modelling, product life-cycle issues, and technology for product implementation.

.*Team Resource*: The knowledge resource for team organisation, including available domain skills of teams and external resources available to teams.

A *Team* can be viewed as a group of individual team resources. Each team is a subset of the team resource, able to take part in an individual design task. It is supported by a blackboard which is based on DAI techniques and modified from a conventional blackboard. This blackboard is divided into 2 parts; a *Controller* and a *Message Board*. The *Controller* contains the activities for managing team interaction, while the *Message Board* contains the communication information for product decisions.

The work flow of the system can be illustrated simply as follows: A project manager, based on a customer need in the modelling workspace, and the product knowledge resource develops a plan for the project, and uses the team resource to form a team with an appropriate mix of expertise. An activity is allocated to a team and the team works through the team blackboard to communicate and negotiate; finally the team give the design results back to the project manager, through the modelling workspace.

Discussion and Conclusion

This paper has reported on the results of studies in the field of design coordination for project management. Design coordination is based on relations between product aspects, design team and activity/process, i.e. design coordination is mostly concerned with (1) product aspects and modelling, (2) design activity co-ordination based on communication between different design teams, and control of the dynamics of product aspects, and (3) product design team co-ordination based on control of the dynamics of team members and their perspectives.

It is foreseen that a virtual design coordination system such as the VTWCs can provide an important and missing level of support for coordinating design. The System proposed in this paper is a concept based on current understanding of design coordination support. Aspects of this architecture are now being implemented to act as a test bed for the ideas.

References

Andreasen M M, Bowen J, MacCallum K J, Duffy A H B, Storm T, Design Coordination Framework, CIMMOD/CIMDEV workshop at Torino, 22-23 sept. 1994

Crowston, K G, Towards a Coordination Cookbook, PhD Thesis, Cambridge, MIT, 1991

Dougherty, D, Interpretive barriers to successful product innovation in large firms, Orhanisation Science, May 1992

Duffy A H B, Andreasen M M, MacCallum K J, Reijers L N, Design Coordination for Concurrent Engineering, Journal of Engineering Design, Vol.4, No.4, 1993, pp251-265

Engelmore R, Blackboard System, Addison-Wesley, 1988

Lesser V K, and Corkill D D, Distributed Problem Solving, In: SC Shapiro (ed.), Encyclopedia of AI, pp245-251, Wiley, 1987

MacCallum, K. J, Liu, X, Model for Computer Supported Design Coordination in CE, CAPE onference, London, Dec 1994

Van de Ven, A H, Central Problems in the management of Innovation, Management Science, Vol. 32, 1986

Decision support re product development

Lisbeth Crabo Ljungman

Vinkelvagen 7A
S-135 50 TYRESO
Sweden

The paper will report findings from ongoing research and propose an outline for structuring early requirements for product development, supporting decision making.

Organizations have to find ways to monitor changing market demands in new ways, interpret and translate those demands into company language and funnel them through an efficient decision process.

Background

Product development is risky business, and to succeed, you need luck or manage information about future customer preferences well.

The time from idea to market launch is often several years. In a large company, several groups ofspecialists have to agree on product development ideas in order to produce good results. They need access to different aspects of the same requirements in order to cooperate and communicate well. Therefore a structure has to be available to handle information regarding future potential markets, customers and their requirements.

Not only must an organization predict the future, it has to do it in a turbulent and dynalluc world. With the quick information and competence transfer over the world, a company can be sure that successful new products brought to the market will be imitated. Soon competitors will benefit from much lower R~D and market introduction costs. Time to recuperate investment has become shorter and failure in the marketplace, either commercial or technical, has become very expensive. First to market is no longer enough, if you don't t have the capital or the organization to scale up and penetrate the market, a company will not reap the benefits of a "successful" product.

A case study - The Company

A case study is being conducted at a certain company, called The Company. The main business of The Company is to develop, and supply communications equipment and solutions to the business society. The Company is represented in some 40 countries around the world, and with R&D units in 7-8 different countries.

In The Company, you find products being developed, that can pass as incremental, substantial transformational product development.

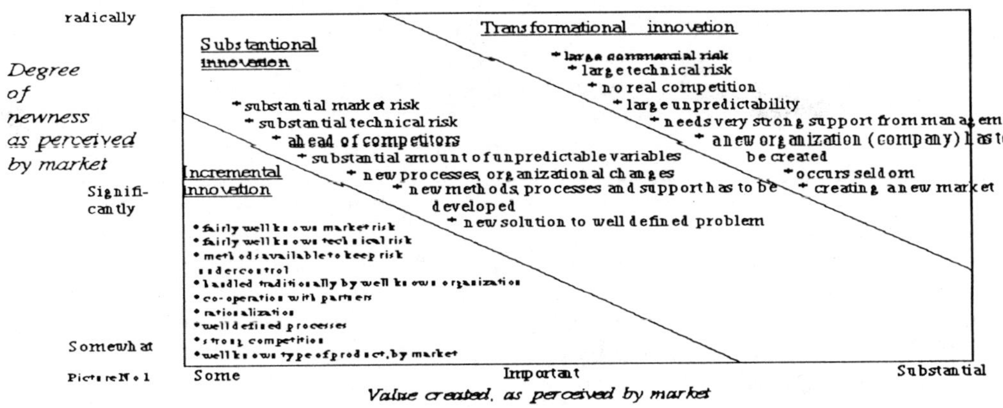

Picture No 1

The communications industry is in a turbulent phase, where old monopolies are dissolved, and where maturing and emerging technologies exist, side by side.

The technique to handle the transition period between technologies is to prolong the life of products, belonging to the aging technology (cash cows), by specializing. Specialization here means adapting products to suit specific customer needs. A smaller part of the R&D capital is geared towards product-, and service-development to suit various segments and microsegments. A major effort is at the same time made to invest in new technology and competence.

In order to be able to handle these different strategies The Company needs different information, different knowledge, different decision criteria, and different organizational structures.

Hypothesis for study

This paper will concentrate on information structures regarding Company environmental information. The proposal is that these information structures will help decision making.

The hypothesis for the entire case study is; - strategies and tactics for incremental, substantial, and transformational innovations have to be differentiated. - information collection and information management supporting different strategies and tactics has to differ between incremental, substantial and transformational product-development.

Theory

General Systems Theory, together with Ashby's Law of Requisite Variety, provides a good basis for viewing the flow of information and data within a company. General Systems Theory is also useful for predicting the implications that changes in the environment will have on an organization. (Checkland 1993, Bowler 1981, Langefors/Samuelson 1976, Agrell 1991)

Ross Ashby has stated, that if the relevant environment of an organization has a great deal of variety, the organization can reduce the environmental effects only by increasing the variety of its actions at the interface. He has also said that as the complexity of the environment increases, the organization must become structurally and functionally as complex as the environment in order to deal with it successfully. (Ashby 1964)

Methods and tools for handling Company external flow of information, and for verifying that it is valid and requisite, must be found. A method for structuring of data is suggested below.

This paper suggests that the early stages of information collection for product development follow two principles;

- according to a Piaget inspired principle
- according to a principle inspired by Popper's schema Pl TT EE P2.

According to Piaget knowledge is built on personal or organizational interest, creativity and activity. It is not necessarily structured according to a conscious and visible structure. Individual models are built. These cognitive models change with learning and experience. The search for information is a continuing process. (Ginzburg/Opper 1969, Popper 1994)

According to Popper; "... knowledge starts from problems and ends with problems (learning spiral)." Popper's schema is applicable to knowledge in the objective and in the subjective sense.

A systemic flow of information

A systemic view of The Company shows organization boundaries the incoming flow of information and matter, the company internal information and matter processes and the output.

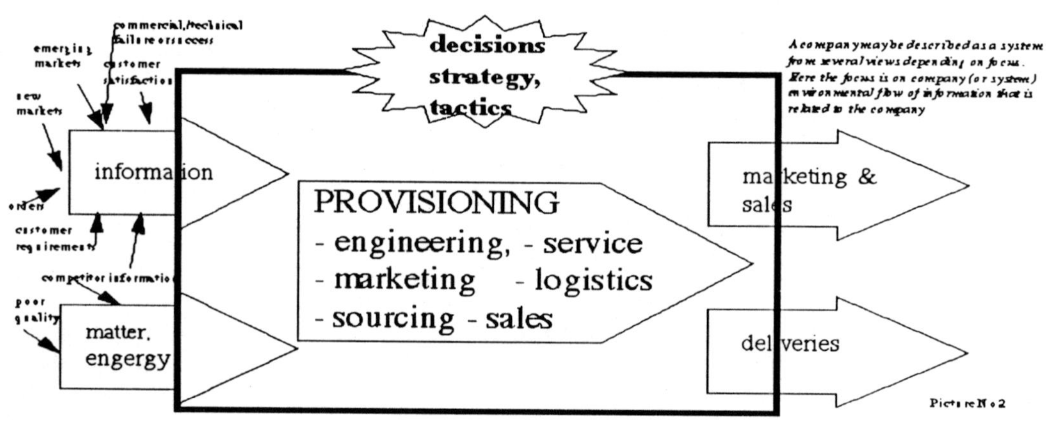

Picture No 2

Ashby states that a company must have relevant knowledge to judge whether a certain piece of information is useful or not (Ashby 1964).

According to the Piaget principle for information collection a broad and varied interest for the environmental informationflow must be promoted in an organization, if the organization is to accept it in its knowledge base.

According to the Popper principle, any information, that contributes to the answer to a problem is included in the organization's knowledge base.

The Company has to be well structured to handle the same complexity as its environment.

Requirement structure for product development

Market and customer requirements arrive at The Company in many shapes and forms. The information is stored in binders, people's heads or various databases, little is used systematically in specific situations.

The Company has not an overall picture of the present or future market/customer requirements. There is no systematic search for information to answer specific questions regarding what products to develop. It is difficult to handle and manage information that is not structured. If specific knowledge is not available it will not be understood as information, and may well be noted as noise. (Ref. i.e. the Shannon-Weaver Communication Model).

Information and knowledge

It is difficult to value information. In The Company the evaluation is personal. Information in The Company is normally communicated to peers of the same discipline only. (Mintzberg 994)

The Company has been rather successful in integrating different disciplines in its multifunction project model for product development, but The Company is not enough process, project and team oriented. (Hammer/Champy 1993)

It is vital for an organization to collect all relevant information that may affect it. If an organization does not know that a change has occurred, it cannot respond. If an organization finds out about a relevant change in the environment after two other changes have already occurred, the reaction will be to the wrong change. The organization will be reacting to what it thought happened, when in reality the relevant environment is at a different state.

Time-to-market-penetration

The measurement starts when an idea is formulated in the marketplace or in the developer's organization. This paper deals with information collection and decision making from the point at which an idea is formulated - until a decision is made to start a conscious study of the product idea. The first point is not easily measured, but valid, since it is a reality, and since it is necessary for The Company to identify and act on market opportunities early. In The Company, time between product idea and study start has to be shortened.

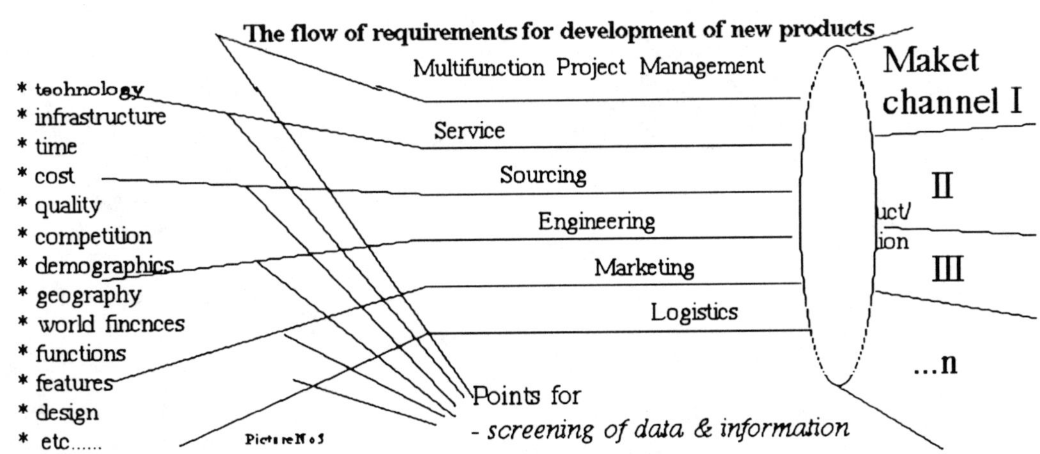

It is necessary that decision criteria are well known and that decision makers are able to ask the necessary questions. Decision criteria must allow for new ways to be explored.

One way to test new ideas is to involve customers who are leaders of their branch.

Customer Involvement

Customer involvement to assist evaluation of early market requirements has been used with success.

However, concerning transformational innovations, customers are not aware that they will become customers two - five years ahead. Other factors, such as top management involvement, technology break-through, and "gutfeel" are essential.

Information collection

Regarding information collection in connection to product development, this paper proposes two different ways, irrelevant of discipline The way information is collected depends on whether the product is to be;

1 incrementally/substantially new or
2 substantially/transformational new.

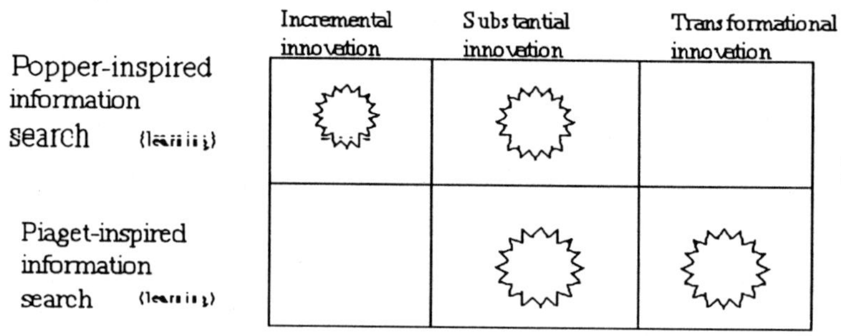

The principles of information searching are different when searching for information regarding incremental, substantial or transformational product development.

It is necessary that all disciplines have a responsibility to actively gather environmental information and act proactively with regard to product development.

Suggested definitions for management of market/customer requirements

low

product, solution maturity

demographics	potential markets	potential customers	segments	segments of segments
emerging competition	new competitor constellations	possible competitor	several competitors	fierce competition
emerging technology	technology uncertainty	known technology simple applications	complex applications	complex, networked customized applications
	exclusive marketing	heavy broad marketing	channeled marketing	micro channeled marketing
		possible partnering	change of/deepening of partner rel.	change of deepen of partner rel.
		personalized distribution	specific distr. channel	specialized distr. channel
		personalized service, exclusive	own service org spec. competence	own/outsourced service org/ personal service
		personal selling	sales efficiency focus	automated sales/ personal selling
		emerging processes	rationalization	mass production/process driven, efficient, high quality

high

In order to facilitate management of environmental information, it is necessary to use a common structure and common definitions. In pictures No 5 and 6 a structure and a set of definitions as well as a suggestion for information structure in connection with product development, are displayed. The suggested definitions and structure can be used independently of newness of product.

The information gathering principles are different. For

- incremental innovations - the Popper inspired principle should be used
- substantial innovations - the Popper or the Piaget inspired principle can be used

- transformational innovations - the Piaget inspired principle should be used. (see picture No 4 above)

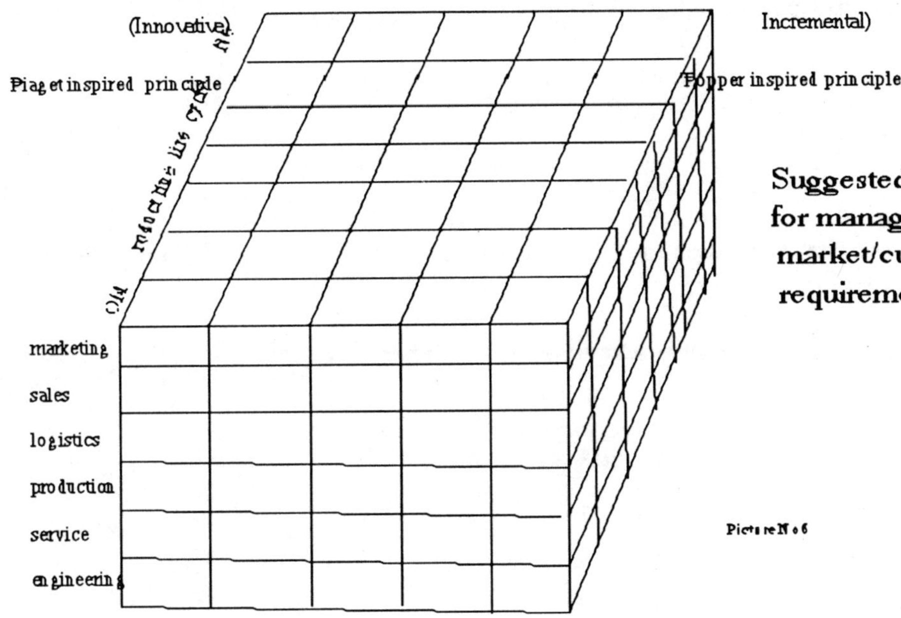

Suggested structure for management of market/customer requirements

Picture No 6

With the fast aging of information and with companies organizing themselves around the b informationflow, and decision process, decisionmaking must be integrated in the informationflow and not be seen as hindering the informationflow. The decision process is an important part of information management concerning product development.

Conclusion

In a dynamic and turbulent world it is extremely important to have a communicated and well known decision process in the early stages of product development.

It is important to find tools and methods whereby it is possible to benefit from engagement and still not let strong sentiments blind people, to ease and to quicken de-learning and learning.

Prestige is often at stake as new information and knowledge is making its way and as old knowledge and old myths will have to go. Failure to systematically manage environmental information will result in collapse. (Argyris 1992)

The next thing to be done in this research work is to test presented method and structure of information collection and management in the early stages of product development. The decision process in connection to product development will further be studied.

References

Peter Checkland, “Systems Thinking, Systems Practice”, John Wiley & Sons, 1993

T. Downing Bowler, “General Systems Thinking, its Scope and Applicability”, North Holland, 1981

Borge Langefors, Kjell Samuelson, “Information and Data in Systems”, Petrocelli/Charter, 1976

Per S Agrell "Systems Theory for Systems Practice. A simple framework for complex studies", Informatics and Systems Science Dep. Stockholm University & Royal Institute of Technology, TRITA-IS 5437/FOA D 10229, 1991

W. Ross Ashby, "An introduction to Cybernetics" London, Methuen, 1964.

Herbert Ginzburg and Sylvia Opper, "Piaget's Theory of Intellectual Development" Prentice Hall, Englewood Cliffs, New Jersey, 1969

7.) Karl R. Popper, "Knowledge and the Body - Mind Problem", Routledge, London 1994

Henry Mintzberg, "The Rise and Fall of Strategic Planning" Prentice Hall Intnl. (UK) Ltd., 1994

Michael Hammer and James Champy, "Reengineering the Corporation, a Manifesto for Business Revolution", Harper Business/Harper Collins, N.Y., 1993

Chris Argyris, "Interpersonal Barriers to Decision Making ", Harvard Business Review, Harvard College, 1992

Managing a Virtual Hospital

Hannu Jaakkola
+358 39 6272712

Tampere University of Technology (Pori)
P.O.Box 30
FIN-28601 Pori, Finland

Pekka Loula
+358 39 6272740

Tampere University of Technology (Pori)
P.O.Box 30
FIN-28601 Pori, Finland

This paper will discuss the concept of virtual hospital and telemedicine as a part of daily hospital operations. The concept of the virtual hospital is analyzed by using business, application and technology point of views. In addition to the hospital oriented operations the operational level activities, i.e. mobile hospital, independent living, care of wellness and treatment chain, form an essential part of it. A common feature of all of these operations is the central role of telecommunication to implement the operative systems. The virtual hospital may not only be seen as a collection of new technical tools at the hospital; above all there is a question on the adoption of new operations model to achieve effectivity and better quality from both the hospital and the patient point of view.

Introduction

Virtual hospital is a virtual organization which provides medical services by the means of information and telecommunications technology. It is a global, distributed and flexible network of co-operating organizations providing highly qualified hospital services to the patients. In addition to the medical diagnoses (*telemedicine*), there is a question on the new kind of arrangements in the administration, workflow management, laboratory operations etc. as a part of the hospital's daily activities.

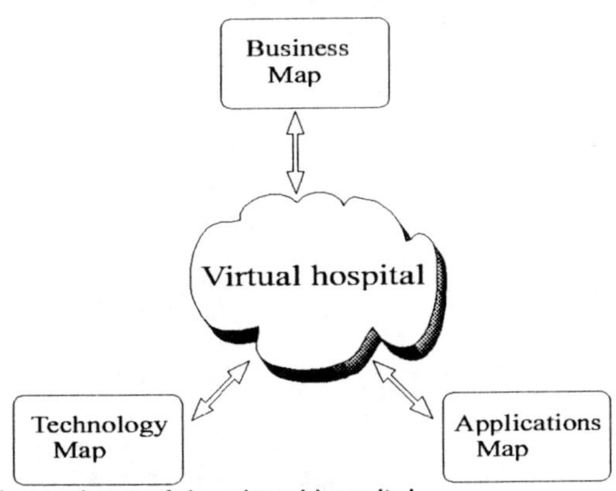

Three views of the virtual hospital

To clarify the "virtual hospital" concept, different viewpoints are needed. Figure 1 is a modified top level description of the *BITMap Reference Model* (Jaakkola 1995; Haikala and Jaakkola 1996). It points out three views (maps) worth discussing in the analysis and management of the utilization of information technology (IT) in industrial companies. This approach will be used to analyse the "virtual hospital" concept in this paper. The *Business Map* contains factors derived from the business (the hospital is seen as an enterprise) activities of the hospital. The *Applications Map* (originally Information Systems Map) studies applications and information systems of the virtual hospital. The key point is the integration of applications using global level interrelations between them. The *Technology Map* considers the problem from

0-7803-3552-X/96/$5.00 © 1996 IEEE

technical solutions point of view. The following chapters follow this map structure. At the end of the paper some experiences in the adoption of telemedicine applications in Satakunta region will be discussed.

Business View of the Virtual Hospital

Telemedicine is one of the major ways to increase effectivity and quality in the hospital. The major benefits in re-engineering of hospital activities by the means of telemedicine are

- *better quality of the diagnoses (highly qualified experts available in all nodes of the virtual hospital);*
- *regional equality in medical care;*
- *faster diagnoses;*
- *decreasing need to move the patients;*
- *easier follow-up of the patients;*
- *decreasing need to revisit the doctor;*
- *improved access to the national data bases;*
- *increasing effectivity in the hospital operations.*

The virtual hospital works in an efficient manner by providing services when they are requested. Telemedicine helps to reduce unnecessary and redundant work by connecting *contents producers* (e.g. laboratories, hospitals, health care centers, different specialities etc.) and *service users* (e.g. communities) together, shown in Figure 2. The virtual hospital has the right to choose suppliers. The virtual hospital may acquire the requested services wherever they are produced with enough good quality and reasonable price. The main goal of the business management is to ensure continuous services with reasonable time delay. The virtual hospital makes the contracts with different type of service providers or provides them itself. If the selected contents producer or service provider is overloaded or not accessible, the virtual hospital may use the other suppliers.

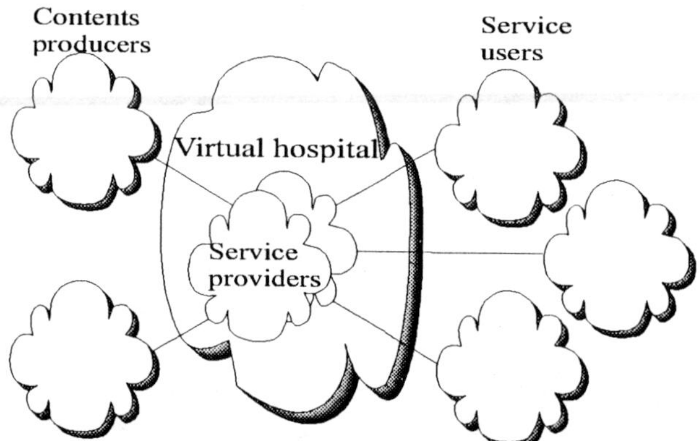

Connectivities of the virtual hospital

The virtual hospital is based on the use of networks, computers and databases. This causes costs in building, updating and maintaining the hardware and software. The spectrum of the provided services of the virtual hospital is a balance between costs and benefits including also the "soft" values i.e. equality and quality aspects. In some cases, it is very hard to measure and compare all the benefits, e.g. in estimating a quality of life aspects.

Applications View of the Virtual Hospital

The information flows and systems related to the virtual hospital are various. Figure 3 shows the basic four elements of the future virtual hospital namely: *care of wellness, independent living, mobile "hospitals" and traditional treatment chain* in a primary and specialized health care. The elements are not independent parts of health care delivery processes. They are integrated and connected to each other very strongly.

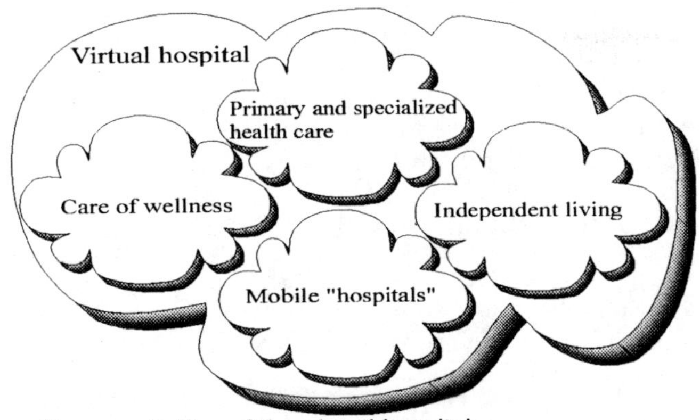

Focus activities of the virtual hospital

The care of wellness concept provides an extension to the concept of prevention. It contains the aspects related to the maintaining of a good health status, but it includes also the development and utilization of modern technology for education, exercise and monitoring physiological variables (Saranummi and Kivisaari 1996). The independent living concept makes possible the use of health care telematics services at home - not only to manage the environment and equipment. It means that the patient, disabled or elderly person, can manage his/her illness or well-being using multimedia platform at home. Simultaneously the health care organizations can provide services, e.g. call-centers and medical education programs, which improve health care delivery at home. The concept of mobile "hospitals" includes the existing emergency services, e.g. ambulance and helicopter services. Furthermore, it also provides non-emergency services e.g. home visits.

The management of the treatment chain in all points in primary and specialized health care processes is essential in the virtual hospital. The management requires transparent information in different phases of the chain, e.g. booking function, resource reservation, patient record information, lab tests, images, statistics, etc., must be available in an electronic form to ensure information for every phase. It is essential that there are developed access mechanisms to the real-time databases. This requires standard interface protocols for exchanging information.

On the other hand the patient is playing a key role in the virtual hospital. The patient is in the center of the virtual hospital, shown in figure 4. It means that the patient will come into one of the contact sites, e.g. health centers, where the services will be provided by the means of modern telecommunication technology. The unnecessary patient transports and booking operations in specialized health care are tried to avoid in order to minimize the total delivery cycle in the treatment chain.

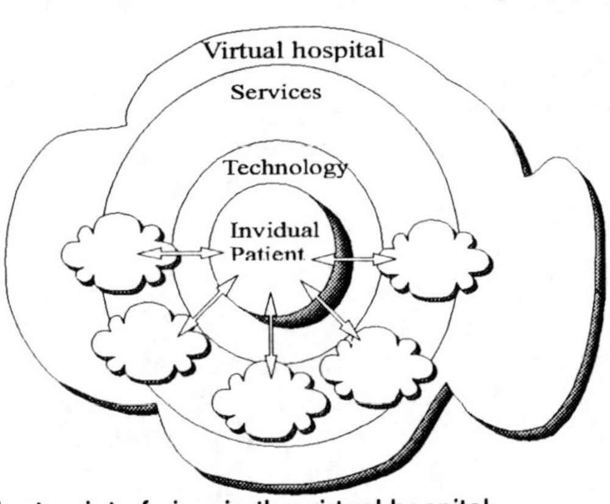

The patient point of view in the virtual hospital

Telemedicine has been used in various application areas (Johnson Ellen et al 1995) such as: *cardiology, dermatology, emergency medicine, family practice, gastroenterology, neurology, obstetrics/gynecology, oncology, ophthalmology, pathology, preventive medicine, psychiatry, radiology, surgery, urology, general telemedicine, education, nursing.*

Different application areas are in the different phase of their life cycle. Teleradiology and telepathology are the most studied application areas. For example, the use of teleradiology consultation services is possible in practice almost in every country.

Technology View of the Virtual Hospital

Technology produces the smallest problems to implement *BITMap Reference Model.* The technology is available and ready to use. The transmission can be realized by using ATM (Asynchronous Transfer Mode) to ensure large bandwidth (155 Mbit/s) as well as ISDN technology (nx64 kbit/s) to minimize the expenses. It is also possible to use fixed WAN (Wide area network) solution with various bit rates or modems including also high bandwidth modems i.e. HDSL and ADSL modems. The selected solution depends on the need and the demand for interactivity.

In Finland, the network infrastructure is very advanced e.g. all the Universities are connected together with ATM, and the number of GSM-phones per inhabitant is the biggest in the world. The hospitals are also well equipped and computerized, e.g. every University hospital has made ATM trials and every hospital uses some kind of distributed network for electronic patient record system. The nurses and doctors are used to handle computers in their daily work.

The applications which used EDI protocol are also implemented. The laboratory requests and resource allocation applications are typical cases. The communication between different applications is still problematic. The software developers are not, in all cases, able to settle the standard interface protocols or make the tools to handle this problem. There exists an ambitious project which tries to specify the aspects related to the implementation of new softwares to the existing *Integration Platform* frame. It specifies the platform itself, interfacing protocols and the structure for databases. Some standards are already standardized internationally and ready for use, e.g. DICOM 3.0 for storing CT, MRI, PET, x-ray images.

Legal aspects set also some criteria for technology. Certainly, the security and privacy should be provided, but the level of security and privacy is not clearly defined anywhere. There exist a lot of possible solutions to ensure security, e.g. smart cards can be used as a key to a database, encrypt algorithm can be used for data transmission. When the patient gives his/her card to the clinician and the clinician is also justified for data then the data can be fetched.

Discussion

In Satakunta province, the west coast of Finland (population 240.000), we have launched a health care development program (Loula 1996). The objectives of this program are to develop networked services which improve quality of care including equality, availability of medical competence, independent living and prevention of illness aspects. The program will be arranged with various local pilot projects. The outputs of these pilots will demonstrate how to use efficiently resources and how to develop the health care delivery processes by the means of virtual hospital concept.

The program will focus on developing telemedicine services including centralised archiving, database capabilities, on-duty arrangements, education, security and consultation aspects. Satakunta central hospital will provide telemedicine services for Noormarkku, Harjavalta and Kankaanpää health care centres, and Rauma area hospital using different types of transmission techniques from ATM to ISDN. The provided services will cover at least radiological, clinical neurophysiological, pathological and ultra sound consultations and cost/benefit analysis at the first phase.

The virtual hospital is a reality now, e.g. the US army used mobile virtual hospitals in the Gulf war with success. The attitudes of the people vary a lot among citizens. This and the security aspects may be the top questions in order to build the virtual hospital. The cultural differences, e.g. insurance system, should be also noticed.

References

Haikala Ilkka and Jaakkola Hannu. *BITMap - Three Views of Information Systems.* Tampere: Tampere University of Technology. Hypertext material and unpublished manuscript, 1996.

Jaakkola Hannu. "IT Trends and Their Effect on the SW Engineering Culture". *Proceedings of the Fifth International Workshop on Experience with the Management of Software Projects.* Karlsruhe: University of Karlsruhe and GI/GM/IFIP/IFAC (Sept. 27-29, 1995).

Saranummi Niilo and Kivisaari Sirkku. "HealTH - Health Technology in the Home". *Proceedings of the Nordic Telemedicine Conference.* Stockholm, Sweden (Jan 30-31, 1996).

Johnson Ellen, Debold Vicky, Chuang Dennis, Tolbert Jennifer, Cameron Andrew, Miller Edward. "Telemedicine: An Annotated Bibliography: part I, part II and part III". *Telemedicine Journal.* 1: 155-165 & 257-292 & 377-394, 1995.

Loula Pekka, Lummevaara Timo, Rantanen Pentti. "Telemedicine development program in satakunta region: C-Medi". *Proceedings of the First Nordic Congress of Telemedicine.* Kuopio, Finland, (Jun 17-19.1996).

Intelligent Reference Models for Designing Virtual Factories

Julio Macedo

Institut de Stratégies Industrielles
229 Forest, Pincourt, P.Q., J7V8E8, Canada

A virtual factory is a temporary partnership of factories created to offer the right portfolio of products to changing target markets. A new artificial intelligence based tool that helps to select the partners of a virtual factory is presented. This tool generates the organization that the manufacturing system of a candidate partner must have in order to be incorporated into the virtual factory.

Introduction

A virtual factory is a temporary partnership of real factories created to offer the right portfolio of products to changing target markets (Davidow and Malone 1995).

All products inherit managerial attributes from the organization of the factory that makes them. Production delay, percentage of defective products, overhead production cost and product variety are all examples of managerial attributes. These managerial attributes must satisfy a target market. Hence, the factory partners of a virtual factory must make products with managerial attributes desired by the target market of the virtual factory.

Whereas the definition of virtual factory is clear, there have been no tools created to select its partners until now (Goldman, Nagel and Preiss 1995). A new tool that helps to do this selection is presented in this paper. This new tool is a set of intelligent reference models that generates the organization that must have the manufacturing system of the candidate partner in order to satisfy the target market of the virtual factory.

Intelligent Reference Models

An intelligent reference model is a mathematical structure composed of an energy function and of a qualitative simulation model (Figure 1). When the managerial attributes desired by the virtual factory target market are expressed as binary values and are introduced in the reference model, and once this model is solved, the control variables get binary values. These values of the control variables thus represent the organization that a partner of the virtual factory must have.

0-7803-3552-X/96/$5.00

Energy function: $\underset{\mathring{p}_{j-1},\, \mathring{p}_j}{\text{Min}}\ Z = \int_0^T m_n [y_n - y_d]^2 + m_{n-1} [y_{n-1} - y'_d]^2\, dt;\ -1 \le \mathring{p}_{j-1}, \mathring{p}_j \le 1$

Qualitative simulation model:

$$\begin{bmatrix} \mathring{x}_{i-1} \\ \mathring{x}_{i-2} \\ \mathring{x}_i \\ \mathring{x}_{n-1} \\ \mathring{x}_n \end{bmatrix} = \begin{bmatrix} \mathring{S}(x_{i-1}) & 0 & 0 & 0 & 0 \\ 0 & \mathring{S}(x_{i-2}) & 0 & 0 & 0 \\ 0 & 0 & \mathring{S}(x_i) & 0 & 0 \\ 0 & 0 & 0 & \mathring{S}(x_{n-1}) & 0 \\ 0 & 0 & 0 & 0 & \mathring{S}(x_n) \end{bmatrix} \left[\begin{array}{ccccc:cc} 0 & w_{i-1,i-2} & 0 & 0 & 0 & w_{i-1,J-1} & 0 \\ 0 & 0 & 0 & 0 & 0 & 0 & w_{i-2,J} \\ w_{i,i-1} & w_{i,i-2} & 0 & 0 & 0 & 0 & 0 \\ 0 & 0 & w_{n-1,i} & 0 & 0 & 0 & 0 \\ w_{n,i-1} & 0 & w_{n,i} & 0 & 0 & 0 & 0 \end{array}\right] \begin{bmatrix} \mathring{x}_{i-1} \\ \mathring{x}_{i-2} \\ \mathring{x}_i \\ \mathring{x}_{n-1} \\ \mathring{x}_n \\ \hdashline \mathring{p}_{J-1} \\ \mathring{p}_J \end{bmatrix}$$

$$p_{j-1}(t) = \int_0^T \mathring{p}_{j-1}\, \mathring{s}(p_{j-1})\, dt;\quad p_j(t) = \int_0^T \mathring{p}_j\, \mathring{s}(p_j)\, dt; \qquad \begin{bmatrix} y_n \\ y_{n-1} \end{bmatrix} = \begin{bmatrix} x_n \\ x_{n-1} \end{bmatrix}$$

Point values selected to solve the equations:

$x_{i-1}(0) \in [x^l_{i-1}(0); x^u_{i-1}(0)]$; $x_{i-2}(0) \in [x^l_{i-2}(0); x^u_{i-2}(0)]$; $x_i(0) \in [x^l_i(0); x^u_i(0)]$

$x_{n-1}(0) \in [x^l_{n-1}(0); x^u_{n-1}(0)]$; $x_n(0) \in [x^l_n(0); x^u_n(0)]$

$p_{j-1}(0) \in [x^l_{j-1}(0); x^u_{j-1}(0)]$; $p_j(0) \in [x^l_j(0); x^u_j(0)]$

$w_{i-1,i-2} \in [w^l_{i-1,\ i-2}; w^u_{i-1,i-2}]$; $w_{i-1,j-1} \in [w^l_{i-1,\ j-1}; w^u_{i-1,j-1}]$

$w_{i-2,j} \in [w^l_{i-2,\ j}; w^u_{i-2,j}]$; $w_{i,i-1} \in [w^l_{i,\ i-1}; w^u_{i,i-1}]$

$w_{i,i-2} \in [w^l_{i,\ i-2}; w^u_{i,i-2}]$; $w_{n,i-1} \in [w^l_{n,\ i-1}; w^u_{n,i-1}]$; $w_{n,i} \in [w^l_{n,\ i}; w^u_{n,i}]$; $w_{n-1,i} \in [w^l_{n-1,\ i}; w^u_{n-1,i}]$

m_n, m_{n-1}: Constants known from training; y_d, y'_d : Introduced constants

Reference model outputs: $p_{j-1}(T)$, $p_j(T)$ (point values)

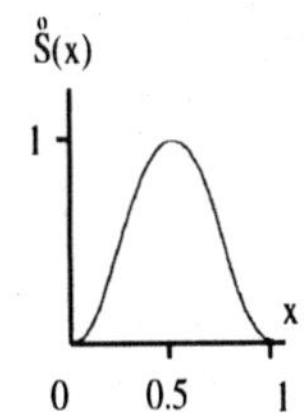

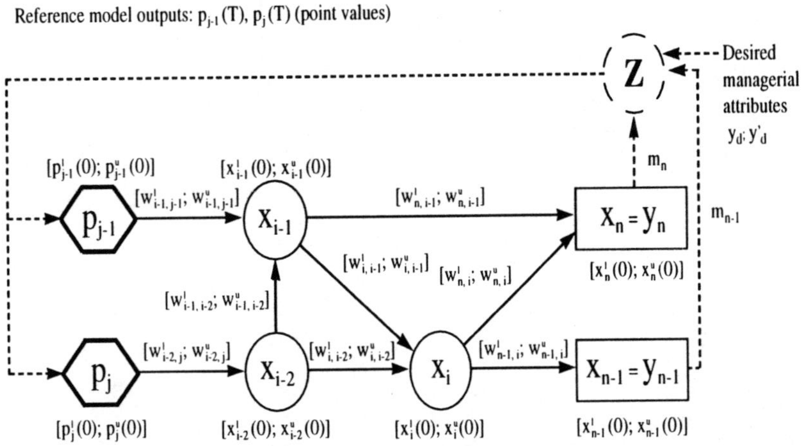

Figure 1. Structure of an intelligent reference model which is composed of an energy function and a qualitative simulation model. The state variables are in circles, the control variables in hexagons and the managerial attributes in boxes. The superscripts l and u denote lower and upper bounds respectively. The cause-effect weights are between -1 and 1 and the initial values of the state and control variables between 0 and 1.

As shown in Figure 1, the qualitative simulation model is the decomposition of the managerial attributes in terms of statc variables and of control variables related by cause-effect relationships. The rate of variation of each state variable is represented as the weighted sum of the variables that influence it, multiplied by a bell shaped function. The latter pushes the final value of the state variable to 0 or 1. As for the energy function, this is an optimization function that minimizes the deviations between the current managerial attributes and the ones desired by the target market.

At this point note that the values of the control variables produced by the solution of the intelligent reference model are robust to small variations in the values of the cause-effect weights and in the initial values of the state and control variables. This robustness is used to construct the intelligent reference model in the following way: First, a tentative reference model is postulated; second, known multivariate points of the type "desired managerial attributes, initial values of the state variables, initial values of the control variables" are introduced in the tentative model; finally, the reference model is solved and modified until the known final values of the control variables are generated (Macedo and Ruiz Usano 1995). At this point note also that the energy function is directed to self-organize and to adapt the control variables of the reference model to structurally different desired managerial attributes (Macedo 1994).

In order to select a partner for a virtual factory, the intelligent reference models are operated in the following form. First, the managerial attributes desired by the virtual factory target market are expressed as binary values and entered into the computer. Second, the current values of the state and control variables of the candidate partner are expressed as intervals in a scale 0 to 1 (or as linguistic values) and entered into the computer. Third, the cause-effect weights of the logics used for organizing the manufacturing system of the candidate partner are expressed as intervals in a scale 0 to 1 (or as linguistic values) and entered into the computer. Fourth, the computer identifies the stored intelligent reference model whose cause-effect weights, initial values of state and control variables (expressed as linguistic values) and desired managerial attributes (expressed as binary values) match the ones of the candidate partner. Fifth, the matched intelligent reference model is solved by using as point values for the initial values of its state and control variables and its cause-effect weights any values in their corresponding intervals of robustness. The final values of the reference model control variables represent the organization that the candidate partner must have.

Application

An intelligent reference model limited to the human factor was constructed by using the procedure above (Figure 2). In this construction, the managerial attributes desired by five target markets named craft, world, timer, quality and cheap were used along with known values of cause-effect weights and of state and control variables.

$$\text{Min } Z = \int_0^{40} [DEF - DEF_d]^2 + [DEL - DEL_d]^2 + [COST - COST_d]^2 + [VAR - VAR_d]^2 \, dt$$

$$-1 \le \dot{PROP}, \dot{PAL}, NSUP, \dot{SIG}, \dot{ELOI}, \dot{METH}, \dot{EXC}, \dot{POS}, \dot{PROC}, \dot{MAIN}, \dot{ERG}, \dot{EQUI}, \dot{DIF}, \dot{ZERO}, \dot{NAUTO}, \dot{DPDE} \le 1$$

$[DEF_d, DEL_d, COST_d, VAR_d]$ = [0, 1, 1, 1] Craft ; [0, 0, 0, 1] World ; [1, 0, 1, 0] Timer ; [0, 1, 1, 0] Quality ; [1, 0, 0, 0] Cheap market

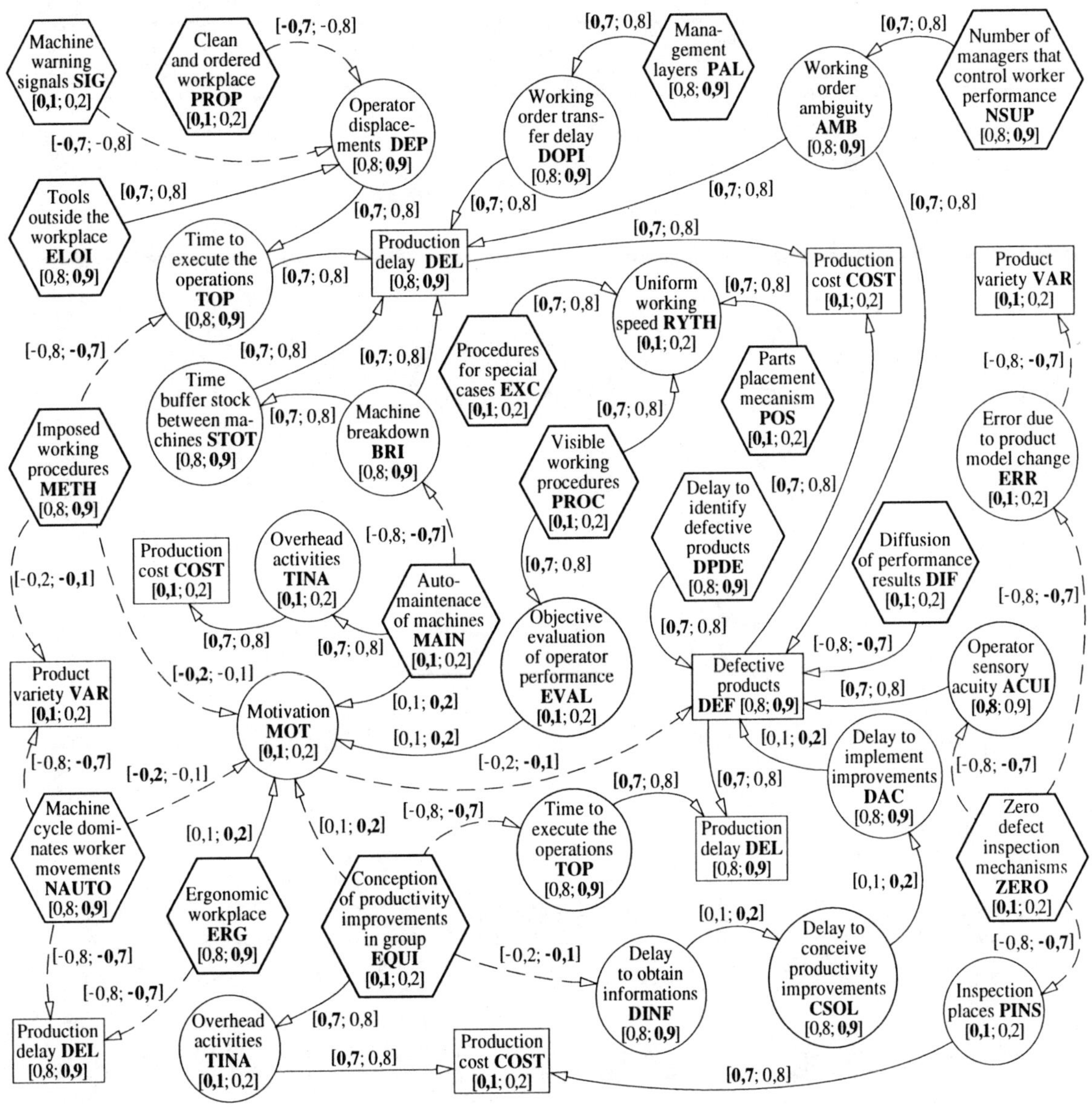

Figure 2. Intelligent reference model of the human factor. The intervals of robustness of each variable or cause-effect weight are in brackets. The numbers in bold have been selected as point values to solve the reference model.

In order to illustrate the use of the reference model in Figure 2, it must be assumed that a virtual factory intends to feed a world class market and wants to know if a candidate partner should join it. It must also be assumed that the characteristics of the candidate partner match the reference model of Figure 2. When this model is solved, the control variable values in the third column of Table 1 are then generated. These values indicate the organization that the candidate partner should have in order to join the virtual factory. At this point note that the other values included in Table 1 were obtained by solving the reference model of Figure 2 with the desired managerial attributes in the upper part of Figure 2. These values indicate the organization that a candidate partner must have in order to join the virtual factory when the latter feeds structurally different target markets.

Control variables	Final value of the control variable for a target market				
	Craft	World	Timer	Quality	Cheap
PROP	0	1	1	0	1
SIG	0	1	1	0	1
ELOI	1	0	0	1	0
PAL	1	0	0	1	0
NSUP	1	0	1	0	0
METH	0	0	1	0	1
EXC	0	1	1	0	1
PROC	0	1	0	0	1
POS	0	1	1	0	1
MAIN	0	1	1	0	1
DIF	1	1	0	1	0
NAUTO	0	0	1	0	1
ERG	0	1	0	0	1
EQUI	1	1	1	1	1
DPDE	1	0	1	0	1
ZERO	1	1	0	0	0

Table 1. Final values of the control variables obtained by solving the intelligent reference model with managerial attributes desired by different target markets.

References

Davidow, W. and M., Malone. *The Virtual Corporation. Harper*-Collins Publishers: New York, 1992.

Goldman, S.; R., Nagel and K., Preiss. *Agile Competitors and Virtual Organizations.* Van Nostrand Reinhold: New York, 1995.

Macedo, J. and R., Ruiz Usano. "Generating manufacturing system organizations using expert-neural models". *Intelligent Engineering Through Artificial Neural Networks.* C. H. Dagli, ed., p. 1001-1006. ASME Press: New York, 1995

Macedo, J. "Reference model recognition system for designing the manufacturing system organization". *Intelligent Manufacturing Systems.* P. Kopacek, ed., p. 205-210. Pergamon Press: Oxford, 1994.

Using Information Technology to Leverage Research Productivity

Larry A. Mallak
(616) 387-3780

Department of Industrial & Manufacturing Engineering
Western Michigan University
Kalamazoo, MI 49008

New information technologies provide tremendous leverage to research, especially as practiced in a traditional teaching university seeking to become a research institution. This paper reports several applications of fax machines, teleconferencing combined with taping and transcription, databases, and the Internet as powerful and inexpensive tools for achieving greater research output with minimal increases in resources. A survey on ISO 9000 implementation conducted via fax machine generated a 74% response rate and required only a $300 budget. Other examples of using information technology to leverage research productivity are offered.

The Need for Increased Productivity in Research

Several current pressures and trends in academia are forcing the need for increased research productivity, especially in universities seeking to move up in their Carnegie institutional categorization. Productivity is defined as the ratio of outputs to inputs. In research, the outputs are completed experiments and action research projects, publications, and funded grant proposals. Common inputs to the research process include researcher time, research assistants' time, money to fund development, experiments, equipment, computers and computing resources, facilities, phone and fax charges, office supplies, etc. The time inputs are becoming increasingly scarce as demands on faculty and R&D staff time escalate. This paper reflects the perspective of a faculty member in a regional state university in the Great Lakes conducting research in engineering management.

Some of the research pressures facing faculty members today include:

- Teaching *and* research—historically, regional universities have prided themselves primarily on their teaching quality and class size. On the research side, however, some more progressive regional universities are seeking to raise the stakes on attracting research funding and, correspondingly, on faculty to do the research and graduate programs to facilitate the research-teaching linkages.

- Expectations have changed greatly, infrastructure has changed much more slowly. The research expectations in many regional universities have increased greatly with proportionately smaller changes in infrastructure to support research. Departmental support staffs are stable or shrinking while faculty rolls are increasing.

- Increased competition for research resources—as academia turns toward research dollars to fill in where state monies leave off, competition for grants, contracts, and other awards becomes very keen. New players must build their networks and deal with much lower probabilities of grant proposal being funded than was the case five or ten years ago.

- Bottom line: do a lot more with the same or a little more. In an era with increasing expectations for output and nominal increases in input (no graduate or undergraduate assistants, full teaching loads, bare-bones support staff, commuter graduate program), a recommended way to increase productivity is to produce more output while basically holding input constant.

0-7803-3552-X/96/$5.00 © 1996 IEEE

This discussion of trends is not meant as an excuse for not achieving the desired research goals—it is merely a reflection of current trends sweeping academia, industry, and government. These trends are here to stay and this paper offers methods for individuals to make creative use of information technologies to increase their research productivity—to do more with less. Information technology provides the opportunity to rethink how we conduct research and what kind of research we can do.

A variety of tools based on information technology can form the basis for leveraged innovation. By no means is this paper's coverage exhaustive—it offers examples and ideas on how personal research productivity can be increased using information technology. Certainly, new information technologies and new applications of these technologies will suggest many new ideas in the years to come.

Information Technology: The Raw Materials for Leveraged Innovation

Pay the Upfront Price

New technologies come with a nonfinancial price tag: the time it takes to make them work for you. You'll no doubt spend hours figuring out how to make these technologies work. Look around your organization for someone who has proven his or her computer skills. This is the person you'll want to have help you. Often, the lack of time or comfort with a known product leads to an endless spiral of never installing or using the new software and realizing the potentially time-saving capabilities. At some point you made the switch from typewriter to computer. That type of switch needs to happen again and again to keep up with technology.

Example Applications of Leveraged Research Using Information Technology

The many functions and features of information technologies such as computers, software, the Internet, and fax machines are well known and can be learned elsewhere. This paper focuses on specific applications of these technologies for enhancing your research productivity. Here are some applications to get you started thinking about how your research productivity can be improved.

ISO 9000 Survey

Two of my colleagues and I embarked on a research study to examine the relationships between ISO 9000 certification and organizational culture. We decided a survey of ISO 9000 management representatives would form the basis for our data collection. Rather than send the survey through the mail (with its "good" response rate of approximately 30%), we opted to fax the survey to these management representatives after contacting them by telephone and gaining commitment to participate in our study. We sought $300 of internal funding to cover phone calls and fax charges. We sent the surveys out directly from the PC (mostly after 5 p.m. to save on phone charges and to avoid interfering with other PC duties, such as preparing for classes and writing papers and grant proposals) and request they also send their responses via fax.

Out of 110 surveys sent, we received 85 usable surveys, for a response rate of 77%. This figure was more the double the rate expected for a conventional mail survey. We analyzed these data and have produced two publications—one conceptual/overview piece published in *Total Quality Review* (Mallak, Lyth, & Bringelson 1995) and an empirical piece currently in review (Mallak, Bringelson, & Lyth, in press). For $300 and the time required for three people to make approximately 50 3-minute phone calls each, we were able to complete a small study and publish two different papers.

Tools Used	computer-based fax and traditional fax machines, telephone
Assessment	Achieved 77% response rate; respondents were interested in participating.
Costs	Approximately $300 in phone and fax charges. No postage required.

Joint Venture for Management Book Research and Writing

A recent project with Virginia Tech's Center for Organizational and Technological Advancement (COTA) led to a joint project with COTA's director, a graduate student in COTA, and a principal of a North Carolina-based leadership institute. This joint project concerned the conceptualization, research design, data collection, and publishing of results in a popular business book format. Our initial meeting was held in person in Blacksburg, Virginia, and we planned to hold subsequent meetings as teleconferences.

All four project members have busy schedules of teaching, research, administration, consulting, training, continuing studies, and other projects and the reality of collaborating across three states. Part of the graduate student's funding involved the logistics for these virtual meetings. Our meetings were scheduled approximately once every 4-8 weeks, depending on progress and schedules. Each meeting was initiated through a Commonwealth of Virginia operator in Richmond, who called all four parties, connected them to the conferencing network, taped the meeting, and transcribed the conversation. The transcribed material was then sent to the graduate student who prepared clean copies, sorted by our working table of contents, and distributed these documents either through electronic mail or U.S. mail.

The conference call, while not cutting edge technology, does preclude the need for all parties to give up one or two days to attend a one-hour meeting. The total cost of a typical 60-minute call involving four parties runs about $312, less than the cost of one mid-week airplane ticket, and that figure includes taping and transcription. The taping and transcription ensure a detailed record is kept of these meetings so we can move forward each time. In-person meetings would be superior, on the surface, because we would be able to have longer meetings, use flip charts or other devices to share ideas, distribute handouts at the meeting, and experience the full communication of a face-to-face meeting. However, the conference call meetings have the following distinct advantages: no travel is required to attend the meeting, thereby saving time and money; members can attend the meeting from any location; members have all the relevant materials (books, articles, etc.) with them that they wouldn't likely bring to an in-person meeting; and members aren't likely to be interrupted during a conference call.

Tools Used	telephone conference call, recording and transcription, e-mail and file transfer
Assessment	Effective for maintaining progress in absence of in-person meetings; a useful augmentation for a working group and not a replacement for in-person meetings.
Costs	Conference call fees for a 60 min. call and 4 parties, including recording and transcription: $92; Graduate student time (20 hrs/calling event @ $11/hr = $220) for a total of $312 per meeting.

Text-Retrieval Systems

The old models of trudging off to the library to research companies, practices, management, or industry trends are obsolete when you have access to full-text retrieval systems. Systems such as Lexis-Nexis offer the user a variety of searching capabilities and the ability to screen results by keyword in context, by full text, and by focused searches to further refine the search. Depending on the access fees, this can be a very expensive method for obtaining information and may lend itself to research productivity only when time and not money is considered as the primary input.

Don't rely on the World Wide Web for very much full-text retrieval yet. Even journals targeted to researchers and practitioners such as the *Harvard Business Review* limit their World Wide Web sites to publishing abstracts of current and back issues; *Business Week* has its cover story and table of contents posted at its Web site. Most sources tend to publish a sampling of their content, making the Web route to searching an incomplete approach since most of the needed content is both missing and unsearchable across publications.

Text-retrieval systems are limited, however, to the text contained in them. Most academic journals must still be hunted down in person and do not exist electronically. As a result, text-retrieval systems are good for obtaining company-specific information, industry profiles, financial information (such as SEC filings and annual reports), and examples or case studies of various management concepts and techniques. For example, a researcher investigating the turnaround strategy at Apple Computers could find articles in *Business Week*, the *New York Times*, *Fortune*, *Forbes*, *PC World*, Reuters, Associated Press, CNN, and others by using a text-retrieval system. However, the need for information on the underlying management practices would not typically be met through text-retrieval systems. One would need to search via ABI/Inform or other databases to identify articles and then get those articles from the library.

Tools Used	Lexis/Nexis, Business Database Plus
Assessment	Effective for current popular periodicals; company, industry, product, and financial information. Most academic publications are not included.
Costs	Access fees vary based on contract, sample range: $40-100/hr for Lexis/Nexis and $2.95/hr for Business Database Plus (through CompuServe)

Research Grant Information

With research grants quickly becoming the ticket for institutions to gain clout and fiscal power simultaneously, competition for research grants has intensified, but without the requisite budgets for seed studies and trips to sponsors. What you need is to expand the set of potential grantors. While there is no substitute for someone who does these searches for a living, the ability to run a few searches on your own, changing your key words each time without a sigh, and not incurring hourly fees to break the bank is worth investigating. Several grant sources for federal funding are available in searchable form on the Internet, including the Catalog of Federal Domestic Assistance (CFDA). The CFDA requires some assistance to understand its listings, but information is available on how much funding the particular program received in current and previous fiscal years and the average size of grant. (See the box at the end of this section for some Internet addresses to get you going. These addresses were current as of this writing. Please note Internet addresses change often. You may wish to use a search engine to find the particular Web site if the listed addresses don't work.)

The ubiquitous Internet is not the answer to all your grant problems. Nothing beats talking to the program officers on the phone or in person to determine their needs and tailor your proposal accordingly. However, several searchable databases on the Internet can help you locate funding sources, types of programs, abstracts of funded projects, and much more to get you started. As suggested earlier in this paper, the unwieldy Internet may offer more hype than hope at this stage of the game; but having the exact addresses of searchable government sites (where full information is typically available) should keep you on track.

Tools Used	World Wide Web sites: Catalog of Federal Domestic Assistance (http://www.sura.net/main/members/gsa.shtml), FEDIX (http://web.fie.com/htdoc/fed/all/any/any/menu/any/index.htm), IRIS (http://carousel.lis.uiuc.edu/~iris/option1.html/), National Science Foundation (http://www.nsf.gov), Commerce Business Daily (http://cos.gdb.org/repos/cbd/cbd-intro.html), Federal Government sites (Defense, Energy, Labor, Commerce, Education) Search engines: Open Text (http://opentext.com:8080/), Excite Netsearch (http://www.excite.com/), Yahoo (http://www.yahoo.com/), DEC's Alta Vista (http://altavista.digital.com/), Lycos (http://www.lycos.com/)
Assessment	The unwieldy Web is highly popular, but contains about 95% noise with respect to whatever you're looking for. Its relatively low costs are often offset by the professional time consumed in a search. Web content, as opposed to library content, is often selective based on who has paid a sponsor fee. Content is often incomplete, serving to tempt users to obtain a larger product. Access to information is unparalleled and a Web search often turns up information you're unlikely to encounter elsewhere.
Costs	University-based researchers usually have free unlimited Internet access. Most other services provide access for approximately $3/hr. Important cost to consider is professional time and searching effectiveness. If the information source is known, a phone call is usually more effective; however, the serendipitous discovery of other related information will be eliminated. The price to pay for this serendipity is lost time and distractions for the variety of sites returned from a search.

Conclusions

Conducting high quality research effectively demands access to resources and data. Researchers must become more productive, that is, produce more research using the same or slightly more resources. Several information technologies such as the fax machine, telephone with taping and transcription, and Internet can help improve research productivity if they are used effectively. The barriers to effective use of information technologies typically involve the learning curve.

Each new information technology carries a nonfinancial price with it—how do you get up and running on the new technology when you still have all your other activities crowding your work schedule? A certain portion of time should be set aside, despite all other crises and demands, to learn the essentials of the technology. As mentioned earlier, find a person in your organization (or outside of it) and have them walk you through the steps to accomplish your desired task. Also, software support lines have improved in quality and may be worth the effort, but these are a second choice. The ability to stay on task is essential when wandering the Internet. Testing out the various search engines on your own and then pointing your browser to your favorite one should reduce the noise when you're hunting for information on the Internet. Try the search engines mentioned earlier in this article to get you started.

We now have new tools in our tool box. It's always fun to use the new tools, but the old tools are still there and useful. Don't use a power saw when all you need is a pair of scissors. If a phone call will be more effective, use the phone.

With fewer funds and more researchers and the moving target of expectations, it's essential to learn methods to make ourselves more productive. The applications presented in this article are a small slice of how information technology resources can be leveraged to improve personal research productivity. In time, many more applications will become apparent and affordable. The researcher should be aware of the possibilities offered by information technology so the questions changes from "How can I possibly do research with this budget?" to "What do I want to do today?"

References

Mallak, L. A., Bringelson, L. S., and Lyth, D. M. (in press). Cultural Values Critical to ISO 9000 Certification.

Mallak, L. A., Lyth, D. M., and Bringelson, L. S. (1995). ISO 9000: Cultural Factors Impact Registration. *Total Quality Review*, Sept./Oct., 53-57.

Information System support for House of Quality

Heloisa Martins SHIH — Department of Industrial Engineering and Engineering Management
+852-2358-7100

Li Chung SHIH — Department of Information and Systems Management
+852-2358-7733

Stanley S. L. KWOK — Department of Industrial Engineering and Engineering Management
+852-2358-7100

The Hong Kong University of Science and Technology
Clear Water Bay
Kowloon
Hong Kong

In this paper we propose an information system to support the building and implementation of House of Quality in the companies. The system provides: (a) a general guide to build House of Quality step by step; (b) a user friendly interface for each step; (c) a data storage and retrieval system throughout the process; (d) a data repository for future House of Quality developments. The system contains also an Analytic Hierarchy Process-based procedure to support the prioritization of customer attributes. The proposed features and a prototype of the system are presented in the paper.

Introduction

House of Quality or HOQ (Hauser and Clausing 1988), also known as Quality Function Deployment (QFD), supports the design of products and services aiming concurrently at optimal deployment of technological and financial resources and maximum customer satisfaction. HOQ is an organized planning matrix that relates the attributes considered important by the customers with the technical characteristics of the product or service.

The building and implementation of House of Quality are usually complex and time consuming processes. They include survey of customer preferences and perceptions, technical tests of own and competitor's products, and identification of relationships among technical attributes and characteristics perceived by customers. One of the most challenging tasks in the development of a HOQ is the evaluation of the relative importance of the customer attributes. Aswad(1989) stressed that inconsistency in quantifying expert judgments is a common weakness in using QFD techniques.

The building of a HOQ is integrative and iterative, as the House is continuously reviewed during its development. The complexity of the process is enhanced by the diverse background and different perspectives of the HOQ development team, often working in different divisions of the business, sometimes in diverse geographical locations and time zones.

An information system could greatly facilitate the important, time-consuming and complex task of building HOQ. Such system would provide a structured, user-friendly way to store, retrieve and modify the necessary information throughout the process. The system could also include assistance for prioritizing customer attributes.

In this research, we aim to develop an information system to support the building of a basic HOQ. The system will provide user-friendly interface and seamless management of the database containing the house's characteristics and attributes. An Analytic Hierarchy Process-based support for the assignment of relative weights to customer attributes will also be included.

0-7803-3552-X/96/$5.00 © 1996 IEEE

Background

House of Quality

Figure 1 outlines the general organization of HOQ. The house contains several sections or submatrices. Sections related to the requirements of the customers, or customer attributes (CAs), appear in the horizontal dimension of the house. They include: (a) the list of CAs, (b) the relative priorities or weights of CAs, (c) the customer competitive assessment of the product under evaluation, when compared with competitors' products. The matrices related to the technical or engineering characteristics (ECs) are displayed in the vertical dimension: (a) the list of ECs, (b) the correlation among ECs, also called the roof matrix, (c) the technical competitive assessment of the product under evaluation, when compared to competitor's products, (d) the target values for each EC, (e) other technical information like cost or difficulty of modification. The center of the HOQ contains the relationship between CAs and ECs; that is, how the customer attributes may be translated to technical characteristics of the product. The matrix of column weights appears at the bottom of the HOQ. It combines the relative importance of each CA and the relationship between CAs and ECs. The overall column weights are utilized to set priorities among the technical characteristics of the product. Hauser and Clausing (1988) contains a good guideline for development of HOQ.

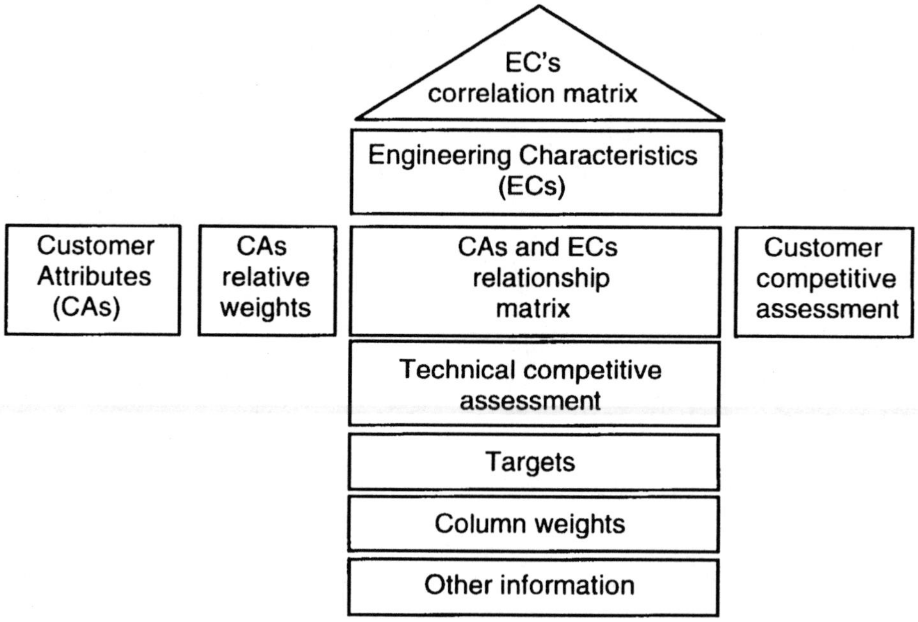

Sections and submatrices of HOQ

One of the steps in building the HOQ is the assignment by the development team of relative weights to CAs. Adding to the difficulty in quantifying the relative importance, there is the challenge to maintain consistency in the judgment. Aswad(1989), Akao(1990) and Armacost et al (1994) proposed Analytic Hierarchy Process (AHP)-based approach to facilitate the process and to ensure consistency in prioritizing CAs.

Analytic Hierarchy Process

AHP, proposed by Saaty(1980), is a multicriteria decision making technique, particularly useful for evaluating complex multiattribute alternatives including subjective criteria. AHP decomposes a complex, multicriteria problem hierarchically. The priorities of the elements at each level are determined in separate, and then combined to determine the overall priorities of the alternatives under evaluation. In order to assess the relative importance of the elements at each level, the evaluator compares those elements pairwise, to estimate their relative importance or preference.

Saaty developed a nine point scale, where the evaluators express relative preference of each pair of elements as equally, moderately, strongly, very strongly or extremely preferred. These preferences are respectively translated to pairwise weights 1, 3, 5, 7 and 9. Weights 2, 4, 6 and 8 are used as intermediate values. The priorities of the elements of each level are calculated using Saaty's eigenvector method, and then weighted by the priorities of the higher level elements. After the process is completed, an overall weight or priority of each of the decision alternatives is obtained. See Saaty(1980) for the mathematical basis of AHP.

One of the advantages of AHP is its embedded notion of consistency. When comparing elements pairwise, if element X is evaluated as four times more important than element Y, and element Y is twice more important than element Z, then a consistent judgment will consider X eight times more important than Z. The eigenvector method permits a quantitative assessment of consistency. Saaty recommends a consistency ratio of 0.1 for acceptable evaluation.

In the development of HOQ, the issue is not to evaluate decision alternatives, but to set relative importance for the CAs. By using pairwise comparison of CAs and the eigenvector method, we may get the relative importances, and, at the same time, provide a consistency ratio of the comparisons.

System development

In this project, we aim to develop a basic information system support for HOQ. The main requirements of the system are: (a) user-friendly interface, (b) database management transparent to user, (c) AHP-based routine for prioritizing customer attributes, (d) facilitated later enhancement of the system.

Software and Hardware

The user interface was developed using Microsoft Visual Basic 3.0, while the database was created using Microsoft Access 2.0. The communication between the two softwares is realized within the Visual Basic program.

The system runs in Windows 3.1 environment, demanding a 486 or larger PC, with at least 8M RAM. The requirement for hard disk space depends on the size of the HOQ, i.e., essentially it depends on the size of the database, that is a consequence of the number of CAs and ECs in the house.

System Prototype

A fully operational prototype of the proposed system was developed. The system is menu-driven, with fill-in forms for each of the submatrices of the HOQ. Usage aid features, like pull-down menus, on-screen explanations and facilitators for data manipulation, were employed in all interfaces.

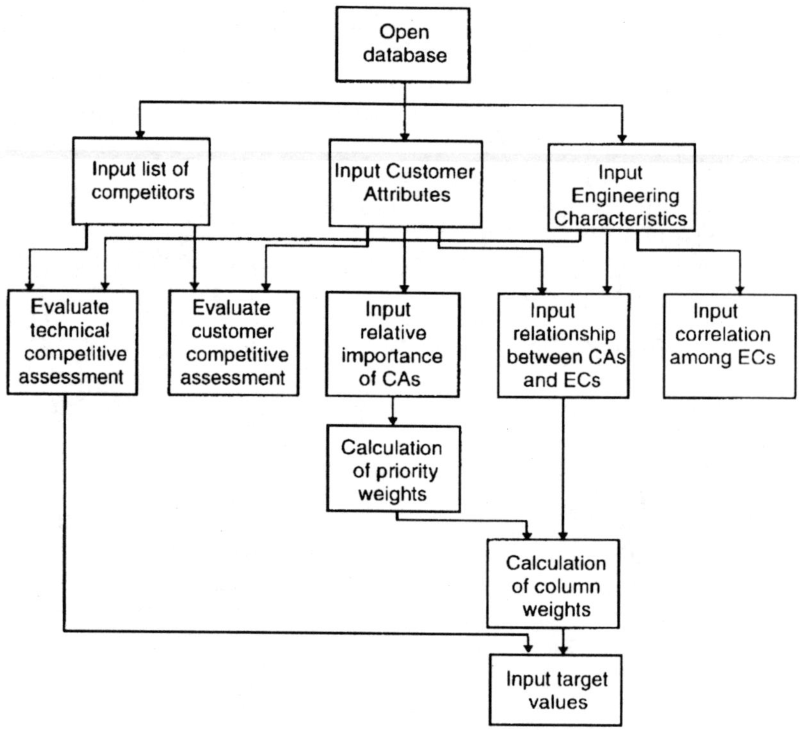

Flowchart for using the proposed system

Figure 2 contains the flowchart for building HOQ with the developed system. For a new HOQ, the first step is to select a database name to store all attributes and characteristics of the house. Then three sets of independent variables may be entered: (a) customer attributes and their relative importance, (b) engineering characteristics and their correlation and (c) list of competitors.

In general, the development of HOQ starts with the CAs, as this choice will drive the later selection of related ECs. When the list of CAs is finished, the AHP-based procedure for prioritizing the CAs may be used. The user is asked to input the relative importance comparison for all pairs of CAs, using Saaty's nine point scale. The system calculates the priority weights for the CAs, as well as the consistency ratio of the pairwise comparisons.

After selecting and entering the three groups of independent variables, the interrelationships may be tackled. Customer competitive assessment uses the list of CAs and the list of competitors. The engineering characteristics and the list of competitors are used for the technical competitive assessment. The interrelationship between CAs and ECs and the priorities of CAs are used to calculate the column weights, an indicator of relative importance of the ECs. The column weights are calculated automatically by the system, using the HOQ procedure (Hauser and Clausing 1988). The column weights and the competitive technical assessment are utilized by the HOQ team to set the target values for the ECs.

For modifications in existing HOQ or for phased data input, additional data or modifications may be entered in any of the submatrices, provided that the basic data for that particular procedure is already available in the database. The prototype contains limited aid to ensure consistency in modifications. For example, if one EC is added, all the submatrices utilizing the list of ECs must be modified.

Conclusion

The internal testing phase of the prototype is concluded. The next step is testing with actual HOQ developers. After the testing of the prototype is complete, we intend to develop a full scale system. In addition to the features already covered in the prototype, the complete system will include: (a) full error handling procedures, (b) on-line help to guide HOQ development, (c) enhanced database features to support cross-reference to related tables in HOQ of similar products, (d) cost or other information considered important.

We believe that the proposed system will greatly facilitate the development of HOQ by providing a structured, reliable and user-friendly support to store, retrieve and modify the necessary information throughout the process. The system may act as a communication facilitator and data repository for the HOQ team. Moreover, the databases of the complete HOQs may become important reference in the development of other houses in the company.

References

Akao, Y. Quality Function Deployment: Integrating Customer Requirements into Product Design, Productivity Press: Cambridge, MA, 1990.

Armacost, R. L. et al. "An AHP Framework for Prioritizing Customer Requirements in QFD: An Industrialized Housing Application," IIE Transactions 26(4): 72-79, 1994

Aswad, A. "Quality Function Deployment: A Systems Approach," Proc. IIE Integrated Systems Conf., IIE: Atlanta, GA, 27-32, 1989.

Hauser, J. R., and Clausing, D. "The House of Quality," Harvard Business Review 66(3): 63-73, 1988.

Saaty, T. L. The Analytic Hierarchy Process, McGraw-Hill: New York, 1980.

Economic Decision-Making in the Virtual Firm

Thomas W. Mason
812-877-8155

Rose-Hulman Institute of Technology
Terre Haute, IN 47803

Microeconomics has contributed much to the foundations of management practice from investment and financial structure, to strategic analysis of markets and the organization of activity. However, rapidly changing technology and the increasing importance of knowledge and communication represent significant challenges to static theories of the firm. This paper explores shortcomings of the traditional model suggests that the model of the entrepreneur is more appropriate for the virtual enterprises of the 21st century.

Introduction

The model of the firm began over two centuries ago when Adam Smith outlined principles underlying much of the industrial revolution. Later scholars have contributed frameworks for management thinking about almost every aspect of business. Many of these intellectual developments were not really intended as practical guides to managers' thinking, but as analyses of how the overall economic system worked to meet with human needs with scarce resources. Nevertheless, the models of economic analysis have become important components of business conventional wisdom. Now, those components must be rethought in view of the growth of the virtual enterprise and its processes.

Some Contributions of Economics to Engineering Management

The model of consumer behavior has established the notion of responsiveness to variables like prices of products, prices of related products and consumers' tastes and incomes. While public policy may have helped, it was probably the belief that consumers would respond rationally to rising real prices of energy that led to investment in more energy efficient products. The same understanding of consumer behavior probably led to the development and implementation of wonderful biomedical innovations with inadequate attention to their costs. Those responsible for funding the research and development of new products and processes will continue to apply such economic principles in the developments of the future, but that application is getting increasingly complex.

Production thinking has also been heavily rooted in economics since Smith's notions about specialization and division of labor in the eighteenth century. Economies of scale and, more recently, scope have been important considerations of the design of processes in many industries. Mathematical optimization models and the growing power of computers have made economic ideas of optimal resource allocation a day to day reality. moreover, economics has contributed significantly to the analysis and allocation of costs and decisions about investing in machinery and equipment and other facilities. Economics also provided the basis for determining the size of the firm and its boundaries.

Economists have also explored relationships of size of R&D operations and their productivity and the connection of time and development costs. They have provided research on the conditions for explaining patent output and the advantages of parallel development efforts. Moreover, economists have established the importance of technological change in improving productivity and living standards. This helped to bring about government action to stimulate research and development.

0-7803-3552-X/96/$5.00 © 1996 IEEE

Finally, economics has made contributions to strategic thinking in management decisions. The ideas of barriers to entry and strategic competitive advantages have helped to guide managers in their selection of markets. Within markets the research on oligopolies and game theory have suggested approaches to maintain or improve competitive positions.

Impacts of the Virtual Enterprise

The concepts of the virtual corporation as described by Davidow and Malone (1992) represent a significant challenge to the economic model of the firm. Complications have always existed in applications, and the abstraction of models is methodologically advantageous because of the power to isolate major relationships. The question is whether the implications of models are sufficiently misleading to cause more problems than they solve. This appears to be the case in a number of important respects. If economics is to contribute to good decisions about the operations of firms, then the implications of all of the revolutions in production, distribution and organization will need to be incorporated.

Technological Change

Typically, economists assume that technology is given and focus on other decisions like price, quantity and investment decisions. In the virtual corporation, the status quo is change, particularly with respect to technology. Obsolescence of production facilities is far more important than their physical life. The optimal decisions for the firm may be driven by the need to react to and even drive the constant changes in technology that define its markets. Viewing firm decisions with static models may ill equip people to make decisions in 21st century industries.

Aggregate models of technological change certainly indicate its importance to modern societies, but they do not contribute much to management decisions. Research on firm level technological change has been done, but it usually emphasizes explanations of patent output. Patents are an indication of successful research operations, but they are the wrong focus. Technology is changing so fast in many industries that patent protection is not enough incentive to overcome the disadvantages of disclosure. Moreover, a lot of the technological changes in virtual firms are not patentable hardware breakthroughs, but flexibility and creativity applied throughout all aspects of the business.

Managers of virtual enterprises have to continually implement technology before there is evidence of its cost effectiveness. Extrapolation of past cost savings trends or understanding of the technology cycle play more of a role than objectively measurable rates of return. In addition, much of the intellectual property assets of the virtual corporation will not be in patents or copyrights, but in the knowledge residing in employees' heads. This means that human capital investment is critical and employees may be the most important part of the firm's capacity, rather than just another resource.

Barriers to entry

The pervasive model of industry structure in the 20th century has been a market dominated by a few firms with large economies of scale, cumulative advantages of years of product differentiation and possession of insurmountable technological advantages. As the century ends, the evidence mounts that thinking on barriers of entry to new competitors needs significant revision. Mini-mills like Nucor have had profitable growth at the expense of major steel firms who were locked into old production approaches. Alternate access suppliers like MFS have been enormously successful in taking the best parts of local telephone markets from Regional Bell Operating Companies. In the world of virtual corporations, large legacy systems and established ways of doing business are likely to be burdens, rather than insulation from competition. The software industry has already provided lots of examples of how short industry life cycles can be when new entrants can leapfrog established technology.

Boundaries of the firm

In most economic models, the firm and the product are well defined, and the focus of attention is on other concepts. The implications of rapid technological change and vulnerability to competition have significantly altered the transactions costs relationships that have been used to define the boundaries of the firm. Like other aspects of the world of virtual enterprises, the organization has become a much more dynamic and vaguely defined concept. Information management is now the key to production and distribution, and businesses use opportunistic strategies to maximize the competitive advantages from generating and using information. The idea that such things as the make or buy decision can be permanently made on the basis of internal costs verses the costs of negotiating and implementing arrangements with vendors does not appear to be very useful when technology and business conditions are changing so rapidly. It is not uncommon now for firms to be collaborating on one deal and competing on another. Other firms may be customers, suppliers, partners and competitors at the same time. Strategic alliances are formed and dissolved rapidly depending upon pragmatic, ad hoc arrangements that seem appropriate for a particular time, place and deal. Because things have to be accomplished so rapidly, production may be a result of the work of teams made up of representatives of several firms working together as one organization.

The time required to work out the formal boundaries of all of these relationships is probably longer than the goal for the cycle from design to delivery. Although there are probably some legal pitfalls to all of the sharing of knowledge and facilities, the benefits of flexible arrangements seem to far outweigh any risks from ill-defined firm boundaries. However, the fact that key design and production planning activities are the result of cooperation makes it increasingly important for each employee to understand the unique resources that give her firm competitive advantages. Since the strategically competitive firm resources are likely to be knowledge that is protected primarily by treating it as trade secrets, it is increasingly important that the definition of where the multi-firm team stops and the individual firm begins must be in terms of the uniquely valuable advantages that will give the employee's company a long-term competitive position.

From the consumer to the customer

Another major change in the traditional model of economic decision-making is that the virtual corporation does not deal with a homogeneous group of consumers prepared by advertising and other marketing to buy a standardized product. Instead, the business considers each individual customer's desires and then attempts to configure the combination of features in its product to those desires in a way which will maximize its revenue. This introduces a new level of complexity to decisions about everything from design and production to pricing and distribution. Most models of economic decision making have both consumers and producers responding to relationships involving a quantity (q) of output which is meaningful. If output is customized to individual buyer desires and rapidly changing over time, the variations in each unit may make analysis of those relationships far more complicated.

The entrepreneur as a "new" model

A model of entrepreneurial behavior is now more important than the implications of static theories of the firm. The entrepreneur is a driver of disequilibrium, a term which seems to characterize most industrial markets of the late 90's. The entrepreneur is the agent of change who combines or recombines resources in innovative ways. This is the model which should be at the core of economic decisions within the virtual firm. This approach to thinking about economic decisions puts the decision maker into the position of selecting the most promising innovation and finding ways to make it succeed. The entrepreneur has the vision of unfolding technology, rather than optimizing the status quo. Entrepreneurs understand sustainable competitive advantages that are central to prospering in changing marketplaces so they can define boundaries of their organization. And they view the complexity and change as sources of new opportunities rather than aberrations to be controlled. This kind of model may not be as elegant as the traditional theory of the firm and needs more development, but it should help more to foster the kind of decisions which will lead to competitive success in a world of virtual enterprises.

References

Davidow, W.H and Malone, M.C. *The Virtual Corporation.* New York: Harper Business, 1992.

Creating a Virtual Enterprise in Manufacturing to Manage the Risk in Discontinuous Innovation

Christopher McDermott
(518) 276-4861

The Lally School of Management and Technology
Rensselaer Polytechnic Institute
Troy NY 12180-3590

This paper examines the extent to which elements of the manufacturing function are outsourced in discontinuous innovation. Preliminary observations indicate that a large portion of early manufacturing tasks are performed externally to reduce risk. Firms whose core lies in fabrication and assembly appear to perform more manufacturing tasks externally than do firm's whose core is in materials processing.

Incremental vs. Discontinuous Innovation

A growing body of literature has emerged which emphasizes the critical importance of identifying and developing a balanced portfolio of new products - focusing on incremental changes and extensions to existing product lines while at the same time spending energy and resources on projects which are completely new to the market. However, the uncertainty and volatility of these more discontinuous innovations (Sloan Proposal, 1994) presents a very different environment from that of incremental innovation. They are typically longer in duration, and often deal with uncertain and evolving technologies which are being created for markets which may not yet perceive a need for the product. In addition, because of this high degree of uncertainty, market analysis becomes fuzzier and traditional financial measures fall short in their ability to capture the potential of these new products in a meaningful way. Discontinuous innovations may also utilize process technologies which are revolutionary, and often unstable. Traditional process control methods (design of experiments, SPC) may not be easily applicable in these situations.

Incremental innovations typically consist of extensions to current offerings, and other small steps into newness that are less risky than the ones closer to the other end of the spectrum. However, the argument has been made that in many industries, the development of these is not enough for the long term success of firms (Gaynor, 1993; Sloan Proposal 1994; ,Morone, 1993). While incremental innovations can sustain an advantage, it is the innovations on the more discontinuous end of the scale that provide the initial competitive advantage in the first place. Firms that focus only on incremental innovation are minimizing risk, but at the same time are missing tremendous opportunities which may negatively impact the long term success of the firm.

Given the differences between these two types of innovation and the environments in which they occur, it is not at all unreasonable to expect that successful practices associated with new product development may be significantly different for discontinuous and incremental projects. The ideas and conceptual models presented in this research are based on the preliminary findings of a larger, ongoing, multi-year study on the management of discontinuous innovation which is supported by a grant from the Sloan Foundation. In addition to the support provided by the Sloan Foundation, the Industrial Research Institute (IRI) has also proactively become involved in this project, and its Research-on- Research committee has adopted this project as its lead study, providing evidence of the importance and relevance of the research goals to managers and practitioners. The firms participating in this study are members of the IRI and by their voluntary involvement in this study, are demonstrating their keen interest in the study of management practices in the volatile

0-7803-3552-X/96/$5.00 © 1996 IEEE

environment surrounding more discontinuous innovation. The observations presented here are based on the study of eight ongoing discontinuous new product development projects within six large firms in a variety of product and market environments.

Manufacturing and Discontinuous Innovation

Much needs to be learned about how manufacturing can foster innovation in environments of discontinuous innovation. While much has been written about the role of manufacturing in more incremental environments (e.g. Rosenthal, 1992; Gerwin, 1993), it is unclear what practices firms follow with respect to manufacturing in cases of more discontinuous innovation, much less which of these practices are most appropriate.

For manufacturing, the costs associated with setting up the facilities and equipment required to produce a completely new line of products are usually quite substantial. Given the volatility and uncertainty of environments of more discontinuous innovation, this becomes especially problematic. The intended customers for the product may be either unknown or currently non-existent and the ability to accurately make long-term manufacturing decisions such as plant location, staffing requirements, not to mention production equipment design and purchase, becomes extremely limited, increasing the risks for manufacturing. Thus the risks for manufacturing in an environment of more discontinuous innovation are much higher than in more incremental settings.

Benefits of Outsourcing

As a result of this heightened risk, firms may turn their attention to sharing this risk with suppliers and partners. The potential benefits are many. Firms commonly look externally for support during the development of innovative projects because the realization of the product itself requires a capability which is outside of the firm's area of expertise. Working with a firm more familiar in this area offers the potential to save both time and money, as there is less effort attempting to learn about existing - but external - manufacturing capabilities. By forming partnerships, firms find that they don't need to "reinvent the wheel".

There are also potential risk-sharing benefits to working with a partner. Rather than shouldering all of the responsibility of early manufacturing involvement (e.g. prototyping, designing facilities) for the project alone, aligning with other firms presents the opportunity to share some of the financial risk with a willing vendor and/or partner in exchange for the potential of a long term relationship. Further, such inclusion of other firms can also result in a sense of validation of project through a "critical mass" of parties involved: many parties involved in a project might signify its importance to upper management.

Potential Penalties

However, there are potential pitfalls to this approach. In some sense, these tactics may only result in an exchange of risks: trading risk of failure and financial commitment for the risks associated with a loss of control of the project. This is potentially dangerous if the firm attempts to outsource a technology or capability which is at the core of what makes the new product unique. Whereas loss of control of a peripheral manufacturing process which is not the major source of the innovativeness of the product might be relatively easy to overcome, close control might become more critical as the outsourced task is more greatly responsible for the core benefit and innovative content of the new product. A miscommunication or shortcoming here can derail the whole project. In addition, there is also the potential risk of choosing a supplier early in the initial phases of product development only to find out too late that they are not capable of providing the expertise and capabilities that the evolving product requires.

Findings

This study explored these and other questions in eight projects within six different firms. Although seven of the eight projects observed in this study do have external partners or significant suppliers involved in their early manufacturing efforts, there is a great deal of variation in both the number of external relationships and the extent to which these relationships work in areas which are core to what makes the new product unique (see Table 1). At one of the ends of this continuum is P2, which is currently developing a large mechanical system. The leaders of this project view their firm as a "system integrator". Very little, if any, of the detailed product design and manufacturing is done in house, and the firm relies almost exclusively on its numerous partnerships and supplier agreements to build the new product. Even the component which is agreed to be most critical to the success or failure of the project is being developed externally. Their focus is on the coordination and final assembly of the product.

At the other end of the continuum are P4 and P8. These projects are based at their core on *process* innovations which the respective firms developed in-house. However, the specific *product* applications for which these innovative new materials are targeted reside in areas outside of the firms expertise. As a result, each of these projects has bring in other firms who are established in production in these areas and markets. While, arguably, the technological "leap" in these projects is in the development of a material, the targeted application/product is outside of the firm(s) area of expertise and so each project has chosen to work with outside firms.

Table 1

	P1	P2	P3	P4	P5	P6	P7	P8
External Manufacturing Involvement?	Yes	Yes	Yes	Yes	Yes	Yes	No	Yes

Looking at the sample as a whole it appears that a considerable amount of what might be considered the core product technology is developed in-house (with the exception of P2 and perhaps P1). This is consistent with what one would expect (Prahalad and Hamel, 1990). The innovating firms build on their known strengths and expand with new products keeping control of what they feel to be central to the innovation. P2 and P1, which both appear to be relying on outside firms to design and develop the critical technology their project innovation have clearly chosen a different path. It remains to be seen if this will be successful for them.

There is noticeably less concern among all the projects regarding where the development and location of the required process technology occurs - in the majority of the projects studied, at least one core piece of production is being coordinated with an outside firm. Given the level of secrecy surrounding many of these discontinuous projects, it was somewhat surprising to observe the large number of outside firms involved in the development of the new product. These external manufacturing relationships are usually established once the project technology has been targeted for a specific product application, and an appropriate external source thought to be more familiar with the manufacturing requirements for this particular product/market can be identified. These new partners may help in using this newly created "component" into a new product.

By extension, the preliminary observations of this study suggest that firms whose core manufacturing competence is in fabrication and assembly are able to create a "virtual enterprise" - working closely with vendors and other component manufacturers - to help share the design and manufacturing risks of more discontinuous innovation, while firms whose core lies in materials processing find it more difficult to do so because of the intrinsic link between product and process in their businesses. The exception, as noted above, occurs when a materials processing firm sees an opportunity to use one of its new materials in an unfamiliar product environment.

Conclusions

Creating a manufacturing "virtual enterprise" in an environment of discontinuous innovation present unique opportunities and challenges for the firms within our sample. There are clearly advantages found in terms of tapping into outside expertise which can help the firm to better provide the benefit of the product. Additionally, some of the risk associated with developing and manufacturing components is shared as they are assigned to outside partners and suppliers. There are, however, risks associated with this tendency to look outside. Within the context of this study, it too early to tell whether the risks outlined above will end in substantial penalties as a result of with undertaking virtual manufacturing in an area of discontinuous innovation.

References

Gaynor, G. H., Exploiting Cycle Time in Technology Management, New York: McGraw Hill, 1993.

Gerwin, D., "Integrating Manufacturing Into The Strategic Phases Of New Product Development" California Management Review, pp. 123-136, Summer 1993.

"Managing Discontinuous Innovation: A Proposal to Investigate New Business Creation in Established Firms." Research Proposal Submitted to and accepted by the Sloan Foundation by The Lally School of Management and Technology, Rensselaer, 1994.

Morone, J., Winning in High-Tech Markets, Harvard Business School Press, 1993.

Prahalad, C.K. and Hamel, G., "The Core Competence of the Corporation, Harvard Business Review, pp. 79-91, May-June 1990.

Rosenthal, S.R., Effective Product Design and Development, Business One Irwin, 1992.

Does the Parallel Approach Make Sense in the Development of Discontinuous Innovations?

Christopher McDermott

The Lally School of Management and Technology
Rensselaer Polytechnic Institute
Troy, New York
12180-3590

Robert Handfield

The Eli Broad Graduate School of Management
Department of Management
Michigan State University
East Lansing, MI
48824-1121

While the success of parallel new product development programs such as concurrent engineering is well documented, their focus on low cost and speed to market has meant that the bulk of what we know about this approach relates to their effectiveness in the development of products which focus on incremental product improvements. Little is known about their utility in projects which involve more "discontinuous" innovation. This paper outlines several potential weaknesses of concurrent engineering practices when applied to more discontinuous innovations, and highlights a number of propositions which map out future research questions in the area.

The Parallel Approach

A central outcome of the increased focus on new product development processes in the 1980s is the rise of concurrent engineering, simultaneous engineering, design for manufacturability and other programs which encourage a parallel process in which multiple business functions work together with the goal of bringing goods which satisfy or exceed customers needs/wants to the market more quickly (Clark, 1989; Gerwin, 1993; Rosenthal, 1992; Van Dierdonk, 90). The implications for manufacturing and purchasing were significant. No longer was manufacturing tapped after the product design was complete, but instead, it would become involved throughout the entire process (Cooper, 1990). Similarly, including suppliers early in the process, with purchasing acting as a liaison, helped to bring products to market faster (Cooper, 1990). Including manufacturing and purchasing early in the NPD process is appealing for a number of other reasons as well. Continuous communication and cooperation between the functions, joint problem solving, a move from functional to project-based management, and improved supplier involvement all point toward an improved final product, both in terms of speed to market and product specific capabilities and attributes (Van Dierdonk, 90).

Incremental vs. Discontinuous Innovation

At the same time, a growing body of literature has emerged which emphasizes the critical importance of identifying and developing a balanced portfolio of new products - focusing on incremental changes and extensions to existing product lines while at the same time spending energy and resources on projects which are completely new to the market. However, the uncertainty and volatility of these more discontinuous innovations (Sloan Proposal, 1994) presents a very different environment from that of incremental innovation. They are typically longer in duration, and often deal with uncertain and evolving technologies which are being created for markets which may not yet perceive a need for the product. In addition, because of this high degree of uncertainty, market analysis become fuzzier and traditional financial measures fall short in their ability to capture the potential of these new products in a meaningful way. Discontinuous innovations may also utilize

0-7803-3552-X/96/$5.00 © 1996 IEEE

process technologies which are revolutionary, and often unstable. Traditional process control methods (design of experiments, SPC) may not be easily applicable in these situations.

Incremental innovations typically consist of extensions to current offerings, and other small steps into newness that are less risky than the ones closer to the other end of the spectrum. However, the argument has been made that in many industries, the development of these is not enough for the long term success of firms (Gaynor, 1993; Sloan Proposal 1994; Morone, 1993). While incremental innovations can sustain an advantage, it is the innovations on the more discontinuous end of the scale that provide the initial competitive advantage in the first place. Firms that focus only on incremental innovation are minimizing risk, but at the same time are missing tremendous opportunities which may negatively impact the long term success of the firm.

Given the differences between these two types of innovation and the environments in which they occur, it is not at all unreasonable to expect that successful practices associated with new product development may be significantly different for discontinuous and incremental projects. Much that has been written relating to the management of innovation either fails to recognize or differentiate between these two very different types of innovation, or acknowledges that different types exist and deliberately focuses on more incremental innovation. Many reason that this narrow focus on the management of incremental innovation is because truly "radical" innovation rarely if ever occurs. While this may be true, it is our contention that this view incorrectly categorizes innovation as something that must fit neatly into the states of "radical or "incremental". Perhaps a more accurate scheme is one which follows Gaynor's (1993) model and views the extent of "newness" in a product on a continuum. A product in the development process might thus be viewed as less incremental than its predecessors, while still not being radical.

It may very well be that many of the "rules" that have been established for the management of incremental innovation may not only be unhelpful in less certain, riskier environments, but may in fact actively inhibit goals of this type of innovation. Although concurrent engineering is intuitively appealing in traditional, more incremental innovation environments, it is unclear whether it is at all appropriate in the environments of high risk and uncertainty which often surround more radical innovations. There is growing support for this line of thinking. Handfield (1994) studied defect rates in make-to-order environments and found that breakthrough MTO products developed using concurrent engineering have more defects than breakthrough MTO products developed with sequential processes. Aitsahlia and Johnson (1994) use rudimentary math models to show that uncertainty "erodes the benefits of concurrent engineering and in some cases may be worse than the serial engineering". The central thesis of the study is that the parallel approach to new product development may not be appropriate in generating discontinuous innovations. Based on a set of case studies conducted with a number of managers in the U.S. and Japan, we identify several propositions regarding the new product development process that may help to promote discontinuous innovations.

CE and Discontinuous Innovation

In more discontinuous projects, the intended customers for a product may be either unknown or currently non-existent and the possibility of making long-term manufacturing decisions such as plant location, staffing requirements, and size become extremely limited, dissolving much of the argued benefits of the parallel approach. Similarly, there is a danger associated with committing to an outsourcing decision too early in the process.

Even if there is a known, clearly identified market for the innovation, it is often the case that the design of the product itself is in such flux that manufacturing and/or suppliers may be unable to move forward with the identification, development or purchase of the appropriate production equipment. Thus, an unknown product form can negatively affect the viability of the parallel approach. For instance, the process technology may be too unstable or unreliable to meet the required product specifications.

According to Cooper (1990), predevelopment activities such as preliminary market study and technical assessment are important to successfully bring new products to market quickly. This involves a close integration of marketing and manufacturing to carry out customer tests of the prototype or sample. In many industries, the development of new technologies may be occurring at an alarming rate, with new suppliers appearing on the scene monthly or even weekly! High tech component life cycles in the introductory stage of the product's life cycle tend to be small and short lived, as the technology itself is unstable and changing (Handfield and Pannesi, 1994). Some components at this stage may not make it to general availability, or may even be pre-empted or aborted before they begin the life cycle. Not unexpectedly, purchasers at this stage must pay particular attention to competing sources of supply and alternative technologies. The increased risk associated with uncertain technologies thus acts as a force which encourages the postponement of decision making in both manufacturing and sourcing areas. Thus

Proposition 1

As firms move more toward more discontinuous innovation, the manufacturing/marketing/outsourcing interface becomes more difficult and cumbersome to integrate, and firms will tend to "decouple" these functions, returning to a more serial approach.

Research and practice have suggested that the automation of design processes through CAD, CAM, and CAE is promising with respect to shorter design cycle times (DeMeyer and Van Hooland, 1990). Carter and Baker (1992) argue that such technologies as CAD/CAE/CAM aid in concurrent engineering because of their ability to support design integration through shared product and process models and data bases. However, more discontinuous innovation often occurs in an environment where the level of uncertainty relating to the very basics of product design (rough product shape, possible materials, etc.) and market (how large will the market be, will the product be made custom or standard) are so high that the utility of such technologies for projects is questionable. For instance, Chrysler Corporation created a design facility that enabled all of its design engineers to work in a single building. The level of integration and synergy required to develop new models such as the Dodge Ram and the Chrysler Neon was enabled through this strategy. While few would argue that a completely new model of auto constitutes a "radical" new product, the development process contained enough uncertainty to render computer-based linkages through e-mail, CAD/CAM, and other strategies ineffective. These environmental characteristics associated with discontinuous innovation suggest:

Proposition 2

Because of the inherent uncertainty and volatility of the NPD environment, the advantages of technologies such as CAD/CAE/CAM disappear as firms move toward discontinuous innovation.

One of the key factors underlying success within the stage-gate model is the quality of execution of technological activities such as technical assessment, in-house product or prototype testing, and test/pilot production. However, the ability of firms to accurately assess the technologies associated with each of these activities in a discontinuous innovation is significantly decreased by the high uncertainty of the new technology itself.

For instance, a large computer manufacturer noted that in the case of basic research at the lab stage, informal meetings with key suppliers are very common. Supplier's top managers are approached initially by the R&D group, and asked if they are willing to work on development for a future product. This is much more of a trust-based approach, with a noticeable lack of formal contracts. Suppliers are asked to share ideas on the technology, in hopes that synergies achieved will result in a radical new product. In such meetings, R&D leads the discussion, while purchasing personnel may not even be involved. Discussions are primarily of a technical nature, and often focus primarily on the supplier's technological capabilities and expertise. Once R&D determines that the supplier is capable, purchasing and legal personnel may become involved in developing a non-disclosure agreement for all future joint development.

For incremental products, however, a stage-gate system wherein each function must approve the decision is more common. In such cases, the manufacturing and purchasing groups are primarily involved in the selection and negotiation with suppliers, with design taking on a secondary role. Specifications may already be well-known. The primary benefits of early supplier involvement in such cases often focuses on cost-reduction strategies derived by "fixing" the design at an earlier stage, and performing value engineering where appropriate. These scenarios suggest that:

Proposition 3

Compared to more incremental projects within a given company, projects incorporating discontinuous innovation will tend to follow the stage-gate process less.

It has been argued (Gaynor, 1993) that manufacturing and purchasing are averse to anything new and different. This is most likely because of manufacturing's traditional goals and performance metrics focusing on low costs were at odds with any type of variety and newness. For instance, the lead designer involved in a more discontinuous project had the following to say about his interactions with manufacturing when he proposes design changes "He (the manufacturing person) will give me a list of things that I absolutely do not want to do in an order like this 1) You are offending God, 2) You are offending the Saints, 3) You are offending the Pope ... And in that kind of decreasing order of magnitude, you figure out what you might be able to do."

Both manufacturing and purchasing are functions which are simplified by the reduction of design changes. By extension, it is proposed that

Proposition 4

Early manufacturing involvement in discontinuous innovation will reduce the likelihood that the final product is significantly different from existing product lines or families.

Conclusions

There is clearly a need for more research into the appropriateness of concurrent engineering in discontinuous innovation. The propositions outlined in this paper provide a tentative framework for considering the management of discontinuous innovation. They all point to the difficulty of dealing with parallel or stage gate development processes in pursuing more radical new products. The traditional benefits of concurrent engineering may not necessarily accrue when the technology itself is unproven, and customer requirements are uncertain. In such environments, a longer development period may be necessary to "iron out" these difficulties. As one engineer we interviewed noted: "Concurrent engineering just doesn't allow you to carry out all of the steps in the process when you're working on a breakthrough technology." In such cases, it may be the case that a little patience in extending deadlines can result in a major product technology.

References

AitSahlia, F., Johnson, E., "Is Concurrent Engineering Always a Sensible Proposition", *IEEE Trans. Eng. Mgmt*, 25(5-6), pp. 166, 1995.

Clark, K., "Project scope and project performance: The effect of parts strategy and supplier involvement on product development", *Management Science*, vol. 35, pp. 1247-1263, 1989.

Cooper, R., (1990), "Stage-Gate Systems: A New Tool For Managing New Products", *Business Horizons*, pp. 44-54, May-June 1990.

De Meyer, A. and Van Hooland B., "The Contribution Of Manufacturing To Shortening Design Cycle Times", *R&D Management,* 20(3), pp. 229-239, 1990.

Gaynor, G. H., *Exploiting Cycle Time in Technology Management*, New York: McGraw Hill, 1993.

Gerwin, D., "Integrating Manufacturing Into The Strategic Phases Of New Product Development" *California Management Review*, pp. 123-136, Summer 1993.

Handfield, R., "Effects of Concurrent Engineering on Make-to-Order Products", *IEEE Trans. on Eng. Mgmt.*, 41(4), pp. 1-11, 1994.

Handfield, R., and Pannesi, R., "Managing Component Life Cycles in Dynamic Technological Environments", *International Journal of Purchasing and Materials Management*, pp. 20-27, Spring 1994.

"Managing Discontinuous Innovation: A Proposal to Investigate New Business Creation in Established Firms." *Reseaerch Proposal Submitted to and accepted by the Sloan Foundation* by The Lally School of Management and Technology, Rensselaer, 1994.

Moffat, L. K. and Gerwin, D. *Concurrent Engineering Project Teams: An Examination of Critical Success Factors* - A Report Prepared for the IMS Global CE Test Case, Carleton University, 1994.

Morone, J., *Winning in High-Tech Markets,* Harvard Business School Press, 1993.

Rosenthal, S.R., *Effective Product Design and Development*, Business One Irwin, 1992.

Van Dierdonk, R., "The Manufacturing / Design Interface", *R&D Management*, pp. 203-209, 20(3), 1990.

Reengineering Team Processes in Quality-Focused Software Development Environments

Eugene G. McGuire
(202) 885-3268

American University
Computer Science and Information Systems
4400 Massachusetts Ave., N.W.
Washington, D.C. 20016-8116

Anita J. La Salle
(202) 885-3305

American University
Computer Science and Information Systems
4400 Massachusetts Ave., N.W.
Washington, D.C. 20016-8116

Most software development improvement programs to date have emphasized process and/or technology and not people. In many organizations people and team management practices are not addressed in a systematic and structured manner. In addition, most managers are untrained or inadequately trained in implementing corrective solutions once people and team problems have been identified. This paper examines people and team issues in relation to the implementation of software process improvement initiatives within organizations.

Software Process Improvement

Part of the reason for the recent attention on software process improvement initiatives is that empirical research consistently shows that many software development projects suffer from a lack of proven and well established methods for dealing with process, technology and people issues. Well-known process improvement models such as the Capability Maturity Model (CMM) developed by the Software Engineering Institute (SEI) have gained acceptance by a wide variety of organizations (Curtis & Paulk 1993).

Most improvement programs like the CMM, however, emphasize process or technology and not people. People management practices in many organizations, despite a significant amount of literature addressing people-related issues (Kraut 1995; Pressman 1995; Rasch & Tosi 1992), do not address people issues in a systematic and structured manner. In addition, most managers are untrained or inadequately trained in implementing corrective solutions once people problems are identified (Rettig & Simmons 1993; Statz 1994). Also, organizational factors, while widely cited (Constantine 1993,1995; McIntyre 1992,1994] are often not systematically analyzed for their effect on process improvement efforts.

The characteristics of the CMM levels show that information systems professionals who work on software engineering projects must be capable of being highly productive in complex, team-oriented environments with strong emphases on process control and overall quality (Walz et al 1993; Zultner 1993]. These requirements are critical but may not always be present in current information systems professionals. The lack of appropriate team and process control skills can greatly contribute to internal organizational volatility.

The CMM breaks down the software engineering capabilities of organizations into 5 maturity levels from Level 1 (Initial) to Level 5 (Optimizing).

The Initial Level (level 1) is characterized by ad hoc activities. The successful execution of these activities depends upon the people involved in them, who, although skilled, typically succeed due to "heroic" efforts. There are no well-defined procedures and management. Roles of the design team members are not well defined. Project and design information is not stored or communicated effectively.

0-7803-3552-X/96/$5.00 © 1996 IEEE

The Repeatable Level (level 2) is characterized by activities whose tasks are accomplished in accordance with a set of procedures. Successful execution of these activities depends upon the availability of procedures (e.g., project management procedures) for accomplishing them. The organization uses standard practices for managing development activities including: staff estimating; cost estimating; scheduling; change and version control on requirements, specifications, and design data; and, managerial reviews. The managers are trained in these management practices and software engineers are trained on the use of tools and procedures.

The Defined Level (level 3) is characterized by activities that are accomplished consistently across all projects and business functional areas. Successful execution of these activities depends upon organizational policies and a standard process, which formalizes and organizes task procedures. The standard process focus provides the structured environment needed for communications and transfer of information among organizational elements, which fosters an integrated view of the process in the context of its relationship to other processes. The software development organization is multi-disciplinary and a Software Engineering Process Group (SEPG) is established to lead process improvement.

The Managed Level (level 4) is characterized by activities aimed at predictability of the standard process (and associated products). Successful execution of these activities depends on establishing the metrics necessary to understand cost, schedule, and quality aspects of the standard process and associated so they can be predicted. The senior managers are involved in the product and process management. The organization typically bases its operating decisions on quantitative process data, and conducts extensive analysis of the data gathered during reviews and tests.
The Optimizing Level (level 5) is characterized by activities aimed at continuously improving the standard process. Successful execution of these activities depends on a institutionally established process for managing change. The focus is on controlling change and assessing the impacts of change in the process, their products, and associated technologies, in relation to the achievement of enterprise-wide goals. The organization conducts analyses of the metrics data gathered during the design process and use these for planning improvements.

Skills and Focus Areas for Software Engineers Under the CMM

The CMM is one of the most commonly encountered software process improvement models today. As such it represents many of the newer process-focused and team-driven modes of operation that software professional are encountering in today's organizations. The following are skills and focus areas that software engineering professionals should be aware of as they adapt to development environments that strive to be in compliance with the CMM.

Team Orientation — This is perhaps the most commonly cited characteristic of successful software process improvement efforts, particularly those based on the CMM. A team orientation is often necessary because organizations have become leaner and less hierarchical to remain competitive. Ramifications of this structural change in organizations include requiring that software professionals not adopt insulated work modes but instead develop business, managerial, and political skills to successfully negotiate with multiple constituencies and integrate the software development effort with the strategic plans of the organization. Software professionals must also develop stronger communication, coordinative, and collaborative skills to successfully satisfy the key process areas in the CMM.

Process Focus — Software development professionals also need to develop a process orientation. Software development teams with a process focus work to prevent crises, monitor user satisfaction and the quality of their software products, and manage and improve their software processes based on quantitative information. As the focus on process and process maturity increases, institutionalization of the software process is achieved via policies, standards, and organizational structure.

Quality Focus — Software is increasingly becoming a mission-critical resource for all types of organizations. If software is unreliable in any way, organizations may face disastrous consequences. Improved quality is the reason behind software process improvement efforts and the CMM in particular. The CMM is based on the work of Deming, Juran, Crosby and other pioneers in the quality movement.

Interdisciplinary — Effective teams have members who know the tasks of others and can provide backup and feedback capabilities to the other members of the team. This means that tasks are not compartmentailzed and integrated only at the end of the development effort. In addition, software development teams must interact with many other external teams within and outside the organization to coordinate all aspects of the CMM-based process.

Continuous Learning — Software development professionals must continually update their skills and expertise not only in technical areas but also in organizational, business, managerial and teamwork areas. The CMM specifically requires that all individuals receive mandatory training in teamwork and asks if sufficient resources are allocated to this activity (in the Level 3 KPA of Intergroup Coordination).

Change Agent — Instituting a process-focused software development environment requires much change in the way software has traditionally been developed in many organizations. Change management techniques must be utilized to overcome individual and group resistance to the new ways. Benefits of the change must be clearly articulated and reward systems implemented that recognize the shift from individual effort to team- and process-driven effort.

Awareness of Purpose — Each software engineer must always know why they and their skills and competence have been allocated to a certain team or function in the development process thereby giving each person's work more meaning in the overall organizational perspective. Team and functional assignments should be presented in the correct context both in terms or organization and required results in order to give a clear purpose for the assignment and team composition.

Goal Driven — A team must have common goals and objectives in order for the team members to achieve maximum performance. This can be assured by having team members together plan for the execution of the assignment. Together the team members develop a plan where they show how to reach the goals of their assignment and agree on how each team member will contribute.

Commitment — Successful team work is based on trust. Trust is based on the belief that all parties will complete their work according to previously defined methods and schedules and that any difficulty with these methods or schedules will be communicated and resolved. In a team-based environment there exists both internal commitments within the team among the team members as well as external commitments between the team and external parties.

Empowered Actions — In order to maximize performance and to benefit from each team member's unique experience and skills, teams and team members must be empowered to make decisions regarding their work. Each team member must be empowered to make decisions regarding their work in order to best contribute to the team's common assignment plan and goals. Hence, the team leader's responsibility shifts from supervising to facilitating.

Accountability — To be accountable means that a person is held responsible for the results (good or bad) and is given recognition for the success. From a team member's perspective the team is together accountable for the delivered results since it is the team that commits to delivering according to the plan. This means that from an external perspective all deliverables are team deliverables and there exists no individual ownership of the deliverables.

Intellectual control — Intellectual control of a team assignment means that all team members at all times know exactly what to do, why it has to be done and who is supposed to do it. The prerequisite for intellectual control is to have a clear, visible and feasible assignment plan within the team. Another benefit of having a clear and visible plan is that the team will have the possibility of performing sequence analysis and replanning when changes occur.

Software Engineering Educational Issues

As software development practices continue to evolve and skills sets change it becomes apparent that current software engineering education and training may not be adequate for industry needs. To address this need as well as needs in IS curricula, a joint task force representing the Association of Information Systems (AIS), the Association of Computing Machinery (ACM) and the Data Processing Management Association (DPMA) as well as other relevant associations and industry representation has developed a model curriculum (IS '95) and guidelines for undergraduate degree programs in information systems.

The applied focus of IS '95 draws upon the linkage between theory and practice. The primary requirement of IS '95 graduates is to be able to implement and deploy information systems within an organizational context. The model also emphasizes the need for an understanding of fundamental organizational principles and highlights the requirement for graduates to be able to adjust to rapid change and continue their learning process independently.

IS '95 includes emphasis on the following characteristics:

Communications
- IS professionals must communicate in a variety of settings using oral, written, and multimedia techniques.

Problem Solving
- IS professionals must be able to choose from a variety of different problem solving methodologies to analytically formulate a solution.
- IS professionals must think creatively in solving problems.
- IS professionals must be able to work on project teams and use group methods to define and solve problems.

Organization and Systems Theory
- IS professionals must be grounded in the principles of systems theory.
- IS professionals must have sufficient background to understand the functioning of organizations since the information system must be congruent with, and supportive of the strategy, principles, goals, and objectives of the organization.
- IS professionals must understand and be able to function in the multinational and global context of today's information dependent organizations.

Quality
- IS professionals must understand quality, planning, steps in the continuous improvement process as it relates to the enterprise, and tools to facilitate quality development.
- As the IS field matures, increasing attention is being directed to problem avoidance and to process simplification through reengineering. Error control, risk management, process measurement and auditing are areas that IS professionals must understand and apply.
- Is professionals must possess a tolerance for change and skills for managing the process of change.
- Given the advancing technology in the IS field, education must be continuous.

- Is professionals must understand mission-directed, principle-centered mechanisms to facilitate aligning group as well as individual missions with organizational missions.

Groups
- IS professionals must interact with diverse user groups in team and project activities.
- IS professionals must possess communication and facilitation skills within team meetings and other related activities.
- IS professionals must understand the concept of empathetic listening and utilize it proactively to solicit synergistic solutions in which all parties to an agreement can benefit.
- IS professionals must be able to communicate effectively with a changing work force.

Conclusion

As software development practices continue to evolve, software engineers have new responsibilities placed on them that require them to acquire new and updated skill sets that emphasize team and quality practices. This paper has examined some of the new skills and focus areas that are required of software engineers and examined educational changes that have been developed to accommodate these new requirements.

References

Constantine, L.L. (1993). "Work organization: Paradigms for project management and organization," Communications of the ACM, Vol. 36, No. 10, October 1993, pp. 35-43.

Constantine, L.L. (1995). Constantine on peopleware. Yourdon Press, Englewood Cliffs, NJ.

Curtis, B. & Paulk, M. (1993). "Creating a software process improvement program," Information and Software Technology, Vol. 35, No. 6, June/July 1993, pp. 381-386.

Kraut, R.E. & Streeter, L.A. (1995). "Coordination in software development," Communications of the ACM, Vol. 38, No. 3, March 1995, pp. 69-81.

Longenecker, H.E., Feinstein, D.L., Couger,, J.D., Davis, G.G., and Gorgone, J.T. (1995). "Information systems '95: A summary of the collaborative IS curriculum specification of the joint DPMA, ACM, AIS task force, Journal of Information Systems Education, Winter 1994-95, pp. 174-186.

McIntyre, S.C. (1992). "Overcoming ad hoc development: Enabling/Disabling factors in software quality maturation," Proceedings of the 25th Hawaii International Conference on Systems Sciences, IEEE Computer Society Press, Vol IV, January 1992, pp. 624-633.

McIntyre, S.C. (1994). "Organizational issues underlie software quality," Proceedings of the 27th Annual Hawaii International Conference on System Sciences, IEEE Computer Society Press, January 1994, pp. 820-821.

Pressman, R.S. (1995). "A people-centered approach to software process improvement," Software Process Improvement Forum, Vol. 2, No. 2, March/April 1995, pp. 8-12.

Rasch, R.A. & Tosi, H.L. (1992). "Factors affecting software developers' performance: An integrated approach," MIS Quarterly, September 1992, pp. 395-409.

Rettig, M. & Simons, G. (1993). "A project planning and development process for small teams," Communications of the ACM, Vol. 36, No. 10, October 1993, pp. 45-55.

Statz, J.A. (1994). "Training effective project managers," American Programmer, V. 7, No. 6.

Turner, A. & Tucker, A. (Eds.) (1991). "Computing curricula 1991 — Report of the ACM/IEEE-CS Joint Curriculum Task Force," Communications of the ACM, Vol. 34, No. 6, pp.68-84.

Walz, D.B., Elam, J.J., & Curtis, B. (1993). "Inside a software design team: Knowledge acquisition sharing and integration," Communications of the ACM, Vol. 36, No. 10, October 1993, pp. 63-77.

Zultner, R.E. (1993). "TQM for technical teams," Communications of the ACM, Vol. 36, No. 10, October 1993, pp. 79-91.

Tools for Engineering the Agile Enterprise

Laura Meade
Adrien Presley
K.J. (Jaime) Rogers

Automation & Robotics Research Institute
The University of Texas at Arlington
7300 Jack Newell Blvd. S.
Fort Worth, TX 76118-7115
Email lmeade@arri.uta.edu

This paper describes the work of the Automation & Robotics Research Institute which has as its objectives the development of a set of business process templates and enterprise design methodologies for transforming enterprises to agility and forming virtual enterprises. Business processes in three categories are being studied: (1) those processes which transform external constraints into internal constraints, (2) those processes which acquire and prepare resources, and (3) those processes which use resources to produce enterprise results. Research in business process templates as well as supporting methodologies in three areas is being conducted: metric development, strategic justification, and agile business process configuration. This paper presents issues related to engineering the enterprise, relevant research results accomplished to date, and implications for future research.

Introduction

Agility is defined as the ability to thrive in a competitive environment of continuous and unanticipated change, to respond quickly to rapidly changing markets driven by customer based valuing of products and services (Iacocca Institute, 1991). The ability to adapt to todayís changing business environment will require that enterprises be able to rapidly and effectively reconfigure their business processes. Increasingly, companies will concentrate on those processes which reflect their core competencies - those things which they do well and differentiate them from competitors. These core processes will then form the contribution that each enterprise will make as it aligns with partners in a new form of venture - the virtual enterprise. The virtual enterprise is a temporary alliance of companies coming together rapidly to take advantage of a market opportunity and then disband when the opportunity has passed. The ability to participate in the virtual company will require the ultimate levels of agility (Barnett, Presley, Johnson, and Liles, 1994; Nagel, 1993).

The Agile Templates Group at the Automation & Robotics Research Institute (ARRI) of The University of Texas at Arlington (UTA) has as its mission to conduct research into the development and deployment of advanced methods, models, and tools to assist enterprises compete in the new agile manufacturing paradigm. Specifically, the group is developing a set of business process templates useful for enterprise analysis and design, and developing enterprise design methodologies for transforming enterprises to agility and for forming virtual enterprises. Together, the templates and methodologies form tools for engineering the virtual enterprise.

The research being conducted by the group will define three things. First, what are the activities that an enterprise performs? Second, how should these activities be performed? And finally, how should the enterprise be constructed? An architecture is being developed which will: 1) Identify the essential processes performed by a virtual company (what), 2) Describe how the virtual company and the agile enterprises involved in the virtual company will perform these processes (how), and 3) Prescribe methodologies for the rapid reconfiguration of the virtual enterprise.

The what and how questions will be addressed through the development of the business process templates described section 2. Section 3 will discuss the methodologies under development. In each of these sections an overview of the research and significant results to date are presented.

Business Process Templates Research

The representation of processes is being accomplished through the development of ìbusiness process templatesî. These templates, which will be useful for both conventional business process reengineering efforts as well as for the formation of virtual enterprises, provides a standard approach for defining and modeling a process. The templates will be prescriptive in nature, defining the characteristics that a process will need to be considered ìagileî. The prescriptive nature of the template is especially important for deployment purposes in that companies can use them as an archetype from which to design their own processes.

Two related modeling approaches are being explored for process modeling. The first, employs an object oriented scheme to model business processes (Jacobson, Ericsson, and Jacobson, 1994). In this scheme, the enterprise is made up of processes and the information and tangible things (which are referred to as "products") interacting with the processes. All representations are in terms of objects as opposed to the very different entities used in traditional methods. From this, the user can extract the view or information of interest. This method also supports the specification and implementation of information systems for automating the modeled processes (Barnett, Hari, and Liles, 1995).

The second approach, described in more detail in another paper in this volume, uses IDEF5 (ontology capture method), IDEF0, and IDEF3 to develop an integrated multi-view model. IDEF5 is used to capture information about the enterprise and its environment. From this an IDEF0 activity model is extracted. A resource view is identified by assigning potential resources (partner companies in a virtual enterprise setting) to the activities of the IDEF0 model. This in effect defines the agents in the model and related activities. For each agent, we would then identify the organization view and business process views.

To define each of the process templates in terms of agility, the project is specifying the required characteristics of agility. Once these are finalized, a set of essential business processes will be identified and described. These essential processes are defined as those which all enterprises must perform well to become agile. This research is based on the proposition that an enterprise can be described as a set of business processes which are logically arranged to produce outputs for external customers. Within the enterprise, processes interrelate through the passing of information and materials to each other. Outputs of one process serve as inputs, constraints, or mechanisms to related processes. By describing what is required by a process for it to function and the outputs and performance characteristics desired of that process, the process based configuration and management of the enterprise is facilitated. "Should-Be" business process templates will be created reflecting the characteristics required for agility.

Methodologies

The architecture and templates alone represent a significant contribution to the development of agile and virtual enterprises. This work is further enhanced by research being conducted in methodologies to assist companies utilize the architecture and templates in guiding their transformation to agility. The project is currently conducting research into methodologies in the following areas:

- Strategic Process Metric Development
- Strategic Justification
- Agile Business Process Configuration

The following sections describe the work in these areas in detail.

Strategic Process Metric Development

The metrics development methodology is based on four important themes related to performance measurement systems that are consistently supported in the literature (Adams, Sarkis, and Liles, 1995). 1) Performance metrics should be derived from strategy. 2) Performance metrics should be developed for enterprise processes. 3). Performance metrics should be dynamic, keeping pace with changes in strategies, processes, technologies and the competitive environment. 4) A participative approach should be used to develop metrics.

The methodology develops performance metrics for business processes and activities. The methodology develops metrics derived from and, therefore, linked to the strategy of the organization. It is expected that performance metrics developed with such a tool would be suitable for deploying strategy, as well as for the control and improvement of enterprise activities. The methodology serves as an instrument for institutionalizing the process of performance metric development. The methodology is being developed and documented using the IDEF0 system design technique literature.

Strategic Justification

The Automation & Robotics Research Institute in conjunction with the National Center for Manufacturing Sciences (NCMS) has developed a pervasive activity based business case tool (NCMS, 1994). This methodology considers both traditional financial and strategic benefits of the technology under consideration. Furthermore, the impact of systems transition or implementation path is specifically addressed. The methodology is organized into five phases shown in Figure 1: Identify System Impact, Identify Transition Impact, Estimate Costs and Benefits, Perform Decision Analysis, and Audit Decision. The analysis in this methodology is accomplished through a series of documents and matrices in which the pervasive impact of the technology on the enterprise is determined.

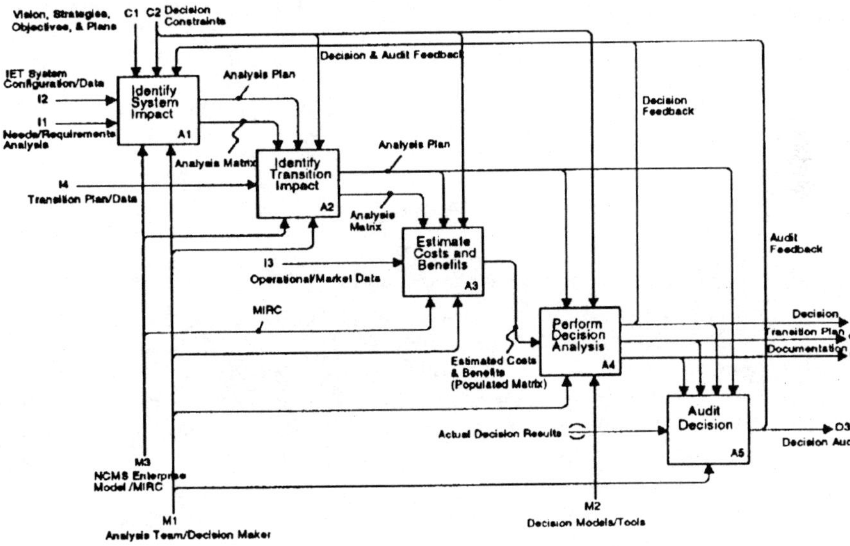

Figure 1. Business Case Methodology

In the *Identify System Impact* phase, the technology is linked to the enterprise. This linkage uses an activity based approach to estimate financial costs and benefits. The impacts of the new technology to the strategies of the enterprise are also identified. In much the same way, the *Identify Transition Impact* activity identifies the impacts of the process by which the new technology is implemented. The *Estimate Costs and Benefits* activity estimates the actual magnitude of the strategic and financial impacts.

In the *Perform Decision Analysis* activity, the analyst(s) performing the justification arrives at a decision recommendation. The *Audit Decision* phase allows for an evaluation of the justification process so that improvements can be made to the methodology itself (Meade, Johnson, Sarkis, and Liles, 1995).

Advanced techniques such as the Analytic Hierarchy Process (AHP) and Fuzzy Set Theory are being researched as possible additions to the SJET to add increased mathematical validity to the analysis. New and novel applications of the SJET are also being developed, for example in such diverse areas as environmentally conscious manufacturing and business process reengineering.

Agile Business Process Configuration

The proposed research will examine business processes within the enterprise as follows: 1) Develop a methodology for assisting the enterprise in determining its critical or most valued processes, and 2) Create a holistic evaluation model which considers strategic and operational factors in configuring agile business processes.

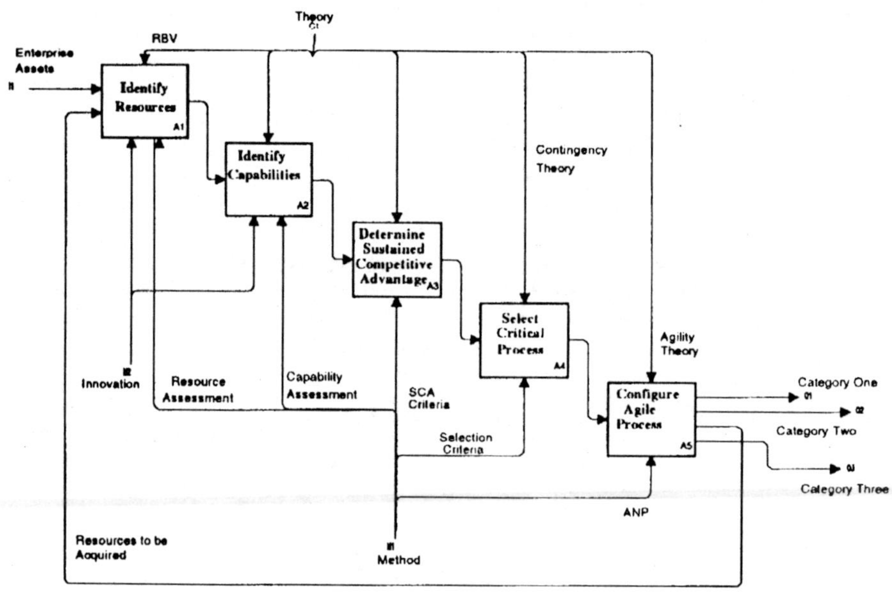

Figure 2. Agile Business Process Configuration

The architecture for configuring agile business processes takes a systems view of an enterprise in which an enterprise is seen as a system which takes in inputs and produces outputs under some set of environmental constraints (Schoderbek, Schoderbek, and Kefalas, 1990). An Analytical Network Process (ANP) model is utilized to aid the decision maker in considering all the multiple criteria involved in transforming a business process into an agile business process.

It is proposed that business processes naturally fall into three categories: (1) those processes which transform external constraints into internal constraints, (2) those processes which acquire and prepare resources, and (3) those processes which use resources to produce enterprise results (ARRI, 1990 and Presley, Huff, and Liles, 1993). Each one of these categories as well as subsets of business processes in these categories will be developed. A separate ANP model is created for each category of business processes.

Conclusion

This paper presented some of the research being conducted by the Automation & Robotics Research Institute to facilitate the engineering of the agile enterprise. The work will provide the tools and framework which will greatly assist the formation of virtual enterprises by providing for the rapid and efficient integration of the business processes of participating firms. In addition to this application, the work will also be applicable to any enterprise engineering effort.

Acknowledgments

Research for this paper is funded in part by the National Center for Manufacturing Sciences, the State of Texas Advanced Technology Program Grant 003656-036 and by the National Science Foundation sponsored Agile Aerospace Manufacturing Research Center.

References

Adams, S. M., J. Sarkis and D. Liles (1995). ìThe Development of Strategic Performance Metrics.î Engineering Management Journal 7(1): 24-32.

Automation & Robotics Research Institute (1990). "Operate a Small Manufacturing Enterprise". Fort Worth, TX, Automation & Robotics Research Institute.

Barnett, W., H. Hari and D. Liles (1995). "Object-Oriented Business Process Modeling". In 4th Annual Industrial Engineering Research Conference, Nashville, TN, Institute of Industrial Engineers.

Barnett, W., A. Presley, M. Johnson and D. Liles (1994). "An Architecture for the Virtual Enterprise". In IEEE International Conference on Systems, Man, and Cybernetics, San Antonio, TX.

Iacocca Institute (1991). "21st Century Manufacturing Enterprise Strategy: An Industry Led View". Bethlehem, PA, Iacocca Institute.

Jacobson, I., M. Ericsson and A. Jacobson (1994). The Object Advantage: Business Process Reengineering with Object Technology. Workingham, England, Addison-Wesley Publishing Co.

Meade, L. M., M. E. Johnson, J. Sarkis and D. H. Liles (1995). "A Tool for the Strategic Justification of Strategic Alliances and Partnering". In 4th Annual Industrial Engineering Research Conference, Nashville, TN, Institute of Industrial Engineers.

Nagel, R. (1993). "The Agile Opportunity". Bethlehem, PA, Iacocca Institute.

National Center for Manufacturing Sciences (1994). A Methodology for the Strategic Justification of Integrated Enterprise Technologies. Ann Arbor, MI, NCMS.

Presley, A., B. Huff and D. Liles (1993). "A Comprehensive Enterprise Model for Small Manufacturers". In 2nd Industrial Engineering Research Conference, Los Angeles, CA.

Schoderbek, P. P., C. G. Schoderbek and A. G. Kefalas (1990). Management Systems: Conceptual Considerations. Homewood, IL, BPI Irwin.

A Mathematical understanding of the success of multidisciplinary development teams

Bart R. Meijer
(+31 15) 278 6876

CIM Centrum Delft
Delft University of Technology
Landbergstraat 3
NL-2628 CD Delft, The Netherlands
email: b.r.meijer@wbmt.tudelft.nl

Today a product development problem has all the characteristics of a complex multi-criteria optimisation problem, with constraints that are often interdependent and sometimes conflicting. Until quite recently, a functionally organised development organisation would take 3-7 year to solve such a problem. Each step in the development process is doing a sub optimisation over a limited set of constraints and criteria. Thus many iterations were necessary to find a solution that satisfied all constraints and criteria. A "modern" multi-disciplinary, project-oriented development team can do the job much quicker. Depending on the type of product, development lead-times have come down to less than a year.

In this paper an attempt is made to explain this phenomenon by showing the similarity between a traditional development organisation and "conventional" optimisation algorithms such as linear programming. In a multi-disciplinary development team the interaction between criteria and constraints is more immediate. This resembles the behaviour of optimisation algorithms such as genetic algorithms. Genetic algorithms applied as a search technique are known to find good solutions in a limited amount of time. However these algorithms cannot be guaranteed to find the optimum.

Introduction

The demand for higher quality and lower cost products with shorter development lead-times has forced industry to consider various new product development strategies. The introduction of the product life-cycle concept has extended the notion of quality beyond functionality and durability, perceived mainly by its user (Alting 1993). Quality has become a complex function of all relevant aspects during the entire product life-cycle, including environmental issues during production, distribution, use and disposal.

In a traditional functional organisation this is a though challenge. Functional organisations are built around the similarity of processes performed in each department. Seemingly the product of the company as a whole has no significance. The departments priority is a high productivity, not in terms of products produced, but in terms of services carried out. Undoubtedly the division and specialisation of work that has been progressing since the industrial revolution has brought increased productivity, higher functionality of products and better quality, but a price had to be paid (Lindberg 1993). Comparing to the pre-industrial craftsman, a threefold information problem was introduced:

- Communication problem between sales, design and manufacturing.
- Control problem, to make sure the quality of the product is maintained throughout the process
- Synchronisation problem, to plan and synchronise all necessary activities.

Decreasing lead-times in a functional organisation, while leaving the organisation itself intact requires adding extra co-ordination and authority to change priorities within each department. An additional difficulty is posed by the life-cycle aspects that are seldom the responsibility of a single department.

Concurrent or simultaneous engineering (SE) is being recognised as a way to speed up the development process. However, despite all research effort into technical support for SE, Molina concludes that the importance of team working and a cultural change in the implementation of SE is undiminished (Molina 1995).

The organisation of this paper is as follows. First a problem solving framework is presented as a reference for solving design and development problems. Then genetic algorithms and conventional optimisation techniques are described to show the similarity between these algorithms and the way a development process behaves. Finally some conclusions and suggestions for further research are presented.

Problem solving and the Design process.

In a review of Yoshikawa's general design theory, Reich defines design as a process: given a description of a desired function and constraints, called specification, provide a description of an artefact that produces the function and satisfies the constraints (Reich 1995).

Studying the behaviour of this process, the more general process of solving problems can serve as a reference. Hesselberth studied the behaviour of expert problem solvers in industrial research and development (Hesselberth 1994). Although Hesselberth limits the scope of his definition of the word "problem" to existing processes or products. By doing this the importance of creativity in problem solving is lessened with respect to new products or processes. However the major part of an industrial development project can be characterised as redesign or redevelopment of existing products or processes in a new context. This does not differ very much from Hesselberth's problem definition.

Hesselberth distinguishes three types of problems with increasing difficulty. The most simple is the "follow your nose" problem or "apply your knowledge" problem. These problems are relatively straightforward in that they usually involve only a few variables. They are often solved by interpolation or extrapolation from a known point or the current situation. The second category is "define the space and find the optimum" problems. The process associated with these problems usually consist of taking a set of variables, defining the relationships between them often using a set of statistical experiments and then finding an optimum set of values to achieve the best balance of factors like cost, yield, product properties, productivity, etc. Recent advances in statistical experiment design and increases in computation power have added greatly to the power of this problem solving technique. The third category are the "really tough problems". This category of problems involves too many interacting variables to yield the simple follow your nose analysis, nor can it be solved writing the problem out as a body of equations and find the optimum, even if we had the super computers of today.

According to Hesselberth the expert problem solvers that are capable of solving the really tough problems, do at least six things that are different from the other problem solving processes:

- break the problem down into pieces, yet they can still think about the whole system.
- use the "scientific method" to gain understanding.
- look for clues or comparisons that allow them to gain understanding.
- often break away from the equipment or process they are trying to make work and design very specific experiments to answer specific questions.
- seek help from a variety of experts in specialised areas
- have a unique ability to leap beyond the current state in structuring a solution to the problem.

While "define the space and find the optimum" may represent a simple design problem, the more general design problem resembles the category of really tough problems. This becomes clear if one considers the design process phase models such as the VDI model (VDI 1973). Although 7 steps are to be distinguished in this model, basically it comes down to 4 phases: problem analysis, conceptual design, embodiment design and detailed design. The "define the space and find the optimum" class of problems are mainly found in the detailed design phase.

Optimisation techniques

A comprehensive overview of conventional optimisation algorithms is presented by Press ea.. (Press 1986). What a conventional optimisation algorithm comes down to is the following. There exists a solution space spanned by the design variables. The shape of this solution space is determined by the criterion function and interdependencies (equations). The boundaries of the space are determined by constraints (inequalities).

From a starting point a finite step is taken and the criterion function is determined again. The "heuristic" for the direction often comes from the maximum local gradient direction. Because these algorithms lack a helicopter view over the search space, they can be very inefficient travelling down the slop in a narrow valley. In a narrow valley the slopes along the valley represent the steepest gradients. Thus the valley is crossed many times before the bottom or the top is reached.

Optimising along the steepest gradient means developing the locally most important variables. The solution path through the solution space is a serial continuous path. Thus this process strongly resembles a functional design organisation. Only one limited set of related variables at a time are being developed. Other directions are explored only after local optimisation of these variables is finished means that the result of a partial design problem is not passed to the next department before the local optimisation is finished.

If we consider again the expert problem solver, none of his bag of tricks have an analogy in the conventional optimisation technique described above. Apparently the really tough problems can only be solved by decomposition and structuring, yet keeping an overview over the entire problem in order to avoid the pitfall of local optimisation, while the "search" through the solution space is guided by association with already known phenomena in other fields. Critical parts of the solution are covered by very specific experiments.

Although the match is not a perfect one, aspects of this problem solving process can be found in genetic algorithms. Genetic algorithms have first been introduced by Holland (Holland 1975) and have found many applications since (Wellman 1995).

There operation is as follows. The design variables are combined and encoded into a finite string (chromosome). Each element of the string (the genes) represents a partial solution to the problem. We have a population of strings were each individual represents a possible solution to the problem, but probably not a very good solution. Looking for this solution means finding a chromosome that is good enough. For this we need two things: a criterion function to value the chromosome and in some cases the value of a gene (partial solutions) and operands on chromosomes to built a new population of chromosomes. The chromosomes are valued and sorted in order of decreasing value. A fixed number (e.g. the top 100) of chromosomes are used to produce a new population of chromosomes that are valued and sorted again. After a number of generations the "top" chromosome is a representation of a very good solution.

The classic genetic operands are crossover, mutation and inversion. Crossover combines two chromosomes into a new one. Crossover is the combination of genes of different chromosomes into a new, hopefully better chromosome. The mutation serves as a random change of a gene or a set of connected genes (allele). This random influence is necessary to avoid local optimisation of alleles. The inversion changes the order of genes within the chromosome. This operation is necessary to give good genes (good partial solutions) a better chance of survival into the next generation than bad genes.

The reason for success of genetic algorithms is that they don't operate searching along a track to move towards an optimum, instead they reshuffle the search space each generation in such a way that the best solutions are put closely together. One other aspect of the success of genetic algorithms is that all genes (design variables) are developed simultaneously. The randomisation and the influence on partial solutions while the chromosome still represent the entire problem is considered similar to the structuring of the expert problem solver. Also the crossover operand could be seen as the association processes he expert problem solver uses.

Another analogy between brainstorming and genetic algorithms has been presented to me by van Bruggen (Bruggen 1995). A good brainstorm session satisfies the following properties:

- not to many participants (~15)
- participants are not all experts
- evaluation of ideas is forbidden until after the session
- all ideas are available to all participants
- although not necessarily, new ideas can be based on old ideas or a combination of old ideas

The way new ideas are generated is similar to the crossover operand. The role of the non-experts is that of mutation and introducing entirely new ideas. The collection of ideas can be seen as a collection of chromosomes, that become better as the session proceeds. Evaluation afterwards is to make sure that all ideas (genes and chromosomes) get a chance to develop into new ones.

Taking this analogy even further to a development team it becomes clear that they influence each other from all possible angle of the design problem simultaneously. Even though they call themselves expert this is not a problem because they are unlikely expert in all the relevant areas. As a team they can also bring about the association powers that are essential to the individual expert problem solver. What they lack yet is a way to value and rank all their ideas. Morphological methods such as value engineering or quality function deployment could be sufficiently accurate to do the job.

Conclusions and Future research

In this paper the similarity between the behaviour of mathematical optimisation algorithms and the behaviour of development teams has been made plausible. Future studies of these similarities will be guided towards more quantitative aspects of genetic algorithms. If one can relate the speed of convergence to the number of aspects and the quality of the initial solution, it may be possible to use this understanding in the organisation of development teams as well.

Acknowledgements

Part of the research reported here has been carried out while the author was still with the Management and Organisation section of the Mechanical Engineering faculty. The author wishes to express his thanks to Marnix van Bruggen who introduced him to genetic algorithms and who's responsible for the analogy between brainstorming and genetic algorithms.

References

Alting L., Jørgensen J. "The life cycle concept as a basis for sustainable industrial production" Annals of the CIRP 1(42):163-167,1993.

Bruggen M. van, "Line-Balancing with genetic algorithms" ,Delft University of Technology, Faculty of Mechanical Engineering and Marine Technology, report FPA95.007, 1995

Charniak E., McDermott D.,"Introduction to Artificial Intelligence", ISBN 0-201-11945-5, Addison-Wesley. 1987

Hesselberth J.F., "Problem-solving in research and development", Int.J. Technology Management, 2(9):253-260,1994

Holland J.H.,"Adaptation in natural and artificial systems", University of Michigan Press, Ann Arbor, 1975

Lindberg L., sponsored by Solenius G. "Notes on Concurrent Engineering" Annals of the CIRP 1(42):159-162,1993.

Molina A., Al-Ashaab A.H, Ellis T.I.A., Young R.I.M, Bell R. "A review of computer-aided simultaneous engineering systems" Research in Engineering Design 7:38-63,1995

Press W.H., Flannery B.P., Teukolsky S.A., Vetterling W.T., "Numerical recipes; the art of scientific computing", Cambridge University Press, 3rd edt.,1986.

Reich Y. "A critical review of general design theory" Research in Engineering Design 7:1-18,1995.

VDI 2222, "Konzipieren technischer Produkte; Konstruktionsmethodiek", Düsseldorf: Verein Deutscher Ingenieure, 1973.

Wellman M.A., Gemmill D.D., "A genetic algorithm approach to optimisation of asynchronous automatic assembly systems", Int.J.of Flexible Manufacturing Systems 7:27-46, 1995

Impact of Owner Involvement on Innovation: Lessons from Power Plant Projects

Roger Miller
Tel: (514) 987-4256
Fax: (514) 987-4705
e-mail: program@imec.login.qc.ca

Hydro-Quebec Chair in the Management of Technology
Université du Québec à Montréal
Pavillon Ste-Catherine West, Room X-7330
C.P. 11008, Montréal, Qc.
Canada, H3C 4T9

Information asymmetry between suppliers and buyers constitutes a major barrier to innovations. Incorporating novel solutions into engineering projects increases the problems of calculating long-term costs and benefits. Industrial marketing suggests that overcoming these barriers is easier when owners are knowledgeable and involved in the design process. Our study examines whether the same proposition holds for innovation in complex engineering projects by focussing on the relationships between innovativeness and owner internal capabilities. Hypotheses derived from agency and organization theories of innovation are tested. Results indicate that projects where owners have internal capabilities are generally less innovative than projects where owners lack internal capabilities.

The dominant view of innovation takes its inspiration from research on families of products competing to create markets through discontinuous innovations. (e.g. Utterback 1995, Schumpeter 1949). Examples of such products are computers, calculators, snowmobiles semi-conductores, etc. The success of innovators is a function of their abilities to interpret ex ante *the calculus of buyers. (e.g. Rogers & Shoemaker 1971). The main challenge is to overcome consumers' limited imagination or lack of knowledge through superior designs.*

In the case of industrial goods, however, potential buyers have both the experience and the knowledge to make assessments of new technologies. Projects involving innovative and uproven features make costs and benefits difficult to estimate. The question that confronts suppliers in these circumstances is how strongly should they press innovative ideas on their clients?

Current research on innovation provides two different perspectives on this question. The first perspective comes from agency theory which argues that owners with in-house expertise are more favorably disposed towards innovation, whereas owners with less information are more likely to put constraints on their agents' ability to implement innovative solutions (e.g. Aoki, 1990; Aghion and Tirole, 1994). The other perspective, coming from organizational theories, suggests that owners with internal capabilities are more likely to institutionalize past experience into practices and are therefore more likely to evaluate innovative solutions within a conservative framework. By contrast, owners that lack internal capabilities are forced to give their agents greater freedom and are also less likely to interfere in project design and implementation (e.g. Stinchcombe 1990, Von Hipple 1994).

The first part of this paper summarizes these two perspectives as they relate to innovation in complex engineering projects. The second part details the data and measures used to test propositions derived from these two perspectives. The data comes from a larger study of electrical power projects in North America between 1988 and 1994. The third part of the study describes the analysis and the results. Implications are discussed in a fourth section. To conclude, suggestions for future research are made.

Theoretical Overview

Innovations in industrial goods are often the result of problems that exist in the operations of owners. Owners unable to solve these problems internally look to firms that are in the business of developing and producing capital equipment for appropriate solutions. Given sufficient time, firms in the capital goods sector respond to their customers' needs. However, for builders of complex engineering projects that are in need of rapid solutions, this process may be too slow and uncertain. The most

0-7803-3552-X/96/$5.00 © 1996 IEEE

common method to ensure effective problem solving is the project organization linking firms that provide consultative expertise, supervisory services, and equipment.

Owners occupy a central position from which they exert influence on the analysis of problems and solutions. However, their ability and their willingness to exert such influence is constrained by two factors: (i) the degree to which they have requisite internal expertise; and (ii) their willingness to actively become involved in the complex and detailed work of developing and implementing solutions. Both factors represent types of information asymmetries that restrict owners from accurate evaluation of solutions. In the absence of internal expertise, owners are forced to rely on other organizations to develop solutions.

The question of interest in this paper is to what extent having or lacking these prior investments in capabilities influences the innovativeness of the solutions that are eventually incorporated into the project. Owners are in an agency dilemma with respect to the organizations on which they depend for effective solutions. Controlling their agents and ensuring that they perform adequately requires prior investments in expertise. There are essentially two schools of thought on this subject: (i) agency theory in economics; and (ii) organization theory and empirical studies of innovation.

The agency theory argues that principals delegate decision-making to agents who presumably have the competence and the superior information to attain the goals stipulated in the agreement (e.g. Stiglitz, 1987). Information asymmetries may interfere in this relationship. Principals must have sufficient information to evaluate the competence and probity of prospective agents, and they must devise supervisory systems to monitor agents. Principals that strive to reduce one or both of these information asymmetries are more likely to see their agents develop and implement innovative solutions to problems (e.g. Aghion and Tirole, 1994).

Organization theories suggest that firms approach new problems with the experience and competence acquired during previous problem-solving activities (e.g. Cohen and Levinthal 1990). Owners that exercise their oversight perogative at every phase and at every level of the project are likely to discourage contracting organizations from pursuing innovative solutions. Intense involvement on the part of owners is likely to increase the costs of discussions, analyses and rework. Agents thus adopt solutions that are more conventional, and more in line with the experience and views of owners.

Data and Measures

To test the propositions sketched in the previous section, we turned to data collected during a research project on electrical power plants. A list of electrical power projects from 1988 to 1994 was compiled from the DOE and McGraw-Hill yearbooks. The final sample included 178 projects owned by 134 organizations. Most of the projects were undertaken by publicly-owned utilities (municipal, state, provincial, federal) co-operatives, investor-owned utilities, and non-utility affiliates, independent power producers, and cogenerators.

The measurement instrument was tested in meetings with industry executives, and was subsequently pre-tested with 15 randomly selected projects. Fifty-eight organizations representing sixty-seven projects completed questionnaires. A majority of the individuals who responded were vice-presidents of power supply development, the remainder were project engineers, or plant managers.

The main dependent variable was the innovation content of the project. Objective measures were based on a classification of projects according to types of awards or mentions received from industry publications or other institutions. Subjective measures were obtained by asking the respondents to rate the innovativeness of projects. The two measures of innovation were correlated at 0.3281 (p=0.009). The questionnaire also contained questions that dealt with dependence on external organizations, and the degree of owner's control of the project.

Analysis and Results

Two statistical techniques were employed to examine the hypotheses: (i) logistic regression and (ii) correspondence analysis. Logistic regression showed that engineering capabilities impacted negatively on innovativeness, while project management capabilities impacted positively on innovativeness. Stepwise logistic regression was used to find which of the relevant variables had the highest impact on innovativeness. The parameter estimates for the significant variables plus the capabilities that significantly impacted on innovativeness are shown in Table 1.

Table 1. Summary of logistic regression results (*<0.1, **<0.05, ***<0.001).

IMPACT ON PROJECT INNOVATIVENESS	PARAMETER ESTIMATES	STEPWISE PARAMETER ESTIMATES
Engineering Design	-3.2973***	
Construction Management	-.88434	
Project Management	3.5671**	
Turnkey Projects	0.344	
Engineering Design * Construction Management	24.49	
Engineering Design * Project Management	-26.49	-2.7411***
Engineering Design * Turnkey Projects	-11.52	
Construction Management * Project Management	1.94	2.29**
Construction Management * Turnkey Projects	-13.07	
Project Management * Turnkey Projects	24.9	
Owner selected major equipment suppliers	0.6239	
Owner selected major equipment suppliers * Engineering Design	-0.5016	-1.1787**
Owner selected suppliers relying exclusively on competitive bids.	-0.730***	
Owner selected suppliers relying exclusively on competitive bids * Engineering Design	-0.4512*	-0.4839***
Owner did scoping, organized bids, but subcontracted engineering and construction.	-0.989	
Owner did scoping, organized bids, but subcontracted engineering and construction * Engineering Design	-1.311	-1.6354**
Owner did everything internally, except equipment, manufacturing, and some construction	-0.83	-1.335*
Owner invited supplier/contractor to become part owner	0.619	
Owner invited supplier/contractor to become part owner * Project Management	0.4512*	0.701**

The graphical results of the correspondence analysis which are reproduced in Figure 1 show innovativeness as polar opposites close to the origin. The other variables are represented in relationship to innovativeness. All other things being equal, variables that are closest to dimension 1 have the greatest impact on innovativeness.

Figure 1
Innovativeness and Relationships with Consultants and Suppliers

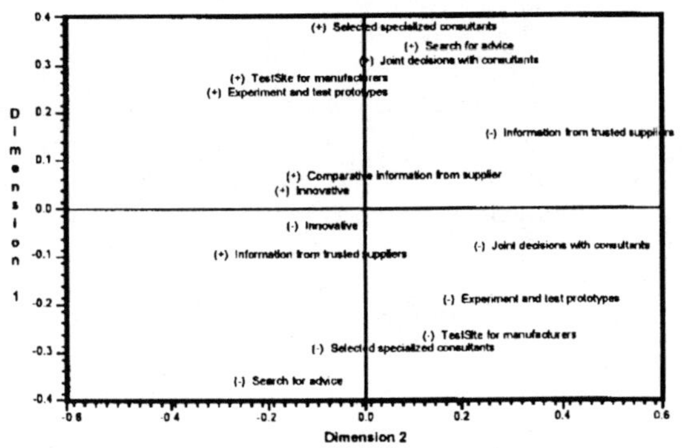

Discussion

Our results strongly indicate that internal capabilities and active involvement on the part of owners tends to reduce the innovativeness of complex engineering projects. Projects where owners had only project management capabilities were on the whole more innovative.

To fully evaluate the costs and benefits of innovative solutions it is not only important to see how such solutions impact the overall project, it is also necessary to have an understanding of the specific problems involved in their implementation. Construction management capability enhances the owner's ability to evaluate implementation claims make by contractors. Furthermore, the fact that construction management alone does not impact innovativeness reinforces the view that its main contribution to innovative problem solving is evaluative rather than active participation in the construction process.

Of the variables describing the relationship between owner's capabilities and those of external organizations only one seems to make positive contribution to project innovativeness. The positive influence of project management capability on innovativeness is the only major support we found for an agency perspective on the relationship between owners and external organizations.

The rest of the analysis leans strongly towards theories advanced by organizational researchers. Owners that had engineering design capabilities tended to have less innovative projects. It seems that wherever engineering and project management capabilities exist, there is a tendency for technical staff to use past experience and authority to dominate the evaluation and implementation of innovative solutions. Correspondence analysis provides a more complete picture of the relationship between innovativeness and an owner's control of a project by suggesting that owners who marshall fresh organizational resources to deal with potentially difficult problems are more receptive to innovative solutions.

Figure 1 shows the relationship between innovativeness and an owner's dependence on consultants and equipment suppliers. Owners that rely and make use of consultants and long-term relationships with equipment suppliers use joint problem-solving at an early stage of projects. The very fact that owners are willing to engage an external network of experts in formal and informal discussions suggests greater openness towards innovative solutions. Obtaining advice from engineering consultants, analyzing and making decisions jointly with consultants, selecting specialized consultants are issues clustered together at both extremes of dimension 1. All these activities involve intense collaboration and exchange with external organizations that premise their strategy on the ability to work closely with owners.

Conclusion

Owners of engineering projects are similar to other buyers: they want the best at the lowest possible cost. However, unlike the buyers of finished goods, owners of engineering projects can not verify the quality and performance of products prior to purchase. They must award the contracts on the assumption that the quality and performance of the completed project will meet their expectations. For their part, contractors, equipment suppliers, and engineering firms must deliver the quality promised at the stipulated price. Reconciling the conflicting demands of quality and price often leads providers of equipment and engineering services to economize on resources by adopting solutions that are inferior to the best possible solutions. Owners who are especially concerned with this problem are likely to draft contracts that seek to deter opportunistic behavior on the part of agents (e.g. Williamson, 1985). Such contracts, however, may also constrain the agents from implementing solutions that are at the cutting edge.

Allaying owners' concerns about innovative solutions is easier when owners lack their own internal design capabilities. At first sight this conclusion seems counter-intuitive. In most market transactions, sellers find it much easier to explain the nuances of technical issues to knowledgeable buyers than to buyers that have little or no competence to judge such details (e.g. Bettman, 1979; Nelson, 1974). In complex engineering projects, however, the expertise of the owner is embedded in technical staff whose primary task is the maintenance of operations, not the exploration of technical frontiers. Owners that lack the internal capabilities to subject the claims of agents to rigorous scrutiny are more likely to go the heart of the matter and ask the basic question: Do our contractors, designers, and equipment suppliers have the necessary competencies and experience to implement proposed innovations? The answer to this question is usually based on factors that go beyond the details of the proposed solutions (e.g. Williamson, 1993). In particular, top management is likely to put considerable weight on such factors as reputation for reliability, innovativeness, previous experience with similar problems, and wherever pertinent, trust arising from past relationships.

References

Aghion, P., and Tirole, J. (1994). Opening the black box of innovation. **European Economic Review**. Vol. 38, pp. 701-710.

Aoki, M. (1990). Toward an economic model of the Japanese firm. **Journal of Economic Literature**, Vol. 28, pp. 1-27.

Bettman, J. (1979). **An information processing theory of consumer choice**. Addision Wesley, Reading, Mass.

Cohen, W.M., and Levinthal, D.A. (1990). Absorptive capacity: A new perspective on learning and innovation. **Administrative Science Quarterly**, Vol. 35, pp. 128-152.

Hipple, E. von (1994). "Sticky Information" and the locus of problem solving: Implications for innovation. **Management Science**, Vol. 40, pp. 429-439.

Nelson, P. (1974). Advertising as information. **Journal of Political Economy**. Vol. 82, pp. 729-54.

Rogers, E.M., and Shoemaker, F.F. (1971). **Communication of Innovation: A Cross Cultural Approach**, 2nd Edition, The Free Press, New York.

Schumpeter, J. **Capitalism, Socialism and Democratism.** Harper and Row New York, 1947.

Stiglitz, J.E. (1987). The design of labor contracts: The economics of incentives and risk sharing. In H.R. Nalbantian (Eds), **Incentives, Cooperation, and Risk Sharing: Economic and Psychological Perspectives on Employment Contracts**, pp. 47-68. Rowman & Littlefield, New Jersey.

Stinchcombe, A.L. (1990). Organizing information outside the firm: Contracts as hierarchical documents. In **Information and Organizations**. pp. 194-239. University of California Press, Berkley.

Utterback, J. **Radical Innovations**. Harvard Business School Press, Boston 1994.

Williamson, O.E. (1985). **The Economic Institutions of Capitalism**. Free Press, New York.

Yang, Y.L. (1995). Degree of supervision, moral hazard, and hierarchical control. **Journal of Economic and Organization**. Vol. 26, pp. 93-105.

Matching Team Management Strategy With The Organizational Culture

Dragan Milosevic
(503)725 4660

PhD, Portland State University, Engineering Management Program
PO Box 751 Portland OR 97207-0751

Tugrul Daim
(503)725 4660

Portland State University, Engineering Management Program
PO Box 751 Portland OR 97207-0751

This work studies the question of "Does different types of organizational culture exhibit different team management strategies?" For a sample of 29 team leaders in product development, manufacturing and construction, the answer is basically a "Yes."

Organizational culture can promote or impede the firm's strategic actions. A recent study found that companies with "strategically appropriate cultures" outperformed selected other corporations with less appropriate cultures [Kotter & Heskett]. The point is that the culture of a firm should match - and be supportive of - the firm's strategy [Wright, Pringle, Kroll, p. 229]. Is such a relationship between the culture and strategy equally valid when it comes to lower level strategies? Specifically, should there be the match between team management strategy and the culture too?

Responding to competitive pressures, corporations are searching for ways to deliver more cost-effective and higher quality products, including the use of teams. The hope is that widespread deployment of teams will result in becoming better, faster, cheaper. [Zenger]. The essential elements in a definition of team, according to Katzenbach, include small number, complimentary skills, common mission, common goals, common working approach and mutual accountability [Katzenbach-HBR][General Foods].

Significant interest has been expressed by corporate leaders in adopting the team approach. In fact, a recent survey of 1,811 companies indicates that 66% have actively used teams [Wall Street Journal]. There is a growing descriptive and prescriptive literature regarding the team approach in organizations. However, no systematic evidence exists as yet to demonstrate the relationship between the team management strategy and organizational culture.

The present study addresses this gap by examining the experience of 29 team leaders. The research focuses on the team management strategies' fit with organizational culture types.

Team management strategy

Few teams would dispute the need to plan and control their projects. The trick is to apply the available tools and methods in a manner that will lead the team members toward the "signed - up" goals within the given time and resource constraints. Effective team management properly defines the work, budgets, and schedules, and then keeps team on track [Thamhain] [Meyer]. Simultaneously, it is engaged in getting the team members to play together and keeping them focused on the goal, while leaving them space to "breathe" and accomplish their non-team activities. It is also to harmonize the technical and human aspects of a project with other company day - to - day business. If all this is the scope of a team's operations, what would be its strategy? Strategy is defined as the framework which guides those choices that determine the nature and direction of an organization, in our case team [Tregoe]. Three elements of Katzenbach's definition - common mission, common goals, common working approach - were found to fit such a definition of team management

0-7803-3552-X/96/$5.00 © 1996 IEEE

strategy. However, for the practical research that we wanted to conduct they were not specific enough. Consequently, the literature search was conducted to break the three elements down into more specific management approaches as components of a team management strategy. The approaches in Figure 1 were suggested by the search [Slevin] [Thamhain].

The culture - team management strategy linkage hypothesis

Culture is defined as the values, beliefs, and norms of an organization. Since teams have been around hundred of years (Katzenbach), it is not unreasonable to assume that teams have been successfully applied in a variety of organizational cultures. Based on the underlying values of organizational members, Denison (1990) [Daft] used two dimensions, needs of environment (change/flexibility vs. stability/direction) and strategic emphasis (external vs. internal) to define four culture types. The types are: an adaptability culture characterized by focus on the environment through flexibility and change to meet customer needs; a mission culture reflecting the values and norms associated with a shared vision of organization purpose, but without need for rapid change; an involvement culture emphasizing the involvement and participation of the organization's members; and a consistency culture based on norms and values associated with an internal focus and a consistency orientation for a stable direction and environment.

Given the emphasis that different cultures put on change, flexibility, stability, direction, external and internal focus, it is felt that teams in such cultures also will use different management strategies. Thus, the first hypothesis:

Hypothesis: Different organizational culture types will exhibit different team management strategies.

Methodology

Sample

Data were collected from 29 team leaders. They were chosen for their knowledge about various aspects of their team projects and their organization. Their task was to assess the relationship between the team management strategy and organizational culture, and to provide data for hypothesis testing. The sample description is given in Table 1.

TABLE I
Sample Description

Organizations		
-Industries	15(Manufacturing) 17(Service) 7(Construction)	
-Annual gross revenue	mean: $37M	range: $5 - 50 M
-Organization's market	17(International) 5(Regional) 7(Local)	
Team leaders		
-Number	29	
-Number of years leading teams	mean: 6.3 yrs	range: 1 - 25 yrs
-Formally accountable for team results	21	
Team Projects		
-Size	mean = $2.2M	range: $50K-33M
-Duration	mean = 11.7 months	range: 2 - 24 months
-Number of team members	mean = 9.9	range: 2 - 20

Table 1

Data Collection Procedure

A mailed questionnaire was a primary vehicle for collecting data. Each respondent was assured of confidentiality. Three team leaders with substantial experience pretested the instrument through interviews that focused on content validity and item wording. The follow-up requests were made through phone interviews and face-to-face meetings with the team leaders.

Measures

Measurement of organizational culture used Denison's typology. The survey asked respondents to select one of descriptions of what constitutes the four culture types that best described their organizational culture. The Cronbach's alpha was computed at satisfactory 0.78.

The difference in team management strategies was defined as difference in the importance that various management approaches have in their team management strategies. On a Likert scale (5 being very high, 1 being very low) the respondents rated the importance. The Cronbach's alphas for the approaches were satisfactory from 0.74 (the lowest) to 0.80 (the highest).

Responses of all team leaders were divided into four culture groups. Then, management approaches were compared for the groups. Since only one respondent characterized her organizational culture as the mission culture, her response was discarded, leaving us to compare the three culture types: adaptability, involvement and consistency. To test our hypothesis, ANOVA was used.

Results

Our hypothesis addressed the relationship between organizational culture types and team management strategy. The results are illustrated in Figure 1. For each culture type, the means of importance of management approaches to team management strategy are reported in a bar chart format.

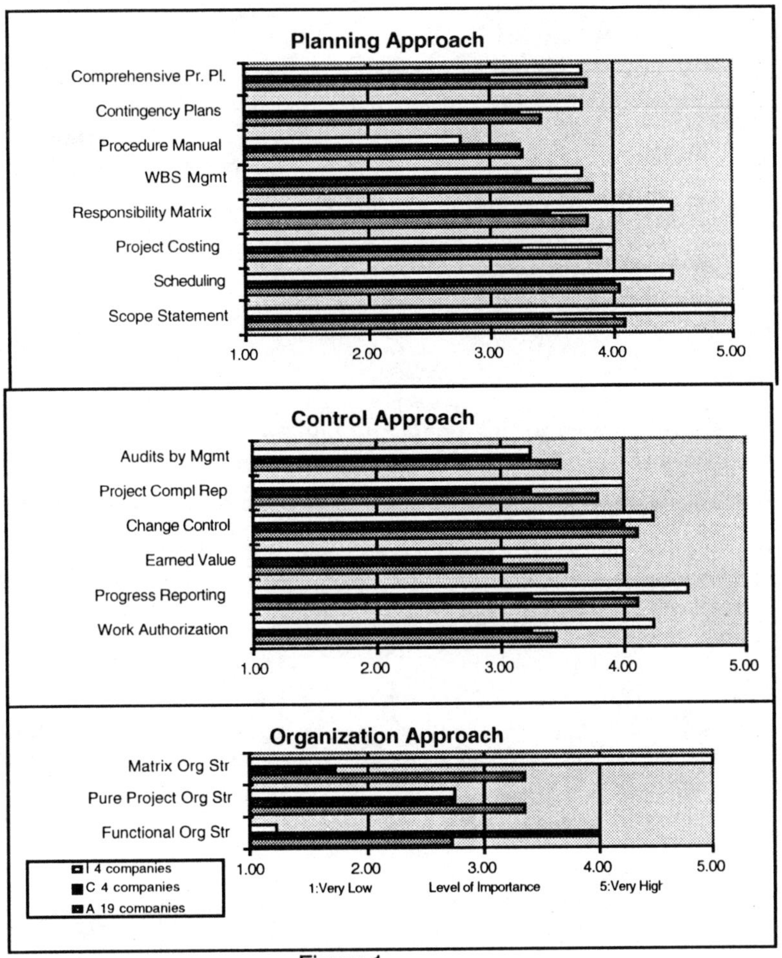

Figure 1

Planning Approaches

ANOVA showed no differences in the importance of management approaches. between the adaptability, involvement and consistency cultures. However, as Figure 1 shows, the three cultures placed different emphasis on the planning approaches. Specifically, the involvement culture had the highest rankings on 5 out of eight approaches. Each of the eight approaches had higher importance in the adaptability culture than in the consistency culture. And the consistency culture featured the lowest importance in seven of the eight approaches. It appears that the culture types that highly value change and flexibility consider the planning approaches more important than the consistency culture that appreciates stability and direction.

Control Approaches

The findings about the control approaches resemble those of the planning approaches. Again, ANOVA did not show significant statistical differences between the culture types. However, in five of the six approaches, the involvement culture had the highest importance rankings. Each of the six approaches was important in the adaptability culture than in the consistency culture. And again, the organizational cultures focusing on change and flexibility placed a higher value on the control approaches than did the consistency culture preferring stability and direction.

Organizational Approaches

As can be seen from Table 2, when it comes to the matrix and functional approach, there were significant differences between the culture types. The involvement culture had placed the highest importance on the matrix, followed by the adaptability culture.

That functional organization would be favored by the consistency culture was no surprise, because this organization is typically found in companies whose cornerstones are internal strategic emphasis and stability - those of the consistency culture.

Anova Results		
	F	p
Functional Org Structure	7.46	0.030
Matrix Org Structure	7.20	0.035

Table 2

Discussion

In the planning and control area, the involvement culture indicated the highest importance in ten of the fourteen approaches. In contrast, in the thirteen approaches, the consistency culture indicated the lowest importance. Further, mean scores of the adaptability and involvement culture for the matrix structure were greater than those of the consistency culture. On the contrary, the consistency culture demonstrated higher preference for the functional organization. The importance of these two organizational approaches was significantly different. All this supported our hypothesis - different organizational cultures exhibited different team management strategies..

This conclusion calls for a caution. Many factors affect team management strategy and our sample is small. Therefore, we are not suggesting that a specific team management strategy is "fundamental" to a specific organizational culture. Quite to the contrary, an organizational culture type may have a variety of subcultures having different team management strategies. Further work is necessary to fully explore this relationship.

However, this study provides some valuable lessons. First, the culture types whose pillars are change and flexibility place comparatively higher importance on the planning and control approaches. Second, the importance of the matrix and functional structures is significantly different for different cultures.

Implications and conclusions

The information that we gathered may provide a useful yardstick for managers responsible for the design and implementation of team management strategy.

Our effort also supports some of the previous writings. Specifically, cultures that feed on change and flexibility put more emphasis on planning and control approaches, and matrix organization. This is certainly important for managers whose job is to evaluate team management strategy.

There are two questions that deserve special attention.

1. Do subcultures in an organizational culture exhibit different team management strategies? Since researchers found that different departments (e.g. R&D, manufacturing) with different subcultures use different organizational designs, it may be that they also use different team management strategies.

2. Do teams whose management strategies are more culturally compatible exhibit higher team success? It is not unreasonable to ponder that if a team's strategy is more aligned with its organization's cultural norms, that its success is more likely.

Some believe that different cultural norms in organizations engender different team management strategies. After this study, we can not claim that they are right. However, it is not unreasonable to claim that we found some evidence that they are not wrong.

References

Teams Become Commonplace in U.S. Companies. The Wall Street Journal, November 28, 1995, p.1.

Bassin, M. Teamwork at general Foods: New And Improved. Personnel Journal, May 1988, pp. 62-70.

Cotter, J.P and Heskett, J.L. Corporate Culture and Performance. New York: The Free Press. 1992.

Daft, R.L. Organization Theory and Design, 5th edition, West publishing Company, Mineapolis (MN), 1995.

Katzenbach, J.R. and Smith, D.K. The Discipline of Teams. Harvard Business Review, March-April 1993, pp.111-120

Meyer, C. How The Right Measures Help Teams Excel. Harvard Business Review, March-June 1994, pp. 95-103.

Slevin, D.P. and Pinto, J.K. Balancing Strategy and Tactics in Project Implementation, Sloan Management Review, Fall 1987, pp. 33-41.

Thamhain, H.J. and Wilemon, D.L. Criteria For Controlling Projects According To Plan, Project Management Journal, June 1986, pp. 75-81.

Tregoe, B.T., Zimmerman, J.W., Smith, R.A. and Tobia, P.M. Vision in Action. Simon and Schuster. New York. 1989. p. 33.

Wright, Pringle, Kroll. Strategic Management. Allyn and Bacon. Boston. 1994.

Zenger, J.H., Musselwhite, e.. Hurson, K. and Perrin, C. Leading Teams. Irwin. Homewood. 1994, p. 13.

Strategic Enterprise Realization and Transformation Processes in the Virtual Organization Era

Benoit Montreuil[1],
Johann Vallerand[1,2]
and Diane Poulin[1]

[1] *SORCIIER* Research Center on the Network Enterprise, Pavillon De Sève, Université Laval, Ste-Foy, Québec, Canada G1K 7P4; e-mail : benoit@osd.ulaval.ca

[2] Department of Management, Moncton University, New-Brunswick, Canada & CERAG Center, IAE Grenoble, Pierre Mendès France University, France

In this paper we introduce a process model of the enterprise in the virtual organization era, having at its core a fundamental ever-active enterprise realization process. This core process is composed of three main subprocesses: the enterprise transformation process, the organizational learning process and the value creation process. Where the transformation process was in the past a big-event project, it becomes an ever-active strategic process itself decomposed in three main subprocesses: the identity rethinking process, the process reengineering process and the network reorganization process. We describe these processes and emphasize their role in helping the firm thrive in the forthcoming global digital economy.

Introduction

As well described by Goldman et al. (1994), Poulin et al. (1994), Tapscott (1995) and many other authors, we are entering a new era of networked enterprises and virtual organizations, in a new global digital economy. For enterprises to thrive in this new era, they will need to learn to manage on a global scale an intricate ongoing interplay of value creation, organizational learning and business transformation. In this paper we describe an integrative process model for enterprise realization which should permit managers to grasp the scope of the changes lying ahead and to get a handle on how to tackle these changes.

The strategic processes of the enterprise

In this section we focus on introducing what we consider the most fundamental and critical processes of enterprises in the new virtual organization era. We do so by describing a process model which focuses first on an holistic process that we term *enterprise realization process*. In our model, the enterprise realization process is the fundamental process underlying all the actions undertaken by an enterprise.

At the most aggregate level, the mission of this process is to continually sustain and promote the realization of the enterprise given its current state and its action potential. In the most generic terms, realizing itself means for the enterprise to become what it can best become for satisfying society's ever evolving needs. Pragmatically, rarely do managers of an enterprise nowadays think in terms of the firm's impact on the entire society, they rather focus on a subset of society defined by the set of its main stakeholders. These include the enterprise's leaders, actual and potential customers and markets, financial organizations and shareholders, personnel, suppliers, government, and so on. Realizing itself then becomes for the enterprise to continually evolve so as to best position itself in terms of needs satisfaction for the various stakeholders, prioritized in function of their impact on the enterprise's viability and prosperity, seeking as much as possible to avoid win-lose compromise situations and to rather hit for win-win relationships.

From this enterprise realization process are derived and justified all the other processes in which the enterprise is engaged at a given time, from customer order delivery to next-generation product development. This is illustrated in Figure 1 which shows the realization process at the top, recursively decomposed into lower-tier

0-7803-3552-X/96/$5.00 © 1996 IEEE

business processes. What are these processes that the business launches and manages is a key strategic decision of any enterprise.

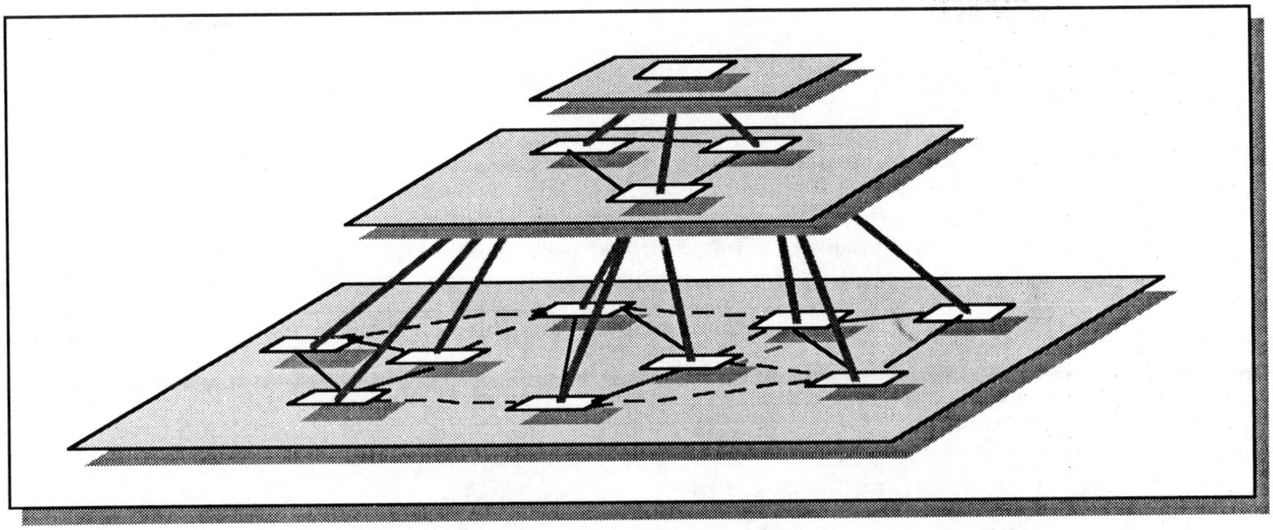

Figure 1. Recursive decomposition of the enterprise realization process

In this fast evolving era, driving us toward a global digital economy, we believe that for many enterprises, explicitly decomposing the business realization process into three main subprocesses may be a key to their survival and prosperity. As shown in Figure 2, in our model these are labeled the *enterprise transformation process*, the *organizational learning process* and the *value creation process*.

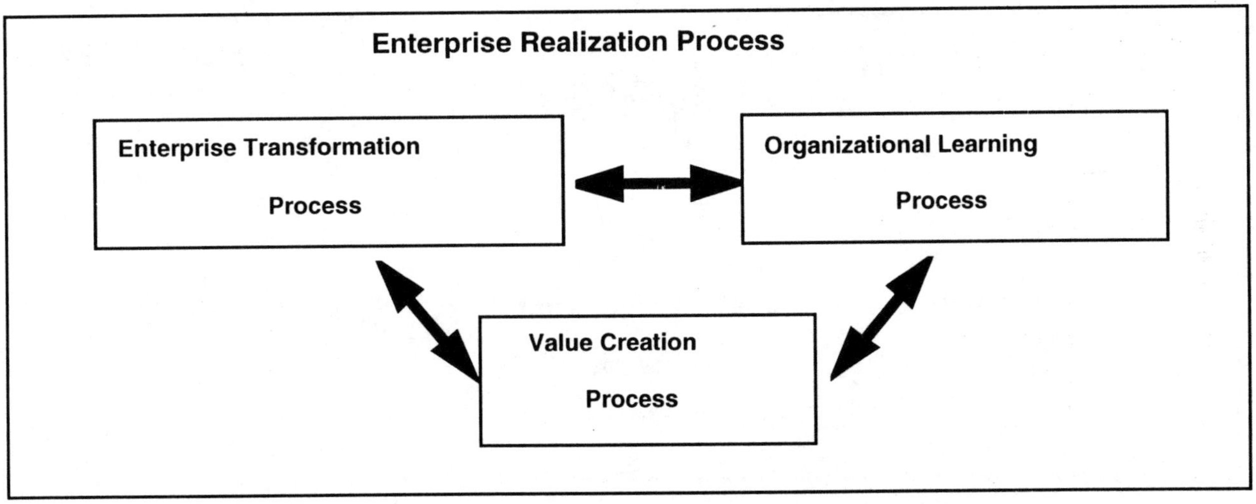

Figure 2. The enterprise realization process and its main enterprise transformation, organizational learning and value creation subprocesses

The enterprise transformation process involves all the activities achieved so as to adapt the identity of the firm, to reengineer its processes, and to reorganize its networks. This process aims to put the stage, the infrastructure, best adapted for performing the value creation process. In parallel, always based on the current results of the transformation process, the value creation process attempts to exploit the firm's identity, to activate and manage the various business processes, and to deploy and manage the enterprise's networks, so as to maximize the value created for the key stakeholders. The organizational learning process continuously attempts to gather information and develop knowledge that has the potential for improving both the transformation process and the value creation process. This involves learning about identity, process and network fitness and know-how, and their intricate interrelationships: learning about new processes, better ways to perform existing ones, learning about potential partners and how to best profit from existing ones, finding ways to leverage knowledge about technology as well as about competition, and so on. From a transformation perspective, it involves learning about the impact of identity, process and network transformations, gaining transformational know-how, and so on.

In a virtual organization mode, none of these key processes may be taken for granted. Each must be tackled from a proactive explicit perspective. Let us take, for example, an identity perspective. In the virtual organization era, a firm is involved with many other firms in many networked projects and ventures, while product life cycles shrink and technological development accelerates. In such a setting, simply stating what is the firm's mission, vision, core competencies and markets is an on-going challenge which must be tackled with in order to insure cohesiveness, and velocity of the firm's actions, and for developing and maintaining a sense of belonging, sharing and commitment among its actors. In the remainder of this short paper we focus on the enterprise transformation process in the context of virtual organization.

The enterprise transformation process

As shown in Figure 3, the *enterprise transformation process* can be decomposed into three key subprocesses: the *identity rethinking process*, the *process reengineering process* and the *network reorganization process*. The prefix *re* is used here to emphasize the continuous transformation from the current status.

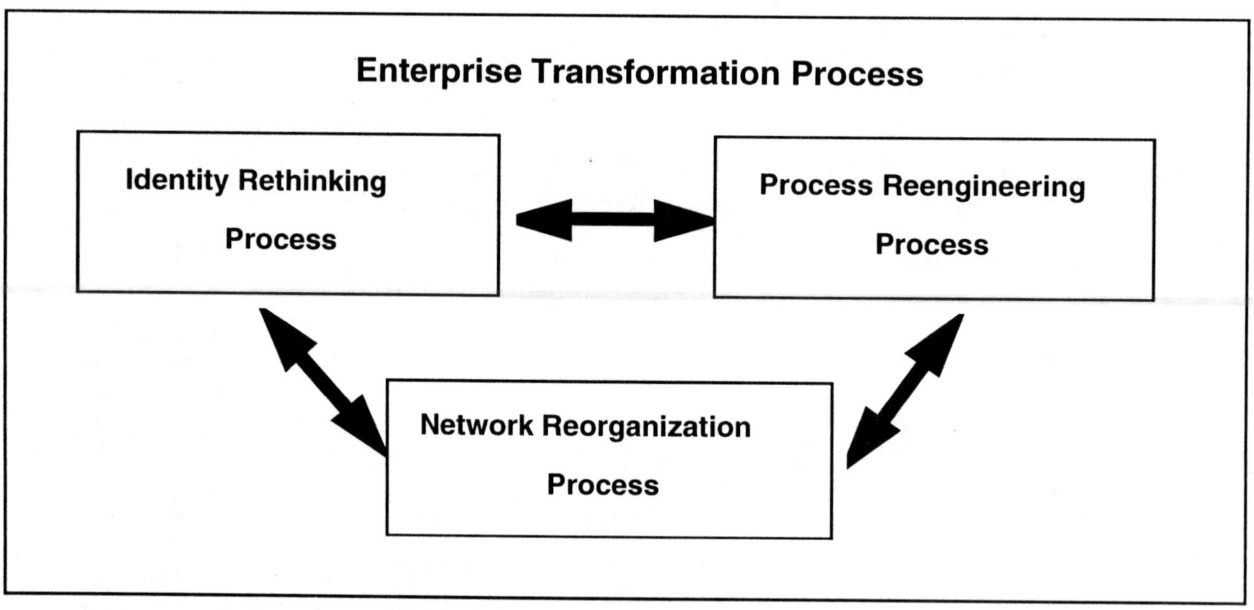

Figure 3. The enterprise transformation process and its main identity rethinking, process reengineering and network restructuring subprocesses

Identity rethinking process

Identity rethinking is a strategic continuous process. Keidel (1994) defines it to generate a revised statement of the identity, values, mission, vision, practices and experiences of the enterprise. Simply put, it means to describe what the firm has been since its infancy, what it currently is, and what it is envisioning itself to become. This implies a deep collaborative understanding of the enterprise and its stakeholders.

Restating the identity of the enterprise deals with the soul of the enterprise. Such a deep level process serves two key needs : rationalization and socialization. These needs relate to the multi-faceted perceptions of the firm's identity by the various stakeholders. The firm has to recognize that these stakeholders mold the firm's identity by virtue of their association with the enterprise being linked to how well it actually satisfies a target set of their needs, and how they perceive the evolution of the satisfaction of these needs. Identity setting has direct impact on survival and prosperity. The identity of the firm must therefore be rationalized.

The firm's mission and vision serve as the blueprint and inference engine from which decisions are to be based. These decisions are to be taken by the firm's network of actors, according to the ongoing business processes. Each of the actors (personnel, vendor, customer, business partner, etc.) must understand clearly the relationships between its actions/decisions and the mission and vision of the enterprise. Each must approve and encourage the development of the firm's identity. Socializing the firm's identity through proactive dissemination and interactive win-win adjustment is a key process in the virtual organization age where hard hierarchical relationships do not prevail.

Process reengineering process

In our enterprise model, we clearly differentiate between verbs and nouns. The verbs are the activities, operations and projects : what we generically refer to as processes in this paper. Each process can be recursively defined in terms of interacting subprocesses. A firm launches, manages and participates in many types of processes. There are hard manufacturing processes, like alloy melting processes in casting factories; order delivery processes, encompassing all activities from the moment a need is expressed by a customer to the time at which this need is fulfilled; product development processes, from idea generation to market penetration; product life cycle management processes; financial resources management processes; partner selection processes; project management processes; and so on. In general processes are not about who does it but rather about what is being done.

The nouns are the actors : humans, resources, processors, computers, software, virtual agents, machines, cells, departments, divisions, enterprises, organisms, and so on. A set of interacting actors is termed a network. Depending on the perspective, the firm deals with many types of networks. Along a hard material/information flow perspective they range from geographical transportation networks to internet/intranet based web networks. From a value creation perspective, there are processor networks within a production center, factory networks, project networks, supply networks, distribution networks, and so on. There are also business networks involving multiple enterprises, from long-term partnerships to virtual webs. Poulin et al. (1993, 1994) characterize these networks according to a variety of factors: stable Vs dynamic, rigid Vs flexible, formal Vs informal, hierarchical Vs heterarchical, open Vs closed, local Vs regional or global, etc. In our model each actor can always be perceived as a single entity or as a network of lower-tier actors. As shown in Figure 4, responsibility for launching, operating and managing processes is assigned to actors/networks.

Business process reengineering has now been high fashion for a few years (e.g. Hammer & Champy 1994, Manganelli & Klein 1994). The emphasis has been on a radical reconception and redefinition of business processes that can bring drastic improvements in performance such as cost, quality, service and efficiency. This stage has been the infancy stage of business process management and engineering. For many firms, the reengineering project has been the first time ever they have taken a hard look at their processes and questioned their appropriateness, specially in the presence of fast-evolving information technologies.

For enterprises to thrive in the virtual organization era, we believe that they have to expand drastically their capabilities and activities related to process *re*engineering since from now on processes and networks have to be understood as distinct yet highly interrelated. As new business opportunities are deciphered, process learning pays off, technological breakthroughs are achieved, the firm has to reengineer its processes. In project management, this is a well known paradigm. For every project, the common practice first emphasizes defining the project scope (identity), then the work breakdown structure and project activity network (process), followed responsibility assignment to project teams and subcontractors (project network), and finally activity schedule planning and management (switching to the value creation process).

Decoupling processes from networks has the clear advantage of forcing the enterprise to define the mandate of the process, its informational and physical input requirements, its planned outputs, its performance metrics, its planned life cycle, the capabilities required by actors for process responsibility assignment, and so on, before the network of actors is actually determined precisely. When actors take the responsibility for a process, they know precisely what they have to expect and to deliver. This is key when actors may range from an internal team to a company on the other side of the planet, and when processes evolve rapidly.

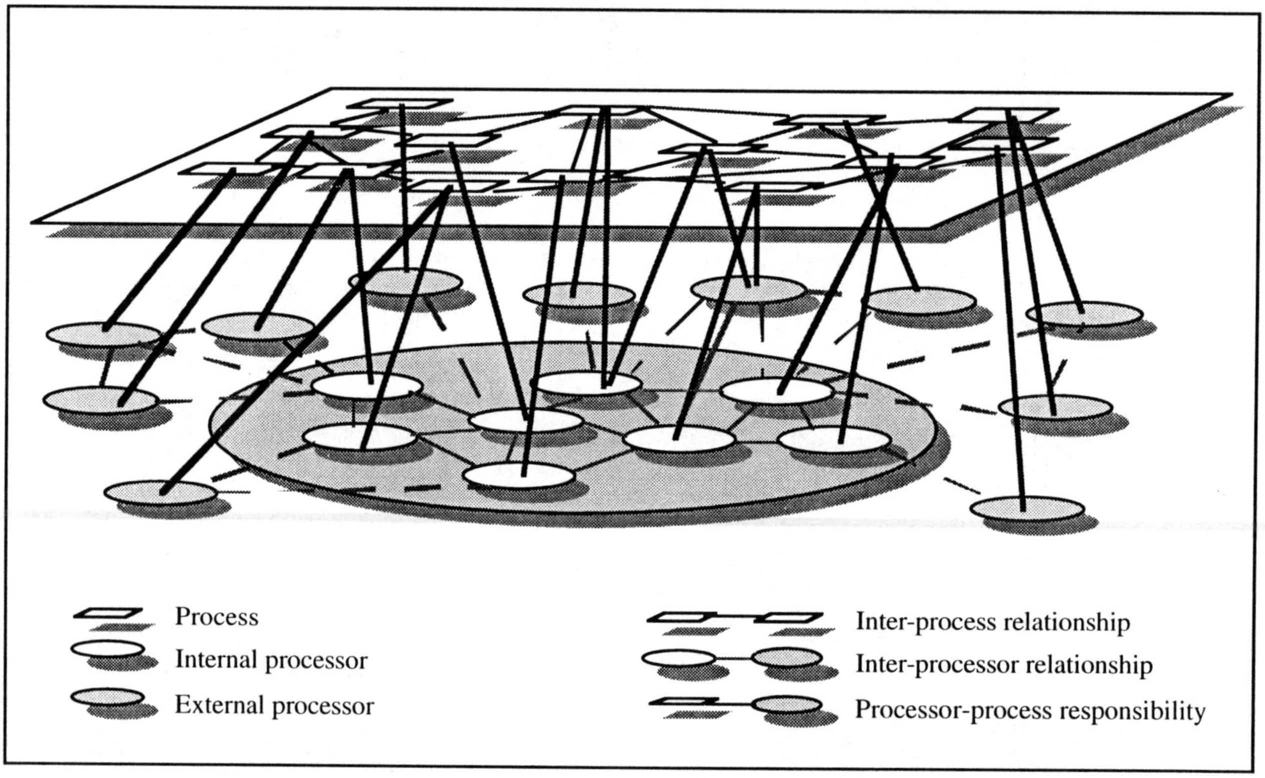

Figure 4. Processor responsibility for a set of processes

Network reorganization process

Dynamically organizing the firm in terms of networks including both internal and external actors has also been a strong trend for the last decade (e.g. Miles & Snow 1986 & 1992, Child 1987, Montreuil 1991, Montreuil et al. 1992, Poulin et al. 1994). Industry has witnessed a rapid growth of alliances, consortiums, coenterprises, multi-enterprise business networks, supply chains, industrial clusters and districts, electronic commerce and more recently virtual webs (e.g. Goldman et al. 1995).

In a generic way, Figure 4 represents a partial instant image of a firm's global network. Enterprises have now to become masters at dynamically transforming their set of networks. They constantly have to strategically manage their set of actors/processors/partners, manage their relationships with each of these, manage the overall configuration of their networks (Poulin et al. 1994). They have to develop strategic capabilities for understanding and deciding when and how to develop mutually rewarding long-ranging relationships, and when and how to take advantage of internet-style access to global networks of customers, vendors, subcontractors and partners.

This networking has to take place at all levels of the enterprise, from product packaging to strategic deployment and financing of major projects. For every process, the firm has to decide whether it does it itself using its internal network, has it done by its external network, does it together with partners in synergy between its internal and external network. For each actor in each of its networks the firm has to dynamically engineer and manage its relationships so that both the actor and the firm get as high as possible a degree of satisfaction of their targeted needs, both gaining value for their stakeholders through the relationships. The challenge is to put together the infrastructure for rapid opportunity based value creation, for continuous innovation through networked intelligence, for proactive durable enterprise realization.

Conclusion

Enterprises are now facing tremendous pressure for concurrent performance in satisfying the ever refined needs of their stakeholders, including customers, financial partners, leading personnel, and so on. In this paper we have introduced a process model of the enterprise having at its core a fundamental ever-active *enterprise realization* process. This core process is composed of three main subprocesses: the *enterprise transformation process*, the *organizational learning process* and the *value creation process*. Where the transformation process was in the past a big-event project, in our model for the enterprise operating in the forthcoming virtual economy it becomes an ever-active strategic process itself decomposed in three main subprocesses: the *identity rethinking process*, the *process reengineering process* and the *network reorganization process*.

Our model is recursive in nature. Everywhere we have written *enterprise*, one can replace it by business unit, cell, and individual. In the virtual organization, every actor will have to effectively manage its realization process. Each will have to continually reflect, decide and act on the development of its identity, processes and networks.

Acknowledgments

Thanks to the Human Sciences Research Council of Canada for partially funding this research through a strategic grant on the international competitiveness of the network enterprise.

References

Child J. (1987). Information technology, Organization and the response to Strategic Challenges, **California Management Review**, v30, no1, pp. 33-50.

Goldman S.L., R.N. Nagel & K. Preiss (1995). **Agile Competitors and Virtual Organizations: Strategies for Enriching the Customer**, Van Nostrand Reinhold.

Keidel R.W. (1994). Rethinking Organizational Design, **Academy of Management Executive**, v8, no4.

Hammer M. & J. Champy (1994). **Reengineering the Corporation**, Harper Business.

Manganelli L.R. & M. Klein (1994). A Framework for Reengineering, **Management Review**, June, 10-16.

Miles R.E. & C.C. Snow (1986). Organizations : New Concepts for New Forms, **California Management Review**, v28, no3, pp. 62-73.

Miles R.E. & C.C. Snow (1992). Causes of Failures in Network Organizations, **California Management Review**, v34, no4, pp. 53-71.

Montreuil B. (1991). The Industrial Microcosm ..., **Proc. of Internat. Conf. on Computer Integrated Manuf. ICCIM 91**, Singapore, World Scientific Pub., pp. 9-13.

Montreuil B., P. Lefrançois, A. Ramudhin & S. D'Amours (1992). A Conceptual Introduction to Symbiotic Manufacturing Networks, **Proc. of Internat. Conf. on Economics, Management & Information Technology**, Tokyo, Japan Society for Management Information, pp. 505-508.

Poulin D., B. Montreuil & S. Gauvin (1994). **L'Entreprise-Réseau : bâtir aujourd'hui l'organisation de demain**, Publi-relais, Montréal, Québec, Canada.

Tapscott D. (1995). **Digital economy : Promise & Peril ... Age of Networked Intelligence**, McGraw-Hill.

Instrumenting Product Development Activities in the Virtual Enterprise

Jeffrey A. Morrow
(206) 463-1855

Industrial Engineering - FU-20
University of Washington
Seattle, Washington 98195 USA
jfmorrow@u.washington.edu

This paper reports on contributions to the methodology of product development (PD) research, specifically that for investigating certain aspects of PD in the virtual enterprise. Means of characterizing consensus states of a PD group using personal computer-based versions of cognitive anthropology methods are described. Described are "drag and drop" interactive pile sorts and triads tests. *Keywords: product development research: product development processes; virtual enterprise, pile sort; triads test; cognitive anthropology*

0. Introduction

This paper reports on contributions to methods for understanding product development (PD), specifically that for investigating certain aspects of PD in the virtual enterprise. For the purposes of this paper, we construe "virtual enterprise" as follows: most PD resources, including human resources, are typically not "owned" by the PD entity -- the developer entity grows and shrinks in size over the PD cycle, adding and shedding quasi-independent resources as needed. In the virtual enterprise, while PD work can occur in a decentralized fashion (often spread around the globe and around the clock), the developers consider themselves a "team," at least for the duration of any particular product development cycle. Finally, it is assumed that the enterprise is "well connected," with high enough PC to staff ratios, good email, reliable Internet access, and computer literate staff.

We place few restrictions on the nature of the product developed: almost any collective effort to analyze market demand and synthesize a profitable response is accommodated. To do so, an appropriate (relatively high) level of abstraction of the PD process must be assumed. We take the perspective that the work of PD teams can be split between design accomplishments and execution accomplishments.

With regard to design accomplishments, the PD team must solve a stream of highly variegated constrained design problems during the PD cycle (e.g., from negotiating the PD schedule and resourcing to finding a concept design to debugging the shipping process). Yet the cleverness of any collective design does not suffice that it succeeds: the development team must execute its clever designs with fidelity and energy. A bit of good fortune may be necessary, too.

When designing solutions for mathematical word problems as students, we all experience the special kind of breakthrough that occurs when shifting to a more subtle (now obvious) perspective after having struggled with an inadequate one. In the same vein, Albert Einstein is said to have claimed that "Perspective is worth 20 IQ points." More recently, Page has shown that for a certain class of search problems, any paramount solution that can be reached by exhaustive application of a search strategy can be found by correct rotation of the search space. (**Page, 1995**). There is also a large and growing literature regarding the importance of accommodating the perspectives of all the stakeholders in the PD process.

Further, concurrent or simultaneous engineering clearly brings diversity of perspective to product development if only by having representatives from relevant functional organizations operate as a team. Yet diversity of perspective does not seem

0-7803-3552-X/96/$5.00 © 1996 IEEE

to represent an unalloyed good in product development. Dougherty describes how the various and closed "thought worlds" of functional organizations thwart communication and generally are associated with PD failure in her sample. (**Dougherty, 1994**)

It seems likely that successful PD requires balance between diversity of perspective and ability to execute in an aligned and energetic fashion. Clausing suggests that a prime benefit of expert PD team use of Quality Function Deployment (QFD) comes in the strong consensus generated. From Clausing's perspective, strong consensus yields coordinated and timely PD team execution. (**Clausing, 1994**) Of course, the other prime benefit of expert QFD use is that it sifts and harmonizes great diversity of perspective very reliably.

What we seek, then, is to track the state of consensus of the PD team over the PD cycle. More explicitly, we define "achieving consensus" as reduction in "diversity of perspective" over time. More subtly, we expect that PD teams starting with greater diversity of perspective around *any* design problem, and having greater ability to achieve consensus despite initial diversity of perspective, will more reliably develop good products. If this is so, then the large investment required to become expert at QFD may be warranted.

From the literature of cognitive anthropology we find methods that characterize, among other things, locus and dispersion of cognitive mental models ("schemas," "frames," etc.) held by individuals in a group or culture. (**D'Andrade, 1995**) These methods include the pile sort, triads test, and structured interviews. Each of these methods tends to be very researcher-labor intensive and requires researcher physical presence for administration. This drawback only worsens when, as we require, locus and dispersion must be tracked over time, thus necessitating multiple reassessments. As classically administered, these methods are probably too burdensome to use for investigating product development done by a globally distributed virtual enterprise PD team.

This paper consists of 4 sections. Following this Introduction is Section 1, in which the pile sort and triads test are described and their analyses reviewed briefly. The linkage between the tools' applications in cognitive anthropology and in PD research is also made explicit. Section 2 outlines a proposed schedule of administration of the instruments within a PD cycle and considers a number of analytic tacks. Section 3 summarizes the paper.

1.0 The Pile Sort, Triads Test, and their use in understanding product development

The pile sort and triads test are used in cognitive anthropology to characterize the way a group of individuals thinks about some aspect of shared reality. Cognitive anthropologists hold that individuals within some sort of culture share (mostly pre-conscious) schemas, or mental scaffolding, for understanding the world in which they cohabit. Classic applications of the pile sort and triads test include investigating kinship terms and their perceived relationships, human personality temperaments and their perceived relationships, ethnic diversities and their perceived relationships, diseases and their perceived relationships, and societal risks and the perceived relationships between types and kinds of risk. (**D'Andrade, 1995**)

A typical pile sort (**Boster, 1993**) works as follows: a deck of plain cards, each having one concept written on it, is given to the informant and the informant is instructed to make as many piles of "related" or "similar" cards as she wishes. In the PC version, a screen is presented with the concepts printed in boxes that have been randomly placed in the startup format. Pile production is accomplished by a series of "click-drag-drop" moves. When the informant finishes this first "free" sort, pile membership is recorded (see Figure 1).

PC screen display of PD Pile Sort after first, "free," sort

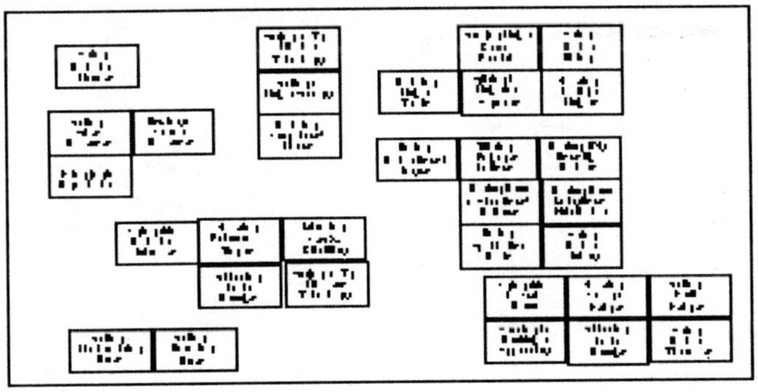

Figure 1

In the PC version, pile membership is written to a file. The informant is then asked to create an additional pile by dividing one of the piles she created in the free sort into two piles. When she finishes, pile membership changes are recorded. The informant continues to split piles in a stepwise fashion, and the researcher to record pile membership, until all cards are in individual piles. Then the piles are returned to the state of the initial free sort. The informant is asked to create one fewer pile by combining two piles from the initial sort. The informant continues to combine piles, and the researcher to record pile membership, until all the cards are in one pile.

From these pile data it is possible to compute Euclidean distances between each pair of concepts in the conceptual schema of the informant. It is also possible to compute the center or locus of the informant's schema. If one administers pile sorts to a number of informants, it is possible to compute mean distances between concepts, a grand locus, and the variances of individuals around those means. It is also possible to project the Euclidean distances between concepts from the n-dimensional space of the schema (where n is the number of concepts) into a 2 or 3 dimensioned space. Of course, in doing so, some distortion of distance occurs (it is an optimization problem). When the distortion is "acceptable," cognitive anthropologists draw inferences regarding the meaning of the principal axes of the projection. We will not need to do this, however, as our interest extends only to characterizing how the variance or dispersion in schema changes over time i.e., in the way diverse perspectives coalesce into consensus via team work.

Because we wish to track change in dispersion of schema over time, we must minimize the so-called "learning effects" due to repetitive administration of the instruments. In the next section we discuss various tactics to minimize learning effects, but a strategic choice is to use two "different enough appearing" but functionally equivalent techniques in some alternating fashion. In this case, we may use a triads test, which also allows estimation of Euclidean distances between elements of schema. We mediate the triads test via PC also. When administered, the informant is presented with a series of schema element triplets and asked to indicate which of the three most differs from the others. Triads tests may be administered as "full factorial" investigations, but such requires a combinatoric explosion (O(n!)) of informant discrimination, and are therefore often administered in fractional factorial experimental designs to a larger sample of informants. PC mediation of the triads tests will support simple but careful data collection and hence allow more complicated experimental designs.

2.0 Administration Protocol Issues

We wish to minimize "learning effects," "Hawthorn effects," and the cognitive burden placed on informants of putting up with relatively frequent and taxing test administrations. We also wish to educe elements of both collective and individual schemata within the PD team and must, therefore, avoid "putting ideas in their heads." Accordingly, the first task of the researcher is to draw from the team the elements of PD-salient schemata. It is proposed that this be accomplished by asking team members to, as individuals, "brainstorm" a list of "All the things we have to do in this PD." Rules for this brainstorming limit informants to only one concept (schema element) per identification, fewer than a dozen words to identify any single element, and require at least one noun and one verb in each brainstormed identification. Where necessary, iterative repair of violations will occur. Almost all of the investigation will be conducted with informants as individuals.

First, each team member will sort his own list of brainstormed schema elements in order of decreasing importance. Then each team member will pile sort his list. Subsequently, an edited condensation of all of the PD team's schema elements (one that spans members' identified elements but removes duplications), will be presented for sorting into order of decreasing importance. Finally, a triads test of the edited condensation elements will be conducted, structured as some suitable experimental design.

Depending upon the compliance of the PD team, the apparent rate of change of diversity of perspective over time, and the PD schedule, it may be appropriate to synchronize this sort of schema-educing and tracking cycle with the PD schedule. For example, if the PD process includes phased activity gated by reviews, review of the team's consensus trends might be of interest. Further, after a gate has been passed, it may be time to reinvestigate what seems important and what seems connected and what seems dispersed.

3.0 Summary

If good tricks solve problems, and good tricks sometimes come from a different perspective, it is well to have some diversity of perspective. If diversity of perspective threatens cohesive, coherent action, effective execution will be hampered. PD teams design problem solutions all the time. Further, PD team members work in non-team settings the majority of their time, even when doing the PD team's work. Members out of the consensus are less likely to take action that furthers the PD team most. Can PD teams have both diversity of perspective and ability to achieve consensus reliably and rapidly--to find clever solutions and execute with discipline and despatch? How might we measure this? We suggest that pile sorts and triads tests may work.

References

Boster, James, "Successive Pile Sort Procedure," course handout from E. Hunn's graduate ethnomethodology class.

Clausing, Don, personal communication, June 1994.

D'Andrade, Roy, *The Development of Cognitive Anthropology*, Publisher: Cambridge University Press, New York, 1995.

Dougherty, Deborah, "Interpretive Barriers to Successful Product Innovation in Large Firms," *Organization Science* V3 no. 2, May 1992, pps. 179-202.

Page, David,"Perspective, Fitness, and Learning," presentation given at the Santa Fe Institute on August 18, 1995

Managing IT Implementation in Virtual Enterprises

Bjørn Erik Munkvold
Tel. +47 38 02 50 55

Telenor AS / The Norwegian University of Science and Technology
c/o Agder Research Foundation
P.b. 2074 Posebyen
4602 Kristiansand
Norway
e-mail: bjoern@ics.uci.edu

The paper presents a framework of factors influencing information technology (IT) implementation in virtual enterprises. The framework is based on previous research on IT implementation, and field studies conducted in virtual enterprises. The aim is to increase the understanding of the implementation process in this organizational context, and the framework is intended as a basis for developing strategies for successful management of this process.

Introduction

The virtual enterprise is one of the new types of organizational forms emerging as strategic options for facing the challenges from global competition and dynamic markets. However, the term lacks a clear definition, and is used differently by different researchers and practitioners (Munkvold, 1995). Here the virtual enterprise is defined as "*...an enterprise created by linking, via IT, businesses, suppliers, customers and/or partner companies to take advantage of specific business opportunities*" (Douglass, 1993, p. 4). This definition distinguishes three vital characteristics of the virtual enterprise: a reliance on use of IT for linking the actors together, interorganizational collaboration, and the temporal nature of the enterprise. IT is used here in a broad sense, comprising both hardware, software and telecommunications technology.

In its ideal form, the virtual enterprise is a streamlined organization, involving seamless integration across geographical and organizational boundaries. However, to reach this stage, an IT infrastructure supporting distributed teamwork needs to be established. Successful management of the implementation of IT for supporting the virtual enterprise is therefore critical. Implementation is defined here as the process of introducing the technology in the organizational context where it is to be used. As will be discussed in this paper, there may be both organizational and technological barriers in this process.

The paper presents a framework of factors that may influence the implementation process in virtual enterprises. In this we draw from our own findings from case studies of IT implementation in distributed organizations, combined with previous empirical research on IT implementation. The aim is to contribute to a better understanding of how the IT implementation process in virtual enteprises can be managed.

Previous empirical research on IT implementation

There are few empirical studies of IT implementation in virtual enterprises. This section gives a brief summary of results from different areas of IT implementation research, that are relevant for understanding the process in this organizational context.

Research on *intraorganizational IS implementation* has identified a number of factors influencing the implementation process. The factors most frequently appearing are management support, championship, planning, resource allocation, training, technical support, users' IT skills and experience, users' expectations, complexity and relative advantage of the technology, and compatibility with existing values, experiences and users' needs (Kwon and Zmud, 1987).

Besides these factors, research on *implementation of interorganizational systems (IOS)* has focused on environmental factors. For instance, high competition intensity in the operating environment of the enterprise is found to be an important driver for adoption (Grover, 1993). *CSCW (Computer supported cooperative work)* research on groupware has illustrated how this type of technology may be more difficult to implement than single-user or mainframe systems, because it needs to be adopted by every member of a group to be successful (Grudin, 1989). A common finding in these studies is that the users need to have some shared understanding of the cooperative nature of the technology, to be able to realize its potential benefits (Grudin, 1989; Orlikowski 1992).

When implementing technology for supporting teamwork, it is also necessary to focus on characteristics of the team, e.g. team members' attitudes towards task, technology and the cooperative arrangement, and the team's previous experience (Applegate, 1991). Research on *computer-mediated communication (CMC)* indicates that users are capable of adapting to using CMC in distributed teamwork, after gaining some experience with the technology.
However, in virtual enterprises the time frame of the tasks will often be limited, leaving short time for building experience with the technology (Munkvold, 1996).

Research on *interorganizational relations* provides a basis for understanding the relations between the actors in the virtual enterprise. This constitutes an important part of the context for the implementation process. The strategic orientation of the cooperation, i.e. vertical or horizontal in the value chain, determines potential competition between the actors. The relative size of the firms may indicate power asymmetries and possible difference in the strategic importance of the cooperative arrangement. Cultural distance between the firms may also act as barriers against implementation (Doz, 1988).

Results from case studies

This paper also draws upon results from case study research of IT implementation in four organizational contexts, as presented in table 1.

Type of enterprise	Number of sites	Technology implemented	Tasks supported	Type of study/ methods used
Network of building constructors	4 (national)	ISDN, e-mail, file sharing	Electronic tendering	14 months field study/ interviews, observation
Network of telecom suppliers for offshore industry	6 (national)	Lotus Notes	Electronic tendering	12 months field study/ interviews, observation.
Corporate R&D in multinational engineering group	8 (international)	Lotus Notes	Process support for research projects	Historical case study/ interviews, doc. analysis
Multinational engineering group	195 (international)	WAN, e-mail, Intranet	Information sharing	Historical case study/ interviews, doc. analysis

Table 1. Case studies of IT implementation in distributed organizations.

The two organizational networks conduct interorganizational collaboration projects of a temporal character, and may thus be defined as virtual enterprises. The corporate R&D department conducts research programs involving all branch offices, independent of geographic location. The large engineering group comprises several independent organizations, with different constellations of these collaborating on each project. Each case includes the implementation of technology for supporting distributed collaboration. The format of this paper only allows a short presentation of the main findings from these cases.

Large geographical distance in combination with task and organizational complexity, i.e. need for frequent interaction and communication (Killing, 1988), were found to be factors facilitating the adoption of communication technology, by creating high user expectations of saving time and money in this process. Several factors were identified as barriers to the implementation process: technological incompatibility, lack of resources for follow-up, lack of management support, lack of champions, key person dependencies, lack of routines, and entry costs. Other factors that proved to be important are the timing of installation and training, local market conditions and relative size of the organizations.

In each case, integrating the technology with existing platforms and software proved difficult, thus creating delays in project schedules. The compatibility factor identified in previous research therefore should also include technical compatibility. The timing of the installation and training was identified as being critical, in that it needed to be coordinated with the project schedule of the virtual enterprise, and other projects in the participating organizations, to get access to the necessary resources. The ideal time for this seems to be before the actual start-up of a scheduled project, when the activity level is still low and time can be spent on learning to use the technology.

Without sufficient resources for close follow-up of the implementation process, there is a risk of each organization prioritizing individual projects. This was found to be dependent on the relative size of the firms, strategic importance of the collaboration, management support, local champions, and the supply of projects in the local markets. The projects studied proved vulnerable to changes in key personnel, both in the participating organizations and the technology vendor, indicating a need for stability in the implementation project organization. Lack of new work routines including detailed allocation of responsibility also hindered the implementation process.

The entry costs of adopting the technology may also serve as a barrier. Due to this, the implementation team in the large engineering group chose to cover some of the installation costs of the communication network, to establish a critical mass of users.

A framework of factors

The previous sections have presented a number of factors believed to be influencing theimplementation of IT in virtual enterprises. By grouping these in main categories, a framework of factors can be formed, that may be used as a basis for developing strategies for managing the implementation process (fig. 1).

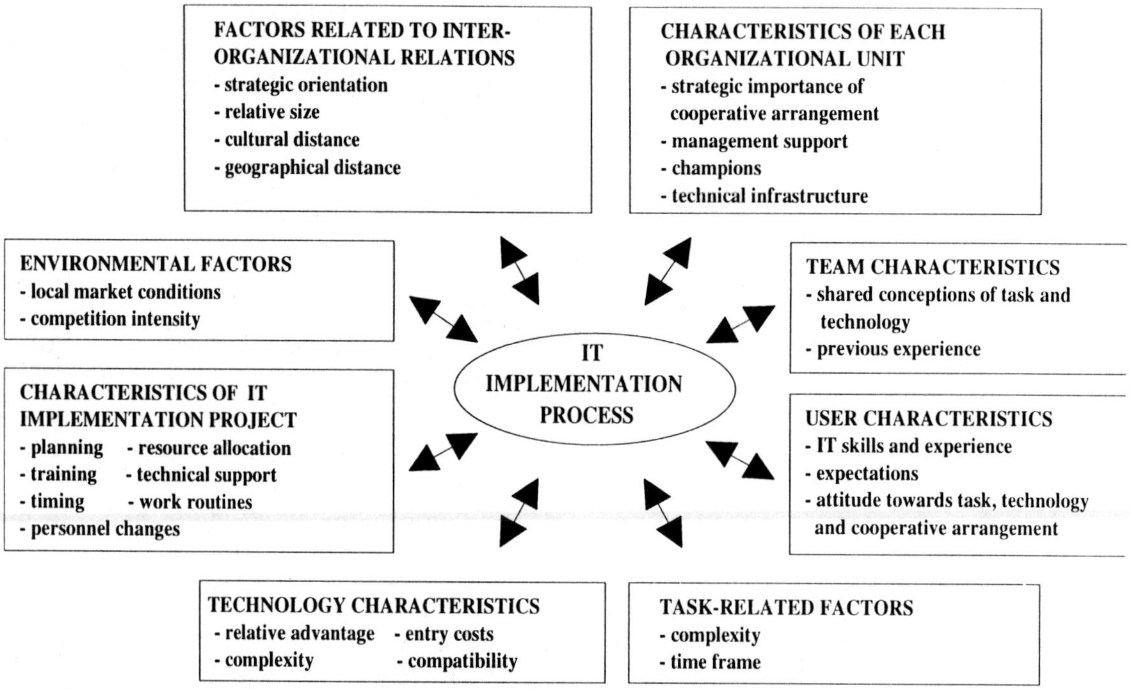

Fig. 1. A framework of factors influencing IT implementation in virtual enterprises

The figure illustrates the reciprocal nature of the implementation process, involving a mutual interaction between the technology and the context in which it is introduced (Leonard-Barton, 1988). Clearly, the degree of influence of each factor will vary between each context. The main purpose of this framework is to draw attention to the different types of factors that need to be considered when planning these projects.

Strategies for managing IT implementation in virtual enterprises

It is not possible to develop a general implementation strategy, utilizing all factors in the presented framework. The strategy needs to be adapted to each context, where the relative importance of the different factors will vary. The following guidelines should be regarded as examples of important elements in an implementation strategy, developed on the basis of the framework.

The implementation team needs to include representatives from each participating organization, to ensure involvement from every actor. The team's mandate must allow for making decisions regarding the implementation process, and effectuating these in each organization. Sufficient resources need to be allocated for close follow-up of the implementation process. A person in each organization needs to be identified as responsible for reporting incidents that may affect use of the technology, e.g. personnel changes, and changes regarding other technology used. If possible, local champions should be identified and used to stimulate adoption in each organization. The internal marketing of the technology must focus on relative advantage and benefits to be expected from its use, e.g. more efficient communication, and reduced time and costs for travels and meetings. However, it is important not to create unrealistic user expectations. In cases where entry costs are hindering the implementation process, steps for lowering these to establish critical mass needs to be considered.

Careful planning of installation and user training is necessary. Because of the dynamic nature of virtual enterprises, the time available for these activities may be limited, thus making correct timing critical. All implementation activities need to be coordinated with the project schedules of the virtual enterprise and each participating organization. A thorough charting of the technical infrastructure in each organization will enable detailed planning of how to link these environments, and may avoid unforeseen problems. Technical infrastructure here includes hardware, network configurations (e.g. LANs), and software used.

The implementation team also should supervise the development of new working routines, including procedures for information flow, and responsibilities for different activities. This includes routines for system administration, e.g. test procedures, back-up, support, security and documentation.

Training should be provided shortly after installation of the technology. In cases with many actors involved, parallel training sessions should be considered to avoid a delay in the adoption between different user groups, thus hindering the establishment of a critical mass of users. Besides functionality and new work procedures, the training may also need to cover some elements of group dynamics and how to work in distributed teams.

Conclusion

The integration of multiple organizations collaborating on temporal projects, makes IT implementation in virtual enterprises more complex than other forms of IT implementation. Careful planning and close, continuous follow-up are critical for succeeding in this process. By identifying different factors influencing the implementation process, the framework presented in this paper may be used as a basis for developing effective strategies for managing IT implementation in virtual enterprises.

References

Applegate, L.M. "Technology Support for Cooperative Work: A Framework for Studying Introduction and Assimilation in Organizations." *Journal of Organizational Computing*, 1: 11-39, 1991.

Douglass, D.P. "Making Alliances and Partnerships Work." I/S Analyzer, 31(10): 1-144, 1993.

Doz, Y.L. "Technology Partnerships between Larger and Smaller Firms: Some Critical Issues". *Cooperative Strategies in International Business*, Publisher: Lexington Books, Massachusetts, 1988.

Grover, V. "An Empirically Derived Model for the Adoption of Customer-based Interorganizational Systems." *Decision Sciences*, 24(3): 603-640, 1993.

Grudin, J. "Why groupware applications fail: problems in design and evaluation."*Office: Technology and People*, 4(3): 245-264, 1989.

Munkvold, B.E. "*The structure of tomorrow's organizations.*" Scientific report 46/95,Telenor R&D, Kjeller, Norway, 1995.

Munkvold, B.E. "An Assessment of Selected Empirical Research on Computer-Mediated Communication, Related to Distributed Teamwork in Concurrent Engineering." *Proceedings of Concurrent Engineering 1996 Conference*, Toronto, August, 1996.

Killing, J.P. Understanding Alliances: The Role of Task and Organizational Complexity. *Cooperative Strategies in International Business*, Publisher: Lexington Books, Massachusetts, 1988.

Kwon, T.H. & Zmud, R.W. "Unifying the Fragmented Models of Information Systems Implementation." *Critical Issues in Information Systems Research.*Publisher: Wiley & Sons, Chichester, 1987.

Leonard-Barton, D. "Implementation as mutual adaptation of technology and organization." *Research Policy*, 17: 251-267, 1988.

Orlikowski, W.J. "Learning from Notes: Organizational Issues in Groupware Implementation." *CSCW' 92 Proceedings*. Publisher: ACM Press, Baltimore, 1992.

IT-Enabled Affiliations as Organizational Forms

Barrie R. Nault
(714) 824-8796

Graduate School of Management
University of California, Irvine
Irvine, CA
92717

Albert S. Dexter
(604) 822-8378

Faculty of Commerce and Business Administration
University of British Columbia
Vancouver, BC, Canada
V6T 2Z2

We have found that information technology allows the creation of virtual enterprises - thereby enabling new forms of organization. This paper addresses the problem where expertise about customers and physical assets are separate. Traditional organizations couple these two aspects, often overlooking the contribution externalities make to incentives. We describe how IT-enabled affiliations can employ the externalities to improve incentives and become more profitable.

Introduction

This work describes the generalization of our existing stream of research on how information technology (IT) can enable new organizational forms. That stream was initiated by our work on franchise organizations in the commercial fueling industry (Nault and Dexter 1994), where we concentrated on local investment (e.g., local customer recruitment) by franchises. Because there are positive externalities between franchises that are not accounted for in traditional franchise organizations, underinvestment at the franchise level occurs (Katz 1989). In our model, IT-enabled transfers between franchises increases the level of franchise investment past the levels in traditional franchise organizations, partially accounting for the inter-franchise externalities. As a result, the franchisor becomes more profitable. We characterize the IT-enabled organization form as "ownership of customers" because those franchises that generate customers through their local investment efforts are rewarded through the transfers when those customers make purchases elsewhere in the franchise network.

We extended that model to include global investment (e.g., national promotion) and found the results on local investment continue to be true, and underinvestment in global investment is also mitigated by IT (Nault 1996b). In addition, we provided conditions under which both the franchises and the franchisor are made better off through this use of IT (e.g., Pareto improvement) (Nault 1996a), and explored the implications of ownership of customers on the location of decision rights in an organization.

The present work focuses on generalizing the types of organizational forms to which our concepts of ownership of customers and inter-participant transfers can be applied. As our analyses suggest, the underinvestment problem that occurs in franchise organizations persists in many other hybrid organizational forms that lie between the extremes of hierarchy and market. In many of those forms, IT-based ownership of customers and transfers can be used to mitigate the underinvestment problem, improving the profitability of the organizational form. This analysis typifies the "move to the middle" hypothesis (Clemons et al. 1993).

The Traditional Organization

Traditional organizations couple two aspects we argue can be separated - the recruitment of customers and the provision of locations from which customers make

0-7803-3552-X/96/$5.00 © 1996 IEEE

purchases. The traditional organization proceeds to recruit customers solely on the basis of selling goods and services to them from its own locations. Therefore, incentives for customer recruitment are only derived from profits made locally. Ideally, an organization prefers to incent customer recruitment from purchases made at a wider set of locations, thereby increasing the incentive for recruitment. In the extreme, the organization wants to incent customer recruitment from profits made across the population of locations, thereby approaching the incentives that arise in a fully integrated operation. That is, these incentives result in levels of investment that are first-best.

Unfortunately, the expertise and specific information required to get the most out of investments in customer recruitment reside with individuals that are not necessarily principals of the organization. Although the principals may be able to estimate the returns to investment in customer recruitment, they cannot write a contract that details how these investments should be implemented. The solution for the traditional organization has been to couple customer recruitment and provision of location in order to partially resolve the investment incentive problem.

However, with the emergence of modern computing and communications technologies, it is now possible to create "virtual enterprises" by decoupling aspects that were previously kept together. IT enables new organizational forms through IT's ability to monitor transactions, thereby providing a basis for the substitution of incentives for hierarchical control mechanisms (Milgrom and Roberts 1992)

IT-Enabled Affiliations

The structure we outline here decouples owners of customers from owners of locations. The former we refer to as "owners" and these are agents that have expertise or specific information that can be used in recruiting a segment of customers. The key to our structure, as we described above, is that principals cannot contract for the use of expertise or specific information held by these owners. Owners of locations, that we call "outlets", own assets related to a geographical location. Owners recruit customers to join the network, and outlets sell goods and services to final customers.

In the commercial fueling industry, trucking firms prefer to purchase fuel across locations from a single provider rather than a series of completely independent fuelers. In that industry, several networks of owners of outlets have evolved. Thus, a truck stop providing fuel to network customers is an outlet, whereas an agent that recruits a trucking firm to use the network of fueling stations is an owner. The network provider tracks purchases by customers at outlets, provides billing information to the owners so that they can invoice their customers, and manages the funds transfers between owners and outlets. Individual owners decide how much investment (or effort) to put into recruiting trucking firms for the network. Individual fuel outlets decide whether to become a network location. On each purchase in the network the network provider sets a royalty from the owner to the provider and a transfer between the owner and the outlet.

We study this structure by modelling its efficiency in three parts. Sequentially, the network sets the level of incentives (royalty and transfer). Then outlets decide whether to adopt, that is, become a node on the network. Finally, owners decide their level of investment in recruiting customers. We approach the solution in reverse order. First, we examine the owner's investment problem conditional on the coverage, where coverage is defined by the number of outlets that adopt. Second we determine which outlets adopt, taking into account the owners' investment levels. Third, we study the network provider's optimization problem choosing levels of the royalty and the transfer.

We identify outlets by potential traffic volume through the outlet and owners by potential volume generated by customers they recruit. Demand at a given outlet from a given customer depends on the potential of that outlet, the potential of the customer, the investment made by the owners in recruiting the customer, and coverage.

The Owner's Investment Problem

A given owner's profit function is the premium per unit of demand, less the royalty and transfer, multiplied by the demand from a given customer at a given outlet, summed over all adopting outlets, less the cost of investment. Each owner optimizes its profits by choosing its level of investment resulting in a set of conditions, one for each owner, that describe a Nash equilibrium in investment. This equilibrium determines the level of investment for each owner as a function of the premium, the royalty and the transfer.

The equilibrium levels of investment are increasing in the premium, and are decreasing in the royalty and the transfer. This is intuitive - the higher the premium the higher the marginal return from investment in customer recruitment, and vice versa for the royalty and transfer.

The Outlet's Adoption Problem

If a given outlet joins the network it receives a transfer per unit on all purchases made by all the owners' customers at that location. It also faces an adoption cost that may be made up of lump-sum and variable components. Should the outlet not adopt, then it is unlikely that it would receive any purchases from owned customers as these customers derive control and convenience from exclusively using the network (Nault and Dexter 1995). Thus, an outlet adopts if its revenues from owned-customer transfers are greater than its costs of adoption. If demands are increasing in outlet potential, then larger outlets adopt and smaller ones do not.

The larger the per unit premium, the greater the coverage because owners invest more in customer recruitment, leading to increased demand at each outlet. The larger the royalty, the lesser the coverage for the opposite reason. Changes in the transfer have indeterminate effects: an increase in the transfer lowers investment in customer recruitment but directly increases the unit revenue at the outlet.

The Network Provider's Problem

The network provider's profits come from the royalty on all demands. The network provider must decide on both the level of the royalty and transfer to maximize profits because the transfer affects profit through the incentive for owner investment. The negative effect on owner investment from increases in the royalty exceed those from increases in the transfer because the transfer has an additional positive effect on owner demand through increased coverage.

Nonetheless, because network provider profits are a direct result of the royalty, it is not possible in general to determine whether the absolute value of the royalty or transfer should be larger. Specific results require additional information concerning the parametric forms which would apply to a particular case.

Extensions and Conclusions

The organization form we described above is isomorphic to a credit card model, and thus our structure and analysis applies there also. Our version of that model has three types of participants: a credit card company (e.g., VISA), banks that offer the credit card (e.g., Chemical Bank) and stores that accept the card for payment. Using the terminology we developed above, the credit card company is the network provider, banks are owners, and stores are outlets. Thus, banks decide how much to invest in recruiting customers to use their card (i.e., local investment). Each bank has some local knowledge of its customer base that it can use in recruitment efforts. That reflects, for example, the credit worthiness of its customers. Next, stores decide whether to be sponsored, that is, to accept the use of credit cards for purchases. Finally, the credit card company has to decide how much to give stores and how much to give banks in order to incent the stores to take the card (adopt the network) and incent the banks to recruit customers.

Our work has shown that the IT-enabled mitigation of the underinvestment problem in franchise operations can also be applied to cases where expertise and information about customers, and physical assets related to location are separate. The basic form of our model is generic - it is generalizable to many other situations. We briefly described its extension to the credit card environment above, and we believe it is applicable to electronic commerce over the internet.

References

Clemons, E.K., S.P. Reddi and M. Row, "The Impact of Information Technology on the Organization of Economic Activity: The "Move to the Middle" Hypothesis," J. MIS, 10, 2 (Fall): 9-35, 1993.

Katz, M. L., "Vertical Contractual Relations", Chapter 11 in Handbook of Industrial Organization, Volume 1, R. Schmalensee and R.D. Willig eds., Elsevier Science Publishers, B.V.: Amsterdam, 655-721, 1989.

Milgrom, P. and J. Roberts, Economics, Organization and Management, Prentice Hall: Englewood Cliffs, NJ, 1992.

Nault, B.R., and A.S. Dexter, "Adoption, Transfers and Incentives in a Franchise Network with Positive Externalities," Marketing Sci., 13, 4 (Fall): 412-423, 1994.

Nault, B.R., and A.S. Dexter, "Added Value and Pricing with Information Technology," MIS Quart., 19, 4 (December): 449-464, 1995.

Nault, B.R., "Mitigating Underinvestment Through an IT-Enabled Organization Form," forthcoming in Organization Sci., 1996.

Nault, B.R., "Information Technology and Investment Incentives in Distributed Operations," forthcoming in Information Systems Research, 1996.

Providing Competence Resources for New Product Development

Jan Österlund
+46 - 888 3 555
S-125 54 ÄLVSJÖ, Sweden

Royal Institute of Technology in Stockholm
Storhagsvägen 41

Demands for more complex product structures, of shortening development times, and higher quality require reorganization of the new product development, NPD, process. Deeply specialized groups have to provide competence resources for operative use in ongoing NPD projects, but in the same time be alert on creating new core competences for future use. This situation demands new organizational means such as a network structure and virtual teams.

Background

More complex products demands use of many various sources of competence to be cooperating in the new product development, NPD, process. Then the design work needs contributions from many deeply specialized groups. We define competence here as *the ability* to do a certain work task with the help of means and support provided by the organization. Because of the need for competence transfer, to take place within the organization, the communication network has to carry extensive flows of information. Management of R&D must focus on the resource structure and the communication system providing competence transfer required for the project process. Figure 1 depicts the trends in management of R&D.

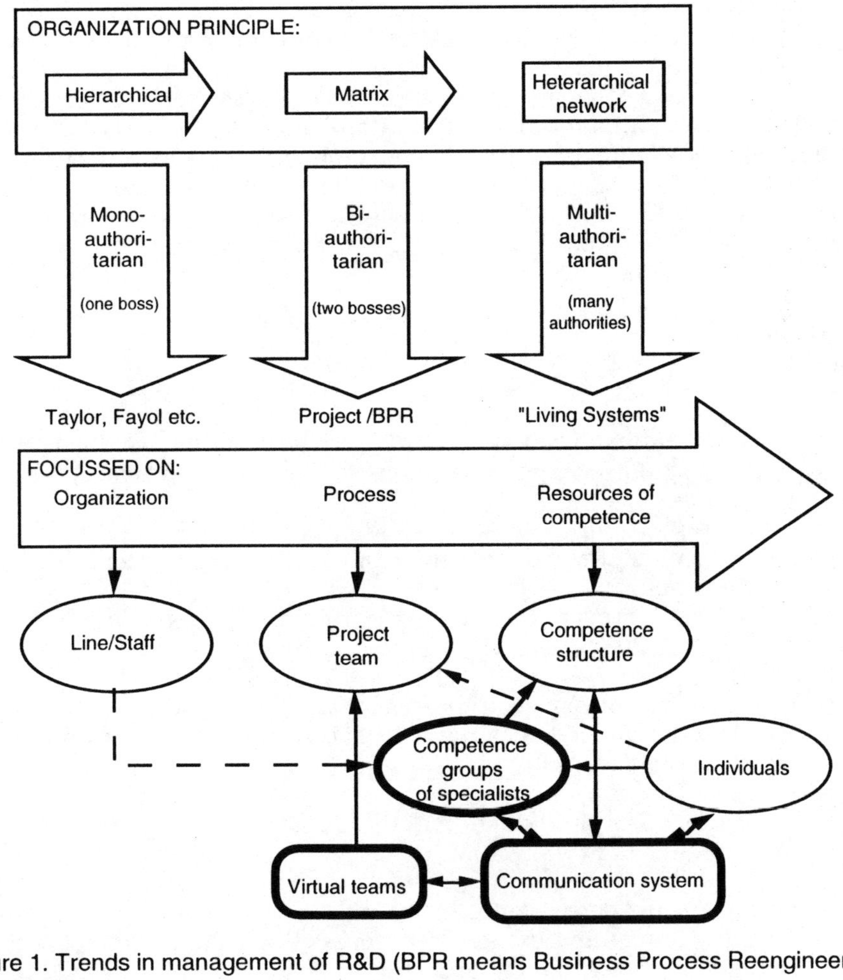

Figure 1. Trends in management of R&D (BPR means Business Process Reengineering).

R. Likert [1961] describes the hierarchical organization as a mean to arrange groups on different organization levels. Managers at each level passage must transfer the information flow between the operating competence groups at the line through transitions. Increasing flows result in information gaps and stress caused by information overload of the managers. From communication viewpoint use of the hierarchical organization is unsuitable for NPD purposes.

The usual way to solve complex design problems by placing the required specialists in separate cross-disciplinary teams causes disadvantages in resource employment according to T. J. Allen and M. S. Scott Morton [1994], and J. Österlund [1994]. The matrix organization has an inherent conflict between the project manager and the line manager, between the process and the resource interests. Who is responsible for what? Who shall do a certain activity? The project manager wants to assign the individual with the best reputation for the actual kind of work. Nevertheless, the resource manager must look to the group as whole, for its competence and efficience. Individual specialists in a project organization without continuous direct contact with the colleagues having same professions as their own, cannot contribute to the project work with more than his or her own limited competence.

The best solution is to remain the specialist working within the competence group of their own competence speciality. The group is then able to provide joint problems solutions based on the competence of all its members in a human system. Then the representatives from all groups involved, transfer their own solution into an overall task solution. Group representatives are participating in a temporary virtual team organized to solve a specific problem. Virtual because it exists on a communication base - they are not physically placed together. The communication can be done through telecommunication means as a complement to meetings. A computer supported network using fullmedia communication can connect the virtual team members together in a way resembling face to face contacts. By appointing Operators collected from different disciplines in problem solving teams as a management tool, they can achieve the benefit of interdisciplinary contacts. M. A. von Glinow and S. Albers Mohrman Eds. [1990] and D. D. Dill and A. W. Pearson [1991] have contributed to this topic. Different authors present general knowledge about interactions in a group [J. A. Shtogren, 1980, E. H. Schein, 1985, and K. D. Mackenzie, 1976].

Problem description

The main problems that R&D management must solve are:

- To create an organization of heterarchical character [Hedlund, 1986] that must be:
  - Flexible in meeting changing demands of competences through self-governing groups.
  - Learning from the process of joint problem solving.
  - Having an effective structure for coordinating use and development of competence resources.
  - Having a method for providing material and immaterial resources to its components (groups).

- To establish a management process that is clear and unambiguous in both operative and in strategical time aspects using the project form assuring core quality design.

- To foster an organization culture that is suitable to the development task and to create a climate of innovation, competence development, and flexibility.

- To support the performance of the individual humans, seen as bearers of the competence required in the NPD process, for them to be creative and innovative in their solution of technical problems.

- To engage and coordinate customers (buyers, deciders, users), suppliers, and Marketing, Production, and User Services (logistics), to cooperate in the NPD process with parallel activities aiming at right product quality and shortest possible time to market introduction of the new product.

- To use an effective communication system providing all corporate members with
  - task information
  - coordination information for project activities and resource administration
  - competence transfer between groups and teams
  - and common management information through Information Technology, IT, tools devoted to use by the human system.

Based on this problem description we divide the forthcoming discussions into following domains:
- The competence resource domain
- The operative domain of virtual teams
- The communication domain

The competence resource domain

The NPD problem-solving and design decisions require groups of monodisciplinary character. Such a group can provide creative solutions through its power of forming a living system that shares its findings to get systems benefits. Group members must have the will and ability to be the best within a single technological specialty by communicating on the same professional language. This specialty can be both of the fragmentary and of holistic character. The required variety comes from differences in personal characters, experiences, manners etc. By supporting each other in a joint dialogue the group creates high technology innovations by the driving force of being *just problem solvers.*

These operative competence groups are self-governed, acting like internal enterprises (interprises). They are direct responsible for fulfilling work packages defined by the project.

A competence resource domain consists of a network structure of interprises supported by groups of experts, both technical and administrative, and external enterprises as subcontractors where the situation requires additional competences. The structure must include resources required for ongoing NPD projects and for strategical development of core competence resources intended for use in future products. I. Nonaka [1994], and J. Österlund [1991] gives examples of competence development.

The operative domain of virtual teams

Before any final design by making drawings can start, making a design concept and a work distribution structure is necessary. A design concept is a breakdown analysis of the functional requirements and corresponding design parameters fulfilling each requirement in an independent solution. This requires problem solving by appointment of a temporary virtual team composed by competences suitable for solving the actual problem. The breakdown analysis proceeds with formulation of new requirements and design parameters level by level, down to the level of the lowest part. The virtual team used to solve a problem must be composed to suit the specific problem. Axiomatic design described by N. Suh [1990] is a useful tool for making this problem solving. Each solution must be documented throughout all levels by answers on the questions of what, how, whom, and why. This documentation forms a tree of requirements and according design parameters. Documentation of all functional requirements will make it possible to follow up the design result against the corresponding requirement assuring a core product quality. This tree makes it also possible to follow up problems that come up later in the design work.

The operative domain of virtual teams consists of a network of temporary problem solving teams composed of representatives from competence groups and other specialists providing competence resources required to solve the specific problems within the structure.

The communication domain

Use of virtual teams in a flexible network organization demand a competence transfer that cannot be performed solely by face-to-face communication. A computer supported telecommunication means can usually replace face-to-face communication by using fullmedia. A fullmedia communication involves voice-, video-, and graphical media besides written text.

Use of the Resource Box model is an approach to meet the communication requirements. J. Österlund [1991 and 1994] describes the principles of the Resource Box model. It separates the communication flows of a competence group – a Resource Box – into three different axes depending on the purpose of transferred information.

Computer support is favorable in information handling of the functional requirement's structure, work breakdown structure, structures for product and process control and quality verification process. The basic question for application of a computer support is:
How to provide the required information in a way that improves the user situation as a human being and increases its information handling capability?
Formulating the answer to the question above is not easy because various factors affect the user situation in many different ways. The communication must be efficient and easy to interpret. It must be unambiguously at distance in the same manner as in direct, face-to-face contacts, and without causing stress by information overload. To be of value, the computer support must fulfill requirements from factors set by the involved individuals, the organization, and the communication system described as the Management by Informatics concept by J. Österlund [1994].

Conclusion

Design of an NPD organization from inside and out, with the help of Resource Boxes, is a systems approach to solve the problems. It demands a resource domain *structure* of core competences appropriate for operative work and competence development for strategic use in future products. Learning by the joint problem solving is an important component in development of deep core competences. It needs also *methods* in the operative domain of virtual teams to provide competence resources into problem solving process achieving right quality level. Effective use of the methods requires support from the communication domain by use of *information technology means* adapted to human use.

In a multidisciplinary systems analysis the problems must be considered from several perspectives.

1. *The systems perspective.* The source of competence resources is individuals in a group forming a social system for learning and support. To get the appropriate competence resource available in a particular situation requires studies of actual individual resources and their behavior in group systems. A favorable organization culture, use of the appropriate driving forces, and strategical goals known by the individuals are essential for development of the right competences.

2. *The biological perspective.* An improved capability of information handling by the individuals is necessary for solving the communication problems. Knowledge of the human information handling must be the base for development of a suitable communication profile of media and means.

3. *The technological perspective.* Human adaptation needs new trends in the development of IT means for communication and documentation purposes. Structuring enables concentration on valuable information contents and choice of valid sources to avoid the risk of information overload.

4. *The organization science perspective.* Obtaining a basis for efficient and valuable information handling requires managerial actions through organizational means. Use of small self-governed competence groups as basic resource units in a network organization offers a favorable climate for using interprises and enterprises in combination forming a virtual corporation.

References

Allen, T. J. and Scott Morton, M. S. eds. *Information Technology and the Corporation of the 1990s.* Oxford University Press, New York, 1994. ISBN 0-19-506806-8.

Dill, David D. and Pearson, A.W., "The Self-designing Organization: Structure, Learning and the Management of Technical Professionals." *IEEE Proceedings of PICMET `91.* Portland State University, Oregon, October 27-31 1991.

von Glinow, Mary Ann and Albers Mohrman, Susan eds., *Managing Complexity in High Technology Organizations.* Oxford University Press, New York, 1990. ISBN 0-19-505720-1.

Hedlund, Gunnar, "The Hypermodern MNC – A Heterarchy?" *Human Resource Management.*Spring 1986, Vol. 25, Number 1, Pp. 9-35.

Likert, Rensis, *New patterns of management.* McGraw- Hill Book Company, New York, 1961.

Mackenzie, Kenneth D. *A Theory of Group Structures.* Gordon and Breach Science Publishers, New York, 1976. ISBN 0 677 05310 X.

Nonaka, I., "A dynamic Theory of organizational knowledge creation." *Organization Science.*Vol. 5, no. 1, February 1994.

Österlund, J., "The Resource Box – Competence Management by Informatics in High Technology R&D Work." *IEEE Transactions on Engineering Management.* Vol. 38, No. 2, May 1991.

Österlund, J., *Competence management by Informatics systems in R&D work.* Royal Institute of Technology, Stockholm, Sweden, 1994. ISBN 91-7170-891-X.

Ramström, D., *The efficiency of control strategies – Communication and decision-making in organizations.* Gebers Förlag AB, Uppsala, 1967.

Schein, Edgar H., *Organizational Culture and Leadership.* Jossey-Bass Inc., San Francisco, 1985. ISBN 0-87589-639-1.

Shtogren, John A. ed., *The Structure of Competence – the theories and facts about managing people.* Studentlitteratur, Lund, Sweden, 1980. ISBN 91-44-00003-0.

Suh, Nam P., *The Principles of Design.* New York, Oxford University Press, 1990. ISBN 0-19-504345-6.

Organisational Networks in R&D

Bernice Pacitti and Alan Pearson
00 44 161 275 6455

R&D Research Unit
Manchester Business School
Manchester
M15 6PB

In all organisations effectiveness and efficiency is of real importance. This is no different with R&D. All organisations are increasingly trying to reduce in-house expenditures and to enhance performance by outsourcing and collaboration. More recently, the concept of networking has come to the forefront and the literature suggests that this is a useful form of organisation under conditions of increasing turbulence in the environment.

In this paper, a review of literature is provided as a background to the presentation of the way two multinationals claim to be moving towards the introduction of a new organisational structure. Activity to date and implications for management are presented.

Introduction

New organisational forms arise to cope with new environmental conditions[17]. In the last century three forms of organising emerged - the functional, the divisional and the matrix[3]. More recently interest in the *network organisation* has arisen[16]. Although many studies have attempted an introduction and a description of activity, few have undertaken an in-depth analysis of the network organisation[13] to highlight how they are started, how they are managed and why they succeed or fail[9]. The first sections of this paper draws on literature relating to the emergence of networks and relevant issues. In the latter sections we will draw on the recent experiences of two multinational companies in moving towards a network structure.

The 'Network Organisation'

Extensive use of the network concept in recent literature has resulted in differing and opposing definitions. This is because authors define firms and networks indistinctly and inconsistently and provide sweeping generalisations to cover a variety of cases that are in many ways dissimilar[21].

The more frequent interpretations of a network describe business arrangements whereby independent firms cooperate in order to achieve a strategic purpose[11]. These collaborations focus on various company stakeholders and range from short-term cooperations to long term strategic partnerships[15]. Networks have been used to describe horizontal relationships between companies operating in similar areas; or as a substitute for vertical integration[20].

Analytical Studies

Analytical studies have focused on communication flows in order to identify, for example, where power lies in the network[19], or to propose a typology by determining information exchange, which is highly connected to organisational structure[1].

Miles and Snow (1992)[18] propose three network structures - the *internal*, the *stable* and the *dynamic* network with each employing varying degrees of outsourcing and market exposure. More recently[18]
they extended this concept and introduced the "Spherical Structure" where the firm partly owns and partly outsources required resources.

0-7803-3552-X/96/$5.00 © 1996 IEEE

The Virtual Organisation

The phenomenon of the "virtual organisation"[4] has emanated largely as a result of the development of computer networks[14] and does not need all or sometimes any people in one place to deliver a service[10]. Prerequisites for success in the virtual organisation, are assumed to include employee trust and loyalty[7].

Chesbrough and Teece (1996)[6] question the desirability of using the virtual form for certain types of innovation. They distinguish *autonomous* from *systemic* innovations, the latter of which should not be organised by the virtual set-up. However, Bidault & Fischer (1994)[2] provide significant evidence that systemic innovations are built on a 'network' rather than the 'market' concept.

Managing the Network Organisation

The management of external relationships is complex and most collaborations fail to yield expected results[5]. Studies which are aimed at understanding the management of the network organisation refer to the general management problems. Very few examine management in depth and most discuss 'partnerships' when, clearly, networks involve multiple members. Nevertheless some recurring issues which are likely to figure in the effectiveness of organisational networks are learning capability, trust, reputation, geographical, social and technological proximity, commitment, power, shared goals, and adaptation[8].

Research

The importance of information flow and external inputs into the innovation process has been well documented in the literature[22]. Collaborations in R&D are nothing new. The use of outsourcing is well established, for example, the use of external test facilities in industries as diverse as oil and pharmaceuticals. Contract research organisations are a well established part of the R&D infrastructure. However, an increasing emphasis on cost effectiveness and downsizing has resulted in a closer look at the balance between internal and external activities[12]. As part of this process, movement to a form of networking is attractive and a number of organisations say they have started to or are seriously considering doing so. This study reports the way in which two such organisations are attempting to do this, the reasons why and the perceived results to date. The companies were chosen because of their expressed interest and efforts made to improve their effectiveness through networking.

Networking at Company A

Company A is a multinational operating in the food industry. In the last fifteen years the group has decreased business interests from nine to the four that remain today. Now it operates a matrix structure which replaced an earlier functional structure. This has enabled the best multidisciplinary team to be accessed more easily for any given project and is seen to have increased the flexibility of the company.

The group sees the future of competitive advantage to lie in Innovation and Radical Change which requires 'technological foresight' and speed to market. They identify "Dynamic Networks" as critical in the realisation of this goal. The company sees networking not only as an objective of R&D but to improve the overall performance of the group - by avoiding re-inventing the wheel, increasing efficiency and dispersing knowledge throughout the company.

They believe the success of networking to be dependent on clear long-term technical plans. From these plans the company can ascertain where external linkages can help. For this, organisational change was needed and networking was envisaged to play a part in that change.

Over the last few years management has encouraged greater interaction across divisions and with outside organisations. To encourage internal networking, the company has 'common interest groups' and management development programs. Employees from across the group exchange and share problems, capabilities and solutions. The objective of this networking is to exploit internal knowledge, skills and experiences.

At present they have no measures to determine the value of these internal networks. An investigation has not been undertaken to identify, for example, who is involved or who has the power in these networks. Possessing such knowledge would enable the dissemination of information much easier and would provide information regarding the level of learning in the company.

The Technology Centre is responsible for the groups' five strategic sciences and group research which provides the enabling technology which operating companies can draw upon to achieve new product development goals. The Director has, in recent years, encouraged scientists to access external knowledge in Universities, Research Associations/Institutes. Due to the nature of their work, the Centre has more external linkages than any other part of the group not only with academic institutions, but with supplier companies and customers in the development of technology for new or improved products and processes. Outside the Centre, external linkages are essentially limited to sub-contractual supplier relationships.

Evidence of an increase in networking within the Centre is provided by the fact that scientists now spend a significant amount of time in institutions across Europe, developing contacts with researchers in related fields, whereas as recently as three years ago they would have been located mostly in their home institution.

Not much has been done to provide incentives for employees in adopting this new role. However, there is a feeling in the company that individuals are self-motivated enough to create and maintain external relationships. On the other hand, decision making in the whole group remains hierarchical and according to some employees, there have been no real changes throughout the company to promote this external networking.

The company has done nothing to measure if the knowledge acquired from external or even internal networks has been disseminated, acted upon or implemented; how valuable it is or even if it is having a positive or negative learning affect. The Centre, being a relatively small operation, might be flexible enough to absorb and act on the information but without any measures and controls, no-one knows. The philosophy at the moment seems to be that anything external is necessarily a good thing.

Virtual Laboratory at Company B

Company B operate in the telecommunications industry. Ten years ago they had a 'classic' centralised R&D laboratory with approximately 350 people. In order to bring closer together the research and development functions and the manufacturing operations (so that their research could be closer to the market) management perceived a need to change the R&D organisation. This was done through the closure of their centralised laboratory and a decrease in permanent R&D staff in the UK from 350 to 20 key people. Meanwhile, they bought manufacturing units in Australia, North America, Italy, Spain and Germany each with their own R&D and technology functions. This enabled them to be closer to their customers. Small technology groups were set up at these international sites which started the move towards a 'virtual' laboratory.

The company downsized to focus on core competences, outsourcing and collaborating with suppliers for non-core activities. For examples the company now contracts for machinery and polymers which, in the past, were manufactured internally. This requires collaboration with suppliers in the development of technologies.

To facilitate the operationalisation of the virtual laboratory key people were based at sites in UK, Australia and America - it was felt that there was a need for at least one or two key people to know each other in order that the collaborative projects could be undertaken with more ease. The company distinguishes between two types of 'virtual' collaboration. The first is that where the collaborating partners are separated by geography but are owned by the same parent and the second is that where the companies are independently owned. 'Virtual' operations are, at present, more focused on the former category.

However, it is felt that the management of the 'virtual laboratory' is not easy. For example, the success factors of collaborations are widely acknowledged but, the experience is that when it 'comes to the crunch' partners back off and are not willing to collaborate wholeheartedly.

Difficulties arise when setting up collaborations with suppliers, especially in the development of technologies. For the supplier effort to be of real value to the company, the supplier needs to know the company thoroughly, i.e. technologies and capabilities. But suppliers often work with competitors, thus creating a risk of opportunistic behaviour.

Difficulties also arise when there is a time difference between collaborating companies and a need to have R&D activities in 'synch'. IT, it is felt, does not resolve this or many of the other problems related with the management of the 'virtual' lab. They find IT to be useful tool for transmitting specific, formal information but do not regard it as being *the* factor in the success of collaboration, a point noted by others[23]. Personal communication is, therefore, encouraged with employees visiting the laboratories around the world in order that at least 'one or two' people know each other.

The one case where the 'virtual' laboratory set up *did* work was in the development of a cutting edge technology in optical telecommunications. There did not exist a single group of people who were capable of undertaking job alone, so the decision and willingness of R&D employees from the various companies to work together was really a 'necessary evil'. In other areas where technology related knowledge are 'older' and more widespread, there is a reluctance to collaborate as individual groups feel that they can achieve the end result themselves, making the management of the collaboration more difficult.

Conclusions

It is clear that the companies studied are concerned about increasing networking in an attempt to increase efficiency. What is also clear is that they have not established any performance measures to help evaluate these structures.

The changes that have been initiated have mainly been to encourage integration between different business units or companies i.e. internal networking. In Company A the main source of external networking has been to Universities. Company B sees more opportunity and cause for collaboration with industrial partners, mainly due to the nature of their technology.

Hence, the degree of external networking now being discussed in literature is not reflected in these two companies. There has been no real attempt to identify specific performance measures which can assess the effects of such changes. This will not perhaps be surprising to those who have been involved with research on R&D/Technology Management as there is little evidence that changes in organisational structure have been analysed against clear performance indicators. This is clearly a deficiency which future research must address.

References

1. Albertini, A. and Butler, J. 'R&D Networks in a Pharmaceutical Company', *R&D Management*, 25 (4): 377-393, 1995

2. Bidault, F. and Fischer, W. 'Technology Transaction: Networks over Markets' *R&D Management*, 24 (4): 373-386, 1994

3. Biemans, W.G. 'Organisational Networks: Toward Cross-Fertilisation Between Practice and Theory', *Journal of Business Research*, 35:29-39, 1996

4. Bird, J. 'Stimulating Simulating', *Management Today*: 62-64 January, 1996

5. Bruce, M., Leverick, F. and Littler, D. 'Complexities of Collaborative Product Development' *Technovation*, 15 (9): 535-552, 1995

6. Chesbrough, H. W. and Teece, D. J. 'When is Virtual Virtuous? Organising for Innovation' *Harvard Business Review*, January-February: 65-73, 1996

7. Crainer, S.'Re-engineering the Carrot', *Management Today*: 66-68, December, 1995

8. Ford, D. (ed) "*Understanding Business Markets: Interaction, Relationships, Networks*", Academic Press, London, 1990

9. Hagedoorn, J. 'Strategic Technology Partnering During The 1980s: Trends, Networks and Corporate Patterns in Non-Core Technologies'; *Research Policy*, 24: 207 - 231, 1995

10. Handy, C. 'A Glimpse of the Invisible Organisation', *Director*, October: p 101, 1994

11 Harrigan K. R.'Joint Ventures and Competitive Strategy', *Strategic Management Journal*, 9: 141-158, 1988

12. Howells, J. R.'Going Global: The Use of ITC Networks in Research and Development', *Research Policy*, 24: 169 - 184, 1995

13 Hussey, D. 'The Virtual Organisation' *Journal of Strategic Change,* 4: 305-306, 1995

14. Kimura, H. 'Computer Launches 'Virtual' Company', *The Japan Times Weekly International Edition,* April 24-30: p13, 1995

15. Levinson, N. S. & Asahi, M. 'Cross-National Alliances and Interorganisational Learning' *Organisational Dynamics*: 50-63, 1995

16. Miles, R. and Snow, C. C. 'Network Organisations: New Concepts for New Forms', *The McKinsey Quarterly*, Autumn: 62-73, 1986

17. Miles, R. and Snow, C.'Causes of Failure in Network Organisations', *California Management Review*, 26 (3): 53-72, 1992

18. Miles, R. and Snow, C. C.'The New Network Firm: A Spherical Structure Built on a Human Investment Policy', *Organisational Dynamics*; 5-18, 1995

19. Oliver, A. L.'Network Analysis as an Organisational Diagnostic Tool: Bringing Structure into Process', *Journal of Strategic Change,* 4: 323-331, 1995

20. Quinn, J. B., Doorley, T. L. and Paquette, P. C.'Technology in Services: Rethinking Strategic Focus', *Sloan Management Review*, 21 (2): 67-78, 1990

21. Robertson, P. L. and Langlois, R. N. 'Innovation, Networks and Vertical Integration'. *Research Policy*, 24: 543-562, 1995

22. Rothwell, R. and Dodgson, M. 'External Linkages and Innovation in Small and Medium-sized Enterprises', *R & D Management,* 21 (2):125-137, 1991

23. Woolride, A. 'Only Communicate. Thoroughly Modern Corporations Know no Borders', *The Economist*, 24 June, 1995

Earned Quality: Improving Project Control

J.P. Paquin, Ph.D.
(819) 773-1741
Dpt. des sciences Adm., Université du Québec à Hull
101 St Jean Bosco, Hull, Québec, Canada

J. Couillard, Ph.D.
(613) 562-5800
Faculty of administration, University of Ottawa
136 Jean Jacques Lussier, Ottawa, Ontario, Canada

R. Paquin, M.Sc., M.B.A.
(819) 773-1742
Université du Québec à Hull
101 St Jean Bosco, Hull, Québec, Canada

D. Godcharles, M.Sc.
(613) 563-2091
AT&T
Ottawa, Ontario, Canada

The concept of earned value has been widely used in project management to measure and control the achievement of time and cost objectives. However, no such equivalent concept has ever been devised to assess and monitor the quality of a project. The goal of this paper is to propose a multiattribute utility theory approach that enables project managers to measure and estimate the monetary value of a project's accumulated quality attributes. To that effect, the concept of earned quality has been devised as a tool for assessing and controlling quality throughout a project's life cycle.

Introduction

Field studies have shown that many successful projects were the result of their quality attributes (Might and Fisher 1985, Pinto and Slevin 1988). Even though many methods have been devised to monitor and control the achievement of a project's time and cost objectives, few exist to monitor project quality effectively and efficiently(Azambre and Audousset 1992, Pinto and Mantel 1990). Therefore, it follows that a simple and effective method to control project quality is needed.

According to the principles of total quality management proposed by Deming 1986, Juran 1988, Crosby 1984, and Feigenbaum 1983, the International Standardization Organization (ISO) has defined quality as the set of properties and characteristics of a product or a service required to meet the explicit and implicit needs of a client. The first consequence of this definition is that quality is achieved only if a project meets the client's requirements. The second consequence assumes that only the client can ultimately assess the quality of a project. Therefore, any method for assessing the quality of a project must take into account explicitly the properties and characteristics of the project, as perceived by the client, that contribute to his/her needs.

Devising a method to control project quality is not an easy task. The method has to identify, measure, and aggregate the numerous quality attributes that a product or service should possess when delivered to the client. Time and cost objectives, in contrast, are more easily measured and controlled by using the earned value concept. Over the last few years, earned value has become quite popular with project managers as a tool to monitor and control cost and time objectives. This is so, largely because of its relative ease of measurement and use (Fleming 1983). However, no such equivalent concept has been developed to monitor and control quality throughout the project's life cycle.

A comprehensive approach to project control should therefore simultaneously take into account cost, time, and quality variances. Accordingly, we propose to complement the concept of earned value with the concept of earned quality.

The present article is subdivided into three parts. The first part deals with the definition of project quality and explains how this concept is related to multiattribute

0-7803-3552-X/96/$5.00 © 1996 IEEE

utility theory (MAUT). The second part provides a definition of earned quality and proposes a methodology to estimate its value. The third part uses a simplified example dealing with an information system development project to illustrate how to implement earned quality to improve project control. Finally, we discuss the applicability of earned quality as an efficient quality control method for the implementation of quality driven projects.

The Quality of a Project and Its Measure

It is generally agreed that the quality of a project depends, to varying degrees, on the achievement of three fundamental objectives, namely time, cost and quality. According to the International Standardization Organization, quality includes all the properties and characteristics that give a product or a service the capacity to satisfy fully the explicit and implicit client's needs. Measuring the quality of a project is a complex task because it generally involves multiple conflicting objectives as well as imprecise and qualitative attributes.

Over the years, methods were developed to take into account explicitly the client's needs. Such approaches include the Quality Function Deployment (QFD) and Value Analysis (VA), which have had successful records (Bossert 1989, Gevirtz 1994, Sullivan 1986). For instance, QFD aims at decomposing the client's needs in relation to technical characteristics and to summarize the relevant information into a matrix named the House of Quality. On the other hand, VA identifies the primary technical functions derived directly from the client's needs, and their support functions (Azambre and Audousset 1992, Gevirtz 1994).

Having identified the essential properties and characteristics of a project, an overall quality measure can then be proposed to effectively control a project. In that respect, the set of attributes identified either by the QFD or the VA methods can be structured in relation to their relative contribution to the attainment of the project's overall objective. This can be achieved by specifying the hierarchy of attributes and objectives, that is by subdividing the overall objective into lower level objectives, thus clarifying the meaning of the related objective. These lower level objectives may thus be considered as means to higher level objectives. Establishing the relationship between objectives and means enables one to build the hierarchy of objectives and attributes. Valuing these multiattributes trees is then performed with the aid of multiattributes utility theory (MAUT) (Keeney and Raiffa 1976), which provides not only an elegant solution to the multiattribute problem but also enables the systematic aggregation of all the attributes into a single quality value for the project (Paquin and Paquin 1991, Mustafa 1991, Saaty 1980).

The MAUT methodology can best be explained using a simplified example taken in the field of information system development. Consider the case where application of the QFD has enabled the identification of the following four quality properties required by the client: speed, reliability, capacity, and user friendliness of the system. These properties are further subdivided into their corresponding attributes I_{ij}, where I_{ij} represents attribute j of property i. Project quality must then be characterized by specifying the relative importance of the contribution of the different attributes to the achievement of their respective properties and, in the same fashion, the contribution of every property to the attainment of the overall quality objective. Elucidating the client preferences could be achieved in a variety of ways; one such method, proposed by Saaty 1980, requires a binary comparison between attributes and properties.

Such an elucidation process enables the establishment of the hierarchical tree of the project's overall quality objective, as given in figure 1.

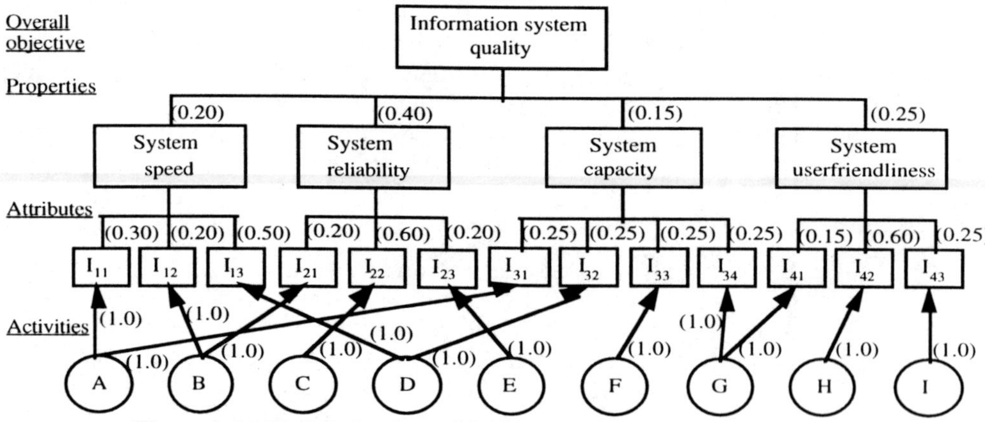

Figure 1: Multiattribute hierarchical structure

Measuring the attributes might prove quite difficult, mostly when dealing with qualitative attributes necessitating subjective evaluations. This should force the project manager and the client to agree on the most pertinent attributes and the appropriate way to assess their value. These attributes retained in the hierarchical structure may be related either in an additive or a multiplicative fashion when interaction exists between them.

In our example, we assume that the attributes interact in an additive fashion with respect to their properties, and so do the properties with respect to the information system's overall quality. The assumption of additivity implies that the attributes and the properties are mutually independent in utility (Keeney and Raiffa 1976). Such an assumption is advantageous for two important reasons. First, estimating the weight of a linear additive utility function is much simpler than those of a multiplicative utility function where interaction terms exist. Secondly, and most importantly, non-linear additive utility functions would be symptomatic of an incomplete or inadequate decomposition of the properties of the information system's quality objective (Keeney 1988, von Winterfeldt and Edwards 1986). Finally, a linear additive utility function means that the marginal rate of substitution between the attributes, or between the properties, is constant.

Once the multiattribute hierarchical structure has been established and validated, the project manager may identify the activities that contribute to the achievement of the selected attributes, and underline the importance of their relative contribution. In the latter case, it is the project manager who conducts a binary comparison of the activities in relation to their relative capacity in contributing to the achievement of each attribute (Saaty 1980). Establishing relationships between the attributes and theirs related activities is essential to validate their pertinence and to ensure quality control.

Returning to the information system example, let us suppose that nine (9) activities were considered essential to the achievement of all of its attributes. The activities' contributions to the quality attributes are described in figure 1. It shows that the project manager has estimated that 30% of the information system's speed was attributable to activity A, 20% to activity B, and 50% to activity D. In this example, each attribute depends on only one activity, thus explaining the contribution rate of 100%. From the hierarchical tree one can determine the relative potential contribution of each activity to the overall quality objective of the information system. For instance, activity A's relative contribution to quality is calculated as follows:

$$(1.00)(0.30)(0.20) + (1.00)(0.25)(0.15) = 0.0975$$

Similarly, the potential contribution or relative importance of each activity to the information system's quality can be determined. The distribution of the activities' weights is given in table A.

Activity	A	B	C	D	E	F	G	H	I
Duration (weeks)	5	6	3	4	4	5	3	2	4
Cost ($ 000)	7	6	8	2	5	10	3	4	4
Potential quality contribution	0.0975	0.1200	0.2400	0.1375	0.0800	0.0375	0.0750	0.1500	0.0625
Approximate weekly contribution	0.02	0.02	0.08	0.03	0.02	0.01	0.03	0.08	0.02

Table A: Potential contribution of activities to quality

The activities are then scheduled using a Gantt chart as shown in figure 2. The Gantt chart was constructed using the latest finish time of each activity. The critical path is given by the activities A, C, F, H, and I and spreads over 19 weeks.

From the Gantt chart, we obtain the planned quality of work scheduled (PQWS) on a specific date, by simply adding the activities' potential quality for each time period. An allocation rule describing the relationship between the (physical) advancement of each activity and the accumulation of quality into the project is needed. Different assumptions are available. We have assumed in the example that the contribution to quality by each activity is proportional to its physical progress. In figure 2, we have retained the assumption of time proportionality, so that the planned contribution of an activity is equal to the ratio of its potential contribution to its scheduled duration. Weekly contributions have been rounded to two decimal places. Thus, activity A, with a total potential contribution of 0.0975 and a scheduled duration of 5 weeks has a potential weekly contribution of approximately 0.02 to the project overall quality.

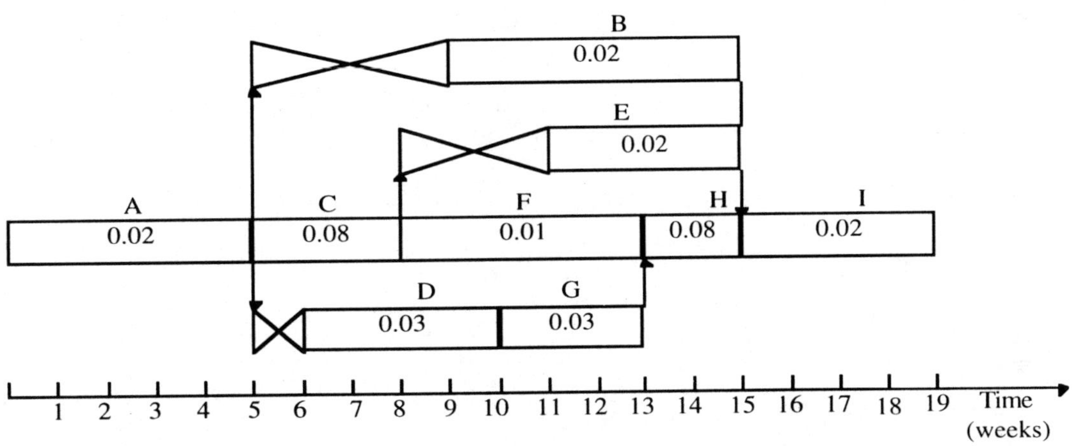

.02	.02	.02	.02	.02	.08	.11	.11	.04	.06	.06	.08	.07	.12	.11	.02	.02	.01	.01	Total
.02	.04	.06	.08	.10	.18	.29	.40	.44	.50	.56	.64	.71	.83	.94	.96	.98	.99	1.0	PQWS

Figure 2: Planned quality of work schedule (PQWS)

Other allocation rules could have been chosen. The allocation of quality could have been specified as proportional to cost, or even only when an activity is actually completed. The accumulated potential contribution of all activities, from the time the project has been launched till a specific date has been reached, provides a measure of the planned quality of work scheduled (PQWS), as shown in figure 2.

Earned Quality

Earned quality measures, at a given date, the planned quality of work performed (PQWP). Such an estimate of planned quality may be compared to the initial plan, given that the same quantity of work has been performed. Therefore, earned quality requires, as in the case of earned value, that the rate of physical progress of all activities be estimated on a regular basis, at least every time a milestone is reached

(Fleming 1992). Such a rate of progress may be estimated in an objective or subjective fashion, depending on the chosen measure.

For instance, as indicated in the example, let us assume that the activities' weekly completion rate has been obtained till week 13. The results have been summarized in table B.

		Activities								
		A	B	C	D	E	F	G	H	I
Potential quality⇒		0.0975	0.1200	0.2400	0.1375	0.0800	0.0375	0.0750	0.1500	0.0625
Date	PQWP									
1	0.0098	0.10								
2	0.0195	0.20								
3	0.0293	0.30								
4	0.0585	0.60								
5	0.0878	0.90								
6	0.0975	1.00								
7	0.1730	1.00		0.20	0.20					
8	0.2485	1.00		0.40	0.40					
9	0.3686	1.00		0.70	0.75					
10	0.4630	1.00	0.20	0.85	1.00					
11	0.5610	1.00	0.40	1.00	1.00	0.10	0.30	0.25		
12	0.6073	1.00	0.50	1.00	1.00	0.20	0.50	0.50		
13	0.6578	1.00	0.60	1.00	1.00	0.40	0.60	0.75		

Table B: Activities completion rates

From these completion rates we can obtain, at each control point, the planned quality of work performed or earned quality. Mathematically, the earned quality is calculated as follows:

$$PQWP_t = \sum_{A_i \in A} \begin{pmatrix} \text{Completion rate} \\ \text{of activity } A_i \end{pmatrix}_t \times \begin{pmatrix} \text{Potential quality} \\ \text{of activity } A_i \end{pmatrix} \qquad (1)$$

The formula explicitly assumes that the potential quality is proportional to its rate of completion. Such an assumption needs to be confronted with the actual accumulated quality at the end of each period, as given in table C.

It has been assumed that each activity is assessed only when it is terminated. As each activity is being implemented, its actual rate of quality attainment is equal to its rate of completion. Only when an activity is completed is its rate of quality achievement actually estimated. If such an assessment would have been performed at the end of each period, whether or not the activity is actually completed, then we would have had avoided these adjustments. This was the case at the end of week 6 when activity A was terminated and consequently assessed: its actual rate of quality attainment suddenly dropped from 90% to 80%. Deciding on the optimal assessment date ultimately depends on economic considerations and on the adoption of a risk management strategy. In the event that such a risk management strategy should lead to assessment dates encompassing more than one period at a time, then an allocation rule would still have to be adopted.

			Activities								
			A	B	C	D	E	F	G	H	I
Potential quality⇒			0.0975	0.1200	0.2400	0.1375	0.0800	0.0375	0.0750	0.1500	0.0625
Date	ACWP	AQWP									
1	1.2	0.0098	0.10								
2	2.5	0.0195	0.20								
3	3.5	0.0293	0.30								
4	6.2	0.0585	0.60								
5	8.4	0.0878	0.90								
6	10.2	0.0780	0.80								
7	14.4	0.1535	0.80		0.20	0.20					
8	19.2	0.2290	0.80		0.40	0.40					
9	24.4	0.3491	0.80		0.70	0.75					
10	28.6	0.4435	0.80	0.20	0.85	1.00					
11	32.5	0.6135	0.80	0.40	1.30	1.00	0.10	0.30	0.25		
12	38.0	0.6598	0.80	0.50	1.30	1.00	0.20	0.50	0.50		
13	41.4	0.7103	0.80	0.60	1.30	1.00	0.40	0.60	0.75		

Table C: Activity's actual rate of quality achievement

Assessing an activity's earned quality poses certain problems. In certain cases, quality is earned only when the activity is actually completed; therefore, earned quality between two assessment dates should be equal to zero, except on the date for which the activity is completed. The assumption of proportionality between earned value and work performed would therefore not provide a measure of the actual increase in earned value but only of its potential increase. Only by measuring the activities' actual rate of quality attainment can one determine the actual quality of work performed (AQWP), which may be calculated in the following way:

$$AQWP_t = \sum_{A_i \in A} \begin{pmatrix} \text{Actual rate} \\ \text{of quality attainment} \\ \text{of activity } A_i \end{pmatrix}_t \times \begin{pmatrix} \text{Potential quality} \\ \text{of activity } A_i \end{pmatrix} \quad (2)$$

Figures concerning the actual cost of work performed need no such calculations for they are provided directly from the accounting system. All these figures allow the project manager a better control over costs, schedule, and quality of the project.

Earned Quality and Project Control

Similarly to earned value, the planned quality of work performed (PQWP) may be interpreted as the earned quality of a project. Any variance between the actual quality of work performed (AQWP) and the planned quality of work performed (PQWP), on a given date, provides project managers with information on the project's capability in achieving its overall quality objective:

$$AQWP - PQWP = (AQ - PQ)\ WP \quad (3)$$

For instance, as shown in table D, at the end of week 13, the project's AQWP is estimated to be 71.03% of the project overall quality objective, thus surpassing by 5.25 percentage points its PQWP originally estimated at 65.78%. The difference represents an equivalent monetary value of $2,572.50, as 100% of the project's overall quality is worth $49,000.

The quality delay of the project may be estimated by the difference between the planned quality of work performed (PQWP) and the planned quality of work scheduled (PQWS):

PQWP - PQWS = PQ (WP - WS)   (4)

At the end of week 13, the quality delay is equal to -0.0522, which corresponds, as shown in figure 3, to a time lag of approximately one week. It is noteworthy to mention that the cost delay and the quality delay should provide the same time lead or time lag with respect to its planned progress.

Date	PQWS	PQWP	AQWP	AQWP-PQWP	PQWP-PQWS
1	0.02	0.0098	0.0098	0.0000	-0.0103
2	0.04	0.0195	0.0195	0.0000	-0.0205
3	0.06	0.0293	0.0293	0.0000	-0.0308
4	0.08	0.0585	0.0585	0.0000	-0.0215
5	0.10	0.0878	0.0878	0.0000	-0.0123
6	0.18	0.0975	0.0780	-0.0195	-0.0825
7	0.29	0.1730	0.1535	-0.0195	-0.1170
8	0.40	0.2485	0.2290	-0.0195	-0.1515
9	0.44	0.3686	0.3491	-0.0195	-0.0714
10	0.50	0.4630	0.4435	-0.0195	-0.0370
11	0.56	0.5610	0.6135	0.0525	0.0010
12	0.64	0.6073	0.6598	0.0525	-0.0328
13	0.71	0.6578	0.7103	0.0525	-0.0523

Table D: Earned quality

Comparing table C and D, one rapidly comes to the conclusion that although the project manager seems to be in control of the project's overall quality, the project is lagging behind schedule and is over budget by $10,550. In summary, the client should be satisfied with the project's quality progress although it will probably not be delivered on time and within budget, unless corrective actions are taken to alleviate the problems.

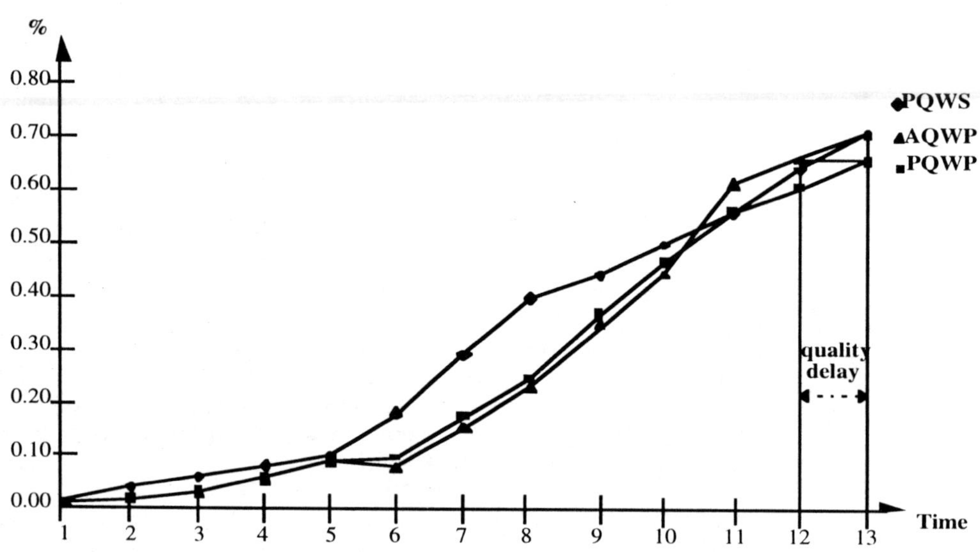

Figure 3: Project quality control

Conclusion

In this article, we have shown that the quality of a project could be addressed explicitly within an ongoing process. This was achieved by adopting the multiattribute utility theory, allowing the decomposition of the quality properties of a project into a hierarchical ordering of their attributes. Such a method enabled us to devise the earned quality concept as a tool for assessing and controlling the quality of an ongoing project in the very same manner that earned value applies to cost.

Implementing such a methodology in project management assumes that the client and the project manager succeed in identifying the pertinent properties and attributes conducive to quality control, and that they be specific on how to measure and assess these attributes when the product or the service is finally delivered. Such a task therefore poses certain practical problems. Those difficulties should not be underestimated. Nevertheless, such an approach should contribute greatly in managing the project towards its overall quality objective. This also means thr project masnager helping the client to elucidate all his needs, implicit as well as explicit, and determine their relative importance to the attainment of the overall quality objective.

In summary, the proposed concept of earned quality should provide managers, given its relative ease of use, an efficient tool to improve project quality control.

References

Azambre, J., and Audousset, J.-G., *Maîtriser une conception de qualité*, Masson, Paris, France, 1992.

Bossert, J., *Quality Function Deployment:A Practitioner's Approach*, ASQC Quality Press, Milwaukee, Wis., 1991.

Crosby, P.B., *Quality Without Tears*, New York, N.Y.: McGraw Hill, 1984.

Deming, W.E. *Out of Crisis*, Cambridge, Mass.: MIT Press, 1986.

Feigenbaum, A.V., *Total Quality Control: Engineering and Management*. 3rd ed. New York, N.Y.: McGraw Hill, 1983.

Fleming Q.W., *Put Earned Value (C/SCSC) Into Your Management Control System*. Publishing Horizons, 1983.

Fleming Q.W., *Cost/Schedule Control Systems Criteria, The management Guide to C/SCSC*. revised edition, Probus Publishing Company, Chicago, Il, 1992.

Gevirtz, C., *Developing New Products With TQM*, McGraw Hill, New York, NY, 1994.

Juran, J.M., *Juran on Planning for Quality,* New York, N.Y.: The Free Press, 1988.

Keeney, R., and Raiffa, H., *Decision with Multiple Objectives: Preferences and Value Tradeoffs*, Wiley, New York, 1976.

Keeney, R., "Value Focused Thinking and the Study of Values" in *Decision making; Descriptive and Prescriptive Interactions*, Eds Bell, Raiffa, Tuersky, Cambridge University Press, NY, 1988.

Might R.J., and Fisher W.A.,"The Role of Structural Factors in Determining Project Management Success", *IEEE Transactions on Engineering Management*, Vol. EM-32, No. 2, May 1985, pp. 71-77.

Mustafa, M.A., "Project Risk Assessment Using the Analytic Hierarchy Process", *IEEE Transactions on Engineering Management*, Vol. 38, No. 1, February 1991, pp. 46-52.

Paquin, J.P., and R. Paquin, "Evaluating a Portofolio of Public Entreprises: a Multicriteria Approach",
Canadian Evaluation Society, in *Evaluation Method Sourcebook*, Ottawa, 1991.

Pinto, J.K., and Mantel Jr., S.J., "The Cause of Project Failure", *IEEE Transactions on Engineering Management*, Vol. 37, No. 4, November 1990, pp. 269-276.

Pinto, J.K., and Slevin, D.P.,"Critical Factors in Successful Project Implementation", *IEEE Transactions on Engineering Management*, Vol. EM-34, 1987, pp. 22-27.

Pinto, J.K., and Slevin D.P., "Project Success: Definitions and Measurement Techniques", *Project Management Journal*, VOL.19, 1988, pp. 67-71.

Saaty, T.L., *The Analytic Hierarchy Process*, McGraw Hill, 1980.

Sullivan, L.P., "Quality Function Deployment", *Quality Progress* 19, no. 6, juin 1986, pp39-50.

von Winterfeldt and Edwards, *Analysis and Behavioral Research*, Cambridge University Press, Cambridge, 1986.

Virtual Corporation Tradeoffs

Robert J. Parden
(408) 554-4984
Fax (408) 554-5474

Santa Clara University
Santa Clara, California 95053

Interest in the virtual corporation has grown because most firms cannot afford to develop high-level competencies in all of the areas required to be truly competitive. Specializing in a narrow area, and outsourcing the rest appears to offer significant leverage. There are tradeoffs, however. Profits earned by external suppliers are lost to the core organization. Value-added experience is lost when the relationship ends, and shared experience becomes available to competitors. It is difficult to accumulate the human capital of the knowledge era. There may not be sufficient momentum to support continuous improvement over the long run. For each gain, there is a loss. Each enterprise will have to decide for itself.

The Search for the Utopian Organization

Charles M. Vest, President of MIT, comments in his annual report, ~..we do not know what the successful organization of the coming decades will look like. Even the most experienced business leaders cannot predict which companies will thrive and which will go under." Concerning information technology, and the new communications tools which are fundamental to the virtual organization, he comments, " We do not know how the vast store of instantly available information can or will be understood and used." (1)

Most of those extolling the virtues of virtual operations, propose that all organizations should get aboard very quickly, or they will miss the opportunity to find gold. Goldman, et. ala., write, " The convergence of computer networking and telecommunications technology is making it possible for groups of companies to coordinate geographically, and institutionally distribute capabilities into a single virtual company and to achieve powerful competitive advantage in the process." (2) Davidow and Malone write:"..we are in the midst of a new business revolution, one as sweeping and as far reaching as its predecessor (the industrial revolution). By the year 2015, the United states will either be a leader in this new business revolution, or it will be a post-industrial version of a developing country. Either a nation of independent knowledge workers, or a colony of economic serfs." (3)

The central premise of this discussion is that there is no "one size fits all" organizational structure. Each enterprise will have to customize its own strategy and execution. The advantages of the virtual corporation should be considered, and after the tradeoffs are weighed, select those features which will strengthen the enterprise.

The Virtual Corporation

Greiner and Metes describe the virtual model as follows. "A lead organization creates alliances with a set of other groups, both internal and external, that possess the best in the world competencies to build a specific product or service in a very short period of time. Such an alliance is "virtual" in that it is really not one homogenous organization, but a hybrid of groups and individuals from different companies, even different industries, that include competitors, customers, and suppliers, whose purpose is not longevity, but bringing a specific, highest quality product or service to market as quickly as possible. Indeed the alliance is temporary, with extremely short concept to-delivery cycles." (4)

0-7803-3552-X/96/$5.00 © 1996 IEEE

This accoolade for the virtual organization makes it appear that there is no other choice. A discussion of the construction management model that follows suggests that there is a spectrum of commitment ot he virtual model, and each organization selects the point at which they feel they are the most competitive, and satisfy their other goals. Longevity may be one of them.

Contemporary Organizational Goals

For an organization to make the virtual decision requires a comparison of their immediate objectives and the ability of the virtual structure to produce better results. Objectives of current interest to many firms include:

1. To make a faster response to rapidly changing market conditions.
2. Lower costs, especially in manufacturing where outsourcing appears to be most attractive.
3. Higher profits for the shareholders.
4. Higher levels of competence to deal with increased complexity.
5. Lower fixed expenses to improve agility.
6. Reduced capital requirements for ease of entry.
7. Rapid utilization of emerging technologies.
8. World class quality.
9. A global presence.
10. The ability to grow, and constantly improve.

Contemporary Management Problems.

While becoming virtual appears to solve some problems, it is not yet clear how many others will be created:

1. Developing the seamless organization will be more difficult with a more dispersed set of suppliers.
2. Understanding of current customer's needs depends on who collects the data, and how well it is fed back to the organization.
3. Anticipation of customer's latent needs, that is, what they will need or want tomorrow, is an extension of the previous concern.
4. High innovation rates, with new products to replace fading products might improve if a more effective R and D group can be outsourced.
5. Creating a competitive urgency throughout the firm, and within an increasing number of outside vendors, will grow more difficult.
6. Communications effectiveness, including the development of more sophisticated information technology, could be outsourced to specialists.
7. Continuous improvement in productivity, reaching higher achievement levels has yet to be widely tested.
8. Achieving interactive optimization among all of the contributors will be more difficult the greater their dispersion.
9. What is the uncertainty factor? How quickly is the industry changing. What are the advantages and disadvantages of becoming essentially a broker of services?

The Tradeoff Process

Matching each of the firm's objectives, and the strengths of virtual operations will help identify the required tradeoffs. The contemporary goals list is used as a checklist. For example, global presence may have no meaning for a local contractor specializing in high-end residence remodeling. When revenue is received for services, not the sale of products, different elements gain priority. A global manufacturing strategy should be the right strategy for global markets, thought it hasn't worked for everyone. Beginning with the highest priority item, what would be the short and long term impact of assuming virtual operations. True, the more specialized products and services may automatically be beyond internal operations because of economies of scale.

These choices become more complicated when it is difficult to place a dollar value on the tradeoff alternatives. In classical make or buy analysis, the pricing of internal production of a good or service is critical to the outsourcing decision. The method of assigning full costs to what may look like a marginal decision, will influence that choice.

An alternate approach may be to assume only a minimum core competence, and an arbitrary outsourcing of everything else. This means becoming a broker of services, with a brokers fee representing the enterprise gross profit. The goal then becomes generating enough value addition to justify the broker's services. Because of the ease of entry, keeping others at bay will be more difficult.

The General Contractors Model

The General Contractor's Model, which has been used for many years in the construction industry, describes some long term experience in virtual operations. Construction for the most part, has a short term, in and out flavor descriptive of the high flexibility/agility proposed for virtual operations. The core group may be composed of one or two people including the owner. Nearly all of the services they provide can be contracted to other groups. They may expand their core group for large projects, then consolidate when activity is slow. They operate by phone and fax, use computer planning, scheduling and control tools, and cellular phones for mobility. Some invest in extensive capital equipment. Others rent or lease. General contractor's have the opportunity to seek lowest bids from many sub contractors, and combine them into a lowest cost project for their clients. Experience with the General Contractor's Model identifies what can happen using a virtual model.

1. Most contractors build relationships with preferred subcontractors. Inevitably, however, the sub-contractors of choice are already busy. A new relationship must be established. The best subs can afford to bid high, because they are world class, which reduces the low cost opportunity. Tradeoff: low overhead, lowest bidders, lowest costs, leads to less project control. Quality may be sacrificed.

2. Qualifying a new sub or contractor can be challenging. They may be winners, in which case you expand your vendor qualification list. Or they may be very uncooperative, lack truly qualified people, and contribute to slipped schedules. By the time you learn, you are half-way through the project and it is too late to replace them effectively. It is not easy to identify world class from a data base. Tradeoff: With your own people, at least you know your competence level.

3. When you do build trust, and go back to the same sub contractor, you worry about losing price competition. This becomes critical if none of your "trusted" subs really want the work, as reflected in their high bids. Tradeoff: partner relationships may not continue to lead to lowest costs, any more than enterprise owned, captive suppliers.

4. It is difficult to decide who to include in the in-house core. Unless you have a continuous, high level flow of work, you cannot hold on to managers and supervisors for very long. If this group grows too large in order to maintain more stability, you are no longer virtual. Tradeoff: unless you can reach certain volume levels, core competency, which now becomes a fixed cost, will challenge even the most virtual of firms.

5. The problem of proprietary information is not often addressed in virtual operations. One of the advantages in using consultants/contractors is gaining the experience they have developed working with other firms, including competitors. When the temporary employee goes back into free agency after a project is completed, some of the experience gained may go to a direct competitor. Tradeoff: proprietary experience is lost, or at least shared with others. In a fast changing industry, this may not be a critical factor.

6. With almost no barriers to entry beyond entrepreneurial vision and leased telecommunications equipment, intense competition will occur almost immediately as others copy and improve. Tradeoff: less opportunity to build value-added competency which will provide synergy. However, some one-person firms with the right strategy, and well controlled execution, do prosper.

7. Everyone would like to employ and retain world-class employees. Everyone would like to employ the unique expertise of the truly innovative. However, these individuals cost more, and are difficult to locate and find available when you need them. Tradeoff: going world-class is a decision to move onto the fast track and pick up speed in all facets of the enterprise. This may stress the core group.

8. An unanticipated consequence of outsourcing is the impact of long-term contracts. In one recent case, a five year contract for supplying information services became a cost overrun as the cost of hardware dropped substantially. Firms are paying more for external services than if they were still handled internally. (5)

The Search Continues

The virtual corporation is proposed as a Utopian organization. Sense the global market, develop a strategy, hire world class competencies, replace the traditional office with a network, find information that you require on the Internet, run in and grab your share, hold on until invaded, and be prepared to fold the tents and move on when necessary.

In the world competition as we know it, virtual corporations are already being tested. We must be careful to insure that the organizational component of virtuality can be isolated for accurate evaluation. The concept might be repudiated for the wrong reasons. There are many reasons for enterprise failure, other than a virtual structure: a flawed strategy; poor market projections; unlucky timing; right strategy, poor execution; leadership faltered; wrong political climate; poor economic environment; even inadequate advertising. It is difficult to know the precise cause of an organizations's success or failure. We cannot replay the match and test by changing one of the variables. It's still up to the participants to weigh the tradeoffs for each enterprise.

References

1. Vest, Charles M., "The President's Annual Report," MIT, January 1996. 2. Goldman, Steven L., Nagel, Roger N., and Preiss, Kenneth, Agile Competitors and Virtual Corporations, Van Nostrand Reinhold, New York, 1995.p. xv. 3. Davidow, William H., and Malone, Michael S., The Virtual Corporation, Harper Collins, New York, 1992. 4. Grainier, Ray, and Metes, George, Going Virtual, Prentice Hall, 1995. 5. Business Week, "Has Outsourcing Gone too Far?/', April 1, 1996, p. 28.

Selection of a Manufacturing Process with Multiple Benefit Attributes

Celik Parkan and Minglu Wu
(852) 2788-8672

Department of Applied Statistics & Operational Research
City University of Hong Kong
83 Tat Chee Avenue, Kowloon, Hong Kong

Process selection is an important issue for manufacturers to increase their products' competitiveness. The present paper applies two different methods to help a semiconductor manufacturer select a process among four alternatives based on their operational benefits. The first one is the technique for order preference by similarity to ideal solution (TOPSIS) for making selection(s). The second is the operational competitiveness rating (OCRA) procedure for gauging the relative performance of operating units.

Introduction

Decision makers in the manufacturing sector always face the process selection problems. Since production process has direct impact on both efficiency and quality which, in turn, determine price and customer satisfaction, and eventually the competitiveness of their products, selecting the best process among a number of alternative processes is very important for the manufacturers. The adoption of a new process usually involves high risk due to its relatively high costs and market changes. So it is a common practice to select and then adopt a process very carefully, always using various methods to reach the final decision.

This paper deals with a manufacturing process selection case from a semiconductor producer. Although process selection belongs to the general field of multiple attribute decision making (MADM), other methods are also used. For example, Khouja 1995 used data envelopment analysis (DEA), a linear programming based efficiency evaluation technique, for technology selection. In this paper, in addition to applying a well-known MADM method called technique for order preference by similarity to ideal solution (TOPSIS) (Hwang and Yoon 1981), we will adopt a recent efficiency assessment procedure called operational competitiveness rating (OCRA) (Parkan 1991 and 1994) to help the manufacturer select a process among four alternatives based on eight operational benefits.

The present paper unfolds as follows: we first give an application based introduction to the TOPSIS method for selecting alternative(s) and the OCRA procedure for measuring performance. We also indicate briefly that the two different methods are both distance based and thus can be used interchangeably. Then, the said case is presented and studied in detail using the two methods. The results are satisfactory as they not only produce a reliable selection, but demonstrate the relationship between the two techniques also.

The TOPSIS method

TOPSIS is an intuitive MADM technique. In TOPSIS we assume that there are K alternatives, $\boldsymbol{A}^1,\ldots,\boldsymbol{A}^k$, each of which takes values on N attributes, $\boldsymbol{X}_1,\ldots,\boldsymbol{X}_N$. For convenience it is assumed that $\boldsymbol{X}_1,\ldots,\boldsymbol{X}_M$ are of "input" or "cost" nature and $\boldsymbol{X}_{M+1},\ldots,\boldsymbol{X}_N$ are "output" or "benefit" attributes. Denoting the value of alternative $\boldsymbol{A}^k$ on attribute $\boldsymbol{X}_n$ by x_n^k, both $\boldsymbol{A}^k$ and $\boldsymbol{X}_n$ can be conveniently expressed as vectors:

$$\boldsymbol{A}^k = (x_1^k,\ldots,x_N^k),\quad k=1,\ldots,K,\quad \boldsymbol{X}_n = (x_n^1,\ldots,x_n^K),\quad n=1,\ldots,N \tag{1}$$

0-7803-3552-X/96/$5.00 © 1996 IEEE

The principle of TOPSIS is that the selected alternative should have the shortest distance from the ideal solution and the farthest distance from the negative-ideal solution (Hwang and Yoon 1981). In TOPSIS, the ideal solution , $\boldsymbol{A}^+$, is composed of all the best attribute values attainable and denoted by

$$\boldsymbol{A}^+ = (x_1^+,\ldots,x_N^+) = (Min_i\{x_1^i\}, \ldots, Min_i\{x_M^i\},\ Max_i\{x_{M+1}^i\}, \ldots, Max_i\{x_N^i\}) \tag{2}$$

and the negative-ideal solution, $\boldsymbol{A}^-$, is composed of all the worst attribute values attainable:

$$\boldsymbol{A}^- = (x_1^-,\ldots,x_N^-) = (Max_i\{x_1^i\}, \ldots, Max_i\{x_M^i\},\ Min_i\{x_{M+1}^i\}, \ldots, Min_i\{x_N^i\}) \tag{3}$$

The distances of $\boldsymbol{A}^k$ from $\boldsymbol{A}^+$ and $\boldsymbol{A}^-$ are defined as the following weighted Euclidean distances

$$\begin{aligned}\rho(\boldsymbol{A}^k, \boldsymbol{A}^+) &= \{\textstyle\sum_{n=1}^{N}[w_n(x_n^k - x_n^+)/\|\boldsymbol{X}_n\|]^2\}^{1/2} \\ &= \{\textstyle\sum_{n=1}^{M}[w_n(x_n^k - Min_i\{x_n^i\})/\|\boldsymbol{X}_n\|]^2 + \sum_{n=M+1}^{N}[w_n(Max_i\{x_n^i\} - x_n^k)/\|\boldsymbol{X}_n\|]^2\}^{1/2}, \quad k=1,\ldots,K\end{aligned} \tag{4}$$

$$\begin{aligned}\rho(\boldsymbol{A}^k, \boldsymbol{A}^-) &= \{\textstyle\sum_{n=1}^{N}[w_n(x_n^k - x_n^-)/\|\boldsymbol{X}_n\|]^2\}^{1/2} \\ &= \{\textstyle\sum_{n=1}^{M}[w_n(Max_i\{x_n^i\} - x_n^k)/\|\boldsymbol{X}_n\|]^2 + \sum_{n=M+1}^{N}[w_n(x_n^k - Min_i\{x_n^i\})/\|\boldsymbol{X}_n\|]^2\}^{1/2}, \quad k=1,\ldots,K\end{aligned} \tag{5}$$

where $\|\boldsymbol{X}_n\| = \{\sum_{k=1}^{K}(x_n^k)^2\}^{1/2}$ is the Euclidean norm of $\boldsymbol{X}_n$, $w_1,\ldots,w_N$ are a set of normalized weights reflecting the attributes' relative importance.

As an alternative with the shortest distance to $\boldsymbol{A}^+$ may not have the farthest distance to $\boldsymbol{A}^-$, TOPSIS considers the distances to both $\boldsymbol{A}^+$ and $\boldsymbol{A}^-$ simultaneously by taking the relative closeness to $\boldsymbol{A}^+$:

$$RC(\boldsymbol{A}^k, \boldsymbol{A}^+) = \rho(\boldsymbol{A}^k, \boldsymbol{A}^+)/[\rho(\boldsymbol{A}^k, \boldsymbol{A}^+) + \rho(\boldsymbol{A}^k, \boldsymbol{A}^-)], \quad k=1,\ldots,K \tag{6}$$

The decision criterion of TOPSIS is that the smaller the value of $RC(\boldsymbol{A}^k, \boldsymbol{A}^+)$, the more preferred the $\boldsymbol{A}^k$, that is, a more preferred alternative is closer to the ideal solution.

The OCRA procedure

OCRA is a non-parametric procedure for operational performance measurement. In OCRA, it is assumed that we want to compare the relative performance of *K* production units (PUs) each of which has *M* inputs and *H* outputs. Representing by vectors $\boldsymbol{u}^k = (u_1^k, u_2^k,\ldots,u_M^k)$ and $\boldsymbol{v}^k = (v_1^k, v_2^k,\ldots,v_H^k)$ the *k*th PU's input and output values respectively, the *k*th PU, denoted by $\boldsymbol{PU}^k$, can be expressed as

$$\boldsymbol{PU}^k = (u_1^k,\ldots,u_M^k, v_1^k,\ldots,v_H^k), \quad k=1,\ldots,K \tag{7}$$

Under general conditions, the OCRA procedure involves the solution of a single linear programming problem to obtain the ratings of the PUs' overall operational efficiencies with respect to the inputs used and the outputs created. However, if all the PUs are assumed to assign the same importance to each input and output activities, the OCRA model can also be solved by the following simple ratio-type expressions:

$$OCRA^k = \textstyle\sum_{m=1}^{M} a_m(u_m^k - Min_i\{u_m^i\})/Min_i\{u_m^i\} + \sum_{h=1}^{H} b_h(Max_i\{v_h^i\} - v_h^k)/Min_i\{v_h^i\}, \quad k=1,\ldots,K \tag{8}$$

$$E^{k^*} = OCRA^k - Min_i\{OCRA^i\}, \quad k=1,\ldots,K \tag{9}$$

where E^{k^*} is the inefficiency rating of $\boldsymbol{PU}^k$, a_m and b_h are called the calibration constants which reflect the relative importance of the *m*th input and *h*th output activities of the PUs and satisfy

$$\textstyle\sum_{m=1}^{M} a_m + \sum_{h=1}^{H} b_h = 1 \tag{10}$$

The OCRA's assessment criterion is that the smaller the inefficiency rating E^k, the more efficient the kth PU. The most efficient PU receives an inefficiency rating of zero.

The relationship between TOPSIS and OCRA

It is noted from (4) and (8) that the two expressions, $\rho(\boldsymbol{A}^k, \boldsymbol{A}^+)$ and $OCRA^k$, have similar mathematical structure. In $\rho(\boldsymbol{A}^k, \boldsymbol{A}^+)$ of TOPSIS, we normalize the values of each attribute by its Euclidean norm and measure the distance between any two alternatives using Euclidean distance. If we define the ideal PU and negative-ideal PU, using the concepts of TOPSIS, as follows:

$$\boldsymbol{PU}^+ = (u_1^+, \ldots, u_M^+, v_1^+, \ldots, v_H^+) = (Min_i\{u_1^i\}, \ldots, Min_i\{u_M^i\},\ Max_i\{v_1^i\}, \ldots, Max_i\{v_H^i\}) \quad (11)$$

$$\boldsymbol{PU}^- = (u_1^-, \ldots, u_M^-, v_1^-, \ldots, v_H^-) = (Max_i\{u_1^i\}, \ldots, Max_i\{u_M^i\},\ Min_i\{v_1^i\}, \ldots, Min_i\{v_H^i\}) \quad (12)$$

we know that $OCRA^k$, normalizing the values of each input or output by its minimum, is a rectangular distance of $\boldsymbol{PU}^k$ from $\boldsymbol{PU}^+$. From this we can conclude that both TOPSIS and OCRA are distance based and thus can be used interchangeably for both performance assessment and selection making. We will examine this point using a process selection case in the next section.

A case study

A semiconductor producer in Hong Kong (the Company) plans to select a more effective manufacturing process to better satisfy the market demand for the electronics industry. The processes under consideration are the existing process and three alternative processes, denoted as $\boldsymbol{A}^1$, $\boldsymbol{A}^2$, $\boldsymbol{A}^3$ and $\boldsymbol{A}^4$, respectively. For simplicity, only the operational benefits of the processes are considered as the selection basis. Identified by the Company's managers and professionals (MPs), these process attributes are: $\boldsymbol{X}_1$: production lead time, $\boldsymbol{X}_2$: operating skills and training required, $\boldsymbol{X}_3$: flexibility, changeover tooling, and set up time, $\boldsymbol{X}_4$: availability of expertise and equipment, $\boldsymbol{X}_5$: equipment maintenance and service support, $\boldsymbol{X}_7$: product quality and reliability and, $\boldsymbol{X}_8$: technological leadership and image.

We use ten-point scale questionnaire to survey the MPs of the Company's Production, Production Engineering, and R&D Departments to obtain relative importance ratings for the attributes and the performance ratings of each alternative process on each attribute. Table 1 summarizes the survey results.

		X_1	X_2	X_3	X_4	X_5	X_6	X_7	X_8
Production Department	A^1	7	5	6	8	5	5	9	6
	A^2	8	4	5	7	5	4	8	7
	A^3	5	6	7	6	7	6	7	8
	A^4	6	7	8	5	7	7	6	9
	Importance	9	7	8	7	6	8	10	5
Production Engineering Department	A^1	4	9	7	7	9	7	6	7
	A^2	7	8	8	7	7	5	6	7
	A^3	6	9	7	7	6	8	7	8
	A^4	10	8	9	9	9	9	9	10
	Importance	8	6	6	6	7	10	9	5
R&D Department	A^1	5	10	7	8	7	10	10	8
	A^2	10	8	7	8	7	6	8	8
	A^3	7	5	10	6	8	8	8	9
	A^4	10	7	9	9	8	9	8	10
	Importance	7	4	6	8	3	9	10	5

Table 1: the survey data

The TOPSIS results

From Production Department's data, the relative importance ratings of the eight attributes are (9, 7, 8, 7, 6, 8, 10, 5) which result in the following normalized weights:

$$w = (0.1500, 0.1167, 0.1333, 0.1167, 0.1000, 0.1333, 0.1667, 0.0833) \tag{13}$$

The ideal and negative-ideal solutions are

$$\boldsymbol{A}^+ = (8, 7, 8, 8, 7, 7, 9, 9), \quad \boldsymbol{A}^- = (5, 4, 5, 5, 5, 4, 6, 6) \tag{14}$$

The distances of the four processes to $\boldsymbol{A}^+$ and $\boldsymbol{A}^-$ are computed by (4) and (5) as

$$\rho(\boldsymbol{A}^1, \boldsymbol{A}^+) = 0.0456, \quad \rho(\boldsymbol{A}^2, \boldsymbol{A}^+) = 0.0613, \quad \rho(\boldsymbol{A}^3, \boldsymbol{A}^+) = 0.0484, \quad \rho(\boldsymbol{A}^4, \boldsymbol{A}^+) = 0.0480 \tag{15}$$

$$\rho(\boldsymbol{A}^1, \boldsymbol{A}^-) = 0.0516, \quad \rho(\boldsymbol{A}^2, \boldsymbol{A}^-) = 0.0446, \quad \rho(\boldsymbol{A}^3, \boldsymbol{A}^-) = 0.0447, \quad \rho(\boldsymbol{A}^4, \boldsymbol{A}^-) = 0.0619 \tag{16}$$

from which the relative closeness to $\boldsymbol{A}^+$ for each alternative is obtained:

$$RC(\boldsymbol{A}^1, \boldsymbol{A}^+) = 0.4691, \quad RC(\boldsymbol{A}^2, \boldsymbol{A}^+) = 0.5786, \quad RC(\boldsymbol{A}^3, \boldsymbol{A}^+) = 0.5200, \quad RC(\boldsymbol{A}^4, \boldsymbol{A}^+) = 0.4369 \tag{17}$$

So the preference order of the four alternatives is

$$\boldsymbol{A}^4 \text{ f } \boldsymbol{A}^1 \text{ f } \boldsymbol{A}^3 \text{ f } \boldsymbol{A}^2 \tag{18}$$

Using the same procedure we can get results for the Production Engineering Department's data:

$$RC(\boldsymbol{A}^1, \boldsymbol{A}^+) = 0.6927, \quad RC(\boldsymbol{A}^2, \boldsymbol{A}^+) = 0.6928, \quad RC(\boldsymbol{A}^3, \boldsymbol{A}^+) = 0.5700, \quad RC(\boldsymbol{A}^4, \boldsymbol{A}^+) = 0.0643 \tag{19}$$

$$\boldsymbol{A}^4 \text{ f } \boldsymbol{A}^3 \text{ f } \boldsymbol{A}^1 \text{ f } \boldsymbol{A}^2 \tag{20}$$

The similar results for the R&D Department are

$$RC(\boldsymbol{A}^1, \boldsymbol{A}^+) = 0.4588, \quad RC(\boldsymbol{A}^2, \boldsymbol{A}^+) = 0.5340, \quad RC(\boldsymbol{A}^3, \boldsymbol{A}^+) = 0.6180, \quad RC(\boldsymbol{A}^4, \boldsymbol{A}^+) = 0.3224 \tag{21}$$

$$\boldsymbol{A}^4 \text{ f } \boldsymbol{A}^1 \text{ f } \boldsymbol{A}^2 \text{ f } \boldsymbol{A}^3 \tag{22}$$

The OCRA results

Applying OCRA procedure to the Production Department's data, the rectangular distances of the four processes to the ideal process are computed by (8) as (0.2900, 0.3864, 0.2953, 0.2133) which produces the following "inefficiency" ratings

$$E^{1*} = 0.0767, \quad E^{2*} = 0.1731, \quad E^{3*} = 0.0820, \quad E^{4*} = 0.0000 \tag{23}$$

This gives the same preference order as that of (18) obtained by TOPSIS.

The OCRA results for the Production Engineering and R&D Departments' data are

$$E^{1*} = 0.4442, \quad E^{2*} = 0.4482, \quad E^{3*} = 0.3615, \quad E^{4*} = 0.0000 \tag{24}$$

$$E^{1*} = 0.1024, \quad E^{2*} = 0.1620, \quad E^{3*} = 0.2128, \quad E^{4*} = 0.0000 \tag{25}$$

which also result in the same preference ranks as those of (20) and (22) obtained by TOPSIS.

The final selection

Applying TOPSIS method or OCRA procedure to the survey data from the Company's three Departments, three different preference orders (18), (20) and (22) are obtained. In order to get a final preference order, we compute for each of the four processes the average preference rating based on the corresponding ratings obtained from the three Departments. From (17), (19) and (21), the average TOPSIS ratings of the four processes are computed as (0.5402, 0.6018, 0.5693, 0.2745). The average OCRA ratings from (23), (24) and (25) are (0.2078, 0.2605, 0.2188, 0.0000). These two set of ratings produce the same preference order:

A^4 f A^1 f A^3 f A^2 which is identical to that of (18) obtained for the Production Department.

According to the above analyses, we can finally conclude that A^4 is the first choice in the Company's process selection. The existing process A^1 ranks as a distant second. A^3 and A^2 are ranked third and fourth, but their rating differences to A^1 are not so significant.

Conclusions

TOPSIS is perhaps the most efficient one among all MADM methods for alternative selection. Its logic is that a more preferred alternative should have a shorter distance to the ideal solution and a farther distance to the negative-ideal solution. The intuitive approach TOPSIS employs leads to a simple computational procedure. On the other hand, The OCRA procedure is a recent one to assess the relative performance of a set of PUs. It is mathematical programming based. If all the PUs assign the same importance to each input and output activities, the OCRA's solution can also be obtained by simple ratio-type expressions.

The TOPSIS method for selecting alternative(s) and the OCRA procedure for measuring performance, despite their different background and construct, are both distance based and thus can be used interchangeably for both areas. We study a semiconductor producer's process selection case using the two methods. The preference ranks of the four alternative processes obtained by the two techniques are identical in all circumstances, not only for each of the Company's three Departments, but for their averages also. This demonstrate empirically the relationship between the two methods.

References

Hwang, C.L., and Yoon, K. *Multiple Attribute Decision Making: Methods and Applications, A State-of-The-Art Survey*. Springer-Verlag, Berlin, 1981.

Khouja, M. "The use of data envelopment analysis for technology selection." *Computers and Industrial Engineering*. Vol. 28, 123-132, 1995.

Parkan, C. "The calculation of operational performance ratings." *International Journal of Production Economics*. Vol. 24, 165-173, 1991.

Parkan, C. "Operational competitiveness ratings of production units." *Managerial and Decision Economics*. Vol. 15, 201-221, 1994.

An Exploration of Technology Intensity Measures

C. Carl Pegels
M.Thirumurthy

School of Management
State University of New York at Buffalo
Buffalo, NY 14260-4000

Introduction

Technological activity is defined for this paper to mean the approaches firms use totranslate research and development (R&D) efforts into advances in their respective product andprocess technologies. These R&D efforts, then, have the potential to provide competitive advantage resulting in improved performance. Several studies conducted in the past have not established any definite relationship between R8D investment (measured in terms of R&D expenditure) and firm performance. One possible reason could be that R&D expenditure per se does not contribute to a firm's success. However, if we could measure the accumulation of knowledge and technological strength resulting from R&D efforts then we might find a determinant of firm performance. This is the connecting link between R&D effort and firm performance.

In this paper we shall first review a variety of papers which report on studies which have attempted to show the linkage between Technological Activity and Technological Intensity withFirm Performance.

We then follow this up with a summary of two studies by the authors which introduce technological intensity measures based on patent activity and patent attributes.

We then conclude this paper with a review of a variety of Firm Performance measures,based on both accounting, market and sales data.

Review of Prior Studies

The relationship between expenditures on research and development (R&D) and firm performance has been an area of interest both for the business press and academic researchers. The business press studies which have had the most visibility for corporate management are largely those reported by Business Week [2] in its annual R&D survey. The results of academic research are found largely in professional journals.

The Business Week surveys over the years have found a consistent positive relationship between R&D expenditures and corporate performance measured on the basis of several dimensions such as return on assets, eamings growth, sales growth, etc. However, theBusiness Week measure used to test the link between R&D expenditure and corporate performance has been the simple correlation measure with its statistical significance. Theweakness of the Business Week survey lies in the fact that statistically significant simple correlation coeffficients can be relatively weak indicators of a relationship between two variables.For instance, a simple correlation measure of 0.158 with a sample size of 100 is statistically significant at the 5% level. However, a simple correlation measure of 0.158 is usually viewed asa quite weak indicator of the relationship between the two variables. As a result, the BusinessWeek results must be studied more closely before taking the claims of the survey too seriously.The results of academic researchers are more mixed. An early study by Parasuraman and Zeren [8] associated R&D expenditures strongly with profit and sales. A 1989 study

0-7803-3552-X/96/$5.00 © 1996 IEEE

byMorbey [5] did not find any direct relationship between R&D expenditures and subsequent profit growth. The study used a sample of 800 U.S. companies. However, a year later Morbey and Reithner [6] reported that they found:

1. There is a strong direct relationship between R&D intensity (R&D per sales) andsubsequent growth in sales.
2. There is a strong relationship between R&D expenditure per employee and subsequent company productivity measured as sales per employee.
3. There is a strong relationship between R&D expenditure per employee and subsequent company profit margin measured as profits per sales.

Morbey explains that he attributes the difference in results between his 1989 and 1990 papers by the fact that the definition of R&D expenditures in his 1989 paper was more inclusive,and included such extraneous items as routine production testing costs, legal costs related topatent applications and the sale and licensing of patents. Morbey's more recent study used R&D data that satisfied a more rigid definition of what R&D expenditures could contain as specified by Compustat.

Other researchers such as Brenner and Rushton [1] found a strong association between R&D expenditure and sales growth in the chemical industry. Franko [3] cites a number of previous studies where a causal relationship between R&D input and sales growth was established.

All of the above studies perceive R&D expenditure as having a direct impact on the performance of a fimm. In fact, this need not be true. R&D efforts definitely contribute to increase in knowledge and innovation, but every experiment may not necessarily result in a successful product.

Mitchell and Hamilton [4] argue that not all the expenditure on R&D would result in successful products as some projects may have to be abandoned after the initial exploratory cost is incurred. No large follow-up investment is made in such projects. Only successful exploratory work which would result in new products and processes is further pursued. They liken this to exercising an R&D option as in American stock option. Hence the real measure of a firm's R&D is not its R&D expenditure or intensity, but how much of this effort is transformed into measurable knowledge, for instance, the number of patents granted to firms. Support for this hypothesis can be found in Pakes and Griliches [7] where they show that the patents are a good indicator of firm differences in advances of knowledge.

A study by Pegels and Thirumurthy [9] found that (1) research and development expenditures were a strong predictor of technological strength; (2) technological strength was astrong predictor of technological cycle time; and (3) technological cycle time was a strongpredictor of fimm performance measured in temms of profitability. The empirical test was done ona sample of 49 large U.S. firms for which technological and financial data were available. The results of this study will be discussed in more detail below.

In a separate paper Pegels and Thirumurthy [10] did a study to evaluate the relationships between technological cycle time, relative technological strength, and, science linkage with firm performance in the pharmaceutical industry. They found that the first two measures were good indicators of fimm perfommance, but not the latter. Again we shall discuss this study in more detail below.

Technology Intensity Model A

The model is described in Exhibit 1. We have focussed on fimm performance as measured by annual operating income (PROF) as the dependent variable. The two control variables are:

a. Assets consisting of plant and equipment (AST)

b. Annual sales revenue per employee (SEMP).

The first variable is a proxy variable for size and the second variable is a proxy variable for employee productivity. The three causal variables used in the model are:

1. Annual R&D expenditures per dollar revenue (RDE).

2. Technology Cycle Time (TCT) is the time that elapsed in years between the current patents and the previous generation of patents. The time separation between earlier related patents and the current patent is defined as the technological cycle time. Since the shorter cycle time patents indicate that a firm is actively pursuing recent technological activity originating from other organizations or itself, it is hypothesized that the lower the cycle time the better the fimm performance will be. This technology measure plus the subsequent ones have been obtained from Chi Research [11].

Exhibit 1 Technology Intensity Model A

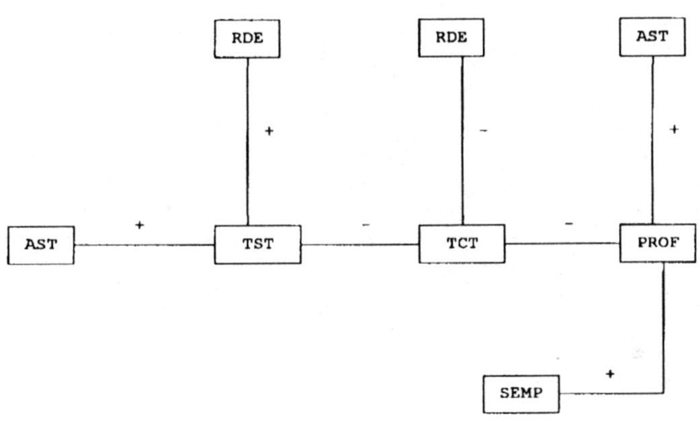

Notes:

	PROF =	Annual Operating Profits
	AST =	Plant and Equipment
	SEMP =	Annual Sales Revenue per Employee
	TCT =	Technology Cycle Time
	TST =	Technological Strength
	RDE =	Annual Research and Development Expenditures per Dollar of Sales Revenue
	PAT =	Number of Patents Issued per Year

3. Technological strength (TST) determined by multiplying a company's number of patents per year by the current year patent impact index. The current impact index is a measure of how important a company's patents are based on how often they are cited in other patents, which shows how frequently they are used as the foundation for other inventions. For example, a company's 1991 index is computed by first calculating the average number of times the patents it was granted in each of the previous five years were cited in new patents granted in 1991. Those figures are divided by the average number of 1991 citations for all U.S. patents in each of the previous five years. That yields a citation ratio for each year. A ratio of 1.0 means that the company's patents were cited as often as the overall average. A ratio of 1.2 means that the company's patents were cited 20% more often than average. Finally, the citation ratio for each of the five years is averaged to obtain the 1991 rating.

Annual operating income (PROF), rather than the annual net income, was selected to avoid the impact of adjustments to net income because of mandated or internal decisions. Only plant and equipment were considered for firm size (AST) because those are the assets used by a fimm to create operating income from operations. Annual sales revenues per employee (SEMP) is not the only measure of productivity but by no means the weakest. Finally, annual research and development expenditures per dollar of revenue (RDE) are obtained from each firm's reported annual R&D expenditures divided by its sales revenues. All data used is 1991 data, except for RDE data which is 1989 data, to include a time lag for this variable. The TCT and TST data were obtained from Business Week [2] and are based on data collected by Chi Research Inc. The financial data were obtained from the Compustat Files.

The model can be evaluated and viewed in two separate ways. First we can view the model as three individual multiple regression equations as shown below.

PROF = f (AST, SEMP, TCT) (1)

TCT = f(TST, RDE) (2)

TST = f(RDE, AST) (3)

Secondly, we can view the model as a three-stage recursive least squares model, where equation (3) feeds into equation (2), and equation (2) feeds into equation (1). The results of the model's two ways will be presented below.

Based on the research results we found that the causal variables RDE, TCT and TST will have the following impact on fimm perfommance (PROF).
1. RDE has a positive impact on PROF.
2. TCT has a negative impact on PROF, because the lower TCT the more effective the firm is in managing its developed technology.
3. TST has a positive impact on PROF.

Separately, for the individual equations we found that,
1. RDE has a positive impact on TST, but a negative impact on TCT.
2. TST has a negative impact on TCT.

The control variables for size (AST) and for employee productivity (SEMP) both have a positive impact on PROF and TST.

Technology Intensity Model B

The model is summarized in Exhibit 2. The predictor variables are measures of technological activity of the firm. The firm perfommance variables consist of two accounting variables, retum on investment (ROI) and retum on equity (ROE) and on one market variable, the ratio of fimm stock price to fimm book value.

Technological activity is measured in temms of three patent-activity and patent perfommance based indicators, as compiled by Chi Research [11]. The three technological indicators are discussed below.

1. Technological Cycle Time (TCT) as defined in Model A. Technology Cycle Time is the time that has elapsed in years between the current patents and the previous generation of patents. The time separation between earlier related patents and the current patent is defined as the technological cycle time. Since the shorter cycle time patents indicate that a fimm is actively pursuing recent technological activity originating from other organizations or itself, it is hypothesized that the lower the cycle time the better the fimm perfommance will be.

Exhibit 2 Technology Intensity Model B

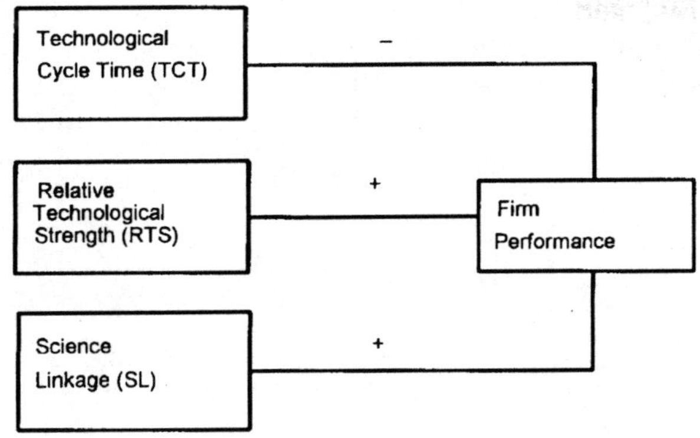

Note: Firm Performance is measured by three measures:

1. Stock Price / Book Value Ratio (P/B)
2. Return on Equity (ROE)
3. Return on Investments (ROI)

2. Relative Technological Strength (RTS) measures how frequently a patent is cited in later patents. A typical U.S. patent is issued with 5 to 6 references to other U.S. patents on the front page of the patent application. These cited patents limit the claims of the patent currently issued. Hence, heavily cited patents indicate that the cited patent contains important advances. The citation count is based on the past five years of patents issued, and is then divided by the expected number of citations to all patents in the corresponding years. RTS is thus an index. Chi refers to this variable as the current impact index (Cll). It is hypothesized that the higher RTS the stronger the firm's perfommance. Note that the RTS measure is different from the RST measures used in Model A.

3. Science Linkage (SL) measures the average number of scientific literature references cited on the front page of a patent. SL measures how close the issued patent is to the scientific research base. The hypothesis that the closer the link of a patent to the scientific research base the stronger will be the fimm's performance was found to bc not statistically significant.

The model was tested on one industry, the U.S. phammaceutical industry. A total of 17 large U.S. publicly listed firms were studied. Both technological and financial information was available on these fimms.

Fimm Perfommance Measures

There are a variety of fimm performance indicators that have been employed in the past in effectiveness studies. Most of these usually have a financial orientation, but many of them also use other measures to evaluate fimm performance. Examples of other measures are units sold, quality level of products/services delivered, customer satisfaction levels derived from customer surveys, market share, etc. Also used as "other" measures is the annual change in any or all of the above measures.

For this paper we shall focus on financial measures based on either accounting or stock market data, or a combination of both.

The two Technology Intensity models discussed above used both. In Model A we used the simple accounting measure of operating profits, as opposed to net profits after ten. Operating profits focuses on the operations performance, instead of on the overall fimm performance which is influenced by levels of debt, quality of debt write-offs and other financial manipulations or effects. In Model B we used Return on Equity (ROE) and Retum on Investment (RIO). These are based on accounting data and are troubled by the concerns expressed above. We did make sure that there were no write offs, and the model was applied to one industry, the phammaceutical industry. As a check we also used the Stock Price to Book Value ratio, a combination of Stock Market and Accounting measures. The book value measure, however, is one of the "safer" accounting measures to use because it is diffficult to manipulate it over the short run. The stock market measure is of course a measure of how investors view the firm. It is independent of accounting data.

In many organizational studies and financial management studies there has been a movement away from using accounting data and instead the use of stock market data is recommended. The reason is that the stock market valuation is considered a better and more realistic performance measure than one based on accounting data. In our study reported for Technology Intensity Model B we found a high correlation between the Price to Book measure and the two accounting measures ROI and ROE. Hence, for a given industry, with fairly homogeneous fimms, as in the pharmaceutical industry, it appears quite realistic to use accounting data as a Fimm Performance Measure

References

1. Brenner, Merril S. and Brian M. Rushton, "Sales Growth and R&D in the Chemical Industry," Research Technology Management, Mar-Apr 1989, pp. 815.

2. Coy, Peter, John Carey and Neil Gross, "The Global Patent Race Picks Up Speed," Business Week, August 9, 1993.

3. Franko, Lawrence G., "Global, Corporate Competition: Who's Winning, Who's Losing, and the R&D Factor as One Reason Why," Strategic Management Joumal, Vol. 10
(1989), pp. 449 474.

4. Mitchell, Graham R. and William F. Hamilton, "Managing R&D as a Strategic Option," Research Technology Management, May-June 1988, pp. 15-22.

5. Morbey, Graham K., "R&D Expenditure and Profit Growth," Research Technology Management, May-June 1989, pp. 20-23.

6. Morbey, Graham, K. and Robert M. Reithner, "How R&D Affects Sales Growth, Productivity and Profitability," Research Technology Management, May-June 1990, pp. 11-14.

7. Pakes, Ariel and Zvi Griliches, "Patents and R&D at the Firm Level: A First Look," !3~. Patents, and Productivity, edited by Zvi Griliches, The University of Chicago Press, 1984, pp. 55-72.

8. Parasuraman, A. and Linda Zeren, "R&D's Relationship with Profit and Sales," Research and Management, Jan-Feb 1983, pp. 25-29.

9. Pegels, C. C. and M. Thirumurthy, "The Impact of Technological Strategy on Firm Performance," Transactions on Engineering Management, forthcoming.

10. Pegels, C.C. and M. Thirumurthy, "Technological Activity and Firm Performance in the U.S. Pharmaceutical Industry," in Management of Technology, edited by R.M. Mason, L.A. Lefebore, and T.M. Khalil, Oxord, UK: Elsevier Advanced Technology, 1996.

11. "Tech-Line-CD Manual," CHI Research Inc., Haddon Heights, NJ.

Maps of Virtual Structures in R&D

Karol I. Pelc
(906) 487–2663

School of Business and Engineering Administration
Michigan Technological University
1400 Townsend Drive
Houghton, MI 49931, U. S. A.

The paper presents evolution of collaborative structures, ranging from traditional pattern of "invisible college" to current virtual structures of R&D projects. Properties of virtual structures in R&D are described and their configurations are presented.

Introduction

In the current era of virtuality it becomes necessary to explore the R&D structures which may fit efficiently with virtual enterprise. This requires, in turn, to review both evolution of such structures and to determine the scope of real and potential benefits they may generate. On one hand, we can view the virtual enterprise as a result of extensive outsourcing, driven by competitive business advantage, on the other hand we may see it in conjunction with the advances of information technology, allowing for dramatically changed modes of communication and control. Drucker (1995) suggests that in the next two decades, "organizations may be outsourcing all work that is "support" rather than revenue– producing." The same author indicates that cooperative arrangements between companies and alliances are expanding for joint activities in research, development and manufacturing.

Virtual structures are emerging. They include research oriented alliances, project and task oriented subcontracts, outsourcing of technology development, dispersed development teams as well as informal research groups, R&D networks reaching beyond boundaries of a company or institution.

Sharing of knowledge, and networking in research, are not new concepts to the scientific community. There is a long tradition of intellectual collaboration, informal contacts and cooperative structures.

The purpose of this paper is to review the evolution of virtual structures in basic, applied research, and development, and to analyze properties and configurations of virtual structures in R&D.

Evolution of Virtual Structures and R&D Networking

Virtual structures exist in almost all disciplines of science and constitute a very important communication mechanism. Scientists working in each problem area are linked together by formal citation network and co–authorship, as well as by informal communications of various types. The latter are considered to be of utmost importance. It may be classified as a spontaneous process which accelerates research through intellectual synergy which takes place at individual, personal level as well as among groups of researchers. Many scientists discuss their ongoing research with other scientists in order to obtain advice or to share knowledge about similar subjects of research. Communication network or "invisible college" links groups of collaborators "through their leaders who communicate with each other and transmit

0-7803-3552-X/96/$5.00 © 1996 IEEE

information informally across the whole field" (Crane 1972). Based on surveys conducted in several disciplines, large proportion of scientists "participate" in the network. Structure of such virtual network is shown in Fig. 1.

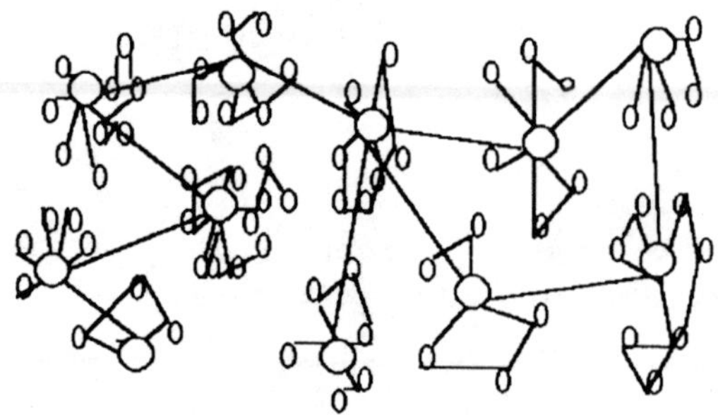

Fig. 1. Structure of an invisible college

Citations and co–authorship allow to track back the evolution of either passive connections (citations) or active contacts among scientists and their groups. It also allows to discover leaders and connections within the groups (Dou 1995). The same approach may be used for analysis of dynamics of inventions in a selected industry and for identification of the core R&D groups. Study of the analog recording technology at JVC shows how the structures of R&D teams/groups were evolving with time and who was the most productive inventor and author in the company (Ijichi, Yoda, Hirasawa 1994).

Format and extent of cooperation and knowledge sharing among scientists differ from case to case. Informal structures are better known in basic and theoretical research, such as mathematics, than in the technology related areas. An example in mathematics follows. In the 1930's, Polish mathematicians of Lwow used to visit the "Scottish" coffee house. They used to keep the record of current mathematical problems in a notebook that was deposited with the coffee house manager and made available to members of the group. This notebook, later called "The Scottish Book," played the role of a bulletin board in mathematics (Steinhaus 1992).

Potential advantage of networking and role of virtual structures in R&D has been recently analyzed in relation with technology projects or in the context of technology strategy of a company. R&D groups which cooperate in technology projects establish ties of different strength depending on the nature of projects. Based on the study of a set of major R&D projects in space engineering , it was possible to determine that typically the nonroutine projects are associated with large weak tie networks while routine projects with smaller strong tie networks (Gales, Tierney and Boynton 1995). In another recent study of R&D networking, focused on experience of a pharmaceutical company , Glaxo in the U. K., the authors (Albertini and Butler 1995) distinguish two basic reasons for R&D networking: "exploratory" in which the company looks outside in a focused way to get good new ideas for the internal discovery process, and "exploitative" in which the company searches outside for complementary knowledge and technology.

Knowledge based industry is becoming dependent on its capability to operate in dispersed mode, which requires coordination of technological capabilities with partners entering at different levels and in different stages of the innovation process. The virtual structures are emerging as a consequence of new strategic concepts. The question arises when and how a company should develop a virtual structure with outside partners, and when should it commit to internal development of technology. The case which qualifies for networked/virtual organization is when there is a definite market opportunity for an *autonomous innovation* while the company does not possess adequate capabilities. Outsourcing of technology development and creating a virtual structure allows to fill the gap in such a case. However, if the innovation is systemic in its nature, the company may consider some form of alliance but should proceed with caution to avoid a longer term negative consequences for its own capabilities (Chesbrough and Teece 1996). Analysis of strategic decisions concerning the virtual R&D lab has been presented by a team of authors (Harris et al. 1996) who distinguished 12 different types of situations as a combination of 4 categories of technology (key, emerging, basic and commodity) with 3 levels of technical capability (weak, moderate, strong) relative to achieving proprietary technical advantage. Conclusion of this study is that the virtual R&D lab is needed in only 4 situations. Such R&D structure can play one of 4 functions. Those functions include: (1) Collaboration and/or monitoring of emerging technology, if own capability is relatively weak, (2)Risk sharing in collaboration with a partner in emerging technology, if own capability is moderate, (3) Outsourcing for a basic technology, if own technical capability is weak and outsourcing is the only viable option, (4) Leveraging technical capabilities in basic technology, when own capability is at moderate level. The virtual R&D structure is never recommended for a key technology or in the situation of own technological strength.

Properties and Maps of Virtual Structures in the 1990's

Business conditions reflected in tendency to outsource as many activities as possible coincide with current advances in communication and information technology. The latter allows for fast transfer, processing and storing of data, information and knowledge on a massive scale at relatively low cost within unlimited networks. Virtual structures that emerged for different type of business activities, including that of R&D, address several needs and create new opportunities for growth of productivity. In case of R&D related virtual structures, those new opportunities include, but are not limited to, the following:

(1) *Just–in–time expertise* delivery system, which allows the company to reduce cost of maintenance of its own technological expertise. Expertise may be obtained from the networked partners or R&D subcontractors at any time.

(2) *Concurrent and synchronous R&D projects*, which can be managed and coordinated in real time across the geographically dispersed organizations.

(3) *Flexible R&D system,* in which a portfolio of projects does not need to be restricted by the internal set of technological competencies. To the contrary, a flexible set of competencies becomes available and is open to extension. The latter feature fits well with the concept of so called *hypertext organization* (Nonaka, Takeuchi 1995).

(4) *Customer oriented R&D system*, in which individual R&D tasks may be easily consulted with and adjusted to local customers through direct networking with them. On the basis of properties described here, it is possible to determine typical patterns of virtual structures in R&D. Those patterns may be mapped in two ways:

(1) According to *organizational linkages and networks*, to reflect different levels of interconnections across the organizational boundaries;

(2) According to *project decomposition* structures, to reflect different possible scopes of distributed tasks and responsibilities.

First of those mapping principles may serve for identification of missing links or potential for establishing new connections among the R&D organizations and companies involved in partnerships.

It is illustrated in Fig. 2, with indication to weak and/or strong ties in the joint structure.

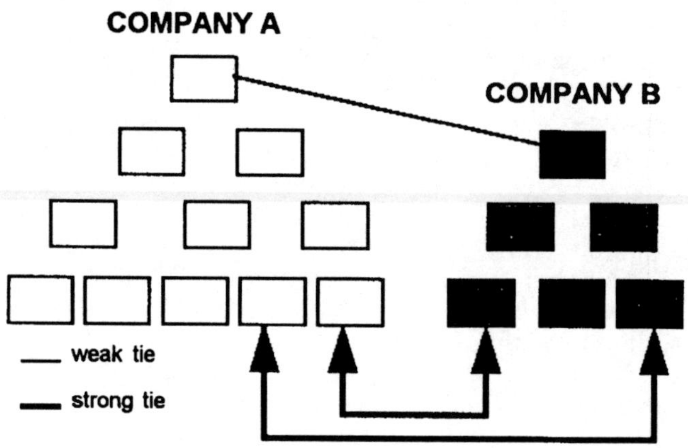

Fig. 2. Virtual structure: Inter-organizational ties

Second mapping convention allows to distinguish two potential patterns of R&D task allocation among partners involved. One of them assumes that the collaborating teams (in partner organizations) work together through all phases of the project with tasks assigned according to the modular structure of the object i.e. each team works on a different module or subassembly of the total design. The map of such module–based virtual structure is illustrated in Fig. 3a. Another possibility is to manage virtual R&D project through distributed phases of the project i.e. each team works on a different phase of the same project. This arrangement is more common in relatively well defined development projects e.g. new software development (Carmel 1996). The corresponding map is presented in Fig. 3b.

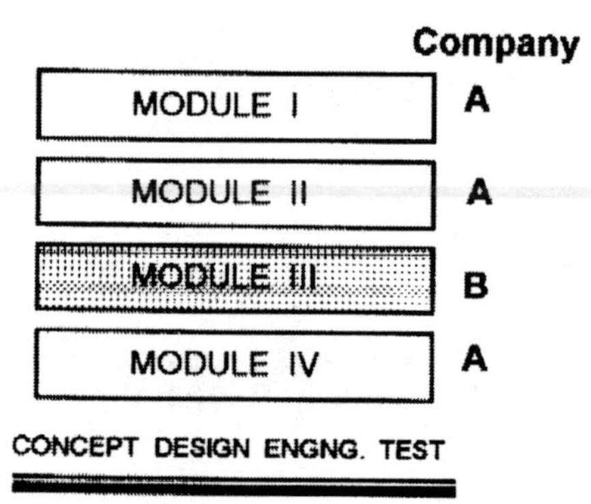

Fig. 3a. Module-based virtual structure

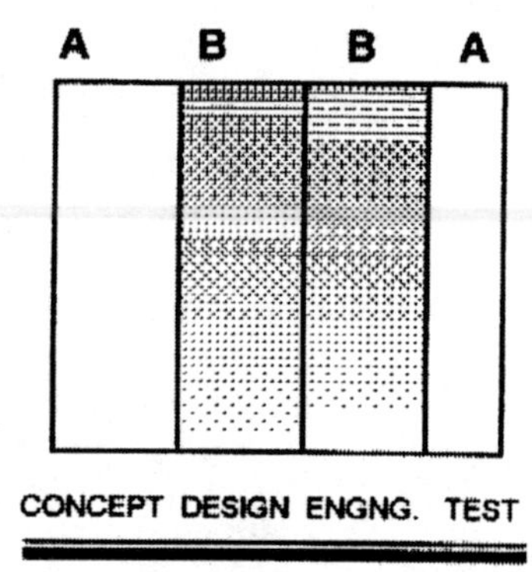

Fig. 3b. Phase-based virtual structure

Several advantages are common to all virtual collaborative systems implemented in the 1990's and integrated via computer networks. They include:

- *Virtual co–location* of individuals and teams participating in joint projects, independently of their distances in physical space.
- *Current research protocols,* which arise through computer recorded communications among participants/coauthors of the project.
- *Shared lab notes,* allowing for up–to–date reporting, sharing and recognition of results.
- *Shared data bases,* representing common reference for all parties involved in cooperation.
- *Interactive evaluation* of results and/or difficulties emerging during the life of R&D project.

Virtual structures are gaining importance for both the industrial and academic R&D environment. However the managerial experience in this domain is still inadequate to develop a well founded set of principles. The author believes that it is a promising field of research in engineering management.

References

Albertini, S., Butler, J. "R&D Networking in a Pharmaceutical Company: Some Implications for Human Resource Management." R&D Management 25(4): 377 – 393, 1995.

Carmel, E. "Globally Dispersed Software Development Teams." Management of Technology V: Technology Management in a Changing World. R. M. Mason, L. A. Lefebre, T. M. Khalil, (eds.). Elsevier Advanced Technology: Oxford, U. K., 1996.

Chesbrough, H. W., Teece, D. J. "When is Virtual Virtuous? Organizing for Innovation." Harvard Business Review:65 – 73, Jan.– Feb. 1996.

Crane, D. Invisible Colleges: Diffusion of Knowledge in Scientific Communities. The University of Chicago Press: Chicago, Illinois, 1972.

Dou, H., Dou Jr., J. M. "Virtual Research Universes and Research Management." Technology Analysis & Strategic Management 7(1): 109–122, 1995.

Drucker, P. "The Network Society." The Wall Street Journal: A4, March 29, 1995.

Gales, L. M., Tierney, P., Boynton, A. C. "The Nature of Information Ties and the Development of Technology." Advances in Global High–Technology Management, vol. 5, Part B, L. R. Gomez–Mejia and M. W, Lawless (eds.). JAI Press: Greenwich, CT, 1995.

Harris, R. C., Insinga, R. C., Morone, J., Werle, M. J. "The Virtual R&D Laboratory." Research * Technology Management 39(2): 32–36, 1996.

Ijchi, T., Yoda, T., Hirasawa, R. "Mapping R&D Network Dynamics: Analysis of the Development of Co–Author and Co–Inventor Relations," Paper presented at the Annual Meeting of the Academy of Management, Technology and Innovation Division, Dallas, TX, August 1994.

Nonaka, I., Takeuchi, H. The Knowledge–Creating Company. Oxford University Press: New York, N. Y., 1995.

Steinhaus, H. Wspomnienia i Zapiski. Aneks Publishers: London 1992.

Reengineering Businesses into 'Virtual Enterprises'

Clive Pereira
Telephone or Facsimile: (905) 270-3964

Principal Management Consultant and Business Architect
1st Reengineers International
1423 Mississauga Valley Blvd. Ste. 208
Mississauga, Ontario, L5A 4A5, Canada
E-mail: clive@icacomp.com
Web Port: http://pages.icacomp.com/~clive/1st.html

This presentation is intended for business architects and executives with vision and fortitude, who seek to reengineer their operations for competitive advantage and leadership, while repositioning their enterprises to leverage the new world of NAFTA, GATT and global commerce for prosperity and growth.
It will deal with:

- ***Phase I - Preparation***
- ***Phase II - Implementation***
- ***Phase III - Management***

Read-on and you will be rewarded with information that is vital for your personal and business success. When you are finished, you will be left with the knowledge that there truly is a light at the end of the tunnel, a straight path to arrive there, and a treasure awaiting the one who collects it first.

Overview

The greatest phenomenon of these times is the discovery of a new frontier as great as the discovery of the Americas. It is a new space we call cyberspace. In a very short time, this new discovery is going to dramatically change the nature of world commerce and social interaction.

It has been clearly demonstrated by leading businesses through the exploitation of this medium, that substantial new wealth can be generated. Business simplification, greatly reduced operational costs, enhanced market awareness, instant market access and substantially improved customer engagement are the desired objectives.

For the purpose of this writing, we shall focus on what it will take to develop and establish a successful presence in cyberspace. Before proceeding, however, you are encouraged to tour a model 'Virtual Enterprise' called '1st Widget Company' on the World Wide Web at **http://pages.icacomp.com/~clive/widget.html,** and wet your appetite for the fundamentals of setting shop for the new marketplace in the emerging new world order.

PHASE I - Preparation

I stress the importance of this phase by drawing the analogy to a commercial real estate re-development project. It would be unthinkable to proceed with the real estate project without an architect or professionally created plans, or to suggest that a plan could be drafted internally as a cost saving measure, perhaps by the current site manager. Likewise, to ignore or minimize this phase in the reengineering effort is to proceed without an optimized and expertly architected vision, and to do so may be fatal.

Business Discovery

The primary objective of a business discovery is to document and model the business functions before implementing change. In the real estate analogy, this is not unlike

0-7803-3552-X/96/$5.00 © 1996 IEEE

the architect calling for the preparation of as-built drawings for the current building and other surveys, prior to production of the rehabilitation blue-prints. In either case, the task is one that calls for the kind of science and professional expertise that is not usually found within the operations group of an enterprise.

The as-is model must represent the business in its' current state and deal with the critical issues for all the business functions. It should clearly identify the customers and the performers, and account for work done before and after the production in each process instance. It should include details of the preparation, negotiation, performance and approval of the process. Each of the four is dependent upon the other and it is important to identify these as an aid to problem determination and resolution. Customer satisfaction is essential. Therefore, the contract terms must be the focal point and clearly identified.

Reengineering Strategy

The greatest military strategists have demonstrated that a bold, offensive strategy is the best defence in war. It is important that business architects and executives accept that commerce is war, they are generals and their businesses are armies. Having implanted this belief, it is then essential to out-manoeuvre the competition and usurp the greater market share, by establishing a capability for intelligence acquisition, responsive change and aggressive growth.

A window of opportunity currently exists, where an out-flanking manoeuvre is possible by the rapid construction of new supply lines through uncharted territory, for access to and winning the favour of the selected target market. This new territory is the high ground of cyberspace. It is necessary to move in swiftly and powerfully, even with risk. It is time to reengineer business into a powerful, intelligent 'Virtual Enterprise'. For those who fail to seize the high ground first, an attempt to gain it later would be a formidable, if not futile task.

'Artificial Intelligence'

Some of the greatest generals won their wars because they had an extra trump in their favour. They had superior intelligence that enabled them to establish where and how the battle would be fought. That was the decisive advantage.

It should be important to weave the components of 'Artificial Intelligence' into the required reengineering model, if the goal is to create an intelligent 'Virtual Enterprise'. Therefore, such technologies should be explored and committed to in the preparation phase.

Software vendors often label their products as intelligent to achieve sales. However, most of these products contain nothing more than hard coded functionality. The best example of intelligence is a biological one. Human intelligence is not hard coded at birth, and nor is it static. It evolves and expands dynamically through most of its life span, and the state of this evolution is dependent upon the accumulation of knowledge and wisdom.

The intelligence required to meet the reengineering objectives can be artificially created by emulating this form. The technologies to create it are available and will be discussed later.

Concept Modelling

A well-conceived model is necessary for the establishment of a powerful vision, so all the horses and all the men may pull in the same direction efficiently. Such a model must be expertly planed and executed to avoid wasted expenditure and time. Savings here can be realized by recruiting professional help. This is certainly not an area to learn as you go.

While most company executives and managers are subjective and trained to think deductively, participants in this process must be objective and think inductively. It must be a drive to push the envelope of capability, by giving serious consideration to concepts and technologies that are ground breaking or futuristic.

Plan of Migration

When a Business Discovery and Reengineered Model are completed, it is possible to develop a plan of migration. Additional research into suitable technologies may be required. Re-education, reassignment and relocation of the business force should be considered. A logical sequence of events should be established in an urgent but realistic time frame.

Special consideration should be given to organizational culture, the cultural change that is required, the motivational incentives needed to effect the change and the time within which cultural change can be effected.

It is then time to appoint a qualified project leader for the execution of the Migration Plan. He should be inspirational, time centric and particularly skilled at communication, diplomacy, coaching and mentoring. Education and experience are no indication of these crucial skills.

The company must carefully draft a 'Vision Statement'. This is where many companies have failed. Their statements have often been prematurely released, without the benefit of models and plans. The input and support expected from their business force has, therefore, been unclear or absent.

It is important to dispel any fear, insecurity or resistance, and stress upon the rewards of an expanded market share, business growth and the value inherent in such an enriching experience. To do this, the statement must contain believable substance transposed from the earlier modelling and planning exercises.

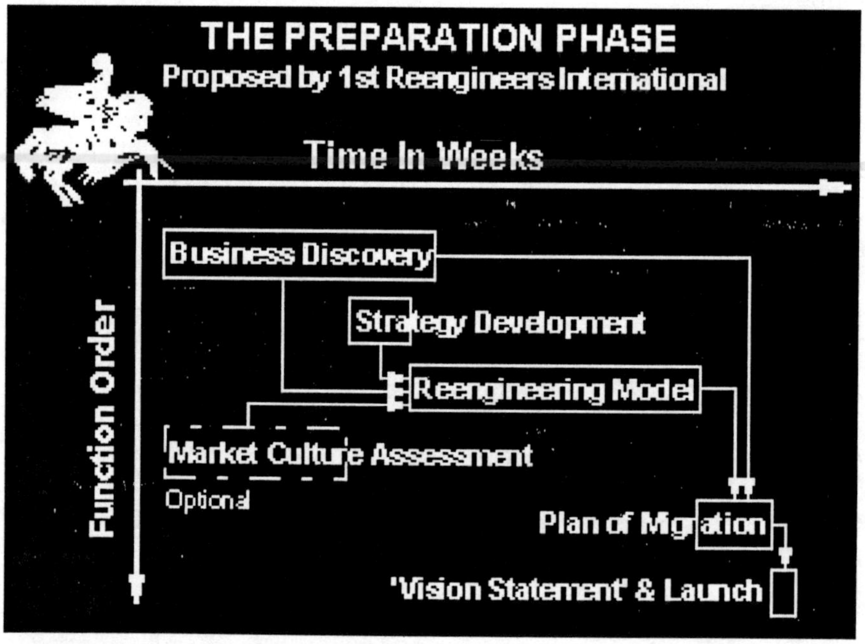

PHASE II - Implementation

It is better if this phase was achieved within 12 months or less, as a limitation of human endurance and to minimize the response time of the competitors. It is expected that professional management and procedures shall be in place and these shall not be discussed here. However, the technologies required to achieve the desired results are an important consideration.

Topology

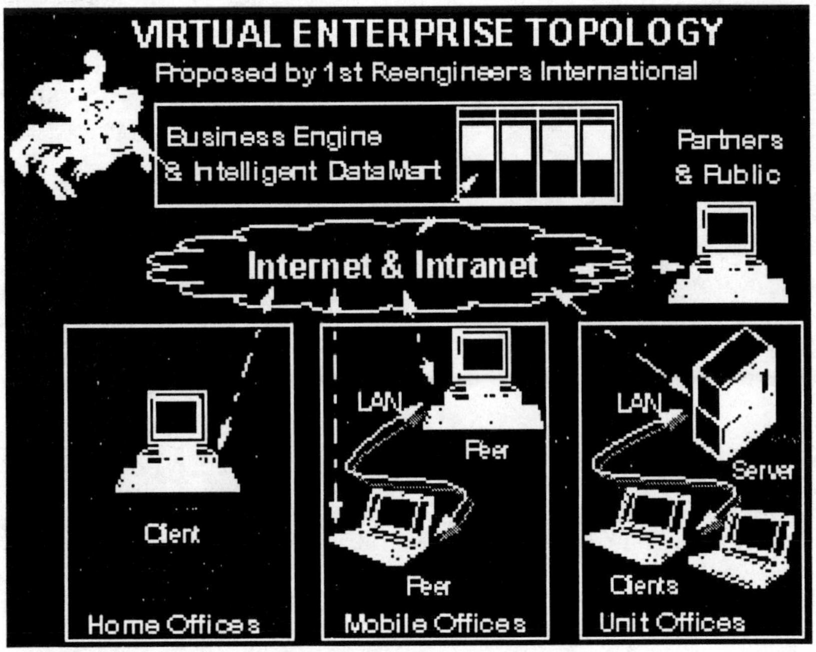

A major operational cost advantage can be gained with the above topology by establishing a home based tele-commuting business force, remote workgroups, labs and production facilities that involve one or multiple companies and business units. The target market can also be rewarded with the ability to cost-effectively integrate access within their systems.

Hardware

Hardware is expected to be an amalgamation of machines that are selected based upon functionality. Compatibility and scaleability must be prime considerations, as well as the manufactures' ability to frequently introduce breakthrough advancements and a migration path that protects and leverages the legacy investments. The company will require a combination of servers. It would be preferable and cost effective if these were of a scaleable parallel architecture.

Software

The primary software technologies considered must be cross platform capable with distributed operations and easy scaleability. The systems should be open and able to share distributed objects.

All server operating systems must be able to share and distribute objects, ad-hoc and relational databases, business applications and workflow. These server systems should also integrate with the Internet, interact with World Wide Web clients and agents as necessary, and respond to a common command language.

The client or peer operating systems should be universal in nature with emulations compatible with the servers, and able to share and support the same objects and databases. These machines should also respond to the same command language as the servers and integrate with the World Wide Web.

Integration should be accomplished in asynchronous mode, with multi-protocol communication redundancy and dynamic protocol encapsulation. Transactional messaging should be asynchronous, with fault tolerant functionality and assured delivery. Also important is the implementation of dynamic configuration and routing, and automation of congestion monitoring and control tasks.

Other technologies requiring consideration include, but are not limited to gateways and replication, digital imaging and conferencing, automation agents for personal productivity and system management, real-time system monitoring, global system management and support, network and data security, a secure log-on co-ordination system, etc. Such technologies go a long way in the recovery of the investment.

Intelligence

Delivery of intelligence would require integration of one or more Cerebrum Systems including a Data Cortex, Dynamic Neural Networks, Rule Induction Technology, Conceptual Clustering Technology, Association Discovery functions, Sequence Discovery functions, on-demand Business Views, desktop decision support tools and suitable data access drivers.

Technology Partnering

It is important to preserve the investment in legacy applications by integration and re-use of existing code, and to ensure that the investment in new software is scaleable. The only best way to do this is to standardize software procurement by partnering with a single software company able to deliver all the tools that would be required. Avoid the compatibility issues and high costs associated with work-arounds caused by a mixed bag environment that, by its' nature, could suppress or delay development of the enterprise.

PHASE III - Management

This phase is specific and dependent upon the nature of the company, it's culture, the authorities having jurisdiction and many more factors that cannot be addressed here. However, we will briefly discuss marketing and administration as it relates to the 'Virtual Enterprise'.

Marketing

It is absolutely vital that the new implementation be fully leveraged to ensure success. To do this would require a strong marketing campaign on all media. This includes press releases, direct information dissemination, liberal distribution of easy connectivity software, partner linkages with others on the cyber frontier, Internet advertising at suitable sites, a professionally designed company image and interface, customer enthusiasm nurturing, etc.

Departmental Administration

Most business today is structured into departmental silos that become the guarded domain of the managers and groups within them. However, business for the most part is a series of iterative processes that must transcend the departmental barriers in an inter-departmental journey during their life span.

Each process iteration has a cost attached to it, and the combined processes as a whole account for the operating costs of the company. The best way to gain competitive cost advantage, therefore, is to manage and execute these processes efficiently.

The problem inherent in a departmental operation is that no single person is empowered and made responsible for the outcome of a process. This neglect results in delays and inefficiencies that translate to higher process costs and a less than competitive business.

Process Administration

It is time to introduce here a new concept that I conceived and descriptively refer to as 'Process Franchising'. The concept lends itself to the management of a distributed operation like the one likely to develop on the cyber frontier, and it lays the foundation for rapid business expansion and diversification at the grass-roots level.

A process can be specified, designed, measured, priced, executed, managed, documented, audited and iterated. A company may, therefore, empower Process Managers and even establish Process Units or Process Teams. These may then be distributed appropriately. Doing so would vaporize most if not all the departmental silos that exist.

Just as it is possible to establish a Process Unit with a manager, procedures and accounting, it is also possible to assign some of the units to Process Owners. The Process Owner would own the Unit under a franchising agreement that is not unlike the franchising agreements of MacDonalds and others.

While the franchising of processes may not be appropriate for all reengineered models, it is concepts like this that will play an important part in the growth and success of many 'Virtual Enterprises' on the cyber frontier.

Summary

Success hinges on the expertise of the business architect and executives who seek to reengineer operations and reposition their company in the new world of NAFTA, GATT and global commerce. It will require a capability for intelligence acquisition, responsive change and aggressive growth. Take courage, however, and forge ahead. Delay or vacillation may play to the competition. The opportunity that presents itself is the light at the end of the tunnel. The path to arrive there has been revealed. The treasure awaits the one who collects it first.

Please direct all inquiries to the author.

The Virtual Enterprise and the Sources of Technology in Discontinuous Innovation

Lois S. Peters
Phone: (518) 276-2977
Fax: (518) 276-8661
peterl@rpi.edu

308 Lally School of Management and Technology
Rensselaer Polytechnic Institute
Troy NY 12180-3590

While there is recognition that technological advance is at the heart of the ability to introduce discontinuous innovation, most studies have not directly addressed the scope or characteristics of the technological contribution, or the management of the sources of the technological contribution. This paper focuses on identifying sources of discontinuous innovation, particularly with respect to internal and external technical alliances and how to manage the formation and productivity of technology partnerships. The initial findings of our study highlight the importance of virtual R&D organization for discontinuous innovation. A working model of virtuous virtual R&D for discontinuous innovation is presented.

Introduction

Technological Advance and Discontinuous Innovation

An appreciation of the temporal and organization sequences of integrating technology into the various states of new product development is fundamental to management of innovation (Van de Ven and Poole, 1990; Tushman and Anderson, 1986). It is a needed domain of research (Loveridge and Pitt, 1990) especially in the relatively neglected area of discontinuous innovation and new product development. Recent work emphasizes that the new technology integration stage is critical in product development (Iansiti, 1995).

When researchers are concerned about the technological contribution to innovation they focus on the degree of radicalness (Green, Gavin and Aiman-Smith, 1995;) and the factors, such as allocation of intellectual property rights, that effect the magnitude of innovation (Aghion and Tirole, 1994). An associated underlying assumption is that a discontinuous innovation is associated with breakthrough technology. Although it is generally recognized that technological innovation comprises many interacting stages and functions internal and external to the corporation, (Daghfous and White, 1994) systematic study of the nature of the technological contribution in these various stages by diverse sectors has rarely been undertaken.

Those concerned with discontinuous innovation have mainly focused on the impact of the radical or discontinuous innovation on the environment. They have documented its ability to change industry structure (Utterback and Suarez, 1991and Tushman and Nadler, 1986). Management research in this tradition has focused on how to manage technology introduction for competitive advantage (Mangematin and Callon, 1995), but not on how to develop technology to bring about industrial change.

Strategic Alliances, Innovation and New Product Development

Consideration of the relationship between environment and technological change has led to studies which illustrate how the process of forming industrial infrastructure facilitates and inhibits technological innovation and provides a framework for understanding the emergence of new industries. Such work points to the importance of strategic alliances in bringing about technological change. Pisano (1990) found that choices in in-house and external sources of R&D are effected when technological

0-7803-3552-X/96/$5.00

change shifts the locus of R&D expertise, but didn't deal with the processes of selecting partners and managing cooperation within the context of discontinuous innovation.

There is also increasing recognition that networks formed from strategic alliances, effect the character of technological innovation and new product development (Kotabe, 1995). The extent of interfirm networking has been associated with the scope of innovations and their degree of incrementality and impact on technological change (Harianto and Pennings, 1994), but there has been almost no research that explicitly addresses the process of managing networks for developing technology in discontinuous innovation projects.

This paper therefore focuses on the following basic questions: What are the technolgical origins of the discontinuous innovation project? What is the role of strategic alliances and technology partnerships in the technological evolution of discontinuous new product development? What is the importance of corporate commitment to technology for discontinuous innovation? When is virtual R&D virtous in discontinuous innovation? Strategic alliances for the purposes of this paper include a broad variety of agreements between firms such as joint ventures and technology exchange agreements or between universities or national labs and firms (see Peters 1995), that are said to be an integral part of the firms new product development strategy. Virtual organization refers to the networks formed by these alliances (Chesbrough and Teece, 1996)

Methodology

The empirical data for this paper is derived from a prospective study of corporate venturing related to the processes involved in the management of discontinuous innovation in established firms. During the pilot phase (Year 1) eight projects were proposed by six participating firms as examples of discontinuous innovation projects, with two firms submitting two discontinuous innovation projects for study. At this stage the sample reflects two primary dimensions: diversity in the technological environment and stage of development. A variety of information gathering techniques are employed, including in-depth interviews, review of project documentation, surveys and interim follow-up phone interviews. This research effort is deliberately designed as an exploratory, theory-building study. At least two visits have been made to all six participating firms. Preliminary observations regarding strategic alliances and the technological contribution to innovation are reported here.

Major Findings

Preliminary Results

Our initial findings are that opportunistic management of an internal technological capability in search of legitimization can be important in the launching of a discontinuous innovation project. Frequently the stage is set by recognition of the convergence of technological advances that can be used in new ways to solve persisting industry problems or perceived customer needs.

In all our cases the initial impetus for the discontinuous innovation project was key technology internal to the company. In seven of the cases the organizational source was corporate R&D and in one case the contribution was from R&D in a SBU and corporate research. During formalization of these projects external organizations as contributors to technology development became involved in all cases. Each of the projects examined involved at least one strategic alliance and in most cases many more were a feature of the projects development. A variety of types of partners in technology development is possible as indicated in the variety of external organizational sources of technology involved in the projects. Partners in our sample included small high tech firms, major established firms, suppliers, government laboratories,universities and independent research laboratories.

Only in three of the projects were the characteristics of the key technology described as a breakthrough and in one of those cases it involved disproving a theory about a familiar technology. In the other two "breakthru" cases the technology had relevance to existing markets rather than new markets. The other projects were built on new applications of familiar technologies or market opportunities provided by convergent or synergistic technological advances.

External sources rarely contributed to core project technological capabilities. As the projects evolved multiple sources of technology were required and external sources of technology became more important. They contributed in varying degrees by project to complementary, synergistic, convergent, development or manufacturing technology. Strategic allianaces also served to keep an eye on competing technologies and provide access to related cutting edge technolgies that could provde entrace into new markets. All projects exhibited multiple alliance strategies involving both technology and market objectives.

Table 1. Alliance strategies

	Proj.1	Proj.2	Proj.3	Proj.4	Proj.5	Proj.6	Proj.7	Proj.8
Acquire key technology								
Acquire complementary technology	X							
Acquire support technology		X						
Joint technology development	X		X					?
Application development		X			X	X		
Product testing				X		X	X	
Market development							X	
Manufacturing	X	X		X	X	X	?	?

A Preliminary Conceptual Model of the Role of Strategic Alliances in Discontinuous Innovation

Three technological conditions seem to provide the foundation for discontinuous innovation projects in established firms - technical breakthroughs, technology gaps, or converging technological advances. The technologies key to the project except for one case all seem to have long company histories going back to the seventies in some cases and the mid eighties for most. Within R&D there seems to be continual product technology interactions that push the technology to the next stage. Sometimes the technology must be shelved until necessary convergent or synergistic technical advances are made. In other cases it is a matter of perceived market readiness. In all cases an outside trigger was an important part of the impetus to formalize the technology development as a discontinuous innovation project. In all cases but one corporate turbulence was a factor. Declining markets were involved in two cases. In four cases it was a matter of job security or restructuring where the technology was viewed as non essential for company business strategy. Two cases primarily reflected company reorientation to a product development focus.

For all cases studied, companies focused on either competing in familiar markets with revolutionary technology or existing markets but new markets to the company with their newly develop technology. No one was attempting to develop totally new industries. Nevertheless it was still difficult for most project to find a company home outside of R&D. A salient feature of our results to date indicate that alliances seem to be important in circumventing either resistance from other parts of the company to new product development or lack of an organizational home. Figure one presents a preliminary conceptual model that depicts the relationship of strategic alliances and external sources of technology to discontinuous innovation projects.

FIGURE 1

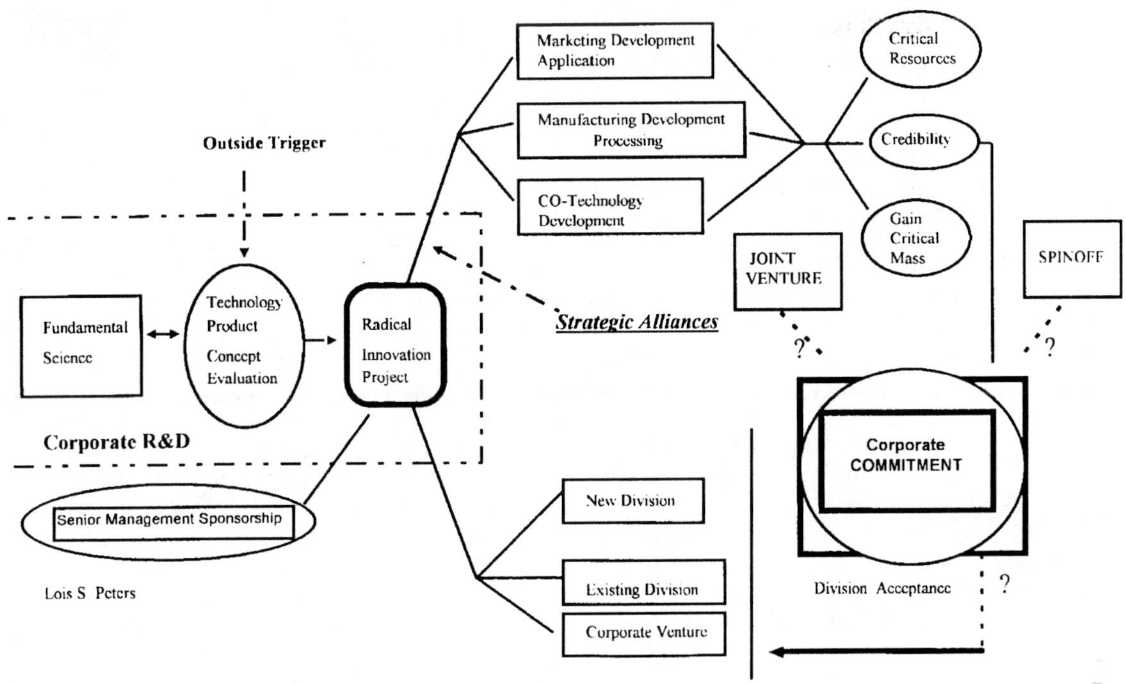

Discussion and Implications

Discussion

This article has attempted to explain some of the features of strategic alliances and the technological contribution to discontinuous innovation. Recent literature points to the importance of strategic alliances in the innovation process. We find here that they are key to the discontinuous innovation process especially after they have been formalized as projects. The literature has suggested that established firms might be better off if they engaged in discontinuous innovation through technological acquisition and strategic alliances that reduce risk. Here we find that they are engaging in strategic alliance to reduce risk but the initial key technology-product concepts are developed by the established firms. The variety of technological contributions and sources appears to increase as the project evolves. Breakthrough technology is not necessarily involved in discontinuous innovation. We are not far enough along in the study yet to determine how the constellation and objectives of partnerships are impacted by technological focus or by phases of project development. As the project gets closer to commercialization it appears that partnerships may become more numerous but this is mediated by technological business objectives. The conceptual model presented suggests strategic alliances brings credibility as well as reduced risk to an established firm's discontinuous innovation project. They leverage the firms technology capabilities to extract maximum value from technology investments and can enable entrance into new markets. The network of technological activity surrounding discontinuous new product innovation appears to be important for gaining corporate commitment to commercialization.

References

Aghion, P., Tirole, J. The management of innovation, The Quarterly Journal of Economics, pp. 1185-1209.

Chesbrough, Henry W. and David J. Teece. When is virtual virtuous? Organiziing for innovation. Harvard business Review, January-February 1996 pp.65-73.

Daghfous, A., White, G.R. Information and innovation: a comprehensive representation. Research Policy, 23, (1994) pp. 267-280.

Green, S.G., Gavin, M.B., Aiman-Smith, L. Assessing a multidimensional measure of radical technological innovation. IEEE Transactions on Engineering Management, 42, 3, (1995) pp. 203-214.

Harianto, F., Pennings, J. M. Technological convergence and scope of organizational innovation. Research Policy, 23, (1994) pp. 294-304.

Iansiti, M. Technology development and integration: An empirical study of the interaction between applied science and product development. IEEE Transactions on Engineering Management, 42, 3, (1995) pp. 259-269.

Kotabe, M., et al. The Role of Strategic Alliances I High-Technology New Product Development. Strategic Management Journal, 16, (1995) pp. 621-636.

Mangematin, V., Callon, M. Technological competition, strategies of the firms and the choice of the first users: The case of road guidance technologies. Research Policy, 24, (1995) pp. 441-458.

Peters, Lois S. The Dimensions of Strategic Leadership in Technical Hybrid Organizational Relationships. Advances in Global High-Technology Management, Volume 5, (1995), Part B, pp. 171-200.

Pisano, G.P. The R&D boundaries of the firm: An empirical analysis, Administrative Science Quarterly, 35, (1990) pp. 153-178.

Tushman, M.L., Anderson, P. D. Technological discontinuities and organizational environments. Administrative Science Quarterly, 31, (1986) pp. 439 -464.

Tushman, Michael L., and David Nadler. 1986. Organizing for Innovation. California Management Review, 28, (3), pp. 74-92.

Utterback, J.M., Suarez, F.F. Innovation, competition and industry structure. Research Policy, 22, (1993) pp. 1-21.

Van de Ven, A.H., Poole, M.S. Methods for studying innovation development in the Minnesota innovation research program. Organization Science, 1, 3, August 1990, pp. 313-335.

Process Modeling to Support Integration of Business Practices and Processes in Virtual Enterprises

Adrien Presley
K. Jamie Rogers
817.794.5900

The University of Texas at Arlington
Automation & Robotics Research Institute
7300 Jack Newell Blvd. S.
Fort Worth, TX 76118

A major issue in the management of virtual enterprises is the integration and coordination of the business processes. The successful attainment of the business goals of a virtual enterprise depends on its ability to align the business processes and practices of partner enterprises. This paper will present a modeling scheme which supports a process centered approach to the analysis and design of the virtual enterprise. An innovative feature of the scheme is the ability to present federated activity, process, business rule, organization, and resource views of a process.

Introduction

Continued competitiveness by North American manufacturers will depend on their ability to participate in an emerging organizational form described as the "virtual enterprise". The virtual enterprise (VE) is a form of joint venture but with important differences. It can be viewed as a temporary alliance of member companies which join to take advantage of a market opportunity. This VE will possess almost no employees or inventoried resources. Each member company will provide its own core competencies in areas such as marketing, engineering, and manufacturing to the VE. Only a small headquarters staff may be required to deal with administrative and management details but the actual work will be performed by geographically separated shareholder joined through electronic networks. When the market opportunity has passed, the VE could be dissolved or reconfigured to address other market opportunities. The VE differs from existing inter-organizational models by the degree of shared accountability and responsibility of the participants and the structure by which companies contribute their competencies through ìplug compatibleî processes (Reid et al. 1994).

A major issue in the formation of the virtual enterprise is the rapid integration of the business processes of the participating companies. While the integration of computer and communication technologies are no doubt critical issues, the successful attainment of the business goals of the virtual enterprise often depends on its ability to align the business processes and practices of partner enterprises. This paper will present a modeling scheme which supports a process centered approach to the analysis and design of the virtual enterprise. The scheme, currently under development, will utilize concepts of object-oriented and agent modeling techniques. An innovative feature of the scheme is the ability to present federated activity, process, business rule, organization, and resource views of the process. The scheme is built around an IDEF0 activity model supplemented with IDEF3 and IDEF5 models to present to the other views. The paper will first discuss the enterprise modeling architecture which the scheme supports. The particular modeling needs of a virtual enterprise are discussed. The scheme is then described in conjunction with the methodology for the creation of the integrated model.

Enterprise Modeling Architecture

An enterprise architecture can be thought of as a "blueprint" or "picture" which assists in the design of an enterprise. An enterprise modeling architecture should assist companies desiring to enter into a virtual relationship by defining the functions and

0-7803-3552-X/96/$5.00 © 1996 IEEE

interfaces of critical business processes, thus allowing for efficient integration of the expertise contributed by each partner. An enterprise architecture must define three things. First, what are the activities that an enterprise performs? Second, how should these activities be performed? And finally, how should the enterprise be constructed?

An enterprise is a collection of enterprise activities organized into a set of business processes which cooperate to produce desired enterprise results (Presley 1993). In the context of this paper, an enterprise activity is defined as any organized behavior which transforms inputs into outputs. This paper proposes that enterprise activities are the basic building blocks of an enterprise and, furthermore, that enterprise activities become useful only when organized into business processes.

An enterprise must be viewed from several perspectives if it is to be fully described and understood (Barnett 1994; ESPIRIT Consortium AMICE 1991). Previous work in the development of architectures by the Automation & Robotics Research Institute (Presley et al. 1993) describes a five view approach.

1. Business Rule (or Information) View: Defines the entities managed by the enterprise and the rules governing their relationships
2. Activity View: Defines the functions performed by the enterprise (what is done)
3. Business Process View: Defines the time sequenced set of processes (how it is done)
4. Resource View: Defines the resources and capabilities managed by the enterprise.
5. Organization View: Defines how the enterprise organizes itself and the set of constraints and rules governing how it manages itself and its processes.

The nature of the Virtual Enterprise raises several issues for the modeling and design of the enterprise. First is the temporary and fleeting nature of the relationship. Companies must find suitable partners and bring together their core competencies rapidly. This implies the need for a representation method allowing for analysis of partner capabilities. The interfaces of processes must be clearly defined to allow for plug compatibility. Provisions must be present for the protection of intellectual properties and core competencies of participants. It has been suggested that the virtual enterprise, or any environment characterized by fast paced change, should be modeled as a series of interrelated component parts (Barnett et al. 1994) or ìblack boxesî. Black box processes permit information and material flow across the virtual enterprise while simultaneously protecting the proprietary nature of member organization business processes. Such an approach would also by definition address the issue of enabling rapid reconfiguration of processes by member organizations.

The Modeling Scheme

In this section, the modeling scheme being developed to address the needs of the enterprise modeling architecture is discussed. The concentration of this research will be on higher level enterprise processes, such as would be encountered when configuring a Virtual Enterprise. It is proposed that at this level, a major issue will be the selection of the mechanisms (partner enterprises) to perform a process. Therefore, a representation method supporting coupling of processes and the agent performing the process is needed. Special emphasis on selection and configuration of processes for the configuration of agile virtual enterprises will be considered. The model is developed by progressively specifying the five views, starting with the business rule view and ending with the organizational view. The methodology being specified identifies the following five highest level steps in creating a model:

1. Define and Plan Project
2. Create IDEF5 Business Rule Model
3. Create IDEF0 Activity Model
4. Identify and Assign Resources
5. Specify Business Process Model
6. Specify Organization Model.

The following sections will describe the steps and the part of the model developed at each step.

Create IDEF5 Business Rule Model

A business rule model identifies the objects of interest in a particular domain and their relationships. In this way, the business rule view is closely related to the concept of an ontology. According to Benjamin, et al (Benjamin 1995, page 225), ìan ontology is a description of the kinds of things, both physical and conceptual, that make up a given domain, their associated properties, and the relationships that hold among them as represented by the terminology in that domain.î. Within the IDEF suite, the IDEF5 ontology capture method (Benjamin 1994) was developed for this purpose. It facilitates the collection of knowledge about physical and conceptual objects along with their associations. It provides facilities for diagrammatic representations of an ontology. In addition, a structured text language for detailed ontology characterization is present.

The ontology model can itself be thought of as the integrated model of the enterprise. Figure 1 shows small part of a model for a generic enterprise. The circles represent kinds (roughly equivalent to class or type) with arrows representing relations. The creation of this model is extremely important as the ontology model will form the basis for the other four views. The model is created through knowledge acquisition methods such as interviews and reviews of existing models and documents. The modeling scheme and method will provide a limited set of predefined objects and relations for creating an enterprise model.

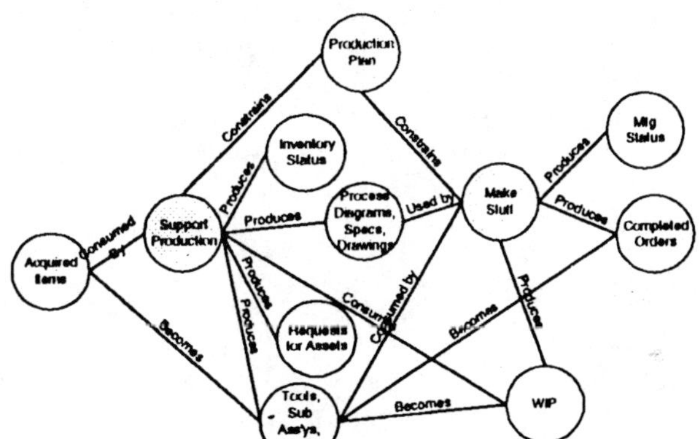

Figure 1. Partial ontology Model

Create IDEF0 Activity Model

An IDEF0 activity model is created based on the business rule model created in the previous step. The use of the IDEF5 model will enforce a level of structure to the standard IDEF0 process. The model is bounded by identifying what is to be included in the model. A process can be considered as a special relationship that exists among objects. In this methodology, however, we define processes as objects. This allows us to capture additional knowledge and make inferences about the processes. The identified processes are used to create a list of candidate activities. The objects and activities are defined in additional detail, with the scheme enforcing some properties which must be defined for the model. This will be done by defining a small core set of properties for activities, mechanisms, constraints, and objects manipulated by activities. The activity model created in this activity will be identical to a "standard" IDEF0 model. At this point however, actual mechanisms are not identified, but rather, capabilities or kinds of mechanisms required are identified. As part of this activity the ontology specification for activities is updated. Properties defined include a description of the activity, and required capabilities.

Identify and Assign Resources

To this point, the model has been partially instantiated. The purpose of this step is to define and select agents capable of performing the activities. This is done by considering the capability requirements previously identified and matching these to potential resources. In the case of a virtual enterprise, the resources are potential partner companies. The assignment is done through user defined selection rules with the aid of available analysis tools. Once the agents to perform the activities are identified, we can define the performance of the enterprise. Only when an activity and a resource are encapsulated can the process, organization, and resource views have meaning.

Specify Business Process View

This step creates the business process view for the agent selected in the previous step. The business process view defines the time ordered sequence in which a process is executed. The reason that the process view is not specified until the agent is defined is that because for the purpose of designing an enterprise, we are interested in the activity view (ìwhat is doneî). That is, knowing ìwhat needs to be done?î for an enterprise, we are now interested in the ìwhat can it do?î when selecting resources. The agent approach taken with this model in which each resource can perform an activity any way it seems fit means that the ìhow does it do itî (internal behaviour) is not of interest until the agent is defined. The behaviour is defined by the set of internal processes used by the holon. For any agent, several process scenarios can be created. Examples include alternate ways of performing a process based on existing constraints, and the tracing of alternate inputs through the agent. The specification of the process model is accomplished by again examining the IDEF5 Business Rule Model. It may be necessary to develop additional IDEF5 models to define information which was not previously available. This process view is specified using the IDEF3 Process Description Capture Method. The method consists of process flow diagrams and elaboration diagrams. One of the appealing aspects of IDEF3 is its relationship to the IDEF5 ontology capture method through the use of object state transition diagrams.

Specify Organization Model

The organization view defines the reporting and constraint structures put in place. Each of these is defined by examining the IDEF5 business rule. A limited set of predefined kinds and relations will be provided within the scheme to assist in developing this view. The modeler would, by identifying the instances of each kind, complete the model. The result of this activity is an organization view specified in IDEF5 along with additional input to the structured elaboration.

Conclusion and Future Directions

A modeling scheme and methodology to support the configuration of business processes and virtual enterprises is presented. The goal of this research is to develop a modeling scheme to support manual and automated analysis and design of enterprise processes. Automated analysis and design would be supported by AI techniques, especially in the evaluation of resources and alternate ways of performing a process. The ontology based approach is being pursued to specifically facilitate the use of automated reasoning tools. The reasoning would incorporate both quantitative and qualitative criteria as defined by the decision maker.

Acknowledgments

Research for this paper is funded in part by the National Center for Manufacturing Sciences, the State of Texas Advanced Technology Program Grant 003656-036 and by the National Science Foundation sponsored Agile Aerospace Manufacturing Research Center.

References

Barnett, W., A. Presley, M. Johnson, and D. Liles. ìAn Architecture for the Virtual Enterprise.î IEEE International Conference on Systems, Man, and Cybernetics, San Antonio, TX, 1994.

Benjamin, P. C., R. J. Mayer, T. M. Blinn, C. Marshall, and R. Tye. ìToward an Environment for Collaborative Enterprise Engineering.î Society for Enterprise Engineering (SEE) Conference, Orlando, FL, 1995.

ESPIRIT Consortium AMICE. Computer Integrated Manufacturing: Open Systems Architecture, Springer-Verlag, Berlin, 1991.

KBSI. Ontology Capture Method (IDEF 5). Knowledge Based Systems, Inc., College Station, TX, 1994.

Presley, A., B. Huff, and D. Liles. ìA Comprehensive Enterprise Model for Small Manufacturers.î 2nd Industrial Engineering Research Conference, Los Angeles, CA, 1993.

Reid, L., D. Liles, K. Rogers, and M. Johnson. ìA Methodology for Engineering the Virtual Enterprise.î Agility Forum Conference, Boston, MA, 1996.

Best Manufacturing Practices... A Uniquely Successful Approach To Teaming And Partnering

Richard L. Purcell
Tel: (301) 403-8100
Fax: (301) 403-8180

Best Manufacturing Practices Center of Excellence
4321 Hartwick Road, Suite 400
College Park, Maryland 20740

There is nothing quite like Best Manufacturing Practices (BMP) in either government or the private sector. BMP is a cooperative effort involving government, industry and universities. For most of its ten-year existence, BMP operated as a "virtual" entity without formal recognition or significant financial support yet was able to effectively harness increasingly greater amounts of people, resources and funding to achieve remarkable success helping to improve the competitiveness of U.S. industry. This paper highlights how BMP developed and implemented successful strategies for teaming and partnering with government agencies, industries, trade associations and academic institutions. The BMP approach is unique in leveraging resources within government to add value and maximize results. It provides a model for how public and private organizations can work together effectively. The BMP approach produced successful strategies for attracting people and resources from a broad range of agencies, institutions and private sector organizations. These are documented along with techniques for leveraging customer focus and delivering products that add value and meet needs. Traps and pitfalls in teaming and partnering and how to avoid them are also addressed.

Best Manufacturing Practices... A Uniquely Successful Approach to Teaming and Partnering Best

Manufacturing Practices (BMP) is a unique government program that has operated successfully and grown through the development and application of effective teaming and partnering strategies involving government, industry, and academic organizations. For most of its ten year existence, BMP operated as a "virtual" program without formal recognition, designation, or a budget. In spite of these seeming shortcomings, BMP was able to attract increasingly greater amounts of people, resources, and funding and apply them effectively to achieve remarkable success. Many of the strategies and methods that BMP uses within and without government to leverage resources, add value, and maximize results have served as models for other agencies to follow for effective teaming and partnering.

Successful strategies for attracting people and resources from a broad range of agencies and institutions have been a key element of BMP. Teaming and partnering are used to leverage these resources and attract funding. The foundation for all of this is unwavering customer focus and the commitment to deliver products that add value and meet real needs. BMP's openness and involvement of a broad range of participants produce a vibrant entity that meets real needs and provides the kind of tools and resources that the industrial base seeks out and uses.

What is BMP?

In 1985, the United States Navy created the BMP Program, bringing together a task force of the best minds in the fields of engineering, computer science, technology and management. Its mission was to help U.S. companies, particularly those in the defense industry, respond to the biggest challenges facing the American economy: decreased defense spending and increased competition from foreign markets. The BMP goal was clear: make U.S. companies stronger, more competitive and more globally minded.

In the past 10 years, BMP has worked to reach that goal with numerous Fortune 500 companies, such as Texas Instruments, IBM, General Dynamics and Boeing, to help

0-7803-3552-X/96/$5.00 © 1996 IEEE

them improve manufacturing practices, benchmark efforts, cut costs, and improve product quality.

Today, the BMP Program has a 30-plus staff of engineering, computer science and management experts. Its funding, which has grown from $50,000 in 1985 to almost $4 million this year, comes primarily from the U.S. Navy. Headquartered in College Park, Md., the BMP Program also operates six satellite offices: Oak Ridge, TN; York, PA; Corona, CA; Oakland, CA; Rockford, IL; and Louisville, KY. And, in September 1994, BMP, the Canadian Department of National Defense and Industry Canada -- an association of commercial companies -- joined forces to launch a Canadian BMP program -- adding best practices from Canadian companies to the existing database of U.S. companies and vice-versa.

How Does it Work?

By identifying the best practices being used in the government, industrial, and academic communities and then promoting the widest possible dissemination of them, BMP has been an engine of progress for the U.S. industrial base.

The benefits generated by BMP begin with a time and money saving way to facilitate technology transfer between companies, thus reducing inefficient, repetitive problem solving. In the process of uncovering today's best practices, BMP promotes excellence in industry, government, and education. In addition, BMP serves as a free source of benchmarking and expert assistance to help American industry become more competitive. Toward that objective, BMP helps companies achieve both incremental improvements and major breakthroughs while avoiding costly mistakes.

The results of the BMP program have been outstanding. Texas Instruments credits BMP for contributing to its winning the Malcolm Baldridge Award; a turbine blade manufacturer realized a 500 percent improvement in productivity and saved $4 million; and one manufacturer was able to avoid a shutdown.

Products and Services

The products and services provided by BMP address real needs in government and industry and add value in areas important to its customer base. These include:

- BMP Surveys - Over 85 surveys have been conducted in which more than 3000 individual best practices have been identified and documented

- Survey Reports - In depth reports that describe best practices and areas of excellence identified during the surveys.

- BMPnet - An electronic network accessible via the Internet and World Wide Web that links thousands of organizations, companies and experts and provides resources such as knowledge bases, communications, expert systems, automated productivity tools, and problem solving aids.

- Program Manager's Workstation (PMWS) - A suite of integrated software tools including the BMP database of verified best practices; KnowHow, an electronic library of expert technical assistance with an intelligent search capability; Technical Risk Identification and Mitigation System (TRIMS), a technical risk management system that may be tailored to the user's needs; and SpecRite, a performance specification development tool. PMWS is available via the BMPnet or on Compact Disk.

- A growing series of "how-to" guidelines documents for selected technical processes and procedures.

- Workshops and seminars.

Leveraging for Success

BMP has never had extensive financial resources or large numbers of full time staff personnel. In order to survive and operate effectively, the program was forced to leverage resources and develop cooperative efforts with other organizations to achieve its objectives. In fact, today there are only two full time government employees assigned to the program. The other members of the core team operating the program are loaned executives from other agencies, government employees who devote part of their time to BMP, and a small cadre of contractors.

BMP is a sub-element of the Navy Manufacturing Technology Program (MANTECH) but, until recently, did not have an identifiable budget line and so maintained a low profile within the Navy. This low level of visibility allowed BMP to evolve relatively undisturbed while its parent, MANTECH, experienced more than a dozen major reorganizations, downsizings, and restructurings. Although the MANTECH program itself remained largely intact and actually experienced substantial growth during this period, it was subjected to frequent budget driven stops and starts and their consequences which are typical in large government programs. By staying small and developing relationships to leverage resources, BMP was able to stay focused on its core objectives.

The Best People - Free!

In order to operate and grow, BMP needed a good supply of highly qualified and experienced technical experts in manufacturing, engineering, management, and specific technical disciplines. Ernie Renner, the Director of BMP, got these people by setting up informal agreements with other government organizations. Agencies such as the Commerce Department, Energy Department, Defense Department, NASA, FAA, Army, Navy, Air Force, national labs, and others loan experts to BMP to serve as survey team members for up to a week at a time. In addition, the agencies cover the costs for salaries and travel. In return the agencies get professional development for their employees, exposure to best practices that can help them improve, and recognition. Often new team members are attracted after a BMP survey at a government facility. They and their management see and understand the value of BMP and want to get involved. Currently there are over 150 people in the pool of technical experts that BMP draws on for survey team members.

Leveraging people resources has been the primary strategy for keeping BMP's actual operating costs low. A typical survey will involve 12 to 15 people at the survey site for 5 days. Salaries and travel costs for a group like this could amount to about $50,000. Annually, with 12 to 15 surveys conducted, this amounts to" in kind" contributions of over $600,000 in value. Other resources from government agencies include assistance in covering printing and distribution costs and providing facilities and support for meetings and exhibits.

Fundamental Principles

The BMP approach to teaming, partnering, and leveraging resources is built on two fundamentalprinciples: customer focus and return on investment.

Customer focus is the most important of these. BMP has a real customer base that uses the products and services and expects to receive value. BMP was built from ground zero on customer relationships. The focus has always been on identifying what the needs of the customer base are and determining how or if BMP can make a value added contribution. Each year BMP gathers its key customers and stakeholders together to review the program and assess whether it is still meeting the needs of its customer base. They decide if the program should continue and what improvements are needed to better meet customer requirements. Customer feedback and participation have been important in keeping BMP from developing a bureaucratic structure and life of its own.

Who are these customers? They are in fact the very people and organizations that have partnered and teamed to make BMP successful. Sixty percent of the customer base comes from industry. These are typically large manufacturing companies with strong representation from major defense contractors, but there is increasingly greater participation from non-defense commercial companies and small and medium size firms. The next largest segment, making up about 30 percent of the customer base, is government. Agencies represented include Energy, Defense, Commerce, FAA, NASA, Federal Laboratories, TVA and others. Colleges and universities account for about ten percent of the customer base and this percentage is growing as much of the knowledge and resources developed through BMP are illcorporated into graduate and undergraduate programs.

The other essential element common to all of the organizations that team or partner with BMP is return on investment (ROI). Most of the contributions that other organizations bring to BMP are non monetary such as people or resources that are already serving some productive purpose. Nevertheless, these organizations seek some sort of return for making their resources available to BMP. BMP in turn provides its products and services and often a cash contribution for which ROI is an important consideration. All the parties expect to gain from the partnership. In every case this return must be clearly definable. The following are specific examples of partnerships that created win-win situations for BMP and its partners.

The Department of Energy has been one of BMP's strongest and most effective partners. They have contributed many technical experts to the BMP survey teams, established a BMP satellite center at the Oak Ridge Centers for Manufacturing Technology, hosted BMP surveys at several of their facilities (including Oak Ridge, Sandia, and Pantex), teamed on exhibits at major conferences and workshops, and provided financial support. In return they have received access to all the BMP products and services, broad exposure throughout the industrial base of their capabilities and areas of excellence, marketing for their work for others programs, benchmarking, technology transfer opportunities, and expanded opportunities for Navy work at their industrial centers.

BMP has established partnerships with many industry associations such as The Aerospace Industries Association, National Center for the Advancement of Manufacturing, National Tooling and Machining Association, Manufacturers Alliance for Productivity and Innovation, the American Iron and Steel Institute, and many others. These relationships provide BMP with a low cost and highly effective means to reach out to companies in specific industry segments. In return, the associations gain access to a whole range of BMP resources they can make available to their members. These resources provide real value to their member companies. Many of the associations are actively promoting and using PMWS. They are putting many of their documents on the BMPnet for wider dissemination such as national skills standards or the NAS-411 Environmental Management Standard. One association provided BMP with $ 100,000 worth of advanced computer equipment and in return BMP set up and now runs the association's Internet home page.

BMP works closely with industry. Industry contributes its best practices and many members of industry contribute their time and effort in helping BMP resolve industry wide problems or by speaking at BMP sponsored conferences and workshops. Some companies such as Litton, Texas Instruments, and Lockheed-Martin have been involved with BMP for many years and have hosted surveys at multiple plant sites. Companies benefit from BMP a variety of ways including, benchmarking, reengineering, consulting, problem solving, new business, and partnerships with other companies.

Each company benefits differently from its involvement with BMP. At Hamilton Standard Electronic Manufacturing Center, BMP helped the company make a successful business comeback. After difficulties in the late 1980's, Hamilton Standard totally reevaluated its products and its means of production. BMP helped the company by providing the means for it to benchmark progress in its continuous improvement effort. Hamilton Standard teams made numerous visits to sites that had been surveyed previously by BMP. The visits provided Hamilton Standard with access to best practices, allowed it to refine its existing procedures, or in some cases, confirmed that it already matched up with the best. When BMP surveyed the company, the survey team found that the Electronic Manufacturing Center had developed one of the best continuous flow manufacturing operations ever documented. After the survey, Hamilton Standard continued to improve and by 1994 applied for the Malcolm Baldridge Award. Hamilton Standard cited another benefit of its involvement with BMP: the ability to demonstrate for customers, such as the Department of Defense, that it is committed to continuous improvement and that a nationally recognized organization (BMP) had confirmed its progress.

Of course, government is the principal sponsor of BMP and it too expects a reasonable return for its investment. In general, BMP has helped to improve the productivity and competitiveness of U.S. industry. This translates to better value and lower costs to the government when it buys goods and services. In addition, BMP has served as a magnet for complimentary activities such as benchmarking, total quality, continuous improvement, and acquisition reform. These and other government initiatives often turn to BMP, with its ten year track record, for credibility and validation.

BMP Center of Excellence

In 1993 BMP took on a significantly broader dimension with the establishment of the BMP Center of Excellence. At that time the U.S. Congress recognized BMP's contributions and directed the Navy to form the Best Manufacturing Practices Center of Excellence. There are three partners in the Center: The Navy; the Department of Commerce's National Institute of Standards and Technology (NIST); and the University of Maryland at College Park. Each partner brings certain strengths and compliments the contributions of the others.

NIST's mission is to assist industry in the development of technology to improve product quality, modernize manufacturing processes, ensure product reliability, and facilitate rapid commercialization of scientific discoveries. The focal point for joint NIST/BMP efforts is smaller manufacturing companies and helping them keep up with the rapid pace of change in the global marketplace.

BMP cooperates with NIST in the latter's Manufacturing Extension Partnership (MEP). MEP is designed to help firms - especially smaller ones - upgrade their equipment, improve their processes, and strengthen their performance. Through MEP, NIST drives its support to the grass root level through 60 MEP centers in 42 states. BMP is supporting this effort by making its knowledge, insight, and experience available to MEP. Solutions to problems of larger companies that have participated in BMP surveys can benefit smaller companies which are able to access them through MEP.

The University of Maryland is taking the BMP experience into the classroom developing accredited engineering curricula based on the BMP knowledge base. BMP also works through the University's Engineering Research Center which seeks to make research and technical knowledge available to local industry. This program shares the same goal as BMP: a stronger, more vibrant industrial base. Another University project is investigating and documenting the best practices in energy efficiency in 15 Maryland companies.

In the case of the BMP Center of Excellence, each of the partner's return on investment was a better ability to meet the needs of its customers. Each partner receives significant added value for the resources committed.

The Center, as with all other BMP partnerships, embodies another principle: simplicity in structure. There were no complicated legal agreements or contracts involved in setting it up. A simple memorandum of understanding was used which amounted to a gentleman's agreement. By keeping the lawyers out of it, the Center has been able to operate effectively and without a cumbersome bureaucracy.

Now, of course, BMP is no longer a "virtual" program. It has a budget and national visibility. The increased visibility and scrutiny present greater risk of budget cuts or cancellation. But the very broad and well established customer base and the favorable reputation the program has developed all work to support its continued growth and development.

Another plus is the BMPnet and the impact of electronic commerce on the information highway. BMP has been on the leading edge of this medium and its effect on the growth of the program has been exponential. The Internet allows the two way exchange of valuable data with fast, easy access. Currently over 2000 people a day are coming to the BMP homepage (www.bmpcoe.org). Most of these are from large Fortune 500 companies, but there is increasing usage from smaller companies, government agencies, and universities. The electronic medium offers a fast way to establish relationships that lead to teaming and partnering.

Traps/Pitfalls/Lessons Learned

The relative informality of many of the BMP partnerships leaves the potential for problems or misunderstandings. This has not happened very often. When it has, it usually involved misuse of the resources or misrepresenting the program in some way. The key here is to be cautious - monitor and police as necessary. It is important to be aware and understand the goals and expectations of each participant. The desired outcome is a win-win for all involved.

Commitment is important. Participants must take ownership of their area of responsibility and perform. A few BMP projects have fallen short because of lack of commitment of key individuals or participating organizations.

The most important lessons learned are the importance of listening to find out what the customer's needs are and responding quickly and effectively. Change and adjust as the needs change. This is the essence of customer focus. It has been the key to BMP providing value to its customers and partners alike. It is an approach that has enabled BMP to get out of its box and break the mold for most government programs. BMP has reinvented the government's role in partnering with government and industry.

Virtual Organizations and Management Control

Lyn Purdy
(519) 679-2111 x 4939
The University of Western Ontario
London, Ontario
N6A 5C2

Frank Safayeni
(519) 885-1211 x 2226
University of Waterloo
Waterloo, Ontario
N2L 3G1

Unni Astad
47 22 85 25 00
Norwegian Computing Center
P.O. Box 114 Blindern
N-0314 Oslo, Norway

This paper will discuss some of the assumptions underlying the management of virtual organizations. In particular the issue of control within the virtual organization will be examined. A number of limitations to the process of controlling virtual organizations are identified and discussed. A preliminary conclusion is that virtual organizations are more likely to be successful where information, control and coordination requirements are relatively minimal.

Introduction

There is a literature which suggests that information technology (IT) can transform organizations from being "traditional" to being "virtual". Further, the literature suggests that organizations can identify their core competencies (Prahalad and Hamel, 1990) and keep those competencies as a part of the "real" organizational structure while the non-core competencies are outsourced to other organizations which may be internationally distributed (Byrne et al., 1993). Finally, if a sufficient number of activities are outsourced then the organization is said to be virtual. Most researchers describe this new emergent form of organization as the result of trying to cope with a market characterized by continuous transformation due to global competition, increasing customer demands, and radical changes in information technology. The markets of the 1990s are complex, competitive, uncertain, and constantly changing (Baskerville and Smithson, 1994; Cravens et al., 1994; Davidow and Malone, 1992; Hedberg et al., 1994; Morton, 1996, etc.).

The purpose of this paper is two-fold. First, some issues will be discussed regarding the concepts of virtual organization and information technology as a communication and coordination mechanism. Second, the problem of controlling the performance of another organization through a virtual link will be examined.

Virtual Organizations

Are virtual organization's new? To answer this question, one must first recognize that the concept of virtual organization is fundamentally vague. For example, other words and descriptions such as "imaginary" (Hedberg et al., 1994), "network of relationships" (Cravens et al., 1994), "edgeless, permeable and continuously changing interfaces" (Davidow and Malone, 1992) have been used to describe the virtual organization. Given the vagueness of the concept, the answer to the above question is both "yes" and "no". Virtual organizations are not new in the sense that one organization depends on one or more others for inputs to their processes. For example, both automotive manufacturers and grocery chains dependent on a number of internationally distributed organizations (e.g., headlights from Spain; tomatoes from Israel). Conversely, virtual organizations can be considered as new in that they are trying to use IT as an effective means of communication between distributed

0-7803-3552-X/96/$5.00 © 1996 IEEE

activities across different organizations and geographical locations. In particular, IT is viewed as the enabler for speeding up the information and coordination tasks necessary for a virtual organization to exist.

IT for Communication and Coordination

Is IT an effective communication and coordination mechanism? The answer to this question from a technological perspective is a definite "yes". Information may be transmitted "instantaneously" from one organization to another. Nonetheless, it is important to keep two issues in mind when assessing IT's role in communication and coordination. First, before IT became popular, there were (and are) a variety of other technologies which allowed for rapid information transmission (e.g., telegram, FAX, telephone). Although these technologies were used in the communication and coordination activities of the organization, they did not lead to the development of the concept of a virtual organization. These technologies have not been envisioned to embody the same characteristics as IT. IT enjoys the positive attributes which have come to be associated with computers -- it is modern, intelligent, and can control a production process, etc. Thus, IT is expected to play a more important role in integrating organizational activities than the more traditional mediums of communication.

Second, communication and coordination are fundamentally related to organizational design. Transmission of information from A to B may only constitute part of the communication and coordination process. A more important aspect of the communication and coordination process involves the actual processing of the information -- what does the information mean? what are the appropriate actions taken in response to the information received?

The responses to these questions are most often related to the design of linkages between various organizational functions. Specifically, functionally designed organizations often have poor systems of coordination and communication, not because they lack information or IT but because functions tend to behave in ways that optimize their own performance. For example, purchasing may buy the cheapest component due to its evaluation criteria while manufacturing requests the highest quality products due to its evaluation criteria. Thus, rapid messages, over IT, from manufacturing requesting better quality components from purchasing do not necessarily lead to cooperation. The relationship between manufacturing and purchasing is constrained by the fact that they are treated as separate units and are evaluated in a manner that local optimization of performance is likely to provide sub-optimal performance for the organization as a whole. These types of communication and coordination problems require organizational redesign and simply adding IT will not solve the problem.

Control in the Virtual Organization

The basic process of control in cybernetics (Beer, 1981) involves a cycle of comparing the system output to the desired or expected level of performance output. Deviations from the expected performance level are then used to develop plans, decisions and actions to generate a closer approximation to the desired level of output. The cycle is repeated until some level of stability is maintained. Thus, to exercise control in organizations there must be: (1) an expected level of performance, (2) adequate information and measurement of the performance, and (3) mechanisms for taking corrective actions. In a virtual organization, all these activities must occur over an IT link.

There are a number of difficulties associated with controlling organizational performance via IT. First, what aspects of organizational activity should be measured or controlled by IT? Theoretically, many activities could be measured (e.g., quality, technology use, costing). Using the logic of the more information the better the control, the solution would be to have information from all possible domains.

Practically, however, this puts considerable strain on the organizations on either side of the virtual link. The one organization would have to gather and input vast quantities of information; whereas the other organization would have to process and take action on this information. As a consequence, organizations must make decisions about the amount and type of information essential to the control process.

Second, in attempting to control the amount of information on an IT link, there may be a tendency to omit or aggregate information. In doing so, information could lose some of its original meaning and, possibly, take on a rather different meaning. For example, suppose the organization under control has been requested to provide information on equipment maintenance schedules. This organization uses three types of equipment with three different schedules -- once every two months, once every two days, once every week. Attempts to simplify the information through aggregation undermines the ability to exert control since the information becomes meaningless in aggregate form (i.e., the average maintenance schedule for all equipment (once every five days) is meaningless with respect to any one piece of equipment). Furthermore, the adequacy of the maintenance schedules is not an administrative control decision but one that requires expert knowledge about the equipment, the throughput rate, and the particular application for which the equipment is being utilized.

The process of control requires expertise about the particular process to be controlled. In a virtual organization, there is a potential loss of various competencies due to becoming virtual. Organizational activities which may, initially, have appeared to be independent (and were outsourced to someone else) because they were handled by different departments may in fact be interdependent. For example, if manufacturing is provided by a virtual partner, to what extent will this impact the design and evaluation of control-related information for manufacturing? The virtual organization may lack adequate expertise to design and evaluate an appropriate manufacturing control system.

Third, there is a problem of the reliability of any processed information from an organization where there is clear self-interest to ensure that a positive picture of the organization is presented. There is considerable evidence to suggest that when output information is used for evaluative purposes, individuals or organizations may exercise selectivity in reporting of the information (Purdy and Safayeni, 1993), as well as overemphasizing information which is likely to be perceived as positive. This overemphasis may vary from simple ways of aggregating data which indicate a more positive picture to the more extreme cases of fabrication of data (Morgan, 1989). The virtual organization is quite susceptible to this problem due to its reliance on indirect information transmitted over IT.

Fourth, another limitation of IT in exercising good control is related to the model that the virtual organization uses for control. Even in cases where the information is accurate, it may well be that the virtual organization still has an inaccurate picture of their networked companies since they may not be requesting the "right" information. That is, the questions asked by the virtual organization and the information sought reflects the model that the virtual organization may have of what is "good". If this model is not a good representation of the process being controlled or of the business environment, then the information collected will not provide a good representation of the situation. For example, in considering supplier-customer relationships (which may be defined as a virtual link), many supplier organizations report difficulties in communicating to their large customer that the customer's model of good manufacturing, although appropriate for a large bureaucratic organization, is not necessarily appropriate for the small supplier organizations. Suppliers have noted that if they have a company of 40 people, many organizational rules and policies are based on implicit assumptions of how things should work as opposed to explicit written procedures for every activity. These suppliers comment that they cannot afford, neither do they need, the overhead associated with these administrative activities (e.g., being expected to write memos for all communication between five managers -- something always handled through mutual adjustment and face-to-face communication) (Purdy et al., 1994).

Fifth, if the control function in a virtual link is carried out by a small organization, measured by volume of business, relative to the one being controlled, then the degree of influence that the controller can exercise will most likely decrease. For example, a furniture operation which buys less than 1% of a lumber operation's product, may have difficulty influencing and controlling the large lumber operation -- they may have difficulty requesting certain types of control-related information or demanding specific delivery schedules.

Concluding Comments

There are examples in the literature which illustrate the effective operation of and control in virtual organizations. The case of a company which markets floral arrangements through a catalog (Craven et al., 1994) illustrates how the communication linkages between marketers, designers, order takers, the distribution system and the customer can be effectively established by IT. This situation is rather unique. The problems of output quality, design specification and, to some extent, service are quite flexible. That is, a good deal of variation in the output can occur and go unnoticed. For example, if flowers were ordered from Vancouver and received in Toronto, the receiver cannot judge the correctness of the design, the quality of flowers ordered, or the time for delivery. From the business perspective, the information that needs to be transmitted is a selected design item number from a menu of possible bouquet designs and an address for delivery. Thus, the situation is characterized by simple information requirements and minimum control requirements.

One may easily imagine situations where informational requirements are more complicated and control requirements are higher. For example, if IT is used between design and manufacturing in an attempt to communicate about design for manufacturability of a new product -- an activity which requires a complex exchange of ideas and negotiations between the parties involved -- then some limitations of IT may become evident. In particular, the role of mutual adjustment in modification, clarification and re-definition of the issues at hand may be decreased because each party may substantially reduce uncertainty on their part prior to sending the message on IT. There is some evidence to suggest that negotiation activities are more likely to be successful when continuous interactive dialogue with interruptions is allowed (Pagerey, 1980).

Finally, as pointed out, there are a number of limitations of the information on an IT link, primarily because the information is indirect. Indirect information is based on reports, charts, or summaries of data regarding product performance or process characteristics. In short, indirect information is information that has been processed by someone else. The organizational literature suggests that there are advantages to relying on direct information -- information that is acquired through the direct observation of the controller or the manager of a situation (Mintzberg, 1979). The primary advantage being that the situation can be seen, experienced and assessed based on its totality as opposed to some select sub-set of information which is supposed to adequately represent it.

This paper has raised a number of issues about some fundamental aspects of virtual organizations; namely, how a traditional business could be deemed virtual and whether IT is an adequate mechanism for dealing with communication, coordination, and control for all possible organizational situations. What is of interest to investigate is the limitation of virtual organizations -- what "types" of organizations could readily go virtual? what types of organizations will experience serious problems if they attempt it? One note of caution is that the literature tends to be indiscriminately optimistic about IT's capability to allow any organization to become virtual -- something practitioners should carefully consider.

References

Baskerville, R., & Smithson, S. (1995). Information technology and new organizational forms. European Journal of Information Systems, 4, 66-73.

Beer, S. (1981). Brain of the Firm. New York: John Wiley and Sons

Byrne, J.A., Brandt, R., & Port, O. (1993). The virtual corporation. Business Week, Feb. 8, 97-103.

Cravens, D.W., Shipp, S.H., & Cravens, K.S. (1994). Reforming the traditional organization. Business Horizons, 37(4), 19-28.

Davidow, W.H., & Malone, M.S. (1992). The Virtual Corporation. New York: HarperBusiness.

Hedberg, B., Dahlgren, G., Hansson, J., & Olve, N.G. (1994). Imaginare organisationer. Sweden: Liber-Hermods.

Mintzberg, H. (1979). The Structuring of Organizations. New Jersey: Prentice-Hall.

Morgan, G. (1989). Crime: The Great Cop Out, The Observer, Sunday 13 July 1986. As presented in Creative Organization Theory, Sage Pub., 11.

Morton, M.S. (1996). How information technologies can transform organizations. In R. Kling (Ed.), Computerization and Controversy, 2nd ed., New York: Academic Press.

Pagerey, P.D. (1980). Communication, control and leadership in telecommunications in small groups. John Hopkins University, Department of Psychology, Technical report #10.

Prahalad, C.K. & Hamel, G. (1990). The core competence of the corporation. Harvard Business Review, May-June, 79-91.

Purdy, L., Astad, U., & Safayeni F. (1994). Perceived effectiveness of the automotive supplier evaluation process. International Journal of Operations and Production Management, 14(6), 91-103.

Purdy, L. & Safayeni, F. (1993). Computerized performance monitoring systems. In A. Lederer (Ed.), Handbook of Human Resource Information Systems, New York: Warren Gorham Lamont.

Project Management and the World Wide Web

Joseph G. Questore
(908) 758-2125
jgq@notes.cc.bellcore.com

Bellcore
331 Newman Springs Road
Room NVC 2Z-453
Red Bank, NJ 07701-5699

Bellcore has created an experimental "virtual project office" web site on the World Wide Web (WWW) for interactive project management of the Internetwork Interoperability Test Plan (IITP) program. IITP is unique in that project team members and labs are from various companies within the telecommunications industry, are geographically dispersed over the United States and Canada and use varied computing platforms. Project management tasks such as effective change control, communication of project information, and forging strong team spirit are particularly challenging in this environment. This virtual project office is an attempt at using the WWW to overcome these obstacles and create a space where project team members can gather to bond and exchange timely project information. This paper discusses the challenges and presents Bellcore's solution and experiences with it.

The Challenge

The Internetwork Interoperability Test Plan (IITP) renders telecommunications service providers and suppliers a cooperative environment for testing their network hardware and software safely with that of interconnected networks. As many as one dozen laboratories scattered about the US and Canada are temporarily connected to each other to form a continent-size laboratory through Bellcore's Interoperability HUB in New Jersey. For each test period--typically three a year lasting 6 weeks each--Bellcore serves as the Project Manager, monitors the connections, collects and distributes data.

There are many challenging factors from a project management perspective:

- Budgets, personnel, and laboratories belong to various companies
- The various laboratories use a mix of computer hardware platforms, application software and test equipment
- Team makeup and laboratories change from test period to test period
- Team members may never meet each other; they generally communicate over the phone
- A large amount of configuration information needs to be shared and understood
- Much of the information is proprietary to one or more of the participants and must be protected

Team-building and communication tools such as a real project office and war room are impractical in this environment because of the physical separation. Collaborative project planning--building a work breakdown structure or risk management plan, for instance--is costly, requiring travel and face-to-face meetings. Quality and progress management are hindered by communications barriers. All of this drives up project risk and cost.

The Solution

Environment

Information presentation and interactivity has essentially been standardized as a product of the almost ubiquitous proliferation of WWW browsers. It almost doesn't matter to the user what hardware platform or application software information is

0-7803-3552-X/96/$5.00 © 1996 IEEE

resident on. Hyperlinks add to the ease of navigating from information point to information point.

Popular WWW technology supports traditional areas of project management at the IITP Web Site, especially Quality, Time, Risk, Human Resources, and Communications management. It is an evolving virtual project office where project members can meet and conduct project activities. It provides security where necessary and it is intuitive to use.

The IITP Web Site is modeled after the office of a sufficiently-funded project (Figure). It consists of a reception area, cubicles for project team members, a War Room, a library, and a pantry.

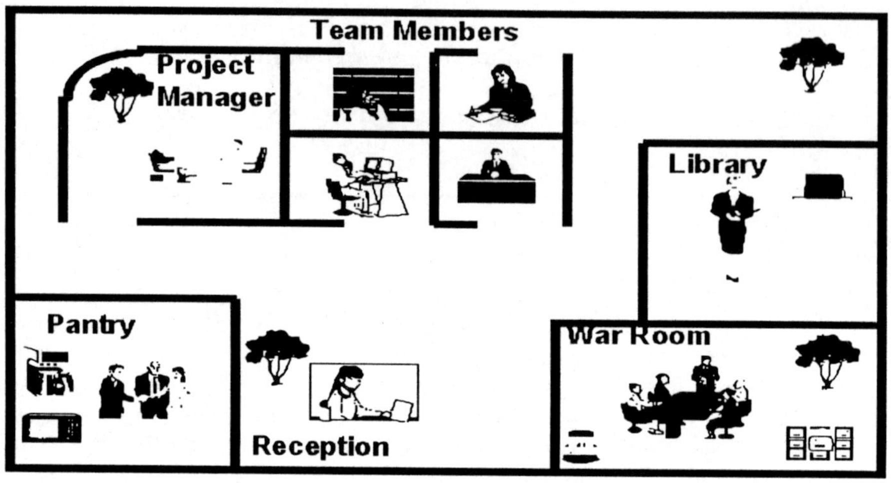

Figure: A Project Office

The reception area is accessible to anyone on the WWW and provides visitors with general information about the IITP such as its mission, background, meeting dates and location. Non-secure documents such as publicly-released reports can be freely downloaded in various application formats or browsed directly in Hyper-Text Markup Language (HTML). The reception area serves as a gate to the rest of the site.

Cubicles provide private workspaces for each of the team members where sensitive communication can occur between the team member and the project manager. A team member's project files can be securely stored and exchanged with the project manager. Sensitive documents contained within cubicles can be given expiration dates, at which time they will be permanently removed. Entry to a cubicle is permitted only to the designated team member and the project manager and is secured with a user ID and Bellcore's S/KEY™ Password Identification System.[1] Transmission of the information from the Web Site to the team member's actual workstation can be encrypted and for the most sensitive documents the actual source files can be encrypted such that only the intended recipient can decrypt and read them.

Team members and the project manager can gather in the War Room to collaborate on project efforts. It also serves as a repository for common project information, e.g., schedules, plans, records, status reports, mailing lists, etc. Entry to the War Room is permitted only to the designated team members and the project manager and is secured with user IDs and Bellcore's S/KEY Password Identification System. Sensitive documents contained within the War Room can be given expiration dates, at which time they will be permanently removed. Transmission of the information from the Web Site to the team members' actual workstations can be encrypted and for the most

[1] S/KEY is a trademark of Bellcore

sensitive documents the actual source files can be encrypted such that only the intended recipients can decrypt and read them.

Supporting references such as Bellcore Generic Requirements, American National Standards Institute (ANSI)/Alliance for Telecommunications Industry Solutions (ATIS) Committee T1 standards, etc. can be accessed through the Library, a set of hyperlinks to the document ordering pages of the appropriate parties. Visitors looking for a reference document need only visit the library to link to the relevant site. The library is open to the general public and is not a secure area.

The office coffee pot often serves as a place where important but informal project communication occurs, and team members bond through casual conversation. The pantry has been built in the virtual project office to house the coffee pot. Informal communication is supported by a bulletin-board system for non-real time communication. Entry to the pantry is permitted only to team members and the project manager and is secured with a user ID and Bellcore's S/KEY(TM)Password Identification System. Late-working team members can click on the microwave oven for access to the menus of local restaurants with delivery service.

Additional Functionality

In some instances, team members need to communicate standard information to other team members. For example, test anomaly reports need to be submitted to the technical writer in a standard fashion to facilitate generation of a final report. In cases such as this, users are presented with a form to complete and the form is submitted electronically to the appropriate party.

In the past, the only team member that could view test data in real time was the tester located in Bellcore's HUB. A War Room feature allows those present in the room to view real-time test data simultaneously. This improves quality since more eyes can scrutinize the data.

Experience

The IITP Web Site described here is currently undergoing beta test; it is expected to be rolled out in September, 1996. The first generation IITP Web Site located at http://carin.bellcore.com:3333 is accessible to anyone on the WWW and provides visitors with general information about the IITP such as its mission, background, meeting dates and location. Publicly-released reports from previous tests can be freely downloaded in various application formats or browsed directly in Hyper-Text Markup Language (HTML).

Experience with the first-generation IITP Web Site is very positive. Some IITP member companies have built links to the IITP site into their intranet pages so that company employees can access the reports directly. This has helped them to address issues more quickly and cost-effectively. We have experienced over 500 hits from all over the world on the Web Site since it was launched in February, 1996.

Conclusion

Management of a geographically scattered Project Team is a challenge under the best conditions. The WWW and its ease-of-use has provided an opportunity to bring the dispersed team together under one virtual roof.

Product Design Using Virtual Environments--A Conceptual Evolutionary Approach

Dr. James. M. Ragusa
(407) 823-5229

University of Central Florida
College of Engineering
Department of Industrial Engineering and Management Systems
Orlando, Florida 32816

Grace Bochenek
(810) 574-7767

VETRONICS Technology Center
U.S. Army Tank-automotive & Armaments Command
Warren, MI 48397

In today's dynamic and sometimes virtual environments, project development organizations must struggle to remain cost effective and efficient. One of the most significant and cost-driving phases of product development is product design. This paper briefly focuses on evolving design methods and technologies, and the need for more robust collaborative, virtual design environments. A conclusion reached is that the application of advanced virtual computer and network technologies has the potential to decrease overall product cost and development time. However, numerous research issues exist and technology challenges remain.

Introduction

Many predict, and present trends indicate, that virtual organizations and their anticipated, supportive work environments are destined to become pervasive for international, national, and corporate enterprises. This evolution is consistent with Tofler and Tofler (1995) who envision that 21st Century "Third Wave" organizations will be more flexible, fast-paced, decentralized, and information-rich. These futurists forecast a new Information Revolution rivaling the Agricultural and Industrial Revolutions of the past.

The evolution and creation of virtual organizations has the potential of becoming reality primarily because of two driving forces: economic needs and enabling technologies. Economic considerations include the emergence of global markets, increasing competition, shrinking product life cycles, and the diversity, dispersion, and availability of natural and human resources. Enabling technologies include: greatly improved computer hardware and software systems, global networks and communications, and importantly their decreasing cost and availability. As this paper will indicate, product design is a key and cost-driving aspect of product development and is, therefore, of critical importance to the overall process of product design, development, and manufacturing.

The purpose of this paper is to describe how a collaborative and distributed virtual design environment under early conceptual development, can be used to accommodate interactive customer/user inputs and eventually improve overall design efficiency and effectiveness. Briefly discussed are the state of product development, the importance of product design, evolving design related technologies, virtual design environments, research issues, and some conclusions.

The state of product development

Traditional product development has been characterized as a sequential, linear process. According to Ulrich and Eppinger (1995) phases of this process include concept development, system-level design, detailed design, testing and refinement, and finally production ramp-up. However, this serial development has also been described by Shu and Flowers (1994) as a sequence of "throw it over the wall" processes where functional groups make contribution to a project independently of other groups. As a result, critical decisions which impact overall product development and design are frequently made without regard to the consequences placed on other functional areas.

But some process improvements are evolving. The need to shorten the development cycle (Vesey 1991; Iansiti 1993; Cooper and Kleinshmidt 1994; Cordero 1991), to achieve high quality (Cordero 1991; Ali, Krapfel, and LaBahn 1995), and to solicit direct customer feedback (Larson-Mogal 1994) have driven changes to be made to traditional product development processes. Now, many new product development efforts include the concurrent use of multi functional teams, the integration of geographically distributed participants, and the application of computer and network systems. Yet more needs to be done--especially within the areas of system-level and detailed product design.

Why is the design phase important?

Product design is considered by some to be the most critical element in the overall product development process because eventual product cost and time-to-market are dominated by decisions made in the early stages of product design. Supporting this idea is a National Research Council Report (1991) which indicates that 85% of the cost of product components is incurred directly due to decisions made before product design release to manufacturing. In a related study, Garcia, Gocke, and Johnson (1994) report similar findings. They found that by the time 10% of total project funds were spent (for design), approximately 90% of a product's development costs had been established.

The contemporary design process

Today, the multifaceted process of product design is being accomplished in a more concurrent manner, offering opportunities for cost and time savings. However, there is a need for additional improvements away from manual, group design review activities. For example, many traditionally product design reviews, conducted in both private and public sector organizations, still require various system and functional representatives to concurrently review hard copy or computer images of drawings in the same work space while someone maintains a record of approved changes. Increasing costs and reduced product life-cycle drivers are, however, motivating responsible managers to find better ways to improve product design methods.

Teaming and people

Product design inherently remains a collaborative process because of its reliant on inputs from functionally, diverse participants. Teams are now composed of members from various elements: product management, development, engineering, and manufacturing (Sansone and Singer, 1992). The basic concept is to create a consensus developing team whose objective is to redesign or develop new products.

Product design can also be considered a social activity because there is a need to maintain a shared understanding among all process participants (Holt 1991). Multi-functional teams proceed by exchanging proposals, arguments, ideas, and seek to form a consensus. Knowledge is transformed into a common pool of information (Mitchell 1994) in an environment where interaction and conflict resolution can take place.

Geographically distributed design

Today, a design team may be in one location or geographically dispersed (Hedberg, 1994), adding an extra burden on the process because of the issues of timing and communications. Exchanging viewpoints and ideas while sharing information is critical to making more informed decisions regarding a product design.

Evolving design-related technologies

Computer-aided design (CAD)

Computer technology has made a positive impact on the product design process. Design automation has resulted from the development of computer-aided design (CAD) tools--beginning in the early 1950's and bloomed in the 1980's (Barfield, Chang, Majchrzak, Eberts, and Salvendy 1987). Importantly, the use of

CAD has been found to save up to 70% of total manual design time (Barfield et al. 1987). Also, CAD technology has made it possible for product designers to develop an assortment of options (and opportunities) for evaluation when designing new products.

Originally, CAD was developed to automate the drafting process of a single designer. However, in today's multi-functional environment this poses a problem because current CAD tools do not allow simultaneous, multiple users design access. Instead, designers are restricted to sequential editing and modification. In addition, current CAD technology does not provide multiple users with the ability to simultaneously view a design from different perspectives.

Virtual reality

As a design environment changes towards more dependence on distributed, collaboration, new design tools need to be developed and new strategies for group interaction investigated. One of the most promising emerging technology which has the capability of impacting the product design process is virtual reality (VR). Burdea and Coiffet (1994) define VR as a high-end user interface that involves real-time simulation and interactions through multiple sensorial channels consisting of visual, auditory, and tactile modalities.

VR allows a person to experience phenomena that appear to be real but exist only in the computer. Thus, the user (or designer) "feels" immersed, as if he or she is actually "in" the three dimensional space. Using VR-based systems, a user or designer can visualize a final product, assemblies, sub-assemblies, or components. The emersion dimension of VR allows the designer or user to go one step further and truly "experience" the design. For example, a car designer could "virtually" sit in the car and visualize its interior, and then "virtually" walk around the car and look at reflections across its surface (Kalawsky, 1993). VR could give a designer or a team a new way to explore reality and make changes (major or minor). While much progress have been made, VR technologies still have a long way to go before they are fully operational and viable for wide-spread commercial design use.

Virtual design environments

As the following figure illustrates, product design has evolved from 2-D drawing to 2-D and 3-D CAD. The next logical evolution is to distributed, multiuser virtual CAD, and eventually to VR supported, collaborative design environments.

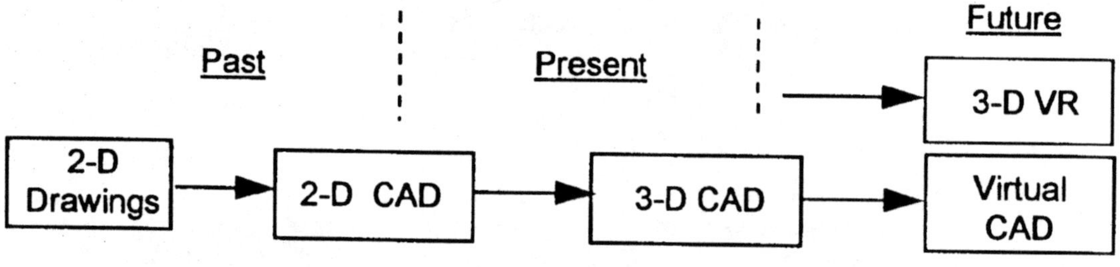

The evolution of product design environments

Army/university concept development

Because of the potential of 3-D CAD and VR virtual environments, VETRONICS Technology Center of the U.S. Army Tank-Automotive & Armaments Command and University of Central Florida researchers are in the early conceptual stage of developing a virtual product design system. Our emerging vision for the future is as follows.

In operation, future, distributed virtual systems (CAD or VR) would use computer-based technologies to simulate and model products or assemblies for review and modification. Virtual mock ups would replace physical prototypes. This "virtual world" environment would allow distributed designers, engineers, production personnel, and customers/users resident to interact simultaneously and constructively in a 3-D product design space.

This unique environment would provide for true team interaction. Design teammates could simultaneously enter a virtual product design world, and jointly evaluate design issues, ideas, and parameters--each from their own viewpoint. People from various functional elements would thus become knowledgeable about all others activities and, therefore, would become an integral part of the decision making process. This unique multi-functional virtual design environment would allow members to share in a concurrent design experience and relate to the perspectives of others.

In addition to participation and perception, using VR for product design could provide design flexibility by allowing the exploration of various options and the opportunity to play "what if" exercise~s where mistakes are less expensive. Significant cost savings could be achieved in system development and production because many of the problems would be identified and corrected prior to the actual physical product construction. The technology would enable developers to refine designs before commitments are made, bring users into the design process much earlier, and allow engineers to solve problems in a collaborative, group setting.

Research issues

Much of the above discussion remains conceptual because of the present lack of stand-alone and integrated technologies. Several research issues can be identified as a result.

Limitations of CAD

As noted earlier, CAD technologies presently do not fully support concurrent user operations in either a stand-alone or networked mode, and variable perspectives are not yet possible. Much R&D remains to be accomplished and many user issues need to be addressed.

Network communications

While great strides are being made, present network technologies and capabilities would only marginally support the throughput requirements of distributed CAD and VR concurrent design environments.

Limitations of VR technology

VR and virtual environment systems and capabilities are presently limited to demonstration systems. Much work remains to develop the capacity to handle the networked, interactive needs of a distributed real-time VR design system. In addition, a basic premise, yet to be validated, is to determine if visualization techniques significantly clarify shared information and improve decision making by multiple team members working in an interactive design environment.

Human and orqanizational behavior

There are a number of human and organizational issues which must be better understood if future virtual design systems ___ __ become accepted. The use of technology, regardless of its promise, does not guarantee that humans working in a virtual design environment will perform better than those using present systems.

Configuration control

Present methods of design revision and drawing configuration control would be compounded in a distributed environment. In addition, electronic signature authority would have to be incorporated into the final virtual system.

Cost efficiency

Whatever systems are developed, positive cost-benefit will be an important part of the determination of organizational use of visual design environments.

Conclusions

Results of this study have lead the authors to conclude that the application of advanced virtual technologies to product design will be the key to improving the efficiency and effectiveness of product design--the major driver of product cost. Near term evolution a~d development of distributed 3-D CAD and eventually VR systems would offer several advantages to private and public sector concurrent and distributed product design groups. While available technologies do not presently support 3-D interactive networked CAD or VR-supported design, a variety of new and improved technologies are emerging. However, numerous research issues and technology challenges remain.

References

Ali, A., Krapfel, R.J., and LaBahn, D. "Product Innovativeness and Entry Strategy: Impact on Cycle Time and Break-even Time." *Journal of Product Innovation Management.* 12:54-69, 1995.

Barfield, W., Chang, T., Majchrzak, A., Eberts, R., and Salvendy, G. "Technical and Human Aspects of Computer-aided Design (CAD)." G. Salvendy, *Handbook of Human Factors,* 1617-1656, John Wiley: New York, NY, 1987.

Burdea, G., and Coiffet, P. *Virtual Reality Technology*. John Wiley: New York, NY, 1994.

Cooper, R.G., and Kleinshmidt, E.J. "Determinants of Timeliness in Product Development." *Journal of Product Innovation Management,* 11(5):381-396, 1994.

Cordero, R. "Managing for Speed to Avoid Product Obsolescence: A Survey of Techniques." *Journal of Product Innovation Management,* 8(4):283-294, 1991.

Garcia, L.A.B., Gocke Jr., C.R.P., and Johnson Jr., C.N.P. *Virtual Prototyping Concept to Production.* Fort Defense System Management College Press: Belvoir, VA, 1994.

Hedberg, S.R. "Design of a Lifetime." Byte, 103-106, October, 1994.

Holt, J.E., and Radcliffe, D.F. "Learning in the Organizational and Personal Design Domains." *Design Studies,* 12(3):142-150, 1991 .

Iansiti, M. "Real-World R~D: Jumping the Product Generation Gap." *Harvard Business Review,* 138-147, May-June, 1993.

Kalawsky, R.S. *The Science of Virtual Reality and Virtual Environments*. Addison-Wesley: New York, NY, 1993.

Larson-Mogal, J.S. "An Immersive Paradigm for Product Design and Ergonomic Analysis." *Virtual Reality World,* July, 1994.

Mitchell, W.J. "Three Paradigms for Computer-Aided Design." *Automation in Construction,* 3(2-3):239-245, July, 1994.

National Research Council. *Improving Engineering Design: Designing for Competitive Advantage.* National Academy Press: Washington, DC, 1991.

Sansone, F.P., and Singer, H.M. "Improving Time to Market in Consumer Products." *AT&T Technical Journal,* 71(3):66-72, May, 1992.

Shu, L., and Flowers, W. "Teledesign: Groupware User Experiments in Three-dimensional Computer Aided Design." *Collaborative Computing,* 1:1-4, 1994.

Tofler, A., and Toffler, H. *Creating a New Civilization.* Turner Publishing: Atlanta, GA, 1995.

Ulrich, K.T., and Eppinger, S.D. *Product Design and Development.* McGraw-Hill: New York, NY, 1995.

Vesey, J.T. "Speed to Market Distinguishes the New Competitors." *Research and Technology Management,* 33-38, November-December, 1991 .

The Importance Of Electronics In Modern Cars

Raimo Ranta
+ 358 31 3652584

Tampere University of Technology
P.O. Box 553, FIN - 33101 Tampere, Finland

Hannu Jaakkola
+ 358 39 6272712

Tampere University of Technology
P.O. Box 30, FIN - 28601 Pori, Finland

This paper deals with the benefits achieved by use of automotive electronics in passenger cars. Because of the magnitude of the topic, only the most important results will be presented. The presentation will be emphasized on the first topic in the list below. Only some general notes concerning other two topics will be presented.

- *Economical aspects*
- *Ecological aspects*
- *Safety aspects*

Introduction

There are a lot of electronic innovations, which have been utilized in passenger cars during the past two - three decades. The first systems in this area were taken in use in early 1960's; since that electronics has been a big influence on the economical, ecological and safety properties in passenger cars. There are several targets like engines, brakes, transmissions, steering, shock absorbing, airbags, etc., in which electronics has penetrated. The invention of the micro processor has made possible an intelligent engine management (Simanaitis, 77). Micro processors control the reliable behaviour of ABS - systems or airbags, too. There are kilometres of cables, tens of electric motors, and tens of electronic units in modern cars (Lemme, 92). Altogether, the innovations related to electronics have become as important as the mechanical design in car manufacturing.

In order to illustrate the importance of modern electronics, a suitable approach for this purpose has to be selected. Because the use of electronics in engine management systems has been very massive during the past two-three decades, its influences on certain events can be quite easily seen. The following paragraphs show, how the economy of cars has developed from 1970's to 1990's. Some notes related to the levels of exhaust emissions will be made, too. Certain connections on the use of electronics in fuel serving of cars will be also presented. The role of safety will not be illustrated deeply, because of the limitations concerning the length of the presentation. Only the future trends of it will be described.

Progress in fuel consumption during the last decades of the 20th century

The role of electricity was changed in 1960's, when a new technology of semiconductors came on the market. At that time, the requirements concerning the cleaner environment got more of interest in USA. The first tight limits for levels of exhaust emissions were set. This meant also the seeking of solutions in order to manage ignition timing and fuel injection. The first modern injection system based on discrete components was introduced in 1967 (Sitterding, 69). In fact, it started a new era in use of electronics in fuel serving.

Increasing of an engine power was the main trend in USA and in Europe during the early 1960's. The engine volumes were growing all the time, and fuel consumption followed, naturally, the same trend. It is easy to make a decision, that an increasing energy consumption means greater exhaust emission values. The more severe limits in this respect were taken at first in use in USA. This meant a very big change in car designing policy. The compression ratios of engines decreased, and catalysators and engine management systems were taken to wide use. The manufacturers in Europe and in Japan had also to follow this development in cars, which they sold on the USA market. In Europe and in Japan the legislation was earlier less severe than in USA. The use of electronics was, however, everywhere a key issue in cutting of fuel consumption and reducing of emissions.

It is typical, that the car manufacturers in Europe and in Japan announce the fuel consumption of their models at the speed of 90 kilometres per hour. American car companies do not use this kind of definition, but many car magazines have measured the fuel consumption for their cars in specific tests. It can be easily shown, that this speed is suitable for reliable comparisons between different types of cars, because the air resistance does not play any remarkable role at that speed. It is practicable to use a unit, that gives fuel consumption as litres per 1000 kilograms. Figure 1 shows, how the fuel consumption of different size European and Asian (primarily the Japanese ones) cars, equipped with SI engines, has developed during the past two decades. The cars have been put into three classes:

Class I:	Engine volume 0, ... , 1.5 l
Class II:	Engine volume 1.5, ... , 2.0 l
Class III:	Engine volume 2.0, ... , 3.5 l

The curves in figure 1 are based on the consumption values of 80 - 90 models per a particular year (totally analysed about 1500 cars). The symbols related to a particular class are: -------- Class I, ––––– Class II, –––––Class III.

Fuel consumption
(l/1000 kg)

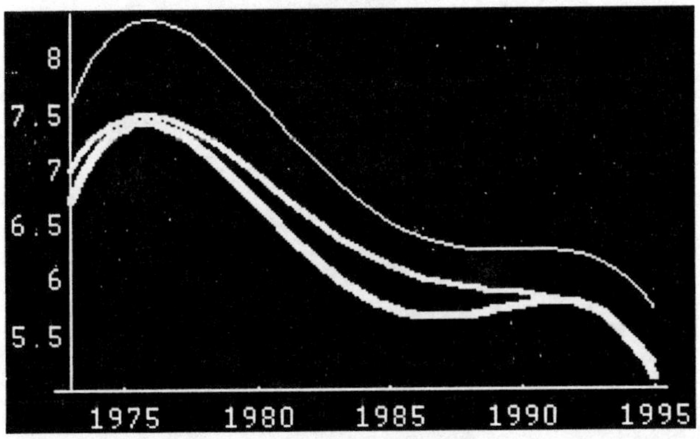

Figure 1. The development of fuel consumption in small and medium size European and Asian cars.

We can see, that fuel consumption was slightly growing before an oil crisis in these types of cars. The main reason was probably, that the tuning degrees of engines were growing. Car manufacturers have equipped their models with fuel injection systems during the past 20 - 30 years.

Figure 2 shows, how big part of models has been equipped with these systems as a standard equipment. The curves in figure 2 are based on the Pearl method (Martino, 93).

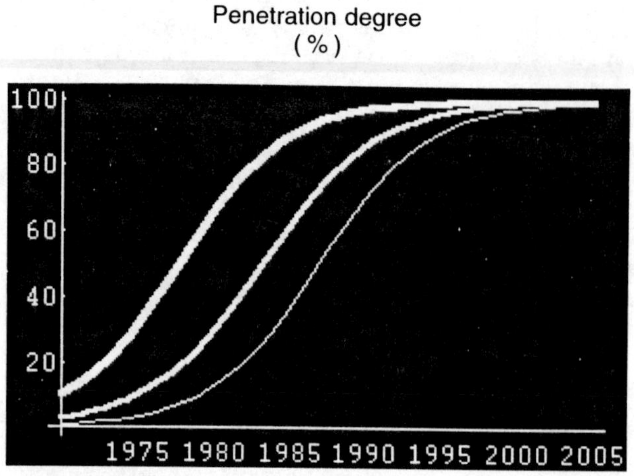

Figure 2. The penetration of fuel injection systems in small and medium sizeEuropean and Asian cars.

It is easy to make a decision, that the growth of fuel injection systems is one of the main reasons for the big progress in fuel consumption of cars. On the other hand, the growth of compression ratios of engines is important as well. That has become possible by means of advanced signal processing controlled by engine management systems. Figure 3 shows the growth of the program length in advanced engine management systems.

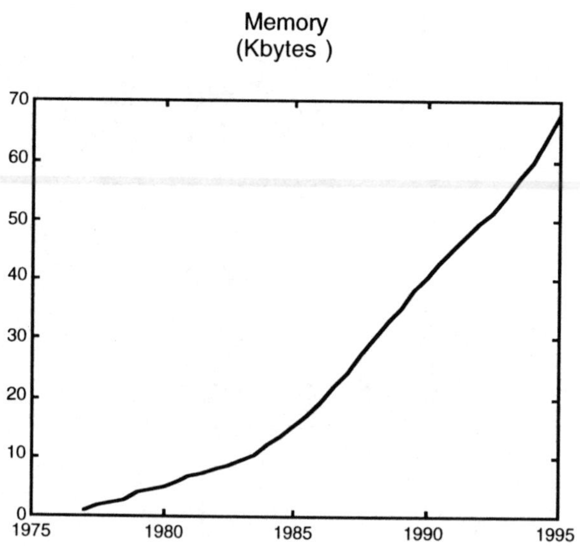

Figure 3. The growth of the program length in engine management systems.

Of course, the speed of systems has also become much bigger since 1970's. When we are considering the complexity of these systems, a good reference is a mobile telephone. Modern GSM phones contain about 200 Kbytes of program memory.

Costs of automotive electronics

Figure 4 shows, how big part of the electronics consumption has been put for car production. The curve is based on an annual statistics of Electronics magazine.

Automotive electronics consumption/electronics consumption (%)

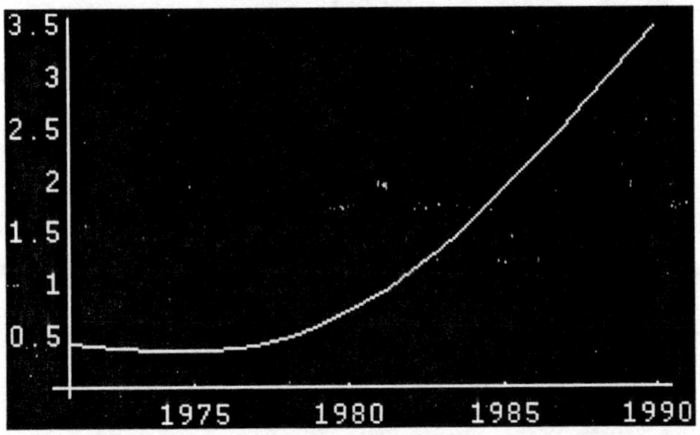

Figure 4. The costs of automotive electronics related to the electronics consumption in USA.

The costs in figure 4 contain the hardware costs of systems and components used in car production. The costs of engine management and exhaust emission systems have been about 50 % of the all costs generated by the utilization of electronics in cars. For example the costs of safety electronics has been about 10 % of that. It will become much more important in the next decade. Electronics has the first two - three decades primarily been used to develop the economical and ecological properties of passenger cars. Safety trends will be more emphasized in the near future. The total costs of electronics will be over 25 per cent of all production costs in passenger cars in the near future (Ehlers, 88).

Conclusions

Figure 1 illustrated the trends towards decreasing fuel consumption in small and medium size European and Japanese cars. We can however find the same kind of behaviour for American cars, too. The growth of the fuel injection systems in American models has followed the similar trends as shown in figure 2.

When we are considering, how much the energy consumption of cars has really decreased during the past two decades, we have to take into account, that an average weight of cars has increased 5 - 10 per cent. It can be estimated, that modern cars use 20 - 25 per cent less fuel, than the cars 20 years ago. The oil production has been approximately 3×10^{12} litres per year during this time period. The production of passenger cars has been about 35 - 40 million units per year. If we make an assumption, that one car has consumed 10 litres/100 km and the yearly distance of driving is 20000 km, we can easily calculated, that an average save of oil has been a couple of per cent. Today, the save in liquid energy is much bigger related to the whole oil consumption.

The upper limits of exhaust emissions are usually depending on the amount of carbon hydrates (HC), carbon monoxide (CO), and nitrogen oxides (NOx). For example in USA, the maximal allowed limits in the year 2000 are 1 % in HC, about 5 % in CO and 5 % in NOx compared to values in the beginning of 1970's, respectively. The European and the Japanese limits differ from the American ones. There are, of course, many measurements, that show, how well these limits can be met. Some car magazines regularly test this kind of values. These tests give the levels of exhaust emissions in the new cars. The emission measurements of almost 100 different types of cars, equipped with SI engines, have been analysed. It can be concluded, that the values of exhaust emissions are about half of the maximum allowed limits. They often fulfil more severe limits to be set at the end of this decade.

The mechanical design is naturally very important, when the progress in energy consumption or in exhaust emissions is evaluated. We have to note, that electronic units are usually acting as a part of mechanical systems. Electronics has, however, an important role in economical, clean, and safe cars.

References

Ehlers Karsten, Sensoren und Stellglieder im Automobil. Elekronik 7/31.3. 1988 pp. 51 - 55.

Lemme Helmuth, Sensorsysteme fur das Auto von morgen: Einige Highlitghts der VDI - Tagung "Sensoren". Elektronik 17/1992 pp. 64 - 66.

Martino Joseph P., Technological forecasting for decision making, third edition. McGraw - Hill, Inc., USA,1993.

Simanaitis Dennis J., MISAR: An Electronic Advance. Automotive Engineering, January 1977 Volume 85, Number 1 pp. 24 - 29.

Sitterding Herbert, Europäische Personenwagen fur 1968. Automobiltechnische Zeitschrift 69 (1967) 11 pp. 397 - 405.

Collaboration in Meta-Organizations: Research Issues and Challenges

Bharat P. Rao
(O) 706.542.3775

Department of Marketing
University of Georgia
148 Brooks Hall
Athens, GA 30602

Alliances and other types of inter-firm relationships facilitate knowledge transfer and generate innovation that is fast and relatively inexpensive - factors that can often spell the difference between successful products and also-rans in the marketplace. In this paper, we argue that alliances need to be conceptualized as meta-organizations that transcend the structures and operational elements of traditional organizations. The implications of strategically managing such innovation in contexts of competition and collaboration, and measuring the results of such management are discussed.

Collaborative new product development

New product development (NPD) has become an expensive proposition given shorter product life-cycles, uncertainties in product-markets, rapidly changing technologies and intense competition (Bettis and Hitt 1995). A number of firms are resorting to collaborative NPD: both with partners from within and outside their own product-market domains and industries (Johansson 1995; D'Aveni 1994; Hamel, Doz and Prahalad 1989). Alliances are formed with the underlying intention of creating strategic leverage given perceived opportunities and inherent capabilities of the allying firms. New products introduced by collaborative alliances are an important way of realizing this leverage (Lele 1992). In the context of NPD, these collaborative alliances and other types of inter-firm relationships primarily facilitate knowledge transfer and generate innovation that is fast and relatively inexpensive - knowledge that may ultimately be put to use to achieve the objectives of the collaborating firms. They also bring with them the overall benefits accorded by alliances: reduced risk and entry costs, increased feasibility of conducting expensive research, utilizing the synergy's created by the innovation process, and expansion into previously uncharted markets and technologies (Achrol and Stern 1990; Sheth and Parvatiyar 1989; Hamel and Prahalad 1994). As Ouchi (1980) points out: NPD, manufacturing and marketing constitute a natural aperture through which inter organizational cooperation could progress. However, there is little research that clearly links the concepts of innovativeness and learning that take place during the process of NPD to broader strategic forces that drive firms to engage in such collaborative activity. Also, collaboration can often take place between firms competing for market share or the same product markets: i.e. the boundaries of collaboration and competition are hard to delineate. Recent work suggests that the emergent paradigm might be one of 'coopetition': a hybrid form where the same set of firms can collaborate at one level and compete at another (Dowling 1993). In this paper, we look at the way firms are dealing with collaboration in the arena of new product development. It is suggested that the collaborative metaphor operates at a meta-organizational level: organizations need to recognize and foster an environment that facilitates genuine collaboration.

The changing nature of innovation

Given the nature of competition in the current decade, the innovation process has changed radically over the last few years. Rapid diffusion of knowledge and slender lead times mean that firms have to be on the alert at all times. There are two basic forces that drive this phenomenon: technology and the marketplace. Increasingly, such an approach is required in all industries: even those that were deemed as being 'low-tech' and 'slow-moving' have had to devise innovative strategies and execute them in order to survive tough competition at the level of the firm and in the

0-7803-3552-X/96/$5.00 © 1996 IEEE

marketplace. There is also a lot of borrowing of technologies, operating procedures and systems and increased access to knowledge that can transform businesses. At the level of new product development, collaboration is fast replacing competition as a viable alternative. As technology fusion gains prominence as a viable strategy, it is to be expected that 1. the need for more sensitive marketplace information will be felt, 2. the number of such collaborations will increase, and 3. designing environments that facilitate collaborative product development will become a key priority for many of the world's leading firms (Kodama 1992). The use of alliances in order to facilitate new product development is not new: semiconductor companies have long used cross-licensing and second-sourcing agreements since the mid-80s. In a matter of time, firms like Motorola were fully exploiting the alliance-based strategy to expand into new areas (Yoshino and Rangan 1995). Semiconductor companies also stepped up alliance activity partly due to the increased presence from Japan and Europe. From a playing field of Motorola, Texas Instruments, Intel, National Semiconductor and Fairchild. in the mid-70s; the semiconductor industry had to deal with competition from Hitachi of Japan, Philips of Holland and Siemens of Germany (Yoshino and Rangan 1995). A number of research consortia were established with the aim of transferring know-how and resources; the MCC (Microelectronics and Computer Technology Corporation) being key case in point (Smilor , et al 1990). In an attempt to bolster US competitiveness and also as a result of lobbying by MCC and its shareholder companies, the US Government also began to push forth a joint-venture/consortia based strategy especially in the area of semiconductors. This ultimately led to the passing of the National Cooperative Research Act by the US Congress in 1984 (Rogers and Valente 1991). By 1988, more than 80 such R&D consortia had been established in the United States, and this in turn led to a proliferation of alliances in the semiconductor industry. As Rogers and Valente (1991) elaborate:

"... Basic technology is changing most rapidly. This rapid rate of change means that a large number of technological innovations are generated each year. Each such innovation provides an opportunity for a new business venture, as an innovation is usually the entry ticket for a start-up firm into an industry......Collaboration, where neither partner has ownership over the other and cannot give orders to the other, is replacing pure competition in the microelectronics industry (and in similar industries). Japan set off the wave of cooperation among US industries, but now US firms cooperate extensively with each other in R&D and even, in some cases, with Japanese and other internationally based firms."

These developments clearly indicate the importance of collaboration as a vehicle to achieve product development and innovation objectives of firms. In the following section, we describe how such collaboration can lead to learning and innovation advantages.

Product innovativeness and inter partner learning

Any idea that is perceived as new by an individual has been described as innovation (Rogers 1962). In terms of products or services, there can be various degrees of innovation: typically characterized by the effect of the innovation on existing patterns and described as continuous, discontinuous, and dynamically discontinuous (Robertson 1967). In order to understand the process of innovation and the degree of innovation, one has to look deeper into the underlying patterns of organizational culture, the evolutionary stage of the industry, etc. This is a complex task, given that it difficult to pinpoint where one form of enterprise ends and another begins, and also due to the fact that organizations themselves are constantly reshaping their boundaries. In an era where products and services seem to be converging, along with the underlying technologies that drive them, there are a myriad of problems associated with pinpointing the exact source or degree of innovation. Bringing together people and organizations with different competencies and skill-sets in an atmosphere that facilitates knowledge transfer and cooperation can be a direct route to achieving innovation efficiencies. In any process of knowledge transfer, innovation and inter-partner learning is also both an objective and an ongoing process (Helleloid and Simonin 1994). Innovation patterns in the firm and industry are triggered by the rate at which firms learn from research and practice. Therefore, it becomes crucial to foster a climate where there is a greater chance of genuine learning between and within partner firms. By doing so, the existing core competencies of partners can be enhanced, leveraged and deployed, leading to competitive advantage in the short and long run. Problems with knowledge transfer in alliances include: ill-defined objectives, hidden agendas or lack of organizational resources. Learning need not always be triggered by differences in knowledge bases of partnering firms: it can be unintended and often serendipitous. Also, there is an underlying process of innovation evolution taking place both at the level of the individual firms and the industry. Therefore, product development teams have to realize that there can be different returns at different points along the evolution continuum.

Collaboration as a metaculture

Using alliances as a vehicle for NPD calls for important changes in the way organizations deal with the process of collaboration. Littler, et al (1995) point out that improved communication is a key to collaborative product development. The presence of a product champion can greatly increase chances of success. This view is supported by Rubin (1992), who points out that the unusual boundary conditions of hybrid organizations require unusual leadership skills that combine visionary encouragement with the politicking and discipline of a project manager. Global alliance managers are faced with the paradox of simultaneously managing cooperation and competition (Kozmetsky 1987). In case of joint product development, it becomes crucial to recognize that there are forces of creative tension that can make or break a collaboration. As pointed out by Friar and Horwitch (1986), the resolution of such tensions can lead to value creation. Focusing on similarities rather than differences, and the establishment of an environment that facilitates experimentation, irrespective of the long-term success or failure of the venture are also frequently cited mechanisms for improving and enhancing mutual learning in such contexts (Rogers 1990). In a loosely structured collaboration, control often emerges through confusion, and winning strategies often occur through 'systematic positioning' and engaging the creative forces at the venture's disposal in a 'total plan of attack' (Wing 1988). As Rubin (1992) lucidly describes, the "ethos which guides such hybrid arrangements is based on multivalence and temporality, rather than on the unitary values and rules of continuity and perpetuation that characterize statutory institutions". It might be useful to think of the collaborating firms as rising to a new metaorganizational form whose main objective is to facilitate knowledge exchange. These two firms with radically different competencies perceive the need for collaboration, then interact in a metaorganizational culture that transcends both the parent firm cultures and conclude their collaboration by returning to their original competitive positions.

Firms competencies can reside in superior technical leads, managerial capabilities, world-class practices or combinations of these. Competence building at the single-firm level requires "a 'strategic architecture' that 1. establishes clear objectives 2. identifies core competencies that need to be built and 3. identifies the technological constituents of those competencies" (Rubin 1992). The building blocks of competence are 1. committing the organization to long term objectives (Hamel and Prahalad 1990) 2. accumulating skills and resources and 3. enhancing the bases of knowledge and skills through cycles of learning. All these building blocks appeal to and reflect logic of strategic alliances. Improved products and processes can result from alliances that 1. capitalize on parent firms competencies 2. are committed to the long term and 3. that engage in dialog and learning (Chiesa and Barbeschi 1994).

A research agenda for the future

The importance of collaboration in today's business world calls for innovative approaches to researching issues of importance to such activity. The structuring of collaborative mechanisms, with the basic purpose of fostering innovation and knowledge exchange, the prime drivers for subsequent value creation, needs substantial consideration. Additionally, normative insights are needed in the areas of team design and performance, the role of technology and the emergence of virtual enterprises. An interdisciplinary approach to tackling these problems is strongly recommended. Just as the fusion of distinct skills with an enterprise can add value, a fusion of research perspectives can provide us with valuable insights into the phenomenon of competition in an age of collaboration.

References

Achrol, Ravi S., and Scheer, Lisa K. and Lou W. Stern (1990), "Designing Successful Transorganizational Alliances," Marketing Science Institute, Report No. 90-118.

Bettis, Richard A. and Michael A. Hitt (1995), "The New Competitive Landscape," Strategic Management Journal, 16, 7-19.

Chiesa, Vittorio and Murizio Barbeschi (1994), "Technology Strategy in Competence - based Competition," in Competence-Based Competition, Gary Hamel and Aime Heene, eds., Strategic Management Society.

D'Aveni (1994), Hypercompetition: Managing the Dynamics of Strategic Maneuvvering. Free Press: New York.

Dowling, Michael J, Barbara A. Carlin, William D. Roering and Joette Wisnieski (1993), "Multifaceted Relationships Under Coopetition: Description and Theory," Working Paper, University of Georgia.

Friar, John and Mel Horwitch (1986), "The Emergence of Technology Strategy," in Technology in the Modern Corporation. New York: Pergammon Press.

Hamel, Gary, Yves Doz and C.K. Prahalad (1989), "Collaborate With Your Competitors-and Win," Harvard Business Review, 67 (January-February), 133-139.

Hamel Gary and C.K. Prahalad (1994), Competing For The Future, Harvard Business School Press, Boston, Massachusetts.

Helleloid, Duane and Bernard Simonin (1994), "Organizational Learning and a Firm's Core Competence," in Competence-Based Competition, Gary Hamel and Aime Heene, eds., Strategic Management Society.

Johansson, Johny K. (1995), "International Alliances: Why and How?," Journal of the Academy of Marketing Science, 23 (4), 301-304.

Kodama, Fumio (1992), "Technology Fusion and the New R&D," Harvard Business Review, July August, 70-78.

Kozmetsky, George (1987), 1987 Annual Report, IC2 Institute, The University of Texas at Austin.

Lele, Milind M. (1992), Creating Strategic Leverage: Matching Company Strengths with Market Opportunities, John Wiley and Sons.

Littler, Dale, Fiona Leverick and Margaret Bruce (1995), " Factors Affecting the Process of Collaborative product Development: A Study of UK Manufactures of Information and Communication Technology Products," Journal of Product Innovation Management, 12 (1), 16-32.

Ouchi, W.G. (1980), "Markets, Bureaucracies, and Clans," Administrative Science Quarterly, 25, 129-141.

Prahalad, C.K. and Gary Hamel (1990), " The Core Competence of the Corporation," Harvard Business Review, 68 (3), 79-91.

Robertson, T.S. (1967), "The Process of Innovation and the Diffusion of Innovation," Journal of Marketing, 31, 14-19.

Rogers, Debra M. (1990), Global Innovation Strategy: Creating 'Value-Added. Masters Thesis, Massachusetts Institute of Technology, published by IC2 Institute, The University of Texas at Austin.

Rogers, Everett M. (1962), Diffusions of Innovations. Free Press of Glencoe: New York.

Rogers, Everett M. and Thomas W. Valente (1991), "Technology Transfer in High-Technology Industries," in Technology Transfer in International Business, Tamir Agmon and Mary Ann Von Glinow, eds., New York: Oxford Press.

Rubin, Michael S. (1992), Creating and Managing Consortia Cultures through Transitional Episodes," in Technology Transfer in Consortia and Strategic Alliances, David V. Gibson and Raymod W. Smilor, eds. , Rowman and Littlefield Publishers, Inc.

Sheth, Jagdish N. & Parvatiyar, Atul (1990), "Towards a Theory of Business Alliance Formation," Scandinavian International Business Review, I (2), .71-87.

Smilor, R., D. Gibson and C. Avery (1990), "R&D Consortia and Technology Transfer: Initial Lessons from MCC," Journal of Technology Transfer, 14 (2), Spring.

Wing, R. L. (1988), The Art of Strategy: A New Translation of Sun Tzu's Classic 'Art of War.' New York: Doubleday Press.

Yoshino, Michael Y and U. Srinivasa Rangan (1995), Strategic Alliances: An Entrepreneurial Approach to Globalization, Harvard Business School Press, Boston, Massachusetts.

The Effect Of Organization Architecture On New Product Development

Tomlinson G. Rauscher
(716) 381-2192

85 Sunset Boulevard
Pittsford, New York 14534

Executives at many companies have recognized the importance of improving the time their firms require to deliver new products to market. Often they address the problem by enhancing the product development process to yield faster time to market. While this approach may yield some improvements, executives can achieve significantly better results by examining the underlying architecture of the organization which effects the product development process.

In this paper, we examine the theory and the practice of this idea. After describing the concept of organization architecture, we examine the role of the organization in the time to market aspects of product development. We then apply the fundamental organization tenets to the activities of a product development team. To illustrate the effectiveness of this approach, we describe a new multifunction digital systems product of moderate complexity (over 300K lines of software) and the organization activities used in its development. The results show a dramatic increase in productivity; the product was delivered to the market in less than half the benchmark time. We conclude with a description of the role of management in organizing a firm for rapid time to market.

The Importance of Time to Market

For some time, industry consultants have been sounding the warning cry that time to market can have a significant effect on a company's business, indeed on the bottom line. In their classic book on the subject Stalk and Hout explain that "The ways leading companies manage time represent the most powerful new sources of competitive advantage." (Stalk 1990) Wheelwright and Clark from Harvard Business School have said "Firms that are slow to market with products that match neither customer expectations nor the products of their rivals are destined to see their market position erode and financial performance falter." (Clark 1993) Paul Allaire, Chairman of Xerox Corporation, has said "What we're aiming for is to put together a company that is much more connected to the customer, a company that can bring technology through our development process to the marketplace in a much more efficient and timely way." (Allaire 1994) It may be difficult to quantify the cost of suboptimal performance in delivering products to market in a timely way because the cost does not appear as a direct line item on a Profit and Loss Statement. However, the opportunity cost of late product delivery can be enormous. Indeed, Xerox reported in early 1996 that time to market inefficiencies in years before undertaking their time to market initiative cost as much as $520 million annually. I am convinced that Time to Market is the battlefield from which the leading firms of the 21st century will emerge.

Organizing for Failure

After a company develops its vision, it should establish an organization to implement the strategy that will realize that vision. For many companies rapid time to market is a critical strategy; thus, the organization should encourage early product delivery. However, our observations of the organizations of most companies lead us to conclude that the opposite is true. Therefore, oversimplifying outrageously, we state Rauscher's Third Law:

The organization encourages late product delivery.

One executive expressed the application of this law to his company by saying, "We are organized to fail."

0-7803-3552-X/96/$5.00 © 1996 IEEE

The last several years have witnessed a revolution in the science of organizations. The following definition provides deep insight into organizational behavior:
It is productive to define an organization as a legal entity that serves as a nexus for a complex set of contracts among disparate individuals. These contracts specify the rules of the game within the organization, including the three critical dimensions:

1. the performance measurement and evaluation system,
2. the reward and punishment system,
3. the system for partitioning and assigning decision rights among participants in the organization. (Jensen 1983)

More recently, the term "organization architecture" (Brickley 1996) has been used to describe this concept to differentiate it from the more general usage. When establishing or changing the organization architecture of a firm, the fundamental tenet in developing the performance measurement/evaluation system and the reward/punishment system is to

Give individuals incentives to act
in a way that maximizes the value of the firm.

Personnel courses may describe this concept with the phrase, "merging individual needs with organization needs."

The fundamental tenet in developing the system for partitioning and assigning decision rights is to

Assign decision making responsibility
to the point in the organization
where the knowledge to use it exists.

A Fortune 100 CEO has expressed this concept by saying, "Decisions should be made at the level where the managers have the expertise to make them."

The Role of Organization Architecture in Time to Market

With this background we can evaluate the effectiveness of organizations used in product development firms. First, consider the traditional functional organization, illustrated in Figure 1, in which program managers coordinate the work of people in the various functional groups who report to the functional managers. Proponents of this type of organization profess that this organizational model will improve time to market for the following reasons:

- Grouping individuals with expertise in a functional area enables the functional manager to bring required expertise to the solution of specific problems, especially when one product development team encounters problems.

- Grouping individuals with expertise in a functional area enables development of standard platforms that can be used in a variety of products. Thus each project will not have to develop its own platforms, but merely pick them off the shelf and integrate them into their particular product.

This type of organization exhibits overwhelming disadvantages:

- The product managers have no decision making responsibility to assign people or other resources to their products, thus suboptimizing the time to market priorities.

- The functional orientation encourages development of functional skills rather than product skills, yet the product should be the output to the customer.

- The standard platforms developed are not ideally suited to any one particular product and require substantial time for rework.

In the project organization, shown in Figure 2, the product managers have some decision making authority, for example, through funding to the various functional groups. However, the dual management by product managers and functional managers inherently confuses decision making authority, thus suboptimizing time to market.

In the business team (or product team) organization shown in Figure 3, each product manager has responsibility for all the various functions that contribute to product development. Thus the product manager can make the necessary decisions to optimize time to market. (Note that while this organization optimizes time to market, it does not necessarily minimize cost at any point in time. However, it does optimize profitability.)

Figure 1. Functional Organization

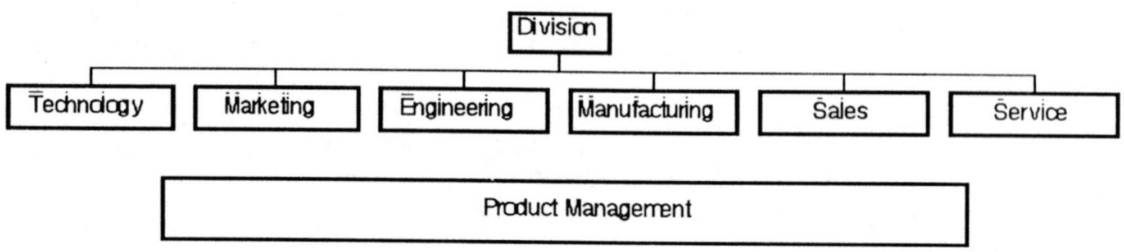

Figure 2. Project Organization

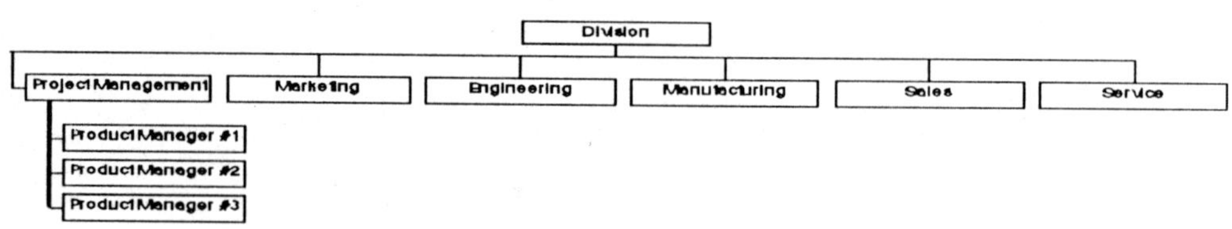

Figure 3. Product Organization

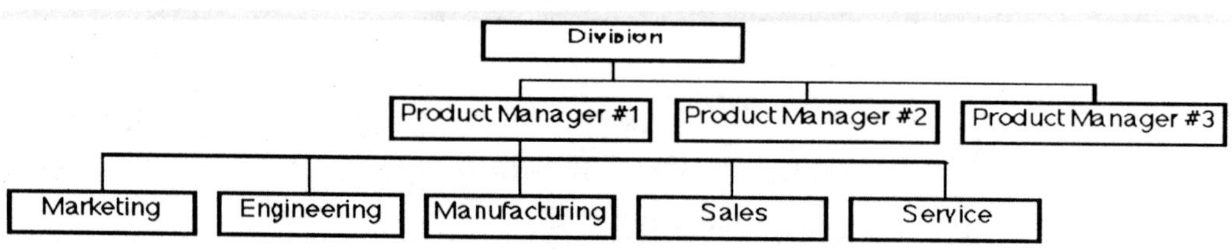

Overview of the Xerox ES8150 Engineering Document System

The Xerox ES8150 Engineering Document System is a mid-volume modular multi-function digital engineering document system. The key words in this appellation have the following meanings:

- Mid-volume - The volume refers to the speed and duty cycle of the system. By mid-volume we mean a system whose input paper scanning speed and output paper printing speed are in the range of 1 - 4 inches per second, and whose output duty cycle is 2,000 - 8,000 square feet of paper per month. The ES8150 has an input speed of 2 inches per second, an output speed of 3 inches per second, and a typical duty cycle of 4,000 square feet per month.
- Modular - The modular architecture enables the system to be
  - configured to meet a variety of customers' requirements
  - upgraded to enhance performance
- Multi-function - The ES8150 provides several document services:
  - Copying (printing a copy of a scanned original document)
  - Plotting (printing a file received from a network client)

- Scanning (creating a digital version of a scanned original document)
  - Document management (storing, retrieving, combining, deleting digital documents both locally and across the network)
  - Finishing (preparing output to facilitate use, e.g., folding, sorting, and rolling)
- Digital - The ES8150 transforms paper documents into digital representations, so that they can be manipulated and distributed electronically
- Engineering Document - The ES8150 handles engineering documents that range from A size (8.5" x 11") to E size (36" x 48" and longer); typical documents are blueprints, architectural drawings, and engineering drawings

Figure 4 shows the copier configuration of the ES8150 Engineering Document System.

The ES8150 provides the following features:

- Touch screen user interface
- Digital image processing
- 25% - 400% reduction enlargement in .1% increments
- Electronic collation
- Network connectivity
- Document storage

The ES8150 effects new digital productivity in several ways:

- Intuitive touch screen user interface with 17" color monitor
- Multiple copies from a single scan
- Electronic collation
- Scan / print concurrency
- Automatic
  - Roll switching
  - Reduction / enlargement
  - Document size sensing

Xerox ES8150 Breakthroughs in Time to Market and Productivity

In developing the ES8150, we paid close attention to the organizational architecture aspects of the product development activity. As a result we were able to make breakthrough achievements in time to market and productivity. Table 1 summarizes these accomplishments.

Table 1. ES8150 Time to Market Breakthrough

Project Start	June 13, 1994
Product Initiation	October 3, 1994
Product Announcement/ Demonstration	March 8, 1995
Launch / First Customer Ship	April 28, 1995
Time from Product Initiation to Launch	30 weeks
Benchmark Cycle Time	77 weeks

The consulting company PRTM developed product benchmarks against which we compared our product development goals. The benchmarks were based on two factors:

- Product complexity measured by number of drawings and/or lines of code. For this dimension, the ES8150, which contained over 300K lines of code, was in the second of four levels of complexity.

- Product newness measured by the amount of new development required to complete the product. In this dimension, the ES8150 was in the second of three levels of complexity.

The ES8150 software development effort also illustrated the concept that breakthroughs in productivity accompany breakthroughs in time to market. The software comprised two major subsystems: user interface / job control software and image management / peripheral control software. In both these subsystems, the ES8150 software developers generated over 20K lines of code per man-year compared to typical industry productivity of 2000-5000 lines/man-year. As another benchmark, we compared the ES8150 user interface / job control software development of 5 man-years to that of another Xerox product with similar functionality which had consumed 120 man-years of effort.

How Did The ES8150 Product Development Team Achieve This Radical Improvement In Time To Market?

Conventional wisdom holds that hard work determines success. Indeed, the ES8150 Product Development Team worked very hard. However, As Xerox Chairman Paul Allaire has said,
"Wishes and trying hard are not sufficient to achieve the new productivity." It was more than hard work that enabled the ES8150 team to achieve a time to market breakthrough.

The ES8150 team followed the unconventional wisdom:

The organization determines time to market success.

Recalling that organization architecture addresses performance measurement, reward and punishment, and delegating decision rights, we can see the effect of the Xerox ES8150 organization activities on time to market:

- The Xerox Engineering Systems (XES) Senior Staff
  - Delegated the product definition and development activity to the Systems Business Unit
  - Supported the team through periodic constructive reviews
- The Vice President of the XES Systems Business Unit
  - Delegated product development responsibility to the ES8150 Product Development Team (PDT) Manager
  - Promulgated a management style of acting as a "police escort" in contrast to the traditional "traffic cop"
- The ES8150 PDT Manager built a dedicated cross-functional Product Development Team with specific knowledge of
  - Customer requirements and marketing process
  - Engineering design process
  - Product development process
  - Manufacturing process
  - Product evaluation and test processes
  - Launch and support processes
- The ES8150 PDT Manager delegated decision making responsibility so that team members at all levels in the group made decisions based on their knowledge
- The ES8150 PDT Manager set aggressive objectives and rewarded the team members for achieving them
- The ES8150 Engineering Manager used his vast technical knowledge to
  - Build on existing technology to reduce effort and gain time to market
  - Develop technology in situations where performance or functionality required advances
- The ES8150 PDT used its time to market knowledge to define an innovative product that would
  - Satisfy or delight customers
  - Serve as an architectural basis for a family of products
  - Set new benchmarks for time to market
- The ES8150 PDT addressed proposed changes in requirements (a major cause of product delay) by
  - Examining the costs and benefits of the proposed changes
  - Making a joint decision to select a course of action

The Role of Management in an Organization (Jensen 1987)

We have seen that the fundamental cause of time to market problems lies in the organization of the firms that develop the products. Generalizing on our earlier research, (Rauscher 1987) we can state that

> Most product development projects that fail to meet their original objectives do so as a result of inadequate organizational management;
> few product development projects fail due to difficult technical problems.

Organization changes must be made to correct the time to market problems, and it is the managers of the firms that must make those changes. In this section, we examine the role of management in an organization to place this task in the context of their activities.

At all levels in an organization, managers have two fundamental activities:

- Decision making or policy making. The manager, using some problem solving process, decides on a desired action and tells someone to do it; i.e., the manager gives specific directions on the operation of the firm. For example, a first level manager decides on the design of a module for a specific product. A high level manager decides on the strategic business areas the firm is to pursue over the next five years.

- Deciding on the organization architecture. Each manager has an understanding (perhaps informal) of the organization architecture, i.e., the rules of the game - the reward/punishment system and the system for partitioning/assigning decision rights. Based on this understanding, each manager decides on an organization structure for the managed group. For example, a first level manager gives assignments to individuals based on the manager's understanding of each contributor's knowledge and capabilities. Similarly, the manager would reward contributors to various degrees. A high level manager decides the groups that should develop certain products and rewards managers for their contributions to corporate objectives.

At higher levels in an organization, managers should

- Devote less time giving specific directions
- Devote more time deciding on the organization structure

There are two reasons for higher level managers to devote more time to the organization structure:

- Individuals have inconsistent objectives. They are self interested. Giving individuals specific directions is not the same as getting it done. If higher level managers do not devote time to develop an organization whose reward/punishment system gives individuals incentives to act in a way that maximizes the value of the firm, i.e., that aligns personal objectives with the firm's objectives, then the firm's objectives will not be met.

- Individuals have incomplete knowledge. Even upper level managers cannot know everything. Although computer technology makes more information available to upper level managers, they cannot comprehend the totality of technical, marketing, financial, legal, etc. information required to operate the firm. If higher level managers do not devote time to develop an organization whose system for partitioning and allocating decision rights assigns decision making responsibility to the point in the organization where the knowledge to use it exists, then incorrect decisions will be made and the firm will perform suboptimally.

The organization of a successful firm must change over time, because the environment in which the firm operates changes over time: technologies evolve, competitors introduce new products, customers' requirements change, financial markets adjust daily, new laws affect product design, and so on. Organizational changes ensure that individuals' objectives and knowledge change with the firm's requirements. A firm that fails to evolve its organization architecture will suffer "Organizational Darwinism," fail to adapt, become competitively disadvantaged, and possibly die. Thus, upper level managers must continually devote substantial time to deciding on the organization of the firm.

Summary

The study of organizational architecture has led us to the theory that the organization determines time to market success. We have demonstrated the theory using a proof by construction; the Xerox ES8150 organization delivered its product to market in less than half the benchmark time.

References

Allaire, Paul, "The New Xerox: An Interview with Paul Allaire," *Xerox Benchmark*, Volume IX, Number 2, 1994.

Brickley, James A., Smith, Clifford W., Jr., and Zimmerman, Jerold L., *Organization Architecture: A Managerial Economics Approach*, Richard D. Irwin, 1996.

Clark, Kim B. and Wheelwright, Steven C., *Managing New Product and Process Development*, The Free Press: New York, 1993.

Jensen, Michael C., "Organization Theory and Methodology," *The Accounting Review*, Vol. LVIII, No. 2, April 1983, pages 325-326.

Adapted from Jensen, Michael C. and Meckling, William H, "Coordination, Control, and the Management of Organizations," University of Rochester, William E. Simon Graduate School of Business Administration, Spring 1987.

Rauscher, Tomlinson G. and Ott, Linda M., *Software Development and Management for Microprocessor-Based Systems*, Prentice-Hall, Englewood Cliffs, New Jersey, 1987.

Stalk, George and Hout, Thomas M., *Competing Against Time*, The Free Press: New York, 1990.

Factors For Success In Acquiring Information Technology

Ali Q. Jawad
968 601 538

P.O.Box 244
Jabroo 114
Sultanate of Oman

Richard Reeves
Tel. 44 1234 754773

R&D Management Centre
Cranfield University
Bedford MK43 0AL, UK.

A large number of acquisitions of IT systems has been examined and 43 different management factors which promote success in such projects have been identified. In addition eight measures by which the degree of success of an acquisition can be assessed have also been identified. The findings are incorporated into a structured approach for managing acquisitions, which is called the Success Map.

Introduction

The acquisition of new Information Technology (IT) systems is expensive and risky. In one case studied in this research, a new computer system purchased to automate an important records system turned out to be an obsolescent model and it was soon found to be impossible to work with, so an investment of $100,000 was almost entirely wasted. In another case a budget allocated for computerising an organisation's administrative operations was all spent on hardware and software without realising the extent of training and support that would be needed, or the necessity of redesigning the operations. The system ended up being used for wordprocessing only.

Such unfortunate experiences occur on a small scale, such as in the purchase of PCs by individuals, and also on the largest scale. The attempted computerisations of the London Ambulance Service and of the London Stock Exchange were well publicised examples of very costly failures.

As well as wasting money, an unsuccessful acquisition means that the expected services will not be provided, or the organisation may become committed to a particular approach to its information management which is inefficient and costly to change. This research examines how organisations can acquire IT systems with greater chances of success. Study of the literature showed that no comprehensive scientific study of this problem had yet been made.

Acquisition and purchase

It is important to distinguish between *acquisition* and *purchase*. When an IT system is required, attention is usually concentrated on the act of purchasing the actual hardware or software, but in many cases this will lead on to an expenditure of time and effort which is worth much more than the purchase price. When the system is finally working or not, the positive or negative impact on the purchaser's business may be worth a great deal more again. The whole acquisition process must be successfully managed, which is the total process of adoption, implementation and use of information technology.

THE METHODOLOGY - success factors

The success factor approach was used. This method has been adopted in different fields by many researchers, including Cooper in his studies of new product development (Cooper 1993). This research was conducted in two main phases. In the first phase, people who had been involved in 33 different IT acquisitions in the public sector in Oman, were interviewed to find what factors were thought by them to

0-7803-3552-X/96/$5.00 © 1996 IEEE

influence the outcome of an acquisition. A list of 43 *candidate success factors* resulted. In the second phase, questionnaires were filled in by 144 respondents involved in different projects of varying degrees of success, to assess quantitatively the extent to which each of the 43 factors was in practice associated with success.

The initial interviews were semi-structured. They were recorded, and analysed by Content Analysis (Gale and Grant 1990) to produce the list of factors. An inter-rater reliability check was carried out and an expert panel reviewed the list and the definitions which had been assigned to each factor. The factors were operationalised to produce questions in the questionnaire, which was pilot tested before use. Rank-order correlation was applied using the SPSS statistical package, to analyse the questionnaire returns.

Analysing The Acquisition Process

Defining Success

During a pilot stage of the interviews, it became apparent that some people would describe a project as successful while others described it as a failure. In fact it turned out that most people, while willing to state whether a project was successful or not, had not previously thought about what they meant by success of an IT acquisition. It therefore became an objective of the interviews to seek *success measures* in addition to the *success factors*.

Players

Standard marketing theory tells us that when something is sold to a large organisation, a number of different people are involved: an Initiator of the purchase, an Authority to sign for it, a User, and a Purchasing official. This research concerns purchasing, not selling, but it was felt that a similar idea could apply. The term *Players* was coined, and three main groups of players were identified from the interviews:

1. **Senior management**: who are involved in either IT decision-making or in using IT for management purposes.

2. **IT staff**: such as IT managers, advisors, engineers, analysts, programmers or technicians.

3. **Users**: who use a terminal as a regular part of their work, as data entry clerks, or to retrieve information.

The comment may be made that an IT acquisition is a team project, but often the players do not realise they are in a project or a team and they do not know how the game is played. Often the team has no captain.

Acquisition lifecycle
It became clear during the survey work that each IT acquisition project went through three main phases: *Initiation*, *Implementation*, and *Service*, as shown in the lifecycle diagram below.

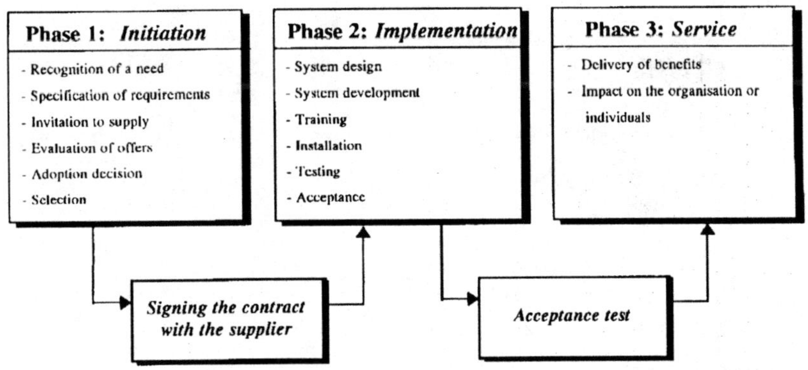

Figure 1. IT acquisition life-cycle.

There are two high quality milestones in the process: signing the contract with the supplier, and the user's acceptance test. Both are very definite in time and signal total commitment to the next phase. Acceptance tests before handover to users are normal practice in Oman.

Results of the research

Results of the interviews

The 43 *candidate success factors* identified in the research are listed in Figure 2. Their relative importances were measured in the following phase of the research. This list of factors is likely to be comprehensive, in that there was an 80% chance of detecting a factor whose likelihood of occurrence in a given case was as low as 1 in 20.

Each factor had an associated definition. Two of these are illustrated in Figure 3.

	Technological factors		
1	Use of current technology	23	Appointment of a project manager
2	Use of proven technology	24	Use of project management methodology
3	Ease of use	25	Accurate statement of requirements
4	Enhancability of the system	26	Stable user requirements
5	Open system interconnectability	27	Sufficient time for completion
6	Off-the shelf software	28	Considering hidden costs
7	Existence of IT standards	29	Supplier analysis
	Organisational factors	30	Provision of user training
8	Internal communication	31	Provision of technical training
9	Indigenous R & D capabilities	32	Comprehensive contract terms
10	Previous IT experience		**Human factors**
11	Sufficiency of financial resources	33	Top management IT awareness
12	Indigenous skill level	34	IT managers' business understanding
13	Clarity of business objectives	35	IT staff technical capabilities
14	Existence of IT strategy	36	User maturity
15	Existence of IS strategy	37	User acceptance
16	Policies and procedures		**External factors**
17	Presence of IT champion	38	Client-supplier relationship
	Management factors	39	Supplier competence
18	User involvement	40	Use of independent consultant
19	Top management support	41	Flexibility of general regulations
20	Realistic project objective		**Data factors**
21	Feasibility study	42	Data availability
22	Piloting the project	43	Data validity

Figure 2. The candidate success factors. The factors have been grouped under subheadings for convenience.

9	Indigenous R&D capabilities	The capability of an organisation to search for, assess, adapt, modify, and develop relevant technology.
13	Clarity of business objectives	The business/strategic objectives of an organisation are clear at all levels.

Figure 3. Definitions of two of the success factors, as examples.

Eight *Success Measures* were identified from the interviews, as follows:

1. Business feasibility
2. Technical feasibility
3. Completion on time
4. Completion within budget
5. System performance
6. Impact on the organisation or individuals
7. User satisfaction
8. Amount of actual use of the system

Analysis of the questionnaire results

The questionnaire measured the degree to which each of the 43 success factors was present in each case, plus the degree to which each of the 8 success measures was achieved. Correlation analysis revealed the degree to which each of the 43 factors was associated with each of the 8 measures.

Nearly all factors correlated at 95% or higher confidence level with one or more success measures. This indicates that IT acquisition is a complex process and that nearly all factors identified are important. Interdependencies among the factors themselves would be expected, and to clarify the situation, a partial correlation process was applied. This discriminated factors which directly correlated with success measures from those whose correlation was indirect. Factors with direct correlation with success measure were called *Superfactors*. Fifteen of these were identified.

Synthesis - The success map

The success map provides a means of bringing the above ideas and findings together and showing where in the acquisition process each factor is important. In Figure 4 the eight success measures have been allocated to the three phases in the acquisition lifecycle. Each measure is associated with 2 to 4 of the 15 superfactors, and the figure shows which superfactors occur in each phase.

Initiation Phase ⇒	Implementation Phase ⇒	Service Phase
success measures • Business feasibility • Technical feasibility	success measures • Completion on time • Completion within budget • System performance	success measures • Impact • User satisfaction • Actual use
Superfactors 1. Stable requirements 2. Project management methodology 3. Realistic project objective 4. Clarity of business objectives	**Superfactors** 1. Supplier competence 2. Off-the-shelf 3. Client-supplier relationship 4. Ease of use 5. User acceptance 6. Considering hidden costs 7. Enhancability of the system	**Superfactors** 1. Indigenous skill level 2. Ease of use 3. User training 4. Project objective 5. IS strategy 6. Supplier competence 7. Proven technology

Figure 4. The simplified Success Map-superfactors only

Figure 4 is derived from a more detailed map which shows which superfactor promotes which success measure, and which of the remaining factors are associated with each superfactor.

Discussion

The importance of the different measures of success varies from one project to another, and it follows that the importance of the different success factors also varies from project to project.

This research was carried out in Oman. It would be worthwhile repeating it elsewhere to check whether similar lists of factors, measures and correlations would be obtained.

A quick guide to managing IT acquisition
1. Identify the players, and bring them together.
2. Decide which success measures matter most in your particular project.
3. In the light of this, use our correlations to prioritise the required success factors.
4. Identify the degree to which the required success factors are present, preferably before you start the project. We have developed a questionnaire for doing this.
5. Appraise the results and manage accordingly.

References

Cooper, Robert G. Winning at New Products. 2nd ed. Addison-Wesley, 1993.

Gale, Rodney and Grant, Janet. *Guidelines for change in postgraduate and continuing medical education*. The joint centre for educational research and development in medicine, London 1990.

An Integrated Management Model for Virtual Enterprises: Vision, Strategy and Structure

Ms. R. Leigh Reid
(214) 266-0942
Northrop Grumman Commercial Aircraft Division
P.O. Box 655907 M/S 49R-LR
Dallas, TX 75265 USA

Mr. James B. Tapp
(214) 266-4468
Northrop Grumman Commercial Aircraft Division
P.O. Box 655907 M/S 49R-03
Dallas, TX 75265 USA

Dr. Donald H. Liles
(817) 272-5902
Automation & Robotics Research Institute
7300 Jack Newell Blvd. South
Fort Worth, TX 76118 USA

Dr. K. J. Rogers
(817) 272-5900
Automation & Robotics Research Institute
7300 Jack Newell Blvd. South
Fort Worth, TX 76118 USA

Ms. Mary E. Johnson
(817) 272-5919
Automation & Robotics Research Institute
7300 Jack Newell Blvd. South
Fort Worth, TX 76118 USA

Three distinct, but related concepts are discussed in this paper: a six step virtual enterprise life cycle, enterprise processes organized into three management functions and an organizational model known as integrated decentralization. When combined, the three conceptual models form an integrated enterprise management model that is equally applicable to single firms operating in extended enterprises and to multi-firm virtual enterprises. The structure and operating model of Northrop Grumman's Commercial Aircraft Division is described to provide an example of a firm that conducts business using the integrated enterprise management model.

A Virtual Enterprise Life Cycle

The virtual enterprise is a temporary relationship with two or more participants which is formed, operated, and dissolved to accomplish specific short term goals. It differs from existing inter-organizational models by the degree of shared accountability and responsibility of the participants and the structure by which companies contribute their competencies.

0-7803-3552-X/96/$5.00 © 1996 IEEE

A *virtual enterprise life cycle* having six elements is shown in Figure 1 and described below.

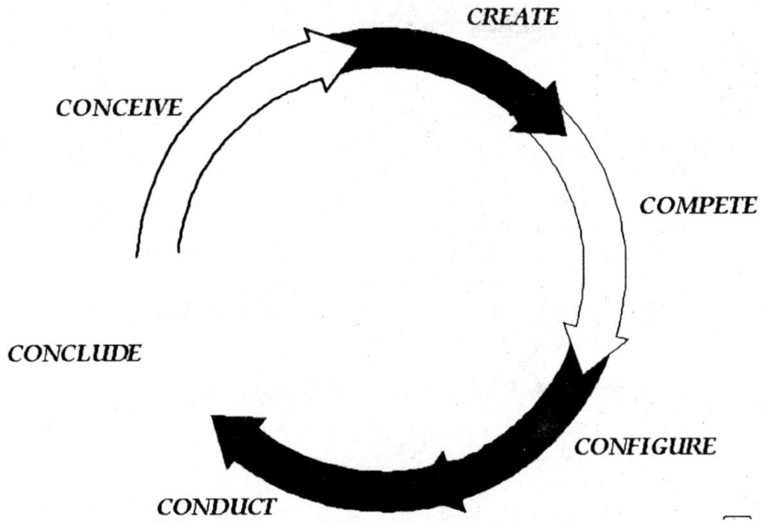

- A virtual enterprise is *conceived* when a need is recognized in the marketplace and a business objective (or set of objectives) is established. This step requires understanding of the customers' expectations/needs and what it will take to satisfy them. The enterprise that is required to meet the need is visualized, and a transformation/migration strategy is articulated. This activity can be accomplished by a single firm or by an existing virtual enterprise. This step is essentially the conceptual design of a new enterprise.

- The enterprise is *created* when relationships are established that will eventually bring together the requisite competencies, when a strategy is crafted and a "product" is "designed" to meet the identified need. At this stage, the firms that comprise the enterprise will likely develop and implement new or improved processes and systems to prepare for the next stages of the cycle. Activities in this stage constitute detailed design of the new virtual enterprise and complete preparation for implementation.

- The virtual enterprise *competes* when the "product" is offered in the marketplace. This activity may be accomplished in several ways. The enterprise may offer new or alternative solutions to previously unmet needs, or it could identify, pursue and capture a defined opportunity to produce and deliver its product. Finally, the enterprise could secure new customers for existing products.

- After competing, the enterprise is *configured* as assets and competencies are acquired and the requisite processes and infrastructure are deployed to accomplish the objectives of the enterprise. The assets, processes, and procedures are acquired or developed, and integrated as specified by the enterprise design to produce and deliver the required product. These activities comprise the actual implementation step for the new virtual enterprise.

- The virtual enterprise then *conducts* operations to produce, deliver and support the "product" and to maximize stakeholder value.

- It *concludes* operations when the objectives of the enterprise are satisfied, by terminating the relationships and by re-deploying and/or disacquiring assets.

The activities of the virtual enterprise cycle are accomplished by processes that are owned and operated by individual members of the virtual enterprise or shared processes that are "owned" jointly by the enterprise as a corporate entity. Whether the processes are individually or jointly owned is largely predicated on the objectives of the enterprise and how they are to be accomplished. This concept is discussed more fully in the following section.

Enterprise Processes Organized Into Management Functions

Enterprises operate with various business and production processes, which can be divided into three categories:

- *Those that mold external constraints into an internal constraint structure (Category 1)...* These processes are generally *jointly owned* in virtual enterprises.
- *Those that acquire and make ready the resources used by the enterprise (Category 2)...* These processes are most often *jointly owned* as well, since their focus is assets and resources that are used to accomplished shared objectives.
- *Those that transform inputs into the desired outputs or results (Category 3)...* Enterprise members generally retain *individual ownership* of these processes and contribute the competency and/or the results of the process to the enterprise to accomplish specific objectives.

The process categories can also be modeled as high-level *management functions*, each with its own set of objectives and operating characteristics. *Category 3 processes* comprise the function of *Value Chain Management.* The objectives of the firm's value chain managers are to satisfy the firm's customers, to maximize profits and to win new business and grow.

Category 2 processes are organized into the *Asset Management* function. The objectives of a firm's asset managers are to optimize the performance of the firm's assets by continually improving operating efficiencies, to provide capabilities and services that enable the effective operation of the value chains and to position the firm for the future through continually evolving the asset "mix" and improving the capabilities and functionality of the individual assets. In virtual enterprises, this function could be performed by the individual owner of the contributed asset (or capability), but it is more likely to be shared among the enterprise members as they try to accomplish common objectives. Four categories of assets are typically managed by enterprises: *human* assets, *financial* assets, *physical* assets (plant, property and equipment) and *intellectual* assets.

Category 1 processes comprise the function of *Strategic Management.* The primary objective of the firm's strategic managers is to maximize the value delivered to all the firm's stakeholders. This objective is accomplished by establishing and maintaining a unifying vision and strategic direction and by reconciling the conflicting objectives of the asset managers and the value chain managers within that context. In many firms, and in virtual enterprises, this function is generally accomplished by an *integrated management team* made up of the asset managers and the value chain managers and led by the chief executive. Figure 2 illustrates the three management functions interacting to operate the firm (or the virtual enterprise). Decisions made in the central circle, by the *integrated management team*, drive the virtual enterprise cycle (see Figure 1).

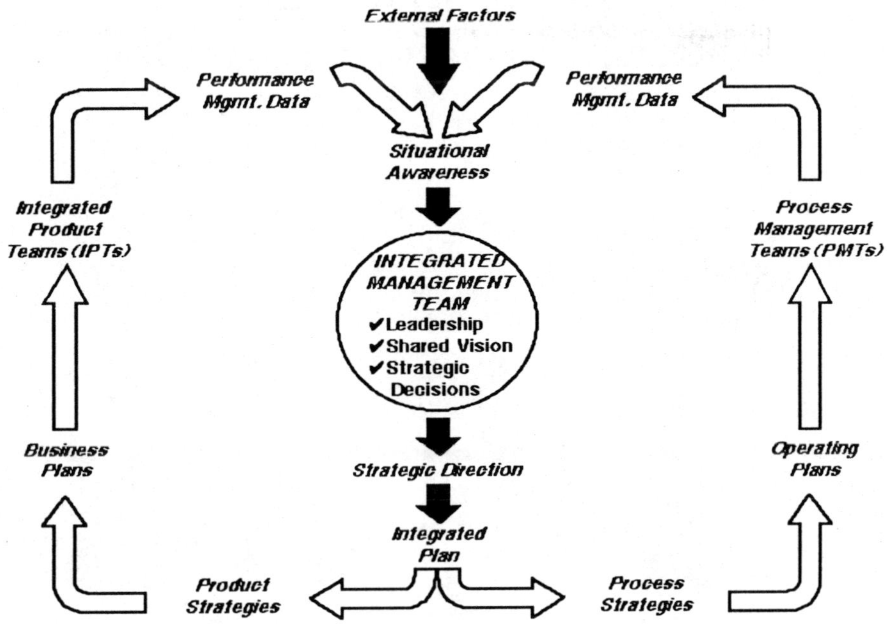

Agility and Integrated Decentralization

Agile firms typically exhibit all (or most) of the following operating characteristics or skills:

- *Sensing,* or the ability to perceive environmental conditions; gather useful information from the system and readily detect changes; also the ability to anticipate changes...
- *Learning,* or effectively modifying organizational behaviors and beliefs through experience; the ability to use information (see "sensing") to improve the organization...
- *Adaptability*, or the ability to effect changes in systems and structures in response to (or in anticipation of) changes in environmental conditions...
- *Resilience,* or robustness to diversity and variability, the ability to recover from changes...
- *Quickness*, or the ability to accomplish objectives in a short period of time; pace at which changes are accomplished; the rate of movement within organizational systems...
- *Innovation,* or the ability to generate many solutions to a problem...
- *Flexibility*, or the ability to do a number of different things with the same systems and structures; having processes, systems and equipment adaptable to many different uses...
- *Concurrency*, or the ability to effectively perform related activities at the same time, with the same assets...
- *Efficiency*, or using minimal resources to accomplish desired results...

Firms that are fully *integrated*, yet operate in a *decentralized* manner are much more likely to demonstrate these characteristics. The effects of *integrated decentralization* are observed at the level of the three management functions described previously. Decentralized, relatively autonomous, *value chains* permit firms to produce many different "products" (or solutions) for multiple customers, who may have widely divergent requirements and expectations. The decentralized value chain manager is also better able to anticipate and rapidly respond to changes in the market and operating environments. On the other hand, integrated management of the firm's *strategy* and *assets* maintains a unified vision and direction and enables synergy and increased efficiency among the operating elements of the firm.

Figure 3 is an example of the three management functions operating in an *integrated decentralized* firm (Northrop Grumman Commercial Aircraft Division - NGCAD). It is presented to illustrate a typical organizational structure within a single firm, although many of the features are likely to be found in virtual enterprises made up of a number of different firms.

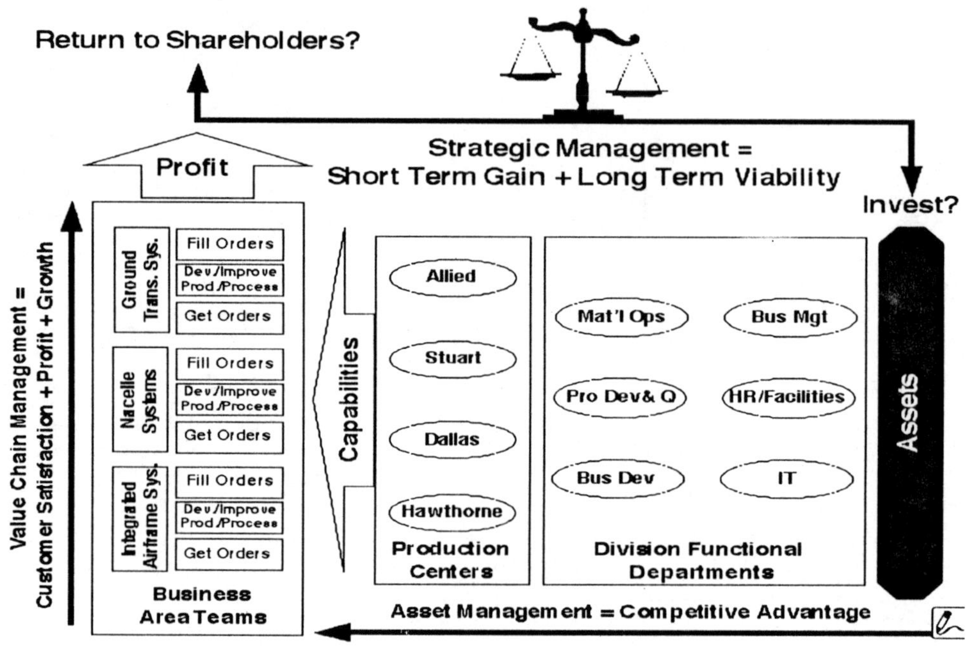

In this example, the function of *Value Chain Management* is performed by decentralized, relatively autonomous *Integrated Product Teams (IPTs).* The IPTs have the accountability and authority to operate the value chains to return the desired business results; namely, customer satisfaction, profitability and growth. They take complete responsibility for the quality of the delivered product, and for cost and schedule performance, as well as for acquiring new customers and orders, and for developing and improving products and value chain processes.

The capabilities and assets of NGCAD actually reside in four geographically dispersed *production centers*. The senior managers who "own" and operate these production centers have a dual role. They are, first and foremost, *Asset Managers*, responsible to provide best-in-class production capabilities to the value chains and to optimize the division's shared production assets. They are also responsible for *filling orders* - actually producing and delivering products on behalf of the business areas - and for developing and improving many of the *value chain processes.* In these respects, they operate as *Value Chain Managers.*

Asset Management is mainly performed within the division's "functional" departments. The objectives of NGCAD's asset managers are to transform "raw" assets - money, people, facilities and equipment, information, technologies, etc. - into *capabilities* that will provide a sustainable competitive advantage in NGCAD's market niches. They provide these capabilities to the value chains through the production centers. Together, the production center managers and the heads of the functional departments also take actions to continuously improve the other (non-value -chain) enterprise processes to maximize their efficiency and effectiveness. They provide support services to all business areas, acting as internal suppliers to the value chains and to each other.

The General Manager and his direct reports (the Business Area IPT leaders, functional department heads and production center managers) form an *Integrated Management Team* to perform the *Strategic Management* function. This team acts to position the firm for the future and to maximize shareholder value. It provides a unifying vision and strategic direction to its subordinate teams and manages the overall performance of the enterprise. Through a constant series of decisions and actionable plans, this team balances the desire for short-term performance results with the need to invest in capabilities and assets that provide long-term viability.

Virtuality and Uncertainty in the Domain of Discontinuous Innovation

Mark P. Rice
(518) 276-8398

304 Lally School of Management & Technology
Rensselaer Polytechnic Institute
Troy NY 12180

Established firms engaging in the discontinuous innovation process often establish virtual enterprises as a mechanism for coping with uncertainty. The virtual enterprise is assembled from a set of disparate and dispersed technical, human and financial resources, some internal and some external, to achieve one or more milestones, then reconfigured as some of the original resources are shed and new resources are acquired to achieve the next set of milestones, and so on. This paper will explore the role of virtuality and uncertainty in the pursuit of opportunity associated with discontinuous innovation.

Introduction

Virtuality, Uncertainty, Discontinuous Innovation, and Entrepreneurship Ettlie et al. (1989) suggest that discontinuous innovation occurs when the magnitude or cost of change required to address discontinuity (in product or market) is large, compared that required for an incremental innovation. Souder and Moenaert (1992) identify four sources of uncertainty: (1) consumer, (2) competitor, (3) technological, and (4) resource. The first three relate to the concept of an opportunity, i.e. the fit between a gap in the marketplace and a product/ service that can fill that gap (Timmons 1994). The linkage between opportunity and resources is embodied in the classic definition of entrepreneurship by Stevenson et al. (1994), i.e. "the pursuit of opportunity without regard to resources currently controlled." Uncertainty also often exists with respect to organizational and leadership dynamics. Uncertainty reduction is a prerequisite to converting an innovation into a successful new product or venture (Souder and Moenaert 1992).

According to Chesbrough and Teece (1996), the combination of rapid response and incentives defines "what's special about virtual." Firms need to respond quickly to opportunities associated with discontinuous innovations, not only to minimize cumulative investment in aborted ventures that fail in a test of a critical go -- no go assumption (Block and Macmillan 1993), but also to seize the market opportunity ahead of competitors when a discontinuous innovation is pursued all the way to commercialization. The opportunity to achieve an attractive return on investment by capturing higher margins associated with a monopoly that may occur during the initial stages of market development for a discontinuous innovation is a powerful incentive.

In the pursuit of opportunities related to discontinuous innovations, all eight projects in this study have employed some form of virtual enterprise involving the integration of various internal and external business components based on a variety of linkage mechanisms, e.g. communication, computer and energy technologies.

Methodology

This research project employs the case study methodology. The complexity of case study research, and the high level of interpretation that is necessary, create an advantage for the use of research teams. Multiple investigators can bring a variety of experience and complementary insights to the research. A mix of different perspectives can increase the likelihood of discovering novel insights. Convergence of opinions from various researchers can enhance confidence in the findings; conflicting views can keep the research from premature closure (Eisenhardt, 1989).

The research reported in this paper, which is part of a three year prospective study of corporate venturing related to the processes involved in the management of discontinuous innovation in established firms, employs a multi-disciplinary team of ten faculty and several Ph.D. students. During the pilot phase (year 1) six firms, all members of the Industrial Research Institute, have hosted two site visits each. Typically, two to three members of the research team typically participate in each interview of each company representative, who undergoes multiple interviews by several subgroups of the total team. There are a total of eight projects in the sample, all of which have been formally established and none of which have reached the point of commercialization. At this stage the sample reflects two primary dimensions: diversity in the technological environment and stage of development.

In accordance with the suggestion of Eisenhardt (1989) that the case study method is especially appropriate for focusing on "how " or "why" questions concerning a contemporary set of events, this research paper focuses on the following question: “How do established firms cope with the high degree of uncertainty associated with managing the pursuit of an opportunity based on a discontinuous innovation?”

Findings: Strategies for Coping with Uncertainty

Partnering to Gain Access to External Competencies

In order to respond as rapidly as possible to opportunities and to minimize expenditures associated with testing the viability of their opportunities, firms in this sample have all formed internal virtual enterprises to access competencies that don’t exist within the enterprise but which are required to pursue the opportunity. In five of the eight projects included in this research project, product development involves a variety of internal and external sources, reflecting sets of technical competencies that are being integrated to achieve a whole product. In six of the eight projects external manufacturing competencies are being accessed, to support the virtual enterprise in some or all of the following: technology development, product prototyping, and full scale production (planned). In three of the eight projects, the discontinuous innovation is a component that will be provided to an external original equipment manufacturer (OEM), and the virtual enterprises intend to rely on the OEMs for marketing and distribution. These competency-related partnering relationships are summarized in Table 1 below.

Partnering to Access External Financial Resources

In all eight cases corporate R&D (CRD) has provided funds at start up to assess the viability of the technology in order to make the go -- no go decision regarding the pursuit of funding required to ramp up the effort from (1) sources internal to the corporation but external to the corporate R&D division, and/or (2) external to the corporation altogether. In five of the eight projects partnering with government R&D agencies for financing of continued product development has been important to sustaining the virtual enterprise. The pressure for short term profitability is particularly significant for operating business units, and in only one of the eight projects was significant financial support provided by a business unit, and in that case only in the earliest stages of the project. External financing partnering relationships are also summarized in Table 1 below.

Table 1. External Partnering Relationships

Functional Domain	P1	P2	P3	P4	P5	P6	P7	P8
Technology Development / R&D	P	P	P	I	P	P	I	I
Manufacturing	P	P	P	P	P	I	I	P
Marketing / Distribution	I	P	I	I	I	P	I	P
Project Funding	P	P	P	P	I	I	I	P

Key: P1, P2 ..... P8 refer to the eight projects. At this stage of the research study, the research team has agreed to cloak the identities of the corporate parents of the individual projects and the underlying innovations until protocols for confidentiality have been resolved with all participants.
I = solely through internal activity; P = partnering plays a significant role in the activity.

Internal Partnering

In most cases partnering with internal organizations that are independent from the virtual enterprise is not significantly different from partnering with organizations that are external to the corporation. In the extreme example (P4), although one of the business units will be the distribution channel, the pressures on the business unit to maximize short term profitability are so severe that the senior manager of the business unit not only refuses to provide any financial support but even requires that the virtual enterprise provide funding to access its people for their competencies and knowledge of the market. In spite of these kinds of obstacles, internal partnering is critical to the success of the virtual enterprises in this sample.

The original home for the project team in all cases has been the corporate R&D (CRD) division and remains so for projects P3, P4, P5 and P6. The current "home" for projects P1, P7 and P8 is a business development group operating within CRD. For P2, a separate corporate venturing organization has been created. The CEO of the corporation believes the viability of the new venture requires independence from the existing business units. With the exception of P2, the virtual enterprises in this sample have a relationship with one or more business units, although the importance of that relationship varies significantly across the sample. In P1 the senior management of the partnering business unit has been a key champion of the technology and the formation of the virtual enterprise. In the other six cases (P3 - P8), the virtual enterprises have accessed the business units primarily for marketing and manufacturing expertise, but in some cases also for access to parts or fabrication capabilities necessary for developing prototypes. Regardless of whether the virtual enterprise has its own identity independent of corporate R&D (CRD) or at this stage continues to be a project group within (CRD), the virtual enterprises continue to benefit from their relationship with CRD, particularly with respect to accessing technical competencies, equipment and occasionally funding.

The role of champions in the success of virtual enterprises

The importance of champions in corporate venturing is well established (Block and Macmillan 1993), and has been reaffirmed in this project, not only with respect to the traditional roles related to financing and organizational issues, but also with respect to the critical issues facing virtual enterprises: control, coordination and communication (Chesbrough and Teece 1996). An inventory of champions is provided in Table 2 below, where Y = yes, there is a champion at the specified level, and N= no, the project does not have a champion at that level. NA = not applicable.

Table 2. Champions for Virtual Enterprises

Level of Champion	P1	P2	P3	P4	P5	P6	P7	P8
senior corporate management	Y	Y	Y	Y	N	N	N	N
senior technical management	Y	N	Y	Y	Y	Y	N	N
senior business unit management	Y	N	N	N	N	N	N	N
senior scientist	Y	N	N	Y	N	Y	Y	N
business development group mgr	Y	Y	NA	NA	NA	NA	Y	Y

With respect to championing, persistence and intensity are important aspects, given the high degree of uncertainty and the inevitable mistakes and setbacks involved in pursuing discontinuous innovation. Five of the eight projects have died and been resurrected as the result of the efforts of one or more champions. Persistence and intensity are also important for the corporate entrepreneur heading up the virtual enterprise. In projects P2, P4, P5 and P6 the corporate entrepreneur has demonstrated a powerful commitment (either in years or in the extent to which his/her career is on the line) and in P2 and P4 the CEO of the corporation has been the primary senior management champion. P2 and P4 are clearly closest to successful commercialization of all eight projects.

The importance of continuity of project leadership

With the exception of project P8, these projects have been in development for 5 to 10 years. P1, P2, P4, P5 and P6 have benefited from the continuity of championing. By comparison, while the lead champion for P3 was off on temporary assignment for a couple of years, P3 died and would not have been resurrected without substantial financial support from a government R&D agency. With his return, the project benefited from his championing, but he recently retired and the commitment of his successor to sustaining the project is unclear, as is the future of the project. Project P7 has gone through three generations of champions and is currently on life support because of the interest of a senior scientist and the beneficence of the director of the business development group. The senior scientist indicates the project may not survive much longer.

The use of special linkage mechanisms

In addition to special organizational mechanisms, e.g. the use of business development groups and championing, the corporations in this sample have developed other coordinating mechanisms. For example the corporate parent of P1 has developed a virtual board of directors including representatives of key internal partners and the single most important external partner in order to ensure through a variety of communications technologies that the champions of each of these entities remains current on the progress of the project with respect to milestones. In the case of P2 the virtual enterprise has relied heavily on electronic mail to keep the members of its project team informed, and to ensure confidentiality, has developed the capability of preventing the reproduction of any material transmitted through its e-mail system.

Conclusions

The development of a virtual enterprise that partners with internal and external entities is one strategy established firms use to manage the uncertainty inherent in the process of pursuing a perceived opportunity based on a discontinuous innovation. The use of the internal virtual enterprise organization form allows the firm to shorten the technology / business development process and minimize the investment of its financial resources that must be expended to test the technical, market and organizational assumptions reflecting the high degree of uncertainty. Through the virtual enterprise the firm gains access to financial and non-financial resources, knowledge and expertise from other internal entities and from external entities. Even so the investment is typically considerable and deploying a virtual enterprise, while benefiting the effort in some respects, can introduce an additional set of challenges and uncertainties, particularly related to control, coordination and communication. Chesbrough and Teece (1996) have suggested that successful virtual enterprises (1) develop and zealously protect special competencies that are the basis for their competitive advantage, and (2) exhibit staying power by establishing themselves at the center of the network in a way that leverages there special capabilities, competencies and resources, and all the firms in this sample have pursued these two strategies. As this research project follows the progress of the ongoing discontinuous innovation projects in the sample, and as some projects enjoy greater success than others, it is hoped that additional insights can be gleaned to illuminate practices that enhance or diminish the performance of virtual enterprises in this domain.

References

Block, Zenas, and Macmillan, Ian C. Corporate Venturing. Harvard Business School Press: Boston, 1993.

Chesbrough, Henry W. and Teece, David J. "When is Virtual Virtuous? Organizing for Innovation." *Harvard Business Review*. January-February: 65-73, 1996

Eisenhardt, Kathleen M. "Building Theories from Case Study Research." *Academy of Management Review*. 14(4): 532-550, 1989.

Ettlie, John E., Bridges, William P., and O'Keefe, Robert D. "Organizational Strategy and Structural Differences for Radical versus Incremental Innovation." *Management Science*, 30(6): 682-695, 1989.

Souder, William E., and Moenaert, Rudy K. "Integrating Marketing and R&D Project Personnel within Innovation Projects: An Information Uncertainty Model." *Journal of Management Studies*, 29(4), 485-512.

Stevenson, Howard H.; Roberts, Michael J.; and Grousbeck, H. Irving. New Business Ventures and the Entrepreneur. 4th ed., Irwin, Boston, 1994.

Timmons, Jeffry A. New Venture Creation. 4th ed. Irwin, Boston, 1994.

Process improvements in distributed computing support in a large corporate environment

Michael E. Richerson, P.E.
(316) 526-9316

Boeing Information & Support Services
P.O. Box 7730, MS K79-24
Wichita, KS 67277

This paper describes process improvements made in supporting computer workstations in a large corporate environment. Over 10,000 computer workstations are supported. Many are in engineering organizations, but workstations in manufacturing and business management organizations are also supported. A central support function was restructured to provide support in geographic zones. Other issues addressed include downsizing, product changes, increasingly more complex technology and new company initiatives.

Computer workstation support had been provided from central organizations. Each organization specialized in one or a few functions and provided those services to all users. For example, one organization performed all of the network installations and another organization performed problem determination after trouble calls were placed. A centralized service approach provides for some concentration of skills and economy of scale. However, it does not lend itself to addressing requirements specific to individual organizations and projects. Also it does not provide for job enlargement among the employees.

Representatives from end user organizations were asked about service requirements. Requirements for more organizational and project specific support were identified. Other developments within the company were also taken into account. The company was restructuring into responsibility centers. Other service functions outside the computing area were restructuring to support the responsibility center concept. Five geographic service zones were defined. These zones closely corresponded to responsibility centers. In some cases more than one responsibility center was within a zone. There were not enough support staff available (in terms of skills and coverage) to define one zone per responsibility center. Additionally, requirements coordination for telephone services were also included in the zone support concept.

Computer workstation problem determination staff and telephone requirements staff already were operating in zone concepts. However, those functions had zones separately defined from each other and separately defined from the responsibility center concept. The initial five zones included computer workstation installation staff, computer workstation problem determination staff and telephone requirements coordination staff. The new zone boundaries were not drawn without a certain amount of pain and controversy. However, certain compromises were reached and the initial zone teams were established.

Other functions were then assigned to the zone support teams. These included process administration and support of certain cabling and network components. For the time being, other functions have been kept in centralized organizations. These include system design, help desk functions, server administration, scheduling of large equipment moves, coordination of contractor services and acquisition functions.

0-7803-3552-X/96/$5.00 © 1996 IEEE

Some challenges were encountered in the implementation of the zone support teams. These included:

- assigning employees to new organizational structure under different management
- developing an attitude of teamwork among employees on each zone team
- handling peak load situations (where staff had to be "borrowed" from other zones)
- handling backup within a zone when someone was absent
- breaking the paradigm of only performing one specialized function and helping the
- team achieve service goals where needed
- allocation and scheduling of common support resources

This resulted in a broader scope of job assignments for the zone team members. Also the zone team members developed a closer working relationship with end users in their respective zones. The zone team concept then resulted in additional types of service requests. End users asked for "walk up" service from the teams in their areas. Also in some cases they asked the teams to assume the roles of requirements coordination and submitting service requests (those roles are normally performed by organizational computing focals). This creates the challenge of being asked to provide additional services without additional staff. The challenge is to find the balance between value added services and affordable overhead service costs.

Another challenge in locating the service providers closer to the end users is maintaining balance between crisis response and process integrity. There are times when a computer workstation needs to be moved to another location on short notice (and in advance of the normal service request process). However, in a large enterprise with thousands of networked workstation, some configuration control is also necessary. For example, if the network node used for a short notice move is not logged, then that node may still appear in the data base as unassigned and later be assigned to another user and interrupt the first user's service.

The zone team concept also required more self directed activity on the part of the teams and the individual team members. Under the central service concept, move and installation work was scheduled by a central coordinator. After the teams were implemented, each team became responsible for scheduling the small moves. (Large moves and moves between zones were still scheduled by a central coordinator.)

Implementation of the zone support team concept is an ongoing and developing process. End user issues, employee issues and management issues need to be balanced. Paradigm shifts are not always easy to achieve. It is a step in the direction of letting business operations drive support services rather than being constrained by a support service structure.

Support of networked workstations in a large enterprise

Michael E. Richerson, P.E.
(316) 526-9316

Boeing Information & Support Services
P.O. Box 7730, MS K79-24
Wichita, KS 67277

This paper describes support of over 10,000 networked workstations in a large corporate environment. The initial environment was terminals and PCs with terminal emulation networked to mainframes. The current environment includes both mainframes and very large local area networks. Several hundred engineering workstations are also attached to the current network. Many of the supported workstations are used by engineering organizations. However, manufacturing and business management organizations are also supported. Issues described include: network transition, staffing issues, organizational issues, technology, logistics and lessons learned.

The initial large installed base of networked workstations consisted of terminals attached to cluster controllers in order to deliver mainframe data center services. Intelligent devices such as personal computers and engineering workstations then became part of the installed base. At first, these used terminal emulation products for network attachment. Local area network products then came into use. At this site, token ring based products were used in the first major local area network implementations. Now most of the token ring based products have been phased out, and ethernet based products are used for local area network implementation. There is still a large installed base of terminals attached to cluster controllers. However, future business applications will be delivered from distributed computing servers; so a migration away from mainframe oriented terminals to personal computers and other workstations is expected.

This evolution of network attached workstations has also caused changes in support labor skills, support concepts, logistical support and organizational structure. Reasons for these changes include more technically complex workstation products, different network products and migration from mainframe hosts to servers.

Initially, the more technically challenging tasks were in the network and the mainframe host. Installing, moving and troubleshooting terminals was not as technically complex as supporting the other parts of the delivery system. Personal computers with terminal emulation capability increased the technical complexity of the network attached workstations. Workstation support now included a processing environment, operating systems, network software and applications software. The network and host support remained technically challenging as new products increased capacity and speed. Local area networked personal computers changed all three areas of support.

During the course of this development, the workstation support concept changed considerably. A variety of software and hardware options became available for personal computers. Not only was the workstation more technically complex, but frequent upgrades were available. Additionally, a variety of peripherals became available to attach to the workstation. Workstation support now involved both hardware and software skills. These workstation features then required various supply items, such as diskettes and printer cartridges. These developments then generated the need for corresponding support structures.

As local area network technology matured, many applications were hosted on servers rather than mainframes. This architecture resulted in continued requirements for

0-7803-3552-X/96/$5.00

strong technical skills to support the host and network. It also increased the technical skills needed to support the workstations. Providing applications software from servers reduced the number of individual software packages purchased for each workstation.

A network supporting several thousand workstations requires configuration and capacity management. Management of node assignments is necessary to assure requested services and avoid unintentional reassignment. Network addressing standards are needed for a common installation and support process across the enterprise. Network capacity and workstation locations must be monitored to assure controllers, concentrators, wiring and other network components are in place to support the workstations in each building.

This architecture provided increased computing power and function for the end user. However, support costs are also incurred. The management challenge is to provide a greater benefit from the computing power (applicable to real business needs) than the support costs to provide that capability.

Organizational structures also changed as the computer delivery system changed. Initially separate organizations supported mainframes, networks and workstations. As products changed, supporting workstations required software skills and network skills. As servers hosted large applications, the system administration and configuration management skills of operating a data center now applied to something that looked like a workstation. Problem determination now required skills across the delivery system.

Even though technology and products have changed, and the computer support organization has responded to those changes, the installed base contains everything from ten year old mainframe attached terminals to current LAN attached workstation products. Due to a process where the end user organizations drive equipment acquisition and upgrades, the support organization gets spread very thin supporting a ten year (or more) spread in technology. This also creates the challenge of maintaining support skills for older equipment while learning to support new equipment.

Other "user driven" issues also present management challenges. Personal computers have become "commodity items" and "appliances." End users have personal computers at home and develop various brand loyalties and personal preferences. This can conflict with company standards (for volume purchase agreements) and economy of scale support strategies. Also users with home PCs usually do not have network connectivity beyond a modem and telephone line. However, PCs in the corporate environment often have more sophisticated network capability with access to multiple hosts and servers. Providing network access to meet company requirements usually adds some acquisition and support costs beyond that for home PCs.

In the large corporate environment, end user computing support is now focused much more on distributed computing and the workstation. The workstation is the interactive delivery device for computing services, information access and access to e-mail and information services outside the company. If the employee moves to another location in the company, often the workstation moves with the employee. There is a trend towards one computer workstation for every office employee. This raises the issue of using only one workstation to access all necessary services and applications. This is particularly important in engineering organizations where applications require considerable processing power and often high resolution graphics.

Providing a single workstation for engineering applications, popular PC software, access to company data as well as company and external e-mail is both a technical and management challenge. In the past, computer workstation hardware and software suppliers have tended to specialize in certain ranges of capability rather than offering a "one size fits all" solution. Management issues include acquisition cost, maintenance costs, office space and furniture costs, training costs and operating costs. Providing multiple workstations per employee tends to increase all of those costs. As engineering and other technical employees become more and more dependent on computer based applications for job performance, the cost and support of the appropriate computer delivery system becomes more important to management.

In conclusion, networked computer workstations have undergone considerable changes in product offerings and technology. This has then caused changes in the support structure. Decisions concerning workstations are important to both in users and management. The workstation is increasing important in performing job functions in large corporations. Management is challenged with providing necessary services while controlling support costs.

Managing contracted services in a large corporate environment

Michael E. Richerson, P.E.
(316) 526-9316

Boeing Information & Support Services
P.O. Box 7730, MS K79-24
Wichita, KS 67277

Contracting service functions is increasingly considered in today's business environment. Contracted services have the potential to decrease costs and increase flexibility. However, contracting brings management challenges different from using company employees to perform those services. In order to achieve company objectives, contracted services must be managed not just purchased and "hope for the best." This paper describes service contracting challenges, management responses and references real life examples from contracting software development, computer installation and computer maintenance.

In today's business environment, large corporations use contracted services for many reasons. These reasons include:

- lower cost (particularly for low to medium skilled tasks)
- cost advantages from economy of scale (particularly when using a contractor who specializes in a service and has other customers so fixed costs can be spread over a larger base)
- acquisition of necessary services outside the organization's primary business emphasis
- handling peak loads without hiring/layoff of regular employees
- short term needs for highly specialized skills
- reduce capital and other long term investment costs

However, these advantages do not just happen automatically. Analysis, decision making and effective management are necessary. Contracting must be treated like other business alternative decisions.

This paper describes experiences and observations with contracting computer support functions. In particular, contracted support of computer workstation installation, computer workstation maintenance and software are addressed.

First some general discussion about contracting services in a large corporation. The negotiations prior to selection of a contractor are extremely important. Detail and flexibility are both important parts of requirements. Anything that is critical should be specified in detail. If a performance target is necessary or performing a task in a certain way is necessary, then specify it. On the other hand there are always some uncertainties and unknowns in a large operation in a changing environment. Both the customer and the contractor need some flexibility to respond to these situations. Specific targets and procedures tend to increase the cost of the contract. Allowing the contractor options and innovations can sometimes decrease the cost of the contract. There are trades in both cost and value received in the rigidity or flexibility of the contract.

Part of the negotiations also involve assessing the potential contractors. The contractor needs to have the capability and capacity to perform the requested service. Can the

0-7803-3552-X/96/$5.00 © 1996 IEEE

contractor make a realistic bid? Can the contractor respond to growth and changes? What level of risk is associated with a particular contractor?

It is also important to designate responsible management from both the customer and the contractor. Someone on the customer side needs to monitor the contractor's performance. The customer needs to know who in the contractor organization can make decisions and authorize activities. On the other hand the contractor needs to know who to approach about out of scope requirements and alternative approaches to service delivery. Each side needs contacts to resolved day to day operational issues. Such a working arrangement is much preferable to frequently appealing to high level executives or resorting to legal action to resolve issues.

Multiple site contracts are another area where cost vs. service requirements decisions must be made. Large corporations often need the same services at multiple sites. A common service contact can allow for greater economy of scale, reduced administration and a lower price. However, each site may not have exactly the same requirements. Requirements from each site need to be analyzed and decisions made on the costs and benefits of site variation. Allowing for some individual site variation may incur small additional contracting costs but may provide for increased benefit in specific site operations. The total picture needs to be assessed before finalizing the service contract.

Also it is important to recognize contracting services is a business transaction. Each side is interested in maximizing their own benefit from the transaction. If a contractor is paid a fixed price, the contractor has an incentive to perform as few services as possible. If the contractor is paid on a per incident basis then the contractor has an incentive to perform as many service tasks as possible. The customer needs to be aware of how the contractor maximizes profit and monitor performance accordingly.

Determination of what services to contract is also a business decision. There can be considerable risk in contracting out the core competencies of the business. On the other hand it is generally a low risk decision to contract routine services where there are many service providers in the market and market pressures keep the price competitive and service readily available.

Computer workstation installation and maintenance are services frequently contracted out by large corporations. There are benefits to contracting such services, but some management challenges need to be addressed in order to realize the benefits. Benefits include:

- the contractor stocks repair parts and manages parts inventory
- the contractor provides test equipment, tools and vehicles for installation and maintenance
- the contractor is responsible for ongoing training of support staff
- ∞ often the contractor can offer economy of scale due to larger base including other customers
- often the contractor handles warranty and related issues

Examples of management challenges include:

- getting the contractor to handle "peaks and valleys" in the workload (it is easier just to have a fixed staff and work to that capacity)
- keeping the contractor informed of new products and product variations to be included for support

- assuring the contractor is reasonably accommodating to end users without allowing the end users to generate requirements with excessive cost increases

- vigilance to assure the contractor meets targets and does not "cut corners"

Managing the contracting of computer workstation support services can be not only challenging, but also exasperating and amusing. This point is made with some anecdotes of actual experiences. One anecdote concerns a contractor hired to perform maintenance on computer workstation equipment. At the start of the contract there was considerable discussion about the need for the contractor to provide all the necessary tools to their employees and to assure their employees were fully trained. As time went on, various concerns were expressed to contractor management about this. Each meeting, the contractor assured us they were addressing this and we did not need to be engaging in such oversight of their work. One day some of the contractor's employees were observed repairing a board by using a pair of scissors to pull a chip instead of a chip puller. When this was brought to the attention of contractor management they made improvements in tools and training.

In another case, a contractor was used to deliver and install computer workstation equipment. Some surplus terminals were acquired from a company site in another city. The terminals had been stored in a warehouse where birds had got inside. When the shipment was received, it was found the terminals were covered with bird droppings. There was a provision in the contract about cleaning surplus equipment before delivery. However, there was some difference in interpretation as to whether that applied to bird droppings. After much discussion and management attention to the issue, the contractor finally agreed to clean the terminals as a gesture of "good will."

Contracting software development also has benefits and challenges. Benefits include:

- economy of scale cost reduction if the contractor has prior experience and other customers for similar applications

- economy of scale cost reduction for ongoing support if the contractor has other customers for similar applications

- the contractor supplies specialized skills not readily available within the company

- the contractor is responsible for the computing systems necessary for the development of the software

- the contractor is responsible for training and/or training materials

In many ways the challenges of contracting software development are similar to the challenges of developing software in-house. These include:

- development of reasonable, specific and firm requirements

- software development schedules tend to be overly optimistic

- getting high quality documentation on time along with software

- compatibility with full range of necessary hardware, software and networks

- portability to new hardware, operating systems, data base managers and networks

- dependencies on off the shelf software, particularly new releases

- consideration of user interface issues

- development of adequate test scripts
- availability of qualified support staff for maintenance, upgrades and support
- configuration management

Software contracting is also a business transaction. Even when using experienced contractors with good history, the customers need to look out for their own interests. Separate low difficulty requirements and high difficulty requirements into separate modules. Allow more development and test time for high difficulty modules. Require interim reviews to assure progress and avoid surprises at the end. Require acceptance tests to verify software meets requirements for both function and performance. Require acceptance tests to verify necessary compatibility and portability. Require acceptance tests to verify documentation matches software functions.

Contractors can be a cost effective means of providing critical software expertise. However, contractors must be managed and someone has to look out for the customer's interests. Sometimes a very aggressive management approach is necessary.

With effective management, contracted services can be cost effective and beneficial. In today's business environment, many large corporations are using or considering contracted services. It is increasingly likely that engineers and engineering management will encounter contracted services as part of their projects. In the past, engineering education has not given much attention to a role for contracted services in project management or other areas of engineering. In the past, large corporations tended to perform many of the technical tasks with regular employees and engineers were often only concerned with the functions performed by just their own organizations. The making of business decisions concerning contracted services and the coordination and management of such services will be increasingly important skills for engineers and engineering management.

Quality and productivity improvements in support of computer workstations in a large corporate environment

Michael E. Richerson, P.E.
(316) 526-9316

Boeing Information & Support Services
P.O. Box 7730, MS K79-24
Wichita, KS 67277

This paper describes efforts to improve the quality and productivity of computer workstation support in a large corporate environment. Methods included both statistical quality control and process analysis. Over 10,000 workstations are supported for engineering, manufacturing and business management organizations. Metrics and analysis are important aspects of managing this activity. Analysis, results and management decision making processes are described.

Computer workstation support in a large corporation is a significant operation. It involves hundreds or thousands of acquisitions, installations, moves and network address assignments. Measurements and analysis are important aspects of managing this type of operation. It is important to understand such factors as work mix, capacity, work load and flow time.

This paper presents some examples of such analysis applied to computer workstation support. Examples include use of statistical control charts. Traditionally control charts have been used to analyze manufacturing operations. However, this technique can also be applied to other processes.

The first example involves "small" move requests (where only one or a few devices are moved on a single service order). A Pareto diagram was constructed to analyze the relative frequency of different categories of requests. Next, p-chart statistical control charts were made to analyze how consistently each category represented a certain percentage of the total requests. Figure 1 shows the Pareto chart and Figure 2 an example of one of the p-charts. Even though p-charts are usually used to track defects, this technique can also be used to track percentage of occurrences other then defects. This helps determine if the process involves a consistent mix of work.

In many areas the number of service requests performed per unit time are tracked. In these cases, x bar control charts can be constructed. This can help determine how consistent the work load is from time period to time period. Figures 3 through 6 show control charts of some network support activities. Figure 7 shows a control chart for the number of devices moved per time period during "large" moves. The "large" moves are usually organizational relocations involving several pieces of equipment and other facilities issues.

The control limits define an expected range for a stable process. The control limits then need to be compared to management targets to determine if the process has the desired "capability." Finding assignable cause for out of control points and runs (above or below the mean) can help identify process limitations, changing customer requirements and capacity constraints.

The next example is from an effort to reduce the flow time in acquiring certain types of computer equipment. The process included:

- end user initiating the request
- budget approval

0-7803-3552-X/96/$5.00 © 1996 IEEE

- processing approved request (including any technical review for nonstandard equipment)

- ordering and receiving the equipment

- delivery, installation and network connectivity
-
- updating equipment inventory data base
-

Two flow time measures were used. One without approval time, to measure the operational part of the process. The other was the total time (including approval), to measure the total process. Figure 8 shows the time after approvals. Figure 9 shows the total time. During 1995, some additional approval steps were added for cost control reasons. This caused some increase in flow time after some reductions had been made the year before. This data was also examined using x bar statistical control charts. Figure 10 shows the chart for the total flow time. Figure 11 shows the chart for the flow time after approval. In Figure 11, a process shift is seen. This indicates a statistically significant change from the baseline.

Supporting computer workstations in a large company has many of the same management issues as other operational organizations (such as manufacturing or transportation). Data collection and analysis are an important part of the management process.

Due to various reorganization and restructuring activities, a number of computer workstation support functions were consolidated. These functions were either assigned to a common organization or identified as part of processes managed by that organization. Functions either directly managed or processed managed included computer workstation hardware and software acquisition, computer workstation installations and moves, computer supplies ordering/distribution, network node assignment and process administration.

This restructuring took place in phases. As each organizational change was implemented, management tasks have been to identify functions performed, available measurements, current performance and opportunities for improvement.

Initially three management actions were taken. These were

- support of ongoing company quality initiatives to document processes

- spending time with employees to understand details of work performed

- evelopment of volume of work measurements and in some cases work flow time measurements

Measurements are very important in understanding service and capacity.

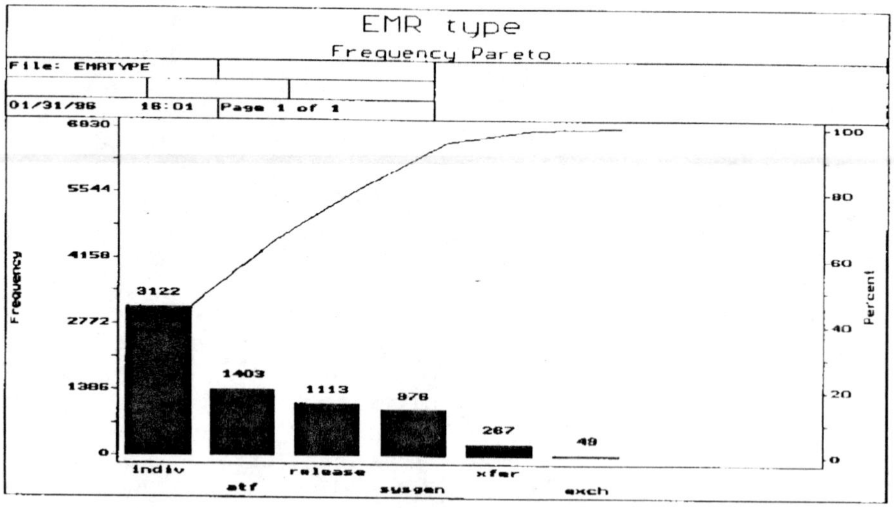

Figure 1. - Request type

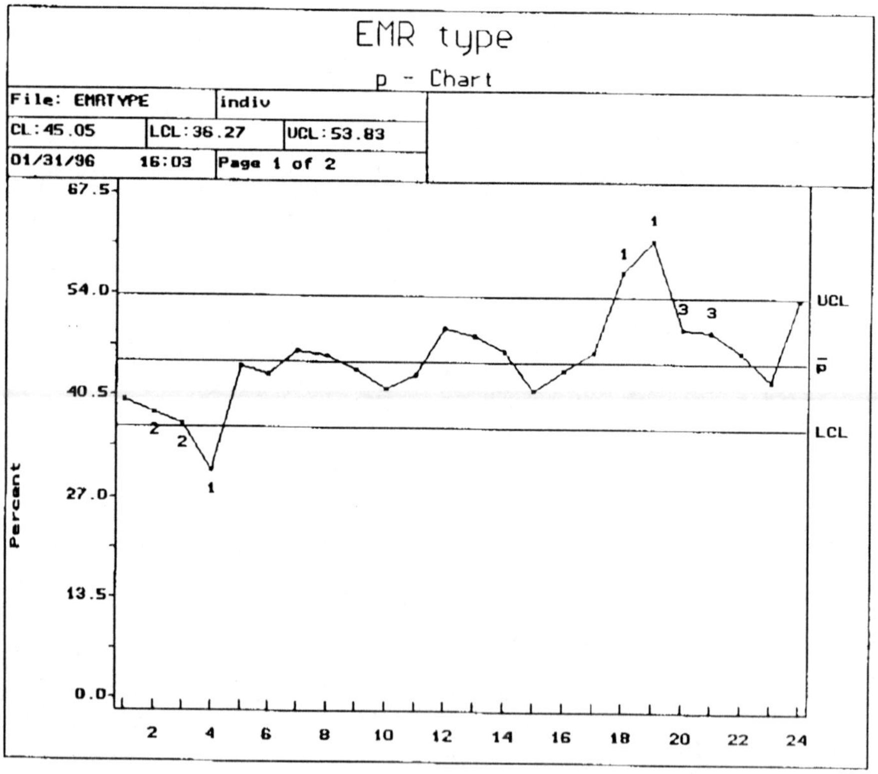

Figure 2. - Individual moves

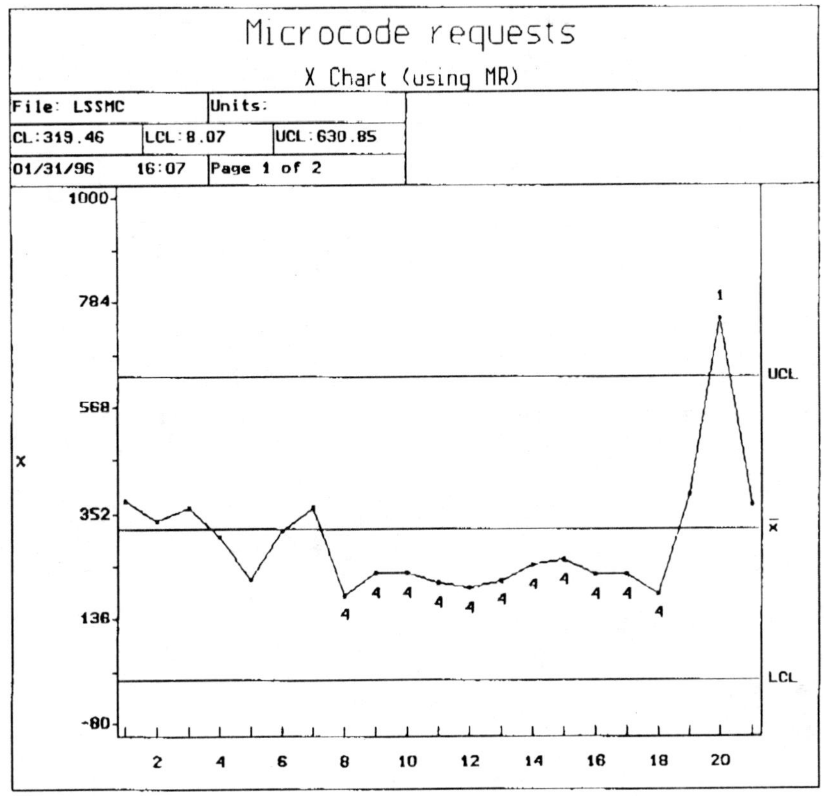

Figure 3. - Microcode requests

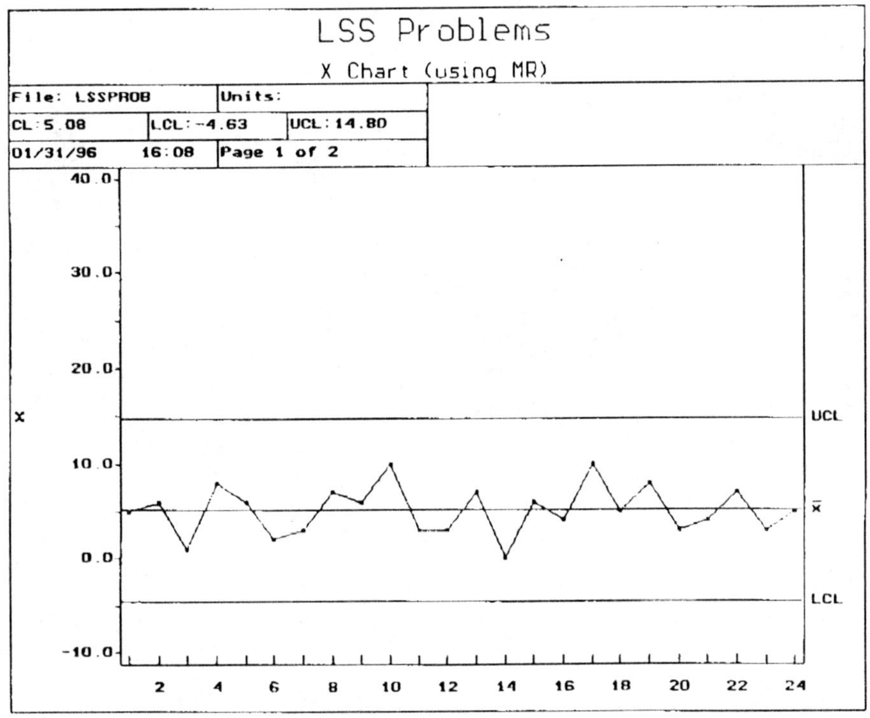

Figure 4. - Problems

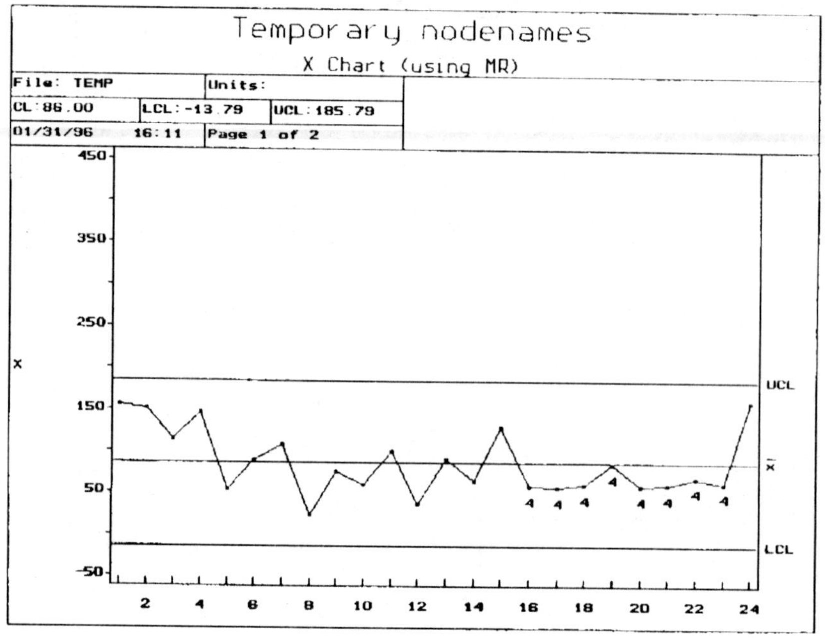

Figure 5. - Temporary nodes

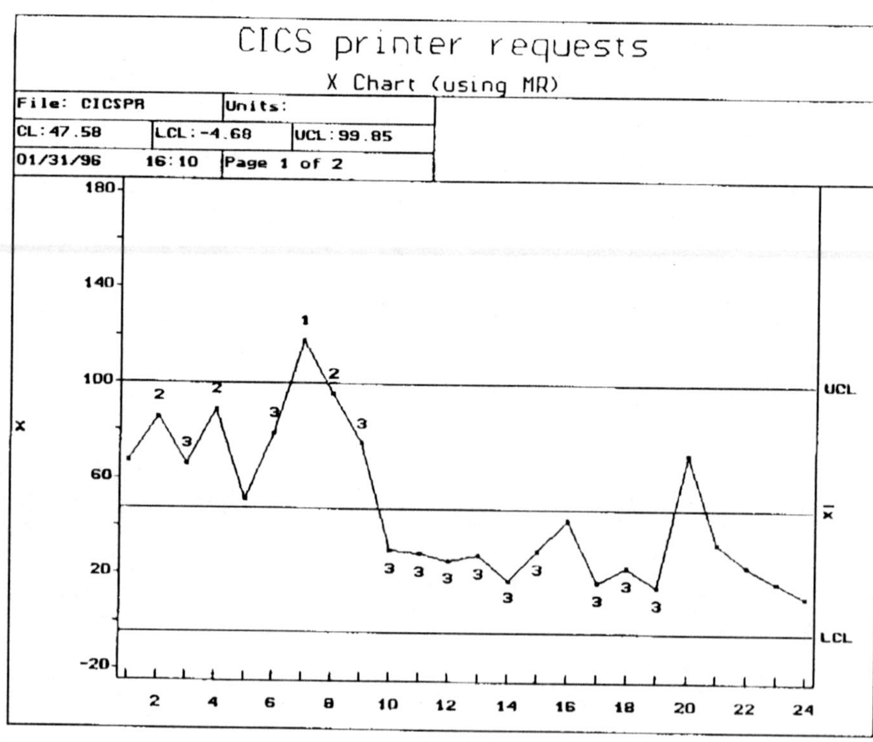

Figure 6. - Printer nodes

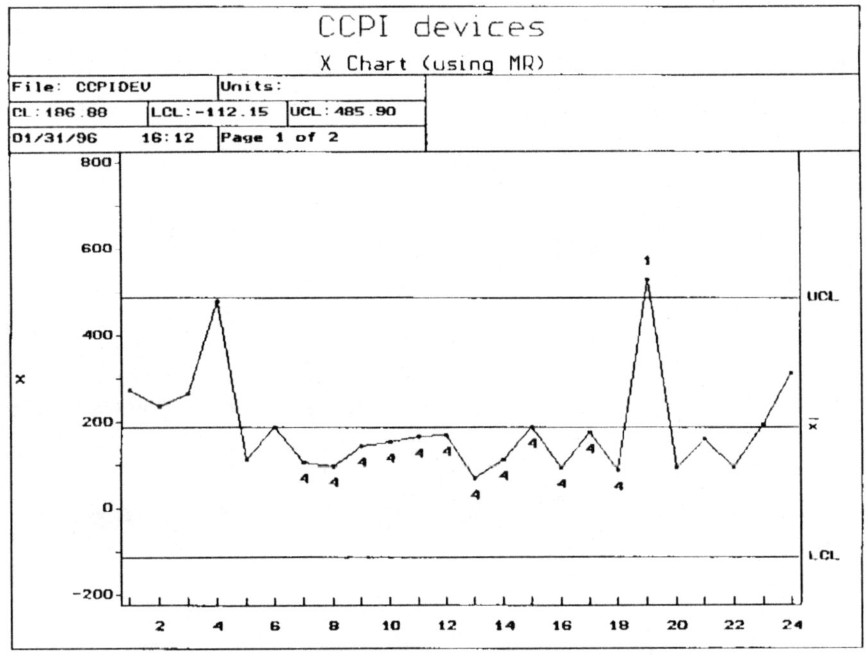

Figure 7. - Devices moved

Average flowdays excluding approvals

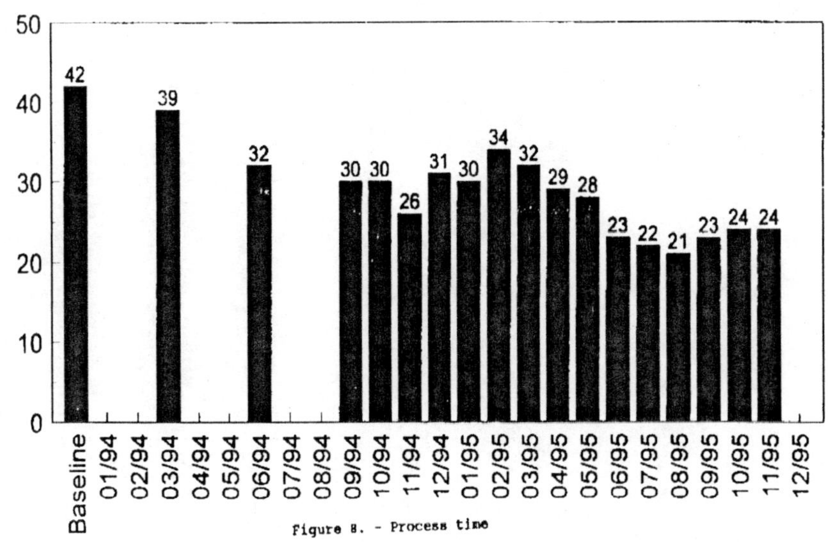

Figure 8. - Process time

Total Overall Average Flow Days

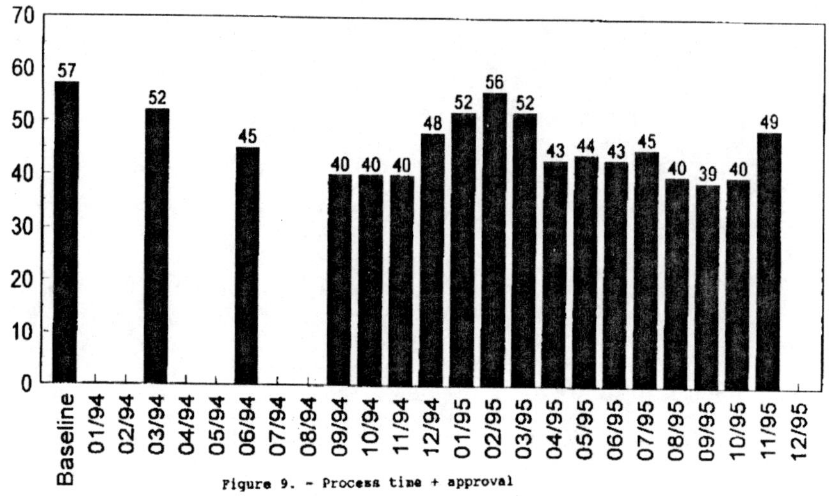

Figure 9. - Process time + approval

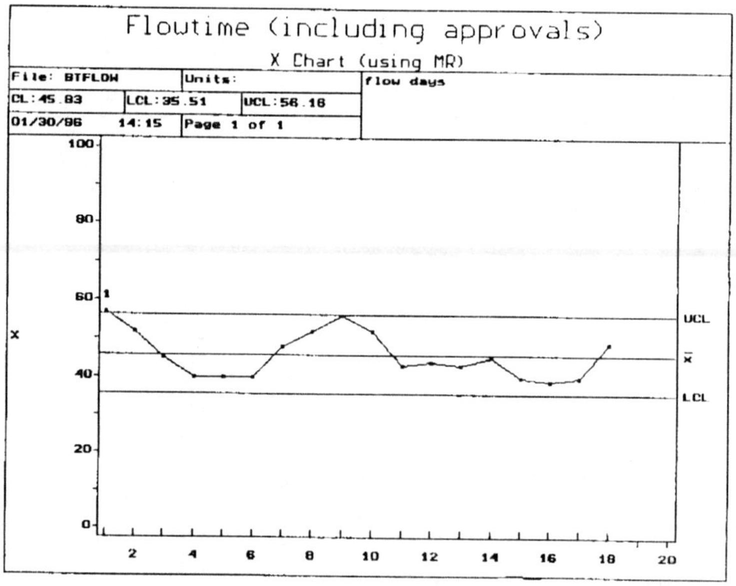

Figure 10. - Process time + approval

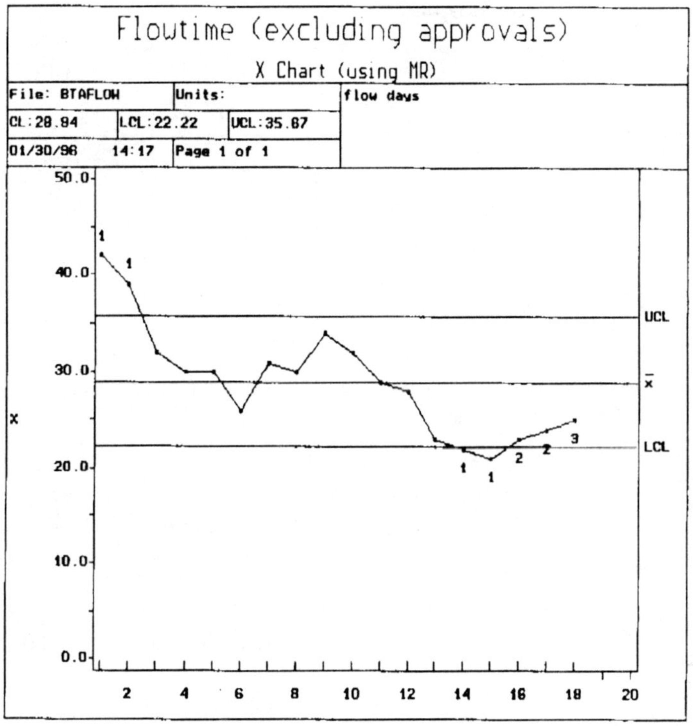

Figure 11. - Process time

Laptop Computers & The Ultimate Virtual Enterprise - Education

A.T. Roper
(812) 877-8324

Rose-Hulman Institute of Technology
5500 Wabash Avenue, Terre Haute, IN, USA

Education is, and always has been, the ultimate virtual enterprise. Students appear in class, but the work on which most of their learning depends takes place in their homes, residence halls, fraternities and sororities. Thus, there are parallels between the roles of the professor and the traditional manager in their respective virtual workplaces. This paper describes the virtual classroom process, lists problems related to it, and describes solutions to some in the context of the computing environment being built at Rose-Hulman. It then deals with unresolved issues facing both academia and industry.

Introduction

Think about it. Neither the computer revolution nor the information highway fostered the idea of "homework."[1] Yet homework, or something very like it, is a central theme in the virtual workplace. What attracts us to the virtual workplace is not that it is new, but that we now have tools that can make it spectacularly productive.

Education is every bit as much a virtual activity as the business of business is becoming. I do not mean that virtual activities in education are identical to those in the workplace. There are obvious differences. Perhaps the most basic are that educational work products are internal, individual centered, and non-material while those of the workplace tend to be more external, group directed and product oriented. However, we should dwell on the similarities while keeping the differences in mind. There are many parallels between the roles of the professor and the traditional manager in their respective virtual workplaces.

Education As A Virtual Enterprise

The really important "value addeds" of education are personal knowledge, understanding and expertise. It is an educational myth that we produce them in the classroom. Since the death of the apprentice system, the most important venue has been the time between classes when students are dispersed and work alone or in small groups. In today's jargon, the most important venue is the virtual classroom. Professors introduce, clarify and sharpen perceptions through, lectures, discussions, laboratories, and other group experiences in the physical classroom. They also test individual development there. Recently, the physical classroom has become an important arena for practicing and refining group skills such as teaming, inter-personal relations and active communications. However, the real value is still added in the virtual classroom.

Education's traditional way to foster student development in the virtual classroom has been through homework (e.g., readings, themes, exercises, problem sets, papers, case studies, programmed learning materials, etc.). Assignments are made by the professor. Then students prepare responses in the virtual classroom which they feed back to the professor. The
professor adds information designed to improve student mastery and returns them. Later he or she tests how effectively students have internalized the knowledge, understanding and expertise implicit in the assignment.

Simplistically viewed, the professor sends a signal which (supposedly) requires growth in the student's knowledge, understanding and expertise to interpret. The student receives the signal, interprets it, and responds. The professor compares this response

0-7803-3552-X/96/$5.00 © 1996 IEEE

with a range of desired responses, notes deficiencies, and feeds the information back to the student. Finally, the professor tests to measure the growth produced by the process. Not a bad system! However, control, motivation and transference have always been uncertain.

Virtual classroom problems

One weakness of the traditional process is that the original signal may not be crafted carefully enough to produce the growth that is intended. Moreover, the range of acceptable responses may be too restrictive, the feedback uninformative, or the testing inadequate. These problems are not unique to virtual activities. There are other weaknesses, however, which are more amendable to modern technology. For instance:

- The process of stimulus, response, feedback and testing is too slow. Often it may take weeks to complete.
- There is usually only one feedback loop when many may be needed.
- Students often must perform repetitive, time-consuming tasks that add little to their growth.
- Even carefully crafted signals may lack the complexity that students need to transfer the mastery they gain to the "real world."
- Students usually have no way to discover how knowledge, understanding and expertise they gain in one area relate to those they need in another.
- The process is perceived as irrelevant by many students because they lack an overview of context.
- The process often fails to motivate students raised on the "gee whiz" visual technology of computer games, television and virtual reality.
- Many students succumb to information overload caused by a bewildering influx of apparently unrelated signals.
- Virtual and physical classroom activities often are not well coordinated.

There are undoubtedly more, but public confession is a pastime easily overdone. For many of the above, you can simply think "traditional manager in the virtual workplace," instead of "professor in the virtual classroom."

The solution context

Solutions to most problems depend upon context. Thus, I must tell you something about Rose-Hulman to set the scene. We are a private, four-year engineering and science college. Our enrollment is about 1350 highly qualified, mostly undergraduate, men and women. Since 1983 we have planned and implemented programs designed to harness computers to pull our educational wagon (Roper, 1995). Our goal is a seamless curriculum in which computers play important, though transparent, roles.

Our residence hall rooms, offices, and most classrooms are connected to the campus computing network and to internet. Faculty have used classrooms with fixed-base (i.e., desk top) computers to develop new approaches to teaching calculus, differential equations, dynamics, and controls coursework. They also have developed an integrated first year curriculum that employs team teaching and discovery-based education to integrate information across traditional disciplines. These programs draw heavily on computer-based tools, especially graphics, symbolic manipulation, and simulation. Their goals are to increase student productivity, to mirror real world complexity, and to increase each student's understanding of basic concepts and techniques.

By 1991 our curricular development had come to a standstill because of three major problems:

- *Over utilization* limited student access to computer-based classrooms for homework assignments. This prevented full realization of the potential of our curricular changes.
- *Obsolescence* demanded major expenditures to replace fixed-base equipment.
- *Diffusion* of curricular advances required major resources to increase the number of fixed-base, computer classrooms.

To solve these problems we needed to revise our approach.

In 1995, we began a program that requires entering students to buy a network-ready laptop computer designated by the Institute. This program resolves each of the three major problems listed above. *Over utilization:* Since laptops are portable, all students have personal computing power and access to networked facilities from their virtual classrooms. *Obsolescence:* Each entering class buys the newest technology and each year the oldest technology "graduates." *Diffusion:* We can install network and power-ready laptop classrooms for less than it costs us to replace three obsolete fixed-base computers. *Bonus:* The laptop solution lets us focus resources on network infrastructure and support which are the most permanent elements of the computing environment.

The computing/curricular environment we are building has let us attack some, but not all, of the virtual classroom problems enumerated in the last section. Some of our approaches are outlined in the section which follows.

Solutions to some virtual classroom problems

The key to achieving the potential of the virtual classroom is a supportive computing/curricular environment. For business, simply replace /curricular by /working and you'll have it. This environment demands mobile, personal computing capability, a solid network, powerful software, and continuous user training and support . . . *but these are not enough.* We must study how we educate (i.e., do business) and modify it to leverage the opportunities provided by modern computing technology. I will describe a few of the ways we are trying to attack virtual classroom problems in the following paragraphs.

The traditional virtual classroom process depends on hardcopy and hand carried feedback. This dependency greatly increases the time it requires and limits the number of feedback loops that are feasible. However, we can now work in a networked environment in which each student has mobile computing power. Thus, we can use software, such as Norton Textra™ Connect (Tuman, 1995), to speed the process, make feedback more immediate, and provide time for additional loops.

Textra™ Connect allows assignments to be made, completed, turned in, graded, returned, revised and returned again over the network. It also allows the professor to organize work group discussions and join them. Professors can use this last feature to improve coordination between activities in the physical and virtual classrooms. This year two of our professors are using Textra™ Connect to teach writing skills and we expect that their experience will foster its use in most courses having a writing component. The software and upgrades are free to the Institute. Student cost is less than that of a traditional writing handbook which the software includes on-line.

No software has done more to reduce repetitive, non-productive tasks than word processing.[2] It improves the quality of student writing as well (surely, a violation of the Second Law of Thermodynamics). Perhaps no one tool has done more to increase student and faculty productivity. When word processing is enriched by spreadsheets, CAD packages and graphics programs, students have the means to produce professional (real world) quality in their reports. Moreover, using Textra™ Connect, these reports can be submitted, corrected, returned and re-submitted electronically. Modern symbolic manipulation packages such as Maple™ allow similar reductions in non-productive tasks for engineering and science students. However, to realize the benefits from these software tools, students must be able to use them in the virtual classroom. Thus, networks and mobile personal computing are critical.

Our students learn to use the Working Model™ analysis package in dynamics. In addition to the usual analytical tools, this software uses modern graphics and animation to make the principles of mechanics more understandable to today's "visually motivated" students. Its power also allows professors to assign homework that reflects real world complexity [see Binder (1996) for an industrial application]. When coupled with Cadkey™, Maple™, word processing and a spreadsheet, Working Model™ allows students to climb several rungs up the complexity ladder without forfeiting sound understanding of basic principles.

The integrated first year curriculum gives a quarter of our first year students an integrated view of basic mathematics, science and engineering principles. This curriculum depends heavily on the tools described above and on "homegrown" simulation software to introduce basic principles in a team-based, integrated, discovery mode. Thus, it provides both integration across disciplinary boundaries and a context within which to understand specific assignments. Because the course is team taught, it places a premium on coordinating virtual and physical classroom activities. Simulation packages play an important role because they allow professors to quickly coordinate laboratory-like experiences with topic development.

Finally, the mere universality of the computer applications described above gives a tool-based context for the myriad of topics and issues with which students are confronted. This in turn, provides indirect assistance in relating topics in one course to those in others, and in the real world.

Unresolved issues of the virtual workplace

Both the good and the bad news are that unresolved problems remain. Bad, because these could seriously reduce the benefits we realize from the virtual classroom or even destroy its credibility. Good because ... well ... without problems, there would be no jobs. Many of these problems have implications for the virtual workplace in business.

Perhaps the most vexing problem is how to continuously train and retrain professors to use computer tools. Even mature software may change each year, and those who are competent with the last version are not instantly productive with new. This problem is compounded by the variety of packages with which personnel must be familiar and the rising tide of new software. Busy users can drown without a well conceived training program. Yet these programs cannot detract from traditional professional development activities. To my knowledge, no one in education has found a satisfactory solution to this problem.

One problem enumerated for the virtual classroom is information overload. Computer technology can intensify this problem by shortening the assignment/submission/correction/ feedback cycle. Professors like to say that understanding requires "soak time": time to reflect. Without care, we could easily pack activities so tightly that students have no soak time. Industry has increased productivity by using portable computers, e-mail, voice mail, pagers, cellular phones and FAX, to remove the voids from the work week and to appropriate time from the non-working week as well. Eventually this could reduce productivity, promote burn-out, and produce badly considered decisions. We all need time to breathe, to think, and to relax. But our instinct is to eliminate activities which produce no material gain.

The technology that promises to make the virtual workplace so spectacularly productive is new. There always is a lag in the diffusion of a new technology from initial adoption to mainstream acceptance. In education, this lag has produced what William Geoghegan (Geoghegan, 1994) styles a "chasm" between early adopters and the rest of the educational community. This chasm has stalled the diffusion of educational technology and left us "stuck at the barricades." As Geoghegan notes, the chasm "... is less a matter of aversion to technology than it is an aversion to risk, a low tolerance for discontinuous change, inadequate vertical support and perhaps the lack of an absolutely compelling reason to buy into a new and relatively disruptive way to go about one's work." This mind set is not unique to education. Both education and industry must find ways to bridge the chasm.

Finally, we are in the early stages of applying educational technology. Naturally, we have tried to use it to better accomplish what we do. The real quest, and a more difficult one, is how to use it to accomplish what we *wish* we could do .... and, to do it more cheaply.

But for now, I am virtually out of time.

References

Binder, John D. "The Knowledge Revolution: A Working Model For the 1990s." Aerospace America. American Institute of Aeronautics and Astronautics, Inc.: Washington/D.C., January 1996.

Geoghegan, William. "Stuck At The Barricades." AAHE Bulletin. American Association for Higher Education: Washington/ D.C., September 1994

Roper, A.T. "Computer-Aided Education Doesn't Work Without Planning." Proceedings of the 1995 EPMESC Conference. Macau, 1995.

Tuman, Myron C., Ann Arbor Software. Norton Textra[TM] Connect for Word For Windows: Instructor's Guide. W.W. Norton and Company: N.Y./ N.Y., 1995.

[1] I was convinced for years that my 6th grade teacher, Ms. Wetzel, created homework to punish us for being sixth-graders.

[2] Remember the old days when, late at night, you had finally "hunt and pecked" out the lab report only to find that you had left out a paragraph on the first page?

If Technology Is Such an Important Strategic Issue, Why Do the Top Executives Play a Minor Role in Its Decision Process?

Gerald H. Rosenfelder, PH.D. and
Guy H. Gessner, PH.D.
(716) 888-2609

Management/Marketing Department
Canisius College
2001 Main Street
Buffalo, New York 14208

This research examines the roles of the Chief Executive Officer (CEO) and the Chief Technology Officer (CTO) regarding new product technology in U.S. multi-product organizations.

This exploratory research study examines data collected from Fortune "500" companies which listed their annual R&D spending in the firm's Annual Report and SEC's 10K. Two related, but different questionnaires were used, one addressed to the CEO, the other to the CTO.

The research examines the following questions: (1) If new technology development is one of the key sustainable competitive advantages, is it best acquired internally or externally, and who plays the major role in those decisions. A number of firms are using CTOs to provide leadership in this decision process. (2) Do firms make use of a Science Advisory Committee (SAC) to assist in this most difficult task and is there a pattern of use of a SAC that differs depending on the educational background of the CEO?

These findings offer insight into new strategic approaches to technology decisions.

Introduction

Has the pressure to bring new products to market impacted the strategic issue of new technology decisions, and has it impacted the role of the Chief Executive Officer (CEO) and Chief Technology Officer (CTO) in this decision process? A recent article in *Research Technology Management* by R.S. Jonash (1996) cited the need for "Getting more with less... and faster," and placing the focus on more reasonable values and returns from interval research development and from engineering; thus, indicating that new technology decisions are becoming more of a short term issue and less of a long term strategic focus. Deb Chatterji's (1996) writings reflect on the new interest of capturing external technology by U.S. Firms as another means of bringing new products and new technology faster to market at reduced costs.

In general, technology is still considered the key to the long term advantage of the U.S. Multi-Market Firm, but why are we seeing this changed attitude of shorter development reflected toward such an essential part of the firm's future strategy?

There is one major concern when dealing with new technology decisions, how do CEO's that are not technically educated or trained deal with this problem? According to Bhargava and Jespersen (1993) in "Portrait of a CEO," *Business Week,* in the most recent study, only about 20 percent are technically educated and trained from a list of 1,000 U.S. corporations. Do the non-technically educated and trained CEOs use the CTO in a more dominant fashion to make up for their deficiency in training, or do they avoid technology decisions whenever possible?

Another author, Peter Gwynne (1996), observes that the role of the CTO is being de-emphasized in the U.S. corporations, especially if he/she does not possess prior or current Profit and Loss (P&L) responsibility. Thus, creating a concern that this is contrary to the observations of the European and Japanese firms.

0-7803-3552-X/96/$5.00 © 1996 IEEE

On the other hand, Gordon Edge (1995), Executive Chairman of The Generics Group, indicates that the management of emerging technologies is a complex senior management task which can not be delegated because of the strategic implications and its impact on the competitive advantages of firms.

Since technology, and especially its rapid change, has such a profound impact on the future of today's firm, there is a continued need to better understand the roles of the CEO and top technology decisions. This exploratory research study is focused on the data collected from Fortune "500" firms which listed their annual R&D spending in the firm's Annual Report and/or SEC's 10K. The study is presently being conducted using surveys sent to CEOs and CTOs of 400 of the Fortune "500" companies to better understand their tasks, backgrounds, and relationships regarding new technology decisions.

Data

This research has developed two questionnaires, one sent to the CEO and a similar, but different one to the CTO of Fortune "500" companies in the United States. Using 1995 the 10K and/or Annual Reports of the corporations that listed their annual R&D spending as the basis of being included in this study, the questionnaires were mailed during the Spring of 1996, with the expectations to obtain an adequate sample of data by this summer.

The initial part of the questionnaire for the CEO is focused on his/her educational background, and role in the new technology R&D process. Included here is the frequency of attendance in new technology briefings and the importance of personal awareness of new technology projects. The next part examines the CEO's relationship with the CTO, if such a position exists within the firm. The final part is concentrated on the use of a Science Advisory Committee (SAC), and the role, if any, that this tool may play in the CEO's new product technology decisions. One of the major purposes of the study is to better understand the role the CEO plays in new technology decisions as applied to the overall corporate strategy and the perception of the CEO of that role.

A similar questionnaire was mailed to the CTO, or equivalent position in the organization, with its focus on the role of that position plays and the relationship with the CEO, especially regarding new product technology decisions. The CTO's questionnaire also addresses the process of acquiring new technology internally or externally, the use of a Science Advisory Committee, and the committee's possible make up. The CTO is also queried as to his/her authority of allocating R&D funds and reprogramming such funds under special circumstances.

Issues Being Examined

With the concern for new product technology as a major issue for the management of the U.S. Multi-Product firm, it was hoped that this study would provide a better understanding of the role of the CEO regarding such decisions. Does the CEO actually participate in new technology decisions, or is that delegated to the CTO, and at what level in the organization is that decision actually made?

Does technology, being a major weapon in defeating competition, influence the role the CEO plays in the future of product technology? Is his/her participation limited to budget considerations only? The issue becomes particularly complex in the multi-divisional firm, where staying current on new technologies is particularly difficult. The strategic importance, however, is just as great in selecting the right new technology as in a single core technology firm.

With the U.S. firms focus on the short term financial impact does this mean that new technologies are acquired externally rather than developed within the firm?

The Chief Technology Officer's (CTO) position, a relatively new function, could play a major role in support of these strategic decisions, but does he/she participate in the process and at what level in the management structure?

Do firms make use of Science Advisory Committee (SAC) as part of new product technology decisions, and what role(s) do they play, including participation in actual technology decisions, review of technology decisions, and audit of technology functions?

Expected Results

A better understanding of the relationships between the Chief Executive Officer (CEO) and Chief Technology Officer (CTO) and the expectations from each other regarding new technology decisions. It is projected that the technically trained and educated CEO will play a greater role in participating in strategic decisions regarding new technologies. Past research of these authors indicate such findings, but show substantial variations between firms and the overall focus on technology.

It is hoped to obtain insight as to what the role of the CTO is in the selection of new technologies, and if his/her role is more as an advisor or a decision maker.

Since CTOs in Japan and in European firms are generally part of the corporate board of directors, does the smaller representation on U.S. boards indicate a lesser influence of the CTO in the U.S. firm? Is this organizational structure changing?

Are U.S. firms being driven to acquire new technology because of the pressure to bring new products to market quicker? These results may indicate that U.S. companies are reducing their effort to pursue long range R&D projects, especially at central R&D laboratories. A number of recent articles have reflected on the change of performance of R&D at the product business unit rather than at the corporate research laboratory.

The Science Advisory Committee (SAC), a tool to assist in new technology decisions, was not well implemented in most U.S. firms. The study is trying to detect why it is not used and what are the conceptions of the use of such an organization. In an earlier study, it was hypothesized that the use of a SAC would be most desired by a CEO without technical education or training in the assistance of new technology decisions. Prior research was not conclusive, but indicated that the technically educated and/or trained CEO would be more likely to use such an organizational tool.

With the pressure to improve new technology decisions in order to provide better strategic guidance to the firm, the results of this research could provide new insight and understanding to this complex process.

References

Bhargava, Sunita Wadekar, and Fred F. Jespersen. 1993. "Portrait of a CEO." *Business Week*. 11 October: 66-106.

Chatterji, Deb. 1996. "Accessing External Sources of Technology." *Research Technology Management*. March-April: 48-56.

"The Corporate Elite (Chief Executives of Business Week 1000)." 1993. *Business Week*. 11 October: 66-106.

Edge, Gordon. 1995. "Thinking About the Technology Future." *R&D Management*. April: 117-128.

Gwynne, Peter. 1996. "The CTO as Line Manager." *Research Technology Management*. March-April: 14-18.

Isaacson, Robert, ,Robert Mitchell, and Lee Star. 1994. "Getting the Most from a Scientific Advisory Board." *Research Technology Management*. March-April: 33-37.

Jonash, Ronald S. 1996. "Strategic Technology Leveraging: Making Outsourcing Work For You." *Research Technology Management*. March-April: 19-25.

Roberts, Edward B. 1995. "Seeking Global Technological Advantage." *IEEE-*

Real-Time Resource Tracking System For Semiconductor Industry

Saengpongpaew, Pichit
Sirinaovakul, Boonmark, Asst. Prof. Dr.

Ph.D. Program at Assumption University
Bangkok
Thailand 10240

In the high growth of Electronic industry, manufacturer needs to maintain the resource reliability and maximize the critical resource utilization. Anyway how manufacturer can define or track the actual resource downtime and cause of those resource problems. Since the real time resource tracking system (RTC) ;increased the responsive time to more control the resource reliability and utilization tightly is the main systematic tool to analyze the utilization factors, and system reliability in order to gain more fast return on those resource investments in a fast paced high tech world.

The data, collected from Integrated Circuit Operations through outcome from the real-time resource tracking system, generally support the frame work and are used to draw conclusion which is useful for industrial organizations and technological strategists.

Introduction

RTC (Real-time resource tracking system) provides custom engineering system and system automation to support the manufacturability to response quickly to the comprehensive data acquisition, monitoring and control options, reliability of performance and versatility, including ease of use and integration.

The case for improved strategic manufacturability, gained through RTC, that allows a manufacturer more successful control is well supported. However, the effect of RTC on the ability to fully resource effective goals has not been adequately addressed in the literature. The difference being emphasized here is that while overall system effectiveness principle allows manufacturers to measure the resource utilization in an its environment to achieve the delivery on time or short cycle time, high quality and high productivity fashioning a fast return on investment, often unique, that achieve the business goal.

In today's competitive semiconductor manufacturing environment, it is important to achieve resource control over the bottle neck machines. Equipment must be operated within customer requirement and satisfaction for the manufactured product to meet performance and reliability specification. In time consuming and expensive procedures, such as the utilization of final functional electrical testing , manufacturers run their equipment to ensure they are operating within output rate. However, despite such low yield, system down and no material ontime in the equipment cause the delay delivery and quality issues. These problems are often not detected until many days problem feedback from production control. With testing machine processed the electrical testing per hours between 10k and 20k, equipment is capable of quickly producing large Kunit amounts of throughput. In response to these issues, real-time resource tracking system (RTC)was designed for use on semiconductor manufacturing equipment. In the wafer sawing operation, RTC analyzes real-time sensor data from the equipment to immediately detect the presence of an event. The signals of interest are automatically collected while the equipment is processing by an operator at a personal computer which has preset recipe parameters; meanwhile, a microprocessor based controller operates the complex of blades, safety devices, and sensors in a cleanroom; every half-second a data center queries each of more than 100 sensors on cutter mechanism throughout a specific die size and adjusts their operation to extend their availability.

0-7803-3552-X/96/$5.00 © 1996 IEEE

In the company under study, the nature of its manufacturing process is characterized by following 1. Mass continuous production environment, 2. Manual data logging for each production steps, and 3. New and high turnover operators. Furthermore , the nature of its product is characterized by complex assembled product and the need of high quality and reliability. This situation leads to problems ;

a.) how to assure every downtime recording to be transferred and calculated in % equipment utilization report accuracy as well as compared with the goal.
b.) how to assure that technician and engineer can identify the root causes of each equipment reliability and know the system effectiveness status accuracy,

This paper focused on resource tracking and measuring and improving ; the means by which primary transformations are carried out at process operations. Thus in the following sections we will first introduce the conceptual design then the overall equipment effectiveness study, common on the RTC system, will be described and categorized. The final sections will describe the result comparison between the goal and actual results from RTC usage.

Conceptual design

Production needs to maintain or improve the machine reliability and maximize the machine utilization. Anyway production does not have these data recording so RTC will provide the computerized system to record, analyze, and report these strategic information Benefits from RTC system The main information of this system is % resource utilization and jam factor for capacity planning and control and system effectiveness and reliability for machine performance improvement.

% Machine Utilization (Operating hours) is defined as the one minus the sum of the utilization factor percentages calculated using the operating hours of the equipment. The total time in operative state (Uptime) is the sum of time spent in active use and in the standby state. And the downtime is the sum of the time spent under active repair and calibrate as well as waiting for spare parts, QC buy-off, machine set up, etc. Please see figure #1 for data flow of RTC system.

Jam factor is the factor used to include incidental jams that do not warrant technical assistance (that are cleared by the production operator). The jam factor accounts for the time from when a jam occurs to the time when the machine resumes normal operation.

Figure #1: Data flow of RTC system

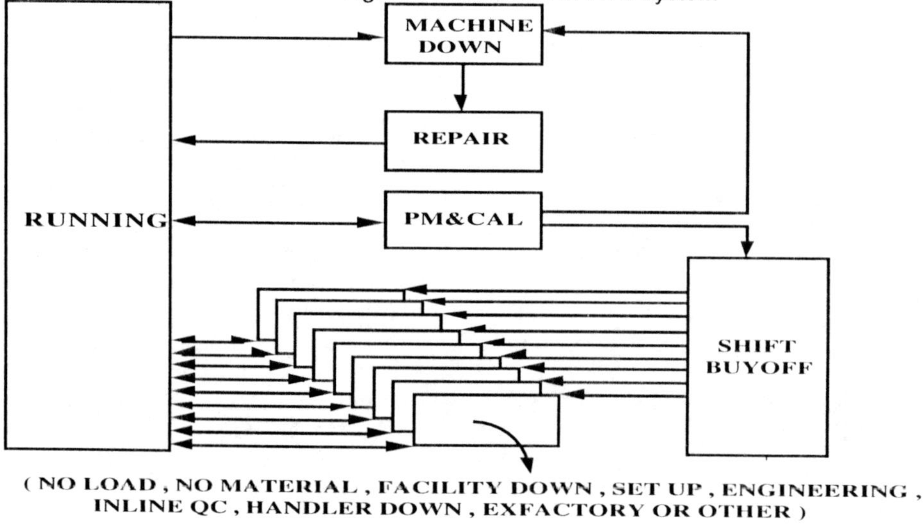

System effectiveness is another vision to improve the productivity. Manufacturer needs to estimate of the reliability of the individual components and combine these known reliability to estimate the overall reliability of controlled system. Applications of such a model include to evaluate the actual results against goal, compare alternative designs and evaluate trade off among availability, reliability and capability. Figure #2 is the example to model the system reliability measurement.

Figure #2: System reliability model

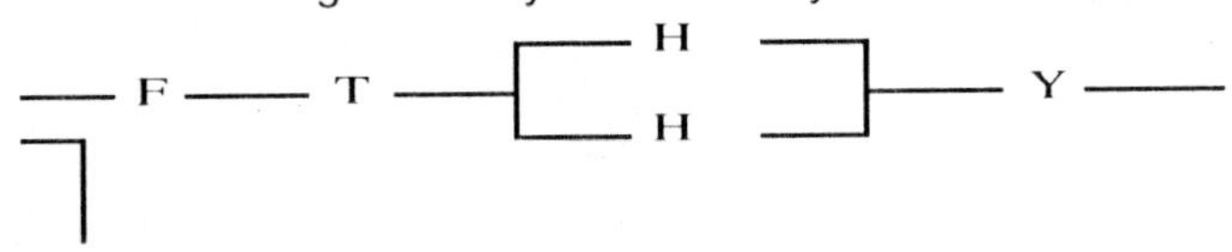

$R\ tester = (Rf) * (Rt) * (Ry) * [1 - (1 - Rh)^2]$; *Time in Week*

$R\ handler = Rj1 * Rj2 * Rj3 * ...$; *Time in Day*

$R = e \wedge -rt$

R = *Component reliability or probability of failure free operation for a time period to or greater than t.*

t = *Specified time period*

r = *Failure rate*

An overall equipment effectiveness (OEE) study

An OEE study was carried out on the final electrical testing in the cleaning room. From the record, the average OEE of these equipments was low at 75%. In order to meet the world class manufacturing standard, the average OEE needs to be improved to 85% or above. The experimental design was suggested with the objectives of identifying the source of variation that caused the low OEE, improving the OEE, and increasing the capacity.

A completely randomized experiment was carried out with three machines number#1,#2 and #3. Five runs were tested on each machine, in a randomized order of testing for the runs. The response was the OEE report from the RTC system. Each experimental run took one and a haft hours. The data was collected by recording the time when a stoppage occurred and the time when the machine started to operate again. The length of time for each stoppage was recorded under the particular causes for stoppage. The total time for the stopgaps is the time for idling, system down and system jamming, which is one of the factors for calculating the OEE. The OEE is calculated based on the equation:

OEE = Availability x Performance Efficiency x Rate of Quality. Based on the NOVA table of the experiment (see Table#1), it was identified that there were significant differences among the OEE of the machines at 95% confidence level.

To reduce the variation of OEE among these machines, manufacturer can analyze the causes of the machine down, idle and jam stoppages. Based on the Pareto diagram, and Reliability report from RTC tracking system, manufacturer can increase the OEE by reduce the Mean Time Between Assist (the time needed to assist the machine when machine asked for jam clearing, system hang, etc.) taked care by operators and line technicians and increase the Mean Time Between Failure of each nature of tooling with effective tool life preventive system. By implementing autonomous maintenance, set up, and repair, waiting time can be reduced near zero, and unplanned downtime can be reduced to a minimum (see figure#3).

Table #1 The ANOVA Table for Completely Randomized Experiment

Source of variation	Sum of Squares,SS	df	Mean Squares,MS	F-ratio	Sig. Level
Between Groups	452	2	227	4.992	0.0265
Within Groups	551	12	46		
Total (Corrected)	1003	14			

Figure #3 Effect of Total Productive Maintenance (TPM)

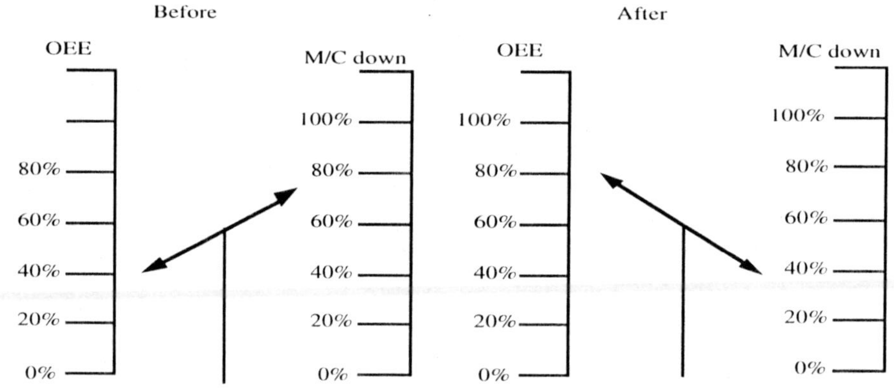

Conclusion

Manufacturing system and equipment are operated and controlled by microcomputers programmed with equipment control software. This type of software requires: real-time operation, concurrent activities (multitasking/processing), high reliability and performance, extensive data communication, intensive requirement but friendly operator interface. All of these technologies are extremely useful in the high competitive environment with more closely monitoring, controlling and decision making. And from this technology, manufacturer can use it to be the strategic weapon to improve the OEE of major resources with the cross-functional team in developing the smart maintenance program preventing unscheduled stoppages, breakdowns, instability, and unreliability.

References

[1] Joseph Shigley and Larry Mitchell, "Mechnical Engineering Design", New York:McGraw-Hill Book Company, 1983, p 14.

[2]Richard Wilson, "Analyzing the dairy risks of life",Technology review, Febuary 1979, pp10-18.

[3]Yefim fasser and Donald Brettner "Process improvement in the electronic industry", 1992, pp6-8.

[4]L.C.Morrow,ed.,Maintenance Engineering Handbook,New York:McGraw-Hill Book Company,1957,pp1-4.

[5] Hussein M. Reda, "A Review of Kanban-the Japanese Just-in-time production system", Engineering Management International, 4,1987, pp.145-146.

[6] Eliyahu M. Goldratt and Jeff Cox, The Goal: A process of ongoing improvement, rev, ed., Croton-on-Hudson,NY:North River Press, Inc.,1986.

[7] W. Edwards Demming " Quality, Productivity, and Competitive position", Cambridge University press, 1982, pp30-31.

[8] A.J.Duncan "Quality control and index trial statistics", Rich chard D. Irwin, Inc,1989, pp423-429.

[9] James Robert Taylor "Quality control system" , The free Press, 1988, pp327-334.

[10] Phillip R. Thomas " Competitiveness through total cycle time", McGraw-Hill Publishing Company, 1990, pp155.

[11] J.M. Juran "Juran on Planning for quality", The free Press, 1988, pp327-334.

The Iterative Component of Design

Michael J. Safoutin
Robert P. Smith
206-543-4605
206-685-3072 (fax)

Contact Information:
Industrial Engineering, Box 352650
University of Washington
Seattle, Wash. 98195
smith@ieng.washington.edu

Iteration is a common thread that runs through most definitions of the design process. In spite of its importance, it lacks precise definition. Iteration exists in many varieties that play different roles in design. Efforts to improve design by manipulating iteration, or to develop automated methodologies that include iteration, should be preceded by an understanding of iteration and its role in the design transformation. This paper develops a classification of design iteration based on observations of existing design practice and common approaches to automated design methodologies. This classification suggests directions for future research in design automation and modeling.

1. Introduction

Iteration is one of the most pervasive features of design. Traditional definitions of design, whether rigorous or casual, provide for iterative feedback loops that span several classical phases of the design process (Suh 1990, Eder 1989, Suri 1989, Shigley 1995). But beyond this introductory concept, many distinct varieties of iteration can be perceived in design practice. The term "iteration" is commonly used to refer to loops that are broader in scope than the traditional notion, such as the successive model releases of a product, or narrower, such as the iterative mathematical techniques used to solve small engineering optimization problems. Iteration varies in its purpose and role as well, as suggested by the prevailing attitude that iteration is necessary yet often disruptive or costly. The fact that all of these varieties appear in design literature, usually under the same name, suggests that they hold enough in common to be relevant to the study of design.

The variety and pervasiveness of iteration suggests that it must exert an influence on cost, time, quality, and other concerns of design. For example, designers often enforce the use of engineering change orders (ECOs) to document changes that take place during design and manufacturing (Hegde et. al. 1992), which can become a burden that is costly and potentially disruptive to the design process. Many researchers have focused on reducing iteration time (Eppinger 1991) and making it more efficient (Thurston 1990, Baya et. al. 1992). But in order to improve design by manipulating iteration, or to develop new design methodologies with the same intent, it is necessary to recognize all of the varieties of iteration and fully understand the roles they play in achieving a design. Otherwise it is difficult to support any changes in iterative activity with proof that the net effect is beneficial.

Most literature mentions iteration only peripherally and anecdotally, leading to confusion. Even when studies involving iteration offer a precise definition (Eisenhardt 1995), (Smith 1993), the type of iteration under scrutiny is only rarely contrasted with other types (Smith 1993). Iteration is conflictingly portrayed as a costly problem that should be avoided, a useful means of improving a design, or even as a catalyst for innovation (Eisenhardt 1995). This makes it difficult to be confident that an observed instance of iteration exerts a net positive or negative effect. Efforts to reduce iterations and their costs should account for the effects of iteration on other aspects of the process.

The uncertainty is compounded as design methodologies shift toward automation. A common approach in automated design methodologies is to begin with a pragmatic

0-7803-3552-X/96/$5.00 © 1996 IEEE

attitude and emulate typical design practice, for instance, catalog-based design or design by modification of existing designs (Joskowicz, 1990), which are iterative. As if to suggest that design and iteration are inseparable, these methodologies often inherit an iterative approach, probably to take advantage of easily codable algorithms. But when no other objective justification for an iterative basis is provided, it suggests that common design practice, being iterative, might exert a bias toward making automated methodologies iterative. Ideally, the role and reason behind traditional practices, including iteration, should be understood in their original context before they are given new life in automation.

Classification schemes or taxonomies (Ullman, 1990) are useful for revealing commonalities and facilitating discussion. There is no taxonomy of iteration that would allow us to interpret casual observations of iteration and gain insight on their role in effecting a design, their commonality with other observed iterations, or their nativeness to human or automated design. Here we will describe some types of iteration common to conventional design processes and propose a simple classification organized by the scope of the activity and the apparent role behind it. Then we will categorize the types of iteration currently seen in automated design methods, and comment on future directions for design automation research.

2. Classification

Several varieties of iteration are visible in design. All involve repetition with a goal of improvement, but each is distinguished by differences in the scale of the iteration with respect to the overall process, or by differences in the cause or purpose of the iteration. Sometimes these differences are subtle but could have strong implications on their role in the design process.

2.1 Iteration in Conventional Design

"Conventional design" means any design process conducted by human designer(s), characterized by mental or communicative processes within or among team members. This contrasts with "automated design", in which a design process is conducted solely or primarily by a computer algorithm. Conventional design may include the use of automated design tools, but the design methodology employed by the tool is an example of automated design. For example, a chassis designer is practicing conventional design when integrating a wheel and axle that were separately designed by an automated algorithm.

2.1.1 Scale

Iteration refers to a cycle of proposal, testing, and modification of an evolving design or portions of it. These components can span any portion of the design effort, from a mathematical optimization algorithm to an entire product design cycle. The "proposal" aspect can be anything from an intermediate mathematical result to a product release. "Testing" could include the evaluation of an error term in a mathematical algorithm, planned testing of mockups or prototypes, or the use of the product by the consumer. "Modification" could be a repetitive mathematical recomputation, a routine revisitation of an earlier stage of design, or a costly product recall.

Micro-scale iteration. The smallest scale of iteration is the mathematical iteration at the heart of many iterative mathematical techniques that may be used in solving some of the low-level design problems that make up a design task. It is an error-driven process with little or no meta-level oversight. The plan is to repeat a sequence of computations until an error term has been minimized

This type of iteration may seem insignificant to the broader scope of design, but it can be seen as a simplified model of the greater design process. Realistic design processes can be described as the minimization of a more complex error term -- the difference between the existing state of the design and the design goal. The similarity is superficial because iteration in this context is more costly and time consuming. Also, humans do not rely solely on iterative repetition because they "reason" given domain knowledge and knowledge of the design goal. But micro-scale iteration could take on greater significance in automated design, where iteration is less costly and "reasoning" difficult to imitate.

Meso-scale or PTM iteration. A more interesting type of iteration is the familiar proposal, testing, and modification cycle that links distinct stages of the design process. Since the scope of this type of iteration lies between the small and the large, we call it meso-scale iteration. A more descriptive name would refer to the proposal, testing, and modification cycle, "PTM iteration". This type is more meaningfully characterized by its purpose or cause, discussed in greater detail later.

Macro-scale iteration. Macro-scale iteration refers to iterations of entire products. For example, a software product or an automobile model may be said to have gone through several "iterations" if its design was refined with each annual release. This iteration takes place between several distinct design processes, where the design process is brought to a climax with the completion and release of the product (even though in reality the product may undergo constant design). Macro-scale iteration does not take place within the day-to-day cycle of designing.

2.2 Iteration in Automated Design

Some common types of iteration in conventional design were classified above. By "conventional design" we mean design practiced by humans as a mental or communicative exercise. It would be interesting to see if computational implementations of design (automated design models) favor the same varieties. If the ultimate goal of an automated design model is the same as in conventional design, one would expect that all varieties of iteration might have a place in automated models. However, the realities of coding an algorithm of any kind would suggest a preference for a more algorithmic, microscale variety. A good number of computer-directed design algorithms have been developed, and most of them are based on an iterative process. Two approaches dominate: design as a hill-climbing search, and design as a nonsearch process of reasoning, with the former being by far the most pervasive (Wang 1994, Chandrasekaran 1989). Some use an evolutionary approach (LaFleur 1991) that is inherently iterative because it depends heavily on variation and selection, while others generate alternatives that still must be filtered iteratively.

3 Implications on Current Design Practices

We have seen that human design is constrained by the high cost of iteration and limitations on information processing ability (cognitive capacity). For this reason, humans approach design in a "forward" manner, planning as much as possible by applying domain knowledge toward the anticipated solution until testing and iteration becomes more efficient than further study. Computers potentially have advantages over human designers in both speed of computation and accuracy of information storage and retrieval. This suggests that properly applied they could design more efficiently than humans via better information processing, cheaper iteration, or both. To date, most automated techniques take implicit advantage of the speed of algorithmic iteration rather than information storage and retrieval ability. In many of these algorithms, iteration has been retained and made more fundamental, reduced to a rapid iterative search of a large database of previous designs, or a methodical or random iterative variation of design parameters, strategies that are relatively easy to theorize and code. The overall approach embodied by these automated methods shifts away from the "forward" meta-level control practiced by humans and instead emphasizes a more blind, error-driven process -- often a simple feedback loop. Although these algorithms are iterative, they are fundamentally different from the way in which human design is iterative. The processes they most rely on are not practical for human agents to carry out, so they seem ill suited to offer insight toward improving human design processes.

There are good reasons for these early algorithms to rely on iterative error-term evaluation. For one, a process that is driven directly by focusing domain knowledge toward the design goal would be domain-specific, in the same way that a human designer trained in a particular field is only qualified to design in that field. Error-driven routines merely use an objective evaluation function that measures the difference between goals and current state without regard to the domain of the problem. Second, simple algorithms are easier to encode and do not require a sophisticated information storage technique or artificial intelligence. However, there are reasons to search for an approach that is more than an exercise in micro-scale iteration. First, the information storage and retrieval techniques developed for iterative algorithms may not be useful for human design because humans do not find it convenient to design in the same mode as the algorithms. The information techniques employed support only a micro-scale, error-driven search process that calls up previous designs or parametric variations, not the knowledge-driven process conducted by humans. Only a small portion of creative design involves searching an existing library of previous designs, and many designers would consider it poor practice to attempt to optimize a design by simply methodically varying parameters until the best design is found, especially if the parameters interact in complex ways and the variation is not being performed from within a computerized algorithm. Also, an algorithm that relies on preexisting designs or parameter variation is of limited applicability. Design parameters, components, and/or preexisting designs must exist already, because this approach does not generate them. In many design problems, established parameters and previous designs do not exist. In these cases initial designs must somehow be generated in the absence of a searchable domain or a computable error term. Here, iterative error-driven search algorithms seem to offer little help in formulating or solving such ill-defined problems.

4 Conclusions

All of the causes and purposes of iteration mentioned above could theoretically be prevented by more efficient application of existent information, although often, in human design, proceeding with design activity is a more practical alternative. If the information ultimately becomes represented by study or iteration, why is this distinction significant? Because if computational implementations of design are not constrained by the same things that constrain human design, iteration could become less important or even unnecessary.

Is design possible without iteration? One should not be led to believe that iteration is so inherent to design that automation of design will necessarily involve the automation of iteration. Reliance on iteration to elicit information seems mutually exclusive with the ability to retrieve the information via study. That is, increased study prior to the advancement of a prototype should lead to fewer or more effective iterations. If the problem is sufficiently understood at the beginning of the process, and design knowledge is optimally applied at exactly the time and place that it is needed, and there are no mistakes or oversights, iteration should be left without a role in the design transformation. This leads to the conclusion that iteration, though in typical practice indispensable, stems from the needs of the agent rather than the needs of the design transformation the agent is charged with carrying out.

While this conclusion is moot for human-directed design, it may be important for those developing automated methodologies for design. Previous automated models have retained or increased a reliance on iteration by structuring the design problem in such a way that the cost of testing is minimized. A computer algorithm can evaluate millions of iterations at practically no cost. But if iteration is seen as a coping mechanism by human designers rather than an elemental component of the design transformation, the door is opened to design automated methodologies free of an iterative aspect. They would be driven directly by a "forward" application of existing domain knowledge rather than a "backward" minimization of an error term. If there are ways of organizing domain knowledge in such a way that this is no longer a difficulty, broadly applicable techniques that are driven directly by domain knowledge become possible.

Some designers question the automation of design tasks, feeling that design is an art that should not or cannot be algorithmically reproduced. We do not seek to promote or discourage the development of automated design methodologies, nor comment on the effectiveness of such methodologies. But fully automated methods will continue to be developed and become more sophisticated. As this occurs, it is important that they not merely imitate existing practices but proceed from a more philosophical basis and examine their root causes. This paper suggested that iteration is a need of human agents but not necessarily of automated agents nor of the design transformation proper, making it relevant to seek approaches that are not inherently iterative.

References

Baya, V., Gevins, J., Mabogunje, A. et. al., "An Experimental Study of Design Information Reuse", DE-Vol. 42, Design Theory and Methodology, ASME (1992).

Chandrasekaran, B., "A Framework for Design Problem Solving", Research in Engineering Design, v 1, p. 75-86 (1989).

Condoor, S. S. et. al., "A Cognitive Framework for the Design Process." DE-Vol. 42, Design Theory and Methodology ASME (1992).

Eder, W. E., "Information Systems for Designers", Proceedings of the Institution of Mechanical Engineers, International Conference on Engineering Design, v. II p. 1307-1319 (August 1989).

Eisenhardt, K. M., Tabrizi, B. N., "Accelerating Adaptive Processes: Product Innovation in the Global Computer Industry", Administrative Science Quarterly, v.40, p. 84-110 (1995).

Eppinger, S., "Model-based Approaches to Managing Concurrent Engineering". J. Engr. Des., v.2 n.4 (1991).

Hegde, G., Kekre, Sham and Kekre, Sunder, "Engineering changes and time delays: A field investigation". Int. J. Prod. Econ., v.28 (1992) pp. 341-352.

Joskowicz, L., "Mechanism Comparison and Classification for Design", Research in Engineering Design, v 1, p. 149-166 (1990).

LaFleur, R. S., "Evolutionary Design Theory Using Dynamic Variation and Thermodynamic Selection", Research in Engineering Design, v 3, p. 39-55 (1991).

Ramachandran, N., Langrana, N., Steinberg, L., and Jamalabad, V., "Initial Design Strategies for Iterative Design". Research in Engineering Design (1992), v. 4, pp. 159-169.

Shigley, J. E. and Mischke, C. R., Mechanical Engineering Design, 5th edition. New York, McGraw-Hill (1989).

Smith, R. and Eppinger, S., "Characteristics and Models of Iteration in Engineering Design". ICED '93, August 17-19, 1993.

Suh, N. P., The Principles of Design. Oxford University Press (1990).

Suri, R. and Shimizu, M., "Design for Analysis: A New Strategy to Improve the Design Process", Research in Engineering Design, v 1, p. 105-120 (1989).

Thurston, D. L., "A Formal Method for Subjective Design Evaluation with Multiple Attributes", Research in Engineering Design, v 3, p. 105-122 (1991).

Ullman, D. G., "A Taxonomy of Mechanical Design", 1989 ASME Technical Conference: 1st International Conference on Engineering Design Theory and Methodology, Montreal, September 1989, pp. 23-36.

Wang, J. and Howard, C. H., "Recording and Reuse of Design Strategies in an Integrated Case-Based Design System." AI/EDAM, v.8 (1994) p. 219-238.

Organizational Visions of Virtual Manufacturing: Sociotechnical Aspects of Adopting Technology Computer-Aided Design Based Manufacturing Process

Samar K. Saha
(408) 429-4731

Silicon Systems Inc.
2300 Delaware Ave.
Santa Cruz, CA 95060

This paper describes a virtual manufacturing system consisting of technology computer-aided design tools for semiconductor manufacturing companies. This virtual factory is shown to be very efficient for development of advanced process technology, evaluation of process manufacturability, and improvement of existing fabrication processes. By using computer aids, the design and manufacturability of the target technology can be evaluated prior to actual silicon processing, and a company can establish a competitive edge in manufacturing. The technical limitations of computer-aided design tools and the managerial responsibilities in adopting a virtual manufacturing system for integrated circuits manufacturing are also discussed.

Introduction

It is well-known that semiconductor design and manufacturing companies face a highly competitive global market. In the semiconductor industry, the product life cycles begin every three to five years, and new technology emerges in every three years [Saha et al. 1995]. Therefore, cost-effective manufacturing and rapid introduction of high quality new products in the market place are crucial for the success of the semiconductor companies. In order to establish a competitive edge in integrated circuits (ICs) manufacturing the application of Technology Computer-Aided Design (TCAD) tools are crucial. TCAD point tools include process and device simulation, compact model parameter extraction, interconnect and reliability simulation, and circuit simulation. TCAD tools offer manufacturing and development engineers vast improvements in flexibility, innovation, efficiency, and customer responsiveness [Majchrzak et al. 1989, Saha et al. 1995]. However, due to the lack of systematic methodologies, the full potential of TCAD tools are not realized in technology development and in manufacturing. In this paper, a systematic methodology to apply TCAD point tools for semiconductor fabrication process development as well as manufacturing are presented. The implications of TCAD-based manufacturing process are also discussed.

Conventional ICs development

The conventional methodology of ICs manufacturing relies primarily on several iterations of silicon processing and measurement of device characteristics until the target specifications are obtained. As shown in Fig. 1, the conventional trial-and-error experimentation is neither efficient nor elegant to introduce new products for the "strategic window" of marketing [Saha et al. 1995]. Though some TCAD tools are used in technology development, their usage in manufacturing is limited because of several reasons such as inaccurate simulation results due to limitations of physical models, lack of expertise and unfamiliarity with application methods. But the aggressive technology schedule, global competition, and shortage of world-wide fab capacity offer a significant role to TCAD tools.

0-7803-3552-X/96/$5.00 © 1996 IEEE

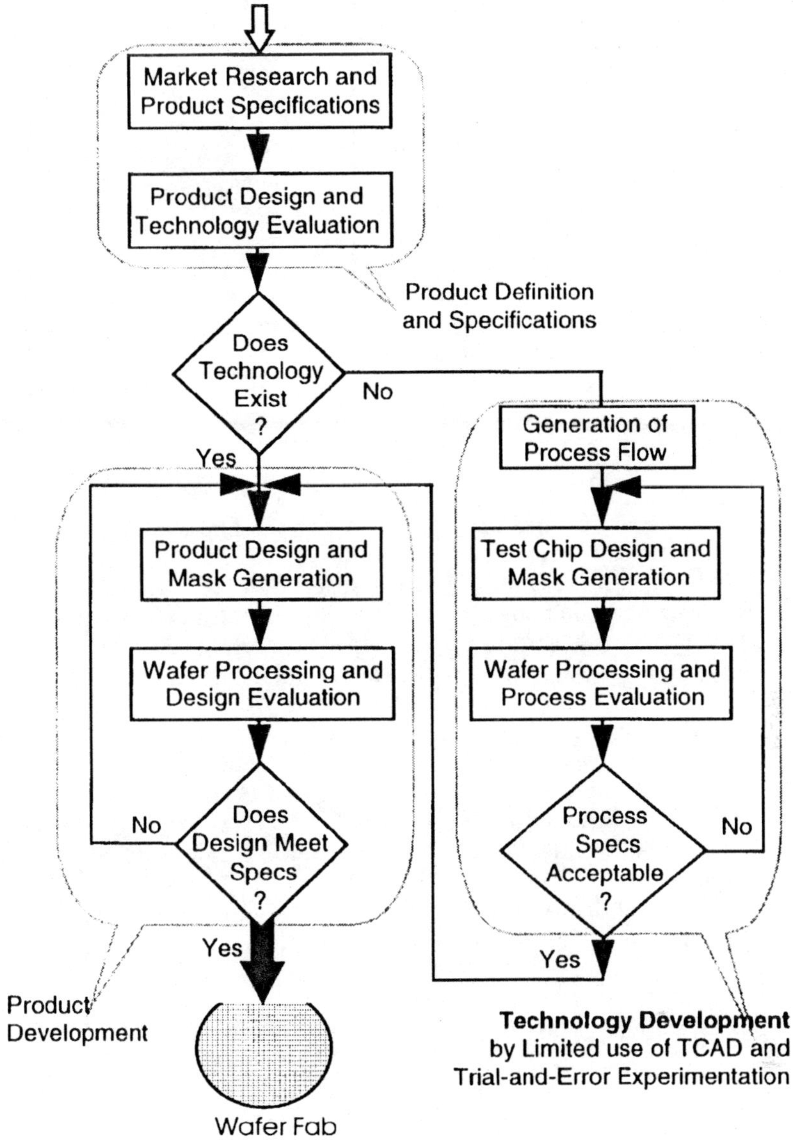

TCAD-based manufacturing

A Virtual Manufacturing System (VMS) is an ensemble of TCAD point tools integrated to a system for semiconductor product development and manufacturing directly from the input of market research. As shown in Fig. 2, the input of this manufacturing system is the product specifications defined by market research and the final output is an optimized manufacturable fabrication process.

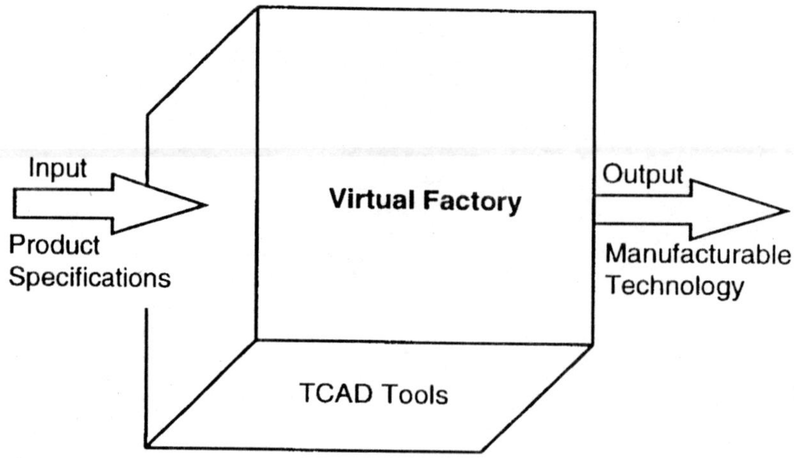

The details of VMS are shown in Fig. 3. The system consists of four major phases: (1) generation of device specifications, (2) generation of technology specifications and process optimization, (3) evaluation of process manufacturability, and (4) design optimization and process control. In the semiconductor industry, product development is strongly tied to new technology development. Therefore, the VMS described in this paper is focused on the entire process of technology development and release of the fabrication process to manufacturing. After market research and decision of product introduction, the use of VMS begins with circuit simulation to generate transistor level device characteristics, and subsequently device simulation using analytical process profile to generate a workable process flow. Then the process, device, and device model parameter extraction programs are used to optimize the process and generate device models for product design and manufacturing. After process optimization, TCAD tools are used for evaluation of process manufacturability. Finally, TCAD tools are used for design optimization and manufacturing process control.

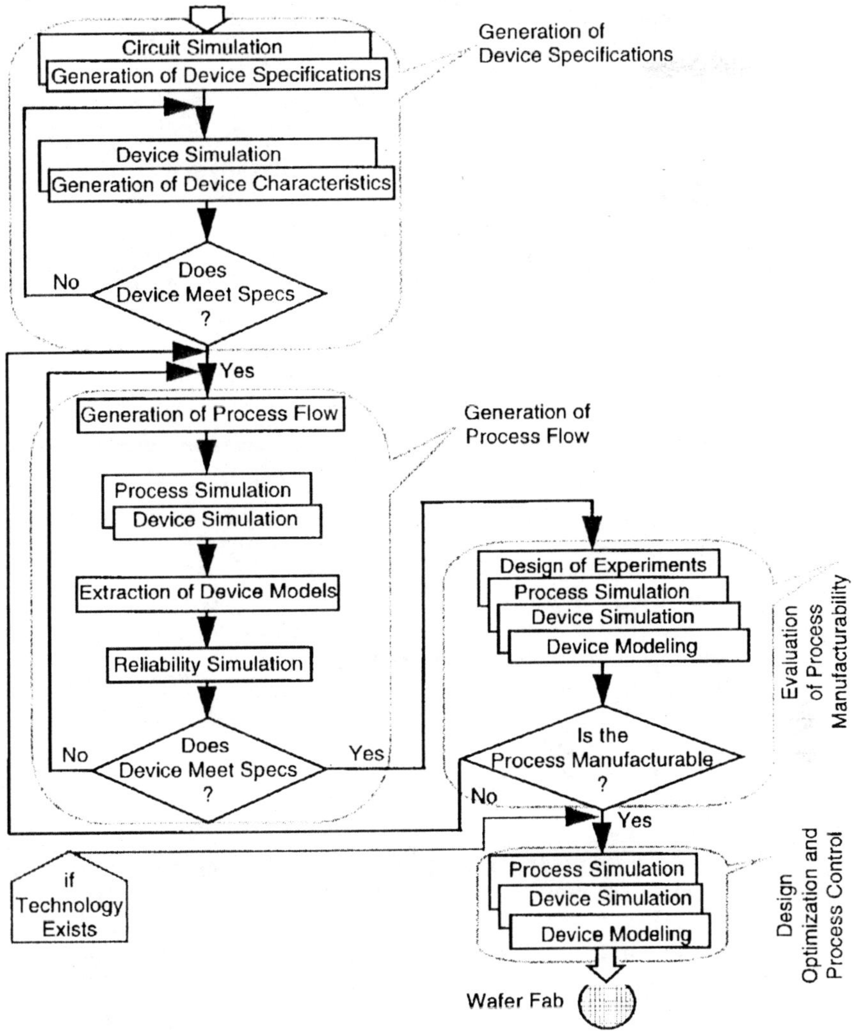

Implementation issues of VMS

With the present state-of-the-art of TCAD tools and the available physical models the VMS described in the previous section has several technical limitations, and therefore, must be clearly identified for realistic managerial expectations.

Over the years, the capabilities of TCAD point tools have significantly improved. However, due to fundamental technical reasons simulation results are not 100% accurate and reliable. Accuracy of simulation results relates to the underlying physics and numerical discretization issues. The predictability issue can be addressed by calibration of process and device models for numerical simulation. TCAD point tools must be calibrated [Saha 1993] with known technology, and can be applied to new process development with higher degree of confidence. In addition, a limited number of wafers for the target new technology should be processed concurrently to correlate the simulation results and update the previously calibrated models. Thus due to technical limitations of TCAD tools, VMS is not a complete replacement of fabrication process, but only a complement to full fab use and an efficient and effective aid to silicon processing.

Though the accuracy of simulation results can be significantly improved by proper calibration of physical models for numerical simulation, the use of TCAD remains limited due to the lack of systematic methodologies and expertise. Effective calibration requires expert users with knowledge of software capabilities and understanding of underlying physics. Therefore, skill level of engineers is critical for adopting VMS, and management must develop proper training and development agenda to train existing engineers.

Another major implementation issue of VMS is simulation throughput [Saha et al. 1995]. In order to optimize process technology and map the process control variables to variations of device characteristics, a large number of process and device simulations has to be performed. With the present hardware and software capabilities, this volume of computation takes an enormous amount of time. However, simulation efficiency can be dramatically increased by providing uninterrupted computational time to each engineer, and adopting new simulation methodologies [Saha et al. 1995]. Therefore, managers implementing VMS must address the computational needs of the organization through effective technology planning.

Organizational issues

In a semiconductor organization, the different functional groups are organized as independent departments. But as seen from the conceptual VMS that the same group of engineers is involved from design optimization to semiconductor manufacturing, and therefore, team environment is crucial for the success of VMS. VMS requires that an individual engineer must be skilled in design, device engineering, and technology. Thus new hires and existing engineers must possess the skills to perform a wide variety of technical job functions, and the organization must be skill-based. Management must create proper training and development plans for re-deployment of existing staff. The core competency of the organization must be re-evaluated. Thus for successful implementation of VMS, the organizational structure, procedures, training programs, job content, and possibly the entire organizational culture must be changed.

The major technical concern of managers implementing VMS is the predictability for new process design. If TCAD tools have to be calibrated for use in design of experiments, "straight out of the box" tools would not be usable to develop new generation of technology. Therefore, top managers may have strong and valid reservations of allocating human and capital resources. However, the predictability issue can be addressed by concurrent experimentation to update calibrated models, and VMS can significantly reduce the development cycle time and cost. And, in manufacturing, the models developed during the development phase can be successfully applied for process control and process manufacturability. Therefore, managers implementing VMS must set clear direction within the limitations of VMS, and inspire the organization by articulating their vision and expectations for the future benefits to be achieved from VMS.

With the present technical limitations, the full benefits of VMS will not be achieved if VMS is regarded as a direct capital-for-labor replacement. Moreover, the full benefits of VMS are not achieved when TCAD tools are purchased to replace the activities previously done by trial-and-experimentation. Rather, the full benefits of VMS are most likely to be achieved when VMS is considered as a tool complementing and enhancing human and fab capabilities throughout the entire development and manufacturing process. Thus in the development phase, consideration should be given to ways to optimize process technology, analyze process control variability, and evaluate manufacturability of the target technology by using VMS.

Conclusions

A systematic methodology to apply a VMS in technology development and manufacturing is described. With proper calibration of the process and device models, VMS offers efficient and cost-effective technology development, optimum product design, design for manufacturability, and manufacturing process control. The technical limitations of VMS and the methodologies to overcome these limitations are described. Knowledge of the limitations of TCAD tools allows management to set realistic goals and expectations from VMS use. Successful implementation will involve training and re-deployment of existing human resources, changes to training programs, job content, organizational structure, procedures, and possibly the entire organizational culture.

Acknowledgments

Grateful thanks are due to Mr. Harinder Singh for critical review of the paper and valuable comments and suggestions. Thanks are also due to Mr. Dominic Massetti, Mr. William Nehrer, and Mr. Nitin Parekh for valuable suggestions and encouragements.

References

Majchrzak, A., and Salzman, H. "Introduction to the Special Issue: Social and Organizational Dimensions of Computer-Aided Design," IEEE Trans. Eng. Manag., EM-36, 174-179, 1989.

Mar, J. "Technology CAD at Intel," Technology CAD Systems, Springer-Verlag: Wien, New York, 1993.

Saha, S. "MOSFET Device Analysis Using MEDICI (PISCES): Substrate Current Modeling," CAD for IC's-Process, Device, Circuits, and Virtual Factory-Industrial use of TCAD, Stanford University, Stanford, California, 1993.

Saha, S., Lindorfer, Ph., and Gadepally, B. "Productivity Increase in Process and Device Simulation Throughput for Integrated Circuit Manufacturing Companies," IEEE Annual IEMC, Singapore, 408-411, 1995.

Economic Benefits in Using Electronic Data Processing in the Administration and Services in the Finnish Cities

Jarmo Salmi
+358 31 2196111

The City of Tampere
P.O.BOX 87
FIN-33211 TAMPERE

The objective of this study was to find, whether the use of IT has given any economical advantage in municipal administration and service production in Finland. The study was concentrated on the development of personnel costs that, on the one hand, is the largest cost group in municipal administration and, on the other hand, may bring the largest economic benefits through the use of IT. The research material consisted of a wide range of reports on IT investments and use, and was collected from Finnish cities during 1980's. Financial investment in each city was proportioned to the operational costs and the number of users was proportioned to all personnel. Averages and growth rates for the whole of the 1980's were calculated using the material. Information about the status of cross section year 1990 was also used.

Annual change indices were calculated for several years to describe the development of personnel costs in each city. The amount of costs in marks were proportioned both to the total amount of operation costs and population of the cities. The economic collapse that took place in the turn of 1980's and 1990's allowed the change indices to be divided into two natural groups: 1987-1990 and 1990-1992.

On the basis of the tests it can be seen that there is no close connection between financial investments and growth of personnel costs. However, a clear connection exists between the extent of the use of IT and the reduction in the growth rate of personnel costs. The bigger the share of personnel with the opportunity to use IT, the more it is possible to reduce the growth of personnel costs.

Introduction

The growth of the use of information technology (IT) in municipal administration in Finland increased in 1980's. New applications were built especially in the service sector. The most important of these were patient information systems for health centres and hospitals, library circulation systems, social welfare systems (subsistence benefit, child home care benefit, childrens' day-care system), administrative systems for schools and institutes, technical service systems (building control, real estate management, CAD applications, location information systems), as well as customer information systems for energy plants, and water and sewage works.

International URBIS and OECD projects were the first (and by the middle of 1980's the only) empirical projects studying the economy on the use of IT in the administration of middle-sized and big cities (King 1985). It was found that in URBIS cities 69% of the cities did not gain any personnel savings, 26% gained some savings, and 5% gained significant savings. As to cost savings, 80% did not gain any savings, 15% gained some savings, and 5% gained significant savings. It was estimated that in 15% of the cities the effectiveness of operation increased a little or not at all, in 54% the effectiveness increased some, and in 31% notably. In OECD cities 20% did not gain any personnel savings, 33% gained some savings, and 47% gained significant

0-7803-3552-X/96/$5.00 © 1996 IEEE

savings. 46% of the cities gained no cost savings, 31% gained some cost savings, and 23% gained significant cost savings. The study revealed that IT does not usually bring personnel or cost savings, but it may help to avoid creating new posts and thus reduces costs.

Lately the connection between computers, telecommunication and other IT as well as productivity has been widely discussed. Though the use of computers, e.g. in USA has grown more than two decades since 1970, productivity (especially in the service sector) seems to have remained unchanged. Interest in productivity and the "productivity paradox" has grown, but so far it has been just a deepening mystery. Nobel prize-winning economist Robert Solow said: "We can see computers everywhere but in productivity statistics."

In his paper Brynjolfsson summarised what we do and do not know about the productivity of IT (Brynjolfsson 1993a). Brynjolfsson states that the closer you analyze the data in the studies on the benefits of IT, the more it seems that measurement errors are the central issue of the productivity paradox. Fast innovation has made IT intensive industries especially sensitive to problems connected with appreciation of quality changes and new products. The increased number of variations, improved delivery times, and more personal services are all additional advantages that are not properly represented in statistics. These all are connected quality factors especially with IT.

Investments in IT have contained five main objectives: labour saving, improved quality, diversification of assortment, better customer service, and faster response time. Excluding labour saving, these factors are difficult to describe as aggregated productivity or yield statistics. Though we see computers everywhere, they only add up to 1% of the costs in companies. Therefore, it is difficult to tell the difference between IT and ordinary occasional fluctuations in normal operation. Brynjolfsson and Hitt (Brynjolfsson 1993b) used the same models as earlier studies, but using more detailed data from more companies.

It was noticed that computer capital and information systems personnel were positively and significantly correlated with the growth of production. It was found that a 1% increase in the use of computer capital amounted to 0.00518% growth in production when other factors remained constant. Since the value of computer capital is less than 1% of the value of annual production, the researchers drew the conclusion that ROI of computer capital (the increase in production in dollars when investing 1 dollar) is appr. 54.2% per year, when other factors are constant.

Objectives and Leading Principles of this Study

The objective of this study is to get more information about whether the use of IT has already been of benefit in the administration and service sector of Finnish cities. As municipal service production is largely based on human labour, the savings are mainly seen in personnel costs.

The following principles have been observed in the making of this study:

1. Use as much comprehensive material as possible. As to the use of IT, the research period is the whole 1980's (1982-1990), and as to economic effects, the second half of 1980's (1987-90) and the beginning of 1990's (1990-92). The data was gathered from all cities selected for the research.

2. The information about the utilisation of IT in cities is proportioned to make it comparable by cities. As far as the use of IT is concerned, the information is proportioned to total operational costs and total number of personnel posts.

3. The use of IT by cities is described with a single index number in order to sort the cities by size. To avoid being dependent on only one examination, several descriptions of the status of use of IT are used. They are:
   - the average of the use of IT during the whole of 1980's
   - the growth rate of the use of IT during the whole of 1980's
   - the use of IT in the cross section year 1990

4. The test subject is the change of personnel costs. Economic benefit is found by comparing annual city-specific growth of the variables under study. Comparison is made for two different periods: end of 1980's (1987-90) and beginning of 1990's (1990-1992). The information is classified both by interval and ordinal scale.

Numerical values measured for variables, on the one hand, and score number of the cities sorted by size on the basis of their measured numerical values, on the other hand, are used in the tests of the study. The whole selected material is used for processing using regression analyses and rank correlation tests. Upper and lower quartile cities concerned as small-scale and large-scale IT users are analysed to find out, whether the distribution of their scores are based on the same kind of population, or whether it can be concluded that distributions are clearly different. The hypothesis H_0 for all the tests is that the use of IT does not affect the increase of personnel costs in any way. The tests are used to ensure whether the hypothesis can be abandoned. (e.g. Conover 1971)

Test Designs

Tests are used to find out the possible connections between the use of IT and change of personnel costs proportioned to operational costs or population. City-specific descriptive indices are used in testing. The smaller the change index, the more the use of IT is connected with positive economic growth. Hypothesis H_0 is the same that has been used in the corresponding international studies, i.e. IT has not given perceivable economic benefit: the growth rate of personnel cost has not been reduced. Correspondingly, hypothesis H_1 is that IT has brought economic benefit. The tests are used to confirm, whether hypothesis H_0 can be abandoned and to which degree of certainty this can be done.

Data Describing the Use of Information Technology by Cities and Its Preprocessing

In this study the use of IT is studied from the point of view from both financial investment and spreading of workplaces where IT is involved. The subjects of the study are the average of investment or use during the whole period of 1982-1990, its growth rate during the same period, and the status of the cross section year 1990. The proportion of financial IT investment of the city total operational costs (Ca/C) reveals the share of IT of available operation financing. Proportioning the number of workplaces equipped with IT to the total personnel (La/L) helps describing the diffusion of IT in the city administration and service production.

Data Describing the Changes of Personnel Costs and Its Preprocessing

The most important cost group for studying the economic effects of IT are personnel costs. On the city-level it is possible to get very accurate information about personnel costs in all Finnish cities. Although the main objectives of Finnish municipalities are quite identical, working organizations in cities are different. Thus it is not possible to make direct comparisons of even proportioned costs. For this reason, this study uses a city-specific annual growth index of the cost group examined.

The great turning-point of the Finnish national economy in the beginning of 1990's has significantly affected the development of cities' economy. The old stable growing development is culminated in the years 1990-1991 and this is clearly the discontinuation point. A discontinuation point like this presents an interesting opportunity to study the effects of utilisation of IT. Cities have been forced to cut their expenses drastically. This study attempts to find out whether the possibility to use IT has made it possible to cut (especially personnel) costs.

In this study the variables to be tested are the share of personnel costs of all operational costs and personnel costs per capita. For the quantities to be tested their development in each city first during the period 1987-1990 and then further 1990-1992 are calculated.

Observations and Results of Analysis

The Connection Between IT Investments And Change Of Personnel Costs

In two cases IT investment seems to be connected with the reduction of the growth of the cities' personnel costs. The share of personnel costs of total costs has grown slower at the end of the 1980's in cities where the share of investments has grown faster. In these cities operational costs have been increased by outside payments (e.g. increasing purchases). In cities, where investment level has been constantly high, personnel costs per capita have grown slower during 1990-1992. This is possibly due to cost cuts. The investment status of 1990 does not affect the development of costs, which is understandable as momentary situations do not easily cause fluctuations in the whole.

	average A(Ca/C); N=83	growth rate T(Ca/C); N=63	in 1990 Ca/C 90; N=76
growth of the share of personnel costs CL/C	not found	found during period 1987-1990	not found
growth of personnel costs per capita CL/Q	found to some extent during period 1990-1992	not found	not found

The connection between relative IT investment Ca/C with the development of relative personnel costs

The Connection Between the Share of the Personnel Using Information Technology and Change of Personnel Costs

In four cases there seems to exist a connection between the share of personnel using IT and the reduction in the growth of cities' personnel costs. It is logical that notable savings can only be achieved, when IT spreads to affect as large as possible operational volumes. The effects are mainly targeted to the growth development of the share of personnel costs. This indicates that operational costs have been increased by increasing outside payments (e.g. purchases). The status of the cross section year 1990 affects the cases strongly. This is understandable because the density of workstations is a cumulative quantity growing steadily without drastic fluctuations that would rapidly change the order of the cities.

	average A(La/L); N=78	growth rate T(La/L); N=75	in 1990 La/L 90; N=82
growth of the share of personnel costs CL/C	found	found during period 1990-1992	found
growth of personnel costs per capita CL/Q	not found	not found	found during period 1990-1992

The connection between the relative share of personnel using IT of total personnel La/L and development of relative personnel costs

Evaluation of the Results

Total financial investment (Ca) in the 1980's in each city is based on the reports of cities given during more than one year. In the reports investments and normal operational costs have not been specified. It is very likely that organizations were not even aware of the existence of all indirect costs. Estimates have been made that purchase costs and so-called visible support of workstations only make up one third of their total costs (Heikkilä 1995).

The share of personnel using information technology (La/L) is calculated using the number of posts at the end of the year as the number of personnel. This figure is problematic, as it is not all the same as the number of persons employed. In this study the difference between part-time and full-time workers has not been made. Cities have no uniform practice for this.

The effect of IT is described with the quantity CL describing total personnel costs. This information is taken from official statistics (Central Statistical Office). Thus, their validity is obviously clear. However problems arise from the fact that not all cities provide their services using their own organizations. In those cases costs not are entered in the books as personnel costs, but as, for instance, costs of federation of municipalities or purchased service costs. Because of these problems only the annual change indices of the quantity in question were selected for this study. The rationale being that the operation of cities has been, as presented earlier, under strong outside control and has continued from year to year steadily and logically. The way the functions are arranged is not essential, only the changes are essential.

Another important result of this study is that it is no use to try and find a connection between total financial investments and total savings to be achieved using today's calculation methods. Estimation of relation of IT costs and advantages gained by the use of IT can be attempted for individual applications and limited sub-units. However, the first problem is, whether the limitation can be done accurately enough to find out the costs and benefits. Other factors affecting yields, such as changes in product assortment and quality, must be eliminated. Changes in legislation affect each city equally, but strong structural changes (growth or fall in size) cause changes in service production that are different from those during steady development. Continuous expansion of IT services (cumulatively) blurs and complicates the estimation of an individual sub-unit. The second problem is how to estimate hidden IT costs. In international seminars there are given estimates of the share of hidden costs which differ from each other considerably. Registering of hidden costs is possible in small scale, but still troublesome. In large scale it is very uncertain. As a conclusion we could say that attempts to estimate the hidden costs of IT is not worthwhile, because they are included in the total costs (or personnel costs) of organizations. In analyses these costs in turn are compared to total costs of other corresponding organizations.

Searching for Other Factors Possibly Affecting the Use of Information Technology

In this study all analyses have been made between two quantities, which automatically leads to the question of existence of a third or more variables that simultaneously affect both variables under testing. More detailed studies are required to find out other quantities affecting the change. These could be, for instance, structure and density of population, degree of unemployment, amount of state subsidies, etc.

The connection between population and investments in information technology

Regression analyses were made to find whether big cities invest relatively more in IT than small cities. In these analyses, quantities depending on population were average and growth rate of IT investments and status of the cross section year 1990 were used.

It can be concluded that big cities have invested in IT relatively more than small cities. In cities with a population of less than 25,000 the relative investment varies a lot, though the smallest investments can be found in this group. In fact, the level of investments in bigger cities seems to be approximately the same. Bigger cities have also increased their investments faster than smaller ones. The widest variation can be found in the smallest cities.

The connection between population and diffusion of information technology

It was also studied as to whether big cities have introduced IT wider (a larger percentage of of users of total personnel) than small cities. This was done using regression analyses, in which the average of diffusion of IT in 1980's, the growth rate during the same period, and the status of the cross section year 1990 were studied. It can be concluded that, population does not affect the current extent diffusion of IT.

The connection between the share of personnel costs and the diffusion of information technology

Finally, it was studied as to whether the share of personnel costs of operational costs is bigger in cities, where the diffusion of IT is limited. This was done by using regression analysis, in which the connections between the share of personnel costs (in 1990) and the average and growth rate of the use of IT and the status of the cross section year 1990 were studied.

It is found that, when the share of personnel costs reduces, the relative extent of the use of IT is wider. This verifies the earlier result, according to which the extension of the use of IT to as many workplaces as possible leads to a reduction in personnel costs. The same general result was found when examining the average extent of use also applies to the growth rate of diffusion as well as to the status of the cross section year 1990, which is natural due to the close interdependence of these variables.

References

Brynjolfsson 1993a, Brynjolfsson E. "The production paradox of information technology." *Communications of the ACM*. December, vol.36, N:o 12 (1993) pp. 67-77

Brynjolfsson 1993b, Brynjolfsson E, Hitt, L. "Is information systems spending productive? New evidence and new results." *Proceedings of the Fourteenth International Conference on Information Systems*. December 5-8, 1993, Orlando, Florida, pp. 47-64

Conover 1971, Conover, W.J. *Practical nonparametric statistics.* John Wiley & Sons Inc. (1971)

Heikkilä 1995, Heikkilä, J. *The diffusion of a learning intensive technology into organizations - the case of personal computing.* Helsinki: Acta universitatis oeconomicae Helsingiensis A-104 (1995); Finland

King 1985, King J.L., Kraemer K.L. *The dynamics of computing.* New York: Columbia University Press (1985)

User Participation in Standardization
New or Merely Recycled?

Liora Salter, FRSC
416 736 5560

Osgoode Hall Law School
York University, Toronto Canada

The user problem in standardization will not be resolved by the well intentioned reforms underway in various standards organizations because its cause lies elewhere. It lies with the different situations of the various types of users, and with contradictions in the standards process itself. Even the new forums and consortia, which are widely hailed as the answer toithe user problems, cannot escape these problems. Nonetheless, participation is not likely to diminish because the reasons for participation lie with benefits other than standards.

Introduction

Despite many efforts at reform, the user problem in standardization remains all but intractable. Neither the new high level OECD user committees, nor the newly refashioned European Telecommunications Standards Institute (ETSI), nor the initiatives undertaken by such private sector organizations as the American Committee T 1 to promote user involvement, nor even the newly proposed structure for members of the International Telecommunications Union seems likely to solve the user problem, although all are intended to integrate users into the standards process[1]. Commendable as these efforts are, especially in light of criticisms that standardization has previously been ill-suited to industry's needs, there is much evidence that the causes of user problems lie elsewhere, beyond the capacity of both public and private sector organizations alike to resolve them.

This paper offers some observations about the sources and consequences of the user problem in setting standards for telecommunications and information technologies[2]. It argues that the category of user conflates four groups with little in common with each other. Each group poses its own user problem, which is described here. The paper concludes with an examination of the new industry fora and consortia, because these are now widely seen to represent a solution to the user problem. On the basis of the discussion offered here, some predictions are made about the possibilities for success. The user problem is intrinsic to standardization, it is suggested, and it stems from the contradictory pressures faced by governments in promoting competitiveness and by firms operating in very turbulent markets.

Who is the user?

Given the effort devoted to users in all standards organizations, it is surprising to fInd few definitions of the user. In fact, there are at least four groups to which the label might properly be applied. They are (l) users of the standards; (2) major industrial users of the technologies, especially those with unique or urgent needs; (3) governments who seek to exercise their procurement powers to support and redirect standardization to meet their industrial strategy needs; and (4) consumers of the new communication and information technologies. Each type of user is in a different position concerning standardization. I will deal with these group, but in reverse order.

Consumers of the new communication and information technologies

Only in Australia are the end-use consumers given much recognition in the standards process[3]. There are good reasons for their neglect elsewhere. The end-use consumer is actually the individual customer for the equipment and services marketed by those who deploy standards. The end-use consumer is only indirectly a user of standards.

0-7803-3552-X/96/$5.00 © 1996 IEEE

Put another way, standards are only one ingredient in the product or service being purchased, and only one of many factors influencing purchasing decisions.

By intervening in standardization, end-use customers seek to play a role in the internal decision making of firms, especially their production and marketing strategies. But customers normally exert their influence elsewhere, through their purchasing power. To be sure, a bad choice of standard can result in customer flight, but so can bad choices about prices, product design, product image etc. No one expects the end-use decisions to be involved in decisions about these other matters, and. similarly, no one should expect the end-use consumer to be engaged in discussions about technical standards.

The point seems obvious, but often it is not. In discussions in standards organizations, frequently mention is made of the need for user representation, in this case, involving the end-use customer of equipment or services, and indeed, many standards organizations now have both user and "human factors" committees, where, in theory at least, end-use customer input can be sought.[4] In most standards organizations, the so called user problem is understood as arising only from the end-use customer's lack of technical sophistication. Although it is true that the end use customer is an unlikely member of a technical committee, this is not the source of the user problem. After all, consumer organizations have become quite expert in dealing with the similarly highly technical and complicated matters associated with telecommunications, and often take their place alongside industry experts in regulatory hearings. Standardization is, however, a collective process, a venue for the resolution of technical issues between and among firms. Collective action among competing firms is only possible in standardization only because firms recognize the importance of the common task to be done. Firms either become involved at the pre-competitive stage, as an integral part of their research and development activities, or they use standardization as an integral part of their marketing strategies at a much later stage. In neither case does the end-use consumer have much of a role to play.

For example, in the pre-competitive stage of technology development, the connection between technical standards and the products or services eventually to be marketed is a very tenuous one. Standards organizations simply lay the ground rules for technology development; they are not harbingers of particular products or services. A consumer organization could usefully make representation about the needs of its members or the general public, but only in the most general terms. Once it has been recognized by those developing standards that the customer wants ease of operation, durability and low cost and once the special needs of target markets (for example, those with disabilities) are recognized - there is nothing left to be said from a customer point of view in the absence of an actual product or service. Most often, however, a standard approved by a standards organization is one which already has considerable application. Firms often participate in standards organizations to defend their technical choices about product and service design from competition as well as from being rendered unattractive because of incompatibility. Here too, the end-use customer is of little importance, because the task which commands the attention of participants in one directly relevant only to the firms whose own standards are, or might be adopted as formal standards. In this case, the "user problem" actually is that the end-use customer has little direct interest in the matters under consideration, and no legitimacy in dealing with the tasks which bring their participants together.

Governments, ProcurementandStandards

Until quite recently, governments have been the mainstay of the standards process. Most standards organizations in the communications sector were inter-governmental in nature, and their deliberations concerned interconnection of equipment and inter-operability of services provided by government-owned telephone companies. Even among the private sector organizations, however, governments have always been important contributors to standardization, underwriting the costs of industry participation in some cases[5], and contributing technical committee members in others[6]. In Europe the government still finances some standards development.

Today, everywhere, governments are reassessing their role, and reducing their budgets, yet it would be mistaken to view governments as having withdrawn from supporting standardization. Some governments have adopted a different strategy, one less dependent upon financing and participation in standardization.[7] Their new strategy is based on three premises: that governments are major users on communications and information technologies, that governments should adopt coherent purchasing strategies based on open standards, and that use of governmental procurement powers constitutes a legitimate form of industrial strategy in an era where direct government subsidies and tariffs are no longer possible or desirable. The first premise is undisputable true, but the other two are considerably less so.

Many countries have adopted policies which call for coherent purchasing strategies based on open standards to be employed within departments and across government. In theory, this makes sense. Departments can garner the best price; they can influence product or service design, and they can co-ordinate their services only by rationally planning their information infrastructure and directing purchasing decisions accordingly. Open standards facilitate maximum inter-connectivity, promote global competitiveness on the part of indigenous suppliers, and prevent single vendor lock in. In practice, however, individual government departments have very different needs, and each is likely to be on a different trajectory of technology development. To the extent that government policy permits it, and on occasion even when it does not, each department or agency will seek to control its own information infrastructure development and purchasing priorities often choosing proprietary solutions to its technical problems because this is seen as the easiest or most relevant course of action. Lack of trust for the central agencies, inter-departmental and agency rivalries, overwork (planning and co-ordination take considerable time and resources), departmental accountability measures also introduced by government, deregulation and decentralization all militate against the implementation of centralized procurement strategies regardless of the fact that such strategies are often mandated by policy or regulation. With prospect for coherent procurement strategies based on open standards being dim, the influence of governments as major users is diminished.

Yet, despite considerable evidence that even mandated procurement policies are often not being implemented, governments persist in conceiving of procurement as an important tool in exercising their power as users in standardization. To understand why, it is necessary to look elsewhere, not at the benefits supposedly accruing to governments following coherent procurement strategies based on open standards, but at the role sought by governments in promoting indigenous industrial development and economic growth. In this second context, standards reflect and important prong of politically acceptable industrial strategies. The reasoning is as follows: Manufacturing firms, or consortia, which have established an early lead in the market by virtue of their technological innovation, can sustain their lead by early standards adoption and through government procurement in the first instance.

Governments often speak of their activities in promoting indigenous firms in terms of competitiveness. They see procurement and standards as closely aligned with each other, and both as important instruments in promoting the competitiveness of indigenous industries which undertake innovation. These government's competitiveness strategies are seen to result in greater expenditures on R&D, with important spin offs for other industries. On the other hand, to promote competitiveness, governments need to cut costs. Buying off the shelf, even an inferior product, or using a de facto or proprietary standard, are often good easy to do so. As well, neither governments nor industry wants to be caught out by promoting a product or service which will later prove incompatible with what is already in the market, and this too militates against using procurement as an industrial strategy or early standards adoption.

In theory, then, governments should be major and model users of the new technologies, and their role as users in the standards process should not be underestimated. By linking standards, procurement and industrial strategies, governments often believe they can promote economic growth and comparative advantage for domestic firms or consortia. In practice, even while governments continue to underwrite some costs of standardization, governments face enormous difficulties in operating effectively as major and model users. They are caught by a contradiction inherent in their own competitiveness policies. Their attempts to promote research, development and innovation on the part of domestic firms, in part through procurement and early standards adoption, are undermined by their equally strong commitments to competition and cost cutting and also by their need to adopt and promote technologies compatible with those readily available in global markets. Caught by this contradiction, governments are in no position to function effectively as users in the standards process regardless of how often they express support for standards, and how "user friendly" standards organizations become.

Major Industrial Users of the Technologies

Most of the efforts to bring users into standardization in recent years have concentrated on the major industrial users of the technologies. Standards organizations of all kinds bristle under criticisms that the standards process has been ill suited to the needs of industry, and that it must become industry-led rather than technology-driven. But because the vast majority of the actual participants are middle management engineers professionally committed to finding the best possible solution to complex technical problems, this worthy goal is hard to realize. Either engineers must convince senior management to take notice of standardization, or senior management must become directly involved in the standards process.

Five years ago, efforts were directed to interesting senior management in standards development[8], but faced with little success in this regard, the standards organizations are now seeking the involvement of new participants. To attract senior management, promises are being made to the effect that senior management will now set priorities for and tailor standards work programs to meet industry needs.[9] Is industry likely to be convinced? It is too soon to know but, despite the participation of key industrial users, the prospects are not good. Were it feasible for major industrial users of the technologies to agree upon their needs, and to make these needs known in anything other than the most general terms, the major industrial users of the technologies would indeed represent a significant force in the standards process. Yet in many respects, the situation of the major industrial user is not very different from that of the end-use customer, discussed above.

Despite their importance generally, these firms represent the customer base for the companies developing products, services and networks. They are no more likely to be welcome when R&D decisions are being made (as is the case when standards reflect pre competitive aspects of the technologies), or when marketing strategies are being designed (when standards are developed for existing products or services) than was the end-use customer. Furthermore, their needs cannot normally be specified in advance of product or service development.

Major industrial users are different from the end-use customer because of their market clout, however. Individual firms sometimes have specific needs not being met in the market, and standards developers fail to pay attention only at their peril. There are many examples of individual firms becoming active lobbyists for standardization, even to the point of promoting new standards. From the firm's perspective, however, there are always serious disadvantages to having unique standards. With tailor-made standards comes the possibility of vendor lock-in and technical incompatibility Furthermore, there are many alternatives to formal standards now available. Industry organizations and consortia develop something akin to standards, and their products are less likely to relate to technological infrastructure than the actual product or service configurations actually needed by individual firms. De facto standards are also readily available in many cases, and adopting them reflects a relatively risk free choice.

As users, then, the major industrial users of the technologies do want standards to be developed, if for no other reason than to minimize the risks attendant upon their technology purchases and market strategies. Furthermore, generally speaking, they view formal standards as preferable to de facto standards. That standards organizations make decisions by consensus, and are now industry-led make formal standards more attractive as an option. However. They often propose that formal standards be developed and later further elaborate by industry forums or organizations As users, then, the major industrial users of the technologies are likely to have their most important influence in pressuring the standards organizations only to speed up the development of standards and to co-operate with forums and industry. Beyond this, there is little likelihood that the major industrial users will want to play a significant role as users, or make their presence felt in the day to day activities of standards organizations.

The message about making standards organizations responsive to user needs has been now heard by all the formal standards organizations, and thus the task is all but completed. Paradoxically, this situation renders the major industrial users of the technologies less important as users now, even when more efforts are being made to court their participation.

The Users of Standards

The real users of standards have always been active participants in standardization. The real users of standards are the firms developing products, networks and services based on the standards. These are the customers of the products produced by standards organizations (i.e. the standards), and they have always been willing to underwrite a significant portion of the enormous costs of standards development. Of course, not all firms are equally involved; participation in standards organizations has always represented one among several possible strategies for establishing market influence or dominance. Small firms generally prefer to be free riders, adopting standards only after they have been developed in standards organizations or, preferably, even after they are established in the market. Very large firms also can often afford to ignore the standards organizations, having no need of the consensus process nor the dissemination capabilities of the standards organizations to promote their de facto standards in the marketplace. But if most of the real users are already in the room, it is easy to see why initiatives designed to attract users are irrelevant.

Why then has so much attention been paid to the user problem? The answer in the short term is very different from that of the long term. In the short term, industry participants have found it necessary to exert maximum pressure on standards organizations in order to speed up the process and render its products useful. Drawing attention to the absence of users is a powerful way to make the point that something is amiss with long-standing practices in the standards organizations. It has created an image of an audience, a market, left unsatisfied, and a perception of urgency. Drawing attention to the user problem has mainly now served its purposes, because all of the standards organizations have undergone a major transformation, and their members' priorities have changed. It is fair to say, notwithstanding the continuing importance of government support, that the standards process is now industry-led.

Why then is the so called user problem still described as urgent, indeed so much so that, in the European case, for example, the relationships between the Commission and European standards organizations is being altered to address the user problem? To understand why requires a closer examination of the benefits historically derived from standardization. To put it boldly, rarely, if ever, were the benefits derived from the actual standards. To be sure, to the extent that decisions regarding basic technological choices could be made in advance of investments and that the rules of the game were known, standards have been considered useful. But all too often, historically at least, standards have been adopted only after technical choices had been made by key firms, or standards had so many options as to render them all but useless as common solutions to technical problems.

The benefits of participation lie elsewhere. Standards organizations have always offered a venue for erstwhile competitors to meet, and to share as much information as they think profitable. Participation has permitted firms to act defensively, for example delaying standardization until they were ready to promote their own products and services. The standards organizations also have offer a patina of respectability: a standards-based solution to a echnical problem, originally promoted by one firm and later adopted by others, is designated as the best possible technical solution for everyone. Its widespread deployment by other firms (especially smaller firms) is guaranteed, and the dissemination of technologies based on the standard is facilitated. Thus while the participants in standardization have been traditionally both producers of the technologies and, at the same time, users of the standards, the benefits of participation have fallen mainly on the side of production not use. The regular participants in standardization are producers and marketers first, users only secondarily. Calling attention to the need for user participation, then, reflects the limited role that they have as users in the standards process.

Concluding Remarks

The standards regime of many years standing has undergone a significant transformation in the past few years, and in the process, increasingly more attention is being paid to users. The OECD created a high level committee of user organizations to assess the problems of standards organizations and offer recommendations. The European Telecommunication Standards Institute has been brought under the direction of a new Board, made up in the main of user organizations. All of the standards organizations now have user committees, and lay stress on their responsiveness to user needs. Even in the International Telecommunications Union, significant attention is now being paid to users, and to ensuring that, in effect, industry sets priorities, especially for the allocation of increasingly scarce resources. Meanwhile, governments are withdrawing from active participation in standardization through standards development committees, preferring instead to conceive of themselves as major and model users of new technologies, as promoters of industry-led initiatives to support a global information infrastructure (reliant upon standards) and as implementing their user mandate through their procurement policies and competitiveness strategies.

This paper has argued that, despite all these efforts at reform, all is not well from a users' perspective. The problems do not reflect a failure of political will so much as the contradictions inherent in standardization itself. In practice, the end-use consumer has no legitimate role in the making of standards. Government procurement policies and competitiveness strategies are fraught with problems so serious that they undermine governments' role as major and model user. The major industrial users of the technologies have already achieved their goal, which was to streamline the standards process even while they ensured that alternatives exist. The standards organizations might well be understood themselves as being user organizations, but their participants do not think of themselves as users for the most part

Has the situation of users changed, such that they now constitute a more important part of the standards process? An argument can be made that it has, because users have fashioned the new forums and consortia which now contribute significantly to the standards process by developing something akin to standards - which can be adopted directly by the formal standards organizations, in some cases - and/or by developing standards profiles.

It would be easy to overestimate their potential, however. The contradictions attendant upon participation from each type of user will not disappear just because the organizations involved are less tightly organized, less bound by consensus, and relatively speedy in their deliberation than the formal standards organizations. Even here, there remains little room for the end-use consumer; little potential for governments to act as major and model users, little incentive for major industrial users of the technologies to co-operate beyond a minimum level to support their individual market ambitions, and little reason to believe that the standards themselves are the most important product of forums and consortia. The forums and consortia, whatever their benefits, do not address the problems described here, problems which have traditionally confounded user participation in standardization, and which are likely to do so for the foreseeable future.

References

1. For discussion of recent developments, see Proceedings of the conference sponsored by CEN, CENELEC and ETSI, entitled "European Standards: Obstacle or Opportunity for the Inforrnation Society?" December 1995 and of the International Seminar on Standards Aspects of the Global Information Infrastructure, sponsored by ISO, ITU and IEC, Geneva, Janaury 246,1996, both of which were attended by the author.

2. The paper is based on a four year study of the politics of communication and information technology standardization, funded under a grant by the Social Sciences and Humanities Research Council of Canada. The study has involved interviews and participation at meetings of many standards organizations and forums, including private sector bodies. Further discussion of the issues raised in this paper is to be found in "The Housework of Capitalism" in the International Journal of Political Economy, Vol. 23, No. 4, Winter 199 43. Papers discussing "User Participation in Standards" were delivered at the Americas Telecommunications Standards Symposium in Rio, Brazil, in 1994 and at the Workshop on User Participation in Information Technology Standards sponsored by the Organization for Economic Co-operation and Development (OECD) in Tokyo in 1993. This paper represents an updated version of the analysis offered in these presentations.

3. The Australian Telecommunications Standards Committee has a section dealing with consumer interests. ATSC often contributes papers on user, i.e. consumer, needs in standardization to standards workshops and symposia. See for example, the contribution of the Australian delegation to the Global Standards Collaboration, meeting in Ottawa, Canada, June 1995.

4. See, for example, Minutes of the 24th General Assembly of the European Telecommunications Standards Institute. The meeting, held in November 1995, was attended by the author.

5. For example in Canada, the Standards Council of Canada sometimes supports industry participation in standards organizations.

6. For example, contributions to the private sector Joint Technical Committee 1 of the International Standardization Organization (.ITC 1/ISO) often come from participants employed by government agencies or departments and even in Committee T 1, a private sector body, officials from the U.S. National Institute of Science and Technology (NIST) or from the U.S military will play an important role as experts or contributors.

7. The following analysis is based on information gathered in conjunctions with a contract study undertaken by the author for the Treasury Board of Canada completed March 1996.

8. For example, both U.S Committee T I and ETSI undertook specific initiatives to educate and involve senior management without much success.

9. This is particularly evident in the recent initiatives taken in Europe in creating the ICT Board, which involves user organizations, to oversee the standards process, and in the OECD expert committee involving major users, which has operated in the last two years.

Video-Conferencing in a Classroom

Chetan Sankar, Professor
413 BB, (334) 844-6504
sankar@business.auburn.edu

Department of Management
Auburn University, AL 36849, USA

We introduced case studies and video conferencing-based projects to supplement lectures in teaching a graduate level telecommunications management course. The effectiveness of these methods in improving learning effectiveness was measured by administering questionnaires to students. Results ofthe analysis suggest that lectures were effective in imparting domain specific skills. Written case studies were more effective in imparting higher level cognitive skills, triggering learning interest, improving communications skills, and learning from others. Video-conferencing projects supplemented the other methods and were perceived as a secondary means of learning.

Introduction

Employers of business graduates claim that too many MBAs were great at narrowly focused skills such as accounting, finance, MIS, economics, but dreadful at diagnosing complex business problems, working in teams, or in motivating colleagues (O'Reilly, 1994). Now, they say, MBAs must be inspiring leaders, capable also of applying knowledge from a dozen different disciplines to unravel and solve problems commonly faced in industries. Many MBA students perceive that they learn theories from multiple disciplines, but have not been exposed to examples where these have been applied to solve "real-world" problems.

We teach a telecommunications management course at the graduate level, but, have problems in making it attractive to MBA students. This project's objective was to make the students understand the importance of telecommunications technologies in solving current business problems. The changing nature of the business environment means that managers can no longer rely on technical specialists to provide them with all telecommunications decisions. Business students must therefore understand how telecommunications in general and telecommunications-intensive-information-systems in particular are part ofthe strategic, tactical, and operational decision processes of an organization. The nature ofthe evolving environment means that a knowledge of telecommunications may mean the difference between achieving competitive advantage and ending in failure.

Although we were convinced ofthe need to teach this course at the graduate level, students and other faculty members perceived this as a domain-specific course teaching technical fundamentals relevant to engineering students. They questioned whether the course delivered higher order cognitive skills, such as identifying problems, choosing alternatives, and solving the problems. Therefore, we substituted new methods for the earlier methods of lectures, site visits, and term paper projects. We incorporated written case studies and video-conferencing based projects in the course besides the lectures. Specifically, using case studies and projects, we expected to introduce the following features (Barnes et al., 1994):

1. The primacy of situational analysis
2. The imperative of relating analysis and action
3. The necessity of student involvement
4. A nontraditional instructor role
5. The development of an administrative point of view.

0-7803-3552-X/96/$5.00 © 1996 IEEE

We integrated four written case studies and four video conferencing-based projects in MN602, Telecommunications Management class during Summer 1995. The effectiveness of introducing these methods was evaluated by administering questionnaires to the students. The results of analyzing the questionnaires provided results that could be of value to MIS and other faculty members teaching technology-based courses.

Research Model and Variables

A research model (Figure 1) was developed based on past research (Hingorani, 1995) showing how teaching methods could influence learning achievement. The variables used in the research are described next.

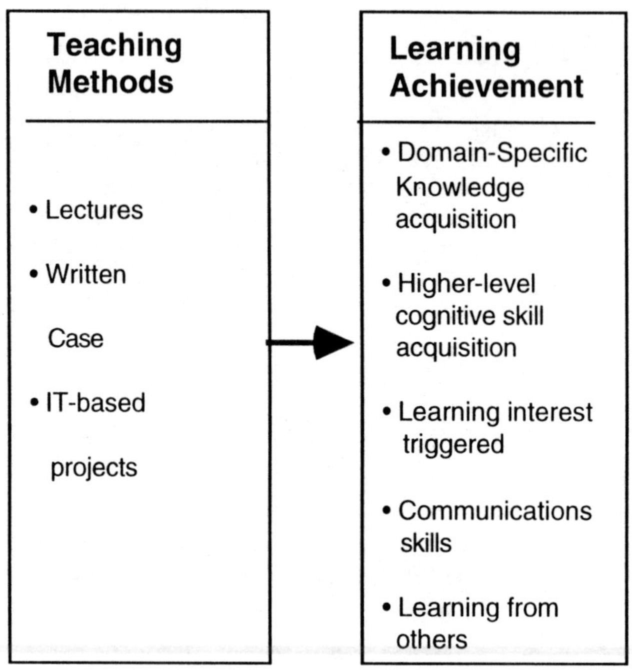

Research Model for Evaluating Teaching Methods

Dependent Variables

The dependent variable for this study was learning achievement. The learning achievement was measured using the variables: domain-specific knowledge acquisition, higher level cognitive skill acquisition, learning interest triggered, communications skills, and learning from others. These variables were chosen from the past hterature (Alavi, 1993; Masoner, 1988; Hiltz, 1988). Based on these past studies, the items under each ofthese variables were identified and used in creating the questionnaires.

Independent Variables

The three methods of instruction: lecture, written case studies, and video conferencing-based projects were the independent variables in this research. We created appropriate instruments based on the past research and validated the instruments using a pilot study conducted with other courses (Hingorani, 1995).

Research Questions

The objective of this project was to examine the effectiveness of the three teaching methods. The specific research questions were:

(1) Which method was more effective in imparting higher level cognitive skills? Compare and contrast lectures, written case studies, and video conferencing-based projects in improving learning effectiveness.

(2) Examine the written case studies and video conferencing-based projects in more detail. Which case studies and projects were more effective? Why?

Research Methodology

The subjects for this project were graduate students enrolled in MN602 - Telecommunications Management, during Summer 1995. There were 11 students enrolled in the course. Two were women and five students had bachelors in engineering. Two survey instruments were designed during past research to measure the effectiveness of the methods. A questionnaire containing sixteen items was created to measure learning achievement. The three methods were rank ordered based on these sixteen items. Another questionnaire containing eight items was created to measure the effectiveness of each case study and project on a Likert scale of 1 to 5. This was used to compare the written case studies and projects.

Design of the Course

Four case studies, Della Steam Power Plant, Credit Reporting Agency, JEI Disaster Recovery, and SHOW Accountants case studies were included as the written case studies. The Video conferencing-based projects were POTS-based video conferencing with a telecommunications manager, ISDN-based video conferencing used by placement services, ISDN-based video conferencing used by building science department, and home pages created for National Engineering Consortium.

Administration

This course was taught during 10 weeks of summer session to graduate students. Four project teams were created and the written case studies and projects were assigned to them. Twenty five percent of the grade was assigned for the written case studies and another 20% for the Video conferencing-based projects. The examinations counted for 40% and homework for the rest 15%. Questions were given to each project team for the case studies and they had to discuss their solutions in the class. The students frequently used Internet to supplement the answers to their questions. The questionnaires were completed by the students at the end of the quarter. Due to the small sample size, non-parametric tests such as the Friedman's tests were run to rank order the methods.

Results

The tables showing the results could be obtained from the author. Ranking the items under the higher level cognitive skill acquisition construct shows that written case studies provided the most value to the students in acquiring these skills individually and in total according to the table. The difference between this rank and the ranks for lecture and Video conferencing-based method was significant as indicated by the Friedman's test statistic. Lectures provided the most value in domain-specific skill acquisition and it differed from the other two methods significantly. Written case studies were effective in learning interest triggered, communication skills acquisition, and learning from others compared with the other two methods. In summary, for this subject population, written case studies were effective in acquiring many skills, whereas, lectures were most effective in acquiring domain-specific skills. It was interesting that the students did not rank the Video conferencing-based projects as the best one in acquiring any ofthese skills.

All the written case studies and projects were perceived to be worthwhile by the students. In particular, SHOW accountants case study was the most liked by the students. The other three shared similar responses. Among the Video conferencing-based projects, ISDN based video conferencing and NCF Home page projects were the best liked. Written comments from the students shows that they liked the mix of lectures, written case studies, and projects.

Discussion

Lectures and written case studies were the most important in improving learning for this course based on this limited sample. Written case studies were found be more effective in imparting higher level cognitive skills, triggering learning interest, improving communications skills, and learning from others. Lectures were effective in imparting domain-specific skills. The Video conferencing-based projects tied with lectures for triggering learning interest, improving communications skills, and learning from others. The finding that written case studies were perceived to be superior to lectures and Video conferencing-based projects for imparting many skills was a surprise. Traditionally, lectures, projects, and short case studies had been used in this course. This finding suggests that full-length case studies on telecommunications management need to be developed and used it in this course.

More specific information about written case studies emerges from the results in Table 6. SHOW accountants was a well-developed case study based on the practices in an accounting firm and has been presented at a conference (Whitman and Sankar, 1994) and is being published by the Business Case Journal. Therefore, the quality of this case study is high. Students preferred such a high quality case study in this class. JEI disaster recovery case was taken from the book by Carr and Snyder (1996). It was not a standalone case study. An improvement to this case study would make it more appropriate for this class. Credit report agency case study was an article in a journal (Gastley et al., 1992). The students were not able to understand the network diagrams in this paper and link those with the business objectives. Della Power Plant was a case study developed to teach vibration related issues to mechanical engineering students. The students wanted a detailed discussion of the telecommunications technology used in this plant and did not relish its current emphasis on vibration analysis. The results indicated that there is a high demand for case studies that integrate telecommunications technologies with business issues. Students prefer well-written case studies that meet the requirements of case study organizations such as the Harvard Business School Publishing or North American Case Research Association.

The students' responses to the Video conferencing-based projects are also important in understanding the role of projects in this class. This group of students perceived the projects to be a secondary means of learning in this class. The students wanted to work on the state-of-the art technologies and were impatient when systems did not work according to specifications. Three video conferencing technologies were used in this class. The students did not prefer the regular phone-based Teleview video conferencing technology due to shaky picture and poor user interfaces. Although this technology was much cheaper than the others, the students did not perceive the use of this technology to be exciting. Data communications using regular phone lines degrade after some time leading to shaky pictures and time differences between voice and picture on the screen. The students found these aspects to be very distracting and were not willing to tolerate the lowered quahty. They were a lot more impressed with the ISDN-based video conferencing systems based on AT&T and Intel products. The clarity of picture and use of current technologies seemed to sway them in choosing these projects as more worthwhile. This was contrary to the results obtained earlier in the use of AT&T Picasso Phones (Kramer et al., 1995; Sankar et al., 1995). In that study, the students preferred the Picasso Phones considerably compared with the written case studies. A reason for this might be that the Picasso Phone technology was simple and it worked without any breakdown. The video conferencing technologies were considerably complex and broke down often due to the need for interface between the computer hardware, video conferencing software, and the telecommunications links. Creation of standalone video conferencing systems such as Picasso Phones might be more appropriate if students had to be taught without getting distracted by technology glitches.

The students also enjoyed learning about the creation ofthe NCF home pages from a distance. Although the technological content was limited, this application seemed to capture the fascination of the students. These results suggest that students are critical in experimenting with new technologies and prefer those that are easy to use, provide better images, and do not fail. This result has deterred us from introducing new video conferencing technologies in the classroom until they are fully tested and found reliable. The results also emphasize that for video conferencing applications, use of digital lines such as ISDN might be essential.

Conclusions and Future Research

Research on use of innovative methods in instruction is not common. Most investments in technology to improve teaching have been made based on gut-feehng, rather than on detailed research studies. This research contributes by using research instruments to analyze the effectiveness of three methods in achieving the objectives of a telecommunications management class. A major limitation of this study is the small sample size. Although we could experiment easily due to the small sample size, the results would be more robust if this were repeated in similar classes. Use of these methods and instruments in other educational institutions would provide ability to interpret results across campuses. A future research issue would be to measure the personality styles of students, their GPA, etc., and connect them to the learning effectiveness.

The research also brings out the need to develop elaborate case studies in use of telecommunications technologies in businesses. Telecommunications is perceived to be a technical subject by many MBA students and use of case studies could make them understand the relevance of telecommunications in businesses. Use of new technologies such as video conferencing met with limited success in this class. Besides resistance to change, many of these technologies are in their infancy and are not robust. The students did not want to experiment with technologies they did not perceive to be state-of-the art and had a difficult time accepting technologies where quahty was lacking. The results indicate that it would take sometime before video conferencing technologies would be common in the classrooms. Further research is essential to understand the potential offered by video conferencing and multimedia in the classrooms and how effective they are in improving learning ofthe students. This study contributes by linking introduction of new technologies in the classroom with systematic evaluation oftheir benefits. It points out that creating and using written case studies in technology based courses in the Colleges of Business are essential. This could lead to improvement in higher level cognitive skill acquisition, communications skills, learning interest, and learning from others. As the Colleges of Business redefine their mission to provide higher quahty education to the students, introduction of new methods and their evaluation becomes essential.

References

Please contact the author for references.

Public-Sector vs. Private-Sector R&D In India: A Comparative Analysis of Two R&D Teams

Jayaram K. Sankaran
0064 9 373 7599, x 7357.

Department of Management Science and Information Systems
University of Auckland
Private Bag 92019
Auckland, New Zealand

V. Suchitra Mouly
0064 3 364 2639

Department of Management
U. of Canterbury
Private Bag 4800
Christchurch, New Zealand

The subject of this paper is a comparative analysis of two Indian R&D teams with similar objectives and activities. The team which we first studied (team A) was located in a public-sector electrical power research institute. The second team (team B) was the R&D unit of a private-sector company which manufactures and sells electrical equipment such as motors, generators, and transformers. Using qualitative methodology, we developed a process model of the ineffectiveness of team A. This model served as an interpretive framework with which to study team B and compare it with team A.

Introduction

The subject of this paper is a comparative analysis of two Indian R&D teams, one of which (team A) was located in a public-sector research organization (Institute X) while the other (team B) was the R&D unit of a private-sector manufacturing concern (Company Y). An earlier paper (Mouly and Sankaran, 1996) presented an in-depth, qualitative case-study of team A. The motivation for the study was to gain rich, qualitative insight into the relative failure of public-sector R&D in India. The major theme that emerged from the study was that a highly centralized administration, excessive bureaucracy, strong authoritarian hierarchy coupled with professional incompetence could seriously damage the quality of life and work of an R&D team. This adverse influence could manifest in severely strained interpersonal relationships, lack of group identity and low self-esteem of the team members, apathetic working lives, and inferior quality of work.

One can conceivably attempt to generalize these findings across public-sector R&D settings. However, the focus of this paper is not on the generalizability of the findings from the study of team A. Rather, it is on a comparative analysis of two R&D teams, one in the public-sector (team A) and the other in the private-sector R&D (team B), whose objectives, nature and scope of activities were similar. During the study of team A, members of the team and other R&D personnel of Institute X had cited team B as a model of a very effective R&D team, comparable with team A. Therefore, we studied team B to try and understand how and why a public-sector R&D team failed where a comparable private-sector R&D team succeeded.

The paper is laid out as follows. We first describe a process model of the ineffectiveness of team A. Then, we present a case-study of team B. Finally, we make a comparative analysis of the various factors that influenced the productivity and quality of work in the two R&D settings.

0-7803-3552-X/96/$5.00 © 1996 IEEE

A Process Model of the Ineffectiveness of Team A

We have developed a process model to conceptualize and understand better the major cause-effect linkages that could lead to the ineffectiveness of a public-sector R&D team such as team A. While we used this model as a starting point for our study of team B, the model is sufficiently abstract and stated in general terms in order that it can be tested across R&D teams located in public-sector institutions.

Prior to an explanation of the process model, it is essential to present a brief description of team A and Institute X. Institute X was founded by the government of India in 1960 to function as a national institute in power research and undertake development projects in the generation, transmission, distribution, and operation of electricity power systems. The R&D personnel were organized into eight divisions. Team A was a group of five R&D personnel that belonged to an R&D division within Institute X. One of the five team members was both the leader of the team and the division. The division had seven supporting staff members. The process model has emerged from the findings of the study of team A by Mouly and Sankaran (1992).

We have identified several factors of ineffectiveness, each of which operates at one or more levels, i.e. at the level of the individual, the group, and the institution. In the model, if two factors are related, then one of them is either an effect or an indicator of the other.

Owing to limitations of space, it is not possible to present the model here. The interested reader is invited to contact the authors for the same.

A Case-study of a Private-Sector R&D Team

We now present a case-study of team B. As mentioned in the introduction, we were motivated to study team B since members of team A had cited team B as a team with which they shared a lot in common, in terms of R&D projects. Further, team B and Company Y were considered very successful by professionals in both academic institutions and industry alike.

Background

Company Y was established in 1946 and undertook the production, manufacture, purchase, and sales of electrical equipment. It worked six days a week (while Institute X worked five days a week). Company Y had a central R&D wing which catered to the R&D needs of all the divisions of the company. The R&D wing had a total of 17 employees of which 5 were R&D personnel, the rest being support staff. The R&D wing was involved in developing new products, improving and testing the quality of materials and providing other services, such as marketing, to various project groups within the organization.

Each R&D project that came to the R&D division went through the stages of R&D innovation such as basic research, applied research, development and design, prototyping, and testing. Since Company Y was a private-sector undertaking where profits were an important consideration, the R&D projects were chiefly motivated by marketability and profit. Taking up all the stages of the project work not only speeded up the process leading to manufacture but also gave the R&D personnel control over their work. Most of the projects were non-sponsored, ie. they were independent research projects pursuing profitable product ideas. A few were specific projects occasioned by requests from the industrial clientele.

The R&D division was headed by a vice-president of Company Y. However, his role was mainly administrative. In fact, during the initial phase of our fieldwork in the R&D division, he was abroad on administrative work. The R&D team in Company Y essentially consisted of the Deputy Manager (who was the head of the team) and four Assistant Managers. The lower cadres of the R&D division comprised the support staff.

Methodology for the Study

Our process model of the ineffectiveness of team A provided us with an initial set of themes to investigate during our study of team B. We found that some of the major influences, either direct or indirect, on the ineffectiveness of team A and of Institute X as a whole, were the project selection process, the personnel recruitment process, the reward structure, the extent of involvement of subordinates in decision-making, the prevailing communication practices, and the external environment. Hence, our study initially focused on these aspects of team B as a group and of Company Y as a whole. We administered an open-ended questionnaire that pertained to the above aspects, to the members of team B. A communications study questionnaire (which was an improvised version of the one that had been developed for the study of team A) was also given to the team members. After this, a three month-long fieldwork was conducted in which we collected data through focused interviews and participant observation.

Again, for reasons of space, it is not possible to list the factors and discuss them at length. The interested reader is invited to contact the authors for a fuller version of the present paper which contains such a discussion.

Conclusion

We now present a comparative analysis of the various influences, at the institutional, group, and individual levels, on the productivity and quality of work of teams A and B.

Institutional-level influences

Since its primary role was to test and standardize electrical equipment, Institute X did not face direct competition. Being a government-funded institution, it did not have to concern with profits and market-position or image. This lack of concern was reflected in a lenient style of functioning as was evident in the recruitment process of Institute X and in the lack of concern of its R&D personnel with quality of work. On the other hand, since Company Y was a private-sector firm, profits and market position were crucial to its very survival. The image-consciousness and concern with productivity of the company was reflected in its recruitment and project selection processes, in its measures to imbibe total quality management (TQM) in its personnel and in its merit-based reward structure for team B members.

Group-level influences

Owing to their involvement in the project selection process, team B members could ensure that they had sufficient time and expertise for the projects that they undertook. Hence, neither time-pressures nor professional inability hampered their work. In team A however, projects were thrust exogenously without regard to the expertise and availability of time of the members. This affected the effectiveness of team A adversely.

In team B, members formed groups by themselves. Thus, members within a group shared interests and could work cohesively. The project leader, even if of a higher rank than his colleagues, was easily accessible to them. To their mutual benefit, project members held informal discussions and brain-storming sessions for "problem-solving purposes". Deadlines were met through peer pressure. Through informal channels of communication and interaction, project members had clarity regarding their respective roles.

In contrast, in team A, members had little in common and yet were forced to work together on all projects. Meetings to discuss project work were rigidly scheduled and regarded as perfunctory rituals. The excessive use of formal communication channels caused delays and loss of clarity among team members regarding their duties and responsibilities, and heightened the dichotomy between the superior and his subordinates.

Individual-level influences

Team B members were aware of their professional aptitude and competitiveness in the job market and were consequently, highly motivated. Their drive and enthusiasm towards work was also fostered by the autonomy granted by Company Y, and the merit-based reward structure and incentives for productivity provided by Company Y. The top-level management of Company Y openly acknowledged the high calibre of its R&D personnel and made them feel wanted for the growth of the company.

On the other hand, team A members were professionally incompetent and their awareness of their lack of competitiveness in the job market led to poor self-images and to apathy towards work. This indifference was aggravated by a total lack of individual autonomy at work and by the reward- and recognition-structure of Institute X which was not merit-based but aligned more with the organizational hierarchy. In fact, the team superior and the director of Institute X held low opinions of team A members, regarding them as mediocre technicians who merely executed what was assigned to them.

We would like to acknowledge that other researchers have identified factors of effectiveness and ineffectiveness similar to those described by us with reference to teams A and B. However, we emphasize that the significance of our comparative study is that it is grounded in in-depth investigations of two comparable R&D settings and not based on inferences from other studies of organizational effectiveness. Hence, for the sake of completeness, we conclude this paper with citations of a few, illustrative studies in the literature, that report findings similar to those from our studies of teams A and B. Through a longitudinal field experiment of job change, conducted over a one-year period with forty-two geographically dispersed railway track maintenance gangs, Pearson (1987) found that those who were engaged in participative goal setting reported higher perceived states of involvement in decision-making and greater job satisfaction. Similar findings were also reported by Morse and Reimer (1956). Coch and French (1948) found that participation improves group cohesion, commitment, and productivity. Similarly, Miles (1975) concluded from a study of R&D professionals that participative goal-setting or MBO may be appropriate for reducing both job conflicts and ambiguity.

References

Coch, L., and French, J.R.P. Jr. "Overcoming resistance to change." *Human Relations* 1: 512-532, 1948.

Miles, R.H. "How job conflicts and ambiguity affect R&D professionals." *The International Journal of Research Management* XVIII: 32-37, 1975.

Morse, N.C., and Reimer, E.E. "The experimental change of a major organizational variable." *Journal of Abnormal and Social Psychology* 52: 120-129, 1956.

Mouly, V.S., and Sankaran, J.K. "An Ethnography of an R&D Team in an Indian Public-Sector Research Organization." Presented at the Second International Conference on Organizational Discourse: Talk, Text, and Tropes, The Management Centre, King's College, London, 1996.

Pearson, C.A.L. "Participative Goal Setting as a Strategy for Improving Performance and Job Satisfaction: A longitudinal evaluation with railway track maintenance gangs." *Human Relations* 40(8): 473-488, 1987.

Infrastructure Requirements for Selected Management Practices

Ahmet Satir
(514) 848-2974

Concordia University
Department of Decision Sciences and M.I.S.
Montreal, Quebec, H3G IM8
Canada

Three management practices (Business Process Reengineering, Total Quality Management and Just in-Time) are revised in the context of a proposed business mission and operating paradigm for production enterprises. The infrastructure requirements for these practices are presented with respect to macro, operational and organizational requirements.

Business Mission and Operating Paradigm

Landscape of production in industrialized countries has been significantly altered especially since the early 1980's, both in terms of hard technologies and management philosophies used. The arena has been proliferated by hardware, software and management thinking / approaches with acronyms for each, such as CIM, FMS, MRP-II, OPT, TQM and JIT. Although naturally open to further developments and innovations, many of these technology and management tools in the marketplace have matured enough to be contained in a framework of an emerging business mission and an operating paradigm for production enterprises.

The rationale inherent in the emerging mission and paradigm proposed is based on three fundamental shifts that occurred over the last fifteen years in the management of (leading edge) production enterprises. These shifts are given in Table 1. It is these shifts and other historical developments in the implementation of hard manufacturing / information and soft management technologies that are instrumental in developing a new business mission and an operating paradigm for production enterprises. The proposed mission and the paradigm are defined in Table 2.

Shift #1:	from "Economies of Scale" to "Economies of Scope", and eventually to "Economies of Integration", with first degree implications in: • organizational structures • marketing / customer order servicing • logistics • costing • technology upgrading and investment appraisal • performance metrics
Shift #2:	from "Hierarchical / Functional" to a more "Participative / Interactive", and eventually to "Fluid / Flexible" networked organizational structure, with first degree implications in: • organizational structure • organizational behaviour and culture • job contents • product / process design • information flow and decision making
Shift #3:	from a "Unit Environment" to a "Chain Environment" concept of production, with first degree implications in: • marketing / customer order servicing • product / process design • supplier relations • logistics • information flow and decision making

Table 1. A summary of fundamental shifts in the management of (leading edge) production enterprises (Şatır 1996)

Business Mission for Production Enterprises The production enterprise strives for creating and exchanging value to stakeholders along its supply segment with the ultimate objective of total customer satisfaction. This is to be achieved at a sustainable wealth creation for its shareholders and, at the same time, meeting the expectations of its societal segment. Operating Paradigm for Supply Chain Management All activities and operations in a production enterprise should be designed for value creation and executed for value addition towards the ultimate objective of "opportunity cost minimization" along the supply chain, rather than "absolute cost minimization".

Table 2. Proposed business mission and operating paradigm (Şatir 1996)

Opportunity Cost

The key phrase in the proposed paradigm is "opportunity cost". In the context of the operating paradigm proposed, it is defined as the cost of operational waste remaining in terms of delays, defects and stocks at various stages of supply chain. The choice of opportunity cost drivers are justifiable. Inflated cycle and lead times as a result of delays, prespecified defect rates built into budgets and, hence into aggregate production plans (and eventually into people's minds) and inventories at all stages acting as "buffers" provide a thick blanket to cover up operational inefficiencies across the supply chain.

Recent management practices like BPR, TQM and JIT have to be approached in this context. These should be perceived as tools of minimizing the overall opportunity cost, not merely reducing the absolute unit cost. In conventional cost accounting, the unit cost is taken to be a function of direct, indirect and overhead costs. As such, it is a rigid measure of control and an indication of performance against budgeted targets. On the other hand, opportunity cost is a function of cycle / lead time delays, defects and stocks. As such, it is a relative indicator of performance measured against a zero base. This shift in emphasis in cost measurement and control requires corresponding changes in cost accounting. Recent developments in the practice of accounting (such as activity-based costing) are encouraging in this respect.

Measurement and reporting of opportunity cost, flow of this information and its use in decision making should be treated as a high priority managerial design issue. The measured levels of this cost in an enterprise would be inversely proportional to the volume of value creation activities. High levels of opportunity cost over a sustained period of time would be an indication of poor performance in value creation. Lack of improvement in opportunity cost, even at the low end, would signal that continuous improvement efforts are fizzling out. In fact, it is this very concept of"opportunity cost" that constitutes the linkage between the business mission and the operating paradigm proposed in Table 2. (Satir 1996).

Infrastructure Requirements for BPR, TQM and JIT

Overall Thrust

Business Process Reengineering, Total Quality Management and Just-in-Time are among the three most frequently reported management practices in dealing with operational waste (opportunity cost) reduction. However, these practices overlap to a large extent in terms of their objectives, domain and tools (Aggarwal 1995). Therefore, prior to discussing the infrastructure requirements for BPR, TQM and JIT, the overall thrust for each one of these practices has been redefined in Table 3. Such a differentiation would also facilitate the grouping of techniques and tools under each practice. The infrastructure requirements for each technique/tool are argued in the next section.

(Process Focused) Overall Thrust of Business Process Re-engineering Radical redesign of core business processes for a significant improvement in enterprise's competitiveness
(Customer Focused) Overall Thrust of Total Quality Management Evolutionary undertaking involving continual and incremental quality driven improvements in enterprise's goods and service offerings to internal and external customers
(Material Flow Focused) Overall Thrust of Just-in-Time Practices Waste elimination along the supply chain through drastic reductions in stocks at all levels

Table 3. Overall thrust of BPR, TQM and JIT

Infrastructure Requirements

Infrastructure requirements that need to be satisfied for a successful implementation of a given BPR, TQM and JIT practice are given in Table 4-6. The requirements are classified into macro (Table 4), organizational (Table 5) and operational (Table 6) issues. In each table, the relevant techniques / tools for each practice is classified in line with the focus associated with that practice in Table 3. The letters "H" (high), "M" (medium) and "L" (low) denote the degree of impact that the existence of a particular requirement have for a successful implementation of the technique / tool in question.

Macro requirements relate to the external environment, whereas the other two classes of organizational and operational requirements are internal. External environment related requirements are taken to be market and supplier based. Demand stability (in terms of range of product mix and individual volumes) and economic stability (in terms of macro balances and price stability) are the two market related requirements. On the other hand, the relative strength of supplier base (in terms of core competencies, technological capabilities and managerial skills) and transportation infrastructure (including the proximity of suppliers) are the two supplier base requirements. The headings for organizational and operational requirements presented in Table 5 and 6, respectively, are self explanatory, and hence, will not be elaborated on.

	MACRO REQUIREMENTS			
PRACTICES	**demand stability**	**economic stability**	**strength of supplier base**	**transportation infrastructure**
BPR Practices (process focused)				
•process identification/mapping benchmarking/design				
TQM Practices (customer focused)				
•design based practices				
•quality function deployment	L			
•concurrent engineering			M-H	
•statistical process control				
•empowerment based practices				
•continuous improvement				
•quality circles				
•source autonomous inspection				
JIT Practices (material flow focused)				
•scheduling based practices				
• lead time reduction	H	M-H	H	M-H
• load leveling	H	M	H	
• Kanban scheduling	H		H	
• set-up reduction				
•procurement based practices				
• stock reduction	H	H	H	H
• supplier certification	H	M-H	H	H

Table 4. BPR, TQM and JIT practices vs. macro requirements matrix

PRACTICES	ORGANIZATIONAL REQUIREMENTS						
	strategic top management committment	cultural infrastructure	networking	fluidity (white collar)	empowerment	multiskill training (blue collar)	cooperative union
BPR Practices (process focused)							
*process identification/mapping benchmarking/design	H	M-H	H	L-M			H
TQM Practices (customer focused)							
*design based practices							
•quality function deployment	H	M		M-H			
•concurrent engineering	H	M	H	H	H	L M	L-M
•statistical process control					H		
*empowerment based practices							
•continuous improvement	H	M-H		H	H	H	
•quality circles	H	H	H		H	H	H
•source autonomous inspection		H			H	M-H	H
JIT Practices (material flow focused)							
*scheduling based practices							
• lead time reduction	H		H				
• load leveling	H	L-M			L	H	H
• Kanban scheduling	H						
• set-up reduction	M				L		
*procurement based practices							
• stock reduction	H		M		M		
• supplier certification	H	H	H	M-H	H		

Table 5. BPR, TQM and JIT practices vs. organizational requirements matrix

PRACTICES	OPERATIONAL REQUIREMENTS					
	effective Layout	**appr. manu-facturing technology**	**established purchasing & stock control**	**established quality control**	**established production planning**	**smooth info. flow & com-munication**
BPR Practices (process focused)						
*process identification/mapping benchmarking/design						M-H
TQM Practices (customer focused)						
*design based practices						
•quality function deployment		M		H		H
•concurrent engineering		M		H		H
•statistical process control		M	M	H		
*empowerment based practices						
•continuous improvement		M		H	M	H
•quality circles		M		H	M	H
•source autonomous inspection		M		H		
JIT Practices (material flow focused)						
*scheduling based practices						
• lead time reduction		H	H	H	H	H
• load leveling		M H	H	M	H	M
• Kanban scheduling		M-H	H	H	H	H
• set-up reduction		H				
*procurement based practices						
• stock reduction		M-H	H	H	H	H
• supplier evaluation					H	

Table 6. BPR, TQM and JIT practices vs. operational requirements matrix

Conclusions

Implementation issues in BPR, TQM and JIT are widely reported in the literature. Among recent publications, one can refer in this respect to (Aggarwal 1995), (Carr and Johansson 1995), (Cordeiro and Turner 1995), (Ramarapu et al 1995) and (Maull et al 1995). In this paper, infrastructure requirements for the implementation of various techniques / tools of these practices are presented from macro, organizational and operational perspectives. In a sense, Table 4-6 provide a checklist for requirements that need to be satisfied in order to successfully implement BPR, TQM and JIT practices.

Among the three practices, JIT is the one most affected by the macro requirements. Hence, one can implement, to a large extent, BPR and TQM based practices even if one does not have the macro infrastructure elements given in Table 4. However, the market and supplier base related macro requirements are essential (with a high degree of impact (H)) for many of the procurement and scheduling practices of JIT.

Strategic top management committment comes out in Table 5 as a top priority organizational requirement for all of the three practices. Empowerment of employees is also a high priority prerequisite especially for successful implementation of TQM practices. Cultural infrastructure in an organization that is conducive to change is another crucial organizational requirement especially for the implementation of TQM practices, among the three management practices studied.

In terms ofthe operational requirements, Table 6 indicates that it is the JIT practices that are highly dependent on the satisfaction of these requirements. Established purchasing / stock control, quality control, production planning and smooth information flow / communication are all crucial in the implementation of, especially, JIT and, to a lesser extent, TQM techniques / tools. The ascertion here is that JIT and TQM are more of perfectionist approaches that require the existence of a certain operational infrastructure as a prerequisite. In that sense, JIT and TQM practices are not the appropriate approaches to implement in order to tackle a faltering quality or stock situation. They rather provide the philosophy / techniques / tools to develop a stronger structure, build upon an already functioning (infra)structure. On the contrary, BPR requires only the operational requirement of smooth information flow and communication within the organization, and not necessarily the other requirements given in Table 6. This is in line with the process focused orientation of BPR practices which are exercised to radically change the quality control, purchasing or any other process that is deemed to be inefficient and ineffective.

References

Aggarwal, S. "Emerging Hard and Soft Technologies: Current Status, Issues and Implementation Problems." *Omega*. 23(3): 323-339, 1995.

Carr, D.K., and Johansson, H.J. *Best Practices in Reengineering*. Mc-Graw Hill, New York, NY, 1995.

Cordeiro, W.P., and Turner, R.H. "20/30 Hindsight: Managers Must Commit to TQM." *Interfaces*. 25(3): 104-112, 1995.

Maull, R.S., Weaver, A.M., Childe, S.J., Smart, P.A., and Bennett, J. "Current Issues in Business Process Re-engineering." *Int. J. Opr. &Prd Management*. 15(11): 37-52, 1995.

Ramarapu, N.K., Mehra, S., and Frolick, M.N. "A Comparative Analysis and Review of JIT "Implementation"Research." *Int. J. Opr. &Prd Management*. 15(1): 38-49, 1995.

Satir, A. "Business Mission and Operating Paradigm for Production Enterprises: A Framework". *Omega*. (under editorial review) 1996.

Impact of Client/Server Computing on the Telecommunications Industry

Salvatore P. Savino
(908) 382-5400, x383

Beechwood Data Systems
100 Walnut Avenue
Clark, New Jersey 07066

Steven M. Queroli
(908) 382-5400, x283

Beechwood Data Systems
100 Walnut Avenue
Clark, New Jersey 07066

Client/Server computing splits a computer application into tasks that are performed using three separate components. These components consist of a client computer, a server computer, and a network which connects the client to the server, where in most cases the client is a personal computer sitting on a user's desk. Client/Server technology touches just about every part of the computer market, particularly systems management. The purpose of this paper is to summarize some of the most popular models of Client/Server computing, and to look at how these models provide solutions for today's business needs, with emphasis on the telecommunications industry. This paper also discusses the benefits client/server technology provides to companies restructuring their management and operations to achieve localized autonomy, with the emphasis on re-engineering business processes that might otherwise be ineffective or impractical to automate.

Introduction

The role of the desktop computer is shifting from the stand-alone personal computing system to that of the "Universal Client". In this model, an intelligent, programmable workstation provides the appearance of a single point of access to an integrated set of information and processing resources. Many companies are migrating from their mainframe-based data centers towards giving processing power to their business units and departments. The tightly-coupled, vertical mainframe application is being replaced by three-tier Client/Server solutions. The migration is intended to empower departments and end users alike and make companies more competitive and responsive to market dynamics.

The model for an effective Information Technology infrastructure is a network computing environment. This environment will result from a convergence of the network system and client/server applications. These applications will empower employees throughout the enterprise by providing instant access to corporate information. As
a result, companies will be able to strengthen their competitive positions, enhance customer service, and streamline staffs. In addition, client/server computing is very adaptable to the profound organizational changes companies are undergoing as enterprises break into smaller, independent business units.

Overview of Client/Server Computing

Client/server computing is a processing model where a single application is partitioned between multiple processors. These processors are comprised of three basic components: the client, typically referred to as the front-end and usually consisting of a programmable desktop computer; the server, typically referred to as the back-end, and consisting of a minicomputer or a mainframe; and a network. The network component, such as a LAN (Local Area Network) or WAN (Wide Area Network), connects the client to the server. These processors cooperate, transparent to the end user, to complete the processing as a single unified task. Figure 1 depicts a typical client/server computing architecture found in most telecommunications companies today. This three-tiered architecture provides for a legacy application and data warehousing layer accessed by the servers. This architecture is recursive; servers can

0-7803-3552-X/96/$5.00 © 1996 IEEE

become clients and request services from other servers on the network or the data warehousing layer. This architecture provides a higher level of processing over the more primitive host-based environment which does not have any distributed application processing capabilities. Host-based application processing, depicted in Figure 2, is performed on one computer system with attached unintelligent, "dumb" terminals.

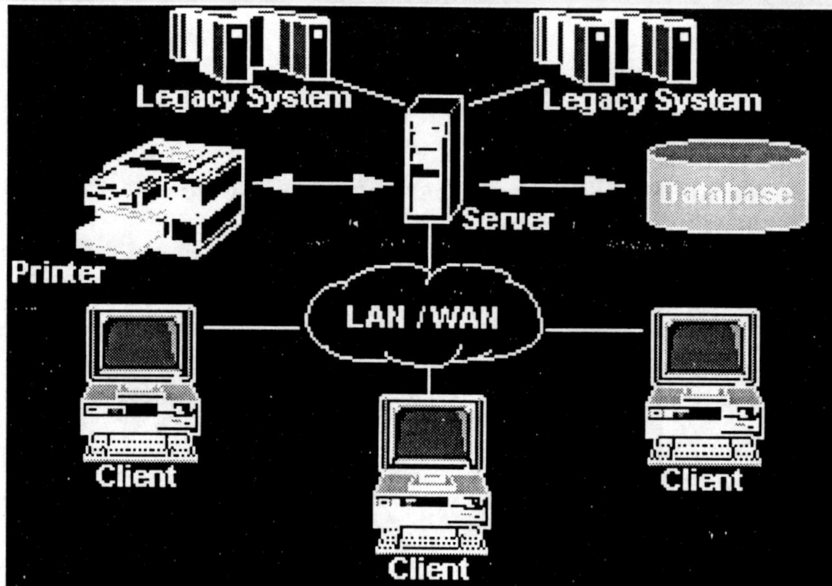

Figure 1: Typical Client / Server Computing Architecture

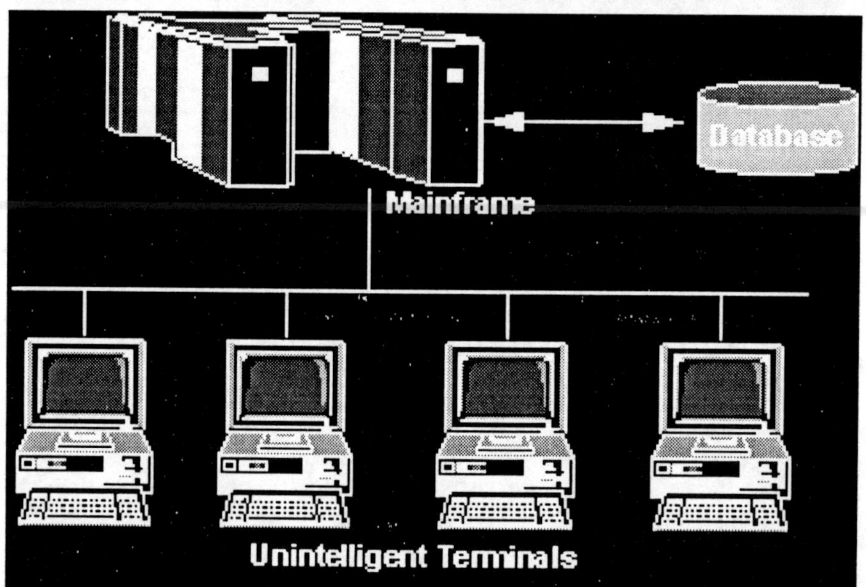

Figure 2: Host-Based Processing Environment

In client/server computing, requests are submitted by a client, or requester, to the server which processes the requests and returns the results to the client. The client software may perform some application logic on the results of the request before passing it on to the presentation component of the software. The presentation component produces the interface with which the user interacts. It is often a graphical user interface (GUI), which provides a graphic-oriented front end to applications. As such, client/server computing creates a network of processors that work together to complete business transactions. This system also permits computing intelligence to be distributed throughout the enterprise at selected scale and functionality levels, and to be unified and interconnected.

Application processing performed on more than one machine in a network may be either distributed computing or cooperative processing. Distributed computing partitions data between two or more computers, which may be geographically dispersed. The user has transparent access to the data. Cooperative processing splits an applications processing between two or more discrete computing systems. Client/server cooperative processing is therefore really a special form of distributed processing, which has emerged as a higher level of shared device processing.

Specialized Client/Server Models

Client/server computing can be grouped into five types of special cooperative processing models: presentation logic functions consisting of distributed presentation and remote presentation; distributed logic; remote data management; and distributed database. These models are briefly described as follows.

Presentation functions are directly related to end-user input/output. Presentation functions perform such tasks as screen formatting, reading and writing of screen information, window management, and keyboard and mouse handling. These functions are typically performed on the client system. The manner in which these functions are separated from the rest of the application determines the two styles of presentation: distributed and remote. *Distributed Presentation* is where a portion of the user interface logic is located on one node, while the rest of the presentation together with the remainder of the application is located on another node. Distributed presentation is especially beneficial when used with personal computers to deliver graphical user interfaces to mainframe applications. In a *Remote Presentation* the presentation part of the application code is placed entirety on one node, while the rest of the application is placed on another node. This style of presentation is best suited for certain types of non-conversational applications where end-user interactions are predetermined.

In a *Distributed Logic* model, the business functions are split so they can be placed on different nodes or systems. The execution of the business transactions is performed as a cooperative effort between all business logic components with the cooperative participation of both presentation and data management functions. Distributed business logic functions are well suited for complex, highly interactive, and database intensive applications.

Data management logic consists of two components: data processing logic (the application code which manipulates data within the application), and the database processing (actual processing of the data). The database processing functions must reside on the same system as the database itself, while the data processing logic may be placed close to the application business logic. When all data and data processing for a given application reside on a single system separated from the application node, it is considered to be a *Remote Data Management*.

Databases are designed to provided independence between applications and data. *Distributed Database Management* deals with databases distributed among multiple nodes. Such distribution provides for higher data availability by placing multiple copies of critical data at different locations, and eliminates single point of failure.

The Fit Between Client/Server Technology and the Telecommunications Industry
Like many other enterprises, the telecommunications industry is undergoing reengineering, or the process of re-thinking how a business operates in an effort to improve its efficiency. Client/server computing is often involved with reengineering projects as a means to quickly implement these new solutions while still protecting the investment in legacy applications. These legacy systems met the need of a specific functional area but did little to provide information for other functional areas. Given

that certain pieces of information were not easily attainable, each system would keep the information it needed to perform it's functions. This practice led to the creation of data and functional redundancies in the telecommunications industry.

The practice of reengineering the business processes attempts to tear down these functional barriers and provide process-concentric view of the enterprise. The most common implementation strategy to meet the reengineered objectives is to not replace the legacy applications, but rather to front-end them with Unix-based servers which implement the reengineered business functions through accessing the multiple legacy systems.

In addition to being able to reengineer the business processes, the introduction of client/server based solutions to front-end legacy systems provides an additional benefit to the user community. No longer is it necessary to train the user on several legacy applications, but rather only on the use of the new, typically GUI-based solution. This solution is typically presented in a model that mimics the business process and not the often arbitrary, function-delineated legacy systems it front-ends. This has resulted in tremendous saving of training and increased customer satisfaction.

Working closely with users and customers to develop business solutions is the basic definition of business-driven application development. Client/server applications forester business-driven development because they facilitate a company's effort to deploy business reengineering, which can lead to orders of magnitude in operational improvements. Examples of such improvements are seen in the following examples.

Customer Service Application

Consider, as an example, the problem faced by a maintenance center service representative. The representative receives a 611 call from a customer who is having trouble with call waiting and they believe it is not working. Prior to the introduction of client/server solutions, the service representative had to determine which legacy systems needed to be accessed, log into those systems, enter information on a number of screens, translate codes (FIDs and USOCs and switch codes) and compare results. At a minimum, the representative would need to access a system to determine if the telephone number that is experiencing the problem is the main billing telephone number, obtain the billing telephone number, access the billing system to see if the user is actually being billed for the call waiting service, access the switch that services the user to determine what parameters are set for the user's line and finally, access the service ordering system to determine if there is a pending order for this change. Each of these accesses would be to a different legacy system, each utilizing different codes and information from one system which would be used as input to the next system. It is not uncommon for a service representative to require 6 months of training before they are able to complete the tasks necessary to diagnose the problem (let alone fix the outage).

A reengineered solution for the service representatives would be to provide a server computer that could navigate each of the legacy systems and a client system that will provide the presentation logic. In this example, the representative only needs to enter the telephone number of the calling party (which can be pre-populated with the Caller-ID number provided with the incoming call). The system would then go out and initiate the request for the proper telephone number. In parallel, the system would also access the billing system, the switch and the service order processor. Once results are obtained from these systems, the client program will display any discrepancies between the bill and the switch and allow the user to point and click at the screen to correct the problem.

The training time for representatives of the reengineered client/server solution was cut from 6 months to 6 hours. The platform that provided this service was then expanded to handle new provisioning as well as maintenance.

Sales Support Application

Major trends in the communications industry, particularly deregulation and competition, are forcing the creation of mechanized interfaces between businesses to drive down expense and push up revenue. The most important challenge is being able to meet both the competition and the demand which presently exists in the communications marketplace.

One way to meet these challenges is to employ authorized sales agents with electronic access to customer service and support systems. The business objectives of the agents is to decrease call volumes to the service provider, improve sales efficiency and increase the general business market sales revenue. In addition, agents are positioning the companies they represent in the marketplace as the telecommunications provider of choice. In order to meet all these objectives, agents need to be in constant contact with their providers in order to obtain customer data , initiate service orders and verify order status, initiate and verify trouble status, and request other types of pertinent information. This dependency on the telecommunication companies has created a significant lapse in time between when a customer initiates a request for service/information, and when the agents can respond back to the customers. The interaction between the telecommunication company and its agents is effectively accomplished by the development of a centralized sales/account database, and a client/server application which allows agents at different locations to gain access to authorized data. This access can be achieved either through the use of client workstations in the agent offices connected to the host via a network, or through dial-in network access to a remote LAN access server. This solution provides an estimated reduction of 15,000 calls per month to the telecommunication companies by the agents. With less follow-up calls and more time to sell, agent revenue is estimated to increase by 15 million dollars over the next several years. In addition, resources at the telecommunication companies, which were handling agent calls, can now be reinvested to other business needs.

Enhanced Customer Service

Another client/server solution was developed to improve the level of customer support. A telecommunication company relied on it mainframes and legacy systems to respond to customer service requests for billing inquiries and the addition, deletion or modifications in the customer's telephone services. However, this level of support could not be provided during the weekends, when the mainframe systems were undergoing maintenance. A client/server application enabled the telecommunication company to become a state-of-art company by providing customer service support to its clients on a 7 day per week, 24 hour basis. This was accomplished by replicating the customer service and billing records to a centralized database in a client/server architecture. The duplicate database was kept current with up to the day updates to the customer records. During the weekend mainframe downtime period, customer service representatives throughout the region serviced by the telephone company would log into the client server application through a graphical user interface on their client workstations, and continue to respond to customer requests, all transparent to the customers. When the mainframe systems were bought back on line, all information processed with the client/server application was uploaded to the mainframe systems. Consideration is now being given to replace the mainframe systems completely with the client/server technology. This application enabled increased end-user sophistication and increased customer satisfaction.

Other Benefits of Client Server Computing

The above examples indicate the magnitude of savings and improved productivity which can be realized by implementing client/server solutions. The software in the Servers in these examples became valuable assets of the corporation. Traditionally, software development had been viewed as a sunk cost and, as soon as the software is released, the annual costs of maintaining it will be viewed as an assumed liability. With the client/server approach, this software is a collection of reusable components of the overall business model and can be used over and over again in different applications. Every time an object is improved through maintenance or enhancements, the value of every other asset or application that uses it becomes more valuable. Additionally, the cost of maintenance can reasonably be amortized across all of the affected assets and applications.

Significant cost savings on hardware expenditures and maintenance costs relative to mainframe-based environments can be identified. Client/server allows organizations to protect current investments by using existing equipment and protecting future investments by using scaleable, expandable products. Client/server applications with reusable software assets can also be developed, implemented and tested in less time than mainframe applications. The users are more involved in the development of the process and in control of the application, once operational. Increased productivity is also a result of the close fit between the system design and the way users actually perform their job.

Conclusions

Client/server computing models, especially when combined with a software asset repository (for reusable client/server components), provides a better way to solve today's business needs. This technology empowers the end user and permits the creation and quick deployment of solutions as needed to solve business problems. By implementing client/server computing, organizations should be able to reduce costs, leverage their personal computer investments, create software assets, deliver valuable services to end users, and improve productivity. Predications are that by the end of the decade, downsized client/server systems will account for nearly half of the applications of major companies. Recent trends in the computing industry, including advances in hardware and software, evolution of local networks into metropolitan area networks and global area networks, and the move away from traditional mainframe applications, all indicate that client/server computing will remain a major force in the computing industry for some time to come.

References

Chivvis, A.M., Downsizing: The Business Decision," LAN TIMES, p. 52. May 20, 1991

Margulies, E., "*Client Server Telephony*". Publisher: Telecom Books, New York, NY August, 1994.

Musich, P., "Client/Server Wave to Continue Unabated, Study Says." *PC Week.* p .52. April 20, 1992.

Pepper, J., "Decentralize? Centralize? Yes!", *Information Week*, p. 34. June 29, 1992.

Scot, K., "Enterprise Computing," *INFOWORLD*, p. S63. February 10, 1992.

The Learning Laboratory: Supporting Learning Organizations with Agent Systems

Detlef Schoder
Ralf E. Strauss
Tom Hummel

Albert-Ludwigs University of Freiburg
Institute for Informatics and Society
Department of Telematics
Friedrichstrasse 50
79098 Freiburg / Germany
email: {schoder|rstrauss|hummel}@iig.uni-freiburg.de

This article discusses how an organizational information market or n:m-communications can be supported through CSCW-applications including e-mail, bulletin board systems and newsgroups. Due to their limitations it is argued that there is a need for new functionalities aimed at supporting n:m-communication. The Information Trader Concept as an actor-based approach is finally introduced, which may serve as an adequate concept for improved support of n:m-communications.

Competitive Advantages in the Learning Laboratory

Programs for continuous improvement and to institutionalize innovation are sprouting up everywhere as organizations strive to improve themselves and gain a competitive edge. Approaches in organization theory such as the "systems-oriented management" or, most recently, the "Fifth Discipline" support this change in philosophy. Both theoretical and practical approaches can be traced back to the fundamental insight that an enterprise has to be understood and designed as a learning organization existing as a network of self-optimizing and self-organizing circles rather than as deterministic (tayloristic) machinery. Some approaches even go so far as to describe the whole organization as a *learning laboratory*, where the factory is changing to a locus of continuous innovation (Leonard-Barton 1992).

The obstacles to defining a conceptual framework start with the notion itself. A clear definition of organizational learning and a concrete concept or framework has proved until now to be elusive (Daft/Huber 1987). As a conclusion out of the different perspectives, most definitions view organizational learning as an organizational process that unfolds over time and links up with knowledge acquisition and improved organizational performance (Duncan/ Weiss 1979). Ongoing empirical experiences and observations change, expand, differentiate, intensify or even completely modify the individually and organizationally stored experience, knowledge, and presentation of the enterprise (theory-in-use). Although individuals are the "learning agents" through whom learning in reality takes place, the process on a macroscopic level is influenced by a much broader set of variables working interdependently, such as the sharing of beliefs, knowledge, assumptions and rules.

As speed to market becomes the central criteroin of competitive success, the need arises to create an organization which is dedicated to knowledge creation, collection, and control – in other words, to realize organizational learning in practice. While in traditional factories any innovation is associated with research laboratories and development organizations far from production, the idea of a *learning laboratory* is that tremendous amounts of knowledge and skills are embedded in physical equipment, processes, and people in the business processes on the floor level. More important than mere technical aspects are the managerial practices and underlying values that constantly renew and support these knowledge bases. According to the interpretation of the organization as a self-organizing system, the learning laboratory cannot be constructed piecemeal, but as an organic whole, which is designed, created, and maintained through the constant communication between the members of the organization and the permanent prepareness to adapt to changing environments.

Leonard-Barton distinguishes 4 critical activities for a learning laboratory: problem solving, internal knowledge integration across functions and projects, innovation and experimentation, and the integration of internal and external information flows (Leonard-Barton 1992). Therefore, communication and information integration is at the bottom of the learning laboratory (Huber 1991, Duncan/Weiss 1979).

A closer examination of the related literatures indicates that much has already been learned about organizational learning (OL) and the contributing processes in the field of organization theory, but still there is a lack in conceptions focusing on the use of information technology. However, the literature concerning information distribution is rich and mature, but an aspect of information distribution that is central to an organization's benefitting from its learning, namely how units that possess information and expertise and units that need this information can find each other quickly and with a high likelihood, is still unexplored (Huber 1991, Constant/Kiesler/Sproull 1994). In terms of information technology, therefore, the relevant (interdisciplinary) research question is how OL can be supported with modern information technology – especially CSCW-applications – , and which new functionalities are needed (Daft/Huber 1987).

Organizational Learning and n:m-Communications

Starting from the standpoint of organization theory according to Huber, one might argue that if organizations are to learn, they must focus on

- lknowledge acquisition,
- torage of information ("organizational memory"),
- information distribution, and
- information interpretation and assumption sharing (Huber 1991, Duncan/Weiss 1979).

Most of these tasks are backed by communication processes between the members of an organization. The central necessity for organizational communication to enhance OL can be put into one sentence: Bring the right information in the right *time* and appropriate *quality* to the *place* where it is needed. Thus, idealistically an *n:m information distribution* or *communication* among organizational members should be achieved. In this vein, the underlying problem of this imperative can be coined in one basic question: How can an *efficient n:m-communication* be supported? The argumentation can be organized as follows: In any given organization there are individuals requesting information and individuals offering information at the same time. This situation can be characterized as the organizational information market. The final article briefly discusses how the organizational information market or n:m-communications can be supported through CSCW-applications including e-mail, bulletin board systems and newsgroups. Due to their limitations it is argued that there is a need for new functionalities aimed at supporting n:m-communication. The *Information Trader Concept* as an actor-based approach is finally introduced, which may serve as an adequate concept for improved support of n:m-communications.

CSCW and n:m-Communications

Four types of relevant interaction relations among members of an organization can be subdivided with different costs arising, 1:1 (known partners), n:1 (individual information request), 1:m (individual information offer) and n:m (organizational information market, Figure 1). For those who need information the adequate contact is often unknown, for those offering information it is additionaly difficult to see who is in need of what kind of information.

Information-source \ Information-seeker	1	m
1	*Well known Partners*	*Individual Information Offer*
n	*Individual Information Request*	*Organizational Information Market*

Figure 1: Communication relationship matrix

If we understand OL as a n:m communication problem, then the question arises, how, to what extent, and which IT-systems may contribute to supporting the organizational information market (Huber 1991)? For various reasons CSCW-applications including e-mail, bulletin board systems and newsgroups will cover n:m-communication situations (Markus 1994). CSCW-applications can generally be defined as computer-based systems that support groups of people engaged in a common task or goal and that provide an interface in a shared environment (Grudin 1994, Johansen 1988).

E-mail is mainly used for 1:1 and 1:m communications where the members of a group are usually known. However, if communication partners are unknown the distribution of informations in a "broadcasting" manner to all members of an organization may be considered. Obviously this leads to a high distribution of knowledge, but increases the "information overload" among the receivers who even may "flame" the sender, regarding his mails as junk mails (Grudin 1994, Constant/Kiesler/Sproull 1994, Sohl/Redding 1987). The implementation of thematical focused newsgroups, the electronic conservation of frequently raised discussions and questions (*"frequently asked questions, FAQ"*), or even the structuring and filtering of incoming informations as in the *"Object Lens System"* (Lee/Malone 1990) may help to solve these problems. In this context newsgroups and bulletin board systems relatively best support n:m-communications. However, these applications have limited functionalities for searching, filtering, and qualifying/assessing incoming data and news. The starting-point to prevent an "information overload", and also for the parallel, automatic, structured, high qualitative information and knowledge acquisition, distribution and adequate presentation may evolve through the use of *intelligent agents* (Chen 1995, Maes 1994).

Supporting n: m-communications: The Information Trader Concept

In any given organization there are individuals requesting information and individuals offering information at the same time. This situation was characterized as the *organizational information market*. A promising approach would be to support the existence of information markets. To be precise, "market" is not the correct metaphor, because there is no real pricing mechanism. There is, however, supply and demand of information which needs to be matched at minimal cost and effort for the individuals affected. A possible concept which might be able to support the n:m communication at low cost could be the *Information Trader*. Figure 2 illustrates the idea of the Trader. Every user is supposed to have his own personal agent which is, however, not tied to the users session.

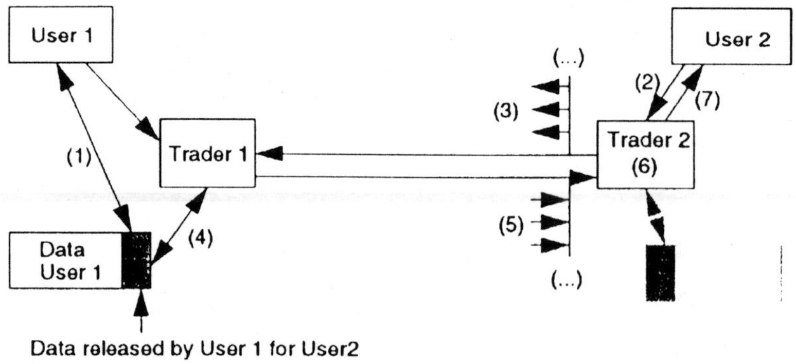

(1) Defined Release of Information through User 1
(2) Search Request for Information by User 2
(3) Request from Trader 2 to other Traders
(4) Trader 1 finds fitting and released Information and passes it on
(5) Other Traders also pass on Information to Trader 2
(6) Preselection / Prefiltering of Information through Trader 2
(7) Proposal of Information through Trader 2 to User 2

Figure 2: The Information Trader Concept

Suppose User 1 releases information to Trader 1 for trading, so that the system can pass on the information to other users autonomously. Normally, a user will not release all his information, thus mechanisms need to be incorporated in the Trader which help to determine what information should be accessible to which person. One way to do this would be a simple access control list. Perhaps more convenient would be an organization model which contains the membership to groups, departments or functions of other individuals in the organization. Information may then be restricted to the specific groups or functions in the organization, e.g. the board of directors or the system administrators.

If User 2 needs information, he can start a search request through Trader 2, his personal agent. Trader 2 sends this request to all the other Traders accessible in the network of the organization. Trader 1, among others, receives the request, searches in the data and may find information which was released for the person who is the owner of Trader 2 (i.e. the person requesting). Trader 1 then autonomously returns all data which fits to the request given by Trader 2. The same happens with other Traders. Trader 2, after receiving all the information, then makes a preselection. Therefore, a valuation component is needed, which can be formed by the user who owns the Trader. The underlying assumption is that individuals in organizations normally have estimates concerning the quality of information received from specific sources in the organization. Information from sources that regularly provide high quality are preferred and in order to filter the information received, an individual then sort the information by quality, normally by using simple heuristics. Consequently, the Trader needs a valuation component to reproduce this action that needs to be flexible for the user' s wishes. Having finished the selection and filtering procedure the Trader can forward the information available to User 2.

The advantages of the concept are:

- it avoids information overload due to filtering which occurs at the agent level not at the user level,
- information seeking efforts are accomplished by the agent, only appropriate or – if specified – ambiguous data is forwarded to the user,

- agents may forward requests to other agents or data bases where the likelihood for finding appropriate information is higher. This helps to avoid the inefficient broadcast of requests to all members associated with a group.

The Information Trader Concept focuses on the exchange of information and expertise in terms of communication and data retrieval functionalities. It can also be extended through artificial intelligence (Johansen 1988). Combined with functionalities for information and expertise capturing the concept may result in computer-resident organizational memories that are superior to the human components or organizational memories and at the same time contributes to OL (Huber 1991, Nunamaker 1991). The implementation of the Information Trader on JAVA concept is currently in the first phase.

References

Bocionek, S. R.: Agent Systems that Negotiate and Learn. *International Journal of Human-Computer Studies*, Vol. 42 (1995), 265 - 288.

Chen, S.: Role of Information Infrastructure and Intelligent Agents in Manufacturing Enterprises. *Journal of Organizational Computing*, Vol. 5 (1995), No. 1, 53 - 67.

Constant, D., Kiesler, S., Sproull, L.: What's Mine is Ours, or Is It? A Study of Attitudes about Information Sharing. *Information Systems Research*, 4 (December 1994), 4, 400 - 421.

Daft, R. L., Huber, G. P.: How Organizations Learn: A Communication Framework. In Ditomaso, N., Bacharach, S. (eds.).: *Research in the Sociology of Organizations*. Greenwich (Conn.): JAI, 1987, 1 - 36.

Duncan, R. B.; Weiss, A.: Organizational Learning: Implications for Organization Design. In Staw, B. (ed.).: *Research in Organizational Behavior*, 1. Greenwich (Conn.): JAI, 1979, 75 - 123.

Grudin, J.: Groupware and Social Dynamics - Eight Challenges for Developers. *Communications of the ACM*, 37 (January 1994), 1, 92 - 105.

Huber, G. P.: Organizational Learning: The Contributing Processes and the Literatures. *Organization Science*, 2 (1991), 1, 88 - 115.

Johansen, R.: *Groupware: Computer Support for Business Teams.* New York, 1988.

Lee, J., Malone, T. W.: Partially Shared Views: A Scheme for Communicating among Groups that Use Different Type Hierarchies. *IEEE Transactions on Information Systems*, 8 (January 1990), 1, 1 - 26.

Leonard-Barton, D.: The Factory as a Learning Laboratory. *Sloan Management Review* (Fall 1992), 23 - 38.

Maes, P.: Agents that Reduce Work and Information Overload. *Communications of the ACM*, Vol. 37 (July 1994), No. 7, 31 - 40.

Markus, M. L.: Electronic Mail as the Medium of Managerial Choice. *Organization Science*, 5 (November 1994), 4, 502 - 527.

Nunamaker, J. R., Dennis, A. R., Valacich, J. S., Vogel, D. R., George, J. R.: Electronic Meeting Systems to Support Group Work. *Communications of the ACM*, 34 (July 1991), 7, 41 - 61.

Stohl, C., Redding, W. C.:Messages and Message Exchange Processes. In Jablin, F. M., Putnam, L. L., Roberts, K. H., Porter, L. W. (eds.): *Handbook of Organizational Communication. An Interdisciplinary Perspective*, Sage: Newbury Park, 1987, 451 - 502.

Comparison of Cooperative R&D Arrangements among Competitive Firms

Jaeyong Sim
+82-42-860-6939

Researcher
Technology Strategy Planning Group
Electronics and Telecommunications Research Institute
161 Kajong-Dong, Yusong-Gu, Taejon
305-350, Korea

Moonsik Bae
+82-42-860-5376

Senior Member of Technical Staff
Technology Strategy Planning Group
Electronics and Telecommunications Research Institute
161 Kajong-Dong, Yusong-Gu, Taejon
305-350, Korea

Thus far, there have been a lot of researches in the field of cooperative R&D and many of them asserted that 'cartelization' dominates all other forms of cooperative R&D. But cartelization can harm both the competition and the consumer by facilitating collusion among the participating firms. This may be the reason why the suggestions of previous researchs could not be readily implemented in a real world.

Our model is built on the R&D classification of Kamien et al. (1992); R&D competition, R&D-cartelization, RJV-competition, and RJV-cartelization, but with the consideration of input sharing in addition to output sharing. Furthermore, to fully explore the effect of competition we use a conjectural variation model instead of a conventional Cournot or Bertrand model. The focus of our analysis is to explore which type yields the best outcome when degree of competition, spillover rate, and input/output sharing rate are simultaneously considered.

In this paper, we show that unlike Kamien et al. (1992), a dominant type of R&D arrangement is not uniquely obtained, rather heavily depends on the parameters(degree of competition, spillover rate, and input/output sharing rate). RJV-competition that have the least desirable outcome in Kamien et al. (1992) may have the best outcome in an appropriate range of input sharing. Also it is shown that R&D-competition dominates all the other R&D arrangements under a low rate of spillover and high competition.

Introduction

One of the characterizing features of a modern economy is the increasing importance of the efficiency of R&D activity for competitive advantage of firms or nations. Cooperation in R&D stage is inevitable to overcome the high risk and enormous investment requirement and to diminish the wasteful duplication of R&D resources.

Thus far, there have been a lot of analytical researches that explore the welfare consequences of cooperative R&D and its various organizational forms. Many of them asserted 'cartelization' is a better organizational form of scarce R&D resources. Here, the cartelization is represented in the model as a joint-profit maximizing behavior of involved firms in R&D stage.

But this result is somewhat contrary to the long-heritaged belief that more competition is better and, therefore, firms should not be allowed to coordinate their production or R&D decisions.

We think that this confiicting view between economic and political sides is largely due to the defects of previous analytic modeling of RJVs (Research Joint Ventures) and if the state of things - in particular, input-sharing of RJV - is appropriately depicted in the model, there will be no hesitation toward R&D cooperation. Our key observation

0-7803-3552-X/96/$5.00 © 1996 IEEE

is that cost sharing can be a useful tool for encouraging R&D cooperation without involving too tight collusion. This will be clearer as one proceeds through this paper.

d'Aspremont & Jacquemin(1988), in their seminal work, considered a two-stage game: in the first stage, two identical firms conduct a process R&D leading to a reduction in constant marginal cost, and the firms are Cournot competitors in the second stage.

Suzumura(1992) extends d'Aspremont & Jacquemin(1988)'s results to the case of more than two firms and more general demand and cost functions. He shows that in the presence of sufficiently large R&D spillovers, neither noncooperative nor cooperative equilibria achieve even second-best R&D levels. In the absence of spillover effects, however, while the cooperative R&D level remains socially insufficient, the noncooperative level may overshoot first- and second-best levels of R&D.

Finally, Kamien et al.(1992) extend d'Aspremont & Jacquemin(1988)'s model to more firms than two, a general concave R&D production function, differentiated products, and Betrand price competition in the product market. Each is modeled as a two-stage noncooperative game with n firms participating as players. In the second stage, the firms are assumed to engage in either Cournot or Betrand competition, while in the first stage they invest in R&D. They show that creating a competitive RJV reduces the equilibrium level of technological improvement and increases equilibrium prices compared to when firms conduct R&D independently. A RJV that cooperates in its R&D decisions (RJV cartelization) yields the highest social welfare under Cournot competition and, in most cases, under Betrand competition.

They assume that in all four cases considered, firms are not allowed to cooperate in the production stage. But in reality, when firms cooperate in the R&D stage, there is a reasonable chance that some cooperation in the production stage might develop as well. Thus, their suggestion that a RJV cartel yields the highest social welfare under Cournot competition and, in most cases, under Betrand competition may lose its appeal in real world R&D activity.

In this paper, we follow Kamien et al.(1992)'s classification of R&D activities (R&D competition, R&D-cartelization, RJV-competition, and RJV-cartelization), but with explicitly considering the input-sharing of participating firms.

We modify their model in two respects. First, rather than a conventional Cournot or Betrand competition model, we use the conjectural variation model to fully exploit the effect of competition on the RJV performance. Second, we suppose that once a RJV is formed, input sharing by a predetermined sharing rate, as well as an output sharing by a higher spillover rate, takes place. This is because in most real world RJV's, output sharing is almost always accompanied by an input sharing.

From these settings, we can find that the superiority of any R&D organization to the others can not be determined uniquely, but heavily dependent on the system parameters, that is, on the degree of competition and spillover rate. Thus, the most desirable cooperative R&D forms in an industry characterized by an intrinsic degree of competition and/or spillover rate can be obtained easily from our findings.

Also, in the case that certain type of R&D organization is given and can not be changed easily, we can find the optimal degree of competition and spillover rate to maximize social welfare. This can give an useful implication for competition policy and science and technology policy.

The Model

Table 1 summarizes the characteristics of four R&D organizations and differences among them with some examples.

Table 1 Comparison of 4 R&D organizations

	Spillover (output-sharing)	Input sharing	Collusion in R&D stage	Example
R&D-competition (scenario *N*)	ϕ	Not	Not	General R&D case
R&D-cartelization (scenario *C*)	ϕ	Not	Collusion	Bad case
RJV-competition (scenario *NJ*)	$\bar{\phi}$	*s*	Not	Competitive cooperation - Japan, TDX(Korea)
RJV-cartelization (scenario *CJ*)	$\bar{\phi}$	*s*	Collusion	MCC

Consider an industry with 2 firms producing a homogeneous product. Let $P = a - Q$ be the inverse demand function, where p and Q denote, respectively, the price and the aggregate output of this product. Let x_i and q_i (i = 1,2) denote, respectively, the amount of R&D investment and the output of firm *i*. The cost of production and that of R&D of firm i are assumed to be $(c\text{-}x_i - øx_i)qj$ and kx_i^2, where $0 \leq ø \leq 1$ denotes the spillover rate from firm *j*'s R&D efforts and *k* is a constant.

The R&D spillovers or externalities imply that some benefits of each firm's R&D efforts flow to other firms without payment. In our specification, the external effect of firm *j*'s R&D is to lower firm *i*'s unit production cost. The cost of R&D is assumed to be quadratic reflecting the existence of diminishing returns to R&D expenditures.

Thus Firm's strategies consist of a level of research and a subsequent production strategy based on their R&D choice. That is, firms are engaging in two stage competition. In the first stage, firms make an irrevocable commitment to a level of R&D, x_i, in full anticipation of the equilibrium that will be established in the second stage, where firms compete in the product market. The second stage strategic variable is assumed to be the level of output, q_i, and the degree of competition is represented by the conjectural variation model.

After all, the profits of firm *i* in the second stage game are defined by

$$\pi_i = [P(Q) - (c - x_i - \phi x_j)]q_i - (1-s)kx_i^2 - skx_j^2$$

where s is a predetermined input-sharing parameter.

We solve the problem by backward induction in order to obtain subgame perfect equilibrium. To begin with, let's consider the second stage. In conjectural variation model, the FOC of product market optimization is given by

$$P(Q)+P'(Q)(1+\beta)q_i = MC_i$$

where $-1 \le \beta \le q_j/q_i$ represents the degree of competition: $\beta = -1$ corresponds to Betrand competition, β=o, Cournot competition, and $\beta=q_j/q_i$ collusion in the product market. Then each firm's optimal production quantity in the second stage game can be given by

$$q_i^* = \frac{(a-c)(1+\beta)+(2+\beta-\phi)x_i+[(2+\beta)\phi-1]x_j}{(1+\beta)(3+\beta)} \qquad (1)$$

Next consider the first stage and analyze four different games, or R&D organizations. As mentioned earlier, we identify 4 scenarios of R&D activity following Kamien et al.(1992); R&D-competition (case *N*), R&D-cartelization (case *C*), RJV-competition (case *NJ*), and RJV-cartelization (case *CJ*). In two cartelization scenarios, firms engage in joint profit maximization and in two RJV scenarios, input-sharing as well as output sharing happens.

R&D-competition (case N) - Firms do not create RJV and do not coordinate their R&D decisions. Instead, each firm simultaneously chooses its R&D investment level to maximize its own production profit net of R&D expenditure. That is, firm *i* solves

$$\underset{x_i}{Max}\ \pi_i = [P(Q)-(c-x_i-\phi x_j)]q_i - kx_i^2$$

where q_i is given by (1) and we can obtain

$$x^N = \frac{(a-c)(2+\beta-\phi)}{k(3+\beta)^2-(2+\beta-\phi)(1+\phi)}.$$

Also, from $\pi_i = (1+\beta)q_i^2 - kx_i^2$,

$$\pi^N = \frac{k(a-c)^2\left[k(1+\beta)(3+\beta)^2-(\phi-\beta-2)^2\right]}{\left[k(3+\beta)^2+(\phi-\beta-2)(1+\phi)\right]^2}$$

Let's use as a welfare criteria a standard social welfare formula that sums consumer and producer surpluses. Then we can obtain the social welfare associated with this scenario as follows:

$$W^N = \frac{2k(a-c)^2\left[k(2+\beta)(3+\beta)^2-(\phi-\beta-2)^2\right]}{\left[k(3+\beta)^2+(\phi-\beta-2)(1+\phi)\right]^2}$$

R&D-cartelization (case C) - Here, as in case N, firms do not create an RJV, whereas they coordinate their R&D decisions so as to maximize their combined profits. That is, they form an R&D cartel while maintaining competition in the product market and each firm solves

$$\underset{x_i}{Max} \sum_i \pi_i = \sum_i \{[P(Q)-(c-x_i-\phi x_j)]q_i - kx_i^2\}.$$

Hence, the optimal R&D investment level in this case is given by

$$x^C = \frac{(a-c)(1+\beta)(1+\phi)}{k(3+\beta)^2-(1+\beta)(1+\phi)^2} \qquad (3)$$

and the associated profit and social welfare are

$$\pi^C = \frac{k(1+\beta)(a-c)^2}{k(3+\beta)^2-(1+\beta)(1+\phi)^2}$$

and

$$W^C = \frac{2k(a-c)^2\left[k(2+\beta)(3+\beta)^2-(1+\beta)^2(1+\phi)^2\right]}{\left[k(3+\beta)^2-(1+\beta)(1+\phi)^2\right]^2}, \text{ respectively.}$$

RJV-competition (case NJ) - Here firms form an RJV. Thus, the spillover parameter ϕ increases to $\overline{\phi}$ and both input- and output-sharing happens. However, as in case *N*, firms do not coordinate their R&D expenditures. Consequently, each firm simultaneously chooses its R&D investment so as to maximize its own profits given the R&D expenditures of another firm. Hence from

$$\underset{x_i}{Max}\ \pi_i = [P(Q)-(c-x_i-\bar{\phi}x_j)]q_i-(1-s)kx_i^2-skx_j^2,$$

we can obtain

$$x^{NJ} = \frac{(a-c)(2+\beta-\bar{\phi})}{k(1-s)(3+\beta)^2-(2+\beta-\bar{\phi})(1+\bar{\phi})}, \tag{4}$$

$$\pi^{NJ} = \frac{k(a-c)^2\left[k(1-s)^2(1+\beta)(3+\beta)^2-(\bar{\phi}-\beta-2)^2\right]}{\left[k(1-s)(3+\beta)^2+(\bar{\phi}-\beta-2)(\bar{\phi}+1)\right]^2}, \text{ and}$$

$$W^{NJ} = \frac{2k(a-c)^2\left[k(1-s)^2(2+\beta)(3+\beta)^2-(\bar{\phi}-\beta-2)^2\right]}{\left[k(1-s)(3+\beta)^2+(\bar{\phi}-\beta-2)(1+\bar{\phi})\right]^2}.$$

As shown in the first equation (4), when s increases, so does x^{NJ}.

RJV-cartelization (case CJ) - As in case NJ, firms create an RJV, except that they also coordinate their R&D expenditures so as to maximize combined profits as in case C. Thus, the firms are cartelized in the R&D stage again (but not in the production stage). Then from

$$\begin{aligned}\underset{x_i}{Max}\ \sum_i \pi_i &= \sum_i \{[P(Q)-(c-x_i-\bar{\phi}x_j)]q_i-(1-s)kx_i^2-skx_j^2\} \\ &= \sum_i \{[P(Q)-(c-x_i-\bar{\phi}x_j)]q_i-kx_i^2\}\end{aligned},$$

we get

$$x^{CJ} = \frac{(a-c)(1+\beta)(1+\bar{\phi})}{kg(3+\beta)^2-(1+\beta)(1+\bar{\phi})^2}, \tag{5}$$

$$\pi^{CJ} = \frac{k(1+\beta)(a-c)^2}{k(3+\beta)^2-(1+\beta)(1+\bar{\phi})^2} \text{ and}$$

$$W^{CJ} = \frac{2k(a-c)^2\left[k(2+\beta)(3+\beta)^2-(1+\beta)^2(1+\bar{\phi})^2\right]}{\left[k(3+\beta)^2-(1+\beta)(1+\bar{\phi})^2\right]^2}.$$

Comparison of results

Comparison of effective R&D investment levels

From the above discussions, we can obtain the relative size of R&D levels associated with four R&D organizations as (2), (3), (4), and (5). Let Xl be the effective R&D levels associated with an R&D organization l; $X^l = (1+\phi)x^l$ when $l=N, C$, $X^l = (1+\phi)x^l$ when $l=NJ, CJ$. That is, X^l means the degree of cost reduction or technological improvement from an individual firm's R&D activity.
The following Proposition 1 summarizes the results of effective R&D level comparison.

PROPOSITION 1 *The relative size of effective R&D levels associated with four R&D scenarios is determined as follows:*

$$X^N \geq X^C,\ \textit{if and only if}\ \phi < \frac{1}{2+\beta},$$

$$X^{CJ} > X^C,\ \textit{always},$$

$$X^{NJ} \geq X^{CJ},\ \textit{if and only if}\ s \geq \frac{(2+\beta)\bar{\phi}-1}{(1+\beta)(1+\bar{\phi})} = s^*,$$

$$X^{NJ} \geq X^N,\ \textit{if and only if}\ s \geq \frac{(\bar{\phi}-\phi)(\bar{\phi}+\phi-\beta-1)}{(1+\phi)(2+\beta-\phi)},$$

$$X^N \geq X^{CJ},\ \textit{if and only if}\ (1+\phi)(2+\beta-\phi) \geq (1+\beta)(1+\bar{\phi})^2,$$

$$X^{NJ} \geq X^C,\ \textit{if and only if}\ s \geq \frac{(1+\beta)(1+\phi)^2+(1+\bar{\phi})(\bar{\phi}-2-\beta)}{(1+\beta)(1+\phi)^2}.$$

It is necessary to pay attention to the third inequality of Proposition 1. This condition says that when input-sharing rate is selected appropriately, RJV-competition can outweigh even the RJV-cartelization.

To see the result of comparison more clearly, let $\bar{\phi}$ be 1 and plot the relative magnitude of effective R&D levels of four scenarios as in Figure 1, 2, 3, and 4. These figures correspond to the case of s=0, 1/3, 1/2, and 3/5, respectively. The horizontal axis represents the degree of competition and vertical axis, the spillover rate.

In these figures, we can see that the relative magnitude of X^l's (l = N, C, NJ, CJ) can not be determined uniquely, but heavily dependent on the spillover rate, degree of competition, and input-sharing rate.

When (and where) the competition is tough, R&D-competition (N) dominates all the other organization modes irrespective of the size of spillover rate (region A and B in Figure 1 and 2, region A in Figure 3 and 4). But when (and where) competition is weak, RJVs replaces top priority (region C, D, E, F, G, and H in Figure 1 and 2, region B and C in Figure 3 and region B, C, and D in Figure 4) but with some room for additional care. That is, when the input-sharing rate is greater than 0.5, RJV-competition (NJ) outweighs RJV-cartelization (CJ) and as input-sharing rate increases, the region of NJ domination enlarges (from region B and C in Figure 3 to region B, C, and D in Figure 4).

Also, in terms of spillover rate, when (and where) spillover rate is high, cartelization becomes more promising than competition, in which firms have a free-riding incentives.

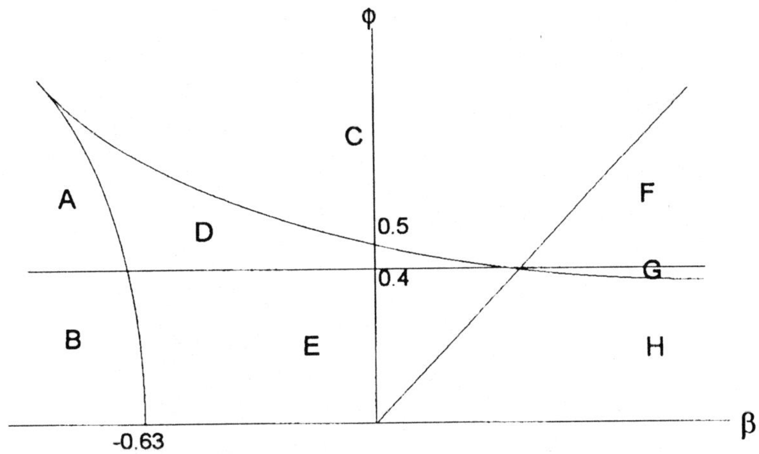

Figure 1 Comparison of effective R&D levels (when $s=0$)

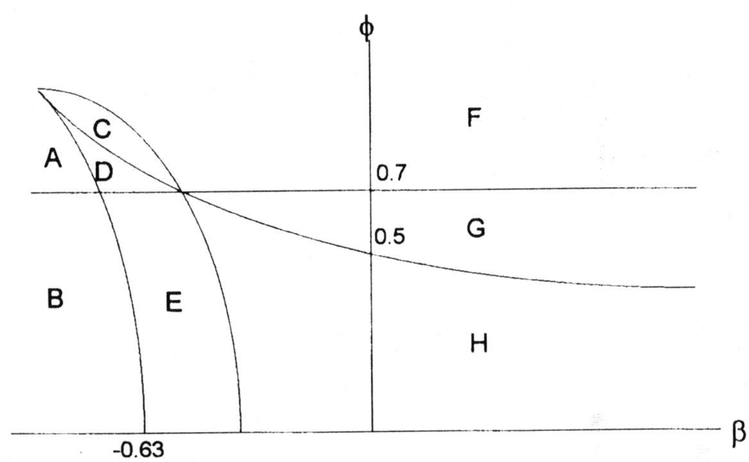

A: $X^N > X^{CJ} > X^C > X^{NJ}$, B: $X^N > X^{CJ} > X^{NJ} > X^C$,

C: $X^{CJ} > X^C > X^N > X^{NJ}$, D: $X^{CJ} > X^N > X^C > X^{NJ}$,

E: $X^{CJ} > X^N > X^{NJ} > X^C$, F: $X^{CJ} > X^C > X^{NJ} > X^N$,

G: $X^{CJ} > X^{NJ} > X^C > X^N$, H: $X^{CJ} > X^{NJ} > X^N > X^C$

Figure 2 Comparison of effective R&D levels (when $s=1/3$)

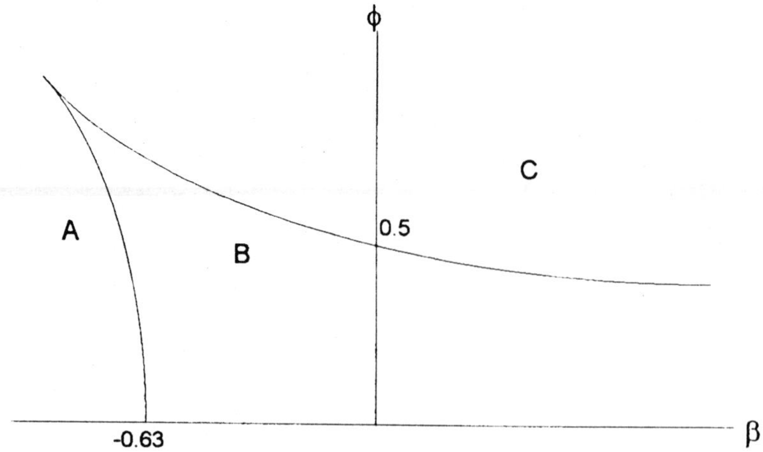

A: $X^{N} > X^{CJ} = X^{NJ} > X^{C}$,

B: $X^{CJ} = X^{NJ} > X^{N} > X^{C}$,

C: $X^{CJ} = X^{NJ} > X^{C} > X^{N}$

Figure 3 Comparison of effective R&D levels (when $s=1/2$)

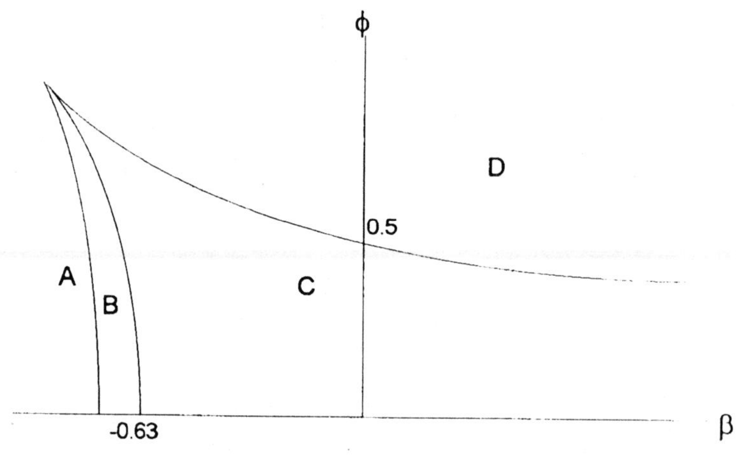

A: $X^{N} > X^{NJ} > X^{CJ} > X^{C}$, B: $X^{NJ} > X^{N} > X^{CJ} > X^{C}$,

C: $X^{NJ} > X^{CJ} > X^{N} > X^{C}$, D: $X^{NJ} > X^{CJ} > X^{C} > X^{N}$

Figure 4 Comparison of effective R&D levels (when $s=3/5$)

Comparison of welfares

The result of welfare comparison is very similar to that of technological improvement comparison, though the exact shape of the figures is somewhat different from one another. That is, when the competition is tough, R&D-competition (N) dominates all the other organization and as competition becomes weak, RJVs - in particular, RJV competition (NJ) when s increases - emerges as an alternative.

Discussions and Policy Implications

In all, the idea of including input-sharing as well as output-sharing is equivalent to creating a negative externality (input-sharing) to overcome the socially undesirable effect - underinvestment in R&D activities - of positive externality (spillover or output sharing). Figure 5 explains this interpretation.

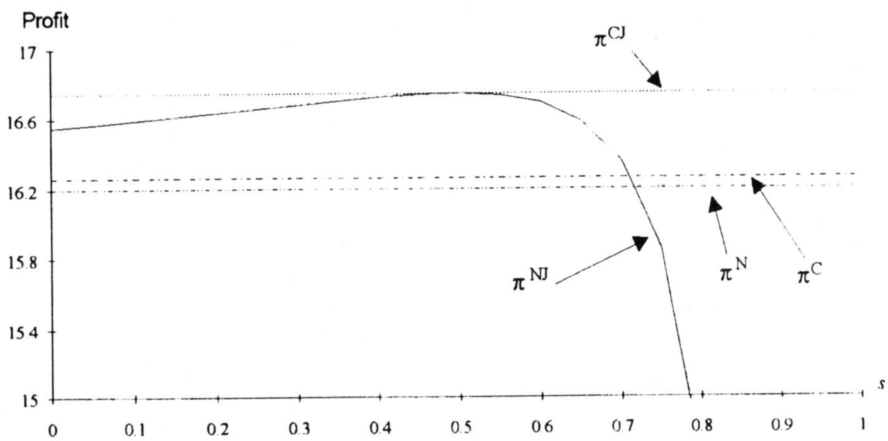

$$(a = 20, c = 8, k = 10, \phi = 0.21, \bar{\phi} = 1, \beta = 0)$$

Figure 5 The effect of input-sharing rate on profits

The role of input-sharing can be evaluated somewhat differently, too. Traditional knowledge asserted that RJV-cartelization (CJ) in highly collusive environment and R&D-competition (N) in highly competitive environment are desirable (making the fear that RJV in highly collusive market makes the problem worse). But, according to our studies, by virtue of input-sharing, RJV can be successfully maintained in highly competitive environment, too, in the form of RJV-competition (NJ) and give the best result.

Conclusion

R&D cooperation is a useful and inevitable mean to avoid wasteful duplication of R&D resources, and to minimize the risk associated with large scale R&D project. Among the various types of R&D organizatins, RJV-cartelization (CJ) may afford the most powerful R&D incentives for firms to participate in the cooperation. But this organizing mode may lower the social welfare, and moreover, may be inhibited by the antitrust authority for extending the collusive behavior to output market competition.

In this paper, we show that this quandrum in R&D cooperation can be settled by explicitly considering the other side of the coin - input-sharing -. Also, we show that the relative size of technological improvement and social welfare corresponding to four R&D organizations (R&D-competition, R&D-cartelization, RJV-competition, RJV cartelization) cannot be determined uniquely, but dependent considerably on the various parameters, that is, on *k*, ß, ø, $\bar{\phi}$, and *s*. Especially, when competition is tough, R&D competition (N) will be superior and when the output market becomes more collusive, RJV-competition (NJ) rather than RJV-cartelization (CJ) will be superior.

Second, firms are not willing to participate in RJV-competition on their own, thus, it will be necessary for government to actively participate in the R&D market.

Finally, the idea of creating another externality of opposite direction to offset the undesirable impact of an externality may be used in other area of policy problems and research questions. For example, it can be applied to environment, telecommunications, and energy, etc.

Further research directions will include (i) more general specification of demand and cost functions, (ii) more than 2 firms, and (iii) integration with empirical data, etc.

References

[1] d'Aspremont, Claude and Jacquemin, Alexis, "Cooperative and Noncooperative R&D in Duopoly with Spillovers", American Economic Review, 78, 1988, 1133-1137

[2] Dasgupta P. and Stiglitz J., "Industrial Structure and the Nature of InnovativeActivity", Economic Journal, 90, 1980, 266-293

[3] Kamien, Morton I., Eitan Muller, and Israel Zang, "Research Joint Ventures and R&D Cartels", American Economic Review, 1992, 1293-1306

[4] Katz, M., "An Analysis of Cooperative Research and Development", Rand Journal of Economics, 17, 1986, 527-543

[5] Quirmbach, Herman C., "R&D: Competition, Risk, and Performance", Rand Journal of Economics, 24, 1993, 157-197

[6] Sinha, D. K. and Michael A. Cusumano, "Complementary Resources and Cooperative Research: A Model of Research Joint Ventures among Competitors", Management Science, 37, 1991, 1091-1106

[7] Suzumura, Kotaro, "Cooperative and Noncooperative R&D in an Oligopoly with Spillovers", American Economic Review, 1992, 1307-1320

Orchestrating Multiple Changes: a Framework for Managing Concurrent Changes of Varied Type and Scope

Albert W. Small, Ph.D., and Elizabeth A. Downey
(301) 946-5858

Downey & Small Associates, Inc.
10103 East Bexhill Drive
Kensington, Maryland 20895 U.S.A.

Accelerating changes in all dimensions of enterprises and their environments present opportunities and challenges. Business environment changes require internal changes that differ in breadth (business unit to virtual enterprise), in range of subject matter (strategy, culture, services, process, information, technologies, organization, etc.), in magnitude (incremental improvement to transformation), and in time frame (immediate to long-term). Competitive advantage accrues when multiple internal changes of varied type and scope are managed concurrently and coherently. This paper proposes an integrated management framework for orchestrating multiple changes.

I. Introduction

The biggest problem for executives today is change--responding to external and managing internal change (Davis 1995). After ignoring implications of technological, social, political, and economic changes around the world until the 1980ís, U.S. business is awake to global competition.

Numerous change-formulas have been attempted. ìDrastic changes are required. The first step in the transformation is learning how to changeî said Deming (1986) who outlined 14 principles to institute quality and continuous improvement. Juran (1988) provided the managerial processes of quality planning, control and improvement. Other authors have addressed ìbusiness processî (Harrington 1991), ìreengineeringî and the significance of ìinformationî (Hammer & Champy 1993), the desirability and difficulty of handling multiple changes (Clemson 1992, Graves 1992, Hall 1993), the importance of vision and strategy (Porter 1980, Robert 1993), and the need to change behavior (Smith 1996).

Dramatic technology advances in digital information processing, storage, and communications have also changed the way people work, and made new types of enterprise possible. Increasingly, work is out-sourced, parts of different companies are brought together to accomplish tasks, employees are becoming contractors and are physically scattered at great distances (Drucker 1995, Peters 1994).

Despite all their effort to respond appropriately, quickly, and coherently to change, the majority of business leaders are disappointed with the results. We believe failure stems from loss of *bearing in the fog* of change, and propose a navigation framework for managing change. The framework essence is provided by a model of the overall process for conducting concurrent changes of all type and scope. In Section II we look at the top level and partial decomposition of the change-model. The model is used to examine scenarios for change categories in Section III, and to determine coordination among overlapping change programs in Section IV.

II. Model of the Change Process

The change-model focuses on the activity *Conduct Changes (A0)*, using the IDEF0 process modeling language discussed by Small and Downey (1992). A model describes a process as a network of activities (boxes) and their interfaces (arrows). Boxes can be activated simultaneously, and feedback loops provide for iteration.

0-7803-3552-X/96/$5.00 © 1996 IEEE

A-1, Context of Conduct Changes

Figure 1 portrays the *Context of Conduct Changes,* denoted A-1 (ìA Minus Oneî), and shows key relationships among three major activities: *Operate Environment* (A-11) includes the operations of customers, suppliers, competitors, governments, and others of interest to the virtual enterprise; *Operate Virtual Enterprise* (A-12) includes all activities of the virtual enterprise; *Conduct Changes (A0)* acts on the operating virtual enterprise to produce a changed virtual enterprise, in response to the environment and virtual enterprise situations. (Although change activities are included in virtual enterprise operations, here they are shown integrated in A0—the focus of this model—to facilitate understanding of the overall change process.)

A0, Conduct Changes

Figure 2 shows the break-out of A0, Conduct Changes, into its six sub-activities.

A1, Manage Changes, receives all current change information and provides guidance to all change activities. Good communication is vital here. A1 also evaluates available resources, and assigns resources for conducting changes.

A2, Analyze Situation & Predict, selects and studies data from the environment and virtual enterprise, and produces current analyses and predictions. These include market research, industry and business projections, new technology assessments, and enterprise effectiveness reports. A2 works closely with A3, responding to analysis needs during re-design in A3.

A3, Re-Design Virtual Enterprise, does the rethinking of the virtual enterprise at all breadths and levels of detail, covering vision and principles, objectives and strategies, products and services, operating processes, organization and resources, for specific periods in time.

A4, Plan to Implement Changes, produces the plan, including timing and responsibilities, for preparing implementation resources and for changing current operations to fulfill the new virtual enterprise design. The implementation plan is critical and must be continuously updated to address implementation issues and ensure coherent and effective transition from current to changed operations.

A5, Prepare Resources to Implement, prepares all the resources needed to implement the changes as detailed in the plan. This includes staffing, off-line training, and facilities and equipment preparation.

A6, Implement Changes, carefully converts the operating virtual enterprise into the changed virtual enterprise, during ongoing operations. A6, which includes on-the-job training, is sequenced to minimize inconvenience to customers.

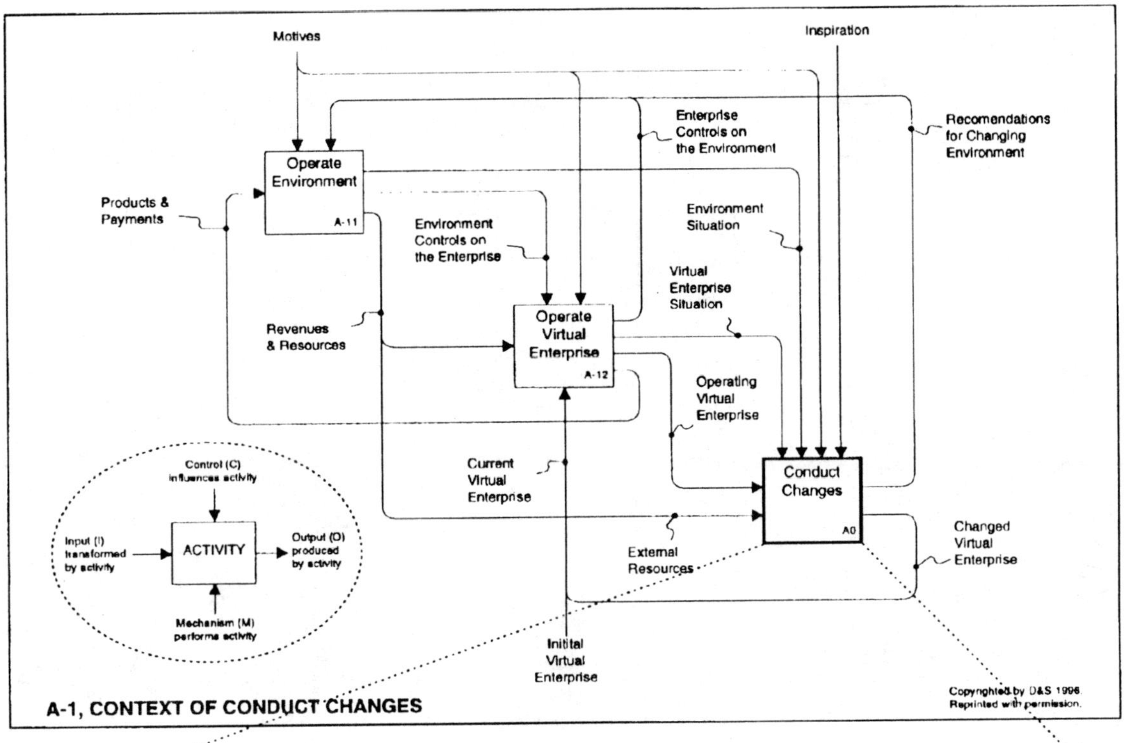

Figure 1. The Change Environment

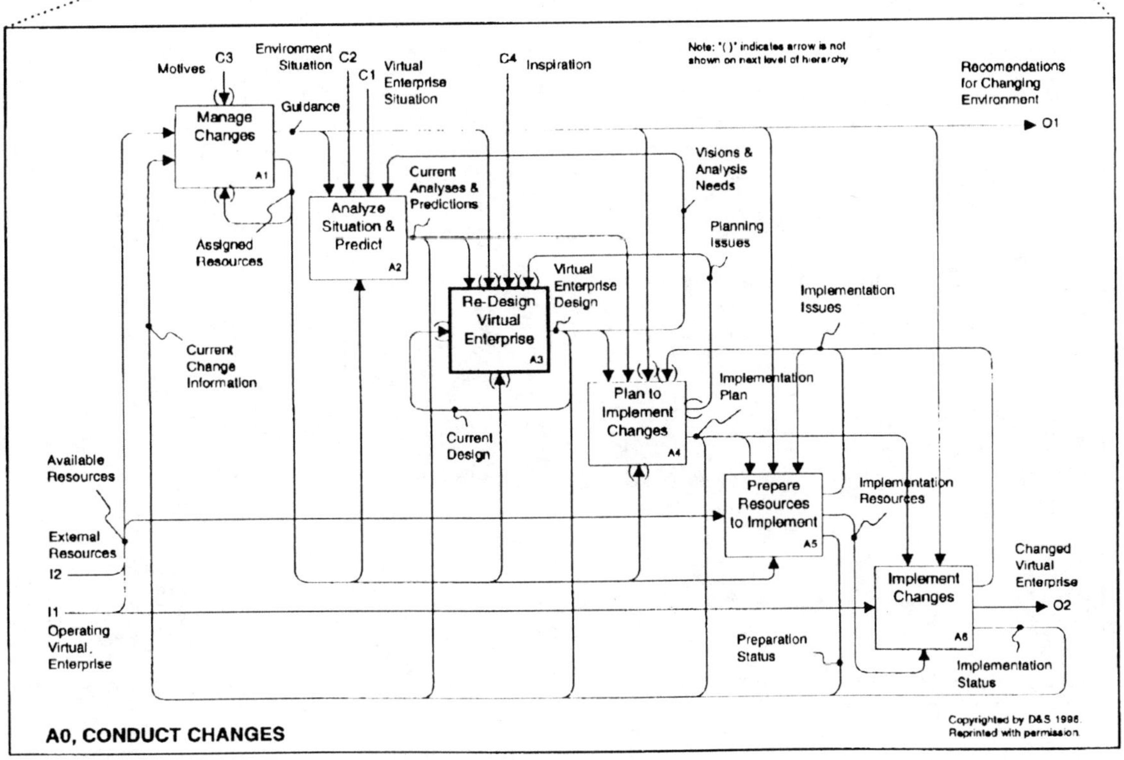

Figure 2. Process for Conducting Changes

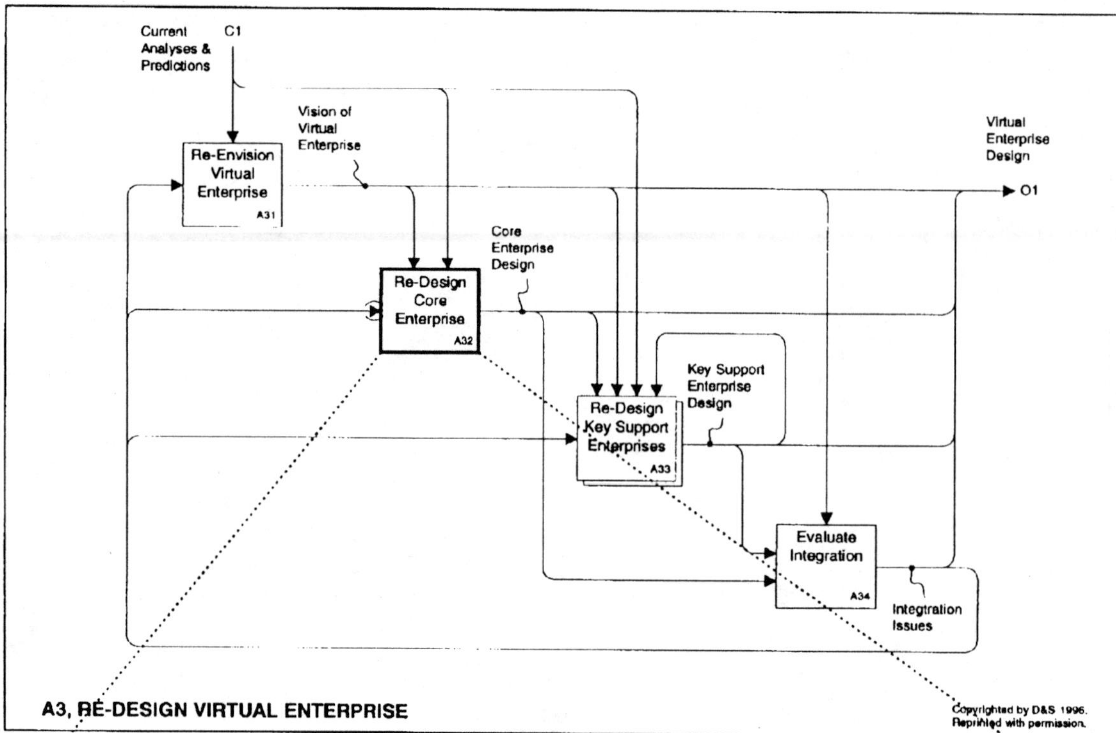

Figure 3. Process for Re-Designing a Virtual Enterprise

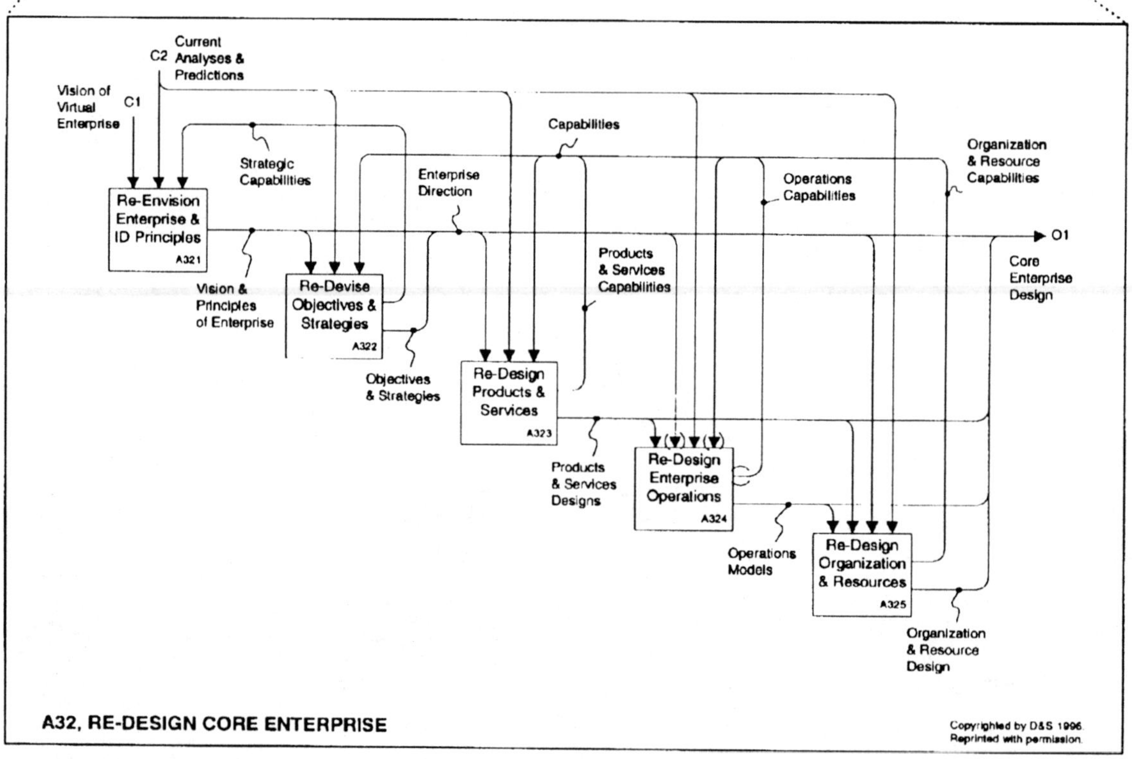

Figure 4. Process for Re-Designing an Enterprise

A3, Re-Design Virtual Enterprise

Changes in a virtual enterprise are specified by the ìre-designî of its properties. *A3, Re-Design Virtual Enterprise,* the third box in Figure 2 with break-out shown in Figure 3, is predicated on the following differentiation between enterprise and virtual enterprise: an *enterprise* is a unit of economic organization or activity that has its own mission; a *virtual enterprise* is a core enterprise plus its key support enterprises, each having a distinct mission. A virtual enterprise may involve one or more companies, or a diversified corporation leveraging the synergies among its related businesses (Ohmae 1982).

A31, Re-Envision Virtual Enterprise, creates the vision of the virtual enterprise that guides the re-design of the core enterprise (A32) and its key support enterprises (A33). Key support enterprise designs must be consistent with each other and with the core enterprise design, as appropriate for the vision of the virtual enterprise. *A34, Evaluate Integration,* evaluates the integration of the evolving re-designs; integration issues are resolved by iteration within A3 and by management guidance from *A1, Manage Changes.*

A key supporting enterprise may itself be a virtual enterprise and require the same re-design coordination as shown in A3; otherwise, its process for re-design is the same as for the core enterprise (A32). Similarly, if the overall enterprise is not considered to be ìvirtual,î then it is re-designed by A32 alone.

A32, Re-Design Core Enterprise

Figure 4 shows A32, *Re-Design Core Enterprise,* broken into five sub-activities.

A321, Re-Envision Enterprise & ID Principles, creates the core enterprise vision, similar to Robertís (1993) strategic profile, consistent with the vision of the virtual enterprise and the current analyses and predictions. This activity also identifies the enterprise guiding principles compatible with the vision.

A322, Re-Devise Objectives & Strategies, devises objectives and strategies to fulfill the vision and principles. The objectives and strategies are detailed hierarchically, consistent with the capabilities of the evolving enterprise re-design.

A323, Re-Design Products & Services, modifies existing designs or designs new products and services, in order to execute existing or new business strategies in response to changes in current analyses and predictions.

A324, Re-Design Enterprise Operations, modifies or designs new enterprise operating concepts (including processes, entity-relations, and scenarios) to produce modified or new products and services, or existing products and services more effectively and efficiently. This activity precedes and iterates with major re-designs of organization or resources, particularly for new technology adoption.

A325, Re-Design Organization & Resources, modifies or designs organization structures and implementing resources (including staff, facilities, and equipment such as information systems and data stores) for continuous improvement or to implement revised or new enterprise operating concepts.

III. Change-Categories and Re-Design Scenarios

Changes require re-design. Several change-categories are differentiated here in terms of their associated ìre-designî activities (A32) discussed in the previous section. Figure 5 shows a sample scenario for the re-design activities of each change-category. The logic of the situation and the change-model determines the sequential/simultaneous nature of the activations in each scenario.

Continuous Improvement--e.g., total quality, technology update

Continuous improvements in products and services, enterprise operations, organizational structure and resources (as by staff training and technology updates) are incremental changes performed continuously with limited coordination, planning, or preparation. Re-designs associated with such changes can occur independently (in A323, A324, or A325). Vision, principles, objectives and strategies are revisited for confirmation but not re-designed continuously.

Reengineering--e.g., new process, new technology, reorganization

Enterprise reengineering is a major re-design of enterprise operations processes (A324), accompanied by appropriate reorganization and resource upgrade (A325). Restructuring opportunities or technology leaps enable consideration of new operations concepts. When new information system technology is driving a re-engineering program, the re-design of operations should be fully explored.

New Products and Services--e.g., new strategy, new technology

Changing conditions may enable or necessitate design of new products or services or major re-design (A323), which should be based upon revised objectives and strategies (A322). Changes in products or services may require partial or full reengineering of operations (A324), organization and resources (A325).

Transformation--e.g., mergers, technology breakthrough

Transformation of an enterprise involves overall change of mission and culture, such as in a major merger or technology adoption. The transformation scenario first re-envisions the enterprise and identifies its principles (A321), develops the objectives and strategies appropriate for the new vision (A322), and then re-designs the products and services (A323), operations concepts (A324), and organization and resources (A325) required by the new enterprise direction.

A32, RE-DESIGN CORE ENTERPRISE		ACTIVATION SEQUENCES																				
	Change-Categories:	Continuous Improvement	Reengineering					New Products and Services						Transformation								
	Time Periods:	1	1	2	3	4	5	1	2	3	4	5	6	1	2	3	4	5	6	7	8	9
A321	Re-Envision Enterprise & ID Principles													X								
A322	Re-Devise Objectives & Strategies							X	X						X	X						
A323	Re-Design Products & Services	u							X	X	X					X	X	X				
A324	Re-Design Enterprise Operations	v	X	X	X						X	X						X	X	X		
A325	Re-Design Organization & Resources	w			X	X	X					X	X							X	X	X
Note: The time periods are not to scale		Key: u, v, w = actions for continuous improvements; X = actions for other change scenarios																				

Figure 5. Sample Re-Design Scenarios

IV. Compound Scenario

The previous section examined fully coordinated re-design scenarios for several change programs. Full coordination is not always possible, particularly when an external change demands immediate response in the midst of a major change program. The change-model can be used as a framework to ensure cohesion and determine the appropriate coordination for programs of different type and scope.

A0, CONDUCT CHANGES		ACTIVATION SEQUENCE																	
	Time Periods:	P1	P2	P3	P4	P5	P6	P7	P8	P9	P10	P11	P12	P13	P14	P15	P16	P17	P18
A1	MANAGE CHANGES	T	Tu	Tu	Tu	Tu	Tu	Tu	Tu	Tv	Tv	Tv	Tvw	Tw	Tw	Tw	Tw	Tw	w
A2	ANALYZE SITUATION & PREDICT	T	Tu	T			u	T		Tv			w				w		
A3	RE-DESIGN VIRTUAL ENTERPRISE			u	Tu	Tu	T	T	T	T	T	T	T	w	w		w		
A31	Re-Envision Virtual Enterprise			.	T						.								
A32	RE-DESIGN CORE ENTERPRISE			u	u	Tu	T	T	T	T	T	T		w	w		w		
A321	Re-Envision Enterprise & ID Principles			.		T	T				.			.					
A322	Re-Devise Objectives & Strategies			u			T	T	T		.			.					
A323	Re-Design Products & Services			u	u			T	T	T	.			.					
A324	Re-Design Enterprise Operations				u				T	T	Tv			w					
A325	Re-Design Organization & Resources				u	u				T	Tv	T		w	w		w		
A33	Re-Design Key Support Enterprises			u	u	u	T	T	T	T	T	T	T	w	w		w		
A34	Evaluate Integration			u	u	u		T	T	T	T	T	T	w	w		w		
A4	PLAN TO IMPLEMENT						u					v	T	T		w	w		
A5	PREPARE RESOURCES							u						T	T	T		w	
A6	IMPLEMENT CHANGES								u				v	T		T	T	T	w
Note: The time periods are not to scale		Key: T = transformation scenario; u,v,w = other scenarios; "." = confirm current design																	

Figure 6. Compound Scenario Example

Figure 6 shows model activation sequences combining several change scenarios in the same time span. A transformation involving a merger of two financial giants is planned. The transformation scenario (denoted by ìTî) starts with activation of change management (A1) and analysis (A2) during the first time period (P1). During P2, ìTî is in the midst of initial analysis (A2) when market research and predictions of surging Internet usage suggest that home banking services be offered before transformation ìTî can be implemented. A1 decides to offer the new service ìuî without delay. During P3, while ìTí is completing analyses (A2), ìuî determines there is no need to revise visions (A31, A321), adjusts core objectives and strategies (A322) to include the service, begins design of the new core and key support services (A323, A33), and starts monitoring design integration (A34). During P4-P6, ìuî completes its design (A3) and planning (A4), and predicts its impact on the market (A2); ìTî re-envisions the virtual, core, and support enterprises (A31, A321, A33), and begins revising core objectives and strategies (A322). ìTî uses results from ìu,î but not vice versa. During P7-P8, ìuî resources are prepared (A5) and service implemented (A6); ìTî begins to re-design products and services (A323), operations (A324), and key support enterprises (A33) -- and to monitor integration of all re-design (A34).

During P9 a new regulation requires immediate change in core enterprise operations that cannot await ìTí implementation; ìTí incorporates the changes in redesign during P10, and ìvî implements in P12 before ìTî implementation. In P12 a critical technology opportunity is recognized; it is decided the opportunity must be taken but implementation of ìTî should not be delayed. Scenario ìwî is initiated but resource preparation postponed until P17 to avoid interfering with ìTísî implementation, and so ìwî is able to take advantage of upgrade in P16.

V. Conclusion

We have outlined a change-model that provides the essence of a framework to coherently manage multiple concurrent changes of varied type and scope. The model was used to study change scenarios. We have included the concept of the virtual enterprise and addressed the additional difficulty it poses.

We propose that enterprises use the comprehensive change-model to create management frameworks appropriate to their situations. The breadth of change activities and their interdependencies shown in the model enable the visualization and communication needed to navigate through the fog of change.

References

Clemson, B. and D. Alasya. 1992. Implement TQM and CIM together. In *1992 International Engineering Management Conference, October 25-28, Eatontown, NJ*. New York: IEEE Engineering Management Society.

Davis, E. 1995. Whatís on American managersí minds? *Management Review,* April, 14-20.

Deming, W. Edwards. 1986. *Out of the crisis.* Cambridge, MA: MIT.

Drucker, Peter F. 1995. *Managing in a time of great change.* New York: Truman Talley/Dutton.

Graves, J. 1992. Leaders of corporate change. Fortune, 14 December, 104ff.

Hall, G., et al. 1993. How to make reengineering really work. *Harvard Business Review,* November-December, 119-131.

Hammer, Michael and James Champy. 1993. *Reengineering the corporation: a manifesto for business revolution.* New York: Harper Collins.

Harrington, H. James. 1991. *Business process improvement: the breakthrough strategy for total quality, productivity, and competitiveness.* New York: McGraw-Hill.

Juran, J. M. 1988. *Juran on planning for quality.* New York: Free Press, MacMillan.

Ohmae, Kenichi. 1982. *The mind of the strategist: the art of Japanese business.* New York: McGraw-Hill.

Peters, Tom. 1994. *The Tom Peters Seminar.* New York: Vintage/Random House.

Porter, Michael. 1980. *Competitive strategy: techniques for analyzing industries and competitors.* New York: Free Press, MacMillan.

Robert, Michel. 1993. *Strategy: Pure & simple: how winning CEOs outthink their competition.* New York: McGraw-Hill.

Small, A., and E. Downey. 1992. Technology support for managing complexity. In *1992 International Engineering Management Conference, October 25-28, Eatontown, NJ.* New York: IEEE Engineering Management Society.

Smith, Douglas K. 1996. *Taking charge of change: 10 principles for managing people and performance.* New York: Addison-Wesley.

A Framework for Optimizing a Virtual Learning System for Minorities and Industry in the Field of Engineering

Paige E. Smith and Brian M. Kleiner
(540) 231-4301

Department of Industrial and Systems Engineering
Virginia Polytechnic Institute and State University
Blacksburg, VA 24061-0118

While many attempts have been made to research and improve the academic achievements of minorities[1], results have been disappointing. This paper takes a unique sociotechnical systems analytical approach to identify important social, technical, and environmental variables. This research has implications for the use of cyberspace for organizational learning; for involving industry virtually in the upstream development of human resources, and for redesigning the overall support system for minorities.

Introduction

Industry is directly affected by a competitive global economy. However, educational institutions, as major suppliers of industrial personnel, have been criticized for producing poorly prepared "products." Universities, in a joint effort with industry, can work together to prepare students with the knowledge and skills they need to immediately contribute to industry upon graduation. With modern information technology, a strategic parternship can be formed to actively involve all interested stakeholders in educating students without having to be collocated.

Virtual learning organizations are characterized by people working together for the purpose of gaining or sharing knowledge although they are (or may be) physically separated by a distance. Virtual learning is "a designed, collaborative, continuous learning process," where people build a knowledge base (Grenier and Metes, 1995; p. 245). The concept of virtual learning originated in distance learning and education. Although distance learning can take on several definitions, within the United States, distance learning can be thought of as "education that takes place when the learner and instructor are at separate locations (Zeller, 1995; p. 123)." Through new and innovative technology, distance learning is enhanced to the point where people can organize together to create a virtual learning organization.

Within predominantly White universities, a disturbing trend is occurring. In general, minority students in engineering fields are not performing at the same level as their white counterparts. For example, African Americans have higher attrition rates and lower graduation rates (National Science Foundation, 1994).

While many attempts have been made to research and improve the academic achievements of minorities in predominantly White universities, results have been disappointing. A unique approach that can potentially help is designing a virtual learning environment with a sociotechnical systems analytical approach. The sociotechnical approach can help to identify important social, technical, and environmental variables upstream (i.e., in high schools and at home), in-process (i.e., university), and downstream (i.e., in industry).

Conceptual framework

According to sociotechnical systems theory, the overall system is comprised of social and technical subsystems with the influences of the environment taken into consideration (Hendrick, 1986). Sociotechnical systems theory prescribes jointly

0-7803-3552-X/96/$5.00 © 1996 IEEE

optimizing the social subsystem and technical subsystem as a function of the demands of the environment as shown in Figure 1 (Pasmore, Francis, Haldeman and Shani, 1982). The dashed lines in the figure represent the open nature of the organizational system. To optimize one of the subsystems would be to suboptimize the other subsystem; hence, guaranteeing the failure of the system as a whole.

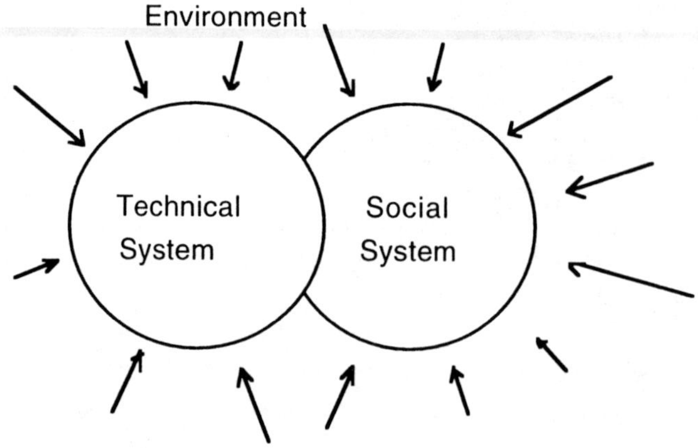

Figure 1. The influence of the environment on the social and technical subsystems

Based on sociotechnical systems theory, a conceptual model is proposed (see Figure 2). The conceptual model shows both the process relationships and the information relationships among the components in the system. The process relationships define where various stakeholders exist within the system and information relationships describe the feed forward and feedback loops of the system.

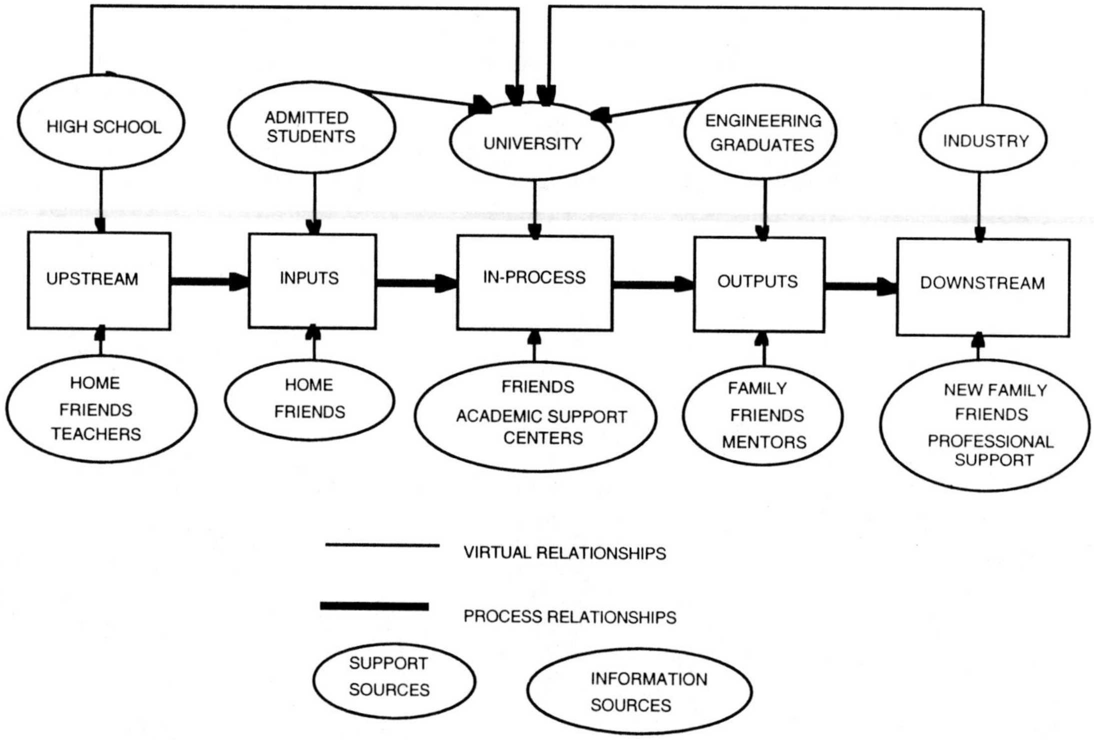

Figure 2. Conceptual model

The social subsystem

The social subsystem of an organization includes the people in an organization and their relationships to one another (Pasmore et al., 1982). Increasingly, people serve on teams. In virtual environments, teams can consist of one time contributors, contributors who follow their tasks through to completion, or combinations of the two types. The individual and group needs are included within the subsystem as well as the demands the group and the tasks place on individuals. The social subsystem combined with the organizational structure create what is defined by Hendrick (1986) as "complexity." He defined complexity as "the degree of differentiation and integration that exist within an organization (Hendrick, 1986; p. 468)." Spatial differentiation describes the geographical dispersion among the members in the team. Therefore, spatial differentiation can be used to describe virtuality.

Minority students have an intricate social subsystem. In addition to academic demands, students must learn to cope in an unfriendly or indifferent institution (Fidler and Godwin, 1994). Typically, minority students turn to their families and friends for moral support in learning to adjust to their new environments (Hughes, 1987). Many researchers point to the need for an increase in positive relationships between faculty and students although this has been historically difficult (Dinka, Mazzella and Pilant, 1980; Hughes, 1987).

The technical subsystem

Skills, knowledge, tasks, and procedures in addition to technology make up the technical subsystem (Pasmore et al., 1982). The purpose of the technical subsystem is to convert input into outputs (Pasmore, 1988). Within a virtual learning environment, two variables should be considered: group process and communication style.

The group process addresses the degree of formalization necessary for the teams to function. In this situation, formalization refers to the degree of standardization of the process. And the group process consists of the team's mission, ground rules for interacting, the roles and responsibilities for each team member and agreement on the tasks. Therefore, if the formalization of the group process is low, teams can interact at their own discretion with no (or very few) guidelines to follow (Hendrick, 1986). Conversely, if formalization is high, then the teams must strictly adhere to the guidelines defined for their team.

Communication style is the other variable of concern. With modern technology, people can communicate either synchronously or asynchronously. Synchronous communication occurs between people "at the same time although (they are) in different places (McGrath and Hollingshead, 1994; p. 15)." Whether the communication is written or verbal, people can share thoughts and ideas in a continuous, dynamic fashion. Asynchronous refers to communication between group members at different time periods (McGrath et al., 1994). Therefore, there is a delay in the process of sharing thoughts and ideas.

Some examples of synchronous communication technology include the telephone, video-teleconferencing, and some group shareware packages. Some problems associated with the technology of synchronous communication include: nonverbal information is lost; and at great distances, there may be lag times in the transmission and feedback of information (McGrath et al., 1994). E-mail is a common example of asynchronous communication. Some other examples of asynchronous communication technology include videotapes and voice mail (McGrath et al., 1994). The largest disadvantage of asynchronous communication is the loss of continuity in sharing ideas. Examining these two styles of communication can help determine which type of communication will work best with virtual teams.

The process by which minority students learn is another component of the technical subsystem. Research supports "a cooperative academic attitude [is] prevalent among the African American students (Jagers, 1992; p. 537)." In addition, "they prefer active involvement rather than passive (Shade, 1989; p. 27)." Another difference is many students of African dissent think holistically as opposed decomposing a problem and then synthesizing the pieces for the final answer (Locke, 1992). Cooperation and holistic thinking contradict the traditional institutional teaching methods. A key to the success of minority students is effective goal setting (Tracey and Sedlacek, 1984, 1985, 1987; Hughes, 1987). The more long range and concrete the goals, the more likely the student will perform well.

The environment

Organizations are open systems. Therefore, the environment in which an organization functions is important to understanding the system. When considering minority engineering students within a university, their environment is relatively isolated. Instructors, peers, and parents influence students. Students have been managing themselves within this environment for many years. A newer dimension in the environment has resulted from diversification. Minorities in predominantly White institutions have additional issues within the university environment that they must understand; and historically performance has been lower for minorities than for white students.

Introducing the virtual component to the environment adds additional complexity to the system. In addition to managing in a new university environment, students are expected to manage when they may not communicate face-to-face. Virtual environments are flexible and adaptable and people rely on cyberspace and other technical links for communication (Barnatt, 1995).

The addition of virtuality to the university environment brings a new paradigm to the university. Grenier and Metes (1995) think of traditional education as one-way interaction from faculty to students. However, in a virtual learning environment, the learning process becomes multi-directional. Figure 3 depicts the contrast between these two paradigms.

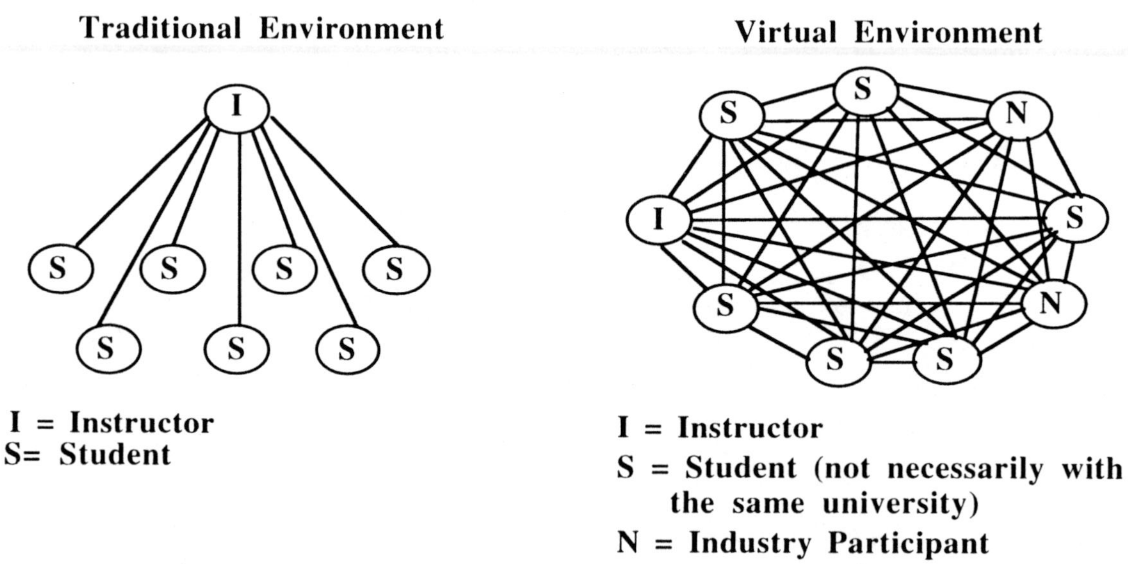

Figure 3. Traditional versus virtual learning environments (adapted from Grenier et al., 1995)

Organizational design

A current trend within universities is the team approach to learning. Teams should be well balanced in terms of team members' skills and abilities (Hilborn, 1994). Team member responsibilities, in addition to completing their tasks, include ensuring that all team members can demonstrate the skills they are responsible for learning in class. In virtual teams, in addition to students, team members can also come from industry or they can be anyone connected with the team via technology.

Conclusions

Through using a sociotechnical systems approach, a virtual learning system can be developed which will have far reaching benefits for both minorities and industries. For minorities, the virtual learning system addresses some of the needs which predominantly White universities currently are missing. For example, minorities in industry can become role models and mentors without having to be collocated with the students. Industry now has the opportunity to help share information students need to become ready for the work place. In addition, high schools can benefit from this parternship. Teachers and counselors, once an important influence on a student, can remain in close contact. In return, college students can help communicate the importance of higher education to high school students. The knowledge that can be shared and gained through a virtual learning system is invaluable.

Future research

Based upon the results of this analysis, sociotechnical design principles, and experimental optimization of design features, a virtual learning system, which integrates suppliers of students and industrial customers, will be designed and validated.

References

Barnatt, C. (1995) "Office space, cyber space, and virtual organizations," Journal of General Management, Summer: 78-91.

Dinka, F., Mazzella, F., and Pilant, D. (1980) "Reconciliation and confirmation: Minorities and Whites at a predominantly White university," Journal of Black Studies, 11: 55-76.

Fidler, P.P. and Godwin, M.A. (1994) "Retaining African-American students through the freshman seminar," Journal of Developmental Education, 17(3): 34-40.

Grenier, R. and Metes, G. (1995) Going virtual. Prentice Hall: Upper Saddle River, NJ.

Hendrick, H. (1986) "Macroergonomics: a conceptual model for integrating human factors with organizational design," Human Factors in Organizational Design and Management - II, 467- 477.

Hilborn, R.B. (1994) "Team learning for engineering students," IEEE Transactions on Education, 37(2): 207-211.

Hughes, M.S. (1987) "Black students' participation in higher education," Journal of College Student Personnel, November: 532-545.

Jagers, R.J. (1992) "Attitudes toward academic interdependence and learning outcomes in two learning contexts," Journal of Negro Education, 61(4): 531-538.

Locke, D.C. (1992) Increasing multicultural understanding, Sage Publications, Inc.: Newbury Park.

National Science Foundation. (1994) Women, Minorities, and persons with disabilities in science and engineering, Arlington, VA. (NSF 94-333)

McGrath, J. and Hollingshead, A. (1994) Groups interacting with technology, Sage Publications, Inc.: Thousand Oaks, CA.

Pasmore, W. (1988) Designing effective organizations, John Wiley & Sons: New York.

Pasmore, W., Francis, C., Haldeman, J., and Shani, A. (1982) "Sociotechnical systems: a North American reflection on empirical studies of the seventies," Human Relations, 35(12): 1179-1204.

Shade, B.J.R. (1989) "Culture and learning style within the Afro-American community," Culture, Style and the Educative Process, Shade, B.J.R. ed., Charles C. Thomas: Springfield, IL.

Tracey, T.J. and Sedlacek, W.E. (1984) "Noncognitive variables in predicting academic success by race," Measurement and Evaluation in Guidance, 16: 172-178.

Tracey, T.J. and Sedlacek, W.E. (1985) "The relationship of noncognitive variables to academic success," Journal of College Student Personnel, 26: 405-410.

Tracey, T.J. and Sedlacek, W.E. (1987) "Prediction of college graduation using noncognitive variables by race," Measurement and Evaluation in Counseling and Development, 19: 177-184.

Zeller, N. (1995) "Distance education and public policy," The Review of Higher Education, 18(2): 123-148.

[1] For the purposes of this paper the term "African American" is not broad enough. As used by Lucinda Roy (Assoc. Dean, Arts and Sciences, Va Tech), the term minority includes African American students and those from the African Diaspora (Afro-Caribbeans, resident Africans, and other members of the "Black" community). The authors acknowledge the ongoing assistance and support of the Virginia Tech Office of Minority Engineering Programs.

Information Security Issues facing Virtual Enterprises

Gerhard Steinke
(206) 281-2377

Seattle Pacific University
350 West Bertona Ave
Seattle WA 98119

Ronald Leamon
(206) 528-3112

Battelle Seattle Research Center
4000 NE 41st St.
Seattle WA 98105

Virtual enterprises require the communication and sharing of information. Only if each participant in the virtual enterprise is assured that the information they bring to the virtual enterprise is secure, will they share it. In this paper we examine some of the security issues facing virtual enterprises and refer to security measures taken by a specific organization.

Introduction

Stiff competition among organizations, in the local as well as the global environment, is common in our society. But the pace of change, the rapid advances in information technology and the cost of capital investments necessary to compete have demanded new organizational forms and alliances. One result has been dynamic cooperation and alliances between firms that might otherwise never collaborate.

When diverse organizations work together to jointly respond to turbulent market conditions and achieve a specific goal in the marketplace, they form what are known as virtual enterprises. These enterprises are event-driven objects that emerge in response to changing market conditions, perform an action or project to meet the market demands, and then allow each individual legal entity to move onto the next activity which may or may not include the same firms.

In this paper we examine information security issues which concern virtual enterprises--issues beyond those facing traditional organizations. The linking and sharing of information between individuals and organizations participating in a virtual enterprise is essential. Each participant brings information to the enterprise. Security of this information is often of critical concern. The virtual enterprise must provide assurance to individual partners that their information is secure and will only be used for the purpose specified by the participating entities.

Information generated during the life of the virtual enterprise must be protected, in particular that individual partners only use the information to accomplish the goals of the virtual enterprise. When the project is over, ownership and distribution of the information and knowledge generated by the virtual enterprise must be determined.

As an example we refer to Battelle, a distributed research organization--a type of virtual enterprise, using our terminology. Battelle partners with researchers and organizations around the world, creating different virtual enterprises. Many of the relationships are defined by specific projects and periods of time. We describe how Battelle deals with some of the security issues facing a virtual enterprise.

Information Security

The dynamic nature of the virtual enterprise as well as the increased use of information technology creates additional concerns for security. Security issues for virtual enterprises can be divided into three areas. First, information which individuals partners bring to the virtual enterprise must receive protection. Second, information

0-7803-3552-X/96/$5.00 © 1996 IEEE

which is generated by the virtual enterprise requires security. And third, upon the dissolution of the virtual enterprise, the information assets of the virtual enterprise must be distributed in a secure manner.

Each entity participating in a virtual enterprise brings information to the virtual enterprise, namely a subset of the individual entityís information. The participating entities will not want to share more of its proprietary information than necessary and agreed upon with the virtual enterprise--and only if it is assured of the security of the shared information.

If a partner in a virtual enterprise shares its own core competence and competitive position vis-‡-vis its adversaries, it risks the erosion of assets which are the core of its competitive advantage. In the worst case scenario, this sharing of information could lead to the dissolution of a participating firm itself. So security of information brought to the virtual enterprise is significant.

The technical nature of how information is varies. For example, an organization may provide a copy of its information to the computer system set up for the virtual enterprise. Alternatively, an organization may permit direct real-time access to specific information on its own computer system. In the latter case the fear is that if you give me access to your computer system, I may be able to bypass your security measures and access more than you intended for me to see.

As participants in the virtual enterprise, each participant has access to the information supplied by others participating in the virtual enterprise. The purpose is to be able to use that information to accomplish the task of the virtual enterprise--not for the benefit of a participating entity. There must be security and audit measures to ensure that information is used for the purpose intended.

The virtual enterprise creates new information and knowledge. Ownership of that new information resides with the virtual enterprise. Security measures must monitor access to try to determine if individual partners use the new knowledge for their own purposes.

Upon dissolution of the virtual enterprise, the information of the virtual enterprise must be distributed or ìdisposedî of. Should each partner have access to some or all of the virtual enterpriseís knowledge? What happens to the information an individual partner gained during the course of the virtual enterprise?

People Issues

Security issues in virtual enterprises must be considered in terms of the human dimension and characteristics that are unique to virtual enterprises. (In a perfect world where one could trust everyone, security measures would be superfluous!)

Virtual enterprises, by their nature, are not permanent organizations. Individuals work for a virtual enterprise for a specific period of time and then move on. There is usually not the time for training and education specific to the virtual enterprise. Loyalty to an organization usually takes a long time to develop. The time frame for involvement means that the commitment of people participating in the virtual enterprise differs from those working for a traditional organization.

If one doesnít have a strong loyalty to the virtual enterprise, the concern for security by individuals may be lower than for individuals working for a ìpermanentî organization. So while the security requirements of a virtual enterprise may be greater than the security requirements of a traditional organization, individuals are probably less concerned about security.

In addition, the work for a virtual enterprise may not be the only work an individual is doing, so the individualís loyalties may be divided--not just sequentially but concurrently. Keeping tasks separate becomes a concern, particularly if there are concurrent conflicts between the requirements of the various tasks one is involved with.

The latter concern means that a thorough check of individualsí involvement in other projects is necessary--something which is relatively rare in the current environment where individuals seldom work for more than one employer at any point in time.

The people issues are therefore complex and difficult to handle. Trust relationships may develop as partners get to know each other after participating together in a number of virtual enterprises. But in addition to trust, security measures must be put into place. Currently, legal documents and agreements are the prime vehicle for specifying and managing the people related information security issues in virtual enterprises.

Battelleís legal department has standard contracts which are designed to cover various consultant and subcontract situations. These contracts vary, depending on whether the client is a state or federal government or an organization from the private sector. Also, in some cases Battelle staff are consultants or have subcontracts from another organization which is the prime contractor. There are also cases in which Battelle staff at one site have consultant agreements or subcontracts with another Battelle component or division.

Battelle assigns the responsibility for security relating to people issues, primarily to the lead on the project, i.e., the leader of the virtual enterprise. Sometimes there are oversight panels who monitor and audit security criteria, especially when a government contract or grant is involved.

Information Ownership

The nature of virtual enterprises results in multiple ownership of information assets. The individuals and organizations participating in a virtual enterprise bring assets to the virtual enterprise. They retain ownership of the information they bring to the virtual enterprise. By virtue of participating in a virtual enterprise each participant receives access (but not ownership) to assets contributed by the other participants in the virtual enterprise.

Ownership of the information generated by the virtual enterprise resides with the virtual enterprise. Upon termination of the virtual enterprise, ownership and distribution of this information must be determined. Often the distribution of assets after the dissolution of the virtual enterprise is agreed to when the virtual enterprise is created.

Legal documents are usually drawn up to specify requirements and ownership of the assets brought to the virtual enterprise, as well as who receives access and ownership of the knowledge generated by the virtual enterprise. If information security is breached, then the issues can be decided by the court system.

Typically at Battelle a client is involved, to whom the virtual enterprise reports and for whom the project is performed. This client is then the owner of the research work. If Battelle does work that is anticipated to result in economic benefit, the client almost always reserves not only rights to the research work but prohibits Battelle from publishing the results. If it is a project devoted more to "pure" research, Battelle is often granted the right to publish the results.

Technological Issues

Advances in information technology and computer networking have allowed rapid access to information, whether it is on oneís own computer or on a computer around the world. (These advances in a sense enable virtual enterprises.)

Often access is not to the computer system of the virtual enterprise, but the computer systems of the individual participating partners of the virtual enterprise. A variety of technical security measures are available to ensure that access is limited according to the information that a partner wants to, or has agreed to share with the virtual enterprise.

The increased use of information technology to communicate, whether it be via email or teleconferencing and videoconferencing, raises further issues in terms of security of the communication and computing infrastructure. This is of particular concern due to the fact that the infrastructure is usually not owned and developed by the virtual enterprise (as it would be for a traditional organization, in part because the traditional organization has a long term concern and commitment to its infrastructure).

Secure access can be provided by a combination of approaches. A firewall can be used to protect the internal network of individual partners or the virtual enterprise from the public network. Depending on what tasks need to be performed by individuals, firewalls can be configured to permit access to the internal network based on specific IP address and protocol. For example, individuals who need only to download or upload files can be given anonymous ftp access to the internal network while individuals who need to log onto an internal server and perform specific tasks can be given telnet access.

Server-based security on an internal network provides great flexibility in terms of specific access permissions. Role or task-based security can be provided by the network operating system or the specific database management system. For example, on an internal network using Novell Netware or NT Server OS, users can be permitted or denied access to particular servers. Once a user logs onto a particular server, access to data can be controlled by the server operating system or by the database management system.

In Seattle, Battelle uses a combination of Novell Netware security and a router-based firewall. Netware allows one to limit access to specific files and directories. In one case a partner who was concerned about network access by other partners was satisfied when a particular data directory was removed from routine backups.

Conclusions

Virtual enterprises are becoming increasingly common. Success of this organizational form will not only be determined if the virtual enterprise achieves it goals, but if security of information is ensured. Only if a participant in a virtual enterprise is assured that the information they bring to the virtual enterprise is secure, will they share it.

We have described some of the security issues. Increased emphasis on providing secure information platforms and architecture is required, particularly for virtual enterprises.

Partnership Management within the Virtual Enterprise in a Network

Zhan Su
(418) 656-2185

SORCIIER
Faculté des sciences de l'administration
Université Laval
Québec, G1K 7P4

Diane Poulin
(418) 656-3105

SORCIIER
Faculté des sciences de l'administration
Université Laval
Québec, G1K 7P4

We propose a new management method who integrates a behavioural approach and an instrumental approach in order to answer the managerial challenges of the virtual corporation.The concept of virtual corporation, developed as an organizational theory in the United States, is gaining more grounds. Communication techniques, who so far where limiting its development, are becoming more powerful and always promising. Virtual corporation has emerged from an answer to new competitive economic constraints, to offer the possibility to grasp opportunities in transitional markets. However, if the flexibility of this new organizational model contitutes the key for corporations survival, its management method is far from easy because it must combine fluidity and control, strong partnership relations and short life expectency, differences and cohesion. Therefore, let's look into this real management method for the new virtual corporations.

Managerial challenges

There is an ambiguity regarding the definition of the virtual corporation. According to Jean-Marie Toulouse (e.g. Boyer, 1994) it is a network of people who all work together on the same project without being part of the same corporation. For Davidow and Malone (1992), virtual corporation is like a molecule that is made, broken and remake each time different atoms are cropped up. As for Poulin, Montreuil and D'Amours (1995), they talked about a network corporation who «organizes and manages its operations in a flexible matter by orchestring a synergy between its inside and outside in order to improve its competitivity».

In this paper, we focused on one type of the virtual corporation which is the temporary association of corporations linked by technological informations in order to achieve a common mission. In fact, virtual corporation uses a strategic networking with others firms to answer growth objectives. Each firm offers its competences in the field where it is more competitive (core competencies) to create a specific product or service after which, the relation does not exist anymore. All the corporation's aspects are influenced by the «virtuality»; from production to sales. We can consider a relative equality among partners specialized in their own field of expertise. Excellence is becoming the key word in order to give to the new formed corporation a maximum of performance at each level, something that a single company could not achieve on its own.

One of the particular elements for the functioning of the virtual corporation is the simultaneity: things are created in parallel instead of linear sequence where processes are rigourously made step by step. In fact, teamwork is done in real time, without geographical contraints and avoid a quantity of reports and meetings normaly seen in traditional organizations. Since information networks help the actors to work together from start to finish until the project is completed. The partnership is based on electronic contracts who permits a faster agreement without any lawyers implication.

Furthermore, in a virtual corporation the physical organization is turn upside down and industrial visits are therefore impossible because of the group and network

0-7803-3552-X/96/$5.00 © 1996 IEEE

phenomena who render useless the concentration of resources in the same place, same town, same country. The workplace is involved with employees, suppliers and distributors, clients and even competitors when legislation permits it. Virtual corporation is a group formed around a joint project. The group can be geographically dirperse, physical liaison between members is done by telephone communication, electronic mail or by computer. Once the project is finished, the group is dissolved permitting to each member to participate into other projects. The lack of borders or precise limits between the different organizations is something new. This new corporate model modifies the outlines of the traditional organizations. Cooperation between competitors, clients and suppliers does not help to define where a firm begins and ends.

All industrial sectors and services can benefit from this new corporative model who is symbiotic, mutually beneficial and reciprocal. But its practice is difficult. In fact, the difficulty resides in forming and preserving a win win relation between each actor in a virtual enterprise; partnership should serve the interests of all parties involved. Furthermore, the risks of an opportunist behavior from one of the partners are real and not very protected since there is no formal agreements between participants.

How to manage and coordinate the relations so that everybody finds its own way, so that all participants can share a co-destiny in a virtual corporation; to make sure there is no opportunist behaviors; and how to have a positive relation? We will address this central problem facing management in a virtual corporation.

Towards a new management principle

The key element in a management problem related to any partnership is control; who contitutes a crucial organizational process. It is more a «pro-active» control, prior to the operation, rather than a «reactive» control subsequent to operations. In other words, it is the capacity to influence the actors' decisions and behaviors, and eventualy refrain the realization of certains operations. Control has become more and more sophisticated and complete as organizational and environmental complexity increases.

Traditional control mecanisms used in other forms of parnership (joint-venture, consortium, etc.), that are the majority shareholding, the allocation of votes and board of directors is no longer justified in a new virtual corporation. Since virtual enterprise is not a new juridical entity, it does not have its own capital structure. Board of directors does not exist, it reduces the potential control from the congruent portion.We think that the specificity of the virtual corporation calls for an instrumental approach as well as a behavioral approach of management, formal and informal mecanisms.

The instrumental approach uses formal tools to guard itself against opportunist behavior from the other partners. There is a definition of the cooperation frame such as contract; the organizational structure choosen for the entity and its verification. Information system management is also a strong tool; eased by the development of the technical informations (computer, fax, modem, e-mail, video conference, etc.). Furthermore, according to Williamson (1985) the principle of «hostage exchange» can be applied. By «hostage exchange» we mean the transfer of specific assets who increases the opportunity cost from an entity of a dismantling entity. The principal indication is the amount of financial resources invested by each party; it can also be the degree of information disclosed by an owner's technology that might be critical for one of the firms. In those circonstances, the value of the assets gives a proportionnal negociation edge to each participant. These negociations edges will have the same equivalence than the value of assets brought in. This situation increases trust who is an informal mecanism of control. The behavioral approach rests on this last mecanism.

Stacked on the formal authority system, informal mecanisms hold an important place in the survival and good functioning of all entity. Essentially it is about trust between partners. In fact, the multiple identity of the entity is such that it adds a weight to the efficiency of the organizational communication. Often, different norms and values must meet, creating a disparate culture who contributes to the ambiguity of communication. Mutual trust that is developed between partners permits to take off that «veil» and secures each attending participant by openness and sincerity.

Formal and informal mecanisms are complementaries. It is indispensable to procede with an integration of both approaches in order to benefit from its advantages and to reduce its mutual weaknesses. But in the virtual corporation, the instrumental approach of control is shaded off by the behavioral approach, who is more informal and based on responsability. If the instrumental approach of control presents the advantage of clarifying responsabilities of the «cocontractant» parties, we have noticed however that the formal rules have rigid and narrow caracteristics (e.g. Hocquart, 1989; Ring and Van de Ven, 1989). That way we avoid rigidity who increases complexity and costs for arbitration. More flexible informal type permits to go further away from functional and divisional boundaries and are able to bring a certain stability to the entity.

Virtual corporation management: the new rules

Trust

Taking initiatives outside the formal rules are favored even incited. It gives a creative development to the entity. Also as mentionned by Perrow (1986), trust permits to avoid rigidity and structural costs for a multifunctional corporation. Macaulay (1964) noticed that transaction costs in the organizations are reduced when there is trust. In a situation of structural, organizational and cultural differences between partners, trust contitutes a flexible way for mutual adjustment capable to give a stable and creative relation. However, if joint-ventures and alliances offer manpower training to help adjust, it is different in a virtual enterprise where everybody comes in with their best assets and where adaptation must be in real time with the help of communication.

Communication

Informal forms of control, like the formal mecanism, do not lack limits. In fact, to set up a trusting climate is a long process who requires tacts for interpersonal relations. We understand the stakes for a good communication. Virtual corporation benefits from information technologies who permit to link people who are not physically together. They give direct ways of communication in real time, and at lower costs. Communication in the virtual corporation contitutes a condition for establishing trust and the success of the organization. Technological progresses in the field of information are not only a communication vector but also contitute at the same time a tool for control. In fact, instant contacts always permit a link and an easy report. Sharing datas and information networks tell us where the project stands, by giving a viewpoint.

Ethics

Loyalty is a must in this new organization where the lack of systematic contracts prevents almost all legal appeals in case of prejudice or the non-respect of commitments. The risk incured is big if the partners have hiden interest and are not honest when playing the virtual game. The sanction for a breach of loyalty is compromise for all future participation in another virtual enterprise.

Co-destiny

It is important to maintain interdependence in order to limit opportunist tendancy of all participants. According to Schmidt and Kochan (1977), even if different partners announce their intention of interdependence, there is a perception of interdependence by the managers of the enterprises' partners. For Ulrich (1983), network is described by the managers as an alternative to competition. In fact, degrees of strategic freedom survive when the behavior of all the network enterprises stick together and each firm has a strategic differentiation inside the system. The strategic process of forming is therefore influenced by the customs to form alliances, and by the perception of the environment from the managers (e.g. Wacheux, 1996). This can be applied for virtual enterprises where each component asserts its differences and brings in what they excel the most. Co-destiny comes from the complementary of togetherness, who could not function in optimal conditions of competences if one of the elements leaves. The overall cohesion is ensured by the complementarity and must be solidified by well managed interactions. Then, the risks for opportunist behaviors are reduced becauce no component by itself can achieve a level of excellence.

New type of control

Loyalty, trust, ethics and communication are necessary, but however, not sufficient for the good functioning of the organization. Control is crucial, not in the sense of sanction but to maintain the organization towards its goal. The virtual firm, formed by expert elements in their own fields, doen not need to be controled as extensively as a multifunctional entity. Although, a focus must be done on critical areas, such as interactions between partners. The notion of focus is the third dimension of control set forth by Shaan and Beamish (1988), after mecanisms and scope. Centering on precise points limits rigidity and permits to the virtual enterprise to keep its lead trump who is flexibility.

Only a rigourous control of the linking zones permits a real synergy, who has an effect on the performance of the system. That synergy depends on the good interaction of the sub-system. To create synergy in an organization where teams and components are autonomous, it is crucial to manage well all the connections and at the same time to delegate the management contained in the components. The frame into which people work must be released but not eliminated, or else this will lead to anarchy without obtaining the fixed goals. To ensure interaction coherence, the context must be integrated in the operation mode of each sub-systems. It is not sufficient to create and communicate a strategic intention; proof is the lack of satisfaction among managers when executing operations. In this manner «communication and hope» do not constitute a substitute from the old managing ways to «order and control». The new battle field is now «context and coordination» key for success for the organizations who have to adapt their design to the permanent evolution of the environment.

According to Haeckel (1995), in firms with variable structures, management is based on principals, models and governing process, who have to be adaptable to changes in the environment.

- governing principles define the soul of the association (the definition of its function) and the founders principles without giving indications on the functioning methods.
- governing method sets an hierarchy of the key competences. To create this model involves a definition of the responsabilities for important stages who intervene in the soul.
- governing process is the way by which governing principles are always up dated and by which they are operational.

These principals must be the starting point of the virtual association. In fact, by giving a functioning frame, clarifying the organizational goals and the responsibilities of its members, they leave total autonomy inside the field of responsibities. The evidence of competences combined in the organization, permits an optimal and exhaustive use. The anticipation of conflits and cautions regarding formality must be translated into open and honest discussions between partners once they have explained their decision. Complications involving lawyers for all contracts are avoided. In fact, they are replaced by a lighter version constituted by aforementioned who are the contracting partners and describe the extent of each parties' commitment. Although penalty clauses can be added regarding the partners' payment, in case of delay, failure or non-respect.

For a good working order and the good performance of the virtual enterprise, the overall management must stay flexible and be shared in order to favour synergies, creativity and the commitment of the partners. Delegation of power and the promotion of autonomy for managers are important. That decentralization generates informal contacts between employees from different parties and tend to develop an informal network of experts as well as a trustful relation.

Conclusion

Virtual corporation constitutes a revolution in terms of its organization and its management. The challenges reside in developing a strong and lasting trustful relation between partners, while the life cycle of the virtual corporation could be short. This paradox implies loyalty between the partners who want to participate in a virtual corporation. You also have to add good communication skills to make sure that all the contact areas between different partners are working together.

Therefore, if autonomy and flexibility are preserved, it is essential to set minimum controls, precisely on the interaction level, since no system can function naturally in an optimal way. Even if the behavioral approach based on trust and loyalty is privileged, it is agreed to define a functioning frame for the virtual corporation, who clarifies its objectives and the competences. Without being a new model of organizational chart, it permits to specify the responsabilities of each actor and becomes contractual since it fixes the involvement and the extent of all the partners' interventions.

References

Boyer, Hélène, «Entreprise virtuelle: tous pour un!», Tendances, Vol 10, No 4, 1 mai 1994, p. 51.

Byrnes, John A., «The virtual Corporation», Business week, No 3304, February 8, 1993, pp. 98-104.

Davidow, W. and Malone M., «Virtual Corporation», Forbes, Decmber 7, 1992, pp. 102-107.

Haeckel, Stephan, «Adaptive enterprise Design: The Sense-and-Response Model», Planning Review, May-June 1995, pp 6-13 and p.42.

Hocquart, C., «Le modèle du chantier», Gérer et comprendre, décembre 1989.

Killing, J.P., «Strategy for Joint Venture success», Croom Helm, 1983.

Macaulay, S., «Non Contractual Relations in Business: a Premary Study», American Sociological Review, 28, 1964.

McQueen, Roc, «Virtual companies learn that the only one reality is action», The Financial Post, Weekly 1, Saturday April 2, 1994.

Perrow, C., «Economic theories of organization», Theory and Society, Vol. 15, 1986.

Poulin, D., Montreuil, B. and D'Amours, S.,«L'Organisation virtuelle en réseau», in Martel, A. and Oral, M., «Les défis de la compétitivité», Publi-Relais, 1995.

Ring and Van de Ven, H.A., Research on the Management Innovation: The Minnesota Studies, Praeger, 1989.

Shaan, J.L., and Beamish, P.W., «Joint venture general manager in LDCs», in F. Contractor and P. Loranger, editors, Cooperative Strategies in International Business, Lexington Books, 1988.

Schmidt, S., and Kochan, T., «Interorganizational Relationships: Pattern and Motivations», Administrative Science Quaterly, Vol. 22, pp. 220-234, 1977.

Scotton, Geoffrey, «First virtual reality: now there's the virtual corporation», The Financial Post, Weekly 5, Saturday September 4, 1993, S13.

Ulrich, D., «Governing Transactions: a Framework for Cooperative Strategy», Human resource Management, Vol. 22, pp.22-39, 1983.

Wacheux, Frédéric, «Alliances: conséquences stratégiques et organisationnelles», Revue Française de Gestion, Mars-Avril-Mai, 1996, pp. 12-24.

Wexler, Joanie M., «Ties That Bind», Computer World, June 28, 1993, pp. 97-98.

Williamson, O.E., «The Economic Institution of Capitalism», Free Press, 1985.

The Design Of An Integrated Circuit Using A Virtual Enterprise

Sherif Sweha
(916)356-2233

Intel Corporation
1900 Prairie City Rd, FM3-99
Folsom, CA 95630

Recent advances in information and communication technologies have enabled the implementation of the Virtual Enterprise in various forms. This paper describes the process followed to successfully complete the layout of a Flash Memory Integrated Circuit project using a Virtual Enterprise. The criteria for project and extended team selection are presented. The team architecture and communication infrastructure are discussed in detail. The paper reports on how today's technology was used to setup hardware and software systems, and how data security issues were resolved.

Introduction

During the late 1970's numerous predictions were made concerning the impact that the technological revolution in telecommunications would have on traditional organizational work arrangements. Forecasts were made of millions of jobs shifting out of the offices right back to the homes (Toffler, 1980). A new paradigm of organization, called the virtual organization, was introduced in the early 1980's and has been refined and elaborated in a number of articles over the past decade (Mowshowitz, 1986, 1994). More recently, the virtual corporation concept has been defined (Davidow, 1992), and the groupings that arise to execute particular episodic tasks has been coined as a virtual team (Hammer, 1993, p.67). A natural extension to the concept of the virtual corporation is the utilization of an external work force. The economics of subcontracting have been analyzed (Radding, 1991), and the advantages and disadvantages of outsourcing on personnel, economic, control, data and organizational dimensions have been studied (Huff, 1991).

In the world of integrated circuits development, the act of subcontracting a small part of the circuit design or chip layout is not uncommon. Recent advances in technology - especially in the fields of communication and personal computers - have made it possible for us to perform all the tasks required for a Flash memory chip layout using a Virtual Enterprise structure. The process we followed consists of six steps, shown in Figure 1, which include the project selection criteria, the selection of the extended team, the requirements for the hardware and software setup, the team architecture, the communication infrastructure and the documentation of the learnings to feed the process forward to future Virtual Enterprises. This paper will report on the details of the process we followed and the implementation on our Flash Memory project.

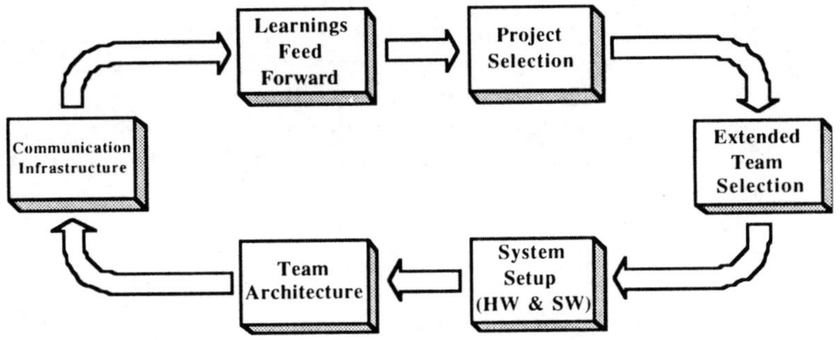

Figure 1: Virtual Enterprise Development Process

Step 1: Project Selection

The project selection step is the key factor to enable the utilization of the Virtual Enterprise. The key characteristics of an appropriate project is the ability to separate the conception and planning of the project activities from the implementation activities. Since conception and planning deal with the abstract model of an activity and their output can be modified without affecting the final outcome of the implementation, they are kept in-house with the core team. On the other hand, the implementation activities can be completed by off-site contractors working in the Virtual Enterprise. The implementation activities must be well defined and the tasks easily separated into sub-tasks so that contingency plans can be developed when sub-tasks are not completed on schedule. This minimizes the risk taken by contracting all the implementation activities on a project. The process of executing the implementation activities must be well defined and understood by the core team. The project deliverables must be defined to meet completion in accordance with a clearly specified set of standards that can be easily communicated to the extended team of contractors through contractual agreements.

In our case of a Flash Memory design project, all the circuit design activities were done in-house by our core team of engineers. As a project, the chip layout was easily broken into conception/planning and implementation activities, which provided a perfect match for our project selection criteria discussed above. All the top level chip planning activities, as well as the project management activities such as scheduling, monitoring and reviewing were conducted by our core team. The implementation activities of the physical design of the chip layout - such as cell layout, section hookup and layout verification - were conducted by the extended team. The milestones of completing cells and section layout in accordance with a set of design rules by a specific date were a natural fit for the criteria of the project deliverables.

Step 2: Extended Team Selection

At the heart of the Virtual Enterprise is the extended team which provides the implementation of the project activities. The process of selecting the extended team covers both the contractors selection as well as the contract setup. With the project tasks and scope well defined, the extended team selection process starts by interviewing multiple contractor companies at the home site where the core team resides. The interview process is expected to evaluate the following:

1. The team size and the level of experience of its members.
2. The hardware and software setup and capability for enhancements and upgrades.
3. The company capability to withstand monetary and personnel setbacks without affecting its output.
4. The communication infrastructure of the company, both internal and external.
5. The training required, if any, to perform the project activities.

Ideally, the selected teams consist of three to six members and a team leader. Each team is given two different sets of project activities to bid on. This limited bidding process provides a competitive environment and provides a check on the bids while not burdening the core team with unnecessary bid review activities.. After the bids are accepted, the formal contract is signed. The contract is the guide for the extended team members activities and is the source for any conflict resolution. The contract incorporates a list of milestones, a payment schedule, and an escalating incentive/penalty scheme for exceeding/missing schedule commitments. The contract also provides the core team with the right to terminate or reassign project activities if certain quality or schedule expectations are not met. All the extended team members are expected to sign a legal Non-Disclosure Agreement to prohibit them from disclosing the project information to any third party.

On our project, the extended team consisted of five contractor companies, each with not less than three members and no more than six members. In addition, one company had three sub-contractors, each with one to three members. The sub-contractors' contracts were set up with the core team and not with the contractor company, to insure our intellectual property protection. The chip layout was divided into three sections and three of the contractor's companies were each assigned a section. The other two contractor companies provided cell level support on the largest section of the chip. Furthermore, this large section was divided into multiple sub-sections that were assigned to the three sub-contractors. Our extended team size grew to a total of thirty members located in more than ten cities across the United States.

Step 3: System Setup

A critical factor in the success of a Virtual Enterprise is standardizing the hardware and software used by the different members. This allows for a high level of flexibility in switching companies or integrating new companies in an efficient manner. On our large project of multiple contractor companies at different cities, standards for hardware, communication protocols and database format were essential.

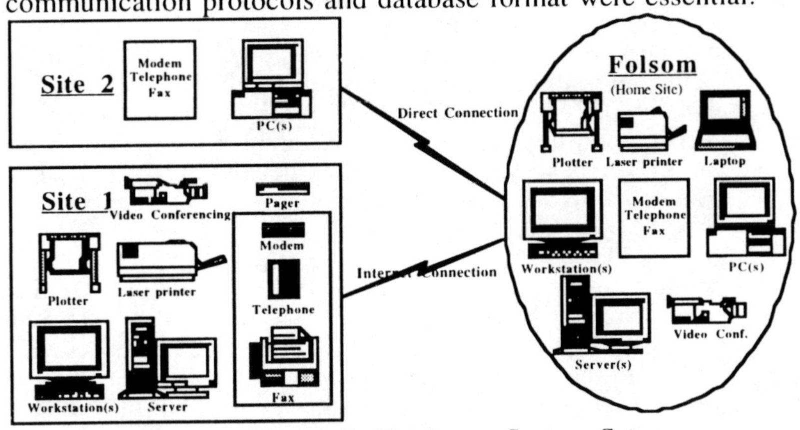

Figure 2: Hardware System Setup

Our hardware setup is shown in Figure 2. At the home site (Folsom, CA), the backbone of the system is the compute capacity which consists of multiple servers and workstations that were used for database creation and verification. Since some of the off-site work was done using PCs, several PCs were provided to enable the core team to use the video conferencing technology to simultaneously review or edit the database with the contractors. In addition, laptop computers equipped with the layout generation software were available for use during travel. Printers and plotters were available to provide plots and schematics for use by both the in-house and the off-site teams. The ideal off-site hardware setup is shown as Site 1. For a company to join the project, the minimum hardware configuration required is shown as Site 2.

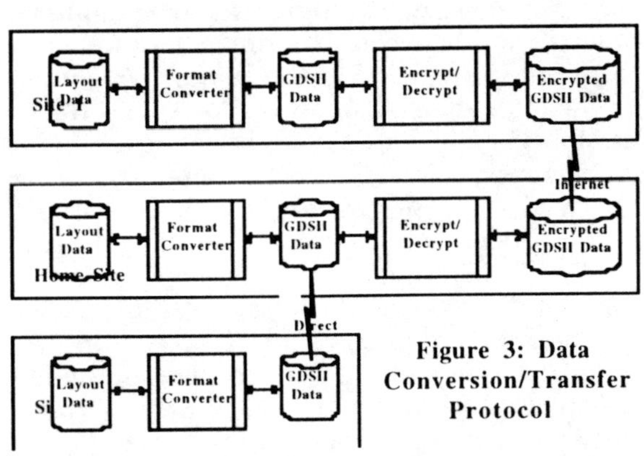

Figure 3: Data Conversion/Transfer Protocol

As it was difficult to standardize all the layout tools across the project, it was much easier to standardize the database format that was used or transmitted to the home site. The protocol used to transmit the database is shown in Figure 3. Before transmission, the data was always converted to an industry standard format (GDSII), which provided flexibility for the contractors to utilize the layout generation tool with which they were proficient. Data transmission was accomplished by using either a direct modem connection or Internet e-mail. In the former case, data encryption was not required as the modem connection provides a secure communication line. In the case when data was transmitted across the Internet, the data was encrypted before transmission and decrypted at the receiving site to guarantee data security.

Step 4: Team Architecture

To maximize the efficiency of the Virtual Enterprise and to provide for a good environment to exercise the traditional management functions, the roles and responsibilities of the various members of the team have to be very clearly defined. The team members fall in one of three categories. They are either Environment Architects, Product Architects or Product Builders. The Environment Architects create and maintain the Virtual Enterprise environment. This includes the key values and guidelines for behavior and communication in the Enterprise. They are the project leaders who set the strategic directions of the project and are responsible for the conception and planning activities. The Product Architects are the Senior Technical members of the Virtual Enterprise. They define the product via the blueprints. Their main role is to optimize the overall product through concurrent interaction and active decision making. They define the problems and drive all the major technical decisions in the enterprise. The Product Builders are responsible for constructing the various elements of the product as per the blueprints. They apply their technical knowledge to solve the problems.

On our Flash Memory project, the Engineering Project Leader and the in-house Layout Leader played the role of the Environment Architects. The leaders of the five contractors companies played the role of the Product Architects as they provided the section level layout planning and supervised the execution of the cell and section layout. In the process, they had to identify all the technical issues on the layout. The rest of the off-site team members, including all the sub-contractors, provided the role of the Product Builders who drew and verified the layout of the chip.

Step 5: Communication Infrastructure

In a Virtual Enterprise, crisp and clear communication is crucial to maintain smooth and constant progress and to minimize conflict among team members. The communication infrastructure of the enterprise has to resolve the barriers created from the competitiveness between contractors, as well as the diverse geographical locations. It has to provide a system for information sharing, coordination and motivational types of communication. The communication infrastructure of a Virtual Enterprise will vary based on the size of the enterprise, number of locations, schedule and nature of the project, as well as the technology available to the enterprise.

In our case, we used a "Single Channel" communication style. This meant that all communication between the home site and the contractors was conducted between the Layout Project Leader and the Product Architect at the contractor company, usually using the telephone. Also, every Product Architect carried a Pager for maximum accessibility. Only Product Architects communicated with Product Builders. Product Architects were allowed direct communication with each other as long as the Layout Project leader was informed, through the use of e-mail on the Internet. Transfer of hard copy documents and plots were faxed or mailed overnight. To finalize the chip plan early in the project and to accelerate decision making and resolution of issues throughout the project, all Product Architects were required to attend a monthly face-to-face meeting at the home site. The meeting was also used to review progress on all sections of the layout with the engineering community on the project. The Environment Architects traveled to the contractor location on an as-needed basis to assess the progress and capabilities of the site, to provide the cohesion across all members of the enterprise and to remove barriers.

Step 6: Learnings Feed Forward

After the completion of the development phase of the project, the Environment and Product Architects document the learnings from the project. On our project, each Product Architect held a meeting with their Product Builders to list the key factors that improved their work efficiency and the issues that hindered work progress. Then, the Product Architects met face-to-face with the Environment Architects to share the findings. The Environment Architects then summarized the learnings and recommended changes to the process for future projects.

Conclusion

In Chinese, the concept of "critical turning point" is represented by two characters. The first means "danger" and the second means "hidden opportunity". The use of technology has enabled us to layout a Flash Memory Integrated Circuit using a Virtual Enterprise, with more than thirty members located in more than ten cities. The full implementation of the Virtual Enterprise in the Semiconductor Industry represents a critical turning point. To find the "hidden opportunity", organization leaders must develop a new culture that exploits the process discussed in this paper. The "danger" in this new paradigm is the high risk incurred by contracting all the tasks on a given project. Anything less than total understanding of the process and development of contingency plans will result in failure.

Acknowledgment

The author wishes to recognize the outstanding efforts and dedication of the core team of J. Avondo, V. Baclig, B. Baltar, M. Chuaprasert, D. Eastman, K. Frary, S. Guliani, P. Lam, K. Loe, J. Petersen, C. Senthilkumar and A. Votino. They made the adventure of developing our Virtual Enterprise a very fruitful experience. The author is grateful to D. Motoda for his invaluable and insightful ideas, suggestions and discussions.

References

Davidow, W. and Malone, M. The virtual corporation. HarperCollins Publishers: New York, NY, 1992.

Hammer, M. and Champy, A. Reengineering the corporation. HarperCollins Publishers: New York, NY, 1993.

Huff, S.L. "Outsourcing of information services." Business Quarterly, Spring: 62-65, 1991.

Mowshowitz, A. "Social dimensions of office automation." Advances in computers 25: 335-404, 1986.

Mowshowitz, A. "Virtual organization: a vision of management in the information age." Information

society 10(4): 267-294, 1994.

Radding, A. "The economics of outsourcing." Bank Management, September: 20-27, 1991.

Toffler, A. The Third Wave. Morrow: New York, NY, 1980.

A Quantitative Framework for Designing Efficient Business Process Alliances

Srinivas Talluri and R.C. Baker
Email: talluri@omega.uta.edu

Department of Management Sciences
The University of Texas at Arlington
Arlington, Texas 76019, USA

The virtual corporations (VCs) are viewed as a solution for quick introduction of a variety of products with high quality and low costs. One of the critical issues in the formation of VCs is the partner selection process. This paper proposes a two-phase mathematical programming approach for effective partner selection in designing a VC. Phase 1 identifies efficient candidates of each business process type utilizing data envelopment analysis. Phase 2 uses an integer (0-1) goal programming model to select an effective combination of efficient partners.

Introduction

Manufacturers are constantly working towards meeting customer oriented performance measures such as high product quality, small lead times, high product variety, and low cost. It is a monumental task for large or small organizations to accomplish all the aforementioned goals in a timely and efficient manner. Large organizations are often very complex and small organizations suffer from a scarcity of resources. VCs are envisioned by many experts as the solution for quick introduction of a variety of products with high quality and low costs (Goldman, 1994; Iacocca Institute, 1991).

VCs are a temporary alliance of independent business processes or enterprises each contributing "core competencies" in areas such as design, manufacturing, and distribution to the corporation (Byrne, 1993; Goldman, 1994; Iacocca Institute, 1991; Presley et al., 1995; Sheridan, 1993). VCs are formed in the event of a market opportunity and would be dissolved when the opportunity passes. A key factor, emphasized by researchers, in forming a VC is the selection of competent and compatible partners (Byrne, 1993; Goldman, 1994; Iacocca Institute, 1991; Porter, 1993). No formal decision models for effective partner selection process have been addressed in the literature. For the successful functioning of the VC, highly efficient and compatible partners are desired.

Problem complexity and solution technique

One of the key issues encountered in the formation of a VC is the selection of partners. For example, consider a scenario in which a VC needs to be formed with three business processes types (A, B, and C). If there are 10 potential candidates for the type A process, 10 potential candidates for the type B process, and 8 potential candidates for the type C business process, then the total number of combinations under consideration is 800. All these combinations must be evaluated and the most desirable one identified. This can be an extremely tedious and time consuming process. Moreover, it is very difficult to incorporate into one decision model both the internal decision variables of candidates of each process type, and the external (compatibility) decision variables among the candidates of different process types.

This paper proposes an alternate decision model for this problem which incorporates two phases. Phase 1 uses a filtering technique based on the internal decision variables of the candidates. The technique identifies efficient candidates of each business process type and filters out the inefficient candidates. This process results in reducing the number of possible combinations that are considered in phase 2 of the analysis.

0-7803-3552-X/96/$5.00 © 1996 IEEE

The technique used in phase 1 is data envelopment analysis (DEA), a fractional programming model developed by Charnes et al. (1978) that identifies efficient candidates by incorporating a range of internal activity and performance measures into the model. Phase 2 utilizes an integer (0-1) goal programming model for selecting an effective combination of efficient candidates (identified in phase 1) to participate in the formation of the VC.

DEA is a fractional programming technique that evaluates the relative efficiencies of homogeneous units in the presence of multiple inputs and outputs. The efficiency is defined as the ratio of weighted sum of outputs to the weighted sum of inputs. DEA has been used primarily to compare the efficiencies of schools, hospitals, shops, bank branches, and other areas where there is a relatively homogeneous set of units.

If there are "n" units with "m" inputs and "s" outputs, then the relative efficiency of a particular unit "p" is obtained by solving the following linear programming problem proposed by Charnes et al. (1978):

$$\text{Max} \quad \sum_{k=1}^{s} v_k \, y_{kp}$$

$$\text{s.t} \quad \sum_{j=1}^{m} u_j \, x_{jp} = 1$$

$$\sum_{k=1}^{s} v_k y_{ki} - \sum_{j=1}^{m} u_j x_{ji} \leq 0 \qquad \forall \, i$$

$$v_k, \, u_j \geq 0 \quad \forall \, k \text{ and } j$$

where: p is the unit being evaluated
k = 1 through s
j = 1 through m
i = 1 through n
y_{ki} = amount of output k produced by unit i
x_{ji} = amount of input j used by unit i
v_k = the weight given to output k
u_j = the weight given to input j

The above problem is solved n times to determine the relative efficiencies of all the units. The unit under consideration each time is referred to as the target unit. The model allows the target unit to effectively select optimal weights that maximize its output to input ratio, but at the same time the output to input ratios of all the n units computed with these weights are prevented from exceeding "one". A relative efficiency score of "one" indicates that the unit under consideration is efficient, whereas a score less than "one" implies that the unit is inefficient. The optimal value of the objective function indicates the relative efficiency of unit "p".

Framework

The decision makers in the framework are top executives acting as brokers. These brokers act as agents in selecting efficient and compatible partners. In the event of an identified market opportunity for a product, the brokers are expected to respond by creating a VC and delivering the product at low cost, high quality, and small lead times. Although, the decision making process is explained by considering only three process types, it is extendible to any number.

Phase 1 is initiated with the identification of the required business process types by the brokers followed by the identification of potential candidates of each business process. The input-output (I/O) measures for each business process are then identified. These I/O measures should be in alignment with the objectives, goals, policies, and procedures of the VC. The inputs should encompass any resources utilized by the business processes, and the outputs should include a range of performance and activity measures (Boussofiane et al., 1991). It is possible in DEA that some units can achieve a relative efficiency score of "one" by indulging in inappropriate weighting scheme. To overcome this problem, weight restrictions can be attached to the DEA model. These weight restrictions should be specified according to the requirements of the decision maker. The general form of these weight restriction structures are:

$$\alpha_j u_1 \leq u_j \leq \beta_j u_1 , \qquad j = 2, 3, \ldots, m$$
$$a_k v_1 \leq v_k \leq b_k v_1 , \qquad k = 2, 3, \ldots, s$$

where: u is the input weight
v is the output weight
α, β, a, b are non-negative scalars

DEA with weight restrictions is used to evaluate all potential candidates of each business process type. Those identified as efficient candidates, with a relative efficiency score of "one", are considered for further analysis, and the inefficient candidates are dropped out. The DEA model reduces the complexity of the problem by minimizing the number of combinations that are to be analyzed in phase 2 of the decision making process. For example, if two candidates are identified to be efficient in each of the three business process types, then the total number of combinations under consideration in phase 2 are eight (2 X 2 X 2).

In phase 2, an integer (0-1) goal programming model is used to select an effective combination of candidates to participate in the formation of a VC. The combinations identified in phase 1 are evaluated with respect to compatibility criteria. Some of the compatibility issues addressed in the literature concerning the linkages among the participating business processes are: effective communication networks, cultural compatibility, frequent interactions, trust, and goals in alignment with the vision of the enterprise (Gilbert et al., 1994; Henderson, 1990; Slowinski, 1992). By incorporating some of these factors into the decision making process, phase 2 selects an effective combination of partners, thereby insuring that the participating candidates are efficient individually and as a group. For example, assume the following goals have been identified by the brokers as important in the formation of the VC.

Goal 1: Minimize the costs associated with the linkage.
These are technical hardware and software costs, standard costs, new equipment costs, new employee costs, and other costs associated with the formation of the linkage.

Goal 2: Minimize the distances among the participating candidates.
To maximize the frequency of personal interactions among the participants in order to respond quickly to the changing needs of the customer, the participating candidates must be geographically close to each other. This assists not only in improving the frequency of interactions but also helps in doing so at lower costs.

Goal 3: Minimize the inception time of the VC.
The inception time is defined as the time taken for a VC to be operational. To capture the market opportunity at the earliest, it is necessary to have relatively low inception times.

The goal programming (GP) model was first introduced by Charnes and Cooper (1961) for dealing with infeasible linear programming problems. The technique has been expanded to deal with variety of problems in areas such as finance, marketing, production planning, corporate planning, medicare planning etc., for more information on GP (see Lee 1972). Although, goal programming reflects complex reality and allows for multidimensionality of the objective function, it has some limitations. Primarily, the decision maker must either set priorities to goals, which in some cases is difficult to do, or assign weights to deviations which in most cases is subjective in nature.

To overcome the above difficulties, conversion factors between goals can be identified. Conversion factors can be determined between cost ($) and the remaining two goals; frequency of personal interactions (distance in miles used as a surrogate measure), inception time (days). These conversion factors are used as weights in the objective function of the goal programming model. The conversion scale between cost ($) and distance (miles) can be identified by estimating the cost incurred in covering a mile of distance. The relationship between cost ($) and inception time (days) can be determined by estimating the cost of lost sales for each day of delay in the inception process.

The integer (0-1) goal programming model is shown below. The combination that minimizes the sum of the weighted deviations from the 'best' goal compatibility measure is selected as the VC.

$$\text{Min } Z = w_1 v_1 + w_2 v_2 + w_3 v_3$$

$$\text{s.t} \quad \sum_{i=1}^{a} \sum_{j=1}^{b} \sum_{k=1}^{c} x_{ijk} = 1$$

$$\sum_{i=1}^{a} \sum_{j=1}^{b} \sum_{k=1}^{c} c_{ijk}\, x_{ijk} - v_1 = c_{min}$$

$$\sum_{i=1}^{a} \sum_{j=1}^{b} \sum_{k=1}^{c} d_{ijk}\, x_{ijk} - v_2 = d_{min}$$

$$\sum_{i=1}^{a} \sum_{j=1}^{b} \sum_{k=1}^{c} t_{ijk}\, x_{ijk} - v_3 = t_{min}$$

$$x_{ijk} = 0 \text{ or } 1$$

$$v_1, v_2, v_3, \text{ and } v_4 \geq 0$$

where: $i = 1$ through a.
$j = 1$ through b.
$k = 1$ through c.
a is the total number of efficient candidates of business process type A.
b is the total number of efficient candidates of business process type B.
c is the total number of efficient candidates of business process type C.
$x_{ijk} = 1$ implies that the combination of ith efficient candidate of process type A, j^{th} efficient candidate of process type B, and k^{th} efficient candidate of process type C was selected for the VC.
c_{ijk} is the cost associated with the formation of combination ijk.
d_{ijk} is the distance associated with combination ijk.
t_{ijk} is the estimated inception time associated with combination ijk.
c_{min} is the cost bound derived from the minimum cost combination.
d_{min} is the distance bound derived from the minimum distance combination.
t_{min} is the time bound derived from the minimum inception time combination.
w_1, w_2, w_3, and w_4 are the weights derived from the conversion factors for deviations from the four goal targets.
v_1, v_2, v_3, and v_4 are the deviations from the 'best' goal compatibility measure (goal target) for the four goals.

Limitations

In general, if the initial data is filtered in an optimization procedure, then the solution to the problem may be sub-optimal. This is also true for the proposed analysis. In the initial screening process using DEA, some units that are dropped out of the analysis due to inefficiency may in fact have better compatibility characteristics when analyzed in phase 2. To alleviate the severity of this problem, the weight restrictions must carefully be specified by the decision maker.

Conclusions

VCs are envisioned as the solution to meet the constantly changing needs of the customer at low cost, high quality, small lead times. In this paper, a quantitative model is proposed to aid the decision making process in the formation of a VC. DEA is utilized to identify the efficient business processes, and a multi-criteria decision making model is utilized to select an effective combination of the efficient business processes.

While the analysis in this paper is performed from the standpoint of brokers, it can also be performed from the standpoint of a lead business process. In such a situation, before becoming involved in the formation of a VC, the lead business process must evaluate its efficiency relative to other similar business processes and implement process improvement techniques to become efficient, if necessary.

In the formation of the VCs, there are other issues to be considered such as VC workers dividing their loyalty and responsibility between this new venture and their parent corporation, which may not be feasible in some types of alliances. Also, the possibility of losing confidential information through the alliance channels is a key deterrent to the effectiveness of the VCs. In spite of all these and other problems, many companies around the world are becoming involved in such temporary alliances to capture a specific market opportunity.

References

Boussofiane, A., Dyson, R.G., and Thanassoulis, E. "Applied data envelopment analysis." *European Journal of Operational Research*, 52, 1-15, 1991.

Byrne, A.J. "The virtual corporation." *Business Week*, February, 98-103, 1993.

Charnes, A., Cooper, W.W., and Rhodes, E. "Measuring the efficiency of decision making units." *European Journal Of Operational Research*, **2**(6), 429-444, 1978.

Charnes, A., and Cooper, W.W. Management models and industrial applications of linear programming, John Wiley & Sons, New York, 1961.

Gilbert, F. W., Young J.A., and O'Neal, C.R. "Buyer-seller relationships in just-in-time purchasing environments." *Journal of Business Research*, 29, 111-120, 1994.

Goldman, L.S. "Co-operating to compete." *CMA Magazine*, **68**(2), 13-17, 1994.

Iacocca Institute. 21st Century manufacturing strategy, Lehigh University, Bethlehem, PA.

Lee, S.M. Goal programming for decision analysis, Auerbach Publishers, Philadelphia, 1972.

Porter, A.L. "Virtual companies reconsidered." *Technology Analysis & Strategic Management*, 5(4), 413-420, 1993.

Presley, A., Barnett, B., and Liles, D.H. "A virtual enterprise architecture." Fourth Annual Agility Forum Conference, 2, 3-12, 1995.

Sheridan, J.H. "Agile manufacturing : A new paradigm." *International Productivity Journal*, 11, 39-48, 1993.

Uncovering Key Variables in Organisational Knowledge Management using Intensive Research Methods

W A Taylor
+44-1274-384424

Bradford Management Centre
University of Bradford
Bradford BD9 4 JL
United Kingdom

This paper argues from a resource based standpoint of the firm, that central to any organization's strategy must be a focus on the way in which it manages its core organizational knowledge base. While there may be an expanding literature in support of this notion there are few case studies of best practice. Using one case study from an MNC over a five year period, the paper shows how the knowledge management concept might begin to be realized in this instance, with particular emphasis on knowledge creation.

Introduction to the problem domain

Consider a typical continuous industrial process, producing high grade product for world markets and operating under conventional process management protocols based upon well developed theories of systems identification, modeling and control. The process had been operating for many years under conditions of infrequently varying demand. It employed a classical control strategy of proportional-integral-differential (PID) structure using torque and viscosity as feedback measurements.

Such control systems rely upon parametric models which fit the input and output data from the plant, in effect a black-box approach in the sense that these models do not rely upon any knowledge of the physical or chemical behavior of the plant. This is problematic in that:

(a) there is noise present in most signals,
(b) most adaptive control techniques assume linear models, which is rarely the true practical situation,
(c) in situations of more frequently varying demand, such linear strategies break down.

Traditionally, control theory had developed a wide variety of techniques to address some of these problems, most recently with the use of neural networks for nonlinear system modeling. Increasingly, reliance upon such 'black-box' approaches is believed to be insufficient. Resource-based theory points instead to the management of knowledge assets as a source of competitive advantage.

The objective of this investigation therefore was to explore what knowledge assets meant in this continuous process context, what the management of such knowledge assets would entail and to what extent the plant's management practices were knowledge based, if at all.

Methodology

A single in-depth case study was chosen as the initial research instrument. While recognizing the various methodological issues concerning case study research, it was chosen because, *inter alia*, case study research "seeks to understand the problem being investigated" (Gable, 1994) and it also allows valuable insights to be gained into new topics emerging in the rapidly changing information systems field, (Benbasat et al, 1987), especially a previously unresearched subject.

0-7803-3552-X/96/$5.00 © 1996 IEEE

The plant was chosen based on its global market position and its willingness to participate. Moreover, this corporation was frequently cited for its innovativeness, technological excellence and receptiveness to adopt new ideas and technologies. To study these knowledge management issues the author gained access to a research project which had been commissioned by one operating site of the corporation.

The aim of the research project was to improve the core manufacturing process beyond its current yield capability by more cost effective methods. No specific objectives were formulated although it was known that the business had been given overall (but commercially sensitive) process yield improvement targets. The management of the operating site committed a budget of $0.5 million to the project, justifying the investment on the basis that any improvements could potentially be applied to its sister plants worldwide and that 'emerging technologies' would be used in ways which would increase staff understanding of the process.

A project team was established, comprising external process experts in control and modeling and staff members of the operating unit from both line management and support functions, especially process control. The project proceeded over a five year period (1990-1995), during which significant process improvements were achieved. Specifically the site achieved yields which were 10 percent better than any of its sister plants worldwide and grew in its reputation as an organizational center of excellence. Neural networks were utilized in the process solutions.

For the whole of the project the author attended project meetings and presentations and received all documentation. What is discussed below are some of the salient observations which arose during this project over the five year period.

Establishing existing process knowledge

Although a long established plant, operational data of trends in system inputs and outputs was not readily available for analysis. There was a dearth of information on system performance even though the computer based control systems had the potential to supply all of the necessary data signals. Management tended to rely instead upon inspection oriented results of final product yield as an indicator of process performance.

Since there was an absence of historical records, data was collected on all system input signals and also on the subsequent behavior of the feedback signals. Analysis using modern identification and signal processing tools identified that there were two key variables for system modeling (viz.; torque and viscosity), a fact which, until that point, was not known widely within the corporation. In effect many of the system measurements being routinely collected were shown not to be necessary for effective management of this technology.

Analysis revealed two other important facts, (a) high levels of noise on one of these two key feedback signals which was actually causing very poor plant performance, and (b) that there was a significant time delay between the two feedback variables. Interestingly these additional factors were also unknown to everyone associated with the management and operation of the process. These were in fact contributing to the perceived problem that the control strategy was difficult to 're-tune' under conditions of changing demand. In addition, it could be shown that the plant was close to instability for most of the time.

Already therefore, basic analysis of data had yielded some additional information about process behavior. However, to explain why these problems existed required additional insights into the process, i.e. to transform the information about noise and time delays into knowledge about why they occur. Moreover, an even more fundamental question was raised as to why this sort of investigation had not previously been carried out at all and why corporate policy did not routinely require such investigations.

Developing a basis for improvement

Building on this new knowledge, the system was redefined as two sub-systems, one modeling plant behavior and the effect of system disturbances, and the other modeling the resultant fluctuations including the newly discovered time delay. Linear models were then developed for each of these sub-systems and improved control achieved through application of predictive modeling using a technique known as Smith Prediction, based on a Neural Network architecture with Multi-layer Perceptrons. System oscillations reduced by as much as 80 percent and more significantly, the plant began to operate at levels 10 percent better than any of its sister plants world-wide.

This significant improvement utilized techniques which were readily published in the research literature but still unknown within the corporation. The techniques could not however be regarded as 'emerging technologies' and to that end were not particularly welcome as a solution within the organization - there were expectations that something more radical and 'leading-edge' must be needed to deal with such a complex and long standing problem. After all, this was regarded as an excellent organization listed in the Fortune 500.

As regards senior site management, once real success was being demonstrated, they began to use the project as a lever to release other funds within the corporation, particularly with respect to new capital investment which went as a matter of policy to the most successful sites. Line management changed their stance from distant skeptics to slightly less distant supporters; in fact, to the rest of the organization they appeared to be even more positively inclined, akin to enthusiastic project champions.

The reaction of the corporation's other sites worldwide was twofold. Firstly there was a move to ensure that this so called improvement was technically valid; consequently a series of site visits took place to conduct confirmatory project appraisals. Secondly, there was a lobby to involve the central resource groupings which should have been generating such solutions. Questions were being formed as to why these central (and expensive) 'experts' were not involved in this project from the outset.

Expediency caused site management to accede to this pressure. Subsequently there was further centralized checking and rechecking of the analysis and the resultant issuing of copious reports, apparently to claim some of the credit. It is perhaps instructive that the real technological improvement was emanating from the periphery of the organization, almost in spite of the organization's efforts to take control of, and credit for it.

The knowledge spiral

The process technology was now being managed more effectively, but there were still inherent problems to be addressed. Specifically the control strategy was not adaptive to changes in flow rates caused by fluctuating demand patterns. So long as demand remained fairly stable with only a few changes per month, it was possible to maintain the process at satisfactory output levels. However, the business environment was increasingly volatile and the site was encountering a six-fold increase in flow rate changes. Therefore, even if the initial strategy had been allowed to remain, the process would have become increasingly unmanageable. It was also discovered that the time delay between the torque and viscosity signals was not constant, but varied with changes in customer demand levels.

Nevertheless the project was regarded as successful by the corporation. It was completed on time and within budget to the satisfaction of the principals. It achieved process improvements exceeding expectations and these were subsequently applied to the other four manufacturing lines within the site being studied.

The new process knowledge uncovered during this project included:

The identification of torque and viscosity signals as the two key signals for the modeling and control of this system.
The demonstration that many of the measurements being routinely gathered were consequently unnecessary.
The identification of noise on one of the key feedback signals.
The presence of a significant time delay between the two feedback signals.
The constant proximity of the system to instability.
The need to re-conceptualize the system as two linear sub-systems, leading to new control strategies.
The identification that the time delay was not constant but varied with demand.

Discussion

This is not simply a report of a successful consultancy project, nor even should the results be dismissed as merely the normal product of research. Rather, the project suggests several important organizational points, including the need to get very close to a problem domain in order to explore the complexities of interactions in socio-technical systems. Moreover, the investigative process became the stimulus for those involved to review and reflect upon what they knew about this continuous process and its behavior, and more importantly to highlight important gaps in their knowledge. Significant improvement was achieved through enhanced understanding and knowledge of the process itself. What is more, knowledge was generated about the ways that such technology was being managed within the corporation and about the organization's behavior. In particular:

The absence of policy for analysis of data, even to produce information, let alone knowledge.
The degree of disengagement of senior management from such fundamental projects.
The reactions of the rest of the corporation, especially centralized groupings with responsibility for such work.

Despite claims of an "organization-wide improvement culture" and practice of the usual raft of fashionable business improvement techniques elsewhere in the organization, it nevertheless seemed as though the internal dominance of the control specialists' mindset and their pre-occupation with stability and control was unconsciously stifling and even blinding them to the possibility of further process performance improvements.

As regards the management of organizational knowledge assets there was very little evidence of this being done. It was clear at the outset that there was little data analysis even in this data-rich environment and without this there was unlikely to be any understanding of the noise on the two key process variables, or of the time delay between them. It seems strange that such fundamental analysis was not routinely done as a matter of policy, yet it was in keeping was the similarly hands-off way that the project was commissioned. While a 'leap of faith' to use new technology can be commended, it certainly does not excuse the lack of project management targets linked to business objectives.

Even more surprising was the reaction of line management to the actual solution. Since it did not meet with their expectations about leading-edge technology there was a sense almost of disappointment, even though the plant was now more competitive than any of the others in the corporation. Senior plant management were however content to use the improvement as a basis for levering more funds from the internal bidding process, even at the price of allowing outside interests to claim some of the credit for this work through their attempts at validation described above. At the policy level, it is worth noting that the process improvements were originating at the periphery of the organization rather than from the centralized resources whose official task this was.

Conclusions

This paper has highlighted the process of knowledge creation and its management. This is but one aspect of a much larger process leading to continuous innovation and competitive advantage (Nonaka and Takeuchi, 1995). This case also suggests that achievement of such knowledge orientated exploration must be a policy mandated activity in organizations, couched in a value system which encourages such deeper level self-awareness and self-knowledge. Related practices for the reward of line management and the basis of inter-site choices for future capital investment may also need to be reviewed in the light of this.

While black-box methods can extract patterns from data, sufficient to enable process modeling and control, deeper exploration is required to generate knowledge assets based upon causality and understanding (Bonn, 1994). The need for information systems to embrace this kind of thinking will continue to grow as (a) firms increasingly compete on the basis of their resources and (b) trends towards outsourcing persist.

While some techniques exist, there appears to be a need for better formed organizational policies and procedures to routinize such knowledge management. The case study presented suggests that technology-intensive organizations may have difficulty adjusting to these ideas, particularly in process industries where there is a dominant mindset of process control. However, this pilot case study of a new and under-researched topic must be complemented by more examples from a multiple case design to see if any of these issues are more generally true.

References

Benbasat, I; Goldstein, D.K. and Mead, M. "The case study research strategy in studies of information systems." MIS Quarterly, September 1987, pp369-386.

Bonn, R.E. "Measuring and managing technological knowledge." Sloan Management Review, Fall 1994. pp61-73.

Collis, D.J. and Montgomery, C.A. "Competing on Resources: Strategy in the 1990s." Harvard Business Review, July-August 1995, pp118-128.

Gable, G.G. "Integrating case study and survey research methods: an example in information systems." European Journal of Information Systems, vol 3, No 2, 1994, pp112-126.

Nonaka, I and Takeuchi, H. "The Knowledge Creating Company." Oxford University Press, New York, 1995.

Managing with technology: a predictive information system for remote supplier development.

W A Taylor and W Doherty
+44--1274--384424

Bradford Management Centre
University of Bradford
Bradford
West Yorkshire
BD9 4 JL
United Kingdom

In technology-based enterprises, the achievement of networked and virtual organisations relies heavily upon the capability to develop suppliers and sub-contractors. While in many cases, suppliers are compliant and will embrace most initiatives in order to maintain contractual relationships, they often do so at additional cost to themselves. Traditional interventions such as second or third party audit, or other criteria-based assessments tend not to achieve genuine and sustained supplier improvement. Moreover, the more recent self-assessment movement is difficult to resource in many supplier organisations because of their (small) size and their remoteness of location.

To this end, the paper details the use of software systems to act in quasi-consultancy mode, by predicting the likely organizational outcome of current management practices. Since the target users are senior managers in these small organisations, it is intended that the reflective style of the system will stimulate genuine learning. The paper discusses the process of developing the predictive system and the validation of predictions to-date in around 100 companies, predominantly employing less than 500 people. The current predictive accuracy of the system is over 80%. If companies were to heed these predictions there would be a marked change in their management practices and in the resultant competitiveness of the majority.

Introduction

"Management is prediction. The theory of knowledge helps us to understand that management in any form is prediction." (Deming, 1993, p104).

Developing the supply chain and facilitating suppliers to move towards improved, and desirably world class performance, is becoming a strategically significant issue. However, when the suppliers are SMEs there are specific challenges which arise. For example, in this size of organization there is a greater scarcity of slack resources and senior managers tend to be more resistant to outside intervention. Moreover, methodologies and improvement techniques developed from work with larger organisations tend to be less easily adopted in SMEs. A variety of alternatives have been proposed, often based upon action learning principles. Learning in organisations, including SMEs can be stimulated by causing people to reflect and analytically assess what they are doing. Harnessing the capabilities of modern software tools can facilitate this reflection and self-assessment, by providing managers with interactive opportunities to explore both their own organisational practices and the implications of these practices for future survival and growth.

One such tool, ProGnosis VI© has been developed using object oriented technology, to help managers explore their knowledge of their own key business processes and the extent to which these processes are likely to be sustainable and to improve the business in the medium term. The approach adopted in ProGnosis VI© is to pose questions to the user about key managerial practices and then to make predictions about whether or not those practices will lead to significant business improvement in the two to three year time frame. In so doing there are two key learning stimuli.

0-7803-3552-X/96/$5.00 © 1996 IEEE

Firstly the manager encounters questions which may or may not be considered important by him, but which have been shown nevertheless by research to be significant in this context. He therefore assimilates better practice by program use. The program also provides back-up amplification of questions and clarification of its lines of reasoning, also with a view to imparting knowledge of better practices to the user. Secondly, the resultant predictions may surprise, even shock the user into re-appraising the consultation with the system and reflecting even more upon the dialogue. Whether or not he agrees with the predictions, he is challenged to face up to a system's assessment of his practices and their implications, based upon a statistically validated model.

The assessment framework has been developed from work done with SMEs since 1988 (Taylor and Trainor, 1991). It was originally focused on traditional business performance measures of cash flow, profitability and sales growth. However, it has now been revised to emphasise process performance and process management practices. In some ways the approach now resembles a Baldrige type framework, except that it has been simplified and adapted for SMEs. However, ProGnosis VI© is certainly not a simple checklist-based replica of a Quality Awards framework. The framework currently covers six fundamental processes, viz.;

- Strategic leadership process
- Market/customer orientation process
- Information management process
- Employee management focus
- Process orientation and management
- Organisational improvement process

In each of these areas of exploration, there is a set of questions to establish the input data to the system, essentially delineating current practices. This, combined with data on business results, enables ProGnosis VI© to make an assessment of the strengths and weaknesses of current practice. These strengths and weaknesses are then fed forward to a prediction module which categorises the likely consequences of current practices into one of the following four output classes:

Dramatic business benefits - there are likely to be significant business benefits with no major weaknesses in practice being found.
Moderate benefits - there appear to be some weaknesses which will cause outcomes to be less than desired.
Marginal benefits - there is evidence of weakness in practices which will result in wasted effort and frustration.
Failure - the initiatives examined are at risk of being discontinued due to management disappointment with the outcomes.

These output categories were chosen to express the expected consequences of current practice on organizational performance. Intuitively, organizations form their own opinions about whether such practices have been successful and based on these opinions they will continue with, or abandon their efforts.

Methodology

The system was applied to responses from senior managers in 113 organizations, gathered during 1992. These organizations were part of a statistically representative sample of 682 predominantly small and medium sized (<500 employee) manufacturing organizations from throughout the Northern Ireland region. This sample represented about 22% of the total regional population of such firms. All of these 113 organizations were claiming to be actively trying to improve organizational performance, primarily through a continuous improvement initiative such as Total Quality Management. This was important since there was little point applying the assessment framework to companies who, by their own admission were not seeking to improve.

Some preliminary analysis of this data had suggested that some of those respondents were not seeing themselves as clearly as they ought. "Many are placing a disproportionate emphasis on internally focused activity .... time will reveal whether these initiatives develop and take root in their organizations in the longer term." (Taylor, 1995). The ProGnosis VI © system predicted outcomes as follows: Dramatic benefits 2%, Moderate 19%, Marginal 52% and Failure 27%. The predictions seemed to support earlier analysis of these organizations, wherein it was argued that "the prognosis seems to be that many of these initiatives will stall or die in the next few years without some significant intervention" (Ibid.). Of course when these comments were written there was no way of exploring the argument further.

The 113 organizations were re-contacted during 1995/96 as part of the validation process for the predictions of the ProGnosis VI© information system. Where practicable, the original respondents were used in this follow-up phase. A total of 92 companies participated in this second phase. The details of the research instrument for this follow-up phase are not included in this paper for sake of space but may be obtained by contacting the first named author.

Predictive accuracy of the system

The predictions for each of the 92 follow-up organizations were first of all compared to their actual claimed outcomes, in order to establish the program's predictive accuracy (at least relative to the perceptions of the respondents). This accuracy level, expressed in terms of the percentage of outcomes forecast correctly, is then discussed in greater detail, to enable some conclusions to be drawn as to the merit of the approach.

Predictive accuracy with four outcome categories

The predictions from the ProGnosis VI© system were presented above and the subsequent (claimed) outcomes for the initiatives in 1995/96 are reported in Figure 1 below.

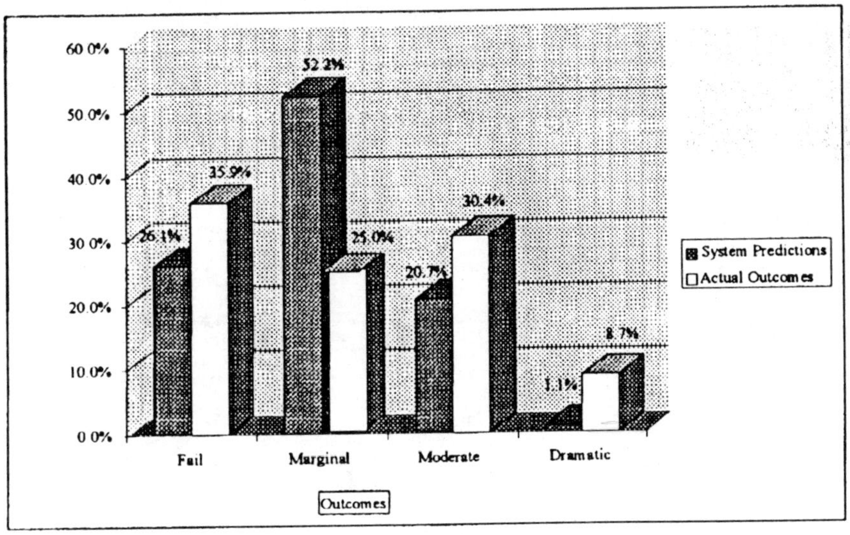

Figure 1: ProGnosis VI System Predictions and Actual 1995/96 Outcomes (n=92)

The actual outcome claimed by each individual organisation, was matched with the predicted outcome and the resultant predictive accuracy of the system, is as shown in Table 1.

Table 1: Accuracy of ProGnosis System Predictions. (Figures are % of total in each column).				
Actual Outcomes (Level of claimed improvement)	Predictions (%)			
	Failure	Marginal	Moderate	Dramatic
Failure	75 (n=18)	23	16	100
Marginal	13	40 (n=19)	5	-
Moderate	13	33	47 (n=9)	-
Dramatic	-	4	32	- (n=0)
Total	100% (n=24)	100% (n=48)	100% (n=19)	100% (n=1)
(n=92)	Key Statistics: τ b=0.46;			

Predictive accuracy with two outcome categories

If Failure or Marginal Benefits are taken to represent *Unsuccessful Outcomes*, and Moderate or Dramatic Benefits are taken to represent, on the whole, *Successful Outcomes* then a more general level of analysis is possible. This approach is a valid one since it can be argued that there is a potential lack of discrimination between such ratings as Moderate or Dramatic, or between Marginal Benefits and Failure. The actual outcomes compared to the predicted outcomes for each case, using this more general two category scenario, are shown in Table 2a.

Table 2a: Accuracy of ProGnosis VI System Predictions for Two Category Scenario. (Figures are % of total in each column)		
	ProGnosis VI Predictions (%)	
Actual Outcome	Failure	Success
Failure	70.8 (n=51)	25.0
Success	29.2	75.0 (n=15)
Total	100% (n=72)	100% (n=20)
(n=92)	Key Statistics: Sig. χ^2 = 0.0002; Cramer's V= 0.39; τ b = 0.39	

Discussion

Based on the two outcome analysis, 71 per cent of unsuccessful initiatives, and 75 per cent of successful initiatives were predicted correctly giving the program a general accuracy level of 72 per cent. This has potential benefits for users, offering a consultative tool whose intent is to stimulate reflection. The majority of system prediction inaccuracies arose from incorrectly forecasting many successful initiatives as failing.

An alternative view of Table 2a can be gained by looking at the data from a different perspective, as shown in Table 2b.

Table 2b: Identification of Actual Outcomes for Two Category Scenario. (Figures are % of total in each row)			
	ProGnosis VI Predictions		
Actual Outcome (%)	Failure	Success	Total
Failure	91.1	8.9	100%(n=56)
Success	58.3	41.7	100%(n=36)
(n=92)	Key Statistics: as for Table 2a.		

Some 91 per cent of unsuccessful initiatives were correctly identified, underlining the ability of the software to expose future troubled implementations. If all of the firms in this study had heeded these predictions then clearly a great many failures could have been avoided. To have identified such areas for improvement when firms were still committed to improvement initiatives is very important, otherwise the all too common disillusionment may set in. In terms of improving the supply chain, especially in SMEs, this must surely have self evident benefits for all concerned.

ProGnosis VI© offers an alternative method for the practice of self-assessment in SMEs. Moreover it offers additional benefits in that it encourages a more honest and open dialogue from the user since a third party is absent. Being computer based it is more likely to fit with the work patterns of busy managers in this size of firm.

The whole area of man-machine interaction was not intended to be part of this particular phase of the research. The core focus was upon establishing the knowledge base and its predictive accuracy. Clearly, there are further steps needed to render such tools useable in the business environment. It is the authors' view that this may be better effected in a commercial rather than a research setting.

References

Deming, W.E. "The new economics for industry, government and education". MIT Press, 1993.

Taylor, W.A. and Trainor, J.J. "Knowledge based diagnosis of business performance". In: Advances in Manufacture. Ed; D.M.J. Harris. University of Ulster. 1991, pp466-479.

Taylor, W.A. "TQM and the need for organizational self assessment: Some empirical evidence." International Journal of Total Quality Management, vol. 6, No 1, 1995, pp3-12.

Managing Technology Transfer For Value Creation And Competitive Advantage: Toward A Contingency-Based, Information Processing Model

Russell W. Teasley
James G. Almeida
Richard B. Robinson, Jr.
803-777-5903
803-777-6876 (FAX)

Internet: rteasley@darla.badm.sc.edu
Internet: jalmeip7@darla.badm.sc.edu
Internet: robinson@darla.badm.sc.edu
Department of Management
College of Business Administration
University of South Carolina
Columbia, SC 29208

A contingency-based information processing model of technology transfer is presented. Effectiveness of this transfer process is dependent upon the fit achieved between information processing demands of the technology and the information processing capabilities of the receiving group. Group characteristics are presented as potential moderators of this relationship.

Introduction

Technology is essential to corporate value creation. Twenty-first century technologies have revolutionized traditional industry structures. Yet downsizing, restructuring, and cost containment constrain companies' internal technology development. Therefore managers increasingly rely on external sources of technology and the transfer of those technologies into their organizations. How can managers reenergize their technology transfer systems to more effectively build value and gain competitive advantage? Technology transfer is the relocation of intellectual property, processes or know-how from a source to a receiver. Transfers are effective for receivers to the degree that they: a) are institutionalized, and b) add value. The receiver-active framework (Kodama and Morin, 1993) argues that technology transfer effectiveness is a function of a receiver's ability to absorb information from its technology supplier and other relevant sources.

Information Processing

Two situational dimensions are useful to describe technologies (Perrow, 1967; Weick, 1990): a) uncertainty, which is the degree that a receiver possesses needed information about a technological situation, and b) "equivocality" which is the degree that a technology is ambiguous to its receiver. Considered together, these two dimensions determine a technology's "information processing requirements". They can be used to order technologies into four unique categories: routine, craft, engineering and nonroutine technologies (see Figure 1).

Managers should consider the informational structure of receiver groups when transferring new technologies. They should structure a group's "information processing capabilities" to fit the information processing requirements of the technologies (Tushman and Nadler, 1978). This is accomplished when the "amount" of processed information matches the uncertainty of a technology and when the "richness" of the information matches the "equivocality" of the technology (Daft and Lengel, 1986). Utilizing planning, incentives and resources managers can influence the amount of information a group processes. They can also influence

0-7803-3552-X/96/$5.00 © 1996 IEEE

richness of the information. The information richness refers to the group's ability to enhance understanding and, therefore, reduce equivocality. Each category of technology will represent unique information processing requirements. Figure 1 shows both the information processing requirements for the four technologies and corresponding information processing capabilities of receiver groups.

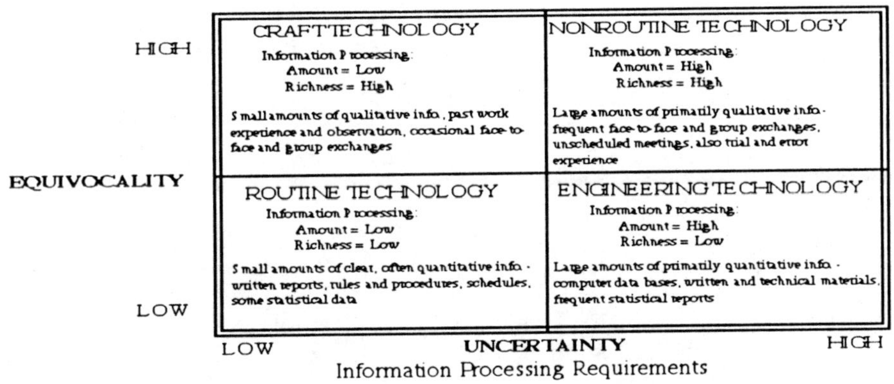

Adapted from Daft and Macintosh (1981)

When receiver groups develop their information processing capabilities appropriately they achieve a "fit" (Drazin & Van de Ven, 1985) with the requirements of a technology transfer (See Figure 2). While fit leads to greater levels of technology transfer effectiveness, misfits create inefficiencies which reduce technology transfer effectiveness. For example, as technology transfer shifts from routine to craft environments, the appropriate response is to increase richness but only in low amounts. Higher amounts of rich information are unnecessary and generate inefficiencies because richer face-to-face interaction is expensive and time-consuming to accomplish. Similarly, as transfers shift from routine to engineering environments, the appropriate reaction is to increase only the amounts of lean, objective data. Figure 2 shows the expected relationships between information processing requirements, information processing capabilities, fit and technology transfer effectiveness. The following propositions are based on the depicted relationships.

Technology Transfer Model

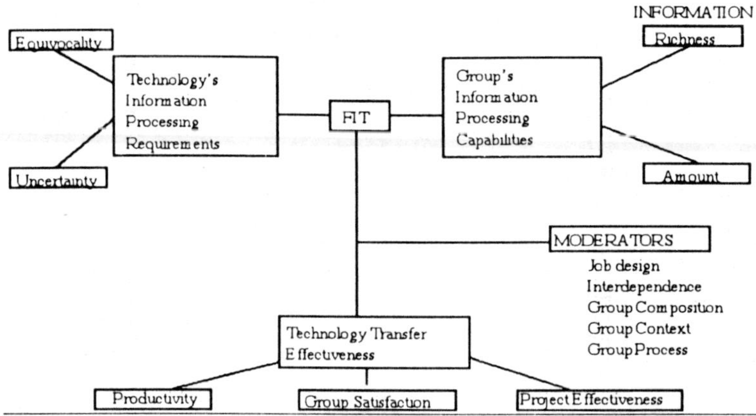

Proposition 1: Transfer of routine technologies is most effective when receivers process minimal amounts of information which is low in richness.

Proposition 2: Transfer of craft technologies is most effective when receivers process minimal amount of rich information.

Proposition 3: Transfer of engineering technologies is most effective when receivers process large amounts of information which is low in richness.

Proposition 4: Transfer of nonroutine technologies is most effective when receivers process large amounts of rich information.

Proposition 5: Deviations from the fit between a technology's "information processing requirements" and of the receiving unit's "information processing characteristics" will reduce the technology transfer effectiveness.

Certain work-group characteristics will moderate the fit-effectiveness relationship. Five group-related factors have been identified from the work group literature (Campion, Medsker, & Higgs, 1993): "Job Design", "Interdependence", "Group Composition", "Group Context", and "Group Process".

Job Design can foster employee satisfaction, escalate motivation and increase group effectiveness. Its characteristics include **self-management, participation, task identity, task variety**, and **task significance.** Self-management refers to the level of autonomy accorded to group members and has been associated with effective work groups (Hackman, 1987). Task variety is associated with making group members responsible for a greater assortment of group work tasks and has been found to motivate workers (Hackman, 1987). Widespread exposure to new and different tasks generates organizational learning, a driving force to enable the absorption of externally-created technologies. Task identity is the degree to which a group completes a whole and separate piece of work. Task significance is the belief of group members that their group work results in significant consequences for its stakeholders. While task significance motivates through a shared vision (Hackman, 1987), task identity has been shown to increase group cooperation and effectiveness (Cummings, 1978). While research supports a linkage between job design and group performance

(Campion et al., 1993), it is likely that a point of diminishing returns exists where groups experience "too much of a good thing".

Proposition 6: Increases of self-management, participation and task variety up to the optimal level will moderate the effects of "fit" on "technology transfer effectiveness" in a positive direction. Increases beyond the optimal level will moderate the effects in the opposite direction. Increases of task identity and task significance will moderate the effects in a linear, positive direction.

Interdependence refers to the structure of interactions occurring among and between the members of a group and is comprised of three dimensions: **task interdependence, goal interdependence,** and **interdependent feedback and rewards**. Task interdependence is the extent to which group members interact and depend on one another to accomplish their tasks. It increases group effectiveness by enhancing a person's responsibility for the work of other group members (Kiggundu, 1983). Goal interdependence is the degree to which individual group members' goals are consistent with the goals of the group and those of the other members. Research has linked group effectiveness with clear goals and interdependent purpose (Guzzo and Shea, 1992; Hackman, 1987). Interdependent feedback and rewards refers to the dependence of individual feedback and rewards on the performance of the entire group. When individual rewards and feedback are aligned with group performance, members are more likely to be motivated towards group goals and increase group effectiveness (Pritchard, Jones, Roth, Stuebing, & Ekeberg, 1988).

Goal interdependence should share a linear relationship with performance, yet increases of task interdependence beyond a certain point will become dysfunctional. Excessive task diversity will deter individual focus and the concentration of group members is likely to suffer. Interdependent feedback and rewards must be mixed with appropriate amounts of individual recognition, else the motivation for individual creativity and problem solving will be compromised.

Proposition 7: Increases of task interdependence and interdependent feedback and rewards up to the optimal level, will moderate the effects of "fit" on "technology transfer effectiveness" in a positive direction. Increases beyond the optimal level will moderate the effects in the opposite direction. Increases of goal interdependence will moderate the effects in a linear, positive direction.

Group Composition contributes to effectiveness through a synergy built from the diverse experiences and capabilities possessed by group members. It is characterized in terms of its **heterogeneity, flexibility, relative size,** and **Not-Invented-Here (NIH) levels.** Heterogeneity is the degree to which group members possess diverse sets of competencies and experiences. Flexibility is the degree to which members are able to perform different job assignments. It enhances technology transfer effectiveness by strengthening members' abilities to complement and fill in for each other's tasks. Groups must be of sufficient size to accomplish their tasks, yet those which are too large can create coordination problems resulting in reduced involvement and subsequent detachment by individual members (Goodman, 1987). NIH levels refers to a group's tendency to discount and/or reject knowledge and solutions from outside the group (Katz, 1982). It hinders the group's ability to absorb and utilize outside knowledge.

Proposition 8: Increases of heterogeneity and group size up to an optimal level will moderate the effects of "fit" on "technology transfer effectiveness" in a positive direction. Increases beyond the optimal level will moderate the effects in the opposite direction. Increases of flexibility and decreases of NIH levels will moderate the relationship in a linear, positive direction.

Group Context affects the group's performance. The effectiveness of technology-intensive groups has three contextual determinants: **training, managerial support** and **technology policy**. Group training programs develop a group's philosophy, enhance its creativity, develop its knowledge-base, and sharpen its decision-making and interpersonal skills. Management support can also influence a group's innovation and technological capabilities. Management should provide not only physical and human resources when needed but also motivational rewards for creative, risk-taking behaviors (Damanpour, 1991). Technology policy is the degree to which technology is emphasized in a company's strategy. It should explicitly define technology transfer strategy, actively assess its entrepreneurial talent, break down functional barriers, and elevate innovative projects to the highest priority (Kanter, 1983). Research has linked technology policy with organizational performance (Ettlie, 1983).

Proposition 9: To the degree that "group context" characteristics are institutionalized within technology transfer projects, the effects of "fit" on "technology transfer effectiveness" will be moderated in a linear, positive direction.

Group Process describes the dynamics occurring within the boundaries of groups. These dynamics can be characterized in terms of **group potency, work-load sharing,** and **group cohesiveness**. Potency is the extent to which a group believes that its work will be both productive and meaningful. It builds positive attitudes and team spirit within a group. Work-load sharing reflects the degree to which group members are willing to do equitable amounts of work and is a measure of individuals' attitudes toward their own productivity. It enhances effectiveness by reducing the incidence of social-loafing or free-riding (Albanese & Van Fleet, 1985). Cohesiveness describes the strength of members' desire to remain in the group and their commitment to it. It has been found to be associated with R&D group performance (Keller, 1986). The process characteristics should influence effectiveness in a continuous manner. It is expected that potency, workload-sharing and cohesiveness will each associate positively with technology transfer effectiveness.

Proposition 10: To the degree that "group process" characteristics are institutionalized within technology transfer projects, the effects of "fit" on "technology transfer effectiveness" will be moderated in a linear, positive direction.

Conclusions and Implications

This model offers insights for managers who are designing technology transfer strategies. These managers must consider the information requirements of each technology uniquely and structure groups to best accommodate reception of the technologies. They should provide groups with information suitable infrastructures. They must also consider group structure and dynamics. Through proactive planning and control they can energize innovation, stimulate technology transfer effectiveness, and gain competitive value for their firms.

References

Albanese, R. & Van Fleet, D. "Rational behavior in groups: The free-riding tendency." Academy of Management Review. 10: 334-365, 1985.

Campion, M., Medsker, G., & Higgs, A. "Relations between work group characteristics and effectiveness: Implications for designing effective work groups." Personnel Psychology. 46(4): 823-850, 1993.

Cummings, T. "Self-regulating work groups: A socio-technical synthesis." Academy of Management Review. 3: 625-634, 1978.

Damanpour, F. "Organization innovation: A meta-analysis of effects of determinants and moderators." Academy of Management Journal. 34(3): 55-590, 1991.

Daft, R., & Lengel, R. "Organizational information requirements, media richness, and structural design." Management Science. 32: 554-571, 1986.

Drazin, R., & Van de Ven, A. "Alternative forms of fit in contingency theory." Administrative Science Quarterly. 30:514-539, 1985.

Ettlie, J. "Organizational policy and innovation among suppliers to the food processing industry." Academy of Management Journal. 26(1): 27-44, 1983.

Goodman, P., Ravlin, E., & Schminke, L. "Understanding groups in organizations." In B. M. Staw, & L. L. Cummings (Eds.), Research in Organization Behavior. 9: 121-173. JAI Press: Greenwich, CT, 1987.

Guzzo, R., & Shea, G. "Group performance and intergroup relations in organizations." In M. D. Dunnette, & L. M. Hough (Eds.), Handbook of Industrial and Organizational Psychology. Consulting Psychologists Press: Palo Alto, CA, 1992.

Hackman, J. "The design of work teams." In J. W. Lorsch (Ed.), Handbook of Organizational Behavior. Prentice-Hall: Englewood Cliffs, NJ, 1987.

Kanter, R. The Change Masters. Simon & Schuster: New York, NY, 1983.

Katz, R. "The effects of group longevity on project communication and performance." Administrative Science Quarterly. 27: 81-104, 1982.

Keller, R. "Predictors of the performance of project groups in R&D organizations." Academy of Management Journal. 29(4): 715-726, 1986.

Kiggundu, M. "Task interdependence and job design: Test of a theory." Organization Behavior and Human Performance. 31:145-172, 1983.

Kodama, F., & Morin, W. Report of the US-Japan technology transfer study panel. National Technical Information Service, US Department of Commerce: Springfield, VA, 1993.

Perrow, C. "A framework for the comparative analysis of organizations." American Sociological Review. 32: 194-208, 1967.

Pritchard, R., Jones, S., Roth, P., Stuebing, K., & Ekeberg, S. "Effects of group feedback, goal setting, and incentives on organizational productivity." Journal of Applied Psychology. 73: 337-358, 1988.

Tushman, M., & Nadler, D. "Information processing as an integrating concept in organization design." Academy of Management Review. 3 (3): 613-624, 1978.

Weick, K. "Technology as equivoque: Sensemaking in new technologies." In P. Goodman, L. Sproull, and Associates (Eds.), Technology in Organizations. Jossey-Bass: San Francisco, CA, 1990.

The landscape of innovation and supply: A graphical model for visualising the interrelationship between the innovation functions and the supply chain

L. Thurlings
tel. +31-40-2744362

Philips Research Laboratories
Prof. Holstlaan 4,
5656 AA Eindhoven,
The Netherlands

The conventional way of presenting an organisation's structure is by means of an organogram, which primarily shows the hierarchy, the functional arrangement and management control within the organisation. However, where it is necessary to visualise the interrelationships between organisational units, particularly the interconnections arising from the supply processes and the innovation processes, organograms are not adequate. The model described in this publication does offer this capability and is therefore eminently suitable for visualising the organisational implications of, say, centralisation versus decentralisation, concurrent en~ineering, technology push, etc.

Keywords: *innovation; value chain; organogram; business process; supply process; product creation process; innovation process*

Introduction

The aim of the model proposed below is to visualise innovation functions such as Research and Development and the activities in the value chain in their respective contexts. Using this model makes it much easier for managers and individual experts to see which neighbouring activities determine the future of their own activities.

The principles of the model

Basically, the model is quite simple. Before a produa can be made, supplies (ingredients), manufacturing tools and instructions on how to manufacture it are needed.

Figure 1 shows the graphical representation: the rectangular block represents the (geographical) production site. The arrow emerging at the bottom shows the outflow of goods and the arrow entering at the top shows the inflow of components, modules, raw materials, etc. The arrow on the right represents the production tools required in the production process, while the arrow on the left shows the manufacturing instructions.

Ingredients, production tools and instructions have to be generated somewhere. These geographical - 'production' sites are depicted in the model by rectangles.

0-7803-3552-X/96/$5.00 © 1996 IEEE

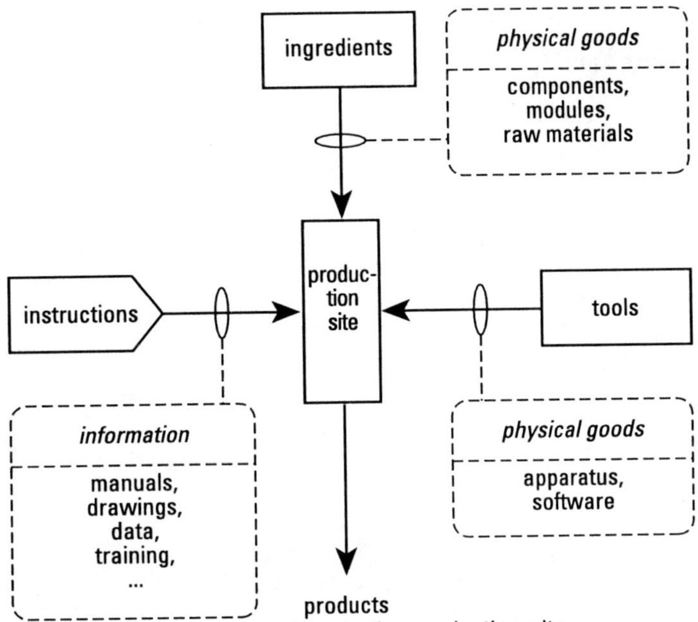

Fig.1: The nature of the deliverables to the production site

The production site for tools will usually be a tool workshop or tool supplier. But the 'production site' in the case of 'ingredients' may be a coal mine, for example. The 'production sites' for the 'instructions' are the research, development and engineering laboratories (called the 'innovation sector'). The output from both the 'tools production site' and the 'ingredients production site' is physical goods. The output from the 'instructions site' is reproducible information such as manuals, drawings, data, training, etc.

The products made by the 'production site' are ultimately supplied to the consumer. This is shown in the model by a horizontal baseline.

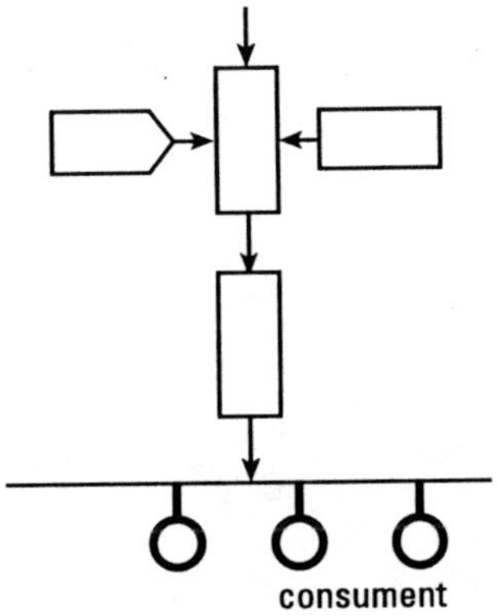

Fig.2: The individual consumer in the consumer market (three of them in the drawing)

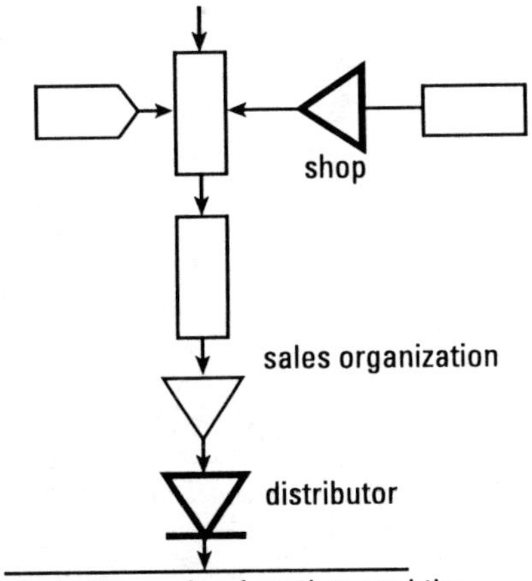

Fig.3: The sales function and the physical distribution function

Below this~line is the large market of anonymous consumers. If a single individual consumer needs to be indicated, this is done by means of a circle linked to the baseline; see Fig.2.

The 'production site' does not normally deliver direct to the customer. This is taken care of by the sales organisation. In the model this function is shown by a triangle pointing in the direction of the goods flow; see Fig.3.

Similarly, the physical distribution of the products is shown by a triangle with a crosswise bar on the point; again see Fig.3. This bar expresses the irreversibility of the supply process (i.e. once the goods have been shipped they are not meant ever to return).

The marketing function is essentially an information processing function. It provides the line of communication between customer needs and the new instructions. This function is indicated in the model by a circle, possibly enclosing the letter M (Marketing), with the connecting lines ending in a crosswise bar at the point of control or of information retrieval; see Fig.4a.

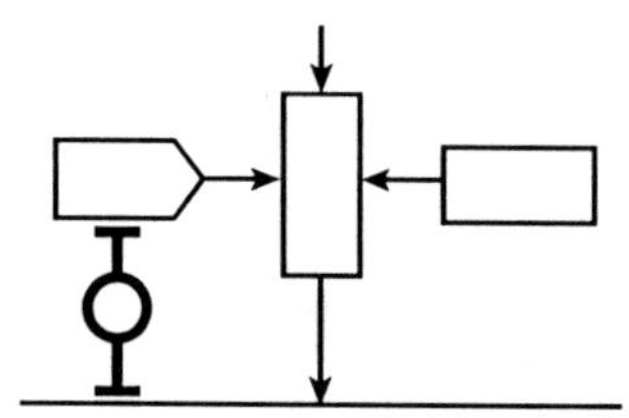

Fig.4a: Marketing interlinks customer needs with the supply of new instructions

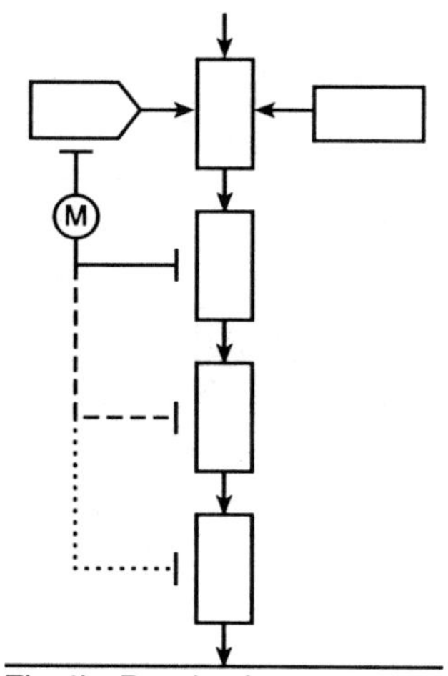

Fig.4b: Depth of penetration of marketing interlinkage into the supply chain

The marketing function is often only in touch with the direct customer, although it will often be the case that the goods m question are to be passed on to a later link in the supply

In this model the connecting line with a crosswise bar at either end is used to indicate that two units are linked and coordinate their activities with one another. In the example given in Fig.10, two development groups are linked in this way, even though they have different tasks in the innovatlon process.

If there is a special unit or official that performs this coordination or control, this is indicated as in Fig.4b by a circle, possibly enclosing one or more letters designating the nature of the control function (e.g. M = marketing, PP = product planning - whatever is convenient).

Finally, there are three other functions that are of great importance. First, 'storage' occurs in both the supply chain and the innovation chain: goods are placed in a warehouse for later delivery and information is stored in an archive. Where this occurs in the landscape it is indicated by the capacitor symbol, as shown in Fig.5.

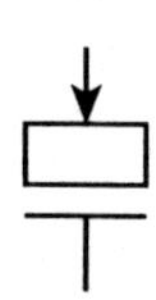

Fig.5: The storage function (archive, warehouse)

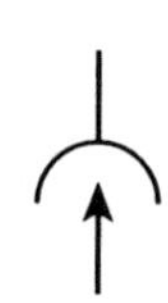

Fig.6: The service function

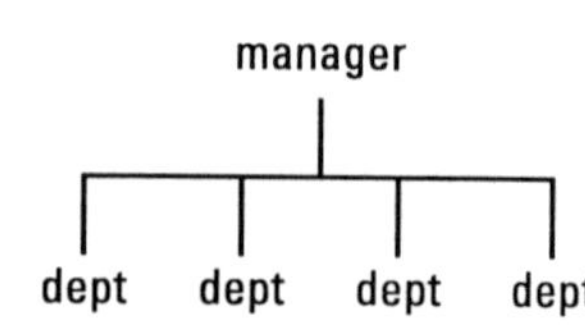

Fig.7: Typical organogram, showing the hierarchy of an organisation

Second, goods are sometimes returned in the supply chain for repair or replacement. Often there is then a (technical) service group that handles the customer's complaint and if relevant sends the market information obtained back to the production and innovation centres. The symbol in Fig.6 indicates this double function: the arrow represents the return flow of goods, while the satellite dish sign indicates the receipt of various types of
Information.

If an organogram is used to represent an organisation (see Fig.7), its purpose is usually to show which functions or departments belong together for some reason or other and which manager they are controlled by. It therefore shows the 'management control', but not the interconnection with the business processes 'innovation' and 'supply'.

In the landscape model this situation is in fact reversed: here the interconnection is paramount - as in a topographical map - and 'management control' is shown last. This can be done in two ways: either by indicating the borders of the organisation with a thin or broken line, as has been done in Fig.8 or, if the units for which the manager in question is responsible are dispersed in the landscape, by hatching or colouring the units concerned. The real-life example shown below will clarify this, see Fig. 10.

These are the conventions used in the model. Conclusively we would like to mention that the arrows in the model point in the direction of the provision of added value. The consideration given in exchange for this - cash and, in many cases, also information - flows in the opposite direction; see Fig.9. These flows will not be elaborated on further in this discussion of the model.

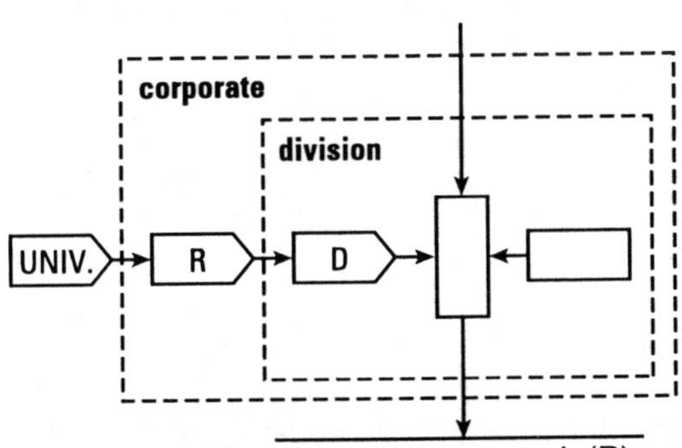

Fig.8: The position of corporate research (R) and divisional product development (D)

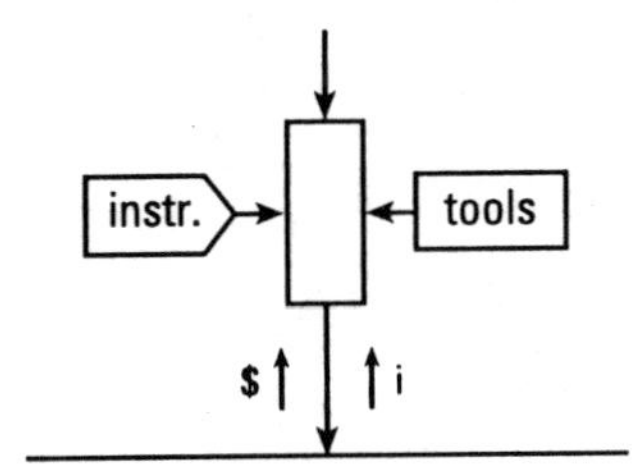

Fig.9: Cash and information as consideration for the added value provided

A real-life example

The following real-life example shows how such a product development group is actually connected - or otherwise! - to other units in the supply chain. The example relates to the former Philips product group "Professional Recording Group".

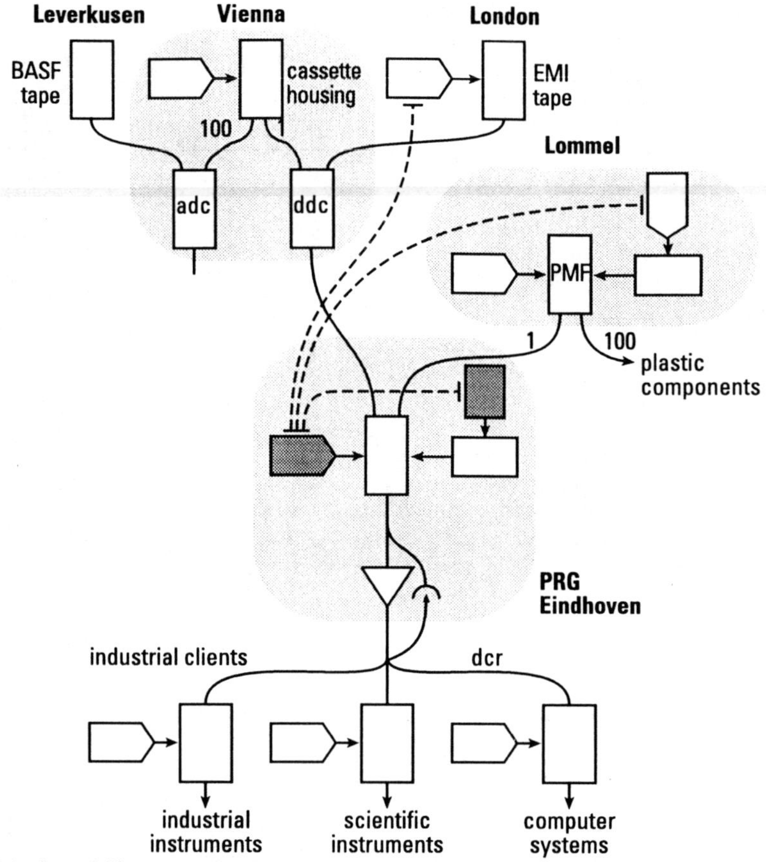

Fig.10: A real-life example: the production of a digital cassette recorder and its most critical components

adc = audio dictation cassette
PMF = Philips Plastics and Metalware Factories, a Philips division
PRG - Professional Recording Group, former Philips product group
dcr = digital cassette recorder

Further research

Practical exercises suggest that it might be possible to define for each basic element of the model the mathematical function as a description of the function performed by that element. Further research is recommended for this approach.

Conclusion

The growing complexity of companies and of their interrelatedness has made it increasingly necessary to visualise this interdependency. The model proposed here can show a wide range of relationships. It gives the physical locations, the link between supply chain and innovation chain and management's actual span of control.

QFD Application to a Software-Intensive System Development Project

Tuyet-Lan Tran
(81 8) 354-61 74

Jet Propulsion Laboratory / Caltech
4800 Oak Grove Dr.
Pasadena, CA 91 109

This paper describes the use of Quality Function Deployment ~QFD3, adapted to requirements engineering for a software-intensive system development project, and synthesizes the lessons leamed from the application of QFD to the Network Control System (NCS) project of the Deep Space Network (DSN).

The selection of QFD started by recognizing that "quality" means customer satisfaction with the product.

The application of QFD to the front-end of the lifecycle was built upon the premise that it is an effective technique for incorporating customer requirements into product design, in such a way that a history of improved product-releases could be achieved and quantifiable relationships could be established between customer satisfaction and the high-level specifications of a communications-and-control system.

Against this background, the adaptation of the QFD technique focused mainly on exploring the measurements of software behavior (or, perfommance characteristics) and on assuring the degree of completeness (or appropriateness) of high-level specifications, while facilitating the analysis and design activities of the product-development team. Where needed, QFD was coupled with other techniques, such as Yourdon's methodology.

Introduction

Developing an infommation-system product has never been an easy task, whether from the grounds up or from a legacy system. From the product-planning perspective, such system development activities range from understanding and capturing customer needs at the highest level to understanding product requirements and controlling at the lowest level those system-perfommance parameters that directly contribute to customer requirements. This paper describes the application of the Quality Function Deployment (QFD) technique to such a product-planning environment (or, pre-project phase) for the Network Control System (NCS) of the Deep Space Network (DSN).

At the time of the NCS pre-project, NASA was planning to evaluate opportunities for agency-wide consolidation and expanded commercialization of space-mission operations. In support of this new NASA horizon, the DSN at the Jet Propulsion Laboratory (JPL) started to evaluate new concepts for more autonomous space operations.

The Business Case

In the current approach to space operations and mission control, an around-the-clock operations team is needed for each DSN communications link to carry out on the ground a variety of functions, such as DSN-wide resource scheduling, spacecraft-commanding, spacecraft tracking and navigation, monitoring the state-of-health of the spacecraft and telecommunications link during a pass, and ensuring correct functioning of all spacecraft and ground equipment, and coordinating failure recovery actions or Uworkarounds." This human-resource-intensive approach is becoming less and less viable in the future due to (1) a desire to optimize the

0-7803-3552-X/96/$5.00

utilization of DSN aperture, (2) a desire to accommodate Principal Investigator (PI)-controlled missions, (3) a desire to accommodate deep space missions with significant round-trip light-time communications delays, and (4) a desire to limit the operations-team and DSN overall operational costs.

In the new concept of DSN operations (to be actualized via the NCS), the stated programmatic goal is to improve its cost effectiveness by 50 % over 4 years of continuous operation, without reducing its overall reliability. The scope of the pre-project phase involves the establishment of customer needs (or requirements) and the conceptual design of a new network-control system that can be rapidly developed and deployed within 1 year of conceptual-design-completed.

The Selection of QFD

After some doses of Total Quality Management (TQM)-rehabilitation, several of us at JPL by now believe that the most important step in product planning is to figure out who our customer(s) is/are. Unfortunately, this is usually harder than it seems at first glance. After a few weeks of system-analysis activities that did not yield any agreeable set of customers or customer requirements, it became compelling that requirements definition is a very risky business - especially when everyone wants to see a "smart system" that controls everything yet is easy to operate but cannot totally displace the worker.

The selection of QFD started by recognizing that "qualityn means customer satisfaction with the product. Additionally, the QFD approach stands out as one that systematizes TQM, when compared to other techniques or methods for requirements definition. QFD introduces quality, indeed, as early as in the requirements phase (or Uupstream") and throughout every product development stage. More importantly, "qualityn is presented from the customer's perspective, in the customer's "language" - as opposed to the designer's or engineering perspective. The goal of QFD is to deploy this customer's perspective (also referred to as Voice Of the Customer, or VOC, or customer requirements) throughout the product's technical requirements and specifications in an explicit and systematic manner. The mapping of design features or essential product-features to those customer requirements becomes a pivotal factor for QFD's role in delivering high-quality products.

For the NCS pre-project phase, a small, cross-functional team was appointed to carry out the QFD process. Specifically, the team's target deliverables are the product-planning matrix/ices, as these represent the most important tenet for estimating the real needs of various customers and capturing their respective requirements, as well as for defining the key requirements of the to-be-developed product. For more detail on QFD, refer to [1] and to Figure 1.

Once a new technique is selected, good software management practices remind us that, in spite of its promised value, a "noveln development methodology represents a significant software risk element to a software-intensive system development project. Accordingly, a risk-mitigation decision was made to adopt QFD only for this pre-project phase. Additionally, the responsibility for "not playing by the book" (as recommended by Professor Yoji Akao himself) and for fine-tuning the technique and adapting it to the NCS was assigned to a dedicated process engineer. The process engineer selected ITI's QFD/Capture as the tool for the process.

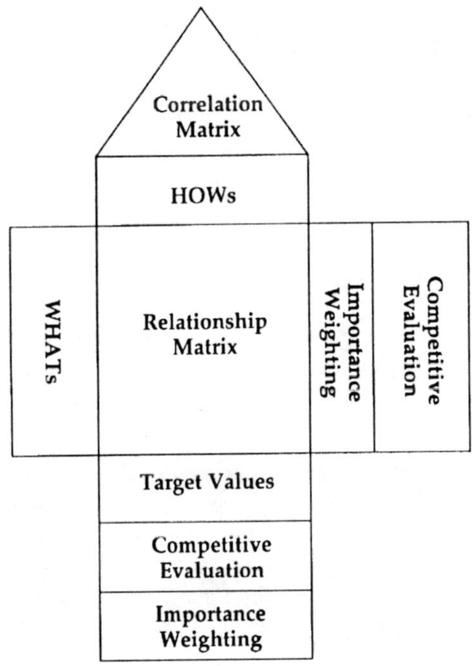

Figure I . Overview of QFD Chart or Matrix

What Made This QFD Application Unique?

In light of NASA's intemal pressures for strategic changes, the political pressure associated with labor contracts in deployment over three different countries, and the complexities of catering to an intemational science community that does not "directly buy DSN data services", the team members who were to include Ops representatives were chosen for a combination of cross-functionality, knowledge and experience with the DSN, and especially respect among their peers and non-bias for their line managers or organizational affiliations. This combination is an absolute necessity for the team to stand a chance with the organizations that might be negatively impacted by the proposed design.

Predictably, the QFD process quickly got the team to get past the usual arguments about what customer requirements are vs. what design is. Instead, the focus shifted to: (a) identifying and agreeing to what the various classes of customers are, and accepting that the extent of overlap in customer-needs should not be used as the discriminating factor in the class partitioning activity; (b) how to level the various customer needs; (c) What the product is vs. What the services are (when the product is used in the Ubackground" as a means to deliver services); (d) how much rigor should be applied to the estimation of customer needs; (e) how to ensure that the QFD matrices are 'bdequate" at capturing the "essential product features" that correlate with the customer needs/requirements; (f) how to ensure the completeness of a minimalist set of product requirements -- whether with or without QFD.

Combining the Yourdon methodology with the QFD process was critical to the successful completion of the NCS conceptual design. The combined process included 5 distinct activities, which are embedded along a spiral model itself.

1. Defining The Context

This activity is similar to the development of the Yourdon's context diagram. The objective is to define the solution space, in temms of a black box that interacts with its environment. The black box is intended to represent the product or services-to-be consumed by the end-user.

This diagram fomms the starting point for developing the conceptual designs of the NCS.
See Figure II for NCS's context diagram.

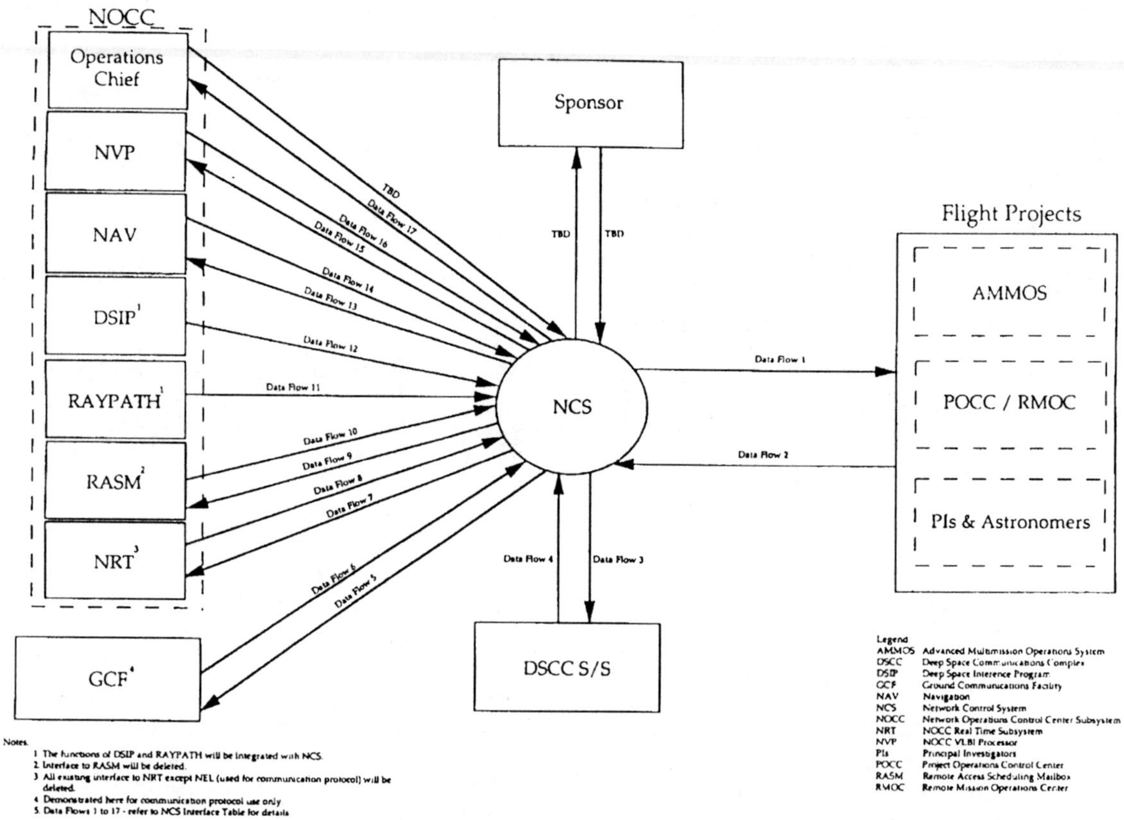

Figure II. NCS Context Diagram (Based on Existing Interface Agreements)

V0 1 February 27 1995

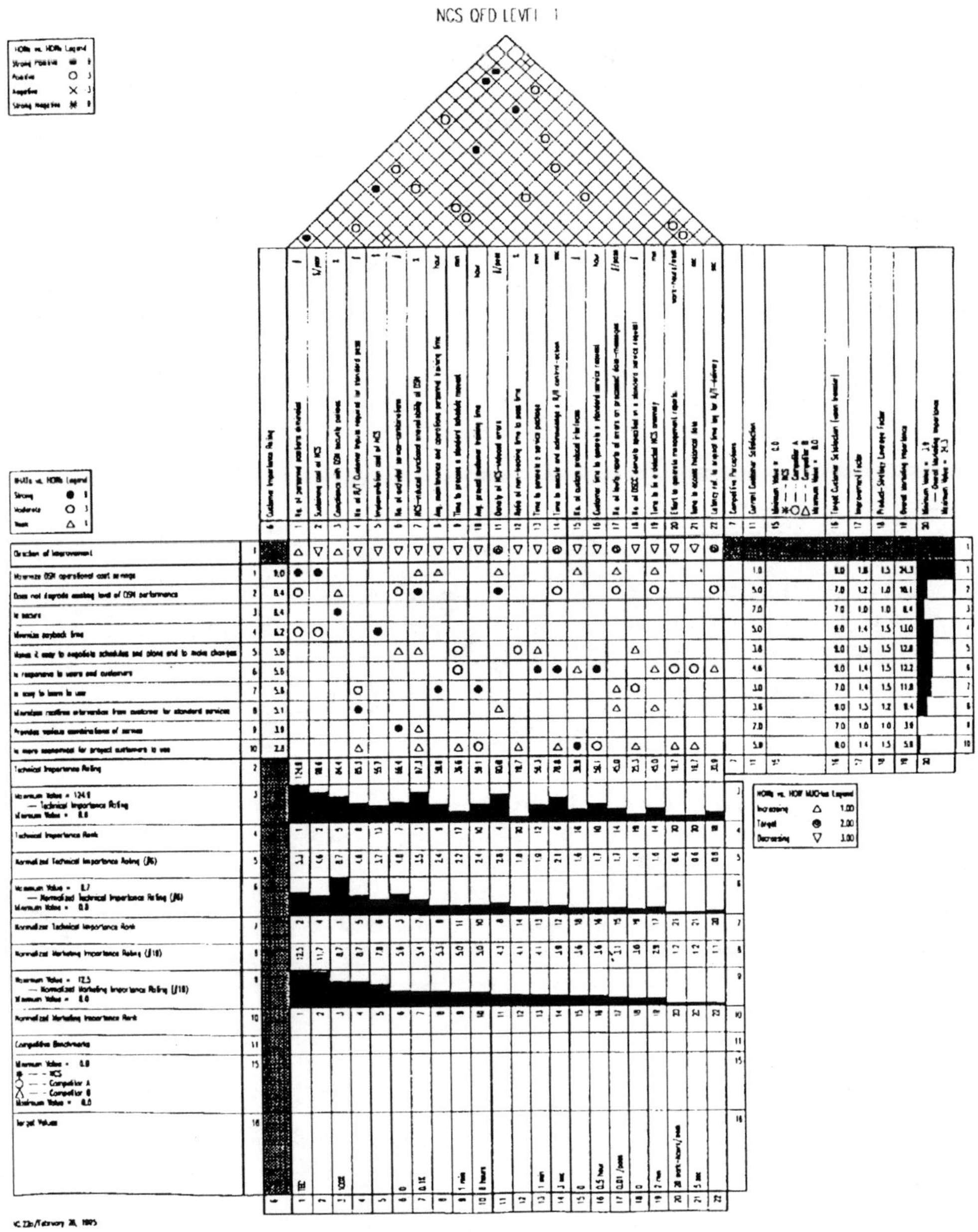

Figure III. Level-1 QFD Matrix for NCS

2. Developing Level-1 QFD Matrix

This consists in eliciting expected customer needs and estimating real customer needs, via use scenarios that are derived from the context diagram and that assume expansions of selected extemal-interfaces. Besides having the context diagram as the input, this activity is very QFD-specific. For more detail, refer to [1 j and [2].

Where the NCS's application is unique is the coupling of the level-1 DFD with the QFD matrix. The level-1 DFD is used, indeed, for the:

- identification of key software processes that make up the product which will solve
- the customer needs;

- specification of performance parameters associated with the primary data flows of

- these essential processes;
- capturing of decision-tree correlations for all "essential perfommance parameters",
- and making technical risk identification explicit.

3. Developing Level-1 Data Flow Diagram

This activity started after the start of the level-1 QFD, but was iterated upon in parallel with the Level-1 QFD matrix.

The objective is to develop a logical model of the system, in temms of processes and data or messages. This model should be high-level enough to allow for consistency check and validation against user requirements and the level-1 QFD; and, to necessitate a next level of decomposition (into a level-2 DFD). Figure IV shows the software architecture of one of the 2 major NCS software end-items. This architecture evolved from an earlier version of software-requirements DFD. The earlier DFD facilitated analysis, and was critical to the verification of completeness of processes. The latter version (i.e., as captured in Figure IV) is a model for further, detailed design.

4. Rapid Iterations for Cost Estimation

The combination of the context diagram, level-1 QFD, level-1 DFD, and associated operations scenario and hardware diagrams were adequately detailed, to support the development of a product-based Work Breakdown Structure (WBS), a level-1 schedule,
and a full cost-estimation effort. The ops scenario and hardware diagrams are, however, not included in this paper.

It is worth noting that the tool environment established and controlled by the process engineer was a key factor in the successful definition and application of the process. The rigor in tum enabled the rapid iterations and refinements of the NCS customer and
product requirements. Finally, the rapid rate at which a cross-functional QFD team could iterate and release the level-1 QFD became an effective catalyst for the problems
of requirements volatility and accurate problem-domain-knowledge acquisition identified
by Curtis et al. [3].

Conclusion

This paper has described how the QFD process has been adapted and mixed-in with Yourdon's methodology, to enable the capture of customer requirements and rapid definition of a software-intensive system 's conceptual design. The author believes that progress toward the goal of having a methodology that enforces rigor in requiremer~ts capture and conceptual design, as well as in their systematic reuse, lies in a development support environment, which includes QFD-based tools ffor the explicit and disciplined incorporation of customer requirements into product design). This would necessitate further bridging between QFD-based tools and other operations-scenario development and simulation tools. The contribution of the QFD based technique is in the front-end of the process, and has most impact in (a) enabling optimal decision making by the right development team; (b) enabling rapid customer-centered model development, checkout, modification and reuse.

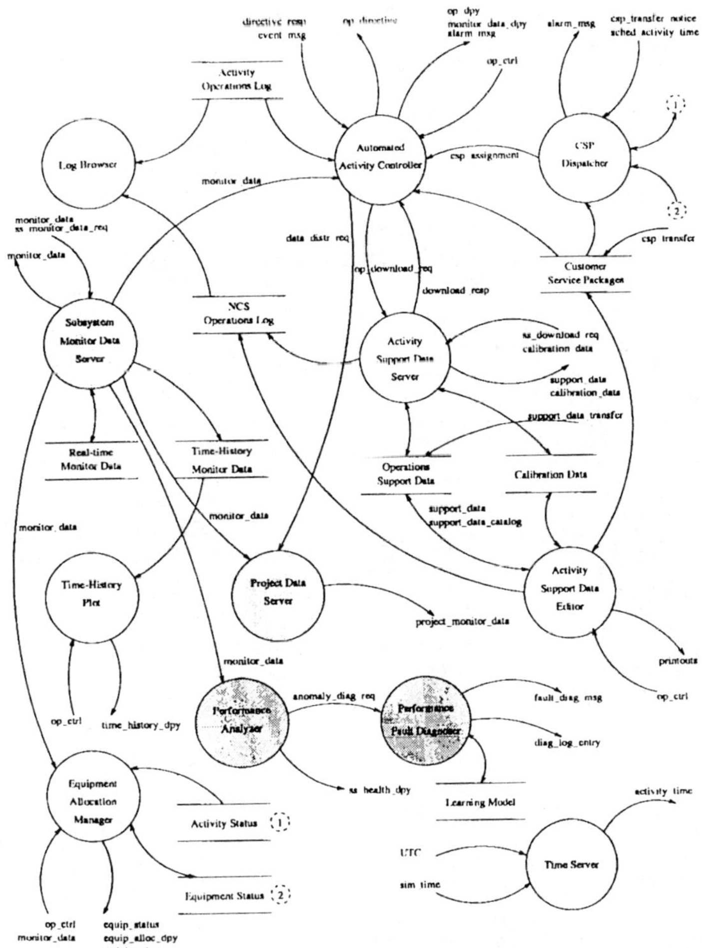

Figure IV EAC Software Architecture

Acknowledgments

The results reported in this paper were carried out by the Jet Propulsion Laboratory, Califomia Institute of Technology, under a contract with the National Aeronautics and Space Administration. Special thanks are given to the entire NCS team; and, in particular to John Leflang, manager of the NCS Pre-project, for having accepted the author's recommendation for using QFD, and to Thang Trinh and Monica Wang for their insightful inputs on what problems they are constantly facing with, when ~applying" QFD or using the QFD/Capture tool.

References

[1] Yoshizawa, T., Togari, H., and T. Koribayashi, "QFD, Integrating Customer Flequirements into ProductDesign,n(Akao, Y. editor), Productivity Press, Cambridge, MA, 1 990.

[2] Tran, T. L.. and J. S. Sherif, "QFD: An Effective Technique for Pequirements Acquisition," Proceedings of Second IEEE Intemational Software Engineering Standards Symposium, pp. 191 - 200, August 1995

[3] Curtis, B., Krasner, H., and N. Iscoe, "A Field Study of The Software Design Process for Large Systems,n Comm. ACM, 31, 1 1, pp. 1268 - 1287, 1988.

Management of Information Technology using Analytic Hierarchy Process

Godwin J. Udo

Department of Decision Sciences
Tennessee Technological University
Cookeville, TN 38505

Information technology (IT) outsourcing has become a common practice among firms who are doing everything possible to stay alive in today's highly competitive business environment. Since some types of information technology services can be used to enhance the firm's competitive position in the industry, firms are faced with the difficult task of deciding which services should be outsourced. In this paper, Analytic Hierarchy Process is proposed as a tool that can effectively handle the complex, qualitative decision variables involved in IT outsourcing decision making. Numerical examples are used to demonstrate that the attributes of this tool can be used to improve the decision quality.

Introduction

The rapid growth of information technology (IT) in industries today is a mix bag of opportunities and problems. On one hand IT presents opportunities in that many companies have used it in ways that help them to gain competitive edge. On the other hand IT presents problems in the sense that the management of IT is a major cause of headache to several companies. Organizations are so overwhelmed by the complexity and need for urgent attention demanded by the challenges of IT management that they are willing and ready to try any technique. Some managers think that since they do not have a good understanding of the technology, there is no hope attempting to manage it. The common practice among companies therefore, is to look and ask around to see how other companies handle their IT challenges in order to copy without asking much questions. For example, since Eastman Kodak successfully outsourced most of its information systems, several other companies have followed suit without a systematic analysis of the consequences of their actions.

The recent trend among firms today is that of classifying their information system needs into two categories: commodity and strategic services. They argue that the commodity services can be outsourced while strategic services should be kept in the house. However, Lacity and Hirschheim (1993) maintain that this type of classification can led a firm into serious problems in the future. First of all, the so-called commodity service may be very critical to the operations of the firm and as such, caution need be exercised in outsourcing such services. Secondly, a strategic service can rapidly become a commodity service in a matter of months. The implication here is that a more vigorous method needs be employed in deciding whether or not to outsource and what aspects of the IT should be outsourced.

In this paper, Analytic Hierarchy Process (AHP) is proposed as a suitable technique for analyzing information systems outsourcing decision. AHP can be used to rank all possible alternatives of outsourcing using several criteria. It has the means of converting subjective assessments of relative importance into a linear set of weights that can be used in ranking the alternatives. The paper presents an example of IT outsourcing decision analysis using AHP. With this techniques, several options are considered in the decision analysis which makes it possible for a company to adequately evaluate and determine whether a particular IT should be outsourced. Since no systematic technique (known to author) has been adopted, the present approach, it is believed, will be helpful to the practitioners who are faced with outsourcing decision making.

0-7803-3552-X/96/$5.00 © 1996 IEEE

Analytic Hierarchy Process

The AHP is a mathematically-based, multi-objective decision making tool which was introduced by Saaty (1990). It uses the pairwise comparisons method to rank order alternatives of a problem that are formulated and solved in a hierarchic structure. The technique has an advantage for being simple and thorough in handling difficult real-life problems. It provides greater utility in applications where information is either incomplete or not available. The AHP approach has be adopted in many applications including resource allocation (Barbarosoglu and Pinhas, 1995), business performance evaluation (Lee et. al., 1995), capital budgeting (Tarimcilar and Khaksari, 1991), and auditing (Yau and Davis, 1993).

The AHP requires that the problem be decomposed into levels, each of which comprises of elements or factors. The elements of a given level are mutually independent, but comparable to the elements of the same level. The structure presupposes that elements of any given level are influenced by elements at the level immediately above them. The process of AHP comprises of the following steps:

1. Structure a problem with a model that shows the problem's key elements and their relationships.
2. Elicit judgments that reflect knowledge, feelings, or emotions.
3. Represent those judgments with meaningful numbers.
4. Use these numbers to calculate the priorities of the elements of the hierarchy.
5. Synthesize these results to determine an overall outcome.
6. Analyze sensitivity to changes in judgment.

A major strength of the AHP is the pairwise comparison where the influence of elements of a particular level over those of a lower level is measured. Since two elements of the same level are compared at any given time to a particular element at a higher level, this simplifies the process to a large extent. Comparisons are provided by means of experienced judgments or from data and experiments (Saaty, 1990). When n elements of a given level are being compared, a total of n(n-1)/2 comparisons are needed. The second half of the comparison matrix is the reciprocals of those judgments lying above the diagonals and usually omitted. The judgments are provided by means of a 9-point ratio scale that ranges from two factors being equally important to one of the factors being absolutely more important than the other. After the expert supplies the ratings, local priorities of each element are calculated by approximating the principal eigenvectors via the geometric mean method. At the end of the analysis, global priorities are obtained by synthesizing each of the local priorities. Transitivity of preference is determined by the consistency ratio and implies that if X is preferred to Y, and Y is preferred to Z, then X is preferred to Z. A consistency ration of 0.10 or less implies a more consistent judgment that is acceptable.

Information System Outsourcing Decision

Information system has become a common place event in organizations across the USA. The cheap and highly-skilled labor pool obtained abroad has led to a sudden increase in global outsourcing. For instance, First Fidelity Bankcorp contracted its computing services to EDS at a cost of $450 million in 1992, and in 1993, John Gigerich, the CIO for Continental Bank Corporation announced the bank's intention to contract the majority of its IS services to outsiders (Fitzgerald, 1991). Also, AT&T announced plans to cut or reassign 4,300 computer workers (57%) because of its merger with NCR (Keller, 1991). In fact, several US corporations such as Citibank, Texas Instruments, New York Life Insurance, and Boeing, have successfully outsourced part of their IS services to other countries, including India, Ireland, the Philippines, and Russia, to name just a few (Shereff, 1989). Global outsourcing is a lucrative alternative that provides capital market gains and cost savings, a larger pool of skilled vendors, and quicker development cycle time. However, its pitfalls include: resistance from IS professionals, poor quality, lack of coordination, and poor communication between the firms and the vendors.

Lacity and Hirschheim (1993) noted that even though information systems outsourcing started over 34 years ago, it has recently taken on a new scope. Initially, IT outsourcing was limited to such tasks as payroll, insurance processing, mailing list, etc., but now it involves multiple systems which may include the transfer of leases, personnel, and assets to the vendor. In a typical IT outsourcing, the vendor will charge a fixed amount for a set of services called "baseline" within a fixed number of years. Within the contract period, any services not included in the baseline has to be purchased for an additional cost. The projected savings are often very attractive and supported by press reports. Unfortunately, failures of IT outsourcing are not adequately reported in the press since no company is particularly pleased to be identified with failure.

IT outsourcing decision making has a strategic importance and so should be systematically analyzed. There are many and complex decision variables involved in this type of decision making situation. Some of the major decision variables for IT outsourcing decision are strategic importance of the services rendered, vendor factors, customer interests, and employee interests. Majority of these variables are qualitative in nature and as such the effectiveness of many of the existing decision analysis tools can be questioned. AHP has been effectively applied by several researchers to decision making problems whose structures are similar to that of IT outsourcing. The following section deals with how AHP can be used in making IS outsourcing decisions.

Application of AHP to IT Outsourcing Decision Analysis

We proposed that IT outsourcing decision should be grouped into four categories as shown on Figure 1. This grouping is not exhaustive but serves as a demonstration of this proposal. Other variables with their corresponding weights can be added to the model based on the particular needs of a given company.

Strategic Importance

This group of decision variables deals with the question of deciding the effect of the IT services on achieving the strategic goals of the firm. The general concept is that the information systems that provide strategic services should not be outsourced but those that provide nonstrategic services could. A weight of .4 is given to this group of decision variables because it is the most important group. Firms pay a great attention to their strategic information systems because in extreme cases, such information systems could as well be the difference between success and failure for the firm. It would be a serious mistake for a firm to outsource its strategic information technology because once the system is duplicated, it fails to play a strategic role. The specific strategic importance variables are grouped into four categories and discussed as follows.

(1) Opportunities to distinguish the firm -- if IT can enhance a firm's ability to distinguish itself, the firm's strategic position in the industry can be greatly enhanced and so the IT can be said to provide strategic services.

(2) Critical service -- if the services provided be a particular IT is critical to the core operations of a firm, it could be said to be of strategic importance. It would not be in the best interest of a company to depend on an outside company to provide the IT services that are so critical to its core business. The idea here is that firms should only outsource those IT that support noncritical operations of the firm.

(3) Being stripped of knowledge -- firms should first decision whether or not the expertise of the laid-off knowledge workers will ever be needed in the future. If the industry is undergoing some growths and progression, chances are the information needs of the firm may change in the future and today's nonstrategic IT can become a strategic hot cake. Outsourcing IT services implies that the firm no longer involves itself with IT staff development and hence it is stripping itself of the knowledge workers that may be crucial in the future.

(4) Internal expertise ineffectiveness -- Firms should not outsource its IT services on grounds of internal ineffectiveness without first providing corrective or remedial programs. The ineffectiveness of the IT department can be corrected with a minor reorganization or policy changes. Outsourcing should only become an option when all remedial programs fail. This decision variable is considered the most important because firms and outsourcing vendors often cite internal ineffectiveness as the main reason for outsourcing IT services.

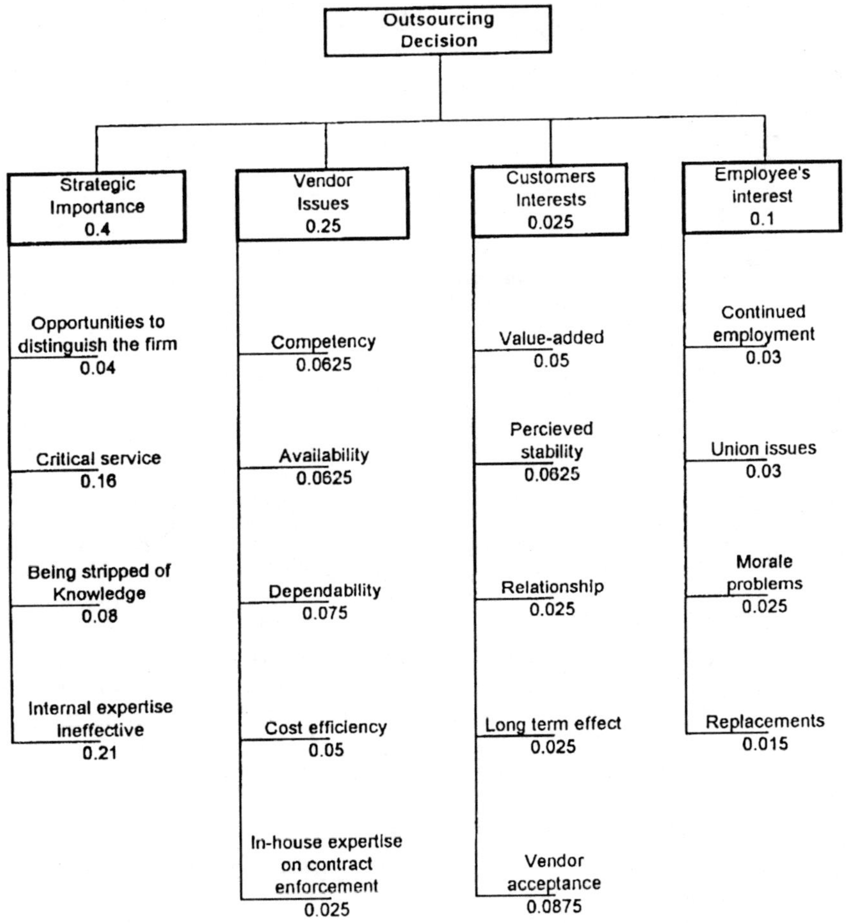

Vendor Issues

This group of decision variables deals with issues relating to the type of vendor the firm should entrust its IT services. If there are no vendors that rank favorably on these issues, then IT services should not be outsourced. The vendor issues to be considered included competency, availability, dependability, cost efficiency, and an effective in-house expertise to enforce the outsourcing contracts. If there are no competent, dependable vendors that are geographically and otherwise close to the firm, IT outsourcing may not be a good alternative. In addition, the vendor has to be much more cost efficient than the IT department before outsourcing can be considered a viable option. There has to be a competent in-house workforce dedicated to enforcing the contract.

Customer Interests

This class of decision variables deals with the possible effects of IT outsourcing on the customer. These effects could be real or imagined but the customer's opinion and perception must be taken into account before outsourcing decisions are made. Specifically, the following questions need be asked in order to determine the customer's interest: (1) Does outsourcing add value to the customers and do customers perceive this decision as adding value to them? (2) Can the firm be perceived by the customers as being stable or will this decision create a notion of bad sense and instability? (3) Will the firm-customer relationship be positively or negatively affected? (4) What long-term effects will the decision have on the customers and the firm's ability to gain new customers? and (5) Will the customers accept the vendor or will the decision create conflict of interests among the customers? A careful consideration of these factors will help a firm decide whether or not to outsource its IT services.

Employees' Interest

This set of decision variables concerns the effect of outsourcing on not only the IT staff but also other workers in the firm. Outsourcing could lead to low morale and negative attitudes among employees, which in turn cold lead to productivity problems. Firms also have moral obligation to account for employees' interest when making a decision on IS outsourcing. The factors to evaluate include:

(1) continued employment -- The firm should endeavor to retrain and employ the replaced IT workers. The firm can also make it mandatory that the vendor continue to employee the IT staff. If no continued employment is provided for, IS outsourcing may not be an attractive option.

(2) Union issues -- Where union exists, firm should first explore its legal possibilities before deciding to outsource. Agreement could be reached with the union if outsourcing can be justified to be beneficial to all parties. The existence of union should not automatically imply that outsourcing can not be considered.

(3) Morale problems -- Firms should analyze the possible morale problems that could result from the decision and so make plans to handle and resolve such problems. If the cost of moral problem is higher than the benefit of outsourcing, the firm should not pursue the option for the sake of productivity and good will.

(4) Replacements -- If employees are difficult to deal with and if it become impossible to strike a deal with the workers, the firm should analyze replacement possibilities. If replacement is more expensive or not possible, IT outsourcing should be discouraged.

Numerical Example

Suppose that there are four criteria for analyzing IT outsourcing as shown in Figure 1. A given company has to be ranked from 1 (being the worse) to 9 (being the best) on each of the subcriteria. Using an hypothetical company (call it company 1) that ranks 5 (midpoint between 1 and 9) in all the subcriteria. The values on Table 1 will be obtained. The top row of Table 1 ranges from 1 to 9, while the bottom row ranges from 1/9 to 1. The center rows are filled in as a moving center. Based on these values, the Eigenvalues are calculation. An Eigenvalue is the 9th root of the product for an entire row. For example, the Eigenvalue for the first row on Table 1 was obtained from (1*2*3*4*5*8*9)^9. This calculation was repeated for the rest of the rows to obtain the eigenvalues given on Table 1. The next step is to convert the eigenvalues into eigenvectors. This done by summing the eigenvalues from the previous step and dividing the individual eigenvalues by the sum. For example, the first value of the eigenvector, 0.308 = 4.147/13.459. The sum of the number in an eigenvector must always add up to 1. The result of this benchmark company can then be used to evaluate any other companies.

1=Worse	9=Best

	Best 9	8	7	6	5	4	3	2	Worse 1	Eigenvalue
Best 9	1	2	3	4	5	6	7	8	9	4.147
8	0.5	1	2	3	4	5	6	7	8	3.008
7	0.33	0.5	1	2	3	4	5	6	7	2.113
6	0.25	0.33	0.5	1	2	3	4	5	6	1.458
5	0.2	0.25	0.33	0.5	1	2	3	4	5	0.999
4	0.17	0.2	0.25	0.33	0.5	1	2	3	4	0.686
3	0.14	0.17	0.2	0.25	0.33	0.5	1	2	3	0.473
2	0.13	0.14	0.17	0.2	0.25	0.33	0.5	1	2	0.334
Worse 1	0.11	0.13	0.14	0.17	0.2	0.25	0.33	0.5	1	0.242
							um of Eigenvalue =			13.459

Rate	Eigenvector
9	0.308
8	0.223
7	0.157
6	0.108
5	0.074
4	0.051
3	0.035
2	0.025
1	0.018
Sum of Eigenvector	1

Table 2 provides the results of two other companies which can be referred to as company 2 and company 3. The values are obtained by replacing the company's scores (scaled 1 through 9) with the corresponding eigenvector. For example, 1 becomes .018 and 5 becomes .074. The eigenvector are multiplied by the corresponding value of the subcriteria. For example, if company 2 scores 9 for the subcriteria "union issues, .03", a score of .002 is obtained by multiple .074 and .03. This calculation is done for company 2 and company 3 as shown on Table 3. The final step is to sum the values for each subcriteria. The grand total is unique to each company and can be used to decide whether or not the company should outsource its IT. The higher the grand total, the more likely the company's IT is suitable for outsourcing. As shown on Table 3, company 2 and company 3 have total scores that are higher than 1, the total score of the benchmark company 1. The implication of this result is that these two companies may succeed if they decide to outsource their information technologies.

COMPANY 1 (standard, note that these cells are formated from page 1)

0.002960	0.004625	0.003700	0.002220	Sum Total for Company 1 (standard)
0.011840	0.004625	0.004625	0.002220	
0.005920	0.005550	0.001850	0.001850	
0.015540	0.003700	0.001850	0.001110	0.08251
	0.001850	0.006475		

COMPANY 2 totals

0.012320	0.004625	0.015400	0.001530	Sum Total for Company 2 (standard)
0.035680	0.003188	0.013938	0.001050	
0.012560	0.002625	0.003925	0.000625	
0.022680	0.001250	0.002700	0.000270	0.14129
	0.000450	0.006475		

COMPANY 3 totals

0.000720	0.004625	0.000900	0.003240	Sum Total for Company 3 (standard)
0.004000	0.006750	0.001563	0.004710	
0.002800	0.011775	0.000875	0.005575	
0.010710	0.011150	0.001275	0.004620	0.08946
	0.007700	0.006475		

Conclusion

Analytic Hierarchy Process has been proposed as a potential tool for analyzing information technology outsourcing decision. The advantage is that AHP has the ability to handle complex, qualitative decision variables involved in the IT outsourcing decision making analysis. The proposed approach is expected to improve the quality of decision making and provides a systematic analysis which is lacking in this particular decision making. The numerical example given in this paper proves that AHP can be used effectively to analyze IT outsourcing decision. The example does not include all possible decision variables and so cannot be used in a real-life case without modifications.

References

Barbarosoglu, G. and Pinhas, D. "Capital rating in the Public Sector Using the Analytic Hierarchy Process." The Engr. Econ. 40(4): 315-329, 1995.

Fitzgerald M. "Bank Considers Outsourcing." *Computerworld*, 48, 1991.

Keller. J.J. "AT&T Plans to Cut or Reassign 57% of Its Computer Workers Due to Merger." *The Wall Street J,* C15, 1991.

Lacity MC and Hirschheim R "The Information Systems Outsourcing Bandwagon." *Sloan Mgmt Rev*, **35**(1): 73-86, 1993.

Lee, H., Kwak, W. and Han, I. "Developing a Business Performance Evaluation System: An Analytic Hierarchical Model." The Engr. Econ. 40(4): 343-357, 1995.

Saaty, T.L. "How to Make a Decision: The Analytic Hierarchy Process." Europ. J. Opt. Res. 48: 9-26, 1990.

Shereff, R. "Creative Clerical Solution: Service Firms Open Up Satellite Global Offices." *Mgmt Rev.* **78**(8): 24-27, 1989.

Tarimcilar, M. and Khaksari, S.Z. "Capital Budgeting in Hospital Management Using the Analytic Hierarchy Process." Socio-Econ. Planning Sc. 25, 27-34, 1991.

Yau, C. and Davis, T. "Using Analytic Hierarchy Process (AHP) to Prioritize Auditing Tasks for Large-Scale Software Systems." J. Sys. Mgt., 26-31, Nov. 1993.

Real-World Challenges: Using the Virtual Enterprise for Successful Product Development Projects

Cinda Voegtli
(415) 966-1650

Calibrated Management
1420 Brookmill Road
Los Altos, CA 94024

By building a "virtual enterprise" to achieve corporate goals, high-technology product development companies can supposedly stay both financially and technologically competitive in today's markets. The company concentrates "in-house" on core competencies, and turns to the outside expertise of organizations and individuals on an as-needed basis. However, if not created and managed properly, the virtual project teams created for such product development ventures can seriously threaten a firm's critical time-to-market goals. The difficulties of actually achieving truly synergistic partnerships and ultimately successful projects using team members outside the corporation must not be overlooked.

When virtual team members are consultants or third-party development organizations, the success of the team depends upon the degree to which these various members are aligned in the following areas: values and goals; priorities; skills, experience, and capabilities; and product development processes. The project manager must understand how to assess and select team members, plan a project in detail considering the specific ramifications of remote members, and manage the project to its successful completion.

This paper explains the "alignment factors" above and illustrates their potential impact, using examples from two projects from the author's experience which made use of third-party development organizations. The paper summarizes the actions project managers must take in planning and managing such projects, including questions for assessing and selecting the virtual project team members.

Definition of virtual team

A project team is a cross-functional group of people working together to develop a product and introduce it to the market. Innovative ways of accomplishing product development are currently in vogue. By building a "virtual enterprise" to achieve corporate goals, high-technology product development companies can supposedly stay both financially and technologically competitive in today's markets. The company concentrates "in-house" on core competencies, keeps its permanent staffing at an optimum level, and turns to the outside expertise and efforts of organizations and individuals on an as-needed basis. A "virtual project team" often includes individuals or sub-groups who are not part of the corporation who owns the product, such as independent consultants or contractors and third-party development organizations.

Critical alignment factors

If not created and managed properly, the virtual project teams created for product development ventures can actually seriously threaten a firm's critical time-to-market and return on investment goals. The success of the team depends upon the degree to which these various members are aligned in the following areas:

- values and goals
- priorities
- skills, experience, and capabilities
- product development processes

0-7803-3552-X/96/$5.00 © 1996 IEEE

Values and goals

These factors affect the commitment given to a project by the team members, both to the project outcome and to the means of getting successfully to that end. The alignment of values and goals must be watched carefully, because virtual team members tend to (and are usually expected to) operate more autonomously, which increases the risk of differing perspectives and misunderstandings which can seriously undermine a project. Mismatches on values and goals most often show up in the definition of the product to be developed, the product development process used to create the product, and the commitment to completing the project on schedule.

It is important to remember that the overall *goal* of developing products is return on investment to the company. Achieving maximum return on investment depends upon a company's ability to correctly balance among four factors when defining a product and planning a project: product performance (P), time to market (T), product cost (C), and development cost (C). The optimum balance among these factors (called the PTCC balance in the Q*R*PD methodology) (Kopelman, 1994; Smith, Reinertsen, 1991) must be determined based on the market window for a product, current and envisioned competition, customers' perceptions of needed products and features, and the monetary size of the market vs. the investment needed to develop the product. All members of a team must understand how the factors contribute to the return goal and commit to making decisions about the product and project that support that goal.

Team members' *values*, the relative worth or importance they attach to different elements of product development, influence how they make PTCC tradeoffs to accomplish the project goal. Incompatible values can sabotage the return on investment of a project. For instance, a developer may be faced with "make vs. buy" decisions for the product element they have been contracted to create. The "not invented here" syndrome is very common among high-technology firms: they value their ability to create a product from scratch, without regard to the time and money consequences. Other organizations exalt the ability to create complex, full-featured, technologically sophisticated products over the expediency of getting to market with a simpler product that would adequately satisfy customers. Consultants and small development firms may value ad-hoc entrepreneurial style over more orderly, process-oriented development, which increases the risk of technical problems and missed schedules.

Priority

"A preferential rating, especially one that allocates rights to goods and services usually in short supply." (Woolf, 1977) Priority of the project with the outside team member is critical. Consultants and third-party organizations often take on multiple simultaneous projects; resources can be pulled from one client's project, or that project left with junior engineers, if another endeavor is of higher priority to the development organization's or consultant's financial success.

Skills, experience, and capabilities

"The ability to use one's knowledge effectively and readily in execution or performance; a learned power of doing something competently." (Woolf, 1977) The ability of the virtual team member to do the job is critical. The hiring of consultants and development organizations is usually accompanied by much less due diligence than the hiring of permanent employee or the acquisition of a company. Team members must be assessed for skill level, experience, and track record in these areas: technological or other functional expertise; estimating, risk management, and other project management skills; and ability to communicate verbally and in writing with other team members. Research into risks in software development projects using outside contractors found that among the most prevalent sources of friction between team members were bidding on projects where the contractor has no skills or capabilities, promising delivery dates that are impossible, and inaccurate and inadequate status reports (Jones, 1994).

Product development processes

A process is "the series of actions or operations conducing to an end." (Woolf, 1977). According to a survey by Pittiglio Rabin Todd and McGrath, a high-technology product development consulting firm, companies determined that only 28% of the tasks they performed during a project were truly new (McGrath, Anthony, Shapiro, 1992). This would tend to imply that most risk in a project is contained within that 28% of new work. However, project teams often repeat mistakes from past projects. As expressed by the Software Engineering Institute's Capability Maturity Model, companies work to gain maturity in their product development processes to ensure repeatability of positive project results. Processes help make the "known" activities more predictable, and the unknown more manageable. A process frames interactions between disparate groups, paces work, reduces uncertainty, and helps build common values. Virtual team members must first understand the value of product development processes; then they must have a defined process and be able to execute to it. Typically, smaller third-party development organizations are less process-oriented than the firms which hire them for outsourced development.

Example projects

In the author's experience, lack of alignment in the above areas has been seen to lead to significant project delays and serious product design problems. The following are vignettes from projects which illustrate the importance of those factors.

Example 1 Multimedia product development.

A publisher of CD-ROM based multimedia education and entertainment titles contracted with a small development firm to develop a new entertainment title. The publisher was new at using outside developers. The development firm was contracted to produce a game based upon a prototype that met the publisher's vision of what their next title should be: a fun flight-simulator for kids, with funny scripts for adults. Payments were to be based on the publisher's receipt of specific deliverables such as documents and prototypes at monthly milestone intervals.

After 4 months the development house communicated that they were approaching a schedule problem. The publisher ultimately realized that although the previous contract milestones had apparently been satisfied, the developers were now late because they had been spending most of their time defining and creating the movies that would be played at different points in the game; the game was becoming a story, not a flight-simulator. The movies were not of high importance to the publisher; the gameplay of the product was, and the publisher realized that the game definitions were almost non-existent. The publisher had an in-house development process that emphasized early definition of game vision and design, but assumed developers would be self-policing in this area, so the early project milestones had not stipulated review of game design documents. The developer viewed the game design as something that would creatively evolve over time and did not have a development process that emphasized early game definition.

This mismatch in values and product development processes resulted in a massive contract re-negotiation when it became apparent that the developers had gotten off-track and were not implementing the type of game desired by the publisher. Schedule time was lost putting effort into the game designs mid-stream and overhauling the scripts to make the product match the publisher's vision. In addition, it became clear that current staffing was not sufficient to make the original schedule, and possibly not even the drop dead date for delivery in time for the next Christmas season. The young development organization simply had very little project management and risk management skill and experience.

Ultimately project management consultants were brought in to help replan the rest of the project and provide the publisher with weekly visibility into progress. The milestones were redefined, much more specifically, to ensure that monthly progress shown was not "artificial". Consultants were brought in to assist with proper detailed game design, and the format and content of a complete game design document was imposed by the publisher. After the final replanning the project was projected to be 5 months late but was expected to finish in time for the Christmas season. The total project cost increased by 50%.

Example 2 Outsourced hardware design

An outside development organization was contracted to create a hardware module for a wireless communications company. The developers' expertise was digital signal processing for radio-frequency applications. The contract was basically handed over to the outside house with little technical oversight. The developer delivered a "complete" design, along with test data, to the company. However, the design proved to not be manufacturable in quantity. It turned out the development organization had never developed hardware for mass production; they were oriented to research and development and in the past had always let someone else improve their designs for volume manufacture. The design verification tests had not been complete because they had not considered corners testing for environmental factors or testing of an adequate number of modules. This R&D orientation also led to a design that Underwriter's Laboratory (UL) would not approve, leading to scrapping of several hundred modules, a re-design of part of the board, and a very late beginning of beta testing. Completion of the project required the company to invest in-house engineering resources to modify the design and shepherd the board through production ramp-up, and in the end this product was a year late coming to market.

Project manager actions

To avoid scenarios like the above project examples, project managers must take the following actions to plan and manage projects consultants and third-party development organizations as virtual team members.

Assessing virtual team members

The virtual team members must be fully assessed against the alignment factors for their suitability for this job. For each factor the project manager must identify possible issues, discuss them with the potential team member, and make a conscious decision at to whether to go forward with the relationship. Note that product development using virtual teams requires an experienced project manager who is willing and able to deal with the complexity and uncertainty of these situations. The following questions can aid the project manager in his assessment:

Values, goals

Does the outside developer understand the "return on investment" goals of this project and the feature, technical and cost tradeoffs that are acceptable? Do they value a structured, predictable product development process? Will they commit to your project schedule and deliver? Will they participate in concrete detailed planning, produce status reports as necessary, attend team meetings if required, and work hours that allow you to reach them quickly and predictably? Do they believe in creating high-quality documentation? Are they interested in a long term relationship with you? Are they interested in participating in your industry in the long term? Are they as an organization or as an individual truly interested in the type of technical work required by this project, or are they taking this project as a stop gap until they find the work they actually want to do?

Priorities

Is the group or individual dedicated to your project or spread among projects? What is the company's workload? Are other projects they're working on higher priority than yours?

Skills, experience, capabilities

Has the person/group done this kind of work before? Do they have the appropriate technical skills? Do they have extensive experience using these technical skills, and a track record of related on-time projects and high-quality products? Do they have experience putting these skills to use *in your industry*? Do they have experience with the development and testing tools you want used on this project? What are the communication skills of the individuals, both verbal and written (will they be able to communicate both technical and management information)? Do they produce usable, complete technical documentation? Do you have a say in what level person works on your project, e.g. junior engineer vs. senior engineer? (For third-party development organizations, the skill level and experience required by the individuals should be defined in the contract)What are the technical and project management skills of the managers in the outside organization and can they be counted on to manage their aspects of the project? Do they know how to estimate project work and can you count on the accuracy of their bid for the project?

Product development processes

Do they understand the concept of using a product development life-cycle for risk management and predictability? Do they have established configuration management processes and systems? Do they know how to define product requirements? Do they routinely perform design reviews? Do they understand design for manufacturability, serviceability, etc. and take a cross-functional, concurrent engineering approach to product development? Do they have defect tracking systems and do they use them during product development? Do they understand and perform all levels of testing, such as unit testing, integration, system testing, usability testing? Do they have significant experience using a development process on different types of projects, especially ones like yours?

Plan project in detail

After deciding to include a consultant or third-party development organization on the project team, the project manager must plan the project in detail, with special attention to minimizing the risks inherent in dealing with these virtual team members.

Agree on the product vision

Both parties must participate in a definition of the vision of the product.Who is the customer? What benefits are being provided? What factors will the customer use to judge the quality of the product (what most dictates the products value to the customer?) What key technologies and features are to be included? What crucial product factors must be considered, such as usability, regulatory certifications, environmental constraints, and user documentation? What are the agreed-upon financial: cost and price targets, perceived market window, allowable development cost, etc.? (Kopelman, 1994)

Negotiate/define the process

The product development process must be aligned between organizations and individuals. The development model (e.g., waterfall, iterative or evolutionary, rapid prototyping, spiral model) must be agreed upon, along with major design review and overall status checkpoints during the project. The deliverables between groups during the development phases must be identified. This effort usually requires cooperation and salesmanship to overcome inertia and resistance to change.

Define deliverables

A project is a continual exchange of information. The team must analyze and plan their interactions to avoid misunderstandings and unnecessary delays. All deliverables must be defined in detail, to ensure that the information is on target and provides value, and to make sure that team members have clear concrete goals for small incremental project progress. What is the purpose of each deliverable? What content must it have to be considered complete, and when in the project should it be complete to minimize risk? On the technical side, the team must pay special attention to defining the content for requirements specifications, design specifications, test plans, and final design documentation. On the management side, the in-house project manager must identify the responsible manager within the outside organization and reach agreement upon the project plan elements, including the level of detail for estimating and planning and the methods for tracking and communicating status.

Plan the work and communication

All members of the team must make accurate estimates of project work, especially considering the possible inefficiencies inherent in working with a "remote" outside consultant or organization. The project manager should analyze communication risks as well as technical risks, looking for areas where the physical exchange of information could break down or be delayed due to tool incompatibilities. The project manager must pay special attention to technical and communication risks and form detailed contingency plans. These plans must address any weaknesses identified in the assessment of the team members against the alignment factors.

Team member responsibilities should be explicitly defined and made visible using an accessible document such as a responsibility matrix. Everyone on the team must be given the big picture, so that they know how their individual work fits within the entire plan. The schedule should be built to provide this information, and to make clear the dependencies between groups.

Manage to conclusion

During the project the manager must continue to treat the outside organization as part of the team, keeping them informed and closely coupled with the in-house team members. He must continue to highlight to the team upcoming dependencies between groups.

The project manager must promote frequent exchanges of information that will show what progress is being made and highlight immediately any misunderstandings regarding project or product goals. Outside organizations or individuals should be included in normal team status meetings. Frequent interim informal reviews should be held on works in progress such as requirements specifications and designs, to uncover misunderstandings as early as possible. Even if an organization is providing a "turn-key" solution, where the company is delivered a final product, the contract should include the right to review engineering specifications. The project manager should also determine whether pre-releases of prototypes from the outside developer would be beneficial for spot checking quality and completeness. In short, the project manager must be especially proactive in promoting communication and identifying and managing risky areas.

Conclusion

The use of outside consultants and development organizations can provide companies with effective avenues for competitive product development. However, any endeavor making use of virtual project teams must be managed appropriately to ensure that the desired return on investment is actually achieved. Attention must be paid to the alignment of the team members' values and goals; priorities; skills, experience, and capabilities; and product development processes. Assessment of these factors will provide the project manager with criteria for selecting outside development organizations, insight into potential problem areas, and an understanding of the special planning and management necessary to make his virtual project team a success.

References

Jones, Capers. *Assessment and Control of Software Risks.* Prentice Hall: Englewood Cliffs, NJ, 1994.

Kopelman, Orion. *QRPD: The Guidebook to Quality Rapid Product Development.* Global Brain Inc.: Palo Alto, CA, 1995.

McGrath, Michael E., Anthony, Michael T., and Shapiro, Amram R. *Product Development: Success through Product and Cycle-Time Excellence.* Butterworth Heinemann: Stoneham, MA, 1992.

Smith, Preston G. and Reinertsen, Donald G. *Developing Products in Half the Time.* Van Nostrand Reinhold: New York, NY 1991.

Woolf, Henry (Ed.). *Webster's New Collegiate Dictionary.* Merriam-Webster: Springfield, MA, 1977.

Teaching Multimedia for Management Through Distance Learning

Laurie B. Waisel
(518) 276–2895

Department of Decision Sciences and Engineering Systems
CII 5015
Rensselaer Polytechnic Institute
110 Eighth Street
Troy, NY 12180–3590

William A. Wallace
(518) 276–2895

Department of Decision Sciences and Engineering Systems
CII 5015
Rensselaer Polytechnic Institute
110 Eighth Street
Troy, NY 12180–3590

A paradigm for distance learning is introduced, and the present state of distance learning technology is reviewed. The development and delivery of a new course, "Multimedia and Visualization for Decision Makers," is presented as a case study. The course emphasizes competence in using visualization for presentation and problem solving, in understanding how multimedia is used in business and industry, and in gaining technical facility with multimedia software.

Distance Learning

Economic forces and technological advances are converging to make distance learning an increasingly popular venue for offering a variety of courses. The emphasis on demonstrable value in exchange for rising tuition has brought competition from "corporate universities" — that is, for–profit corporations that offer academic courses for credit — into the fray. In order to compete and remain financially viable, educational institutions have begun using distance learning to attract students for whom traditional on–campus courses are impractical.

Distance learning has long depended on satellite and videotape recording technology, but now Internet access and small monitor–top cameras (such as Connectix's Quick Cam) provide interactive alternatives to the old–fashioned "talking heads." Both of these newer technologies are inexpensive and rapidly becoming ubiquitous among personal computer users. Students who take a distance learning course through a local institution may have access to higher–end technology such as video conferencing (participants can see and hear each other through a live satellite linkup), computer conferencing (participants can see each other's work in a computer application), two–way compressed audio and video, and distributed and cooperative virtual workspaces.

The trend toward specialization and outsourcing offers another way for universities to get more mileage out of their resources. University departments are stronger in some areas and weaker in others. Rather than offering a full spectrum of courses, perhaps including some that are not of the best possible quality, since no department can specialize in every area, departments from different universities can join to become a "virtual department." For example, if one university's management science department has a large statistics faculty but is weak on operations research, and another university's management science department is in the opposite situation, both institutions would benefit from a joint effort to offer the best of both of their departments to as many students as possible, via distance learning." Tuition banks," in which the credits granted by each university to students of other institutions is tracked, could be established with selected universities to promote intra–university cooperation.

0-7803-3552-X/96/$5.00 © 1996 IEEE

In this paper, we introduce a paradigm for distance learning, one that allows for optimal distribution of educational resources. The production of a single course, from the initial conception through final grades, takes place over four stages: *course design, materials development, course delivery,* and *student certification*. The course designer is an information designer, defining the educational goals and developing a detailed outline of the content to be covered. The outline should include a bibliography and relevant Web sites and suggestions on materials development. The materials developer, a presentation designer, works from the designer's outline and bibliography to create presentation materials for the instructor, integrating computer demonstrations, videotapes, and other media as appropriate. The instructor is responsible for course delivery, using the materials created in the previous step. Student certification is the last step in the process, and can be administered by the student's local institution. The goal of certification is to verify that the student has achieved the goals set by the course designer. Certification may become increasingly competency– based, in a style not unlike video games: one can't move on to Level 2 until one can successfully leap all the hurdles of Level 1.

Developing informative evaluation measures for each step of course production is essential. Is the type and amount of content consonant with the course's goals? Is the content arranged in a logical fashion? Do the course designer's materials provide enough information for the materials developer? Are the course materials informative, well–organized, visually appealing? How well does the instructor present the material? And, ultimately, do the students learn? These evaluations could be done in a number of ways, running the gamut from the traditional form filled out by students to having faculty travel to students' sites for evaluation meetings.

Distance Learning Technology

Teleconferencing options (Barron and Orwig 1995) that can be used in distance learning include audio teleconferencing, audiographic teleconferencing, and video teleconferencing. Audio teleconferencing is the oldest technology. It is simple and inexpensive, but as its name implies, its users are limited to voice communications. Audiographic teleconferencing adds two–way still image transmission to audio teleconferencing. Prepared slides and spontaneously–drawn sketches can be sent back and forth, as well as still images of people. Most audiographic systems connect only two sites; connecting additional sites is considerably more complicated. Image transmission can take up to fifteen seconds and the images themselves will have limited visual resolution. Full motion video teleconferencing is the option closest to actually having a group in one location. Options for distributing a video image include cable television, microwave transmissions, satellite technology, or digital video teleconferencing. Video teleconferencing offers the highest level of information exchange, but it also requires significant financial and personnel resources, which need to be justified. Special physical facilities usually need to be built for video teleconferencing, and instructors usually need some training in order to teach in front of cameras. Computer teleconferencing allows users in different locations to work cooperatively, sketching on a shared whiteboard, working as a group on an shared analysis or design software, and commenting on one anothers' work.

Internet technologies include email, the World Wide Web, File Transfer Protocol (FTP), and Telnet. In distance learning, email may be used for instructors to send out lecture notes, for students to submit assignments, and for students to communicate with one another and with the instructor. The World Wide Web allows quick and easy display of and access to text and graphics. FTP is used to send data files from one location to another, and Telnet allows the user to remotely operate another computer.

A Multimedia Course for Distance Learning

Rensselaer Polytechnic Institute's Department of Decision Sciences and Engineering Systems soon will be offering, for the first time, a course entitled "Multimedia and Visualization for Decision Makers." Multimedia authoring software is a visualization

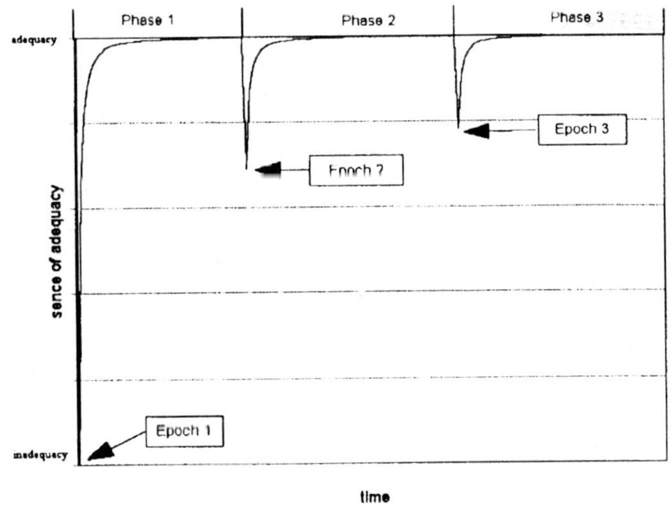

Figure 2: Epochal Transitions and Adequacy (M ,)

Ability. The individual brings an a priori set of knowledge and abilities into the epochal transition which is the accumulation of previous growth. Lawler (1973) implies that ability accumulates with learning where learning is the sum of one's education, training and experience. Mathematically the increase in total ability of an employee is expressed:

$$A_{total} = \int L_t \, dt + C_l.$$

Relevance of ability is deterrnined by the extent of aggregate ability as well as the nature of the new demands. Entry level knowledge can lose relevance as the individual's career progresses laterally or upward in the organization.

The integration of-lea~ning accommodates this with stepwise constants of integration to yield a sum of knowledge ($A_{relevent}$) relevant (Figure 3) to specific demands. A constant of integration, c2 i, exists for each of i positions. Excluding factors such as attitude and organizational support for training, the accumulation of relevant knowledge directly relates to the ability of an individual to produce and contribute.

$$A_{relevant} = \int L_t dt + C_l + C_{2,i}.$$

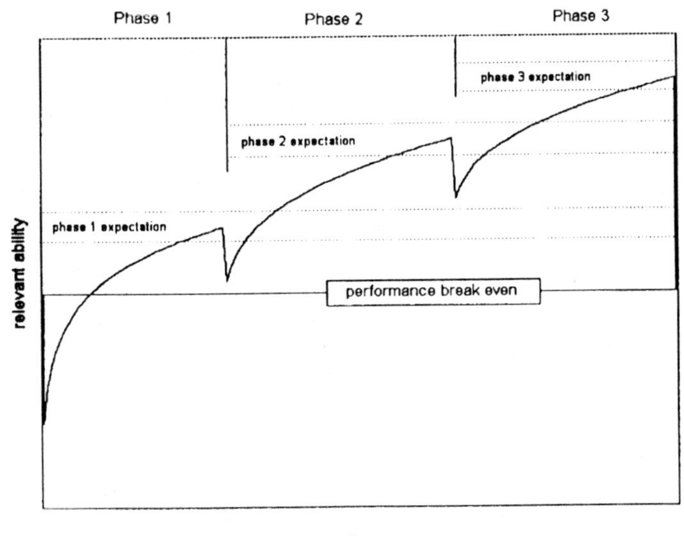

Figure 3: Position Dependent Ability (A_{total})

Since performance relates to position as well as ability, performance must also reflect a stepwise change from one position to the next. Lawler (1971, 1973) suggests that performance is dependent upon aptitude, training and experience:
Performance =∫ (Motivation x Aptitude x (Training + Experience)).

It is been assumed that aptitude and overall motivation remain constant ($k_{aptitude}$, $k_{motivation}$) for any given individual. This is reasonable for aptitude which, like intelligence, is not subject to significant fluctuations under normal conditions. Many components, including those of achievement and competence, comprise one's overall motivation. For simplicity, the career trajectory of such individuals who experience fairly steady growth is considered. This path infers motivational stability at a high level. For these individuals, Lawler's performance function modifies to:

$$Performance = k \bullet (Ability), or P_t = k \bullet A_{relevent}$$

where ability is the accumulation of relevant education, training and experience.

Value. Conceptually, productivity (or performance) may be viewed as one's rate of contribution to business profitability. If job performance, P_t , is measured by contribution to profitability (i.e. dollars per year), then the sum of one's total contribution over the entire period of employment represents the present value, V_t, of the individual's total contribution at zero discount rate. Present value ,V_t of a new hire is simply the integral of P_t.

$$V = \int P_t dt + c_3$$

The constant of integration, C_3, represents initial costs such as search and relocation.

A differential analysis offers additional structure and new perspective regarding the relationship among the attributes of competence motivation, learning, and productivity (MLP). The resulting model separately suggests a learning curve function similar to, but not identical with, the traditionally assumed function of logarithmic decay. The proposed MLP model offers a mechanism to help organizations in understanding and projecting the needs and performance expectations of both the individual and the environment for which they labor.

Learning and motivation are, respectively, the first and second derivatives of productivity. Although ability has the same wave-shape as productivity (Wells/Bailey 1995), its dimensions and initializing constants are not the same. White asserts that an individual will strive for competency by meeting performance expectations. In the proposed extension of the MLP model, the application of White' s competence motive offers differential closure by directly relating productivity to motivation, its second derivative. In control theory, this deficiency among perceived and expected performance is called error - that component of feedback that quantifies deviation. Expected performance, β_e, iS assumed constant for a given position or responsibility. The constant α^2 represents a characteristic of individual responsiveness and allows the equation to be dimensionally correct. Accordingly, White's competence motive is functionally expressed as:

$$M_t = \alpha^2(-\beta_e + P_t) \bullet$$

Without further stimuli, performance remains convergent on β_e the original target. The resulting equation for productivity is expressed as

$$P_t = \beta_e + (-\beta_e + P_0)e^{-\alpha(t-t_0)}.$$

Levy (1965) proposed a semi-logarithmic productivity function which is essentially identical. Levy asserted that the traditional (logarithmic) function was inappropriate at extreme values of time. The first derivative of this equation yields a learning curve function similar to, but not identical with, the traditional learning curve. The second derivative is the function for competence motivation.

A look at the semi-logarithmic learning function plotted in three dimensions against its integral (productivity) and time does not appear significantly different in its trajectory than that of the traditional logarithmic function (Figure 4). In either case, the individual's developmental trajectory proceeds toward the opposing corner of the quadrant, set back periodically by promotion to productivity reductions and elevated learning.

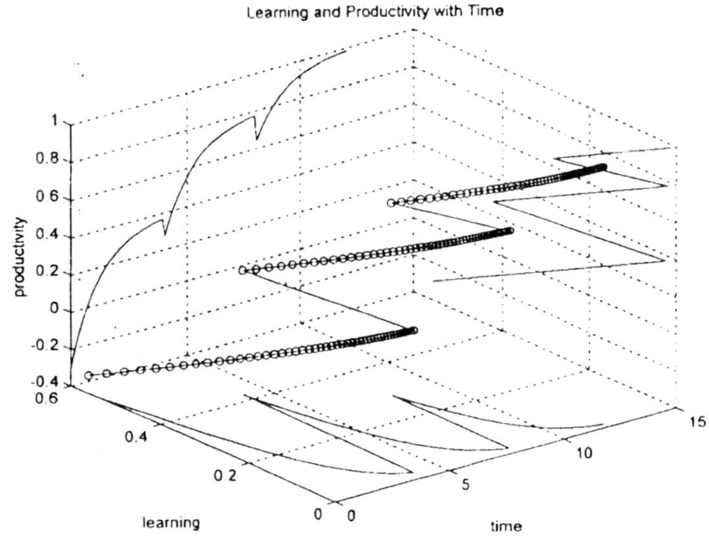

Figure 4: L_t, P_t, vs t: Semi-Logarithmic Decay

The relationships among motivation, learning, ability, productivity, and present value comprise this model. Motivation and rate of learning offer obvious difficulty for quantification, particularly for duties predicated upon analytical and strategic abilities. It may be that rate of learning, and motivation more so, can best be evaluated by their reflection in performance. From the other side, estimating present value is also problematic. Good estimates of present value hinge upon accurate forecasts of developing ability and anticipated duration of employment. Given these, estimates of individual value remain speculative on both fronts.

If a reasonable measure of productivity is made available, then its respective derivatives, learning and motivation become known. An estimate of present worth is still speculative, but the value, and trajectory of one's present value for work completed in the past is available, and a better estimate of the worth of keeping a particular employee on board also becomes available. If one associates dollars per day with individual productivity, then units for the related functions of value, learning, and motivation are set, differing only by constant multipliers.

Market Growth

Business volume is more easily measured and reported than the employee performance; a point confirmed by the significant extent of numerical information available in the form of financial data. Also unlike employee development, business development lacks a tight theoretical construct. In fact, the availability of quantitative measures in business eases the task of market analysis, but complicates the task of producing a detailed theoretical model. A fairly clear response mechanism is inferred, but a functional structure in mathematical terms is not widely discussed. Such a mathematical model, if a definitive and proven one were to exist, may do less to predict the future than it might to explain those structures of causality which cause markets to shift this way and that.

One common characteristic of successful markets is growth itself. The traditional product life-cycle of a new product is comprised of four phases: introduction, growth, maturity, and decline (Figure 5). For example, the initial phase of a new market experiences a low, but perhaps stable, sales
volume. Growth continues, and may even accelerate, until approaching market limits finally become recognized by producers. A market acceleration that continues indefinitely is unlikely and examples of such do not come to mind. Continuing growth, unchecked, must eventually consume its host economy.

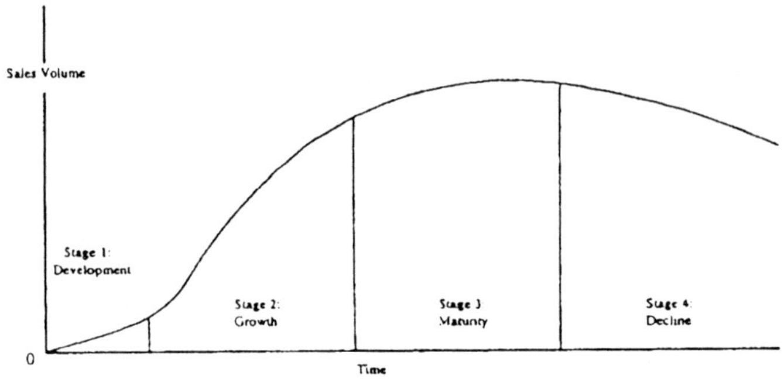

Figure 5: S-Shaped Product Life Cycle (Levitt 1965)

Customers will purchase only the products that match their needs. The difficult tasks for businesses are to determine what those needs are, create a satisfying, distinctive product, and make it available at the precise time. Upon these accomplishments, focus will turn onto growth. Success of an organization coincidentally hinges upon a dynamic understanding of market conditions and customer needs, and the ability of an organizations people to set strategy and implement tactical plans effectively.

Although the S-shaped product life-cycle is widely embraced, it is not exclusive; a plot of product sales could follow an infinite number of trajectories. Nevertheless, this proposed model is based upon the traditional product life-cycle S-curve concept; in particular, the growth stage. The commonly accepted S-curve provides a solid foundation for a unique study of market growth wielding similar techniques and perspectives that were applied to the employee's learning curve.

An Idealized Model for Product Growth. The sales volume trajectories of three products from completely different industries (Plain filter cigarettes 1947-1966, Musical cassettes 1975-1995, and pre-WWII sales for Automobiles 1900-1940) were plotted and non-dimensionalized. In Figure 6, the three product curves are overlaid to illustrate their common s-shape patterns.

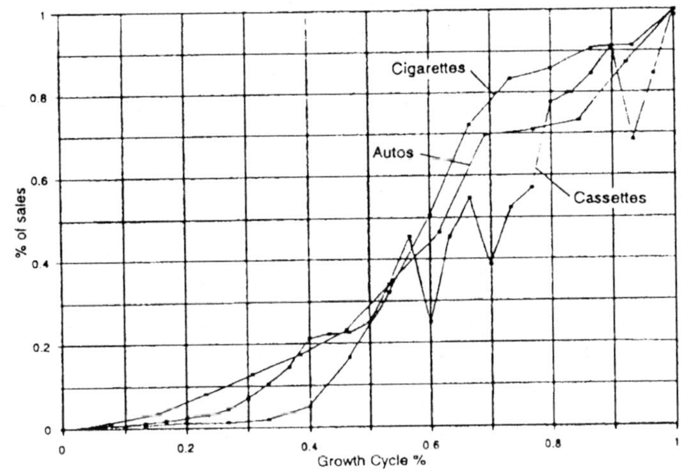

Figure 6: Overlay of Non-dimensionalized Product Curves

A non-dimensionalized fitted data curve was created by averaging this non-dimensionalized product growth phase data. As expected, the fitted data curve strongly resembles an S-shape, and when it is applied to the overlaid curves in Figure 6, it provides a good representation of the product's initiation and growth phases.

The plotted curve (or sections of curve) of numerous mathematical functions exhibit S-shapes: for example, the hyperbolic secant (sech). An attempt at fitting a sech function to the fitted data curve was carried out using the following function [z], where different power and constant values were used to shift and adjust the amplitude of the curve to provide a best fit:

$$z = (\sec h(k_1 * t - k_2))^{k_3} \quad [z]$$

A modifying cosine function was then fitted to a plot of the error; the discrepancy between z and the fitted data curve D. A second modifying cosine function was applied to achieve a tighter fit. This second modifying cosine function (c) is then multiplied to the original sech function, resulting in the following function [y];

y = z * c (Figure 7).

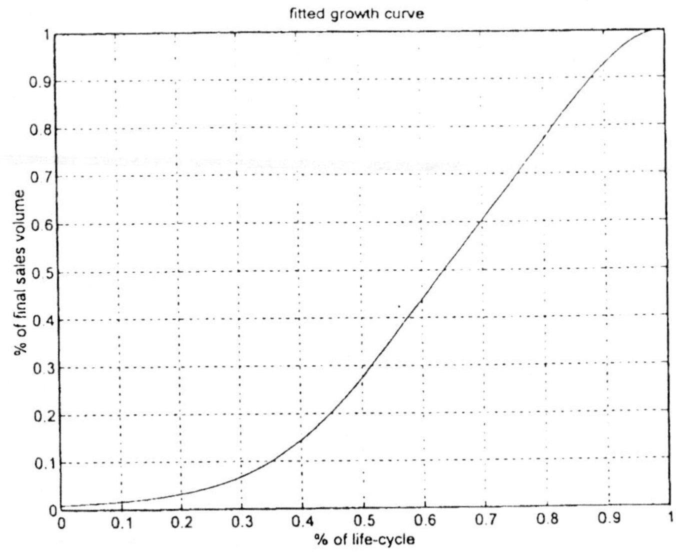

Figure 7: Fitted Functional Curve

At 50% of the life-cycle on the fitted data curve, sales volume had reached approximately 26%. In comparison, at 50% of the life-cycle on the fitted functional curve, sales volume is approximately 27.5%. At 20% life-cycle of both curves, sales volumes are both at 3.5%.

The derivatives of the functional curve also furnish a framework for a differential equation structure. The first derivative is sales growth rate, while the second derivative represents sales acceleration. It was found that the initiation and growth phases of a product life-cycle closely follow the following non-linear differential equation, with M representing demand, and y representing production or current supply;

$$d^2y / dt^2 = (M - y) * (dy / dt).$$

Supporting this relationship is the view that companies advocate increased production when they perceive an opportunity for growth in the market; when (M-y) > 0, and when sales is increasing, dy/dt > O. Producers are most interested in M. In most cases, however, M is unknown and unpredictable. If organizations knew the demand for a certain product, investment strategy and planning would be elementary, and risk-free. As M remains an approximate quantity, it drives the perception of potential growth in the eyes of risk-averse business investors.

It should be understood that a good curve fit does not define structure, but does seem to be compatible with the proposed differential structure (Figure 8). This curve seems to support the differential relationships among market demand and volume, market acceleration, and growth.

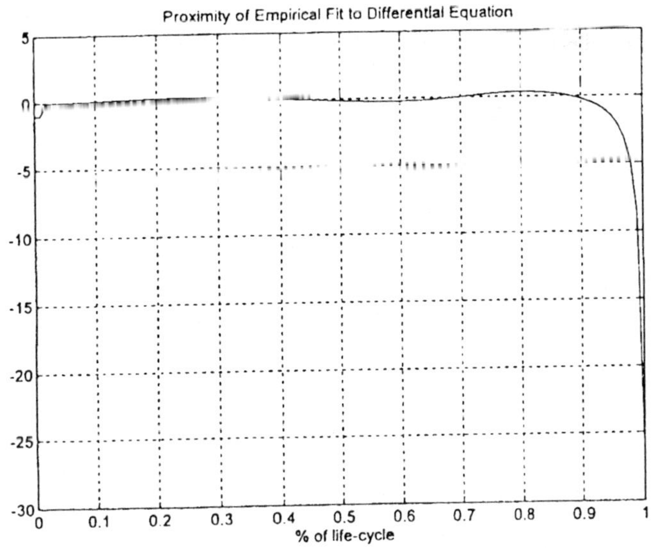

Figure 8: Fitted Data Curve Fit to the Differential Equation

This stable response is present for over 90% of the growth cycle. However, as the growth phase approaches a stage of maturity, the flat response and relationship structure ceases. This is to be expected. In comparison to growing markets, mature products or markets encounter different drivers, structures, variables, and factors.

This non-linear differential equation offers little hope of predictive power. Nevertheless, with its trajectories similar to those empirically driven, the model may provide justification of phase transitions and structural drivers of growth.

Conclusion

The work discussed in this paper does less for quantifying measureables than for clarifying the structural motivators that cause measurements to change. For example, what might be said about the employee growth model? The first point is that employees who develop to a higher level of contribution and who stay employed longer are worth more—a lot more. This means that good employees ought to be supported in their development and provided ample reason to not seek employment elsewhere. The second point is that a dynamic exchange of stresses and contribution exist among employee and employer, and that both stress and contribution, properly applied, yield growth and increased value. Employee and employer need to understand each others role in enhancing growth and value.

Markets are the root driving force for all business, and the maturity of any given market gives context to associated growth opportunities and risk. Life-cycle maturity infers a higher level of competition and increased demands for operational efficiency. Since every product experiences such phase changes, new demands on individual performance propagate through the entire organization. Just as the employee's contract with the organization represents a system of motivation and reward, so also do markets impose an exchange of motivation and reward. A firm's activities in employment and commerce must be managed in order to make best use of limited resources.

A notion of virtual growth does not mean an image of growth, but that growth opportunities are promoted wherever good progress is most attainable and most needed. After all, companies cannot maintain steady-state emphasis on optimized growth with all people and all products. For employees, a natural emphasis on learning and development results from times of promotion and change. For sales, the transitions arise as one product matures and another is initiated. Among these seemingly natural oscillations exist the best opportunities for gaining and disseminating experience, and for putting newly acquired abilities to best service.

References

Haire, M., A New Look at Human Resources, *Industrial Management Review*, pp. 17-23; Winter 1970.

Katz, R., Job Enrichment: Some Career Considerations, in *Organizational Careers*: Some New Perspectives, John Van Maanen, ed, New York: John Wiley and Sons; 1977.

Lawler, E. E., *Pay and Organizational Effectiveness: A Psychological View*, New York: McGraw Hill; 1971.

Lawler, E. E., *Motivation in Organizations, Monterey*, CA: Brooks/Cole Publishing Co.; 1973.

Levitt, Theodore, EXPLOIT the Product Life Cycle, *Harvard Business Review*, Vol.43, No.6, pp.81 94; 1965.

Levy, F. K., Adaptation in the Production Process, *Management Science*, Vol. 11, No. 6, pp. B 136-151; April 1965.

Maslow, A. H., *Motivation and Personality*, New York: Harper & Row; 1954.

McClelland, D. C., Measuring Motivation in Phantasy: The Achievement Motive. In H. Guetzkow (Ed.), *Groups, Leadership, and Men*, Pittsburgh: Carnegie Press; 1951.

McClelland, D. C., *The Achieving Society*, Princeton: Van Nostrand Reinhold; 1961.

Wells, D. J. and Bailey, S., Learning and Productivity: A New Perspective on Employee Development and Worth; 1995, *IEEE Transactions on Engineering and Management*, in review.

White, R. W., Motivation Reconsidered: The Concept of Competence, *Psychological Review*, Vol. 66, 297-333; 1959.

Telecommuting and Consulting Engineering

John Whittaker
(403) 492-4443

Mechanical Engineering Department
University of Alberta
Edmonton, Alberta
Canada T6G-2G8

Tim Cartmell

Mechanical Engineering Department
University of Alberta
Edmonton, Alberta
Canada T6G-2G8

Consulting engineering involves teams sharing information and tasks to create a project. Thus it would seem a suitable area for telecommuting. However, the nature of engineering work creates difficulties with issues related to: payment, team interaction, mentorship, security, legal issues and document control. Consequently it seems that only part-time telecommuting is a viable work strategy for this industry.

Introduction

Consulting engineering, in that it involves the application of physical principles to unique problems, is an industry based on the manipulation, interpretation and presentation of information. Consulting engineering firms traditionally employ a project team approach. This has led to the creation of the contingency work force, a group of workers including contractors, part-time employees, informal workers and telecommuters.

The literature on telecommuting was reviewed to determine the potential advantages and disadvantages. This information was supplemented by interviews and a survey of telecommuting engineers.

Telecommuting defined

"Telecommuting is the use of computer and telecommunications technologies which permits organisational employees to substitute telecommunications for transportation, thereby permitting some or all of a job to be performed at a remote work site (Ramsower, 1985, p.2)." The ubiquity of computers and the internet have provided the technology to make telecommuting a reality.

Terms are such as telecommuting, home work and informal work are applied to the use of remote work sites. "The unifying factor underlying all these labels is that work normally done on an employer's premises is instead performed in the worker's home. Such home work makes its appearance on the contemporary scene as a potential solution to problems as diverse as the energy crisis and the child care problems of working mothers (Lozano, 1989, p.98)".

Telecommuting consequences

In the literature the main advantage identified, beyond the obvious saving of commuting time, is the improvement in employee organisational behaviour as a consequence of the trust implicit in the arrangement (Ramsower). The obvious disadvantage is the loss of face-to-face communication. This both alters communication plus it changes the way workers are judged, evaluated and compensated. Other major consequences are summarised in the following tables.

Advantages of Telecommuting

Topic	Advantages to the employee	Advantages to the Employer
Working Hours	• greater flexibility with which leisure can be integrated with work • more control over how work day is spent	time not spent working not compensated for (including vacations, sick time)
Travel Time	time normally spent commuting now free for other activities	hours saved not commuting often spent working
No Physical Boundaries	no longer necessary to live near the employer	take advantage of labour forces that cannot get to the employer's location (single parents, handicapped individual with restricted mobility)
Increased Flexibility	• can choose which projects to work on • can determine own work schedule- can offer services to wide market	can increase or decrease the size of the workforce as the workload demands
Productivity	greater autonomy leads to greater satisfaction, resulting in higher motivation level and increased productivity	freedom granted to employees results in greater satisfaction, productivity and loyalty
Decrease in Overhead	reduced expenses associated with regular office job (clothes, meals, travel)	• reduced or eliminated cost of benefits • overhead costs (heating, office space) now the responsibility of the telecommuter

Disadvantages of Telecommuting

Topic	Disadvantages to the Employee	Disadvantages to the Employer
Working Hours	workers end up working more hours, inducing more stress into their lives	
Compensation	• contract or piecework compensation results in insecurity, increased working hours and postponed breaks and vacations • control of company over employee formerly expressed through supervision now expressed through deadlines and piecework	
Isolation	• left feeling unnoticed • limited access to guidance and assurance of supervisor • difficulty in gaining access to references and resources	
Intrusion of Home Activities	• no physical separation means telecommuters are always aware of pressing household responsibilities • there is no longer the physical separation that allows the worker to "forget" work	intrusion of home activities can make the employee less productive
Lack of Vision (inability to see telecommuters)		traditional "management by observation" no longer possible

The consulting engineering industry of Alberta

In December 1985, it was estimated that 18,000 persons were employed by Alberta's consulting industry. Most consulting firms were small, averaging 24 employees. Approximately 6 percent employed over 100 individuals, while 70 percent of the companies had ten employees or fewer.

Typically a consultant is awarded a body of work to accomplish a particular function, termed a project. This body of work has a well-defined beginning and end, and will require resources simultaneously from some or all of the technical groups in varying numbers and for varying amounts of time. Thus the nature of consulting engineering work favours a more divisional or market orientation, with teams formed to focus on the requirements of a particular project. Consultants are usually reimbursed according to the number of hours spent on a project. The highly competitive nature of the business requires the company to keep as many individuals as possible performing value added work. The more hours worked, the higher the revenue. Functional grouping allows a company to move people from one project to the next, eliminating down time between projects and enhancing the profits of the company. This dichotomy supports the creation of a "matrix" organisation structure and that is the type of structure that is used by many large consulting engineering companies.

Issue unique to consulting engineering

The problem of piece work

Full-time telecommuting takes the employee out of the sight of the manager, so there is no convenient way of confirming the time spent on the project. In other telecommuting applications, this problem is overcome by making compensation contingent upon results. A piece of work, done to the requirements of the employer will result in payment. The problem for the consulting engineering firm becomes one of determining just what outputs to measure. What "deliverables" can be identified? What value is attached to each deliverable? How should projects be broken into "pieces"?

One author's experience in consulting engineering is that there are many computing tasks that must be done as part of each project, and that the work is deadline driven. Thus superficially, it seems that telecommuting could be easily applied to consulting engineering. However, it is often not known just what tasks will be required, until they are needed. An engineering project is more than just an assembly of different tasks into one whole. It is a body of work that is created as different people provide information about their particular aspect of the work.

Engineering project work is a series of highly interdependent individual tasks. If one task breaks down, the entire project can be affected. So for the engineering firm to provide a deadline and a body of work to a worker is not enough. The cost that incomplete work, uncoordinated work or the wrong work will defeat any savings accrued through informal work arrangements.

Problem of team interaction

Another problem that a consulting engineering firm would have to adjust to would be the variable hours that each telecommuter would keep. Within engineering projects, the perceived advantage of a team effort is the ease of communication between team members. When a crisis emerges, as they frequently do, a member from one functional group often needs an immediate response from a member of another group. It the other person is a telecommuter who has chosen to take that particular time as leisure time, the first person cannot complete their task. While technology takes care of the physical link, if the employee is not available to respond expeditiously to a request the project can experience delays, and therefore cost overruns.

Problem of mentorship

Engineering companies usually have a cross section of skill levels and areas of expertise. This allows assignment of the required resource to a particular project. As a person enters the firm, they are assigned increasingly complex tasks and given reducing amounts of guidance. Thus they develop increasing uncreative skills and expertise, and increase their value to the firm.

A critical element in this process is the ability of the worker to call upon more experienced individuals for guidance. Young employees need to look over the shoulders of the more experienced individuals and learn more about their industry. New employees need a chance to absorb the expectations of the firm regarding productivity levels, competence requirements and communication needs. How will individuals develop, when they are in a remote work site and not exposed to other individuals? How can an inexperienced person get into an industry that does not provide these mentoring activities? Who will replace those workers that leave the industry?
Related to the issue of mentoring is the problem of timely feedback from one's supervisor. The informal worker no longer has the reassurance that comes from a manager's encouragement, or the comfort in the ability to demonstrate how hard one is working. Informal workers are also aware that while the work terms have changed,

often the style of management has not, so the lack of social interaction potentially results in missed promotions (Alexander 1990). "Face time counts. Putting in long hours, even if you are twiddling your thumbs, is still viewed by many bosses as a sign of loyalty (Fierman, March 1994, p.67)."

Problem of security

Many engineering clients are concerned that proprietary information may be released to their competitors. This is not usually much of a concern in the small project/revamp environment, but it can become more of an issue if process information is involved. There is always a question of how secure information is when it is released to an engineering firm.

When informal work methods are added to the equation, the concerns are much more widespread. If informal workers are retrieving such information through modems or other data links, what security measures are in place? How easy is it to get onto the system? This can also be an issue for the engineering company, as anyone gaining unwarranted access to their system has potential access to their own corporate information, can possibly steal long-distance phone time or can sabotage the environment with computer viruses or other malicious means.

Legal issues

Of particular concern to consulting engineers is the issue of legal responsibility. These individuals are offering expertise in many forms including designs, feasibility studies, inspection reports and completion certificates. They are people, and they are not infallible. The issues for most engineers include the circumstances under which one can be sued, and what can be done to protect oneself from that possibility? Consulting engineering firms protect themselves by taking out errors and omissions insurance.

The contract is typically between an engineering firm and an insurance company. The issues that usually arise concern who is insured and under what circumstances. Generally errors and omissions insurance will cover an employee engineer for work done for that firm, while employed by that firm. Work done outside of the firm, on a "moonlighting" basis for example, will not be covered. An employee engineer is covered as long as the employer maintains the insurance policy, even if the employee leaves the firm.

The matter of errors and omissions insurance is a potentially major problem for telecommuters. For those telecommuters employed exclusively by one firm, the insurance policy could extend to the remote worker. When a telecommuter is hired on a contractual basis, the situation becomes complicated. Is the telecommuter still an employee of the firm, and thus covered under the firm's insurance policy? The situation becomes even more involved when the telecommuter does work for several firms, either consecutively or concurrently. What insurance policy covers such an individual?

Errors and omissions insurance is a complicated issue, particularly under non-traditional working arrangements. The one party that stands to lose the most is the telecommuter who works on a contract basis for several firms. This person could find that they are not covered, or could be embroiled in a time-consuming lawsuit to determine responsibility.

Document control

Engineering design work is typically presented on design drawings. Engineered drawings used to be drawn by hand, which meant that only one person could work on them at a time. This relatively simple statement has wide ranging implications in the electronic world.

The consulting work that one author's firm does mainly concerns renovations to existing facilities. Drawings used to describe changes to these facilities are generally existing design drawings. They portray what is there now, and allow the engineer to determine how the item should be altered. In this way drawings can be used many times over on many successive projects.

Because of the nature of this work, most companies have
rigid rules regarding the use of their drawings. For example, only one copy of each drawing is maintained, and all changes go on that drawing. On a particular project, a set of drawings is given to a consultant either to portray new work or to use as reference in completing the project. Changes made to the drawings are carefully noted, so an individual using the same drawing in the future can see what changes were made, when and for what purpose.

Drawings today are almost always consist of a physical drawing and an electronic copy. When a consulting firm has completed a project, changes must be made on the electronic copy of the drawing as well. When this is done in a traditional project setting, it is a cumbersome but manageable task to ensure that all changes are recorded on the electronic drawings.
In a telecommuting application, this problem becomes much more cumbersome. Employees who require a reference drawing to look at will still require this drawing if they work at home. Different workers requiring the same drawing to do their work will want to view and change that drawing. If two persons are working on the same drawing simultaneously, which drawing is now the "original"? Which one will be updated to reflect all the changes? Who will record these changes?

Conclusion

For the consulting engineers, the only viable measure that seem to address these issues it that of part-time telecommuting. That is, telecommuting works in the situation where only one individual is working on a clearly defined task, and long uninterrupted periods are conducive to task completion. Then the remote linkage to information, and the computational ability of a computer, is an asset. However, the person is still an hourly employee, and the employer must still provide both benefits and office space, so many of the potential advantages are negated.

References

Alberta Economic Development and Trade, Consulting Index - Alberta Engineers, Geologists, Geophysicists, (1986).

Alexander, Michael, "Travel-Free Commuting". Nation's Business 78:12 (December 1990), pp. 33-37.

Consulting Engineers of Alberta, 1993 Consulting Services Directory, (1994).

Fierman, Jaclyn, "Are Companies Less Family-Friendly?" Fortune (March 21, 1994), pp. 64-67.

Lozano, Beverly, The Invisible Work Force, New York: the Free Press (1989).

McGee, Bill, "Telecommuting: A Viable Option?" Denver Business Journal 44:52 (September 10, 1993) pp. 18-19.

Ramsower, Reagan Mays, Telecommuting-The Organizational and Behavioural Effects of Working at Home, Ann Arbor, Michigan: UMI Research Press (1985).

System Design for an Integrated Document System and its Impact on Standards Development Efficiency

Swantje Willms
Emile Morse
Michael Spring
(412) 624-9429
swan@lis.pitt.edu

Department of Information Science
University of Pittsburgh
Pittsburgh, PA 15260

The development of information technology standards is slow and expensive. Spring et al. (1994) suggested electronic document development tools can improve the process. CASCADE, a system developed at the University of Pittsburgh, supports collaborative authoring. CASCADE was recently tested in a review of comments made on a mock ballot of POSIX 1003.21 to determine the reliability and accuracy of a parser for mail delivered comments and gather preliminary usability data. Asynchronous electronic commenting and comment review can improve both the quality and speed of standards development.

Introduction: The Standardization Process

Information Technology (IT) standardization has been the subject of numerous studies (Bonino and Spring, 1991; Spring et al., 1994; Oksala et al., 1996; Lehr, 1992; Farrell, 1993; Toyofuku, 1993; Cargill, 1989; Libicki, 1995; Weiss and Sirbu, 1990). Standards organizations operate in an open, voluntary, public fashion, observe due process and operate through consensus (Greenstein, 1992). The time required to develop IT standards is between three and seven years. The cost of an average IT standard is about $10,000,000 (Spring and Weiss, 1994). Standardization involves many individuals and incorporates input from a diverse set of documents. In addition to the standard itself there are comments, minutes, drafts, presentations, briefs, whitepapers, etc. to be managed. Many standards are constructed by the chair with active input from about ten individuals. Large (multipart) standards may have numerous editors; one editor then compiles the components to ensure continuity and clarity of the composite document. Involved parties need to review the document in a timely fashion. All comments and objections must be addressed and a cohesive document must result without changing the substance of agreements.

The study by Spring et. al. (1994) identified two computer-based technologies subjects felt were underused in process management and document development:

- E-mail for document distribution and balloting. (The use of e-mail in the development of IEEE 1012 reduced development time one year.)
- Networked document access for comment annotation.

The standards community is exploring ways to improve the standardization process. These include the IEEE Standards Process Automation (SPA) system, the National Standards System Network (NSSN), and the National Information Infrastructure (NII) standards development effort.

CASCADE

CASCADE (Computer Augmented Support of Collaborative Authoring and Document Editing) is an environment for collaborative authoring designed to support research on augmenting authoring and other document processes through agents using the information available from structured electronic documents.

0-7803-3552-X/96/$5.00 © 1996 IEEE

CASCADE has three goals:

- reduce the cognitive overhead in the authoring of structured documents through tools and agents,
- reduce the complexity of document management through an SQL-compliant DBMS as a repository for workgroup, structured document and hypertext information,
- assure document integrity and availability via database management of document controls and information.

CASCADE seeks to augment human actions, exploit the power of structured distributed documents, and use intelligent assistants in a coordinated fashion. CASCADE represents a structured and controlled approach to the creation and management of network documents. In this way it differs from World Wide Web products that operate with minimal access constraints and are subject to failure, e.g., URL not found. While these systems use an SGML-compliant structure, they do not use this structure optimally to augment the authoring and development process.

CASCADE editorial and review processes are enhanced by having a single copy available for review and annotation. Coordination of reviewers' comments and editors' actions is done by the system. CASCADE makes the authoring process and auxiliary activities of participants easier and insures document and link integrity.

CASCADE will use a CORBA-based client-server configuration with clients using native interfaces for common platforms. The prototype uses X Windows served across a network. The SGML document structures will be managed by an SQL-compliant DBMS. The prototype uses a variety of record files and the UNIX file structure to manage the prototype documents. In what follows, we briefly review the structural architecture of CASCADE. See http://www.lis.pitt.edu/~spring/cascade for more.

The CASCADE document space begins at a node (currently a directory), and fans out as document and comment nodes are added. Within the structure the user can view the main and related documents under the control of a given project. Whenever a document node is accessed, the document contents become visible along with its hypertext links. Given the proper permissions, a document may be directly edited at this point.

In the process of viewing a document in CASCADE, comments can be embedded in the document by positioning the mouse pointer at a location and clicking. This reduces the cognitive effort and number of steps required to make comments, focusing the user on the substance of the comment rather than the processes required to generate the text. Both system- and user-defined links are consistent with the internet Universal Resource Locator (URL) standard. Comments are noted and added to an activity list so that all other users are made aware of them.

The prototype has provided rudimentary agents for:

- activity analysis and reporting
- printing and final form document processing
- mail generation
- mail review and browsing
- mail analysis and parsing
- document and comment review (mural)
- analysis of group participation
- synchronous communication and document review.

Each of these is briefly described below.

Activity reporting

CASCADE maintains a log of document creations and editing, comment insertion and review. At connection time, each user is informed of activity since his/her last connection. The system keeps track of activities the user has reviewed. Ultimately, the user will be able to target an activity agent to watch for given classes or loci of activity.

Printing

CASCADE provides for printing of documents or documents with comments attached as callouts to user defined levels.

Mail generation

E-mail is integrated for communications with commenters and group members. The system provides editable fields which are initialized with selections based on the system's estimation of the user's intent. The system uses preference information and agents to determine how best (i.e. fax, e-mail, pmail, etc.) to deliver mail and communications to users.

Mail review and browsing

A simple mail browser allows multi-filter browsing of system related mail along with a full document vector relevance matching tool that allows user to find related mail notes quickly. This feature will be enhanced to enable identification of relevant comments or document components.

Document and comment review (mural)

Whenever a document is open, a "mural" of the full document displaying all embedded comments is shown next to the vertical scroll bar. For document nodes that are larger than the display window, this allows a preattentive examination of the entire document for given classes of comments based on the color of the displayed comments in the mural. In our preliminary study, this feature greatly facilitated editorial review.

Analysis of group participation

CASCADE tracks usage, connect time, etc. for each group member. The usage pattern, over time, of group members may be visualized. Currently, this information can be used by users to determine the patterns of access, good times for meetings, current access, etc. Ultimately agents will be trained to use this data on the user's behalf.

Synchronous communication and document review

Based on participation information, two synchronous communications capabilities exist. A caucus function connects logged in group members with a shared whiteboard window displaying the currently viewed document component. A communications agent watches for group members or other individuals of interest to a given user. The agent looks for these people using a variety of rules, and spawns a talk session should they log into a system being watched by the agent.

Two additional features of CASCADE, mail parsing and comment analysis, are discussed below in the context of the usability study.

Usability Study

With the permission and support of the editors of the POSIX 1003.21 committee, efforts were undertaken during the summer of 1995 to support the IEEE mock ballot process using CASCADE. (The authors acknowledge in particular the encouragement of Craig Meyers, Pat Place, and especially Bill Pollak of the Software Engineering Institute at Carnegie Mellon University.) The committee used the Scribe Document Processing System, a highly structured, SGML-like, markup language, as well as a mail exploder to share documents and make comments. Because of security firewalls, it was necessary for commenting to proceed via mail ballot. The existence of mail exploders and a structured electronic source document provided an opportunity to experiment with agents to automatically process mail and parse comments into structured source documents. While short of the integrated network access imagined for CASCADE, this situation allowed exploration of how mail might be used to link individuals who did not have full interactive access.

Balloting

In the study, software was written to parse incoming mail into document comments. While interactive commenting is the goal, mail integration may be used to link individuals without full interactive access. CASCADE performs the following steps in ballot processing:

- mailbox files are broken into individual mail notes;
- comments within mail notes are identified, named and stored as individually linkable files;
- comments are mapped, based on final form document locations to revisable form document sections;
- comments are classified by type and
- links are inserted as appropriate.

The automatic parsing of mail notes depended on reviewers complying with a minimal format -- consistent with structured documents. More on the details of this process may be found in Morse et al., (1996).

Comment Analysis

Once comments were located in source documents, they were reviewed by editors. Three tools in CASCADE facilitated this review:

- comments appear in the viewer as a small colored icon (indicating status) with a label that identifies the comment author. (The mural described above allows quick navigation to comments.)
- an interface allows editors to review and modify the classification of comments through a form with radio button selections for features such as type, status, and disposition of comments. The review system annotates comments detailing who made what changes when allowing the editor to add notes explaining the changes.
- a reporting mechanism provides information about the type, status, and disposition of all comments. The reports range in detail from statistical summaries to full text of all relevant comments.

Usability

The usability study provided a variety of data for development: The system was able to identify 1900 out of 1905 comments and determine the type of 1844 automatically. The placement abstracted to sections was satisfactory for all comments. The comments should have been automatically numbered for easier reference. The mural was pointed out as a very useful locator devise for navigating through a document. The editor made frequent use of communications methods such as e-mail as he worked. The overall time spent on an average comment resolution was about two minutes. The editor spent seven seconds of this time to add an annotation averaging 35 characters in length.

Conclusions

CASCADE is still in early stages of development as a grounded research platform for studying and augmenting the collaborative authoring process. A variety of data visualization tools are in place to assist developers in identifying usage patterns that agents can later be trained to look for. The specification of an agent architecture to guide the development of the local and server agents has been completed along with guidelines and methods for inter-agent communication.

Related to the standards development effort, preliminary evidence of the value of such a system is encouraging. The basic assumptions about augmentation appear to bear out. Some specialized features have already been put in place and a number of new features are being considered including:

- commenting agent: As individuals make some comments more than once in a detailed review -- e.g. "awkward wording in this sentence", accelerators will automate repetitive comments.
- comment visualization: Currently, comments appear as a small colored icon with a label specifying the author. This system will be enhanced to show different things about the comments: read/unread comments, comment age, comment type, etc.
- comment analysis: Preliminary analysis of comments by agents will replace the activity report provided at login.
- editorial assistant: We are currently experimenting with software which will prune and graft branches of SGML document trees to allow dynamic granularity in the locking of documents. Versioning will be incorporated to ensure that the most current version of each section of a document is made available, while guaranteeing the ability to review, track, and revert to any previous versions.
- editing agents: The editor responsible for a working document section will monitor the comments on the most recently published version and incorporate suggested changes that gain group approval. This will be especially easy for editorial, factual, or clarification comments.

References

Besen, Stanley M. "The European Telecommunications Standards Institute: A preliminary analysis." Telecommunications Policy, 14: 521-30, 1990.

Bonino, M. and M.B. Spring. "Standards as Change Agents in the Information Technology Market." Comput. Standards Interfaces, 12: 97-107, 1991.

Cargill, Carl F. "Information Technology Standardization: Theory, Process, and Organizations." Digital Press, Rockport, MA, 1989.

David, Paul A. and Shane M. Greenstein. "The Economics of Compatibility Standards: An Introduction to Recent Research." Economics of Innovation and New Technology 1 (1): 3-42, 1990.

Farrell, Joseph. "Choosing the rules for formal standardization." University of California, Berkeley, 1993.

Farrell, Joseph and Saloner, Garth. "Coordination through Committees and Markets." RAND Journal of Economics 19 (2): 235-252, 1988.

Greenstein, Shane M.. "Invisible Hands and Visible Advisors: An Economic Interpretation of Standardization." Journal of the American Society for Information Science 43 (8), 1992.

Lehr, William. "Standardization: Understanding the Process." Journal of the American Society for Information Science 43 (8), 1992.

Libicki, M. "Information Technology Standards: Quest for the Common Byte." Butterworth-Heineman, Boston, MA, 1995.

Morse, Emile, Swantje Willms, William Pollak, Craig Meyers, and Michael B. Spring. "Improving the Standardization Process Using CASCADE". Technical Report. University of Pittsburgh, Department of Information Science and Telecommunications, 1996.

Oksala, Steven, Anthony Rutkowski, Michael Spring, and Jon O'Donnell. "The Structure of IT Standardization." StandardView 4 (1): 1-14, 1996.

Spring, Michael B., Christal Grisham, Jon O'Donnell, Ingjerd Skogseid, Andrew Snow, George Tarr and Piehan Wang. "Improving the Standards Process." Standards Development and Information Infrastructure, National Institute of Standards and Technology, Cambridge, MA, June 15, 1994

Spring, M.B. and M.B.H. Weiss. "Financing the standards development process." In: Standards Development and Information Infrastructure, B. Kahin, Ed., Cambridge, MA, June 15. National Inst. of Standards and Technology, Center for Science and International Affairs, Science, Technology and Public Policy Program, John F. Kennedy School of Government, Harvard University, 1994.

Toyofuku, R.T., III. "Free ridership in the market for 10Base T products." Master's thesis, Department of Information Science, University of Pittsburgh, Pittsburgh, PA, 1993.

Weiss, M.B.H. and M. Sirbu. "Technological choice in voluntary standards committees: An empirical analysis." Econ. Innovation New Technol. 1, 111-133, 1990.

Engineering Research Center: A New Virtual Innovation Organization in China

Qingrui Xu, Jin Chen, Bin Guo
Tel: 0086-571-7990529
Fax: 0086-571-7951358

School of Management
Zhejiang University
Hangzhou 310027, P.R. China

In Western view, Vitural Enterprise (VE) is the attempt to be more competitive in timing. But in China, VE is mainly used to bridge the capability-gap between R&D, manufacturing and marketing. Because ERCs are a brand new style of scientific research organization as well as a new type Vitural "Enterprise" in China, there are inevitably some problems in establishment and operation of ERCs. This paper, based on the extensive case study on the ERCs in China, one in America and Cooperative Research Center in Australia, with the expection to explore the effective establishment and management mode of ERCs in China.

Background

It is well known that America became a world leader of industry innovation after World War II. But by the end of 70s, America has lost its leading position in most industry fields in facing the ever-increasing challenge from Japan and Europe, and suffered the economy recession and productivity decline of manufacturing industry. Some authorities pointed out that one of the causes of this situation is that the American industry sector, especially the manufacturing industry, had failed to transfer the industrial technology results and commercialize them in timely and effective way (Carlof Mazazita, 1991). Then, supported by government, in 1985, the US National Science Foundation launched a vigorous efforts to link universities, industry,and government in an attempt to close the international competitiveness gap. An outstanding one of them is Engineering Research Center (ERC) Program. It started in 1983, involved establishment of ERCs in university campus. It has played an active role in promoting the process for US to regain its leader position in industry innovation (Henry Etzkowitz, 1994). According to a investigation report issued by American Competitiveness Commission in September of 1994, US has regained and strengthened its international competitiveness in most key industrial technological fields.

Australia was started the Cooperative Research Center (CRC) Program in 1990, which is similar to ERC program. It aims to enhance the effectiveness of the overall Australian R&D effort by drawing together researchers from several institutions to create the concentrations of resources which are needed, in many research fields, to keep pace with the rapid scientific and technological progress which is occurring internationally (Shantha Liyanage & Hellen Mitchell, 1995). Fifty two Centers has been approved across six major R&D and industry sectors. When these centers are fully established, this program will represents some six per cent of the total Australian R&D efforts. So it will have profound influence upon the R&D system and industry innovation process.

China has made great efforts during the past 40 years to raise its scientific and technological level and develop the manpower for R&D, however, the scientific and technological structures and capabilities created, in many cases, failed to fulfill their potential. The failure is apparent from the weakness, or lack of an effective network to bridge research and development as well as manufacturing and marketing (Xu Qingrui & Chen Jin, 1996).

The end of the 1980s and nowadays in China witnessed the lauching national support to bridge the gap between above-mentioned "Two gaps", which laid heavy emphasis

0-7803-3552-X/96/$5.00 © 1996 IEEE

on the concept of "National Center" for science and technology development in China. Supported by the World Bank, the State Planning Commission established the Engineering Research Centers in some key universities and research institutes to develop the key engineering technology in industry,which is expected not only to enhance and improve the indigenous technological innovation capability of Chinese large enterprises, but also assist the engineering technology improvement for medium & small-sized enterprises.

Up to now, the ERCs in China have primarily been established with organizational structure improved, but because ERCs are a brand new style of scientific research organization in China, as a new Vitural "Enterprise" in China, there are inevitably some problems in establishment and operation of ERCs. Therefore some more efforts should be made to explore the suitable managerial mode and operation mechanism to improve engineering research efficiency.

The establishment and running of ERCs, are a pioneering project without precedent experiences for China. In period of national economic reformation in China, the problems existing in the establishment and running especially are as follows (Chen Jin et al, 1995):

- heavy burden of loan repayment As ERC is a credit project mainly supported by the loan of the State Planning Commission as well as the World Bank, it must pay the Whole or part of loans by the end of construction period. General speaking, the investment of the construction in ERC is very large, so the difficulty facing by ERC is the heavy burden of loan repayment.

- the serious issue of lack of the talents As the national level centre for science and technology development in China. It is essential to retain a group of high-level engineering research talents with appropriate knowledge structure to ensure the effectiveness and continuity of ERCs. In addition, the young talent in research institutes often go abroad or enter some joint ventures for the low salary and poor housing conditions. The brain drain will weaken the R&D capacity severely if we do not take some strategic measures to change this situation.

- no enough ways to get incomes to cover operating cost ERCs can get incomes by three means: technology transfer, selling their own products or offering technical service. But at present, most of ERCs mainly make money by selling its own technical products, and seldom consider the means of engineering technology and service to the whole industry.

- no clear-cut development strategy on technology If we do not make efforts to correct present tendency, the ERC may probably end in two results: a high profitable and high-tech enterprise or a research institute having no difference with the previous organization. Both results do not meet the aim and mission of ERC. And even more, few of ERCs are at lost on how to invest its capital into research projects because of the lack of clear-cut technology development strategy.

A Comparative View Of ERC/CRC Programs

It is necessary to definite tasks of ERCs after we have a clear awareness of its mission. From the long-range view, ERCs should be a social technology service unit which do lots help to industry to improve their international competitiveness. ERCs should cover the process from applied research to self-reliance development and secondary development in the chain of technological innovation chain (see Figure 1).

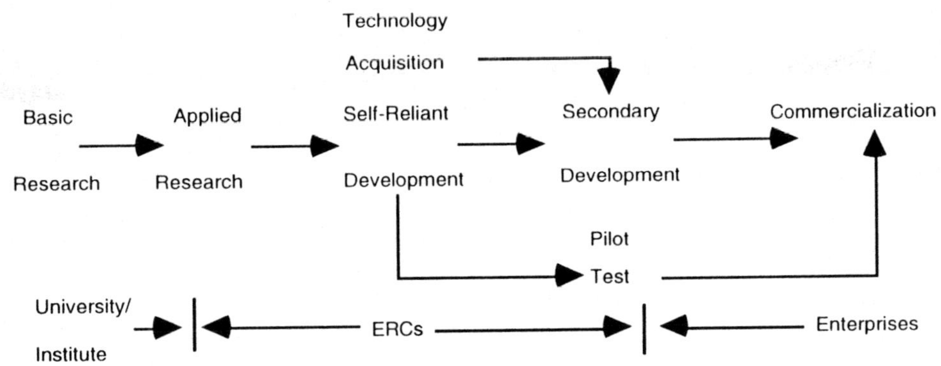

Fig. 1 The Role of ERC in Innovation Chain

And also, the basic task of ERC should be:

- Carry out research activities on crucial basic engineering technology and key comprehensive technology to the development of important industrial branches in national economy, and continuously gain results of innovation on engineering application through interdisciplinary and multidisciplinary research;
- The main source of getting money should be offering key engineering technology and service instead of selling its own products;
- According to the experience of ERC in America, ERC should help to foster the "ERC culture" and cultivate the talents with strong engineering capacity to fit the development demand of industry. At the same time, attach importance and make efforts to carry on re-education and retraining to the engineering personnel;
- ERC also should take the task to explore and monitor the leading edge of international engineering technology development.

The main tasks of ERC can be illustrated as Figure 2.

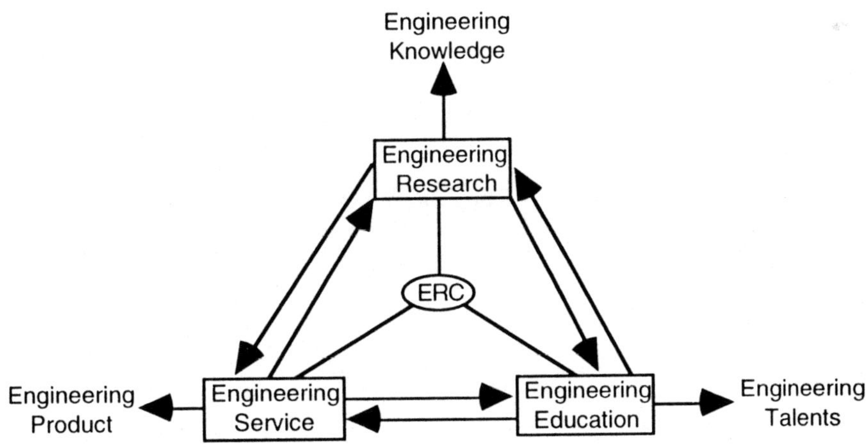

Fig. 2 Main Tasks of ERC

A Practice Perspective of ERC/CRC Program

As to China, one long existing problem in the industry innovation process is the serious interruptions between technology research and production, as well as research and marketing. The reasons are two fold, one is the lackness of enough engineering capacity for industry, the other is the lackness of effective collaboration institution.

Location
In US and Australia, the main operating unit of center is university. So US set up their centers in university, while Australia establish their CRCs in university or nearby. But in China,the ERCs are established in different type of units, including university, institute and enterprises. This makes it very difficult for China to set up effective management system and evaluation system.

Management Mode
S. Liyanage and H. Mitchell divide the management mode of CRC into three type by the investigation of Australian Cooperation Research Program, i.e.,corporation style, research style and integrated style (Shantha Liyanage & Hellen Mitchell, 1995).

In Australia, the most majority of CRCs are research style, secondly the integrated style. For US's ERCs, most of them are integrated style, and lay same emphases on the research function, engineering and commercialization function and education function. Generally speaking, US's ERCs adopt corporate management mode. As to China, the ERCs have a strong tendency to corporate style management, and often lay much larger emphasis upon engineering and commercialization function than education function.

Industry-Academy Relations
The cooperative research transition has a long term influence on the effectiveness of centers. It will influence the attitude and way for industry and academia to work together in the cooperation. For US's ERC program,because of its long running period and the fine cooperation research transition, it is relatively easy for them to set up the framework for collaboration among industry and academia. As regard to Australia's CRC program, although it is started almost at the same time as China and has no long experience yet, they can absorb the running experience from US's ERC for the similarity of its R&D management system and market mechanism. By contrast, the real situation for China' ERC is very different from the others for its special national conditions, the effective management and evaluation system are expected to be shaped.

Commitment and Investment
ERCs in China are also facing the troubling situation of capital lackness. Not only is the average funding amount much lower than US and Australia, but also a large portion of investment of China's ERC comes from the loan of the World Bank. It makes ERCs face a great pressure to pay off its loan debts as soon as possible, so the ERCs have to pay much attention to its for-profit activities than its long term development strategy. This will be harmful for ERC to fulfill its primary aims of promoting the technology innovation and diffusion for industry sectors.

Enthusiasm of Participator
Up to now, there still lack enough enthusiasm for industry to participate into the ERCs activities. Maybe three reasons can explain this situation. Firstly, the industry are not well aware of the importance of engineering technology for its market competitiveness, i.e., from the demand side, a relative low demand for high level engineering technology. So the number of ERC's industry partners in China is much lower than US, and almost the vast majority of industry partners are from large enterprises. The ratio of partners from large, medium and small sized enterprises in US is 32:28:40, while in China 76:21:3 respectively (Chen Jin et al, 1995). Secondly, there exists "culture difference" in collaboration for industry and academia in China. Thirdly,from cooperation research and transfer i9s also a essential factor, which to a great extend destroy the active enthusim for academia to take part in the cooperation research and technology diffusion.

Discussion

By our study, the effective management of ERC lies in the selection of its operating mode. The existing operating mode as followings:

- non-independent operating mode
  In this case, ERC is not a corporate organization, but the personnel and logistic service all depend on the parent unit. The ERC in Panda Electronics Group is such a case. Then the achievements of ERC is significantly related to the management effectiveness and operating level of parent unit.

- semi-independent operating mode
  In this case, ERC is also not a corporate organization, but is adopts secondary level accounting system in financial affairs. The features of this mode is that there is a clear-cut clarity of responsibility, authority and benefit between ERC and parent unit, and thus result in autonomy and flexibility to some extent.

- independent operating mode
  The ERC of this kind is a corporate organization which adopts independent accounting and system of personal responsibility for profit and loss. It has the autonomy to offer technology and technical service to industry in automonous and flexible way. The national Die & Mould CAD ERC is a typical one.

In general, most ERCs in China are market-oriented, and adopt management mode of technology corporation. This may be helpful for ERC to bridge the "Two Gaps" in a short term view. But from perspective of long term view, it will put negative impact on its role in enhancing the international competitiveness and engineering technology level of Chinese industry. Because once ERC becomes a pure profit-pursuit corporation, it is

inevitable for it to lay more attention on money-making than facilitating the engineering technology transfer to industry. In an extreme condition, this may lead to a technology monopoly position. More ever, this disadvantageous tendency has already appeared, to some extent, in some ERCs.

Therefore, ERC should not get involved too deeply in commercialization process, they should lay more attention to precompetitive technology, especially some key generic engineering technology that form the foundation of future competitiveness for industry.

Then, there are two suitable ways for ERC to get its income to ensure its long term development. One is getting industry subsidiary by technology transfer, the other is to spin off a technology company which only has technology relation with ERC, i.e., ERC signs a technology transfer contract with that company to get its income, and does not interpose management and operation of the spin-off company.

Therefore, we conclude that ERC can take the operation mode of profit-making center to ensure the repayment of loan debts during the construction period, under the condition that not give up the mission to serve and meet the technology demand of industry. But it should turn into independent or semi-independent accounting unit progressively from the construction period. At last ERC should become a social service organization to provide technology to the firms. The government department in charge of ERC should ensure such a transition of missions of ERC dynamically.

Conclusion

In general, Vitural Enterprise (VE) in Western Countries is the attempt to be more competitive in timing. But in China, Vitural Enterprise is mainly used to bridge the capability-gap between R&D, manufacturing and marketing. The Engineering Research Centers (ERCs) eatablished in several key universities, scientific research institutes and enterprise is the one of the Vitural Enterprise in China, its organizational structure improved, and operation mechanism are expected to be improved. Most of them have done a lot of work to improve engineering research efficiency.

The ERC program in China should be able to satisfy the need of industry, in some cases ERC have proved to be successful in the industry-academy collaboration research. However, the primary objective of China's ERCs is still far away the industry's need, so following measures should be taken:

- Set up a perfect legal framework for technological innovation and technology diffusion, thus create a proper outside context for industry-academy collaboration in ERCs in China;
- The existing profit-pursuing operating mode of ERC, to a certain extend, have constrained the effective collaboration between industry and academy. ERC itself should have a balance consideration on technology locus, i.e., should integrate appropriateness and advancement in the engineering technology research under a long term strategy for ERC technology development;
- Take some effective measures, especially some policy ones, to facilitate medium and small sized industrial enterprises to participate in the operation of ERC, thus ensure the validity for ERC to enhance the competitiveness of industry;Pay more attention to the education function of ERC, to cover the "cultural difference" among industry, university and institute, and also provide enough amount of research talents with engineering technology background for the industry in China.

References

Carlof Mazazita, " Technology Transfer in the United States: Industrial Research at Engineering Research Centers versus the Technological Needs of U.S. Industry", Technology Forecasting & Social Change, 39: 397-410, 1991.

Chen Jin, Xu Qingrui and Guo Bin, National Engineering Research Centers: Establishment and Operation Perspective", Scientific Research Management, 5, 1995.

Henry Etzkowitz, "Technology Centers and Industry Policy: The Emergence of the Interventionist State in the USA", Science and Public Policy, 2(21), 1994.

Ralph O. Slatyer, "Cooperative Research Centers: The Concept and its Implementation", Research Grants and Funding Symposium Proceedings, Wood F.Q. & Meek V.L. (eds), Anutech Publishers, 121-129, 1993.

World Bank. Staff Apprasal Report: China Technology Development Project. Report No 12814-CHA. Industry and Energy Operation Division, China and Mongolia DEpartment, East Asia & Pacific Regional Office, January, 18, 1995.

Xu Qingrui, Guo Bin and Chen Jin, The Management of Engineering Research Center in China, Proceeding of Multinational Symposium on Management of Technology and Innovation, 1995.

Shantha Liyanage & Hellen Mitchell, "Organizational Management in Australia Cooperative Research Centers", Technology Analysis & Strategic Management, 1(5): 3-14, 1995.

Xu Qingrui & Chen Jin et al, " Paths and Policies for the Improving of Indigenous Technological Innovation Capability in China", Technical Report to International Development Research Center of Canada, Zhejiang University, April, 1996.

Managing Innovation Portfolio: Experiences and Lessons in China

Qingrui XU, Bin GUO, Jin CHEN

Research Center of Management Science and Strategy Development
School of Management, Zhejiang University
Hangzhou 310027, P.R.China

Over the past decade, although people have attach much attention to the innovative activities in firms, the excutives often view the innovation process as a static, linear one, and often limit their strategic consideration on technology innovation in an isolated and seperated way. As a matter of fact, the innovation process is a dynamic, non-linear and systemetic process, which makes the traditional innovation management too simple and imappropriate in a long term view. By reviewing these works and investigation results of some leading firms in China, this paper advances a new paradigm for innovation management, i.e., portfolio innovation, to integrate technological, organizational and institutional factors from a strategic and systemetic view.

Introduction

Over the past decade, many studies and literatures have drawn upon the importance of technological innovation for firms' sustainable growth in the ever changing environment, and also an increasing number of industrial case studies has highlighted the importance of technological innovation for industrial competitiveness. Although people have attach much attention to the innovative activities in firms, the excutives often view the innovation process as a static,linear one, and often limit their strategic consideration on technology innovation in an isolated and seperated way.

As a matter of fact, the innovation process is a dynamic, non-linearand systemetic process (Dosi, 1982). In the development process of a firm, the technology innovations often happen in cluster, and among them exist interaction and integration relations, which makes the traditional innovation management too simple and imappropriate in a long term view.

In recent years, a new innovation management paradigm of how firms' innovative activities interact, which is essential to the field of strategic innovation management, has begun to emerge. Known as the "innovation portfolio", it regarded by some as a coordinative process of product innovation and process innovation in adaption to firm's environmental, technological and organizational change (Weiqiang Wang and Qingrui Xu, 1993; Qingrui Xu, 1995a).

By reviewing these works, this paper advances a new paradigm for innovation management, i.e., portfolio innovation, to integrate technological, organizational and institutional factors from a strategic and systemetic view angle. The essence of this pattern is that , all of these factors must take into consideration in the formulation of firm strategy. These include, portfolio of product innovation and processs innovation, portfolio of incremental innovation and radical innovation, portfolio of visible innovation benefit and hidden innovation benefit, as well as portfolio of technological innovation, organizational innovation and cultural innovation.

Then some case studies are analyzed to form the conceptual framework model of portfolio innovation in this article. As a result, experiences and lessons are concluded to gain some inside looks into the strategic management of technological innovation. Finally, the charateristics, as well as the strategic meaning of innovation portfolio, are discussed. It is pointed out that there should be a paradigm shift in the technological innovation management from conventional technology management strategy to portfolio strategy, i.e., dynamic, systemetic, strategic and portfolio view.

0-7803-3552-X/96/$5.00 © 1996 IEEE

Portfolio Innovation

People nowdays attach more and more importance to portfolio innovation due to the intensive competition in both local and global market. Weiqiang Wang and Qingrui Xu (1993) proposed out that, portfolio innovation is a coordinative process of product innovation and process innovation activities for firm's long term development. This may be correct, but regarding the sustainable competitive advantage for a firm, it is often not enough. In fact, portfolio innovation has a more broader connotation, no matter from therotical or pratical meanings.

From the strategic view of a firm, innovation portfolio innovation can be thought of a systemetic and coordinative process, which is a flow of series of innovative activities under the direction of firm's development strategy. It at least include four dimensions that underlie the nature and process of firm performance, all of which must take into consideration in the formulation of firm strategy. These include, portfolio of product innovation and processs innovation, portfolio of incremental innovation and radical innovation, portfolio of visible innovation benefit and hidden innovation benefit, as well as portfolio of technological innovation, organizational innovation and cultural innovation. The emergence of the new "portfolio innovation" paradigm, has greatly changed the tranditional thinking way of technological innovation management. It urgently asks people to pay more attention to the systemetic characteristics of technological innovation flow of a firm, from the view of portfolio and strategy (See Table 1).

Table 1. Paradigm Shift in Technological Innovation Management

Conventional Paradigm in Technology Innovation Management	Portfolio Paradigm in Technology Innovation Management
view innovationprocess from isolated and statistic way	view innovation process from systemetic and dynamic way
imbalancely emphasize the importance of product innovation	emphasize the coordination between product and process innovation
overemphasize radical innovation	high value the role of incremental innovation
ignore the hidden benefit of innovation activities	trade off the hidden and visible benefit of innovation activities

Portfolio of Product & Process Innovation

For a firm to ensure its long term returns, it is almost impossible to completely depend on product innovation, while without strategic consideration on process innovation. It is essential to take the firm's existing processes into consideration in carrying out product development (Utterback and Abernathy, 1976). So, there exists an interaction and interdependence between product innovation and process innovation. The portfolio of product and process innovation therefore forms the base for the portfolio innovation management. The porfolio of product and process innovation is the basis of realization of innovation returns (Qingrui Xu and Weiqiang Wang, 1993). This can be obviously seen from the the technological development history of Japanese industry, just as Mansfield had pointed out that one of important factors facilitating Japanese economic development is the emphasis on process innovation (Mansfield, 1990). But in China, there exists heavily imbalance between product innovation and process innovation, which became the bottleneck for Chinese firms to close their technological gaps contrasted with international advanced technology level (Xiaobo Wu, 1995).

By tracing back the technology development history in developed countries, we can find that almost all of them, to a varied extend, emphasize the coordinative development between product and process innovation. The different input strategies in product and process innovation have led to a dramatic result. According to the famous research report by MIT Commission on Industrial Productivity in 1990, the ignorance of process innovation is one of the main reason that caused US to lose its industrial leading position and competitiveness (MIT Commission, 1989). By contrast, it is said that the high growth rate of Japanese industry and strong competitiveness of Japanese products have close corelation with its strategic focus on the match between product innovation and process innovation.

Table 2. Comparison of Input in Product & Process innovation in U.S. Japan and Germany

Input in	Product Innovation	vs.	Process Innovation
US firms	2		1
Japan firms	1		2
German firms	1		4.1

Data Source: MIT Commission on Industrial Productivity, *Made in America: Regainning the productivity edge*, MIT Press, 1989.

In China, many firms have a traditional thinking in the management practice of technology, that is, weight product innovation more heavily, while look at process innovation as "a secondary demand" along with product innovation. This thought has led to a absolute dominated position of product innovation in firms. Because of the lackness of integrated consideration of product & process innovation, some serious problem emerged and caused great troubles to firms. First, the mismatch between product & process innovation has greatly hampered the improvement of productivity and high product quality. Second, the backward ststus of process innovation and the level of process base has become a bottleneck to the improvement of technological capability. According to our investigation to acqusition process of 473 foreign technologies in Hangzhou in 1990, it is found that only 0.8% of them had been successfully adapted and created new technology innovation. This to a large exteend came from the mismatch between technologies acquisited and the realatively backward indigenous process level for firms in China.

Portfolio of Radical & Incremental Innovation

Generally speaking, technological innovation can be divided into two group, radical innovation and incremental innovation. Radical innovation and incremental innovation describe two different kinds of technological changes. Radical innovation refer to a revelutionary change that contains high degree of new technical knowledge, whilee incremental innovation represents a renovation and adaption of current technology with low degree of new technical knowledge (Dewar and Dutton, 1986). In conventional technological management, more attention were attached to the economic meaning of radical innovation to a firm, but seldom did people gave the incremental innovation a proper position in determining the technology strategy. In recent years, more and more studies were carried out on the role of incremental innovation. By comparing the return from an initial-radical-process innovation and the return from the incremental changes in petroleum refining proceess, it is found that incremental improvements generated higher returns than from its innitial innovations (Yin,1994). Rosenberg also has noticed the accumulated economic meaning of incremental innovations (Rosenberg, 1982). These findings suggest that firms should integrate incremental R&D efforts into competitive strategies to generate desired benefit from technological progress in a timely fashion.

Many Chinese firms have put too little weight on incremental innovation visus radical innovation. This tendency has greatly influence the R & D resources allocation decision-making in technology project selection. The firm executives often take it for granted that radical innovation would fully ensure their splendid future in the maket. But things are not always this case. Some Chinese leading firms in manufacturing industry were observed a declining competitiveness in maket with both high level of R&D capacity and poor capability in translating technological advantages into maket advantage.

Regarding new product developments, they can be classified into three groups, i.e., radical product development, incremental product development and differential product development according to the technological innovative extend to basic products (Qingrui Xu and Bin Guo, 1995b). By applying a modified model based on J.Z.Yin's Preesent Value Index Model to returns caculating of new product development activities in a SBU of Hangzhou Telecommunication Equipment Works (HTEW), we found that the new products based on technological improvement have higher economic returns than the base products. For incremental product series, the returns are increasing by turn according to its technological development order. The differential product series have the similar tendency, while with much lower economic benefit than incremental ones.

This large economic benefit of incremental products can be attributed to following factors, such as technological factors, manufacturing factors and marketing factors. Referring to technological factors, as known, generally the incremental products are developed on the basis of existing products and relative mature technology, so the cost of trial-and-error is very below average. And also the experience, skill and capacity accumulated in previous product development activities can greatly reduce the resource input, especially capital input (Calantone, 1993). As to manufacuring factors, incremental products often have many common piece part, similar process and facility, thus can lower the manufacturing cost. At last, the marketing factors also play a big role in gaining returns. Incremental products have many similar functions, so are of great substitutability with existing products (Cooper et al, 1993; Bruce, 1994). Thus they can attract users of previous products by taking advatage of the marketing influence of existing products, such as promotion capability, distribution channel, as well as good relationship with suppliers.

Portfolio of Visible & Hidden Innovation Benefits

The benefits of technological innovation can be found in two ways. In one hand, through innovation project, some visible economic returns, which can be measured by some business indicators such as sales and profit and so on, will be brought out. On the other hand, there is also some hidden benefits along with innovation activities, which is very important for the long term development of a firm. Among them, most important is the accumulation of firm's technological capacity that forms the foundation for future competitivity.

Many firms often limit their view on the visible benefit of technological innovation, such as growth in product sales and net profit, while ignore the hidden benefit of innovations. Actually, the hidden benefit in the technology innovation, is very essential to the long term development of firms, especially the accumulation and improvement of oganizational technological capability. These neglected facets must be laied more attention in the technology management of firms.

The hidden benefits of innovation project are imbodied in two perspectives, i.e., capabiliity enhancing and technological paradigm formulation. First, the capacity development is a path-dependent process on the knowledge and skill accumulation in previous innovation projects. Second, by the formulation of technological paradigm, it will help a lot select innovation project with both economic and technical feasibilities in decision making stage, and also can highly raise the R&D efficiency by reducing resources wasting by the trial and error process. Just as Wilson and Hlavacek (1984) has observed that the innovation projects based on firm's storehouse technology have a far great success rate than new venture development projects. Therefore, the hidden benefits of innovation projects come from followings perspectives: enhance efficiency of technological resources usage of following innovation projects; raise success rate of following innovation projects; reduce time usage of technological resources usage of following innovation projects.

And it is just because of the existence of "hidden benefits effect" that make some failure innovation projects become very treasureful and useful for firm's long term development, just as the development history of Hitachi's numeric control machine technology has shown (Hayes and Pisano, 1994) . It is obvious that not all of failure projects will benefit firm.

As a matter of fact, there is some typical features for "useful failure projects". By our investigation into 40 failure innovation projects in Chinese manufacturing and electronics industry since 1993, we found that: firstly, the failure projects should have the same core tchnology with following projects, that is to say, its technology locus should locate in the firm's technology developing trajectory, which can only be promised by strongful technological strategy; secondly, these projects are often of great complexity involved great technological interrelation and participation of many fuctions besides technology department; thirdly, the termination of these projects often take place in the later stages of innovation process; and finally, learning by R&D is a typical learning mode in the development activities.

Portfolio of Technological, Organizational & Cultural Innovation

Numerous researchers have undertaken studies on what factors influence technological innovation. There are lots of factors affecting technological innovation. According to Twiss's categorization, these factors can be divided into five types labeled as corporate objectives, strategy, policies, and values; marketing; research and development; financial; and production. Amongst these factors the organizational culture and structure aspects play the first role and provide a framework within which this consideration can take place (Twiss, 1974).

Most of the empirical research has focused on the basic technological and marketing factors. However, having the similar technological capacity and facing the same marketing condition, the different firm would select the very different project to develop even in the same industry. There must be some other factors besides technological and marketing ones affecting technological innovation. These factors have the identical features that are very different, they are rooted in firm's developing history and operating style, and mainly refer to organizational culture and structure.

Because each project within a firm has at least some linkages both technologically and organizationally with other on-going projects and past projects within the firm, so there exist mutual interaction between organizational, cultural and technological factors. The introduction of new technology often induces the corresponding structure change in organization (Tyre, 1991). And also, organizational structure will greatly influence the performance of a firm, especially the efficiency of R&D function. Thus, organization and culture is the another coin of portfolio innovation. Some of Chinese Firms senced this new trends and got great gain by organizational and cultural innovation.

The typical case was did on one famous telecommunication manafacturer, Hangzhou telecommunication Equipment Works. Established in 1958, Hangzhou Telecommunication Equipment Works (HTEW) is a specialized manufacturer of China National Post & Telecommunications Industry Corporation.Its main products cover radio communication equipment, cellular mobile communication equipment, and SPC, digital communication and laser-typesetting equipment. With sales networks covering all parts of China, HTEW enjoys a remarkable rapid sales growth of its products. Since the 90's, HTEW has gone into the national 500 largest industrial enterprises, and the national 100 largest electronic enterprises. During the process of technological development, HTEW adjusted organizational structure at both firm's and divisinal level as well as R&D structure. Driven form the study of Saleh (1993), cultural at firm could be defined as the mixture of corporate strategy, coopeerative spirit and incentive environment. By using such a framework, we give the detailed matching pattern of technological innovation, organizational innovation and cultural innovation (See Table 3).

Table 3. Matching Pattern of Technological Innovation, Organizational and Cultural Innovation: Case of IITEW in China

Innovation Stage	Technology Acquisition	Assimilation & Adaption	Indigenous Innovation
Firm Structure	Joint Venture	Stockhold System	Stockhold System
Division Structure	Functional Process Oriented	SBU Product Oriented	SBU Product Oriented
R&D Organization	Semi-functional	Integrated: functional and divisional	Highly Intergrated
Corporate Strategy	Defensive	Risk-taking	Innovative
Internal Communication	Set up interfunctional communication	institutionalized communication	Interfunctional & interdivisional
Incentive Environment	Entrepreneurship	Entrepreneurship	Entrepreneurship Intrepreneurship
New Product Value Over Whode Sales(%)	5%	10%	30%
Labor Productivity (RMB yuan/per year)	7000-16000	16000-24000	24000-860000

Discussion and Conclusions

In recent years, people have attach much attention to the innovative activities in firms, with a view change from static linear way to a dynamic, non-linear and systemetic one. This has largely change the conventional thoughts and attitutes toward innovation strategy and management practice in firms.

In this paper, a new paradigm for innovation portfolio has been advanced to integrate technological, organizational and institutional factors from a strategic and systemetic view angle. In fact, portfolio innovation can be thought of a systemetic and coordinative process, which is a flow of series of innovative activities under the direction of firm's development strategy.

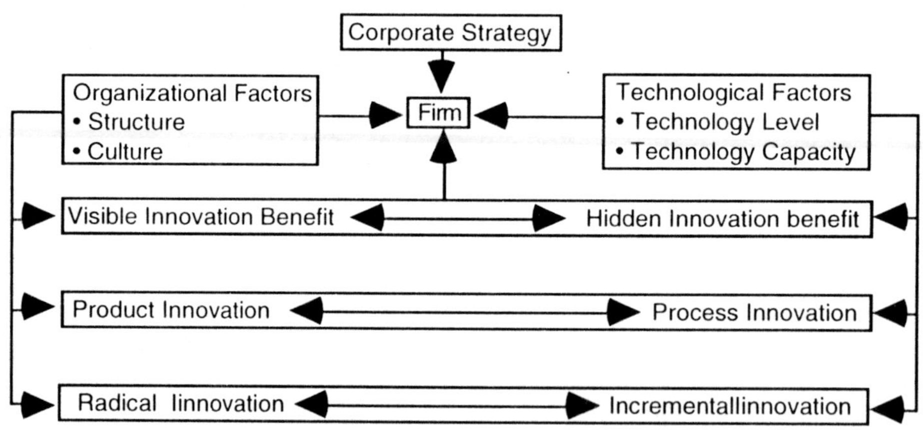

Fig 1. A Conceptual Framework Model of Protfolio Innovation

Because of interrelation and interaction among technological activities in firm, the portfolio innovation has some very different features by contrast with single-innovation-mode: (a) multi-facet characteristic. Portfolio innovation has four dimensions that underlie the nature and process of firm performance, including portfolio of product and processs innovation, portfolio of incremental and radical innovation, portfolio of visible and hidden innovation benefit, as well as portfolio of technological, organizational and cultural innovation. The portfolio of product and process innoation is the basic layer of portfolio innovation, which forms the foundation or platform for profit gainning. (b) systemetic characteristic. The aggregate returns of a appropriate innovation project portfolio is more than the total returns of each project, and also, show a dynamic tendency with the number of projects portfolioed increasing (Robecca Handerson, 1994). This is due to the R&D cost sharing, technical and market risk sharing. (c) dynamical characteristic. As the core heart of portfolio innovation, the technological capability of a firm itself has also a dynamic and acumulative developing process, which make the organizational, technical and financial infrastructure of a firm an indispensable support for benefits realization of portfolio innovation.

As for firms in China, there still exist dominant strategic thinking that view the innovation activities as a static and linear and isolated ones, thus make the technology strategy seperate from corporate strategy. Although many Chinese firms put much resources into R&D and innovation activities, they often found them facing the dilema of less returns virus heavy input, i.e., low efficiency in the translating technology into profit. In order to get out of such puzzle, it is urgent for Chinese firms to fully understand the nature and characteristics, as well as the strategic meaning of portfolio innovation. And also, there should be a paradigm shift in the technological innovation management practice from conventional technology management strategy to portfolio strategy, i.e., dynamic, systemetic,strategic and portfolio view.

References

Bruce G.S. Hardle, et al,°∞The logic of product-line extensions°±. Harvard Business Review., 72(6), 1994.

Cooper, R.G. and E.J. Kleinschmidt,°∞Uncovering the keys to new product success°±. IEEE Engineering Management Review, 21(4), 1993.

Dewar, R.L. and J.E.Dutton,°∞The adoption of Radical and Incremental Innovations: An Empirical Analysis°±. Management Science, 32: 1422-1433, 1986.

Dosi, G.°∞Technological paradigms and technological trajectories°±. Research Policy. 11(3): 147- 162. 1982.

Mansfield, E. °∞Industrial Innovation in Japan and US°±, Chinese Journal of Overseas S&T Policy Mnanagement, 1990.

MIT Commission on Industrial Productivity, °∞Made in America: Regainning the productivity edge°±, MIT Press, 1989.

Qingrui Xu and Weiqiang Wang, °∞Study on the Sources and Pattern of Firm's Process Innovation°±, Chinese Science & Research Management, No.4, 1993.

Qingrui Xu, °∞Theory and Approach of Portfolio Innovation°±, NSFC report, Research Center of Management Science and Strategy Development, Zhejiang University, 1995a.

Qingrui Xu and Bin Guo, °∞Benefit Evaluation of New Product Development°±, Chinese Science Management Research, No.4, 1995b.

Rebecca Henderson, °∞Managing Innovation in the Information Age°±, Harvard Business Review, 72(1): 100-105, 1994.

Hayes, H. Robert and Gary P. Pisano,°∞Beyond World-Class: The New Manufacturing Strategy°±. Harvard Business Review. 72(1): 77-86. 1994.

Calantone, J. Roger. °∞Organisational,technical and marketing antecedents for successful new product development°±. R & D management, 23(4), 1993.

Rosenberg, N. °∞Inside the Black Box°±, Cambridge: Oxford University Press, 1982.

Saleh, S.D. °∞The Management of Innovation: Strategy, Structure, and Organizational Climate°±, IEEE Transactions on Engineering Management, 40(1), February 1993.

Twiss, B.C.°∞Managing Technological Innovation°±, Longman, London, and New York, 1974.

Tyre, J. Merice. °∞Managing the Introduction of New Process Technology: International difference in a multi-plant network°±, Research Policy, 28: 57-76, 1991.

Utterback and Abernathy, °∞The dynamic model of product and process innovation°±, Omega, 3: 639-655, 1976.

Weiqiang Wang and Qingrui Xu, °∞Study on the nature of benefit in technological innovation°±, Chinese Science Management Research, No.6, 1993.

Wilson, T.L. and J.D. Hlavacek, °∞Don't Let Good Ideas Sit on The Shelf°±, Research Management, 27(3): 27- 34.1984.

Xiaobo Wu, °∞Evolution Process of Secondary Innovation°±, Science & Research Management, No.2, 27-35, 1995.

Yin. J. Z., °∞Managing process innovation through incremental improvements: empirical evidence in the petroleum refining industry°±, Technological Forecasting and Social Change, 47: 265, 1994.

New Wave of Managing Innovation

Hiroyuki Yamasaki , Senior Member, IEEE
phone +81 727 84 7328
fax +81 727 80 2685
email yamasaki@lsi.melco.co.jp

ULSI Laboratory, Mitsubishi Electric Corporation
4-1, Mizuhara Itami, Hyogo 664, Japan

Jun'ichi Baba
phone +81 3 5573 4431
fax +81 3 5573 4404
email yoshidas@gad.ark.melco.co.jp

Mitsubishi Electric Corporation
5-2-20, Akasaka Minato-ku, Tokyo 107, Japan

In the computer network era, it is imperative that the policy of "closed-management using principal management resources," be renewed to "open-management using outside resources." This renewal significantly alters corporate identity (autonomy, originality). Administrative behaviors to alter corporate identity are different from those used to maintain the present identity. We define the former behaviors as corporate governance. For innovation in particular, how to manage the forces changing the technologies, organization and corporate system should be considered from the governance viewpoint. Considering this background and the objective to adopt open management, the examples of Mitsubishi Electric are introduced in a new wave of innovation management.

Introduction

Due to advancements in computer, communication and transportation technologies, it is necessary to transform corporate management style from "closed management that exclusively uses management resources within the company," to "open management that effectively makes use of outside resources." This transformation is indispensable to ensure that corporate innovation [1] is conducted in a speedy, effective and efficient manner.

Transformation significantly alters corporate identity, and activities to maintain the present identity are different from those used to alter identity. We define the former as "management in the narrow sense" and the latter as "governance"[2] (Yamasaki and Baba, 1996).

Innovation is necessary to build an everlasting corporation by altering its identity, such as technologies, products, customers, organization and manufacturing and sales systems in response to external changes and forces. How to manage the forces causing change to such areas should therefore be considered from the governance viewpoint.

Innovation Model for Governance/Management

In order to transform the manufacture of power-generation equipment (hardware) into a power-systems business (hardware and software), Mitsubishi Electric invested in human resources and computers to develop system analysis technologies. The end result was the establishment of a factory that specialized in the systems business. Figure 1 shows a flow chart of the processes involved in this transformation. These combined processes represent the innovation model for governance/management.

Starting from the investment in system analysis technologies, it took 20 years to realize the construction and operation of the specialized factory. Based on conditions at the time, it was impossible to forecast what business would be like 20 years in the future. This can be understood by the fact that digital computer business had not yet surfaced in the market. It can be said that creating a new business structure is like

0-7803-3552-X/96/$5.00 © 1996 IEEE

establishing order from chaos. A project advances as entropy decreases, thus indicating that innovation becomes possible only through application of open-style management.

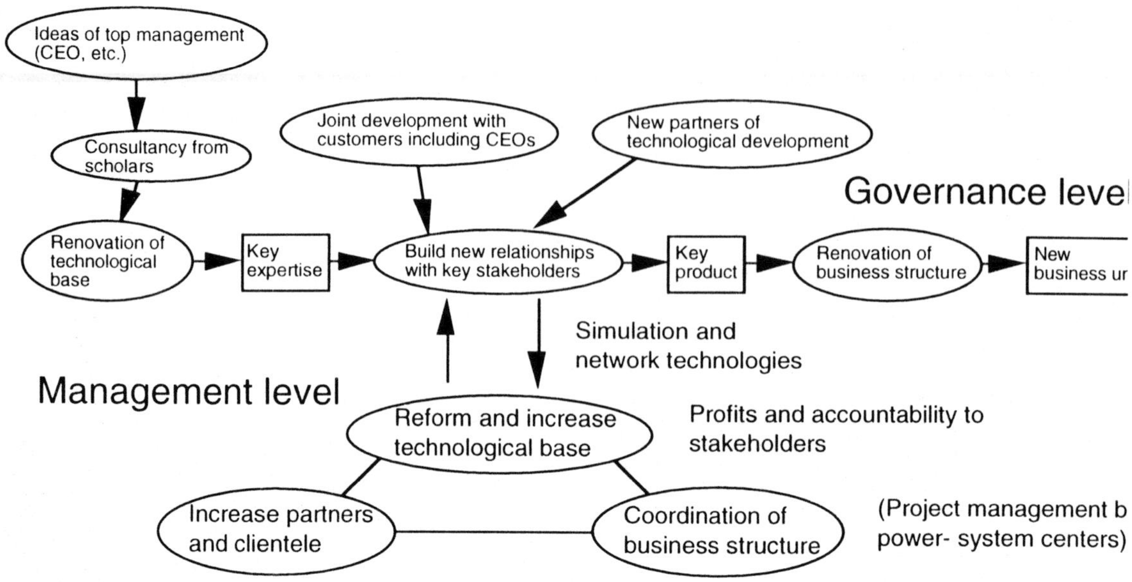

Figure 1: Governance/Management Model of Innovation

New Wave of Innovation Management

In the following we describe what was learned from the processes in the project, what we call "a new wave of innovation management."

1. The ideas and ambitions of top management result in the promotion of new business for which technological and marketing capabilities are yet unknown when development starts. Peoples' emotions yield ideas. The development of emotions begins from childhood, and is greatly influenced by the environment surrounding each person as he/she grows.

2. In order to establish a technological foundation (system analysis technologies), top management made a substantial investment in a simulator. This process is consistent in consideration of morphogenesis for positive feedback in second cybernetics (Maruyama, 1967). In management, deviation from a fixed paradigm is contained by negative feedback. One should be aware that the modes of feedback are completely different for governance and management.

3. Building new relationships with customers and selecting strategic partners for technical development should not be implemented according to prescribed methods (e.g., manuals). Such matters require the person in charge to have tacit knowledge (subjective knowledge, which is difficult to verbalize, and empirical knowledge, which is developed through practical conduct)[3]. In comparison to the management level, tacit knowledge plays a larger role at the governance level as paradigms are not established (Baba and Suzuki, 1993).

Examples of New Wave Management at MELCO

Role of the Strategic R&D Planning Office in Semiconductor Device Development Center

New business is implemented based upon excellent ideas introduced by top management (CEOs), but the planning staff that supports the CEOs is also important.

Mitsubishi Electric has laboratories at both the corporate level and business groups level. Four corporate laboratories are located in Japan, and three are located overseas. The administrative services for these laboratories are provided by the Planning & Administration Department of Corporate Research & Development.

Expenditures during the development stage are higher than those of the research stage due to the increased need of human resources and capital for product development. More capital is therefore invested in the R&D activities of the business groups. Each business group has a development laboratory where research and development aimed at the future business opportunities are promoted. The research budgets for each business group are raised from the profits made by the group.

The Semiconductor Group has three development laboratories, ULSI Laboratory, Optoelectronic & Microwave Devices Laboratory and System LSI Laboratory. It was thus necessary for this group to have a standing organization to correlate laboratory plans, and to see and coordinate research and development activities so that overlaps of wasteful activities and insufficient activities in a specific field could be avoided. For this purpose, the Research & Development Planning Department was established as the planning department of the Semiconductor Group.

In this modern era where information and environmental technologies are of great importance, it is not correct for corporations to act on their own, indifferent to social and economic activities. If a particular company's research and development were to be planned solely taking into account trends in science and technology, the resulting activities would be one-sided and incomplete. Proper research and development require full projection over all areas, which means that movements in the market and economy must also be taken into account. Corporate research and development must now be conducted based on a wider frame of references including living creatures, society, economics, culture and so on. In consideration of the above fact, the Research & Development Planning Department established the Strategic R&D Planning Group to collect and analyze vast types of information, and then draft high-quality strategic plans.

Establishment of a Private Engineering School

Those people who understand the technologies of the entire company are the ones capable of providing positive feedback during the business innovation process. To foster this concept among employees, Mitsubishi Electric has established an Engineering Forum (supported by affiliated engineering societies) and a Technology Training Camp.

Engineering Forum

In a product era where technologies are becoming increasingly complex and fusing together, personal networks are important. How much people "give and take" through these networks is the key as to whether or not success will be attained. Known by the more popular name of "Mitsubishi Electric Institute of Engineering," the Mitsubishi Electric Engineering Forum is contributing to the creation of a company-wide network for engineers.

The Engineering Forum was established in 1978 as a place for engineers in the corporate crossroads to go for self-enlightenment, self-training and exchange. There are presently 14 technical societies and each society has an average of five or six special sub-societies. Activities in each technical society include study meetings, presentations, lectures, the publication of a society bulletin and so on. Recently, the utilization of teleconferencing systems between Tokyo and Osaka is increasing. This allows engineers in each city to give their presentations simultaneously. Furthermore, lectures intended to cover various engineering fields are being delivered via communications satellites on predetermined dates.

In the future, greater effort will be made to exchange information via E-mail. By utilizing E-mail, development propositions can be discussed on the network. It is believed that by utilizing E-mail, more young engineers will be exposed to opportunities of technological exchange, thus more effectively promoting the transfer and development of technologies through generations.

Technology Training Camp

A system in which engineers are trained as managers is well established at Mitsubishi Electric. The company created the "Technology Training Camp," in which the general manager or members at the management level of the company's various works and labs lecture and teach the concepts and philosophies of technology management to the young engineers (around 30 years of age) who will be the future leaders of the works. This is the very reason that Mitsubishi Electric has consecutively turned out presidents whose backgrounds are in technical fields.

The Technology Training Camp is located at the Central Education Center in Kobe, Japan. It accepts only 42 young engineers each year, 21 in the spring and 21 in the fall. Every participant has been recommended by the general manager of works/labs. Upon entering the program, an engineer attends 12 lecture sessions, one each month for a year. Each session is six days long. The participants enter the training center on a Sunday evening and remain there until the evening of the following Saturday. The lectures cover all business fields, from energy conversion and electronic devices to materials.

The number of graduates from the Technology Training Camp now totals 500, and the members have formed an alumni association called, *"Hishi-juku-kai"*. Recently, they frequently utilize E-mail to exchange information.

Awarding of Bonuses to Supereminent Employees

The awarding of special bonuses to employees who achieve excellent results motivates other employees and substantially contributes to strengthening the role of positive feedback.

In 1993, Mitsubishi Electric introduced a "new employee evaluation system," also called the "added-points principle evaluation system." The former bonus system more or less attached importance to: 1) the principle of harmony, 2) wages according to personnel records and 3) the principle of relativity. Contrarily, the new system considers the following three points to be important: 1) the will to challenge (i.e., employees who challenge matters are evaluated positively even if some mistakes are made), 2) serious consideration to carrier development (i.e., fostering creative and innovative employees capable of coping with diversifying and advancing business operations, and 3) employee evaluations oriented more toward absolute evaluation based upon facts rather than the conventional evaluation based upon seniority system and relativity.

"Special points" are given to those employees who make conspicuous challenges and efforts and achieve results. Even if an employee had made mistakes previously, the added-points evaluation offers him/her the opportunity to makeup for the past through efforts and achievements afterward.

There are two types of special points: the "President's Award System," in which an award is presented to an employee (or group of employees) for the development of a superior new product that contributes to corporate business; and the "Special Bonus" system, in which one individual receives an award for his/her excellent performance. The President's Award was also given in the past. However, as of the anniversary date of the corporation's establishment, February 1, 1996, the award amount has been doubled.

Mitsubishi Electric formulated this incentive policy as a straight way of stimulating the employees' willingness to work.

Conclusion

Since the end of World War II, we have learned much about management concepts and methods from the United States, and for that we wish to express our sincere gratitude. Working from within Japanese-style management, it is our wish to contribute something of universal value to the world. In this respect, we believe that the ideas of top management (CEOs), the tacit knowledge of the middle people and the strength of supportive emotions are aspiring factors that will result in advances in business innovation.

Footnotes

[1] The usages of "effectiveness" and "efficiency" herein are similar to that of C.I. Barnard (Barnard, 1938). Effectiveness means achievements of the organizational purpose, and efficiency, satisfying personal motivation.

[2] There are several definitions for "governance." For example, R.I. Tricker (Tricker, 1984) differentiates management and governance, and indicates the former is "running a business" and the latter is "seeing that business is run properly" from a standpoint of the entire company. The authors believe that it becomes simple and comprehensive to define governance from the viewpoint of "innovation of identity" (Baba, Koike and Fukuda, 1988). Of course, this does not contradict Tricker's definition.

[3] Professor Nonaka (Hitotsubashi University) pointed out the importance of tacit knowledge in Japanese corporations at an earlier date (Nonaka, 1990).

References

Baba, J., Koike, M. and Fukuda, T. "The Relationship between Innovation Process and Governance.", *Journal of Science Policy and Research Management* 3(2): 162-171, 1988.

Baba, J. and Suzuki, H. "Development of Power-Generation Systems Business." *New Management for the 21st Century* -Part 5, *Corporate Management Strategies for the New Century, Collective Laws and Ordinances*, Vol. 17, 1993 (in Japanese).

Barnard, C.I. *The Functions of the Executive*. Harvard University Press, 1938.

Maruyama, M. "The Second Cybernetics: Deviation - Amplifying the Mutual Cause Process." *Sociology and Modern Systems Theory*. W. Buckley, 1967.

Nonaka, I. "A Theory of Organizational Knowledge Creation." Nikkei (Japan Economic) Newspaper Co., Ltd., 1990.

Tricker, R.I. *Corporate Governance*. Grower Publishing Company Limited: Vermont, USA., 1984.

Yamasaki, H. and Baba, J. "General Manager's Role and Responsibilities in a Research and Development Center." *Proceedings of the Fifth International Conference on Management of Technology*, 478-487, 1996.

The Virtual Enterprises in Japan

Kazuo Yanagishita
81-762-94-6710
E-mail: yanagist@manage.kanazawa-it.ac.jp

Kanazawa Institute of Technology
Dept. of Managerial Engineering
7-1 Ohgigaoka, Nonoichi, Ishikawa 921
Japan

Japan Electronic Industry Development Association (JEIDA) started 6 year project in 1981 to survey how to work at home (telework). The committee consisted of a dozen leading electronic apparatus manufactures. Teleworking (TW) is not popular so far. Here is an analysis of the reason why it was unsuccessful and as well as recent change. The author was a member of the committee and the chairman in the latter 3 years.

Japan Electronic Industry Development Association (JEIDA)

JEIDA consists of leading electronic industries in Japan. There are more than 20,000 industrial association in Japan, but JEIDA is one of the most active organization and have many committees to develop future businesses.

There are successful committees like factory automation (FA) and office automation (OA), but vegetable factory (VF) and teleworking (TW) were not popular so far.

The aim of the TW committee was to estimate the number of teleworking people and develop electronic apparatuses necessary for them. The committee consisted of a dozen leading electronic apparatus manufacturers of Canon, Fujitsu, Hitachi, Matsushita, Mitsubishi Electric, NEC, Ricoh, Sanyo, Sharp, Sony, Toshiba, and Yokogawa.

Not only monthly meetings held in JEIDA, but also we surveyed TW in the US in 1984 and in Europe in 1986. We published 5 reports and held two open conferences to promote teleworking and one demonstration at electronic product trade fair.

Social Change in Japan

Longevity of Japanese

Japanese lives very long and average age is around 80 years. Rate of aging, i.e. the years between percentage of people over 65 years old becomes 8% to 18% is only 40 years in Japan compared with 160 years in Europe and US, which means Japan is growing old at very fast speed. Pension crises is apparent for fewer young people who pay and many aged people who want to receive, which means higher age to receive and less money to receive. Therefore, even aged people must work, but crowded train is dangerous for aged people. They prefer teleworking.

Centralizaton to Tokyo

There was a strong tendency of concentrating business to Tokyo, for it is necessary to have close connections with government because there are lots of governmental regulations in Japanese business. And more than 10% of population in Japan living in Tokyo, which means poor living conditions and jammed transportation. Many businessmen leaves their family in local cities where living conditions are better and education circumstance is better for their children.

0-7803-3552-X/96/$5.00 © 1996 IEEE

High Real Estate in Japan

Tenant fee of office building in Tokyo is still very expensive compared with Hong Kong and Singapore and many foreign banks in Tokyo moved to them. But real estate price is very low if teleworking is done at 3 hours off Tokyo.

Deregulation

Public communication in Japan has long been monopolized by governmental enterprizes, i.e. postal service by the ministry of postal service and telephone and telegraph by Nippon (Japan) Telephone and Telegraph Public Corporation (NTT). But telephone lines are permitted to use not only by voice but also data communication by facsimile, telex, computer, etc. Then NTT is privatized and there appeared several competitors in long distance communication. Also the division of NTT, like AT&T is now under discussion in order to increase competition, for the monopoly is the biggest hinderance of low cost service. Japanese communication cost is very expensive compared with that of the US, which is also one cause of poor spread of teleworking.

Increasing Girl Students

More and more high school girl students go to 4 year universities. They want to work equally with boy graduates. The equal employment law of male and female guarantees same salary and promotion nominally, but actually quite unequivalent and if they are married, or babies are born they prefer teleworking. Choice of job, marriage or baby is a big decision for them. Japanese husbands are not so cooperative with their wives, and wives say "Ideal husband is healthy and not at home". House wives have less jobs compared with other countries for social conditions such as transportation for pregnant ladies, nursery school, baby sitters are not so convenient in Japan.

Table 1. Accelerating and Decelerating Factors of Telework

Accelerating Factors	Decelerating Factors
Aged society	Communication cost
Less children society	Groupism
Expensive office tenant fee	Large room system
Higher education of women	Office sick
Equal employment law	Small house
Cheap electronic instruments	On the job training
Change in management style	Lack of Social life
Soft economy	Worry of personnel manager
High commuting cost	
Less communicaton cost	
Eqrthquake	

Change in Business

Work Style in Japanese Firms

Most of the employees were not employed as specialists or professionals but general staffs. There is no detailed job description for them. They work in a large room. In some firms one room for one floor of a building. Managers, secretaries and staffs are in the same room. In some companies even executive directors are in one room.

Working in a large room has several merits, i.e. the business trend is very clear, better understanding of colleagues' jobs, on the job training is easy, mutual help among staffs is very easy, mutual monitoring among staffs prevent mistakes, sabotage and doing wrong.

There are demerits too, i.e. no privacy, noisy, unsuitable for thinking deeply, poor keeping secrecy of customers as well as personnel affairs, lack of document storage space.

Japanese Mentality of Work

Most of Japanese employees prefer "When you take shelter, make sure you go under a big tree", while most of American prefer independence when they get good ideas and make their own company and realise American dreams.

Life long employment was a feature of Japanese society. But young people moves from company to company very easily, and elderly people recommended early retirement in this recession. Freeters were very popular, who work at their will on contract basis. Most of them are computer software engineers, industrial designers, and the like whose achievement is quite different by their abilities. Capable persons could get more income in less time than those who work in offices. But after the destruction of economy, many firms reconstructed themselves and many part timers, temporary workers, freeters were fired, and even permanent employees were recommended to retire earlier with increased retiring allowance. The average jobless rate for all ages in Japan increased to 3.4% which is twice as much as before, and the jobless rate of the younsters is twice as much as the average rate, which is quite a new phenomenon in Japan.

Groupism

Very few Japanese executives have strong leadership with indivisuality, but they have coordination power and very good at consensus building. They usually prefer bottom up rather than top down. Therefore, most of their time is spent in meetings, not only in daytime but also at night in restaurants and bars. This is the biggest hinderance for teleworking for senior staffs, even if telephone conference, TV conference or comuter conference are introduced.

Change in Management Style

Flexible time working system is becoming very popular in Tokyo offices for the rush hours in the metropolis is terrible. The system allows employees to choose time of attending and returning freely if they work core time of 10 to 3 o'clock and 168 hours a month, for instance. It means some employees works while manager is not attending. This brought another merit for working at home, i.e. managers evaluate their staffs not by their working attitude but by achievement.

Change in White Collar Job

As the economy become softened many white collar work in market research, business planning, etc. Databases in Japan are expanding in both numbers and volume, although very poor compared with those of the US. There is less need to go to big libraries in Tokyo.

Technological Progress

Progress in Electronics

Electronic apparatuses such as facsimiles, word processors, personal computers are wide spread due to higher performance, easier operation, cheaper price.

Personal Computer Communication

Instead of letters and telephones, personal computer communication using internet become very popular worldwide. Trade off analysis is necessary of instrumentation cost of facsimile, word processor or computer plus comunication cost with tenant cost of office building plus transportation cost of staff plus opportunity loss of commuting time. Commuting time is 3 to 4 hours a day standing in crowded trains and buses in Tokyo area.

Other problems

Office Sick

There are self starters whose jobs are independent ones like lawyers and consultants, but ordinary staffs in ordinary firms feel office sick instead of home sick when they are ordered to telework. Young people needs social life after 5 o'clock.

Worry of Personnel Manager

It is quite difficult to distinguish who should come to office and who should work at home. Besides, if an employee is ordered to do telework and he built a home office in his backyard, but next month there is a reshuffling and it is difficult to tell to work in the office.

Space for Home Office

Japanese houses are rather small compared with other countries and often called "hare's hut", and most of staffs have no study room at home. How can they work at home? Maybe satellite office is one solution.

Future Trends

Satellite Office

Satellite office within 30 minutes from home may be the compromise of long commuting and small house. NEC and Sumitomo Trust and Bank established experimental satellite offices, both in the suburbs of Tokyo and well reputed among their employees, the latter is not only for their own but also TW research group of Fuji Xerox, NTT Data Communications, Uchida and some independent consultants.

Risk Management

After the big earthquake in Kobe on January 17, 1995, risk control is a big concern of Tokyo offices. They realized "Don't put all eggs in one basket". In 1994 a master student in my laboratory wrote "The trigger of start of teleworking will be earthquake" and it is coming true after the earthquake.

The Application of The Principal Component Method in Business Cycle Analysis and Prediction

Tan Yuming* *
Yan Jianmei*

Faculty of Business Administration
**Automation Department
South China University of Technology
Mao zongyuan**
Guangzhou,
PRC, 510641

In the paper, we explain the theory of the principal component method first. And then, we use it in the analysis of business cycle. The result the application of the method are also shown in the paper. We can see that this is a practical and less complicate method in the economic analysis, and it can result reasonable and practicable results.

Key words:
Principal component method, Business cycle, Business fluctuation

1. Foreword

The traditional analysis and prediction of the economic used to be based on the average economic growth rate. This method assumes that the economic is increasing exponentially strictly. But in fact, like other things in the world, the economic always grows fluctuate around its trends. So, it is important to study the fluctuation rules of the economic. Many method had been applied to take out the economic fluctuation, mainly include the economic statistics method and the expansion exponent method. But these methods require rather complicate calculation and so that can't be used in some case. The paper uses the principal component method to analyze the original data, and then uses it as the economy fluctuation analysis target data. We find that it is practical and much less complicate than the traditional method. The paper explains the business cycle analysis and prediction theory and the principal component method theory at first, and then use some county in Guangdong Province, China as example to introduce the application, at last we also give the result of fluctuation analysis based on this method.

2. The basic theory of the economy fluctuation theory

For every economy index, its monthly or seasonal series figures(DF) always contain four fluctuation component: long term trends(T), seasonal fluctuation(S), cyclic fluctuation(C) and random fluctuation(I). So, we must apply some model and method to get rid of the long term trends, seasonal fluctuation and random fluctuation. The formula of the addition model is: $DF=T+C+S+I$ 2. 1

All variations in the equation are absolute figures. This model is very simple and easy to be used, but because of using the absolute figures, different economic index had different measure, it is hard to compare the different index. The formula of the multiply model is: $DF= T\cdot C\cdot S\cdot I$ 2.2

Among them, DF and T are the absolute figures, C, S and I are relative figures. It uses relative figures to indicate the fluctuation, so it can avoid the different measure. In the study of the business cycle, we mainly adapt the multi-indice method. For example, NBER use correlation analysis method to divide the economy indices into three group: leading, coincident and lagging. Leading group can predict the economic trends, and the coincident group indicate the present economic environment, the lagging group be used to check the analysis result of the other two groups.

0-7803-3552-X/96/$5.00 © 1996 IEEE

3. The principal component analysis method

The principal component method is one of the multiple statistics methods. By the algebra point, the principal component is the special linear combination of varieties $x_1 x_2, \ldots, x_p$, by the geometry point, the linear combination is the new coordinate afterrotation of the old one, it makes the new one point to the maximum variation direction. At the same time, it provides the concise expression of the covariance matrix.

Provided the random vector $\mathbf{x}' = \{x_1, x_2, \ldots\ldots, x_p\}$, its covariance matrix is Σ, $\lambda_1 \geq \lambda_2 \geq \ldots\ldots \geq \lambda_p \geq 0$ is the eigenvalue of Σ, consider the linear combination

$$
\begin{aligned}
y_1 &= \mathbf{l}_1'\mathbf{X} = l_{11}x_1 + l_{21}x_2 + \ldots\ldots l_{p1}x_p \\
y_2 &= l_2'X = l_{12}x_1 + l_{22}x_2 + \ldots\ldots l_{p2}x_p \\
&\ldots\ldots \\
y_p &= l_p'X = l_{1p}x_1 + l_{2p}x_2 + \ldots\ldots l_{pp}x_p
\end{aligned} \qquad 3.1
$$

We have

$$
\begin{aligned}
Var(y_i) &= l_i'\Sigma l_i \qquad i=1,2,\ldots\ldots p \\
Cov(y_i, y_k) &= l_i'\Sigma l_k \qquad i,k=1,2,\ldots\ldots p
\end{aligned} \qquad 3.2
$$

Then the ith principal component is linear combination $l_i'X$ It should be confined by the following equations:

$$
\begin{aligned}
l_i' l_i &= 1 \\
Cov(l_i'X, l_k'X) &= 0 \qquad k<i
\end{aligned} \qquad 3.3
$$

At the same time, it makes the variance Var(lj ' X) maximum and Yl 7Y2 7 Yn linear independence.

Consider our application, we conclude the theory to several rules.

Rule 1.

Provided the covariance matrix of the random vector $x' = \{x_1, x_2, \ldots, x_p\}$ is Σ, $\lambda_1 \geq \lambda_2 \geq \ldots \geq \lambda_p$ is the eigenvalue of Σ, and the correspondent normal orthogonal eigenvector is $e_1, e_2, \ldots, e_p$, then the ith principal component is

$$y_i = e_i'X = e_{1i}x_1 + e_{2i}x_2 + \ldots\ldots + e_{pi}x_p \qquad i=1,2,\ldots\ldots,p \qquad 3.4$$

and

$$
\begin{aligned}
Var(y_i) &= e_i'\Sigma e_i = \lambda_i \qquad i=1,2,\ldots\ldots,p \\
Cov(y_i, y_k) &= e_i'\Sigma e_k = 0 \qquad i \neq k
\end{aligned} \qquad 3.5
$$

Rule 1 shows that the principal component of $x' = \{x_1, x_2, \ldots\ldots, x_p\}$, e.g. $Y_1, Y_2, \ldots Y_n$, is the linear combination with the coefficient comprised with the eigenvector of Σ.

Rule 2.

Provided the covariance matrix of the random vector $x' = \{x_1, x_2, \ldots, x_p\}$ is Σ, $\lambda_1 \geq \lambda_2 \geq \ldots \lambda_p$ is the eigenvalue of Σ, and the correspondent normal orthogonal eigenvector is $e_1, e_2, \ldots, e_p$, note $y_i = e_i'X$, $i=1,2,\ldots,p$ as the principal component, then we have:

$$\sigma_{11} + \sigma_{22} + \ldots\ldots + \sigma_{pp} = \sum_{i=1}^{p} Var(x_i) = \lambda_1 + \lambda_2 + \ldots\ldots + \lambda_p = \sum_{i=1}^{p} Var(y_i) \qquad 3.6$$

so, the variance of the kth principal component is the rate of total variance:

$$\frac{\lambda_k}{\lambda_1+\lambda_2+......+\lambda_p} \qquad k=1,2,......,p \qquad 3.7$$

By rule 2, We can know that if the forward principal component variance stand a high proportion in the total variance, they can be used to replace the original varieties, and that will not result in the serious information reduction.

Rule 3.

Provided the covariance matrix of the random vector $x' = \{x_1, x_2, .., x_p\}$ is Σ $\lambda_1 \geq \lambda_2 \geq .... \lambda_p$ is the eigenvalue of Σ, and the correspondent normal orthogonal eigenvector is $e_1, e_2, ..., e_p$, note $y_i = e_i'X$, i=1,2,..,p as the principal component, then:

$$x_k \rho_{y_i, x_k} = \frac{e_{ki}\sqrt{\lambda_i}}{\sqrt{\sigma_{kk}}} \qquad i=1,2,......,p \qquad 3.8$$

is the correlation coefficient of y_i and X_k.

Rule 4.

It is about the principal component of the normalized varieties.
Consider normalized variety:

$$z_i = \frac{x_i - u_i}{\sqrt{\sigma_{ii}}} \qquad i=1,2,......,p \qquad 3.9$$

It can be written with matrix form:

$$\boldsymbol{Z} = \boldsymbol{V}^{-1}(\boldsymbol{X} - \boldsymbol{U})$$

and the $\boldsymbol{V}$ is:

$$\boldsymbol{V} = \begin{Bmatrix} \sqrt{\sigma_{11}} & 0 & \cdots & 0 \\ 0 & \sqrt{\sigma_{22}} & \cdots & 0 \\ \cdots & \cdots & \cdots & \cdots \\ 0 & 0 & \cdots & \sqrt{\sigma_{pp}} \end{Bmatrix} \qquad 3.10$$

So, we know that $E(\boldsymbol{Z}) = 0$, and

$$Cov(\boldsymbol{Z}) = \boldsymbol{V}^{-1}\Sigma \boldsymbol{V}^{-1} = \rho = \begin{Bmatrix} 1 & \rho_{12} & \cdots & \rho_{1p} \\ \rho_{12} & 1 & \cdots & \rho_{2p} \\ \cdots & \cdots & \cdots & \cdots \\ \rho_{1p} & \rho_{2p} & \cdots & 1 \end{Bmatrix} \qquad 3.11$$

The principal component is:

$$y_i = e_i'\boldsymbol{Z} = e_i'\boldsymbol{V}^{-1}(\boldsymbol{X} - \boldsymbol{U}) \qquad i=1,2,......,p \qquad 3.12$$

and we can also know that:

$$\sum_{i=1}^{p} Var(y_i) = \sum_{i=1}^{p} Var(z_i)$$

$$\rho_{y_i, z_k} = e_{ki}\sqrt{\lambda_i} \qquad i,k=1,2,......,p \qquad 3.13$$

According to rule 4, we can use normalized data when the indices have different measure or their values range widely. The method will not affect the application of this case.

In the practice, we often use sample covariance matrix S to replace ~ so that we can apply the theory to the actual problems.

4. The application of principal component method in the economy indices analysis

In the paper, we use a county in Guangdong province, China as example, here name it L county. L county lies in the southwest of Guangdong province -- the most rapid growth province in economic in China. Its population is about 900 thousands, and its GNP is about 40 billions yuan(RMB) in 1991. The GNP is approximate 10 times more than 1978's. We select 24 indices from the county's development data from 1983 to 1992. Each index has 40 seasonal series items. Because the original data have different measure and wide range ,the first thing should be done is to normalize the original data. We can use equation 3.9 to normalize them. Then from the result matrix X we can get the correlation coefficient matrix R. The second step is to get the eigenvalue and eigenvector, then get the contribution rate. Use data from L county, we list the eigenvalue from maximum to minimum, show in table 4.1

table 4.1 eigenvalue contribution rate

19.2004	0.80002	0.80002	0.0218	0.00091	0.99749
2.3794	0.09914	0.89916	0.0209	0.00087	0.99836
0.6547	0.02728	0.92643	0.0097	0.00041	0.99876
0.4911	0.02046	0.94690	0.0079	0.00033	0.99909
0.3990	0.01662	0.96352	0.0073	0.00030	0.99940
0.2559	0.01066	0.97418	0.0041	0.00017	0.99957
0.1705	0.00711	0.98129	0.0032	0.00013	0.99970
0.1269	0.00529	0.98658	0.0029	0.00012	0.99982
0.0986	0.00411	0.99069	0.0020	0.00008	0.99990
0.0647	0.02694	0.99338	0.0011	0.00005	0.99995
0.0394	0.00164	0.99502	0.0010	0.00004	0.99999
0.0373	0.00155	0.99658	0.0003	0.00001	1.00000

From the table, we can see that the two biggest eigenvalue has stand 90% contribution for the whole. We can use it or the first component(contributes 80% for the whole) to represent the 24 indices.

The fourth, we analyze the compose of the two principal component and get the table 4.2. It shows that how the 24 indices contribute to the two principal component respectively. table 4.2 The coefficient of eigenvector of the two principal component

table 4.2 The coefficient of eigenvector of the two principal component

	x1	x2	x3	x4	x5	x6	x7	x8	x9	x10	x11	x12
I1	0.223	0.201	0.224	0.202	0.222	0.22	0.222	0.217	0.217	0.225	0.226	0.164
I2	0.053	0.021	-0.02	-0.008	0.123	0.043	-0.041	0.071	-0.011	0.076	0.063	-0.149

	x1	x2	x3	x4	x5	x6	x7	x8	x9	x10	x11	x12
I1	0.223	0.201	0.224	0.202	0.222	0.22	0.222	0.217	0.217	0.225	0.226	0.164
I2	0.053	0.021	-0.02	-0.008	0.123	0.043	-0.041	0.071	-0.011	0.076	0.063	-0.149

Based on the analysis above and the original data, we can calculate the time series figures of the two principal component as table 4.3. That is the foundation of the economic fluctuation analysis. table 4.3 The time series data of the two principal component

table 4.3 The time series data of the two principal component

	I1	I2		I1	I2
1	15234	1853	21	48580	7255
2	14967	1949	22	51598	7848
3	15462	2125	23	46295	6201
4	16637	2335	24	52426	7821
5	17098	2021	25	56546	8725
6	17003	2343	26	57755	9563
7	18058	2589	27	58908	9886
8	20806	2871	28	58402	9495
9	22174	2581	29	63724	11073
10	22739	2986	30	65096	12014
11	24089	3382	31	70454	13556
12	25552	3598	32	72156	13446
13	28144	3625	33	80694	14683
14	27631	3946	34	81480	15500
15	29532	4421	35	86777	16592
16	31605	4743	36	93811	17508
17	35471	5070	37	109349	20487
18	37084	5756	38	114845	21594
19	41968	6770	39	123264	23516
20	42566	6770	40	130690	24044

5. The business cycle analysis and prediction

In this part, we use the calculation result to analysis the business cycle. For L county, we use the total output as the basic cycle of the economic fluctuation. Base on the calculation of the above, the peaks and valleys are shown in table 5.1 and 5.2 as following. We also graph the economic fluctuation as fig. 5.1. From the table and figure, we can see that the first principle component forward the basic cycle by 1.0 season, the second principle component forward the basic cycle by 0.43 season, and they all have the similar trend with the basic cycle. So, the method explain in the paper is practicable and require less complicate calculation.

table 5.1 turning point comparison (valleys)

Total Output of Industry		The First Principal Component		The Second Principle Component	
year	season	year	season	year	season
1982	4				
1985	2				
1987	1	1986	3	1986	2
1990	2	1990	1	1990	2

table 5.2 turning point comparison (peaks)

Total Output of Industry		The First Principal Component		The Second Principle Component	
year	season	year	season	year	season
1984	3				
1986	1	1985	2		
1988	2	1987	3	1987	3
1990	4	1991	1	1990	2
				1992	2

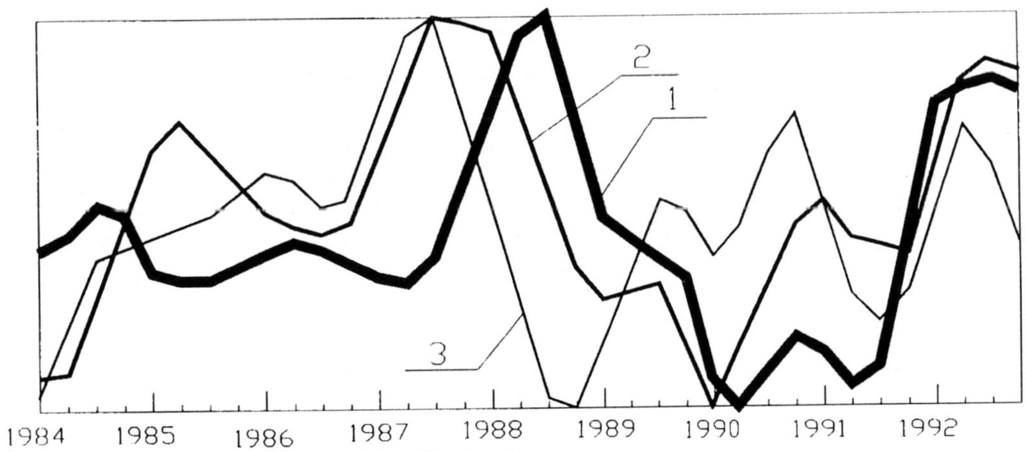

1: the total output of industry
2: the first principal component
3: the second principal component

fig 6.1 the comparison of the industrial total output and the two principal component

6. End

So, from the paper, we can see the method is simple and practical. It can be used when there are too many economic indices to select the group to study the fluctuation of economic. More work should be done to analyze the correlation of the indices, it will reveal more information about the running of economic.

References

1. Wan xueren, Wan Songgui. "Practical Multi-statistics Analysis". Shanghai Technology Press. September, 1990

2. MicQueen, J.B. "Some Methods for Classification and Analysis of Multivariate Observations". University of California Press. 1967

3. Chen Lei, Wu Guizheng, Gao Tiemei. "The Principal Component Analysis and Economic Fluctuation". Research of Statistics Economic and Technology Economic. July, 1993 4. Yu Jing. "Macro Economic Monitor and Prediction System". Statistics and Economic Mathematics. May, 1991

A Model for Analysis Dynamic Characteristics of Strategic Alliance

Sida Zhou

Engineering Management Program, Portland State University
P.O. Box 751, Portland, OR, 97207-0751

Based on the standard theory of the firm, a dynamic model for the study of R&D manage ment in a strategic alliance is presented. The model provides insight into the firm's behavior in a strategic alliance u)ith R&D activity. By using this dynamic model, the optimal level of technology for a firm can be determined from the profit maximization viewpoint. The model indicates that firms u)ith complementary skills and resources in a strategic alliance will be better off in terms of R&D cost, profit and competitiveness.

Introduction

Globalization makes inter-firm cooperative arrangements an essential tool for serving the cus tomers. The organizational forms of these arrangements range from the spot market relationship to complete mergers, due to inter-organizational dependence and strategic impact [1, 2]. High technology organizations, as technology continues to expand and to disperse, team up with their suppliers, distributors, OEM customers and others to form strategics alliance. In recent years, literature on strategic alliances has been substantially enriched in both theoretical and empirical studies of the causes of inter-firm cooperative relationships. Empirical study results indicate that a firm's decision to form a strategic alliance is a function of its characteristics, its competitive position and its strategic choice [3]. Most theories on strategic alliances fall into the following three classes.

- Firms may be motivated to pursue a strategic alliance when the benefits of pooling the resources of cooperating partners outweigh the externalities of these relationships, such as technology dissemination and reputation erosion. The benefits include risk reduction, economies of scale and scope, production rationalization, and convergence of technology
- [3, 1].

- Firms focus on the organizational and contracting efficiency of cooperative arrangements. Factors such as uncertainty, small-number bargaining conditions, and indivisibility of assets make the cooperative arrangement a low cost alternative to appropriate inter-firm synergies and to attenuate the externalities of these relationships [3].

- Firms seek cooperative arrangements to increase market power, to deter entry, and to block competition through strategic corporation. In industries with rapidly changing technologies, strategic cooperation expedites the commercialization process. The cost of sharing the returns of proprietary technology is more than offset by the long-term benefit of first mover advantage and successful pre-emption of competitors [3, 1, 2].

We can clearly see that these theories complement, rather than compete, with each other. Taken together, they provide a complete treatment of short and long-term economic factors that compel firms to form strategic alliances. The next step is to understand the underlying factors for strategic alliances among firms with R&D activity. For this purpose, Zhou [4] has developed a basic(static) model to analyze characteristics of firms with R&D activities within a strategic alliance. This paper will present a dynamic model for studying the strategic alliance for product development.

0-7803-3552-X/96/$5.00 © 1996 IEEE

2 Dynamic Model of the Strategic Alliance

We consider a homogeneous goods industry with demand Q(t) = Q(p(t)), where Q(t) is the in dustry output at time t and p(t) is the price at time t. We assume that firms within the industry may form strategic alliances. For example, firm M in the target industry form a strategic alliance with n firms (i = 1, 2, ..., n), which may or may not be in this industry. These n firms may be suppliers or OEMs of firm M. We also assume that the final product or service is presented to market by firm M. This is the kind of strategic alliance that we will analyze in this paper.

By using the standard theory of the firm, a dynamic model is presented here to study R&D process within a strategic alliance. We assume that firms maximize profit and produce a single homogeneous good under conditions of certainty. Firm M produces a negligible part of the indus try's output, and each of the n firms makes technological contributions to firm M's output (for example, the OEM case). The new twist is that the firms invest a certain amount of inventive inputs—such items as research, development, or engineering—which is expected to increase the firms' technical knowledge and improve their productivity. We also assume that the firms are able to appropriate all the knowledge they produce respectively or obtain it from the other firms.

Each firm in the strategic alliance is divided into two departments: production and research. The production department uses conventional inputs (capital and labor) to produce output ac cording to the technical knowledge of the firm. The research department of the firm engages inventors, engineers, and technical personnel (as well as the relevant capital goods) in knowledge increasing activity, and obtains and/or creates a flow of new knowledge to the firm. This activity need not be research and development; it can be any kind of investment that improves the firm's conventional production function.

For any period of time in a strategic alliance, let LM(t) and Li(t) be labor-hours in production for firm M and firm i respectively, KM(t) and Ki(t) be the number of machine-hours in production for firm M and firm i, RM(t) and Ri(t) be the labor-hours of inventors and engineers invested at time t in knowledge-increasing activities of the firm M and i, where i = 1, 2, ..., n. aM(t) and ai(t) are the level of productivity and have the natural interpretation of the "level of technology" for firm M and i (i = 1,2,...,n), respectively. We should call aM(t) and ai(t) the R&D possibility function for firm M and firm i. In order to develop a simple form of production function, it is assumed that capital and labor are used in fixed proportions, and that there are diminishing returns to scale in K and L. The diminishing returns take the special form that marginal product of conventional factors are constant up to a point and zero thereafter. For firm M without a strategic alliance, we will use the following relationships developed by Nordhaus to represent the present value of firm M. For a detailed derivation and the associated assumptions of this relation please see [5].

$$V_M(T) = \int_T^{\infty} \pi_M(t)e^{-rt}dt$$

Subject to

$$\begin{aligned} \pi_M(t) &= p(t)a_M(t)L_M(t) - W_M(t)L_M(t) - S_M(t)R_M(t) \\ a'_M(t) &= a_M(t)G(R_M(t)) \end{aligned} \tag{1}$$

where p(t) is the price of the output, WM(t) is the price of a bundle of conventional inputs, SM(t) iS the cost of research inputs at time t, r is the market interest rate, GM() indicates the contribution of RM(t) to the level of productivity aM(t). The firm is assumed, at every point of time, to maximize its discounted profits VM(t).

Now let us extend (1) to the environment of a strategic alliance. We assume that R(t) is the total labor-hours of inventors and engineers invested at time t in knowledge-increasing activities of this strategic alliance and that the following relationship exits:

$$R_M(t) = \varphi_M(t)R(t) \quad and \quad R_i(t) = \varphi_i(t)R(t) \tag{2}$$

where $\varphi_M(t)$ and $\varphi_i(t)$ $(i = 1, 2, \cdots, n)$ indicate the fraction that $R_M(t)$ and $R_i(t)$ contribute to $R(t)$ respectively. In summary, we have

$$R(t) = R_M(t) + \sum_{i=1}^{n} R_i(t) \tag{3}$$

Let a(t) indicate the level of technology of the strategic alliance. Due to technology transfer among firms in the strategic alliance, the level of technology in each firm should be enhanced by the strategic inter-firm arrangement. In our model, two kinds of technology transfer are considered. One type of transfers considers the influence of firm i's technology level (i = 1, 2, ..., n) to firm M's technology level; $_a M^{(t)}$ should be enhanced to a(t). The other effect is firm M's technology level's influence on each firm's technology level. This is the influence of aM(t) to each of ai(t) (i = 1, 2, ..., n). If we let $a_{ei}(t)$ stand for enhanced a(t), then we have

$$a(t) = a_M(t) + \sum_{i=1}^{n} \phi_i(t)a_i(t) \quad and \quad a_{ei}(t) = a_i(t) + \psi_i(t)a_M(t) \tag{4}$$

where $\o_i(t)$ (i = 1,2,...,n) indicates the degree of influence that $a_i(t)$ can contribute to aM(t). $\mu_i(t)$ (i = 1, 2, , n) indicates the degree of influence that aM(t) can give to each $a_i(t)$. Therefore, the profit for firm M is as follows:

$$
\begin{aligned}
V_M(T) &= \int_T^{\infty} \pi_M(t)e^{-rt}dt \\
\textit{Subject to} & \\
\pi_M(t) &= p(t)a(t)L_M(t) - \sum_{i=1}^{n} p_i(R_i(t), p(t))a_{ei}(t)L_i(t) - \\
& \quad W_M(t)L_M(t) - S_M(t)R_M(t) \\
a'_M(t) &= a_M(t)G_M(R_M(t)) \\
a'_i(t) &= a_i(t)G_i(R_i(t))
\end{aligned} \tag{5}
$$

where $P_i(R_i(t), p(t)$ is the price that firm M pays firm i for that firm's contribution to the out put. Equation (5) is a direct extension of equation (1), except that a(t) indicates the combined level of technology and the new item $\Sigma_{i=1}^{n} P_i(R_i(t), p(t))a_{ei}(t)L_i(t)$ indicates the total cost for the contribution of the other n firms to the final output.

In the same manner, the profit for firm i (i = 1, 2, ..., n) is as follows:

$$\varepsilon_{p_i}(t) = \frac{\frac{d}{dR_i}(a_{ei}(t))}{a_{ei}(t)} = \frac{p_i(R_i, p(t))}{\frac{d}{dR_i}(p_i(R_i, p(t)))}, \qquad i = 1, 2, \cdots, n. \tag{7}$$

where $P_i(R_i(t), p(t))$ is the price of firm i's output, which is usually decided by negotiation between firm M and firm i. Equation (5) and (6) are our theoretical model of a firm's behavior in the strategic alliance with R&D activity.

3 Analysis of Strategic Alliance

We shall now examine the important aspects of this dynamic model. First, the model indicates a cost reduction for all firms in a strategic alliance. Because of the assumptions concerning tech nology transfer among the firms in a strategic alliance, the level of technology of each firm is enhanced with no cost or lower cost. This means SM(t) and Si(t) can be lower for firms in an alliance than for firms not in one. Technology transfer presented here certainly makes the pro duction more rational since we assume that firms with complementary technology form strategic alliances. Second, the model represents the improved profit and competitiveness, since aM(t) and ai(i) are enhanced through the strategic alliance, and the capacity aM(t)LM(t) and ai(t)Li(t) of each firm is also increased. The economies of scale are therefore improved. All these characteris tics are consistent with the strategic alliance cause model which we discussed above.

Based on the above model, what we are really interested in is how firms in a strategic alliance will behave when R&D is part of their business. One of the most important things is to determine the level of R&D of level of technology. Let us assume the decision for firms in a strategic alliance simply involves maximizing profit with respect to R, RM and Ri, and assume that LM~ Li, SM, Si, WM, Wi are constant. We define the price elasticity of technology level as follows:

$$\varepsilon_{p_i}(t) = \frac{\frac{d}{dR_i}(a_{ei}(t))}{a_{ei}(t)} = \frac{p_i(R_i, p(t))}{\frac{d}{dR_i}(p_i(R_i, p(t)))}, \qquad i = 1, 2, \cdots, n. \tag{7}$$

which measures the responsiveness of the technology level to changes in the price of the output for firm i. Now starting with equation (5), (6) and (7), the first-order conditions for maximizing VM and Vi are:

$$\begin{aligned} \frac{dV_M}{dR} &= \int_T^\infty p(t)\frac{d}{dR}(a(t))L_M - \sum_{i=1}^{n}[(1+\frac{1}{\varepsilon_{p_i}(t)})\frac{d}{dR_i}(a_{ei}(t)) \\ & p_i(R_i, p(t))\varphi_i]L_i + S_M\varphi_i]e^{-rt}dt = 0 \end{aligned} \tag{8}$$

$$\frac{dV_i}{dR_i} = \int_T^\infty [(1+\frac{1}{\varepsilon_{p_i}(t)})p_i(R_i, p(t))\frac{d}{dR_i}(a_{ei}(t))L_i - S_i]e^{-rt}dt = 0 \tag{9}$$

By combining equations (3),(4), (8) and (9), the optimal level of R&D ($a(t)$) for a strategic alliance (as a whole, for firm M and firm i (i = 1, 2, ..., n) can be determined by following equation (10).'

$$\int_T^\infty p(t)\frac{d}{dR}a(t)L_M = \frac{\sum_{i=1}^{n}\varphi_i S_i + S_M\varphi_M}{r} \tag{10}$$

If we know the relationships among pi(t), Ri(t), RM(t) and p(t), we can solve these equations by using numerical analysis methods, or even get an explicit solutions to them. Let us consider a very simple case, at a given time T. The firm chooses a constant research program. As RM and Ri are constant, we have

$$a_M(t) = a_M(T)e^{G_M(R_M)(t-T)} \quad and \quad a_i(t) = a_i(T)e^{G_i(R_i)(t-T)} \tag{11}$$

From equations (3) and (4), we can get:

$$a(t) = a_M(T)e^{G_M(\varphi_M R)(t-T)} + \sum_{i=1}^{n} a_i(T)e^{G_i(\varphi_i R)(t-T)} \tag{12}$$

$$a_{ei}(t) = a_i(T)e^{G_M(\varphi_i R)(t-T)} + \psi_i a_M(T)e^{G_i(\varphi_M R)(t-T)} \tag{13}$$

If we further assume that p'$(t)/p(t) = -h$, and substitute equation (12) and (13) back to equation (8) and (9). After some calculation, we can get a solution for the optimal technology level for firm M and firm i. But for most cases, it is not easy to solve those equations; therefore, a simulation study to understand the characteristics and behavior of the theoretical strategic alliance is necessary. Equation (8) is the usual condition that marginal product equals marginal cost. The left side of equation (8) has two factors that show the determination of the level of R&D. The first factor a'(t)LM shows that the benefits of additional knowledge are proportional to the size of the firm. The second important determinant of the level of R&D is the present value factor. The right side of equation (8) has six factors $(E_{pi}(t), p_i(R_i(t), p(t))a'_{ei}(t)L_i, \wp_i, S_{M}, \wp_{M}, L_M)$ which show the determination of the level of R&D. A detailed analysis of the influence of these factor can be carried out by simulation study.

4 Conclusions

A theoretical model for analyzing the R&D activities in strategic alliance has been presented. Sample cases of solutions to determine the optimal level of technology are discussed. In general, the detailed relationships among various parameters of the model can be carried out by computer simulation.

References

[1] F. J. Contractor and P. Lorange, Coorperative Strategies in International Business. Lexington Books, 1988.

[2] K. Ohmae, "The global logic of strategic alliances," Harvard Business Review, pp. 143-154, March-April 1989.

[3] W. Shan and K. Visudtibhan, "Cooperative strategy in commercializing an emerging technology," European Journal of Operational Research, vol. 40, pp. 172-181, 1990.

[4] S. Zhou, "A model for analysis R&D in strategic alliances," in Proceeding 1992 on IEEE International Conference on Engineering Management, IEEE, 1992.

[5] W. D. Nordhaus, Invention, Growth, and Welfare: A Theoretical Treatment of Technological change. The M.I.T. Press, 1968.

Minimum Distance Method (MDM) For Group Judgment Aggregations

Sida Zhou and Dundar F. Kocaoglu

Engineering Management Program, Portland State University
P.O. Box 751, Portland, OR, 97207-0751

A new aggregation method —Minimum Distance Method (MDMJ) — was developed to support the group decision process and to help decision makers achieve consensus under the framework of AHP. MDM employs the general distance concept, and is proven to be very appealing to the compromise nature of a group decision making. It preserves all characteristics of the functional equations approach proposed by Aczel and Saaty. MDM is based on a goal programming model, which is easy to solve by using optimization soft7uare such as LINDO. Using the goal programming model, MDM provides the weighted membership capability. Sensitivity analysis can be performed to investigate the effect of varying the decision makers ' relative importance in terms of weights in the goal programming model. Sensitivity analysis also allows us to make robust decisions.

1 Introduction

The basic problem of decision making is to choose a best one in a set of competing alternatives that are evaluated under conflicted criteria. In general, the evaluation process is to evaluate de cision elements by quantifying the subjective judgments. The Analytic Hierarchy Process (AHP) provides us with a comprehensive framework for solving such problems. As pointed out by Saaty, AHP "enables us to cope with the intuitive, the rational, and the irrational, all at the same time, when we make multicriteria and multiactor decisions." Furthermore, in most organizations decisions are made collectively, regardless whether the organization is public or private. It is sometimes difficult to achieve consensus among group members, or for all members of a group to meet. Therefore, External rules or methods are needed to derive the "group set of data or preference" from given "individual sets of data or preference". There are two existing methods to aggregate pairwise comparison judgments. One is the function equation approach via geometric mean, which was developed by Aczel and Saaty [1, 2, 3]. The other one is the arithmetic mean, which is the commonly used method. Both of these methods are simple and easy to use. In this paper, a new method will be presented.

2 Problem Definition

The judgment quantification process involves comparisons among decision alternatives according to a given criterion. The individual judgments are made by comparing an object, say C, with another, say B, according to the given criterion. In a group decision process, these individual judgments need to be aggregated into a single judgment. Suppose the pairwise comparison matri ces (Ai) are made separately by each of m individuals in the decision group, we should obtain m n X n pairwise comparison matrices, where n is the number of decision alternatives; The objective is how to aggregate those m n X n pairwise comparison matrices into a single group pairwise comparison matrix. The group priority vector then is derived from this aggregated pairwise com parison matrix.

0-7803-3552-X/96/$5.00 © 1996 IEEE

Suppose $A_G(\ )$ stands for the aggregation method. This approach can be expressed as $\bar{A} = A_G(\{A_i\})$ and $\bar{V} = f(\bar{A})$. More specifically, we have

$$\bar{A} = A_G(\{A_i\}) = A_G(A_1, A_2, ..., A_m) = \begin{bmatrix} \bar{a}_{11} & \bar{a}_{12} & \cdots & \cdots & \bar{a}_{1n} \\ \bar{a}_{21} & \bar{a}_{22} & \cdots & \cdots & \bar{a}_{2n} \\ . & . & \cdots & \cdots & . \\ . & . & \cdots & \cdots & . \\ . & . & \cdots & \cdots & . \\ \bar{a}_{n1} & \bar{a}_{n2} & \cdots & \cdots & \bar{a}_{nn} \end{bmatrix} \tag{1}$$

and

$$\bar{V} = f(\bar{A}) = (\bar{v}_1, \bar{v}_2, ..., \bar{v}_n) \tag{2}$$

where $\bar{a}_{jk}$ is the aggregated group pairwise comparison between decision element j and element k. $\bar{v}_j$ is the aggregated relative weight for decision element j.

3 Minimum Distance Method (MDM)

The distance concept has been used by researchers to aggregate group judgments. Among them are Kemeny and Snell's distance measure to aggregate a set of ordered rankings as indicated in [4], and Yu's general distance approach to solve the group decision problem by using the concept of an ideal solution to describe measurements of compromise in utility space as presented in [5]. The concept of distance between pairwise comparison matrices is also very appealing for the group decision problem defined in section 2 of this paper. This concept can be applied to several aspects of group decision measurements.

- Distance as Accuracy Measurement: Considering n attributes, all of which are mea surable in a common criterion, one would contend that the resultant relative weights of n attributes are correct at the level of significance of the instrument if the attributes are physically measured with a precise instrument. Let a second estimate of relative weights be provided by an estimator that is less precise, a human evaluator for instance. Taking the first estimate as correct, the distance between the two estimates results from errors made by the estimator.
- Distance as Group Disagreement: Another circumstance to consider is the comparison of pairwise comparison matrices supplied by different sources, none of which is sufficiently precise to assure objective correctness. Since no objective standard for correctness exists, the disagreements (distance) D among the judgments of the experts cannot be interpreted directly as inaccuracy. However, some indication of the extent to which the experts agree may be a useful guide to make inferences regarding how well the attributes are known. In general, when group disagreement D is small, the group decision represents better consensus (which does not mean that the estimates are accurate). When disagreement is high, the group decision should be reexamined.

Based on the appealing characteristics of the distance concept for the group judgment ag gregation problem defined in section 2 of this chapter. We also stated that our objective is to minimize the group disagreement (D). Therefore, we call it Minimum Distance Method and have the following expressions:

$$Objective: \quad Minimize\ D \tag{3}$$

and D is defined as:

$$D = \sum_{i=1}^{m} d(A_i, \bar{A}) \tag{4}$$

in which d stands for any distance function between A_i and $\overline{A}$, such as the squared distance, absolute distance and so on. Let's look at following two important aspects, on which the MDM is derived.

First, the weighted geometric mean method is the extension of the geometric mean when an individual has different weights. Aczel and Saaty [3] demonstrated that weighted geometric mean is a robust method for aggregating group judgment. Therefore, the $\overline{A}$ can be expressed as follows:

$$\bar{A} = (\bar{a}_{jk}) = (\prod_{i=1}^{m}(\{a_{jk}\}_i)^{w_i}), \quad and \quad j,k = 1,2,\cdots,n \tag{5}$$

$$\sum_{i=1}^{m} w_i = 1, \quad \bar{V} = f(\bar{A}) \tag{6}$$

where w_i is the weight assigned to person i, and aik takes the form of the weighted geometric mean. Therefore, the aggregation problem defined in expressions (3), (5) and (6) becomes to find the optimal wi so that the group disagreement D is minimal.

There are two significant advantages in assuming that $\overline{A}$ is in the form of the weighted geometric mean:

- Under the distance function defined in (7) the weighted geometric mean form allows us to convert the problem defined in expressions (3) and (4) into a linear programming formulation which can be easily solved.
- The weighted geometric mean form reduces the number of variables to be determined in expression (3) from n(n—1)/2 to m. Due to reciprocal matrix $\overline{A}$ contains n(n—1)/2 independent variables or elements (ajk) and there are only m wi representing the weights to be assigned to the m decision makers.

Second, what kind of distance function of d(A_i, $\overline{A}$) will be the most suitable form for our aggregation problem? There are at least two criteria:

1. $\overline{A}$ should be unique to satisfy the Eqn. (3)
2. $\overline{A}$ should be easy to calculate

Cook and Kress [6] have shown that the unique distance between any two pairwise comparison matrices A and B should be in the following form:

$$d(A,B) = \frac{1}{2}\sum_{j=1}^{n}\sum_{k=1}^{n}\left|\ln(\frac{a_{jk}}{b_{jk}})\right| \tag{7}$$

as long as A and B satisfy three axioms (metric properties, single pair distance, scaling) as shown in [6]. Reciprocal pairwise comparison matrix Ai and A satisfy those axioms given the distance function in expression (7). Given logarithmic form of the distance function (7), the nonlinear relationship between wi and D presented in expressions (5) and (6) becomes linear.

Definition: The consensus pairwise comparison matrix A is that matrix which minimizes the total absolute distance:

$$D = \sum_{i=1}^{m} d(A_i, \bar{A}) = \sum_{i=1}^{m}\sum_{j=1}^{n}\sum_{k=1}^{n} |\ln(\{a_{jk}\}_i) - \ln(\bar{a}_{jk})| \tag{8}$$

and is subject to the following constraints:

$$\ln(\bar{a}_{jk}) = \sum_{i=1}^{m} w_i \ln(\{a_{jk}\}_i), \qquad \sum_{i=1}^{m} w_i = 1, \qquad j, k = 1, \cdots n. \tag{9}$$

where w_i is the weight assigned to decision maker i. The aggregation problem defined in expressions (8) and (9) can be further expanded by using the transformation $\{P_{jk}\}_i - \{N_{jk}\}_i = \ln(\{a_{jk}\}_i) - \ln(\bar{a}_{jk})$ where $\{N_{jk}\}_i \geq 0$, $\{P_{jk}\}_i \geq 0$. The original problem is now equivalent to the following goal programming problem:

$$minimize \quad \sum_{i=1}^{m}\sum_{j=1}^{n}\sum_{k=1}^{n} \theta_i \left(\{N_{jk}\}_i + \{P_{jk}\}_i\right) \tag{10}$$

$$subject\ to \quad \sum_{l=1}^{m} w_l \ln(\{a_{jk}\}_l) - \{N_{jk}\}_i + \{P_{jk}\}_i = \ln(\{a_{jk}\}_i) \tag{11}$$

$$i = 1, 2, \cdots, m, \quad and \quad j, k = 1, 2, \cdots, n$$

$$\sum_{i=1}^{m} w_i = 1$$

$$\{N_{jk}\}_i, \{P_{jk}\}_i, w_i \geq 0.$$

where ϕ_i is relative importance of decision maker i. The ϕ_i $(i = 1, , m)$ in expression (10) can be interpreted as weights on the goals in a hierarchical sense. They tend to sway the aggregated pairwise comparison matrix closer to the judgments of the more important members, and away from those of less important members.

4 An MDM Example

In the above sections, the aggregation method (MDM) has been discussed in detail. Now let's look at numerical example. Suppose there are four estimators (say A, B, C and D) for four decision elements with ~i = 1. The corresponding four pairwise comparison matrices obtained from estimator A, B, C and D are as follows:

$$A = \begin{bmatrix} 1 & \frac{1}{2} & \frac{1}{3} & \frac{1}{4} \\ 2 & 1 & \frac{2}{3} & \frac{1}{2} \\ 3 & \frac{3}{2} & 1 & \frac{3}{4} \\ 4 & 2 & \frac{4}{3} & 1 \end{bmatrix} \qquad B = \begin{bmatrix} 1 & 2 & \frac{2}{3} & \frac{1}{2} \\ \frac{1}{2} & 1 & \frac{1}{3} & \frac{1}{4} \\ \frac{3}{2} & 3 & 1 & \frac{3}{4} \\ 2 & 4 & \frac{4}{3} & 1 \end{bmatrix} \tag{12}$$

$$C = \begin{bmatrix} 1 & \frac{1}{3} & \frac{1}{2} & \frac{1}{4} \\ 3 & 1 & \frac{3}{2} & \frac{3}{4} \\ 2 & \frac{2}{3} & 1 & \frac{1}{2} \\ 4 & \frac{4}{3} & 2 & 1 \end{bmatrix} \qquad D = \begin{bmatrix} 1 & \frac{4}{3} & 2 & 4 \\ \frac{3}{4} & 1 & \frac{3}{2} & 3 \\ \frac{1}{2} & \frac{2}{3} & 1 & 2 \\ \frac{1}{4} & \frac{1}{3} & \frac{1}{2} & 1 \end{bmatrix} \tag{13}$$

By applying expressions (10) and (11), the goal programming model for aggregating the pairwise comparison matrices is constructed as follows:

$$minimize \quad \sum_{i \in \{A,B,C,D\}} \sum_{j=1}^{4} \sum_{k=1}^{4} (\{N_{jk}\}_i + \{P_{jk}\}_i) \tag{14}$$

where $\{N_{jk}\}i$ is the negative deviation from the comparison of elements j and k by estimator i, $\{P_{jk}\}i$ is the positive deviation from the comparison of elements j and k by estimator i, $i = A_i$, D (estimators), j and k = 1, , 4 (the elements being compared. The expressions for constraints are as follows:

$$subject\ to \quad \sum_{l \in \{A,B,C,D\}} w_l \ln(\{a_{jk}\}_l) - \{N_{jk}\}_i + \{P_{jk}\}_i = \ln(\{a_{jk}\}_i) \tag{15}$$

$$i \in \{A, B, C, D\}, \quad and \quad j, k = 1, 2, \cdots, 4$$

$$\sum_{i \in \{A,B,C,D\}} w_i = 1$$

$$\{N_{jk}\}_i, \{P_{jk}\}_i, w_i \geq 0.$$

The goal programming defined in expressions (14) and (15) are solved by LINDO with following results.

$$w_A = 0.5, \quad w_B = 0.0, \quad w_C = 0.15, \quad w_D = 0.35 \tag{16}$$

The corresponding pairwise comparison matrix A can now be obtained by using the expression (5), which are:

$$\bar{A} = (\bar{a}_{jk}) = \Big(\prod_{i \in \{A,B,C,D\}} (\{a_{jk}\}_i)^{w_i} \Big), \quad and \quad j, k = 1, 2, \cdots, 4 \tag{17}$$

Specifically, combining the w_i ($i \in$ {A, B, C, D}) value with the respective pairwise comparison matrices $\{A_i\}$ obtained from the four estimators, the $\overline{A}$, the aggregated pairwise comparison matrix, is obtained as follows:

$$\bar{A} = \begin{bmatrix} 1 & \frac{1}{2}^{0.5} 2^{0.0} \frac{1}{3}^{0.15} \frac{4}{3}^{0.35} & \frac{1}{3}^{0.5} \frac{2}{3}^{0.0} \frac{1}{2}^{0.15} 2^{0.35} & \frac{1}{4}^{0.5} \frac{1}{2}^{0.0} \frac{1}{4}^{0.15} 4^{0.35} \\ 2^{0.5} \frac{1}{2}^{0.0} 3^{0.15} \frac{3}{4}^{0.35} & 1 & \frac{2}{3}^{0.5} \frac{1}{3}^{0.0} \frac{3}{2}^{0.15} \frac{3}{2}^{0.35} & \frac{1}{2}^{0.5} \frac{1}{4}^{0.0} \frac{3}{4}^{0.15} 3^{0.35} \\ 3^{0.5} \frac{3}{2}^{0.0} 2^{0.15} \frac{1}{2}^{0.35} & \frac{3}{2}^{0.5} 3^{0.0} \frac{2}{3}^{0.15} \frac{2}{3}^{0.35} & 1 & \frac{3}{4}^{0.5} \frac{3}{4}^{0.0} \frac{1}{2}^{0.15} 2^{0.35} \\ 4^{0.5} 2^{0.0} 4^{0.15} \frac{1}{4}^{0.35} & 2^{0.5} 4^{0.0} \frac{4}{3}^{0.15} \frac{1}{3}^{0.35} & \frac{4}{3}^{0.5} \frac{4}{3}^{0.0} 2^{0.15} \frac{1}{2}^{0.35} & 1 \end{bmatrix}$$

$$= \begin{bmatrix} 1 & 0.5 & 0.5 & 0.375 \\ 2 & 1 & 1 & 0.75 \\ 2 & 1 & 1 & 0.75 \\ 2.667 & 1.333 & 1.333 & 1 \end{bmatrix} \quad (18)$$

Using the $\overline{A}$ as the aggregated matrix, we can obtain the aggregated priority vector, V, which represents the minimum distance criterion. Several methods as described in Appendix A can be used for this purpose. For example, using the geometric mean method:

$$\bar{V} = (0.13, 0.26, 0.26, 0.35) \quad (19)$$

The individual priority weight are:

$$v_1 = \mathbf{0.13}, \quad v_2 = \mathbf{0.26}, \quad v_3 = \mathbf{0.26}, \quad V_4 = \mathbf{0.35} \quad \mathbf{(20)}$$

$\overline{V}$ the resulting vector obtained after the aggregation process, is the one which minimizes the distance between the A_i ($i \in$ {A, B, C, D} and the aggregated value ($\overline{A}$) in the multi dimensional space.

5 Conclusions

MDM is a goal programming model based on the general distance concept, which is easy to solve by using commercial software such as LINDO. MDM provides the weighted membership capability (i.e. °i) Sensitivity analysis can be performed to investigate the effect of varying the decision makers' relative importance in terms of weights in the goal programming model

References

[1] C. Aczel and T. L. Saaty, "Procedures for synthesizing ratio judgments," *Journal of Mathematical Psychology*, vol. 27, pp. 93-102, 1983.

[2] C. Aczel and T. L. Saaty, "Synthesizing judgments: A functional equations approach," *Math. Modeling,* vol. 9, pp. 311-320, 1987.

[3] C. Aczel and T. L. Saaty, "On synthesis of judgments," *Socio-Econ. Plann. Sci.*, vol. 20, pp 333-339, 1986.

[4] J. G. Kemeny and L. J. Snell, *Preference Ranking: Axiomatic Approach.* Ginn and Co., Boston, 1962.

[5] P. L. Yu, "A class of solutions for group decision problems," *Management Science,* vol. 19, no. 8, pp. 936-946, 1973.

[6] W. D. Cook and K. Moshe, "Deriving weights from pairwise comparison ratio matrices: An axiomatic approach," *European Journal of Operational Research,* vol. 37, pp. 355-362, 1988.

Product Development Strategy for Small and Mid-Sized Air-Conditioning Enterprises in China

Xiuwen Zhu
(022) 747-2344

Dept of Management Engineering
Tianjin University,
Tianjin, 30072 China

John Whittaker
(403) 492-4443

Dept. of Mechanical Engineering
Univ. of Alberta,
Edmonton, AB.
Canada, T6G 2G8

Based on large scale questionnaires the air-conditioner market in China was described and analyzed. Then possible strategies for small and medium sized firms were proposed and evaluated. Facing on strong competitors and increasingly severe competitive environment small and mid-sized enterprises should avoid them, and try to open up the blank area of the existing market by properly selecting product strategies that are the product positioning and the new product development strategies.

Introduction

China is currently undergoing a critical transition from a planned economy to market economy. Against this background China's enterprises, including air-conditioning enterprises, start their own dramatic changing process. China's air-conditioning enterprises are facing very keen competition which will inevitably result in the adjustment of their industrial structure. What is happening and what is its future in China's air-conditioning industry is a special concern for small and mid-sized enterprises. In recent years sale volume of air-conditioners has been growing rapidly. Many air-conditioner firms expanded output, some refrigerator firms turned to produce air-conditioners, many new factories put air-conditioners into production. As a result, China's air-conditioner industry developed too quickly, supply exceeded demand and the result was an "Air-conditioner War".

- There is a major need for accurate, strategic data. Questions needing answers included:
- What is the future of China's air-conditioner market?
- How should China's air-conditioner industry develop under the market
- conomy?
- How should China's air-conditioner industry be adjusted in the next years?
- What is the potential consumers demand?

A questionnaire survey was conducted with both manufacturers and consumers. Completed questionnaires were analyzed.

Based upon the information from the questionnaires and other sources it is possible to analyze the development process, the current status of market and competitive situation of China's air-conditioning industry. The analysis indicates a strategic product direction for small and mid-sized enterprises. The aim of this research is to help the managers of small and mid-sized enterprises to initiate a strategic product plan that is based upon what the consumer wants rather than just on what the enterprise wants to produce or sell. The consumer focus is new in China but it has great significance for product and enterprise future development.

0-7803-3552-X/96/$5.00 © 1996 IEEE

Development background of China's air-conditioning industry

China's air-conditioning industry started in the 1960s but due to limitations of low productivity and state policy it was quite slow to develop. It hasn_t yet achieved economies of scale . Since the economic reforms, the market demand's push has resulted in steady growth in the air-conditioning industry. The key manufacturing equipment for producing air-conditioners was introduced from Japan to Shenyang early in 1982. Then, the air-conditioner assembly lines and manufacturing technology were subsequently introduced by more than 10 enterprises from USA, Denmark, Italy, and Hong Kong. Also the existing enterprises carried out the technical reform. This took the China's air-conditioning industry to a new level. Now it possesses great production capacity, advanced production processes and equipment. The equipment of some leading enterprises are almost the same as that of USA and Japan.

Their products have reached the advanced world level at the beginning of 1990s. The air-conditioners output in China increased from 12,800 units in 1981 to 597,400 units in 1991, increasing by 46 times over 10 years. The 1992's output was about 1.52 millions units, over 3.1 millions in 1993, it broke through 4.5 millions in 1994. There were more than 200 enterprises that produced air-conditioners by the end of 1994 in China. Their annual production capacity is over 8.5 millions units. About 30 enterprises have annual capacity of 100,000 units and the about ten have the capacity of 300,000 units.

In recent years, room air-conditioners have become a consumable "hot point" like colour TV sets, refrigerators, washing machines [C.F.Chen and Z.J.Chen,1994]. The room air-conditioners owned by 100 families in the cities went up from 1.1 units in 1991 to 8 units in 1994, even to 12 units in such large cities as Beijing and Shanghai. After ten years of slow development the market demand increases significantly from 1992. This indicates that air-conditioner market has started up and is into the stage of rapid growth.

Competitive situation in China's air-conditioning industry

The recent past has seen keen competition in China's air-conditioning industry. There was "the advertising war" in 1993, "the price reducing war" in 1994, and "the overall civil war" combining the product quality and the after-sale service in 1995. During that "War" all enterprises discounted the product price by an average of 20% . At the same time the prices of air-conditioner parts went up 10-30%. It was estimated that the advertising expenses of each enterprise increased on average by 80 millions every year as compared with that before the "war", which means that due to keen competition the annual advertising expenses increased 1.6 billions in the air-conditioning industry. Because the name brands of China's air-conditioners are not recognised by the consumers, each enterprise took many sales promoting steps to increase consumer awareness. They all introduced the strategy of offering a more comprehensive services and warranty policy. Consumers enjoy free delivery, installation, product lifetime maintenance and materials such as brackets, ducts, etc were also freely provided.

The stronger enterprises currently produce at less than full capacity. Generally speaking, in order to make a profit an annual output of over one hundred thousands units is required. These stronger enterprises are trying to expand to full capacity production. They hope to achieve a high sales volume by using mark-down tactics Hence it is very difficult that for small and mid-sized enterprises to take one position on the existing air-conditioner market. However, no enterprise will willingly retire from the market. They have all invested hundreds of millions of Yuan of capital and they will fight continuously to realise a return on their investment According to the expert opinion following the air-conditioner war in 1996, 1997 will result in the adjustment of the industrial structure. This will directly threaten the survival of small and mid-sized air-conditioning enterprises. Our forecast market demand for air-conditioner is 3.5 millions units in 1994. The possible annual output is 4.5 millions. Hence, the competition will undoubtedly be more and more intense.

A product strategy for small and mid-sized air-conditioning enterprises

The foregoing outlined the strategic environment facing small and mid-sized air-conditioning enterprises. The following is a product development strategy that small and mid-sized enterprises could take, including strategies of the product positioning, the new product development and the product appearance etc.

The product positioning strategy

Product positioning refers to the place an offering occupies in the target consumers' minds on important attributes relative to competitive offerings. Positioning is a consumer's subjective evaluation of products (Berkowitz,1995), it is how a product is seen through the eyes of the consumers. Through proper positioning, the product could gain a competitive advantage. When considering the positioning of products of the small and mid-sized enterprises there are two possible approaches. One is head to head or direct competition with other firms, the other is to try for product differentiation and avoid direct competition (Berkowitz,1995). Head to head positioning means that small and mid-sized enterprises compete directly with strong rivals on similar product attributes in the same target market. Given their weaker strengths, this would not be a wise strategy. The differentiation positioning seeks to avoid direct competition by stressing unique aspects of the products. An enterprise may position its products on a specific attribute, feature or benefits. From the survey and market information analysis we have learned that most of stronger, larger enterprises tend to manufacture air-conditioners with many functions such as dehumidifying, timing, directional control and remote control. This enthusiasm for multifuctions could be "over quality" solution. According to a theory of value engineering, if a product's functions reduce to a certain degree, product value will greatly increase. The questionnaires revealed that many consumers bought multifunctional air-conditioner at higher price due to their curiosity. In fact some functions are rarely used. Some consumers expressed definitely no need for multifunctional air-conditioner that salesmen recommended. For example, the dehumidifying function is little used in the north of China. Also more complicated the equipment, the higher the cost, and the higher the failure rate. Alternatively, the simpler the equipment, the cheaper it is, and also the more reliable it is. So, although simplifying the functions is contrary to the direction of many enterprises, low price and reliable quality is a potentially competitive strategy. Accordingly, small and mid-sized enterprises should consider positioning their products on the market characterised by the low price, high quality, less functions and easy-to-operate.

New product development strategy

It is very difficult to develop new products: it requires significant capital and creates large risks. Thus a small and mid-sized enterprise must select new product development strategies based on market demand, competitive situation and its own competence. As suggested above, small and mid-sized enterprises could position their products on high quality, less function and easy-to-operate air-conditioners. There are a large number of new advanced products on the air-conditioner market, most of which are multifuctional air-conditioners. Small and mid-sized enterprises may exit and cede these new product market to larger, stronger companies, and foreign and joint venture companies, developing instead easy-to-operate air-conditioners with high quality and less functions by using the following-type (i.e. imitation) strategy. In product competition the imitation strategy is a legitimate market strategy and succeeds frequently. For example, many Japanese companies used the imitation strategy successfully to enter markets and in some cases, ultimately gain market leadership in competition with American companies (Schnaars,1994). The imitation strategy is especially suitable for small and mid-sized enterprises facing many strong rivals. Imitation is less expensive than innovation, on average, its costs are 65 percent of innovation costs, and does not have to slog through the same arduous innovation process as the innovator. Using this strategy small and mid-sized enterprises, that are weaker in the finance and technology, could spend much less money and time on research than larger companies because the product existence and characteristics provide them with a great deal of information that innovator had to obtain through its own research. So imitators can keep costs lower,and by passing those saving along to consumers produce the winning advantage of low price in the market.

The development of the easy-to-operate air-conditioner

The concept of the easy-to-operate air-conditioner came from the easy-to-operate camera, which integrated and automated all necessary functions into a simple point-and-shoot format. Now the functions of air-conditioners are becoming more and more complicated, with controls and functions which are not welcomed by all consumers. If an air-conditioner has some necessary functions and easy-to-operate we are sure that it will win the love of all consumers. Yadu brand air-conditioner has first tested simplifying functions. The Yadu 88 model, introduced in 1994, has only 3 controls: the switch, the high-low wind, and the moisture control. They are advertised as the easy-to-operate air-conditioner. The simplification of the functions made its cost decrease greatly. Although accurate data is not available, interviews with store staff and consumers indicated that the Yadu is finding a large level of consumer acceptance.Also, there is now a National brand air-conditioner with only one button. Interviews revealed that in the air-conditioner market, the air-conditioners offered at a better quality, low price, and simple operation are no doubt attractive. They are especially suitable for the salaried class who seek for real benefits, senior consumer group and rural consumers. These market sectors represent the future potential market of China's air-conditioners. It would seem advisable for small and mid-sized enterprises to concentrate on easy-to-operate air-conditioners by the imitating and modifying Yadu and National brand ones.

To pre-empt the easy-to-operate air-conditioner market the following tactics are worth of studying such as refined improvements, fast product development, exploratory development, constant evaluation of project and ties to international trading companies. The fast new product development tactic often uses parallel product development with multifuctional team. Motorola Inc. is a company that best mastered the concept that fast new product development and fast manufacturing start-up is a sure way to achieve market share. Also, the tactic can be applied to all sizes, even very small firms where there are very little cash and few people to go around. In China this tactic should be practised in small and mid-sized enterprises. They will discover that in long run fast product development is less expensive because they are able to get relevant products to the marketplace faster and hang onto or grab extra market share.

The product appearance strategy

Heating units with the flower pictures printed on the surface sold very well. Air-conditioners currently have just flat, white painted surfaces. The air-conditioners are used for 2-4 months every year, most of time they occupy a part of the room only as a display. So according to their this consumable feature, it is very necessary to design new appearance styles. The styles of current air-conditioners are mostly rectangular shape and pure white colour without exception. With people's increasing enthusiasm for the room decorating, small and mid-sized enterprises should identify this feature and rapidly carry out the product appearance strategy. For example the noble pictures of the mountains, oceans and forests are printed on the outer surfaces with the coating process. Some designs of the special means may also be printed according to the consumer's demand and the calendar and temperature display etc. may also be designed on thc background. If so, while people enjoy cool in the quiet night they admire also beautiful scenery, presenting with the strong living taste. The traditional pure white colour may be broken, brown and black may be tested, brown matching with furniture in the room, black representing the noble style, The new styles of the air-conditioners will certainly be welcomed by the all consumers.

References

Berkowitz,"Marketing," Times Mirror Higher Education Group Inc., 1995.

Chen Chunfang,"Sales of air-conditioners tending to be 'hot' when summer isn't coming yet," Jiefang Daily, 26, March, 1994.

Chen Zengjin,"Air-conditioner being first choice of urban residents in consumer goods in the 90's," Economy Daily,4,May, 1994.

Philip A, Himmelfarb,"Survival of the fittest: new product development during the 90's," Prentice-Hall Englewood cliffs, N.J. 1992.

Schnaars, S.P.,"Managing imitation strategy: how later entrants seize markets from pioneers," New York, Free Press, 1994.